COMPREHENSIVE BIOTECHNOLOGY

IN 4 VOLUMES

COMPREHENSIVE BIOTECHNOLOGY

*The Principles, Applications and Regulations
of Biotechnology in Industry,
Agriculture and Medicine*

EDITOR-IN-CHIEF
MURRAY MOO-YOUNG
University of Waterloo, Ontario, Canada

Volume 2
The Principles of Biotechnology: Engineering Considerations

VOLUME EDITORS
CHARLES L. COONEY
Massachusetts Institute of Technology, Cambridge, MA, USA

and

ARTHUR E. HUMPHREY
Lehigh University, Bethlehem, PA, USA

PERGAMON PRESS
OXFORD · NEW YORK · TORONTO · SYDNEY · FRANKFURT

U.K.	Pergamon Press Ltd., Headington Hill Hall, Oxford OX3 0BW, England
U.S.A.	Pergamon Press Inc., Maxwell House, Fairview Park, Elmsford, New York 10523, U.S.A.
CANADA	Pergamon Press Canada Ltd., Suite 104, 150 Consumers Road, Willowdale, Ontario M2J 1P9, Canada
AUSTRALIA	Pergamon Press (Aust.) Pty. Ltd., P.O. Box 544, Potts Point, N.S.W. 2011, Australia
FEDERAL REPUBLIC OF GERMANY	Pergamon Press GmbH, Hammerweg 6, D-6242 Kronberg-Taunus, Federal Republic of Germany

Copyright © 1985 Pergamon Press Ltd.

First edition 1985

Library of Congress Cataloguing in Publication Data

Main entry under title:
Comprehensive biotechnology.
Includes bibliographies and index.
Contents: v.1. The principles of biotechnology — scientific fundamentals / volume editors, Alan T. Bull, Howard Dalton —
v. 2. The principles of biotechnology — engineering considerations / volume editors, A. E. Humphrey, Charles L. Cooney — [etc.] —
v. 4. The practice of biotechnology — speciality products and service activities / volume editors, C. W. Robinson, John A. Howell.
1. Biotechnology.   I. Moo-Young, Murray.
TP248.2.C66   1985   660′.6   85–6509

British Library Cataloguing in Publication Data

Comprehensive biotechnology: the principles, applications and regulations of biotechnology in industry, agriculture and medicine.
1. Biotechnology
I. Title   II. Moo-Young, Murray
660′.6   TP248.3

ISBN 0–08–032510–6 (vol. 2)
ISBN 0–08–026204–X (4–vol. set)

Printed in Great Britain by A. Wheaton & Co. Ltd., Exeter

Contents

Contents

Preface

In his recent book, entitled 'Megatrends', internationally-celebrated futurist John Naisbitt observed that recent history has taken industrialized civilizations through a series of technology-based eras: from the chemical age (plastics) to an atomic age (nuclear energy) and a microelectronics age (computers) and now we are at the beginning of an age based on biotechnology. Biotechnology deals with the use of microbial, plant or animal cells to produce a wide variety of goods and services. As such, it has ancient roots in the agricultural and brewing arts. However, recent developments in genetic manipulative techniques and remarkable advances in bioreactor design and computer-aided process control have founded a 'new biotechnology' which considerably extends the present range of technical possibilities and is expected to revolutionize many facets of industrial, agricultural and medical practices.

Biotechnology has evolved as an ill-defined field from inter-related activities in the biological, chemical and engineering sciences. Inevitably, its literature is widely scattered among many specialist publications. There is an obvious need for a comprehensive treatment of the basic principles, methods and applications of biotechnology as an integrated multidisciplinary subject. *Comprehensive Biotechnology* fulfils this need. It delineates and collates all aspects of the subject and is intended to be the standard reference work in the field.

In the preparation of this work, the following conditions were imposed. (1) Because of the rapid advances in the field, it was decided that the work would be comprehensive but concise enough to enable completion within a set of four volumes published simultaneously rather than a more encyclopedic series covering a period of years to complete. In addition, supplementary volumes will be published as appropriate and the work will be updated regularly via *Biotechnology Advances*, a review journal, also published by Pergamon Press with the same executive editor. (2) Because of the multidisciplinary nature of biotechnology, a multi-authored work having an international team of experts was required. In addition, a distinguished group of editors was established to handle specific sections of the four volumes. As a result, this work has 10 editors and over 250 authors representing 15 countries. (3) Again, because of the multidisciplinary nature of the work, it was virtually impossible to use a completely uniform system of nomenclature for symbols. However, provisional guidelines on a more unified nomenclature of certain key variables, as provided by IUPAC, was recommended. (4) According to our definition, aspects of biomedical engineering (such as biomechanics in the development of prosthetic devices) and food engineering (such as product formulations) are not included in this work. (5) Since the work is intended to be useful to both beginners as well as veterans in the field, basic elementary material as well as advanced specialist aspects are covered. For convenience, a glossary of terms is supplied. (6) Since each of the four volumes is expected to be fairly self-contained, a certain degree of duplication of material, especially of basic principles, is inevitable. (7) Because of space constraints, a value judgement was made on the relative importance of topics in terms of their actual rather than potential commercial significance. For example, 'agricultural biotechnology' is given relatively less space compared to 'industrial biotechnology', the current raison d'être of biotechnology as a major force in the manufacture of goods and services. (8) Finally, a delicate balance of material was required in order to meet the objective of providing a comprehensive and stimulating coverage of important practical aspects as well as the intellectual appeal of the field. Readers may wish to use this work for initial information before possibly delving deeper into the literature as a result of the critical discussions and wide range of references provided in it.

Comprehensive Biotechnology is aimed at a wide range of user needs. Students, teachers, researchers, administrators and others in academia, industry and government are addressed. The requirements of the following groups have been given particular consideration: (1) chemists, especially biochemists, who require information on the chemical characteristics of enzymes, metabolic processes, products and raw materials, and on the basic mechanisms and analytical techniques involved in biotechnological transformations; (2) biologists, especially microbiologists and molecular biologists, who require information on the biological characteristics of living organisms involved in biotechnology and the development of new life forms by genetic engineering techniques; (3) health scientists, especially nutritionists and toxicologists, who require information on biohazards and containment techniques, and on the quality of products and by-products of biotechnological processes, including the pharmaceutical, food and beverage industries; (4) chemical engineers, especially biochemical engineers, who require information on mass and energy balances and rates of processes, including fermentations, product recovery and feedstock pretreatment, and the equipment for carrying out these processes; (5) civil engineers, especially environmental engineers, who require information on biological waste treatment methods and equipment, and on contamination potentials of the air, water and land within the ecosystem, by industrial and domestic effluents; (6) other engineers, especially agricultural and biomedical engineers, who require information on advances in the relevant sciences that could significantly affect the future practice of their professions; (7) administrators, particularly executives and legal advisors, who require information on national and international governmental regulations and guidelines on patents, environmental pollution, external aid programs and the control of raw materials and the marketing of products.

No work of this magnitude could have been accomplished without suitable assistance. For guidance on the master plan, I am indebted to the International Advisory Board (J. D. Bu'Lock, T. K. Ghose, G. Hamer, J. M. Lebault, P. Linko, C. Rolz, H. Sahm, B. Sikyta and H. Taguchi). For structuring details of the various sections, the invaluable assistance of the section editors is gratefully acknowledged, especially Alan Bull, Charles Cooney, Harvey Blanch and Campbell Robinson, who also acted as coordinators for each of the four volumes. For the individual chapters, the 250 authors are to be commended for their hard work and patience during the two years of preparation of the work. For checking the hundreds of literature references cited in the various chapters, the many graduate students are thanked for a tedious but important task well done. A special note of thanks is due to Jonathan and Arlene Lamptey, who acted as editorial assistants in many diverse ways. At Pergamon Press, I wish to thank Don Crawley for originally suggesting this project and Colin Drayton for managing it. Finally, I am pleased to note the favourable evaluations of the work by two distinguished authorities, Sir William Henderson and Nobel Laureate Donald Glaser, who provided a foreword and a guest editorial, respectively, to the treatise.

MURRAY MOO-YOUNG
Waterloo, Canada
December 1984

Foreword

This very comprehensive reference work on biotechnology is published ten years after the call by the National Academy of Sciences of the United States of America for a voluntary worldwide moratorium to be placed on certain areas of genetic engineering research thought to be of potential hazard. The first priority then became the evaluation of the conjectural risks and the development of guidelines for the continuation of the research within a degree of containment. There had hardly been a more rapid response to this type of situation than that of the British Advisory Board for the Research Councils. The expression of concern by Professor Paul Berg and the committee under his chairmanship, and the call for the moratorium, was published in *Nature* on 19 July 1974. The Advisory Board agreed at their meeting on the 26 July to establish a Working Party with the following terms of reference:

'To assess the potential benefits and potential hazards of the techniques which allow the experimental manipulation of the genetic composition of micro-organisms, and to report to the Advisory Board for the Research Councils.'

Because of the conviction of those concerned that recombinant DNA techniques could lead to great benefits, the word order used throughout the report of the Working Party (Chairman, Lord Ashby) always put 'benefits' before 'hazards'. The implementation of the recommendations led to the development of codes of practice. This was followed by the establishment of the Genetic Manipulation Advisory Group as a standing central advisory authority operating within the framework of the Health and Safety at Work, *etc.* Act 1974 and, later, more specifically within the framework of the Health and Safety (Genetic Manipulation) Regulations 1978. Similar moves took place in many other countries but the other most prominent and important activity was that of the US National Institutes of Health. This resulted in the adoption by most countries of the NIH or the UK guidelines, or the use of practices based on both.

The significant consequence of the debates, the discussions and of the recommendations that emerged during these early years of this decade (1974–1984) was that research continued, expanded and progressed under increasingly less restriction at such a pace that now makes it possible and necessary to devote the first Section of Volume 1 of this work to genetic engineering. Many chapters of the subsequent Sections and Volumes are of direct relevance to the application of genetic engineering.

The reason for identifying today's genetic engineering for first mention in this foreword is its novelty. It was being conceived barely more than ten years ago. Ten years by most standards is a short time. Although in the biological context it represents at least 10^4 generation times of the most vigorous viruses, it is less than one of man even for the most precocious. The current developments in biotechnology, whether they be in recombinant DNA, monoclonal antibodies, immobilized enzymes, *etc.* are mostly directed towards producing a better product, or a better process. This is commendable and is supportable by the ensuing potential commercial benefits. The newer challenge is the application of the new biotechnology to achieve what previously could scarcely have been contemplated. Limited biological sources of hormones, growth regulators, *etc.* are being, and will be increasingly, replaced by the use of transformed microorganisms, providing a vastly increased scale of production. Complete safety of vaccines by the absence of ineffectively inactivated virus is one of the great advantages of the genetically engineered antigen. This is quite apart from the ability to prepare products for which, at present, there is a technical difficulty or which is economically not feasible by standard methods.

A combination of advances in recombinant DNA research, molecular biology and in blastomere manipulation has provided the technology to insert genetic material into the totipotent animal cell. The restriction on the application of this technology for improved animal production is the lack of knowledge on the genetic control of desirable biological characteristics for transfer from one breed line to another.

There are probably greater potential benefits to be won in the cultivation of the domesticated plants than in the production of the domesticated animals. In both cases, the objectives are to increase the plant's or the animal's resistance to the prejudicial components of its environment and to increase the yield, quality and desired composition of the marketable commodities. These include the leaf, the tuber, the grain, the berry, the fruit or the milk, meat and other products of animal origin. This is not taking into account the other valuable products of horticulture, of oil or wax palms, rubber trees and forestry in general. Genetic engineering should be able to provide short-cuts to reach objectives attainable by traditional procedures, for example by by-passing the sequential stages of a traditional plant breeding programme by the transfer of the genetic material in one step. Examples of desirable objectives are better to meet user specifications with regard to yield, quality, biochemical composition, disease and pest resistance, cold tolerance, drought resistance, nitrogen fixation, *etc*. One of the constraints in this work in plants is the scarcity of vectors compared with the many available for the transformation of microorganisms. The highest research priority on the plant side is to determine by one means or another how to increase the efficiency of photosynthesis. The photosynthetic efficiency of temperate crop plants is no more than 2–2.5% in terms of conversion of intercepted solar energy. These plants possess the C_3 metabolic pathway with the energy loss of photorespiration. Tropical species of plants with the C_4 metabolic pathway have a higher efficiency of photosynthesis in that they do not photorespire. One approach for the breeder of C_3 plants is to endow them with a C_4 metabolism. If this transformation is ever to be achieved, it is most likely to be by genetic engineering. Such an advance has obvious advantages with regard, say, to increased wheat production for the ever-increasing human population. Nitrogen fixation as an agricultural application of biotechnology is given prominence in Section 1 of Volume 4. Much knowledge has been acquired about the chemistry and the biology of the fixation of atmospheric nitrogen. This provides a solid foundation from which to attempt to exploit the potential for transfer of nitrogen-fixing genes to crop plants or to the symbiotic organisms in their root systems. If plants could be provided with their own capability for nitrogen fixation, the energy equation might not be too favourable in the case of high yielding varieties. Without an increase in the efficiency of photosynthesis, any new property so harnessed would have to be at the expense of the energy requirements of existing characteristics such as yield.

Enzymes have been used for centuries in the processing of food and in the making of beverages. The increasing availability of enzymes for research, development and industrial use combined with systems for their immobilization, or for the immobilization of cells for the utilization of their enzymes, is greatly expanding the possibilities for their exploitation. Such is the power of the new biotechnology that it will be possible to produce the most suitable enzymes for the required reaction with the specific substrate. An increasing understanding at the molecular level of enzyme degradation will make it possible for custom-built enzymes to have greater stability than those isolated from natural sources.

The final section of Volume 4 deals with waste and its management. This increasingly voluminous by-product of our society can no longer be effectively dealt with by the largely empirical means that continue to be practised. Biological processes are indispensable components in the treatment of many wastes. The new biotechnology provides the opportunity for moving from empiricism to processes dependent upon the use of complex biological reactions based on the selection or the construction of the most appropriate cells or their enzymes.

The very comprehensive coverage of biotechnology provided by this four-volume work of reference reflects that biotechnology is the integration of molecular biology, microbiology, biochemistry, cell biology, chemical engineering and environmental engineering for application to manufacturing and servicing industries. Viruses, bacteria, yeasts, fungi, algae, the cells and tissues of higher plants and animals, or their enzymes, can provide the means for the improvement of existing industrial processes and can provide the starting points for new industries, for the manufacture of novel products and for improved processes for management of the environment.

SIR WILLIAM HENDERSON, FRS
Formerly of the *Agricultural Research Council*
and *Celltech Ltd., London, UK*

Guest Editorial

Since 1950, the new science of molecular biology has produced a remarkable outpouring of new ideas and powerful techniques. From this revolution has sprung a new discipline called genetic engineering, which gives us the power to alter living organisms for important purposes in medicine, agriculture and industry. The resulting biotechnologies span the range from the ancient arts of fermentation to the most esoteric use of gene splicing and monoclonal antibodies. With unprecedented speed, new scientific findings are translated into industrial processes, sometimes even before the scientific findings have been published. In earlier times there was a more or less one-way flow of new discoveries and techniques from scientific institutions to industrial organizations where they were exploited to make useful products. In the burgeoning biotechnology industry, however, developments are so rapid that there is a close intimacy between science and technology which blurs the boundaries between them. Modern industrial laboratories are staffed with sophisticated scientists and equipped with modern facilities so that they frequently produce new scientific discoveries in areas that were previously the exclusive province of universities and research institutes, and universities not infrequently develop inventions and processes of industrial value in biotechnology and other fields as well.

Even the traditional flow of new ideas from science to application is no longer so clear. In many applications, process engineers may find that the most economical and efficient process design requires an organism with new properties or an enzyme of previously unknown stability. These requirements often motivate scientists to try to find in nature, or to produce through genetic engineering or other techniques of molecular biology, novel organisms or molecules particularly suited for the requirements of production. A recent study done for the United States Congress* concluded that "in the next decade, competitive advantage in areas related to biotechnology may depend as much on developments in bioprocess engineering as on innovations in genetics, immunology, and other areas of basic science."

These volumes bring together for the first time in one unified publication the scientific and engineering principles on which the multidisciplinary field of biotechnology is based. Following accounts of the scientific principles is a large set of illustrations of the diverse applications of these principles in the practice of biotechnology. Finally, there are sections dealing with important regulatory aspects of the potential hazards of the growing field and of the need for promoting biotechnology in developing countries.

Comprehensive Biotechnology has been produced by a team of some of the world's foremost experts in various aspects of biotechnology and will be an invaluable resource for those wishing to build bridges between 'academic' and 'commercial' biotechnology, the ultimate form of any technology.

DONALD A. GLASER
University of California, Berkeley
and *Cetus Corp., Emeryville, CA, USA*

*"Commercial Biotechnology: An International Analysis," Office of Technology Assessment Report, U.S. Congress, Pergamon Press, Oxford, 1984.

Executive Summary

In this work, biotechnology is interpreted in a fairly broad context: the evaluation and use of biological agents and materials in the production of goods and services for industry, trade and commerce. The underlying scientific fundamentals, engineering considerations and governmental regulations dealing with the development and applications of biotechnological processes and products for industrial, agricultural and medical uses are addressed. In short, a comprehensive but concise treatment of the principles and practice of biotechnology as it is currently viewed is presented. An outline of the main topics in the four volumes is given in Figure 1.

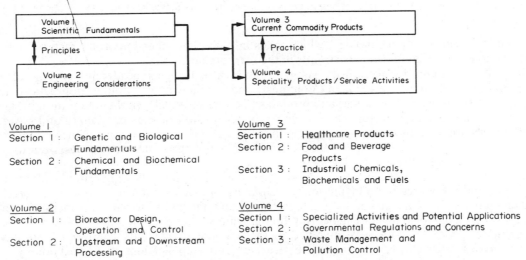

Volume 1			Volume 3		
Section 1 :	Genetic and Biological Fundamentals		Section 1 :	Healthcare Products	
Section 2 :	Chemical and Biochemical Fundamentals		Section 2 :	Food and Beverage Products	
			Section 3 :	Industrial Chemicals, Biochemicals and Fuels	

Volume 2			Volume 4		
Section 1 :	Bioreactor Design, Operation and Control		Section 1 :	Specialized Activities and Potential Applications	
Section 2 :	Upstream and Downstream Processing		Section 2 :	Governmental Regulations and Concerns	
			Section 3 :	Waste Management and Pollution Control	

Figure 1 Outline of main topics covered.

As depicted in Figure 2, it is first recognized that biotechnology is a multidisciplinary field having its roots in the biological, chemical and engineering sciences leading to a host of specialities, *e.g.* molecular genetics, microbial physiology, biochemical engineering. As shown in Figure 3, this is followed by a description of technical developments and commercial implementation,

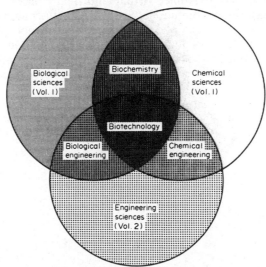

Figure 2 Multidisciplinary nature of biotechnology.

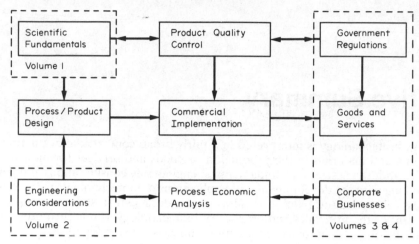

Figure 3   Interrelationships between biotechnology principles and applications.

the ultimate form of any technology, which takes into account other important factors such as socio-economic and geopolitical constraints in the marketplace.

There are two main divisions of the subject matter: a pedagogical academic coverage of the disciplinary underpinnings of the field (Volumes 1 and 2) followed by a utilitarian practical view of the various commercial processes and products (Volumes 3 and 4). In the integration of these two areas, other common factors dealing with product quality, process economics and government policies are introduced at appropriate points throughout all four volumes. Since biotechnological advances are often ahead of theoretical understanding, some process descriptions are primarily based on empirical knowledge.

The four volumes are relatively self-contained according to the following criteria. Volume 1 delineates and integrates the unifying multidisciplinary principles in terms of relevant scientific fundamentals. Volume 2 delineates and integrates the unifying multidisciplinary principles of biotechnology in terms of relevant engineering fundamentals. Volume 3 describes the various biotechnological processes which are involved in the manufacture of bulk commodity products. Volume 4 describes various specialized services, potential applications of biotechnology and related government concerns. In each volume, a glossary of terms and nomenclature guideline are included to assist the beginner and the non-specialist.

This work takes into account the relative importance of the various topics, primarily in terms of current practice. Thus, bulk commodity products of the manufacturing industries (Volume 3) are accorded more space compared to less major ones and for potential applications (part of Volume 4). This proportional space distribution may be contrasted with the expectations generated by the recent news media 'biohype'. For example, virtually no treatment of 'biochips' is presented. In addition, since the vast majority of commercial ventures involve microbial cells and cell-derived enzymes, relatively little coverage is given to the possible use of whole plant or animal cells in the manufacturing industries. As future significant areas of biotechnology develop, supplementary volumes of this work are planned to cover them. In the meantime, on-going progress and trends will be covered in Pergamon's complementary review journal, *Biotechnology Advances*.

M. Moo-Young
University of Waterloo, Canada

Contributors to Volume 2

Ms K. Antonsen
Department of Chemical Engineering, Massachusetts Institute of Technology, Cambridge, MA 02139, USA

Dr. W. B. Armiger
BioChem Technology Inc, 66 Great Valley Parkway, Great Valley Corporate Center, Malvern, PA 19355, USA

Dr H. A. C. Axelsson
Alfa-Laval Separation AB, PO Box 500, S-147 00 Tumba, Sweden

Mr P. A. Belter
Fermentation Research and Development, Unit 1400, The Upjohn Company, 7000 Portage Road, Kalamazoo, MI 49001, USA

Dr D. N. Bull
Satori Corporation, 23 Marion Road, Upper Montclair, NJ 07043, USA

Professor M. Charles
Biotechnology Research Center, Lehigh University, Bethlehem, PA 18015, USA

Professor C. K. Colton
Department of Chemical Engineering, Massachusetts Institute of Technology, Room 66-425, Cambridge, MA 02139, USA

Mr R. S. Conway
Pall Corporation, Scientific and Laboratory Services Department, 30 Sea Cliff Avenue, Glen Cove, NY 11542, USA

Professor C. L. Cooney
Department of Chemical Engineering, Massachusetts Institute of Technology, Room 66-468, Cambridge, MA 02139, USA

Dr R. V. Dove
Bio-Isolates Ltd, Powell Duffryn House, 10 Adelaide Street, Swansea SA1 1SE, UK

Dr C. E. Dunlap
A E Staley Manufacturing Co, 220 El Dorado Street, Decatur, IL 62525, USA

Professor C. R. Engler
Department of Agricultural Engineering, Texas A & M University, College Station, TX 77843, USA

Dr Z. Er-el
Fermentation Unit, Hadassah Medical School, Hebrew University, PO Box 1172, Jerusalem, Israel

Mr F. R. Gabler
Millipore Corporation, 80 Ashby Road, Bedford, MA 01730, USA

Dr R. A. Grant
Ecotech Systems Ltd, Poole, Dorset, UK

Dr R. T. Hatch
BioTechnica International Inc, 85 Bolton St, Cambridge, MA 02140, USA

Professor T. A. Hatton
Department of Chemical Engineering, Massachusetts Institute of Technology, Cambridge, MA 02139, USA

Mr A. H. Heckendorf
Consultant (NEST), 45 Valley Road, Southborough, MA, USA

Professor A. E. Humphrey
Provost, Alumni Memorial Building 27, Lehigh University, Bethlehem, PA 18015, USA

Dr J. A. Howell
Department of Chemical Engineering, University College of Swansea, Singleton Park, Swansea SA2 8PP, UK

Dr S. M. Jain
Ionics Inc, 65 Grove Street, Watertown, MA 02172, USA

Dr F. Kargi
Department of Food Engineering, Division of Biotechnology, Ege University, Bornova, Izmir, Turkey

Professor Dr J. Klein
Institut für Technische Chemie, Technischen Universität Braunschweig, Hans-Sommer-Strasse 10, D-3300 Braunschweig, Federal Republic of Germany

Dr M.-R. Kula
Gesellschaft für Biotechnologische Forschung mbH, Mascheroder Weg 1, D-3000 Braunschweig-Stöckheim, Federal Republic of Germany

Dr M. S. Le
Department of Chemical Engineering, University College of Swansea, Singleton Park, Swansea SA2 8PP, UK

Professor H. C. Lim
School of Chemical Engineering, Purdue University, West Lafayette, IN 47907, USA

Dr R. A. Messing
Consulting Chemist & Bioprocess Technologist, 168 Scenic Drive South, Horseheads, NY 14845, USA

Mr G. D. Moon, Jr
Raphael Katzen Associates International Inc, 1050 Delta Avenue, Cincinnati, OH 45208, USA

Professor M. Moo-Young
Institute for Biotechnology Research, University of Waterloo, Waterloo, Ontario N2L 3G1, Canada

Professor A. Moser
Institute of Biotechnology, Microbiology and Waste Technology, Technical University of Graz, Schlögelgasse 9, A-8010 Graz, Austria

Dr D. E. Palmer
Bio-Isolates Ltd, Powell Duffryn House, 10 Adelaide Street, Swansea SA1 1SE, UK

Professor J. Phillips
Biotechnology Research Center, Lehigh University, Bethlehem, PA 18015, USA

Mr C. W. Rausch
Biopure Fine Chemicals Inc, 136 Harrison Avenue, Boston, MA 02111, USA

Mr P. B. Reed
Ionics Inc, 65 Grove Street, Watertown, MA 02172, USA

Dr M. J. Rolf
School of Chemical Engineering, Purdue University, West Lafayette, IN 47907, USA

Professor K. Schügerl
Institut für Technische Chemie, Technischen Universität Hannover, Callinstrasse 3, D-3000 Hannover, Federal Republic of Germany

Professor M. L. Shuler
School of Chemical Engineering, Cornell University, Ithaca, NY 14853, USA

Dr J. R. Swartz
Genentech Inc, 460 Point San Bruno Blvd, San Francisco, CA 94080, USA

Mr R. S. Tutunjian
Amicon Corporation, 17 Cherry Hill Drive, Danvers, MA 01923, USA

Dr K.-D. Vorlop
Institut für Technische Chemie, Technischen Universität Braunschweig, Hans-Sommer-Strasse
10, D-3300 Braunschweig, Federal Republic of Germany

Mr R. C. Willson
Department of Chemical Engineering, Massachusetts Institute of Technology, Room 66-225,
Cambridge, MA 02139, USA

Dr D. M. Yarmush
Department of Chemical Engineering, Massachusetts Institute of Technology, Cambridge, MA
02139, USA

Dr M. L. Yarmush
Department of Chemical Engineering, Massachusetts Institute of Technology, Room 66-425,
Cambridge, MA 02139, USA

Dr D. W. Zabriskie
Smith, Kline & French Laboratories, 709 Swedeland Road, Swedeland, PA 19479, USA

Contents of All Volumes

Section 2   Governmental Regulations

Section 3   Waste Management and Pollution Control

BIOREACTOR DESIGN, OPERATION AND CONTROL

1

Introduction

A. E. HUMPHREY
Lehigh University, Bethlehem, PA, USA

The commercialization of biotechnology at present is not opportunity limited. Literally hundreds of clones for producing potential bioproducts exist. What is now needed is the translation of these bench-top schemes to large scale commercial systems. The Office of Technology Assessment of the US Congress concluded in its January 1984 report on *Commercial Biotechnology* that 'In the next decade commercialization of biotechnology may depend as much on developments in bioprocess engineering as on innovations in genetics, immunology, and other areas of basic science.'

At present, our ability to perform and control the large-scale culture of cells, particularly plant and animal cells, for the manufacture of bioproducts is at best primitive. Our knowledge of techniques for efficient large-scale recovery of bioproducts from the complex mixtures in which they occur is little more than rudimentary. Moreover, successful manufacture of bioproducts requires the availability of highly sophisticated sensors plus the necessary models for control algorithms so that the bioprocesses may be efficiently operated. Solutions to these problems are needed if the fruits of biotechnology are to be rapidly commercialized.

The material in Volume 2 addresses the state of technology in these areas, *i.e.* bioreactor design, separation and purification, and system operation and control. In Section 1, the focus is on bioreactor design, operation and control. The first four chapters of this section deal with the basic mass and heat transfer principles and their application in the design and scale-up of bioreactors. The next five chapters focus on modelling, monitoring, analysis and control of biosystems. The last three chapters deal with unconventional bioreactor systems, *i.e.* immobilized enzymes, immobilized cells, and mixed enzyme/cell systems. These systems surely will be the bioreactor systems of the future.

With respect to the design and scale-up of bioreactor systems, scale-up occupied first place on a meaningful scientific knowledge base in the mid 1940s due to the research of two sets of investigators. These were Cooper, Fernstrom and Miller from the DuPont Company, and Rushton and Oldshue from the Mixing Equipment Company. Forty years later we still design and scale-up gassed, mechanically agitated reactors on the basis of the correlations developed by these workers. These correlations relate the volumetric oxygen transfer coefficient, $k_L a$, to some function of the gassed power input per unit volume to the reactor, P_g/D^3, and to the superficial gas velocity, V_s, *i.e.*

$$k_L a = K(P_g/D^3)^\alpha (V_s)^\beta$$

Two characteristics of this correlation are still not understood. These are the effect of viscosity on $k_L a$ and the reason for the exponents α and β being scale dependent. The general opinion of researchers in this field is that $k_L a$ is roughly inversely proportional to the viscosity above broth viscosities of 25–50 cP. Also, some researchers feel that the major reason for the scale-dependency of the exponents is due to the difference in surface transfer between small and large vessels and the bulk flow patterns in the vessels. It has been suggested that large vessels with multiple impellers could be better modelled as a series of well mixed vessels rather than one poorly mixed vessel. These problems are addressed to varying degrees in the first three chapters.

In the fourth chapter of this section, Professor Schugerl addresses the problem of scale-up and design of nonmechanically agitated fermenters. Because of increasing energy costs relative to other costs, energy expenditure in bioreactors has become an increasing concern. Typical energy costs now run as high as 35% of the total fermentation associated costs compared to only 15% over a decade ago. It has been found that oxygen transfer is cheaper in a gassed, nonmechanically agitated system compared with a mechanically agitated system, that more kilograms of O_2 are transferred per kilowatt-hour of energy expenditure in a gassed only system. However, nonmechanically agitated systems usually suffer from lower volumetric mass transfer productivity than the mechanically agitated systems simply because they have poorer bulk mixing and lower shear gradients. The challenge in the future is to understand better the gassed only systems and find ways to optimize their volumetric efficiency.

With respect to monitoring and control of bioreactors, the present weakness or bottleneck appears to be due to a lack of meaningful models from which to develop process control algorithms. In the past decade sterilizable sensors for biosystems have emerged. Key sensors include mass flow meters, dissolved oxygen probes, mass spectrophotometers for off-gas analysis of oxygen uptake and carbon dioxide evolution rates, and more recently the multiple internal reflectance IR analysis system which will now let the biotechnologist monitor dissolved carbon dioxide and sugar concentrations in the biosystem.

With the rapidly increasing power of minicomputers and because of a lack of satisfactory bioprocess models, the biotechnologist may turn to on-line process optimization where the systems, through small perturbations of the process, may evolve on-line control algorithms. The knowledge base for on-line control and optimization is covered in the middle chapters of this section.

The final chapters of this section deal with alternatives to the standard dispersed cell system bioreactor. These chapters examine immobilized enzyme and cell systems. Many pharmaceutical companies are beginning to look at immobilized cell systems as a way of overcoming the rheological problems encountered in the highly mycelial antibiotic fermentations. In these systems, cells are entangled in and around porous particles several millimeters in diameter. Typical particles are of vermiculite. The bioreactor is then operated as a fluidized system with the particles suspended essentially in water. These systems are not without their unique problems. A key problem is that of keeping the system viable and active while still producing a product.

The practicing biotechnologist should, therefore, find this section of particular relevance to the commercialization of bioproducts, but a warning is necessary. Since this section covers the research problems which are of greatest concern and in need of attention in order to rapidly commercialize biotechnology, it may well be the section where the knowledge base changes the most in the next decade. Only time will tell!

2
Transport Phenomena in Bioprocesses

F. KARGI
Lehigh University, Bethlehem, PA, USA
and
M. MOO-YOUNG
University of Waterloo, Waterloo, Ontario, Canada

2.1 INTRODUCTION TO TRANSPORT PHENOMENA

The term transport phenomena refers to the various modes of mass, heat and momentum transfer in process operations. As such, the topic is important in virtually all aspects of engineering considerations of biotechnology: feedstock upstream pretreatment, product recovery downstream processing, and the design, operation and control of the bioreactor itself. In this chapter, we will address the basics of transport phenomena, especially mass transfer of nutrient oxygen in fermentation systems, and heat transfer by analogy, especially with respect to bioreactor design. These two types of transport phenomena are inevitably related to momentum transfer in practical systems and the latter is discussed in this context, at least at the elementary level. Elaborations of this treatment are given in the other chapters of this volume.

Transport of reactants to and from the bulk liquid phase to the near vicinity of biocatalysts (*e.g.* microbial cells, immobilized enzymes, *etc.*) can have a significant influence on the performance of biological systems. For example, when the rate of transport of nutrients to microbial cells is slower than the rate of nutrient utilization, the performance of the biocatalysts would be limited by the rate of nutrient transport. Thus, in many biological processes (*e.g.* certain mycelial and polysaccharide fermentations), the rate of product formation can be improved by increasing the rate of transport of a limiting nutrient.

In many aerobic fermentation systems the rate of oxygen transfer to the cells is the limiting factor which determines the rate of biological conversion. The availability of oxygen for microbial use depends upon the solubility and mass transfer rate of oxygen in the fermentation broth and the rate of microbial utilization of the dissolved oxygen. In order to improve the rate of biological conversion, it is sometimes desirable to operate a fermentation process at a high biocatalyst (cell) concentration. However, a high density of active cells may cause rapid depletion of dissolved oxygen in the nutrient medium and therefore oxygen consumption rate may exceed the rate of oxygen transfer. In such cases, oxygen transfer from the gas phase into the liquid fermentation broth is the rate limiting factor and the rate of oxygen transfer needs to be enhanced in order to improve the rate of biological conversion.

In mycelial fermentation processes organisms grow in the form of pellets and the soluble nutrients must diffuse through the pellets to be available for utilization by the organisms growing inside the pellet. In this case, nutrient diffusion and utilization are simultaneous. Since the mycelial cells growing inside the pellet will be exposed to a nutrient concentration lower than that of the bulk fluid, both product formation and microbial growth rates inside the pellet will be lower than in the case of single cell growth. In such cases, the size of the microbial pellet has to be small enough to overcome diffusion limitations on the microbial growth.

The examples of microbial growth coupled with mass transfer can be extended beyond the aforementioned situations. This chapter covers the mass transfer effects in fermentation processes. Physical and physicochemical aspects of nutrient transfer are considered. The transfer and microbial utilization of oxygen are emphasized since oxygen is a sparingly soluble nutrient limiting the rate of microbial activities in many aerobic fermentation systems. Heat transfer in biological processes is briefly covered at the end of the chapter.

2.1.1 Mass Transfer by Molecular Diffusion

Mass transfer in a quiescent liquid or a gas mixture takes place by molecular diffusion. The driving force in molecular diffusion can be concentration, pressure, temperature gradients or external forces, *e.g.* electrostatic or gravitational (forced diffusion). The diffusion due to concentration gradient is known as ordinary diffusion. In a quiescent gas or liquid, if there is a region in which the concentration of component A is higher than that of the bulk fluid, then the species A will diffuse from the high concentration region to the low concentration regions. The rate of mass transfer due to a concentration gradient is defined by Fick's law of diffusion which states that the mass flux is proportional to the concentration gradient, that is

$$J_{AZ} = - D_{AB} \frac{dC_A}{dZ} \tag{1}$$

where D_{AB} is the diffusion coefficient of A in a binary mixture of A and B. Equation (1) is valid for isobaric and isothermal systems and is usually used to determine mass fluxes in dilute liquid and gas mixtures.

2.1.1.1 Estimation of diffusion coefficients

Since the transfer of gaseous nutrients such as oxygen into a fermentation broth involves mass transfer in both gas and liquid phases, the estimation of diffusion coefficients in both gas and liquid phases is of significant importance in microbial processes.

The diffusivity of A in a non-polar binary gas mixture at low densities can be estimated using the kinetic theory of gases (Bird, 1960; Sherwood, 1975; Treybal, 1980). Assuming that gas molecules A and B are rigid spheres of unequal mass and diameter, the following equation can be derived from the kinetic theory of gases (equation 2; Bird, 1960).

$$D_{AB} = \frac{2}{3} \left(\frac{k}{\pi} \right)^{3/2} \left(\frac{1}{2m_A} + \frac{1}{2m_B} \right)^{1/2} \frac{T^{3/2}}{p \left(\frac{d_A + d_B}{2} \right)^2} \tag{2}$$

In equation (2) k = Boltzmann's constant (1.38×10^{-23} J K^{-1}); m_A, m_B = the masses of molecules A and B, respectively; d_A, d_B = the molecular diameters of A and B, respectively; T = absolute temperature (K); p = absolute pressure (atm). Equation (2) predicts that the mass diffusivity, D_{AB}, is directly proportional to $T^{3/2}$ and is inversely proportional to the total pressure, which agrees well with the experimental data up to 10 atm pressure.

For a more accurate estimation of diffusivities in gas mixtures Hirschfelder, Bird and Spotz used Chapman–Enskog kinetic theory and developed the following equation which is only valid for non-polar, non-reacting binary gas mixtures (equation 3; Hirschfelder, 1949).

$$D_{AB} = \frac{1.084 \times 10^{-4} T^{3/2} \left(\frac{1}{M_A} + \frac{1}{M_B} \right)^{1/2}}{p r_{AB}^2 f(kT/\varepsilon_{AB})} \tag{3}$$

Wilke and Lee (1955) modified the Hirschfelder equation by considering a binary gas mixture of a polar and a non-polar gas (equation 4).

$$D_{AB} = \frac{10^{-4}[1.084 - 0.249(1/M_A + 1/M_B)^{1/2}]T^{3/2} \left(\frac{1}{M_A} + \frac{1}{M_B} \right)^{1/2}}{p r_{AB}^2 f(kT/\varepsilon_{AB})} \tag{4}$$

In equation (4) D_{AB} = diffusivity (m^2 s^{-1}); T = absolute temperature (K); M_A, M_B = the molecular weights of A and B, respectively (kg kmol^{-1}); p = absolute pressure (N m^{-2}); r_{AB} = molecular separation at collision = $(r_A + r_B)/2$ (nm); ε_{AB} = energy of molecular attraction = $(\varepsilon_A \varepsilon_B)^{1/2}$; k = Boltzmann's constant (1.38×10^{-23} J K^{-1}); $f(kT/\varepsilon_{AB})$ = collision function (given in Treybal, 1980, p. 32).

The values of r and ε for each gas component can be estimated empirically (equations 5 and 5a; Treybal, 1980)

$$r_A = 1.18 V_A^{1/3} \tag{5}$$

$$\varepsilon_A/k = 1.21 T_{b,A} \tag{5a}$$

where V_A is the molal volume of liquid A at its normal boiling point (m³ kmol⁻¹) and $T_{b,A}$ is the normal boiling point of A (K). The values of ε/k and r for gases are given in Treybal (1980; Table 2.2).

Theoretical estimation of diffusivities in liquids cannot be made as accurately as for gases, since there is no well developed theory for the structure of liquids. However, for dilute, non-electrolyte solutions, the empirical correlation of Wilke and Chang (1949, 1955) can be used to estimate diffusivities in low molecular weight solvents within 10% error (equation 6).

$$D_{AB} = \frac{117.3 \times 10^{-18}(\varphi' M_B)^{0.5} T}{\mu V_A^{0.6}} \tag{6}$$

In equation (6) D_{AB} = diffusivity of A in solvent B (m² s⁻¹); M_B = molecular weight of solvent B (kg kmol⁻¹); T = temperature (K); φ' = association factor for solvent (2.26 for water); μ = viscosity of solution (kg m⁻¹ s⁻¹); V_A = molal volume of solvent at its normal boiling point (m³ kmol⁻¹; 0.0756 for water).

The viscosity of a solution may have a profound effect on the diffusivity of the solute. Assuming that the solute–solvent interactions are not affected by changes in viscosity, equation (7) can be used to correct D_{AB} for changes in solution viscosity

$$\frac{D_{AB}}{D_{AB,ref}} = \frac{\mu_{ref}}{\mu} \tag{7}$$

where $D_{AB,ref}$ and μ_{ref} are the diffusivity of the solute in a reference solvent and the viscosity of the reference solvent (such as water), respectively.

The diffusivity of soluble nutrients in microbial flocs, for example, is significantly less than that in water. The diffusivity of oxygen in microbial flocs varies according to the type and age of microbial cells and also the C/N ratio in the fermentation broth. The diffusivity of oxygen in microbial flocs may be as low as 1 to 2% of the diffusivity of oxygen in water (Bailey, 1977).

2.1.2 Mass Transfer by Convection

The term 'convective mass transfer' is used to describe mass transfer associated with fluid flow and it involves the transfer of mass between a moving fluid and a boundary surface or between two immiscible moving fluids. The mass transfer coefficient for convective mass transfer is defined as

$$k_c = \frac{N_A}{\Delta C_A} \tag{8}$$

where N_A is the molar flux of A (mol m⁻² s⁻¹), ΔC_A is the concentration difference between two points in the same phase or between two phases (mol m⁻³), and k_c is the mass transfer coefficient based upon molar concentration of A (m s⁻¹).

Convective mass transfer may take place in two different forms: forced convection and natural (or free) convection. In forced convection fluid flow is driven by an external force such as a pump or a fan, whereas in natural convection, the driving force for fluid flow is the presence of buoyancy forces which result from the presence of density differences. The cause for density difference can be temperature or concentration differences.

In agitated fermentation vessels, mass transfer takes place by forced convection. Whereas transfer of oxygen from the bulk gas phase to microbial cells in the liquid phase is by forced convection in agitated fermenters, the transfer of oxygen through microbial pellets is mainly by diffusion. Oxygen transfer by natural convection may play an important role in fermentation processes where there are large temperature differences between different regions in the fermenter in the absence of significant bulk mixing. The biological trickling filters used in wastewater treatment or static solid substrate fermentation beds are such reactors where mass transfer by free convection is important.

2.1.2.1 Dimensionless groups

The mass transfer coefficient k_c varies with fluid properties, flow conditions and geometry of the physical system. These parameters are often combined in the form of dimensionless groups. The major dimensionless groups involved in convective mass transfer are:

(a) Sherwood number:

$$Sh = \frac{k_c L}{D_{AB}} \tag{9}$$

where L is the characteristic length of the system. The Sherwood number is a measure of the mass transfer coefficient.

(b) Schmidt number:

$$Sc = \frac{\mu}{\varrho D_{AB}} = \frac{\nu}{D_{AB}} \tag{10}$$

where μ and ϱ are, respectively, the viscosity and density of the fluid. The Schmidt number is a measure of the fluid properties and is only a function of temperature and fluid composition at constant pressure.

(c) Reynolds number:

$$Re = \frac{LV\varrho}{\mu} \tag{11}$$

where L and V are the characteristic length of the system and the linear fluid velocity, respectively. The Reynolds number is a measure of the physical properties of the fluid and flow velocity, and it determines the flow regime. Usually the flow regime is called laminar when $Re<2300$, turbulent when $Re>10^4$ and transition when $2300<Re<10^4$.

(d) Grashof number:

$$Gr = \frac{L^3 g \Delta \varrho_A}{\varrho \nu^2} \tag{12}$$

where g is gravitational acceleration, $\Delta \varrho_A$ is the difference in density of A between two points in the transfer field and ν is the kinematic viscosity of the fluid. The Grashof number is encountered in mass transfer by natural convection and is ac will have access may be particularly important in ensuring accessibility for the disabled. According to circumstances appropriate provision may include:
ler tip speed (V_i) which is proportional to ND_i, where N is the impeller rotation rate (revolutions per minute), and D_i is the impeller diameter. Therefore,

$$Re_i = \frac{\varrho D_i V_i}{\mu} = \frac{\varrho N D_i^2}{\mu} \tag{13}$$

The impeller diameter (D_i) is considered as the characteristic length in equation (13).

The Froude number is another dimensionless number which should be considered in unbaffled fermentation tanks and is defined as

$$Fr_i = \frac{V_i^2}{g D_i} = \frac{N^2 D_i}{g} \tag{14}$$

The Froude number is the ratio of inertial forces to gravitational forces acting on a fluid element. In baffled fermentation tanks, where vortex formation is negligible, the Froude number is insignificant and only Re_i is considered to describe fluid hydrodynamics.

The Reynolds number in agitated fermenters can be based upon the root mean square (rms) fluid velocity (V_{rms}) and average diameter of gas bubbles ($\bar{D}_b$)

$$Re = \frac{\varrho \bar{D}_b V_{rms}}{\mu} \tag{15}$$

The V_{rms} can be related to the power input per unit volume of fermenter broth (P/V) and the diameter of gas bubbles, that is

$$V_{rms} = k_1 \left(\frac{P}{V} \frac{D_b}{\varrho} \right)^{1/3} \tag{16}$$

where k_1 is a constant. The V_{rms} value used in equation (15) should be time averaged since local fluid conditions vary with time in agitated vessels. The rms velocity, $V_{rms} \equiv [V^2(t)]^{1/2}$ is a measure of the average magnitude of the local velocity variations (Bailey, 1977).

2.1.2.2 Mass transfer coefficient correlations

The mass transfer coefficient is usually correlated with the Reynolds number and the Schmidt number in forced convection as follows:

$$\frac{k_c L}{D_{AB}} = a\left(\frac{LV\varrho}{\mu}\right)^b\left(\frac{\mu}{\varrho D_{AB}}\right)^c \tag{17}$$

or

$$Sh = aRe^b Sc^c \tag{17a}$$

where L is the characteristic length of the system.

The constants a, b and c are usually determined experimentally although there are theories which could be used to predict the values of these constants, for example according to the laminar boundary layer theory equation (17) is

$$\frac{k_c L}{D_{AB}} = 0.664\left(\frac{LV\varrho}{\mu}\right)^{1/2}\left(\frac{\mu}{\varrho D_{AB}}\right)^{1/3} \tag{18}$$

Several empirical and theoretical correlations for the mass transfer coefficient (k_c) for various physical systems are given in Welty *et al.* (1976).

In agitated gas–liquid fermenters, the mass transfer coefficient is correlated with the impeller Reynolds number (Re_i) and the Schmidt number (Sc), that is

$$Sh = f(Re_i, Sc) \tag{19}$$

The transfer of sparingly soluble gases (such as oxygen) into liquid media is controlled by the liquid film resistance near the gas bubble and therefore the liquid side mass transfer coefficient, k_L, is correlated with Re_i and Sc for such cases. However, transfer of highly soluble gases (such as NH_3) into liquid media is controlled by gas film resistance and the gas side mass transfer coefficient, k_G, is correlated with Re describing bubble hydrodynamics and the gas phase Schmidt number.

Calderbank (1967) has correlated the oxygen transfer coefficient with Re and Sc for turbulent aeration as follows:

$$Sh = 0.13Sc^{1/3}Re^{3/4} \tag{20}$$

Among other gas transfer coefficient correlations are the equations developed by Higbie and Froessling (Blakebrough, 1967).

Higbie's correlation: $Sh = 1.13Re^{1/2}Sc^{1/2}$ $\tag{21}$

Froessling's correlation: $Sh = 2.0 + 0.552\,Re^{1/2}Sc^{1/3}$ $\tag{22}$

Several theories have been developed for the estimation of mass transfer coefficients. Among these are the film theory (Whitman, 1923), the penetration theory (Higbie, 1935; Danckwerts, 1951), and the surface renewal theory (Danckwerts, 1951). Detailed descriptions of these theories can be found in Astarita (1967), Danckwerts (1970), Treybal (1980) and Welty (1976). The film and surface renewal theories describe two extreme cases of gas transfer into liquid media. At low Sc and low rate of surface renewal (*i.e.* low degree of turbulence) the film theory gives better correlation with experimental data, whereas for high Sc and a high degree of turbulence, the surface renewal theory is in better agreement with experimental findings.

Mass transfer coefficient correlations for natural convection involve the Grashof number (Gr_{AB}) instead of Re and have the following form:

$$Sh = f(Gr_{AB}, Sc) \tag{23}$$

or

$$\frac{k_c L}{D_{AB}} = a'\left(\frac{L^3 g\Delta\varrho_A}{\varrho v^2}\right)^{b'}\left(\frac{\mu}{\varrho D_{AB}}\right)^{c'} \tag{23a}$$

The constants in equation (23) are determined experimentally.

Mass transfer coefficients for gases such as oxygen in agitated fermenters are correlated with the power input per unit volume of fermenter (P/V) and the linear superficial gas velocity (V_s) (Aiba, 1973; Wang, 1979). These correlations for oxygen transfer are considered in Section 2.3.3.

2.1.3   Interphase Mass Transfer

Many chemical and bioconversion processes involve the transfer of material between two phases, such as the transfer of oxygen from the gas phase into the liquid broth during fermentation. In mycelial fermentations in which the liquid medium is not well mixed, the surface of the mycelial pellets is covered by a stagnant liquid film through which the nutrients must diffuse in order to be used by the microorganisms.

In this section, interphase mass transfer is considered from the standpoint of gas–liquid mass transfer since the transfer of oxygen from the gas to the liquid phase has a profound effect on the performance of aerobic fermenters especially at high biomass concentrations.

2.1.3.1   *Gas–liquid mass transfer*

The most widely used theory to describe gas–liquid mass transfer is the 'two-film' or 'two-resistance' theory developed by Whitman (1923). Transfer of gaseous components from the bulk gas phase into the liquid phase involves three transfer steps: transfer from the gas phase to the gas–liquid interface, transfer across the interface into the liquid film, and transfer from the liquid film to the bulk liquid phase. The two-film theory is based upon the following assumptions: (1) there are two films on each side of the interface (gas film on the gas side and liquid film on the liquid side) and the rate of mass transfer is controlled by the rates of diffusion through gas and liquid films; and (2) the interfacial resistance for mass transfer is negligible compared to gas and liquid film resistances. Figure 1 depicts the transfer of component A from the gas phase to the liquid phase by diffusion through gas and liquid films. Since no interfacial resistance is assumed, P_{Ai} and C_{Ai} are equilibrium concentrations given by the system's equilibrium-distribution curve. For dilute liquid solutions these equilibrium concentrations are related by Henry's law as follows:

$$P_{Ai} = H'C_{Ai} \tag{24}$$

where H' is the Henry's law constant.

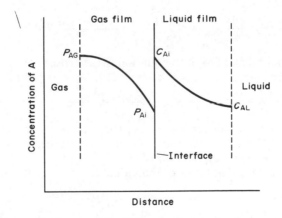

Figure 1   Concentration profiles in gas and liquid films for the transfer of gaseous A into the liquid phase

For the steady-state transfer of component A from the gas to the liquid phase the following equations for the gas and liquid films can be written:

$$N_A = k_G(P_{AG} - P_{Ai}) \tag{25}$$

and

$$N_A = k_L(C_{Ai} - C_{AL}) \tag{26}$$

where k_G and k_L are the convective mass transfer coefficients for the gas and liquid phases respectively; P_{AG} is the partial pressure of A in the bulk gas phase; C_{AL} is the concentration of A in the liquid phase; and N_A is the rate of mass transfer of A per unit transfer area (*i.e.* mass flux of A).

Under steady-state conditions, the flux of A through the gas film is equal to the flux through the liquid film. Therefore,

$$N_A = k_G(P_{AG} - P_{Ai}) = k_L(C_{Ai} - C_{AL}) \tag{27}$$

Equations (25) and (26) are not suitable for calculating the mass flux of A since it is difficult to measure the interfacial partial pressure and concentration of A. In general, it is more convenient to employ an overall mass transfer coefficient based on the overall driving force between the gas and liquid phase compositions. The overall mass transfer coefficient based on the driving force in the gas phase (K_G) is defined as

$$K_G \equiv \frac{N_A}{P_{AG} - P_A^*} \tag{28}$$

where P_A^* is the partial pressure of A in equilibrium with the liquid phase concentration of A, C_{AL}. The overall mass transfer coefficient based on the liquid phase driving force (K_L) is defined as

$$K_L \equiv \frac{N_A}{C_A^* - C_{AL}} \tag{29}$$

where C_A^* is the concentration of A in equilibrium with the partial pressure of A in the gas phase and therefore is a measure of P_{AG}.

For low concentrations of A in the liquid phase the equilibrium relationship is approximately linear since Henry's law is obeyed. Therefore

$$\begin{aligned} P_{AG} &= mC_A^* \\ P_A^* &= mC_{AL} \\ P_{Ai} &= mC_{Ai} \end{aligned} \tag{30}$$

where m is the equilibrium constant between gas and liquid phase concentrations.

Rearranging equation (28) yields

$$\frac{1}{K_G} = \frac{P_{AG} - P_A^*}{N_A} = \frac{P_{AG} - P_{Ai}}{N_A} + \frac{P_{Ai} - P_A^*}{N_A} \tag{31}$$

or

$$\frac{1}{K_G} = \frac{P_{AG} - P_{Ai}}{N_A} + \frac{m(C_{Ai} - C_{AL})}{N_A} \tag{31a}$$

Substituting equations (25) and (26) into equation (31a) yields

$$\frac{1}{K_G} = \frac{1}{k_G} + \frac{m}{k_L} \tag{32}$$

where $1/K_G$ is the overall mass transfer resistance, $1/k_G$ and m/k_L are the gas film and liquid film resistances, respectively. Therefore, the total mass transfer resistance is equal to the sum of the individual resistances of the liquid and gas films.

Similarly, K_L can be expressed in terms of individual transfer coefficients as follows:

$$\frac{1}{K_L} = \frac{1}{mk_G} + \frac{1}{k_L} \tag{33}$$

The relative magnitudes of the individual phase resistances depend on the solubility of the gas. If the solubility of the gas in the liquid phase is high, such as ammonia in water, then m is low and the liquid phase resistance is negligible compared to the gas phase resistance. In this case, the overall mass transfer coefficient is essentially equal to the gas phase transfer coefficient and transfer of the gaseous component A is gas phase controlled. For sparingly soluble gases, such as oxygen and carbon dioxide in water, the value of m is large and the gas phase resistance is negligible compared to the liquid phase resistance. The overall mass transfer coefficient, K_L, is approximately equal to the individual liquid phase coefficient k_L, and mass transfer is liquid phase controlled for such systems. When the gaseous component is moderately soluble in the liquid phase, both resistances should be considered in the evaluation of the total resistance.

The individual coefficients k_L and k_G are dependent on the properties of the liquid and gas phases, flow conditions in the phases, and the nature of the diffusing component. The coefficients k_L and k_G can be determined using the correlations given in Section 2.1.2 and the overall mass transfer coefficient by using equations (32) and (33).

2.1.4 Intraparticle Diffusion

Intraparticle diffusion is particularly important for the cases when biological catalysts (*i.e.* whole cells or enzymes) are immobilized inside a porous catalyst pellet. In a well-mixed bioreactor containing biocatalysts immobilized on porous inert particles, the reactants have to diffuse through tortuous void passages inside the support particles to be available to the organism. In order to describe the diffusion inside the tortuous void passages, an average or effective diffusion coefficient is used which is usually determined experimentally.

Consider a well-mixed reactor containing porous support particles on which microbial cells are immobilized. The reactor is fed by a nutrient solution containing substrate (S) and the effluent contains the product (P). If the reactor is well mixed, the concentration of substrate on the surface of a catalyst particle (C_{ss}) may be assumed to be constant. Substrate diffuses through the tortuous passages in the catalyst and is utilized by the immobilized cells simultaneously. The product P then diffuses back to the liquid medium. Since the concentration of substrate inside the porous pellet will be lower than that in the liquid medium and will vary with the radial distance, the rate of bioconversion will be lower than would be obtained under homogeneous liquid phase conditions. This reduction in the rate of bioconversion is described by the effectiveness factor which is the ratio of the rate of reaction with diffusion to that without diffusion, therefore the reaction rate obtained under homogeneous reaction conditions (*e.g.* a well-mixed liquid medium) should be multiplied by the effectiveness factor to determine the rate in the presence of diffusion.

For a single spherical support particle containing microbial cells inside the pores, the diffusion and utilization of substrate inside the particle is described by the following equation (Bird, 1960):

$$D_s \frac{1}{r^2} \frac{d}{dr} \left(r^2 \frac{dC_s}{dr} \right) = k_1 a' C_s \tag{34}$$

where D_s is the effective diffusivity of the limiting substrate, C_s is the concentration of the substrate, a' is the surface area occupied by cells per unit volume of support particle and k_1 is the first order rate constant. It is assumed that substrate concentration inside the pores is small enough for Monod kinetics to be approximated to a first-order reaction rate. With the boundary conditions of $C_s = C_{ss}$ at $r = R$ and $dC_s/dr = 0$ at $r = 0$, the solution to equation (34) is

$$\frac{C_s}{C_{ss}} = \frac{R}{r} \frac{\sinh (k_1 a'/D_s)^{1/2} r}{\sinh (k_1 a'/D_s)^{1/2} R} \tag{35}$$

where R is the radius of a spherical support particle.

At steady state, the rate of substrate utilization inside the particle is equal to the rate of substrate diffusion through the outer surface of the particle. Therefore, the rate of consumption of substrate in a single particle is

$$m_s = - 4\pi R^2 D_s \frac{dC_s}{dr} \bigg|_{r = R} \tag{36}$$

Using equation (35), we find that

$$m_s = 4\pi R D_s C_{ss} \left[1 - \left(\frac{k_1 a'}{D_s} \right)^{1/2} R \coth \left(\frac{k_1 a'}{D_s} \right)^{1/2} R \right] \tag{37}$$

In the absence of diffusion (*i.e.* if the surfaces of the pores are exposed to the concentration C_{ss}) the rate of substrate utilization would be

$$m_{so} = (4\pi R^3)(-k_1 a' C_{ss}) \tag{38}$$

The effectiveness factor (η) is defined as the ratio m_s/m_{so} and is given by the following equation

$$\eta_s = \frac{3}{\phi^2} (\phi \cot \phi - 1) \tag{39}$$

where $\phi = R(k_1 a'/D_s)^{1/2}$ which is called the Thiele modulus.

Therefore, the rate of consumption of substrate in a single support particle can be expressed as

$$m_s = \left(\frac{4}{3} \pi R^3 \right) (-k_1 a' C_{ss}) \eta \tag{40}$$

The effectiveness factor (η) varies with the Thiele modulus (ϕ). For low ϕ values, $\phi < 0.1$ (*i.e.*

small particle size), the effectiveness factor is approximately equal to one, that is diffusion does not affect the rate of substrate consumption when the particle size is small enough. However, at large support particle sizes (*i.e.* $\phi > 1$) the effectiveness factor is less than one and the rate of substrate consumption is reduced by a factor of η compared to substrate utilization in the absence of diffusion.

The foregoing analysis can be extended to non-spherical particles. For irregularly shaped support particles, the equivalent radius is defined as

$$R_{eq} = 3\frac{V_p}{A_p} \tag{41}$$

where V_p and A_p are the volume and external surface area of a single support particle. In this case equation (40) becomes

$$m_s = V_p(-k_1a'C_{ss})\eta \tag{42}$$

and the effectiveness factor is

$$\eta = \frac{1}{3\phi^2}(3\phi \coth 3\phi - 1) \tag{43}$$

where $\phi = \dfrac{V_p}{A_p}\left(\dfrac{k_1a'}{D_s}\right)^{1/2}$.

A plot of η *versus* ϕ is presented in Figure 2 for support particles of different geometrical shapes.

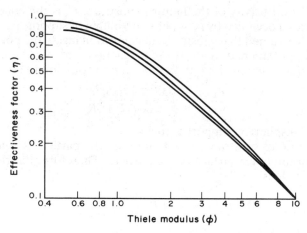

Figure 2   Effectiveness factor *versus* Thiele modulus for porous support particles: top curve, flat particles; middle curve, cylindrical particles; bottom curve, spherical particles

As to the selection of support particles for cell immobilization, particle size should be as small and void fraction inside particles should be as large as possible in order to minimize the adverse effects of diffusion limitations on reaction kinetics. The critical particle radius (below which diffusion limitation is negligible) depends on the kinetic constant of substrate utilization (k_1) and therefore on the type and physiological state of cells as well as on the diffusivity of rate limiting substrate inside the pores. The critical radius should be determined experimentally for a given biological system. The interested reader is directed to a more detailed review of the importance of intraparticle mass transfer in the design of biochemical reactors by Moo-Young and Blanch (1981).

2.1.5   Facilitated and Active Transport Across Cell Membrane

There are three major mechanisms of nutrient transport across cell membranes: passive transport, active transport and facilitated (or carrier mediated) transport. Passive transport is the transfer of nutrients by ordinary diffusion which was covered in Section 2.1.1. Passive transport takes place in the direction of the concentration gradient (*i.e.* from high to low concentration) and does not require external energy. Active transport is the diffusion of nutrients from low to high concentration (against the concentration gradient) and requires external (metabolic) energy. Facilitated transport is the transfer of nutrients across the cell membrane with the aid of a carrier molecule. Figure 3 depicts various mechanisms of membrane transport.

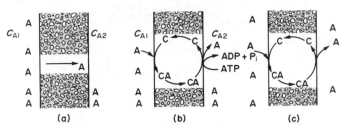

Figure 3 Major mechanisms of membrane transport: (a) passive transport ($C_{A1} > C_{A2}$); (b) active transport ($C_{A1} < C_{A2}$); (c) facilitated (mediated) transport

Biological membranes contain a phospholipid bilayer which is impermeable to polar molecules. This layer prevents ionized, internal metabolites from diffusing out of the cell. In order to transfer certain molecules such as glucose, Na^+ and K^+ ions across cell membranes, biological membranes have specific transport systems (Bittar, 1975; Boos, 1974).

The free energy (ΔG) change in the transport of a solute from a compartment of concentration C_1 to another of concentration C_2 during active transport is given by

$$\Delta G = R_G T \ln \frac{C_2}{C_1} \tag{44}$$

where R_G and T are the gas constant and absolute temperature, respectively. In active transport, $C_2 > C_1$ and $\Delta G > 0$, therefore the system gains free energy. However, in passive transport, $C_1 > C_2$ and $\Delta G < 0$, that is the system loses free energy and passive transport is spontaneous.

If the transferred component is electrically charged, then the free energy change is given by the following equation:

$$\Delta G = R_G T \ln \frac{C_2}{C_1} + ZF\Delta\psi \tag{45}$$

where Z is the number of charges on the transferred molecule, F is Faraday's constant (96 500 C/g equiv) and $\Delta\psi$ is the potential difference across the membrane (V). The first term in equation (45) is concentration (chemical) potential; the second term is the electrical potential gradient and the sum of these two terms is referred to as the electrochemical gradient. The electrical membrane potential is the major driving force in the function of muscle cells and in the membranes of mitochondria. Moreover, cells have an active transport mechanism to maintain proper concentrations of K^+, Na^+ and water within the cell. Usually, Na^+ is pumped out of the cell and K^+ is pumped in against the concentration gradients. This transport mechanism is driven by the energy released from the hydrolysis of ATP. An example of active transport is the transfer of H^+ from blood plasma (pH = 7.4) into the gastric juice of mammals (pH $\approx$ 1). The H^+ gradient is approximately $10^{-1}/10^{-7}$ and the free energy change for the transfer of 1.0 g equivalent of H^+ is $\Delta G \approx 33.61$ J. This energy is supplied from the hydrolysis of ATP (-30.5 kJ/mol ATP). There are mechanisms for the coupling of ATP hydrolysis to active transport of nutrients against concentration gradient (Lehninger, 1975; Stryer, 1975). Various mathematical models have been developed to elucidate the fundamental molecular mechanisms for active transport. Further details on active transport are given in Bittar (1975), Boos (1974), Kaback (1972) and Kotyk (1975).

Facilitated (or mediated) transport of molecules through cell membranes takes place with the aid of a carrier molecule (Figure 4). The transferred compound combines with a carrier molecule on the outside of the membrane and diffuses to the other side where the complex dissociates releasing the carried molecule inside the cell. One of the distinguishing characteristics of facilitated transport is that the rate of transfer exhibits saturation kinetics such as Michaelis–Menten kinetics in enzyme reactions. At low concentrations of the compound transferred, the rate increases linearly with concentration; however, at high concentrations the rate reaches a saturation level, whereas in non-mediated transport the rate increases linearly with the concentration of the compound transferred and does not reach a saturation level. Figure 4 is a comparison of mediated (facilitated) and non-mediated transport.

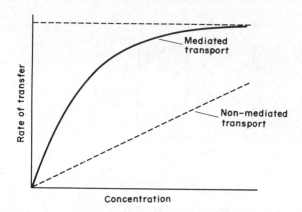

Figure 4   A comparison of mediated (facilitated) and non-mediated membrane transport

Mediated transport has a large degree of specificity for the substance transferred. The erythrocytes of certain vertebrates, for example, have a mediated transport system in their membranes which is specific to D-glucose and other structurally related monosaccharides but has very little activity toward disaccharides and D-fructose. Mediated transport can easily be inhibited by structural analogs of transferred compounds. Because of these properties of mediated transport, it is believed that carrier molecules are proteins which are called permeases or translocases. The substrate–carrier complex undergoes either translational or rotational diffusion within the membrane.

The mediated membrane transport may be either active or passive depending on the criterion used. For example, the mediated transfer of glucose into human erythrocytes requires metabolic energy, however it is not inhibited in the absence of metabolic energy and therefore is called passive mediated transport. The active mediated transport is unidirectional, such as the active transport system of erythrocytes for Na^+ and K^+ which moves K^+ into the cell and Na^+ out of the cell, whereas the passive mediated transport of glucose may be in either direction (*i.e.* it is reversible) and the direction of transport depends on the relative concentrations of glucose on both sides of the membrane. In the absence of metabolic energy, some active transport systems may function as passive transport systems. For this reason, it is believed that those active transport systems contain two components: one couples the action of the carrier molecule to an energy source and the other recognizes and carries the substrate in either direction.

2.2   OXYGEN TRANSFER AND UTILIZATION IN GASSED MICROBIAL SYSTEMS

In aerobic fermentation systems, the transfer of oxygen from the gas phase to the surface of a single microbial cell is of primary importance, especially at high cell densities when microbial growth is likely to be limited by the availability of oxygen in the liquid phase. In many fermentation systems involving high oxygen demand (such as hydrocarbon fermentations) special fermenter designs are required to enhance the rate of transfer of oxygen. The design of a fermenter for optimum oxygen transfer requires a thorough understanding of the transfer and utilization of oxygen in microbial systems. For further details on oxygen transfer in biochemical reactors, the reader is referred to the review by Moo-Young and Blanch (1981).

2.2.1   Mass Transfer Resistances

Oxygen is usually supplied to the fermentation medium by sparging air bubbles underneath the impeller of an agitated fermenter. Oxygen from a rising air bubble is first dissolved in the liquid medium and then transferred to the oxygen utilization sites inside the cell. The process of oxygen transfer may be described in terms of individual mass transfer resistances by considering the film model. Figure 5 depicts a typical pathway for oxygen transfer from a gas bubble to a single microbial cell. There are seven major mass transfer resistances in oxygen transfer to microbial cells (Arnold, 1958) which can be summarized as follows: (1) gas film resistance between the bulk gas and gas–liquid interface; (2) interfacial resistance at the gas–liquid interface; (3) liquid film resistance between the interface and bulk liquid phase; (4) liquid phase resistance for the transfer of oxygen to the liquid film surrounding single microbial cells; (5) liquid film resistance around cells; (6) intracellular or intrapellet resistance (in the case of microbial flocs or mycelial pellets); and (7) resistance due to consumption of oxygen inside a microbial cell.

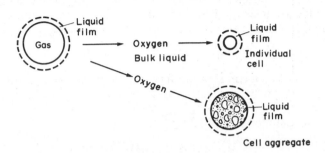

Figure 5   Major mass transfer resistances for the transfer of oxygen from air to microbial cells

The overall oxygen transfer resistance is equal to the sum of the individual resistances. The relative magnitudes of the individual resistances depend on bubble and liquid phase hydrodynamics, the composition and rheological properties of the fermentation medium, the density and activity of microbial cells, and the gas–liquid interfacial phenomena. In general, the following statements can be made regarding the magnitude of certain resistances: (a) the gas film resistance is usually negligible compared to other resistances since the volume of a gas bubble is relatively small and gas phase is well mixed; and (b) the gas–liquid interfacial resistance is also negligible since the concentration of oxygen at the bubble interface is almost the same as that inside the bubble.

Therefore, the major supply side resistance is the liquid film resistance around the gas bubbles and the overall oxygen transfer coefficient (K_L) for oxygen transfer from air bubbles into bulk liquid is then approximately equal to the liquid film transfer coefficient (k_L).

In a well-mixed fermenter, the concentration of dissolved oxygen in the bulk liquid phase is constant and concentration gradients in the bulk liquid are assumed to be negligible. However, in some fermentation processes, perfect mixing may be difficult to achieve. If the fermentation broth is non-Newtonian (such as in mycelial fermentations), there may be significant concentration gradients within the bulk liquid and therefore oxygen transfer resistance in the bulk liquid may not be negligible. One of the major goals in aerobic fermenter design is to enhance bulk fluid mixing to reduce the aforementioned transfer resistances.

Since microbial cells are much smaller than air bubbles the liquid film around a single cell has negligible resistance to the diffusion of oxygen. Although hydrodynamic conditions surrounding the cells and air bubbles are the same, the liquid film resistance around a single cell is much smaller than that around a gas bubble. Another indication that liquid film resistance around a single cell is negligible is the fact that above a critical dissolved oxygen concentration the oxygen uptake by cells is independent of dissolved oxygen concentration (Finn, 1954). However, if microbial growth is in the form of pellets (or cell clumps as in mycelial fermentations) then the liquid film resistance around the cell clumps may be significant since the pellets are much larger than single cells.

Intracellular oxygen transfer resistance is usually negligible compared to other resistances because of the small size of cells and the proximity of organelles inside cells. When the cells form pellets (or clumps), the intrapellet resistance may be important since oxygen has to diffuse through to be available to the cells growing inside the pellet. The size of the clumps should be small enough to avoid anaerobic regions in the interior. The critical size of the clumps depends upon the rate of consumption of oxygen, diffusivity of oxygen and the concentration of oxygen in the bulk liquid.

The rate of consumption of oxygen within the organism may be the limiting factor in microbial utilization of oxygen when the rate of oxygen transfer is higher than the rate of biological oxygen consumption. The specific rate of oxygen consumption is a hyperbolic function of the dissolved oxygen concentration in the liquid medium. If the dissolved oxygen concentration is above the critical value, which is usually 10% of the solubility of oxygen, then the specific rate of oxygen consumption is independent of the dissolved oxygen concentration and is a constant.

2.2.2   Rate of Oxygen Transfer and Utilization

The dissolved oxygen (DO) concentration in a fermentation broth has a profound effect on the performance of aerobic fermentation systems. The dissolved oxygen concentration depends on the relative rates of oxygen transfer and utilization. Assuming that oxygen transfer from gas to liquid phase is controlled by the liquid film resistance around air bubbles, the rate of oxygen transfer is given by the following equation

$$n_{O_2,T} = k_L a(C^* - C_L) \qquad (46)$$

where k_L is the liquid film oxygen transfer coefficient, a is the gas–liquid interfacial area per unit volume of liquid, C^* and C_L are the saturation and actual dissolved oxygen concentrations in the liquid medium, respectively.

The rate of oxygen consumption is

$$n_{O_2,C} = Q_{O_2}X \qquad (47)$$

where Q_{O_2} is the specific rate of oxygen consumption averaged over a volume of liquid and X is the biomass concentration in the fermentation broth. The rate of oxygen consumption can be related to the rate of microbial growth using a yield coefficient, Y_{O_2}, that is

$$Q_{O_2} = \frac{\mu}{Y_{O_2}} \qquad (48)$$

where μ is the specific rate of microbial growth and Y_{O_2} is the ratio of cell mass formed per mole of oxygen consumed. Substituting equation (48) into equation (47) yields

$$n_{O_2,C} = \frac{\mu X}{Y_{O_2}} \qquad (49)$$

At steady state, the oxygen transfer and consumption rates are equal, that is

$$k_L a(C^* - C_L) = \frac{\mu X}{Y_{O_2}} = Q_{O_2}X \qquad (50)$$

The rate of oxygen transfer is maximum when $C_L = 0$ and the maximum rate of oxygen consumption is achieved when $\mu = \mu_{max}$. If $k_L a C^*$ is smaller than $\mu_{max}X/Y_{O_2}$, then oxygen consumption is limited by the rate of oxygen transfer. When the reverse inequality holds, the reaction is limited by microbial consumption of oxygen.

Usually, the specific rate of oxygen consumption (Q_{O_2}) varies with dissolved oxygen concentration according to the Monod equation, therefore equation (50) can be rewritten as

$$k_L a(C^* - C_L) = X Q_{O_2,max} \frac{C_L}{K_{O_2} + C_L} \qquad (51)$$

Assuming that $C_L \ll C^*$, equation (51) can be rearranged as

$$C_L = C^* \frac{K_{O_2}k_L a/(Q_{O_2,max}X)}{1 - C^* k_L a/(Q_{O_2,max}X)} \qquad (52)$$

The dissolved oxygen (DO) concentration, C_L, given in equation (52) has to be greater than the critical value of DO in order to avoid oxygen limitation. Aerobic biological reactors are designed to maximize the value of $k_L a$ in order to keep the value of C_L above the critical level. The critical dissolved oxygen concentration is in the order of 5–10% of the solubility of oxygen. The solubility of oxygen (or the saturation dissolved oxygen concentration) depends on the partial pressure of oxygen in the gas phase, temperature, and the composition of the liquid phase. The total rate of microbial oxygen consumption, $Q_{O_2}X$, is affected by the type and growth rate of microorganisms, pH, temperature, type of carbon nutrients and concentration of microorganisms in the fermentation broth.

In a continuous biological reactor, the cell concentration (X), $k_L a$ and Q_{max} are constant resulting in a constant dissolved oxygen concentration (C_L). However, in a batch culture the total rate of oxygen consumption ($Q_{O_2}X$) varies with time since cell concentration increases during the course of batch fermentation. The specific rate of oxygen consumption (Q_{O_2}) passes through a maximum in the early exponential phase of batch growth although X is small. Figure 6 depicts the variation of Q_{O_2} with time in a batch culture of *Myrotechium verrucaria*.

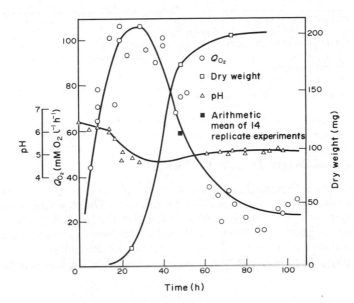

Figure 6 Variation of the rate of oxygen consumption during batch growth of *Myrotechium verrucaria* (Redrawn from Darby and Goddard, 1950)

Several methods have been developed to determine the rate of oxygen consumption (or oxygen demand, $Q_{O_2}X$). The Warburg respirometer is one of the earliest devices used to determine $Q_{O_2}X$. Umbreit *et al.* (1973) used a Warburg respirometer to determine the rate of oxygen consumption of a yeast culture in a shake flask and found that the value of $Q_{O_2}X$ varies with the rate of shaking.

An oxygen mass balance between the inlet and off-gas streams can also be used to determine the rate of oxygen consumption, provided that oxygen consumption is not limited by the rate of oxygen transfer:

$$Q_{O_2}X = \frac{1}{R_G V_L}\left[\left(\frac{F_G P_{O_2}}{T}\right)_i - \left(\frac{F_G P_{O_2}}{T}\right)_o\right] \qquad (53)$$

where F_G is the volumetric gas flow rate, P_{O_2} is the partial pressure of oxygen in the gas phase, T is the absolute temperature, V_L is the liquid volume in the fermenter and R_G is the ideal gas constant. The subscripts i and o indicate inlet and outlet gas streams.

The simplest method used to determine the value of $Q_{O_2}X$ is the dynamic method. This method requires only a dissolved oxygen electrode. A dynamic oxygen balance for oxygen in a batch culture is

$$\frac{dC_L}{dt} = k_L a(C^* - C_L) - Q_{O_2}X \qquad (54)$$

When the air supply is turned off, equation (54) becomes

$$\frac{dC_L}{dt} = -Q_{O_2}X \qquad (55)$$

Therefore, the value of $Q_{O_2}X$ can be determined by following the rate of decrease in dissolved oxygen (DO) concentration with the aid of a DO probe after the air supply is turned off. Section 2.2.3 should be referred to for the details of this method.

There are only a few studies on the locality of gas transfer in stirred vessels. Wilhelm *et al.* (1966) determined the contribution of gas transfer near the sparger–impeller region relative to that from bubbles distributed in the bulk liquid for both sparingly soluble gases (oxygen and chlorine in water) and highly soluble gases (chlorine in benzene). They found that in both cases the largest fraction of gas absorption takes place in the stirrer–sparger region. Table 1 summarizes their results indicating the percent gas absorption taking place in the stirrer–sparger domain.

Table 1   Percent Gas Absorption near Stirrer–Sparger Region[a]

System	Absorption (%)
Oxygen from air into water: 450 r.p.m.	~ 50
560 r.p.m.	~ 40
Oxygen from air into 0.5 N Na_2SO_3 solution: 450 r.p.m.	~ 45
560 r.p.m.	~ 65
Chlorine into water	~ 50
Chlorine into benzene	~100

[a] Gas flow rate: 3 l min^{-1}; liquid depth: 9 in.

2.2.3 Determination of Oxygen Transfer Coefficients

The oxygen transfer coefficient ($k_L a$) is an important parameter which is used to compare the oxygen transfer capabilities of various aerobic bioreactors. The $k_L a$ varies with the intensity of agitation, rate of aeration, gas–liquid interface area and rheological properties of the fermentation broth. Various methods have been developed to measure $k_L a$ values in aerobic fermenters. Each method has certain advantages and disadvantages over the others.

The volumetric oxygen transfer coefficient ($k_L a$) is defined by the following equation

$$k_L a \equiv \frac{n_{O_2}}{C^* - C_L} \tag{56}$$

In order to determine the value of $k_L a$, accurate methods must be developed for the measurement of the volumetric oxygen transfer rate (n_{O_2}). The principal methods used in the determination of $k_L a$ are discussed in the following sections.

2.2.3.1 *Sodium sulfite oxidation method*

This method employs the oxidation of sodium sulfite to sulfate by oxygen in the presence of copper or cobalt salts which act as catalysts. The sulfite oxidation method was originally developed by Cooper *et al.* (1944) to evaluate the performance of gas–liquid contactors.

$$Na_2SO_3 + \tfrac{1}{2}O_2 \xrightarrow{\text{catalyst}} Na_2SO_4 \tag{57}$$

The oxidation of SO_3^{2-} to SO_4^{2-} in the liquid phase is almost instantaneous and the rate is independent of the Na_2SO_3 and dissolved oxygen concentrations. Since the rate of reaction is much faster than the rate of transfer of oxygen, the oxidation rate is mainly controlled by the rate of oxygen transfer to the solution. To find $k_L a$ by this method, air is sparged through a 1 N Na_2SO_3 solution in the presence of Cu^{2+} ions at a concentration of 10^{-3} M and mechanical agitation. The unreacted sulfite concentration in solution is determined by titration with iodine during aeration and the rate of oxygen consumption is calculated using the stoichiometry of the reaction (equation 57). The concentration of dissolved oxygen in the liquid is nearly zero since the oxidation reaction is extremely fast. Therefore,

$$k_L a = n_{O_2}/C^* \tag{58}$$

where n_{O_2} is the rate of oxygen transfer (or consumption) and C^* is the saturation dissolved oxygen concentration in solution.

Sodium sulfite oxidation may be a reliable method of determining oxygen transfer rates in gas–liquid contactors, but it is not a perfect method for determining $k_L a$ in fermenters. The solution of sodium sulfite does not approximate to the physical and chemical properties of a fermentation broth and therefore $k_L a$ values obtained by sulfite oxidation may be significantly different from those in fermenters. Moreover, this method is labor and cost intensive and is also time consuming.

2.2.3.2 Dynamic method

This method is probably the simplest one since it requires only a dissolved oxygen probe. It is mainly based upon the dynamic oxygen balance in a batch culture, which has the following form

$$\frac{dC_L}{dt} = k_L a(C^* - C_L) - Q_{O_2}X \tag{54}$$

where Q_{O_2} is the rate of oxygen consumption per unit mass of cells (mM O_2 g^{-1} h^{-1}).

The various versions of the dynamic method are summarized below.

One method developed by Hixon and Gaden (1950) involves the addition of a lethal agent (such as phenol) into a sample withdrawn from a fermenter and recording the dissolved oxygen concentration *versus* time. For dead cells, the second term in equation (54) will be zero (*i.e.* $Q_{O_2}X = 0$).

$$\frac{dC_L}{dt} = k_L a(C^* - C_L) \tag{59}$$

By integration equation (59) becomes

$$\ln (C^* - C_L) = - k_L at + \text{constant} \tag{60}$$

A plot of $\ln (C^* - C_L)$ *versus* time yields a straight line with a slope of $k_L a$.

The second method, the gassing-out method, was originally developed by Wise (1951). It is an accurate method, but may not be convenient. In this method the fermentation broth is placed in a vessel and gassed-out with nitrogen. Aeration is then initiated at a constant air flow rate after gassing out. The dissolved oxygen (DO) concentration is monitored with the aid of a DO probe. Using equation (60) and a plot of $\ln (C^* - C_L)$ *versus* time the $k_L a$ value can be determined as described in the first dynamic method.

The third dynamic method for measuring $k_L a$ was developed by Humphrey *et al.* Again it is based on a dynamic oxygen balance in a batch culture. Rearranging equation (54) yields

$$C_L = C^* - \frac{1}{k_L a}\left(Q_{O_2}X + \frac{dC_L}{dt}\right) \tag{61}$$

The air supply is turned off at a certain time during fermentation and the variation of C_L with time is followed with the aid of a DO probe. Since the term $k_L a(C^* - C_L)$ becomes zero when air is turned off, the C_L value decreases linearly with time according to equation (54). The slope of the C_L *versus* time curve yields a value for $Q_{O_2}X$. The air is then turned on and the increase in C_L with time is followed as depicted in Figure 7. Having determined the $Q_{O_2}X$ value, C_L is plotted against $(Q_{O_2}X + dC_L/dt)$ as shown in Figure 8. The slope of this plot is equal to the reciprocal of $k_L a$.

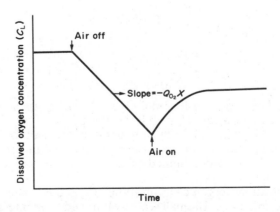

Figure 7 Determination of total oxygen consumption rate ($Q_{O_2}X$) using the dynamic method

One of the problems with this dynamic method is surface aeration when the air is turned off, which can be avoided by purging the surface of the liquid with nitrogen. Moreover, the response of the oxygen probe should be rapid enough to record biological changes.

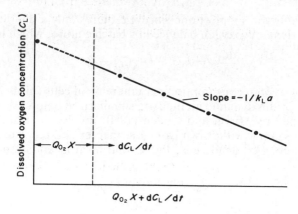

Figure 8 Estimation of the volumetric oxygen transfer coefficient ($k_L a$) using the dynamic method

It is much easier to apply the dynamic method to a continuous culture for the determination of $k_L a$. In a continuous culture, the rate of oxygen supply is equal to the rate of oxygen consumption at steady state, therefore

$$k_L a = \frac{Q_{O_2} X}{C^* - C_L} \tag{62}$$

The total rate of oxygen consumption ($Q_{O_2} X$) can be determined by turning the air supply off and following the linear variation of C_L with time as described previously. Since C_L is constant in a continuous culture at steady state, then equation (62) can be used to determine $k_L a$.

One other dynamic method which was developed recently to determine oxygen transfer coefficients is the frequency response technique. Using a polarographic DO electrode Vardar and Lilly (1982) followed the changes in DO concentration in a fermentation broth in response to sinusoidal air pressure input. With the appropriate transfer functions for oxygen transfer and the electrode, they determined the $k_L a$ value for oxygen transfer in a penicillin fermentation medium containing *Penicillium chrysogenum*. Although the high frequency of the pressure fluctuations may affect the actual value of the mass transfer coefficient in some cases, this method is simple and can be used during fermentation under sterile conditions.

2.2.3.3 Direct method

This method is based upon oxygen balance in inlet and off-gas streams around a fermenter. An oxygen balance for the sparged air yields

$$n_{O_2,T} = \frac{1}{V_L} (n_{O_2,i} - n_{O_2,o}) \tag{63}$$

or

$$n_{O_2,T} = \frac{1}{R_G V_L} \left(\frac{F_i P_i Y_i}{T_i} - \frac{F_o P_o Y_o}{T_o} \right) \tag{63a}$$

where $n_{O_2,T}$ is the rate of oxygen transfer (mol O_2 l^{-1} h^{-1}), V_L is the volume of fermentation broth (l), F is the volumetric flow rate of air (l h^{-1}), P is the total pressure of air (atm), y is the mole fraction of oxygen in air, T is the absolute temperature (K) and R_G is the gas constant (l atm mol^{-1} K^{-1}). The subscripts i and o indicate fermenter inlet and outlet, respectively. The parameters given in equation (63a) are determined using flow meters, pressure gauges, temperature probes and gaseous oxygen analyzers.

The dissolved and saturation oxygen concentrations, C_L and C^*, need to be measured experimentally in order to determine $k_L a$. In small scale well-mixed fermenters, the DO concentration is uniform throughout the fermenter and only one measurement of C_L would suffice. However, in large scale fermenters with non-Newtonian broth characteristics the fermenter may not be well mixed and a number of C_L measurements have to be made to determine an average C_L value.

The C^* value is the dissolved oxygen concentration in equilibrium with the partial pressure of oxygen in the exit gas stream. Again, in small scale fermenters, the exit gas stream can be assumed to be well mixed and C^* can be calculated from the partial pressure of oxygen in the exit gas stream. However, for large scale fermenters, the assumption of a perfectly mixed gas stream may not be valid. In this case, the logarithmic mean of the driving force, *i.e.* $(C^* - C_L)_{log\cdot mean}$, between the inlet and outlet of the fermenter should be used in the following equation to determine $k_L a$

$$k_L a = \frac{n_{O_2,T}}{(C^* - C_L)_{log\cdot mean}} \tag{64}$$

2.2.3.4 Oxygen yield coefficient method

The rate of oxygen transfer can be related to the growth rate of microorganisms using the oxygen yield coefficient according to the following equation

$$n_{O_2,T} = \frac{\mu X}{Y_{O_2}} \tag{65}$$

where μ is the specific growth rate of microorganisms (h^{-1}), X is the cell concentration (g dry weight cells l^{-1}), Y_{O_2} is the yield coefficient of oxygen (g cells per g O_2) and $n_{O_2,T}$ is the oxygen transfer (or uptake) rate (g $O_2\, l^{-1}\, h^{-1}$).

The rate of microbial growth, μX, can be determined by following the growth of microorganisms. The yield coefficient, Y_{O_2}, depends on the microorganisms and the substrate. The values of Y_{O_2} can be calculated from the substrate yield coefficient, and the compositions of the cell and substrate. Mateles (1971) has developed the following expression to estimate Y_{O_2}.

$$\frac{1}{Y_{O_2}} = 16\left[\frac{O'}{1600} - \frac{C'}{600} + \frac{N'}{933} - \frac{H'}{200} + \frac{2C + H/2 - O}{Y_s M_s}\right] \tag{66}$$

where O', C', N', H' are the percentages of oxygen, carbon, nitrogen and hydrogen in the cell; C, H and O are the number of atoms of carbon, hydrogen and oxygen in the substrate, respectively; Y_s is the substrate yield coefficient (g cells per g substrate) and M_s is the molecular weight of the substrate.

In using equation (66), it is assumed that an inorganic nitrogen source is used and the only products of metabolism are CO_2, H_2O and biomass. This method requires an accurate determination of cell composition. Once $n_{O_2,T}$ is determined from equation (65), then equation (56) can be used to calculate $k_L a$ provided that the dissolved oxygen concentration is measured.

Some of the aforementioned methods for the determination of the oxygen transfer coefficient require the measurement of the dissolved oxygen (DO) concentration. Various electrodes have been developed for the measurement of DO (Aiba, 1973; Johnson, 1964). The two major types of steam sterilizable DO electrodes are polarographic (Beckman electrode, Ag anode and Au cathode) and voltametric (Mackereth electrode, Pb anode and Ag cathode) electrodes, which are covered by a polymeric membrane and contain an electrolyte between the membrane and the cathode. The dissolved oxygen in the liquid phase diffuses through the membrane and the electrolyte layer and is reduced at the cathode surface according to the following reaction

$$O_2 + 2\,H_2O + 4\,e^- \rightarrow 4\,OH^- \tag{67}$$

The formation of OH^- ions on the surface of the cathode produces a current between the cathode and anode and the intensity of this current is proportional to the flux of oxygen through the electrode membrane. In general, the flux of oxygen is

$$n_{O_2} = K_o(p_L - p_e) \tag{68}$$

where p_L and p_e are the partial pressures (which are proportional to the concentrations) of oxygen in the bulk liquid and in the electrolyte–membrane surface (it is assumed that the membrane is tightly attached to the cathode), and K_o is the overall oxygen transfer coefficient.

The intensity of current at steady state is

$$i = nFAn_{O_2} \tag{69}$$

where n is the number of electrons transferred in the aforementioned reaction $(n=4)$, F is Faraday's constant (96 500 C/g equiv), A is the surface area of the cathode and i is the steady-state current.

Substituting equation (68) into equation (69) yields

$$p_L = p_e + \frac{i}{nFAK_o} \tag{70}$$

As can be seen from equation (70) the partial pressure of oxygen (or the concentration) in the liquid phase is proportional to the intensity of the current. Therefore, with an appropriate calibration (p_L *versus* i), equation (70) can be used to determine the value of p_L.

In general, the dynamic response characteristics of the dissolved oxygen probe should be considered in the determination of $k_L a$ by the dynamic method. In the absence of microbial cells, the dynamic balance on dissolved oxygen is

$$\frac{dC_L}{dt} = k_L a(C^* - C_L) \tag{71}$$

With the initial condition of $C_L \big|_{t=0} = 0$, the solution to equation (71) is

$$C_L = C^*[1 - e^{-k_L a t}] \tag{72}$$

If the response of the DO probe is first order, the rate of change in probe readings is

$$\frac{dC_p}{dt} = k_p(C_L - C_p) \tag{73}$$

where k_p is the overall sensor constant and C_p is the DO concentration corresponding to the sensor reading. The solution to equations (71) and (73) with the initial condition $C_p \big|_{t=0} = 0$ is

$$C_p = C^*\left[1 + \frac{k_L a}{k_p - k_L a} e^{-k_p t} - \frac{k_p}{k_p - k_L a} e^{-k_L a t}\right] \tag{74}$$

If the sensor constant (k_p) is of the same order of magnitude as $k_L a$, equation (74) should be used to estimate the value of $k_L a$ by measuring the values of C_p as a function of time and using a trial and error procedure provided that the value of k_p is known *a priori*. In some cases, the response of the probe is much more rapid than the rate of oxygen transfer (*i.e.* $k_p \gg k_L a$). In this case, $C_p \approx C_L$ and the response of the probe is given by equation (72). However, for imperfectly mixed fermentation broths (such as mycelial or polysaccharide fermentations) the sensor constant is smaller than the oxygen transfer coefficient ($k_p \ll k_L a$) and the response of the probe is very slow compared to the actual changes in DO concentration. In such cases, equation (74) should be used to estimate the value of $k_L a$.

2.2.4 Oxygen Transfer Coefficient Correlations

In agitated aerobic bioreactors, the magnitude of the oxygen transfer coefficient largely depends on the hydrodynamic conditions around the gas bubbles.

Literature correlations for mass transfer to single bubbles, drops, dispersion of solids and bubble swarms are available in the references listed in Table 2. Since the major resistance to oxygen transfer is the liquid film resistance surrounding air bubbles, the diffusivity of oxygen in the liquid phase is the major factor which affects the mass transfer coefficient. The correlations for the liquid phase oxygen transfer coefficient vary with bubble size. For small rigid-sphere bubbles ($D_p < 2.5$ mm) friction drag is predominant and the mass transfer coefficient is proportional to the diffusion coefficient to the power 2/3 according to laminar boundary layer theory (Calderbank, 1967). For large bubbles ($D_p > 2.5$ mm) form drag is predominant and the mass transfer coefficient is proportional to the square root of the diffusion coefficient (Higbie, 1935). The reader is also referred to the review by Moo-Young and Blanch (1981).

In industrial aerated reactors, air bubbles usually form swarms or clusters. The oxygen transfer coefficient correlations for air bubble swarms are different from single bubble correlations since liquid hydrodynamics around the bubbles are different for each case. Bubble-swarm mass transfer correlations depend on the size of the bubbles. The critical bubble diameter (D_c), which separates large bubbles from small ones, is $D_c = 2.5$ mm.

2.2.4.1 Correlations for small bubbles

Air bubbles with diameters smaller than 2.5 mm are usually encountered in agitated vessels and in sintered plate columns containing aqueous solutions of hydrophilic solutes. Mass transfer

Table 2 Literature Correlations for Mass Transfer to Single Bubbles

System	Ref.
Water–CO_2	Calderbank and Moo-Young (1961)
Water–O_2	Calderbank and Moo-Young (1961)
Aqueous glycerol–CO_2	Calderbank and Moo-Young (1961)
Glycol–CO_2	Calderbank and Moo-Young (1961)
Water–O_2	Coppock and Meiklejohn (1951)
Water–O_2	Hammerton (1953)
Water–air	Kinzer and Gunn (1951)
Water–O_2–CO_2	Sheng-Li Pang (1953)
Water–CO_2	Baird (1960)
Water–H_2	Hammerton (1953)
Toluene–H_2	McDowell and Myers (1955)
Liquid–solid sphere	Steinberger and Treybal (1960)

from small air bubbles in a fermentation broth can be approximated to mass transfer from a rigid sphere. For a small Reynolds number (*i.e. $Re = \varrho DV/\mu_L \ll 1$*) the following equation can be used to calculate the mass transfer coefficient for a small size spherical particle

$$Sh = \frac{k_L D_p}{D_{O_2}} = 1.01 \left(VD_p/D_{O_2} \right)^{1/3} \tag{75}$$

where V is the velocity of the particles (air bubbles) and D_{O_2} is the diffusivity of oxygen in the liquid surrounding the air bubbles. For a solid sphere, the terminal velocity, V_t, is given by

$$V_t = \frac{D_p^2 \Delta \varrho g}{18 \mu_L} \tag{76}$$

where $\Delta \varrho = \varrho_L - \varrho$ and μ_L is the viscosity of the liquid. Substituting equation (76) into equation (75) yields

$$Sh = \frac{k_L D_p}{D_{O_2}} = 1.01 \left(\frac{D_p^3 \Delta \varrho g}{18 \mu_L D_{O_2}} \right)^{1/3} \tag{77}$$

or

$$Sh = 0.39 Ra^{1/3} = 0.39 Gr^{1/3} Sc^{1/3} \tag{78}$$

where $\dfrac{D_p^3 \Delta \varrho g}{\mu_L D_{O_2}}$ is the Rayleigh number (*Ra*).

For large Reynolds numbers (*$Re \gg 1$*), the correlation for single bubbles in a laminar flow field is

$$Sh = 2.0 + 0.6 Re^{1/2} Sc^{1/3} \tag{79}$$

The equations (78) and (79) are applicable to single bubbles and correlations for small size bubble swarms can be derived in a similar way. The following equation is suggested for small size bubble swarms ($D_p < 2.5$ mm; Bailey, 1977; Calderbank, 1967):

$$Sh = 2.0 + 0.31 Ra^{1/3} = 2.0 + 0.31 Gr^{1/3} Sc^{1/3} \tag{80}$$

As the density difference ($\Delta \varrho = \varrho_L - \varrho_G$) gets smaller, the Sherwood number approaches 2.0 and buoyancy effects become negligible compared to pure diffusion.

2.2.4.2 *Correlations for large bubbles*

For bubble swarms of large size ($D_p > 2.5$ mm) the following empirical correlation is suggested

$$Sh = 0.42 Gr^{1/3} Sc^{1/2} \tag{81}$$

or

$$\frac{k_L D_p}{D_{O_2}} = 0.42 \left(\frac{D_p^3 \varrho_G (\varrho_L - \varrho_G)}{\mu_L^2} \right)^{1/3} \left(\frac{\mu_L}{\varrho_L D_{O_2}} \right)^{1/2} \tag{81a}$$

The oxygen transfer coefficient, k_L, varies with the square root of the diffusivity of oxygen for large bubbles, as opposed to that of small bubbles for which k_L varies with D_{O_2} to the power 2/3. This is because of the variation in the shape of gas bubbles and hydrodynamic conditions around gas bubbles. The shape of bubbles is nearly spherical for small bubbles ($D_p < D_c$), whereas bubble shape becomes cap-like or hemispheric for large bubbles ($D_p > D_c$).

2.2.5 Estimation of Interfacial Area

The oxygen transfer coefficient, k_L, may be determined using the correlations given in Section 2.2.4. However, in order to evaluate the volumetric oxygen transfer coefficient $k_L a$, the gas–liquid interfacial area (a) has to be determined. One of the major goals in aerobic biological reactor design is to provide a large gas–liquid contact area to improve the rate of oxygen transfer. In general, the dispersion of gases in liquids is obtained as bubbles by the flow of gases through orifices. Most of the aerobic biological reactors employ spargers underneath the impeller and mechanical agitation to reduce the air bubble size and to increase the gas–liquid contact area.

The value of a depends on the sparger design (*i.e.* orifice diameter), airflow rate, reactor volume and diameter of air bubbles. If the flow rate of air from the sparger orifice is F_a, the residence time of bubbles in the fermenter is t and the average bubble diameter is D_b, then a is given by

$$a = \frac{F_a t \pi D_b^2}{\pi D_b^3 / 6} \frac{1}{V_L} = \frac{F_a t}{V_L} \frac{6}{D_b} \tag{82}$$

where V_L is the liquid volume in the reactor.

The term $F_a t / V_L$ is the total bubble volume per unit liquid volume and is known as the gas hold up H (m^3 gas/m^3 liquid volume). Therefore,

$$a = \frac{6H}{D_b} \tag{83}$$

At low Reynolds numbers, bubble motion obeys Stokes law, that is

$$V_t \propto D_b^2 \text{ or } D_b \propto V_t^{1/2}$$

where V_t is the terminal rise velocity of the bubbles and H is inversely proportional to the bubble rise velocity, that is $H \propto 1/V_t$ or $H \propto 1/D_b^2$. Therefore,

$$a \propto \left(\frac{1}{D_b}\right)^3 \tag{84}$$

For finely dispersed, small air bubbles the gas–liquid interfacial area is inversely proportional to the third power of the bubble diameter. For this reason, vigorous agitation is usually applied to reduce D_b and hence to increase a.

The bubble size at the outlet of the sparger orifice can be estimated by applying a force balance for a bubble leaving the sparger (Bailey, 1977; van Krevelen, 1950). At the outlet of the orifice the buoyant force is balanced by the surface tension force.

$$\frac{\pi D_b^3 g \Delta \varrho}{6} = \pi d \sigma \tag{85}$$

or

$$D_b = \left(\frac{6d\sigma}{g\Delta\varrho}\right)^{1/3} \tag{85a}$$

where d is the orifice diameter, σ is the surface tension at the gas–liquid interface and $\Delta\varrho = \varrho_L - \varrho_G$. However, the bubble diameter in the liquid is usually not the same as that at the outlet of the sparger because of bubble coalescence, bubble size reduction by agitation, and changes in the hydrostatic pressure.

Usually, the bubble size is not uniform and bubble size distribution should be considered in evaluating average bubble size. The surface volume mean diameter or Sauter mean diameter of the bubbles is usually used in equation (83) to determine a. The Sauter mean diameter is given by

$$\bar{D}_{bs} = \frac{\sum\limits_{i}^{n} n_i D_{bi}^3}{\sum\limits_{i}^{n} n_i D_{bi}^2} \tag{86}$$

where n_i is the number of bubbles with diameter D_{bi}.

The bubble size can be measured experimentally by using photomicrographs or high-speed photographs. The pin-dropping technique developed by Rose and Wyllie (1950) has been modified by Calderbank and Rennie (1962) to determine the Sauter mean diameter of gas bubbles from photographs.

As shown in Figure 9, a triangular grid is placed over the photograph of the bubbles and the number of hits and cuts between the lines and particles are counted. The following equations are used to determine $\bar{D}_{bs}$, H and a:

$$\bar{D}_{bs} = \frac{3Lh}{2c} \tag{87}$$

$$H = \frac{h}{2n} \tag{87a}$$

$$a = \frac{2c}{nL} \tag{87b}$$

where L is the length of lines comprising the grid, h is the number of hits (end of line segments of length L in bubbles), n is the number of lines in the triangular grid, and c is the number of cuts (number indicating how many times the lines are cut by the images of bubbles).

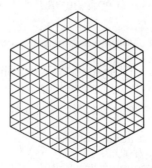

Figure 9   Triangular grid used in pin-dropping technique to determine particle diameter (D_{SM}) and interfacial area (a)
(Redrawn from Calderbank, 1967)

Empirical equations have been developed by several investigators to estimate the Sauter mean bubble diameter (Endo, 1958; Kafarov, 1959). The most widely used equations developed by Calderbank (1958) are the following:
(a) for dispersion of gas bubbles in electrolyte solutions:

$$\bar{D}_{bs} = 2.25\left[\frac{\sigma^{0.6}}{(P/V)^{0.4}\varrho_c^{0.2}}\right]H^{0.4}\left(\frac{\mu_d}{\mu_c}\right)^{0.25} \tag{88}$$

(b) for dispersion of gas bubbles in aqueous solutions of alcohols:

$$\bar{D}_{bs} = 1.9\left[\frac{\sigma^{0.6}}{(P/V)^{0.4}\varrho_c^{0.2}}\right]H^{0.65}\left(\frac{\mu_d}{\mu_c}\right)^{0.25} \tag{89}$$

(c) for dispersion of gas bubbles in pure agitated liquids:

$$\bar{D}_{bs} = 4.15\left[\frac{\sigma^{0.6}}{(P/V)^{0.4}\varrho_c^{0.2}}\right]H^{1/2} + 0.09 \text{ cm} \tag{90}$$

where σ is the surface tension, P/V is the power input per volume of liquid, ϱ_c and μ_c are the density and viscosity of the continuous phase (*i.e.* the liquid), respectively, and μ_d is the viscosity of the dispersed phase (*i.e.* the gas).

The $\bar{D}_{bs}$ values for pure liquids are considerably greater than those obtained in electrolyte or other solutions containing hydrophilic solutes since bubble coalescence occurs more readily in pure liquids.

The empirical equation developed by Calderbank (1958) for the gas–liquid interfacial area in aerated–agitated vessels is

$$a = 1.44\left[\frac{(P/V)^{0.4}\varrho_c^{0.2}}{\sigma^{0.6}}\right]\left(\frac{V_s}{V_t}\right)^{1/2} \tag{91}$$

where V_s is the superficial gas velocity and V_t is the terminal velocity of freely rising bubbles.

Substituting equations (90) and (91) into equation (83) yields a correlation for gas holdup

$$H = \left(\frac{V_s}{V_t}H\right)^{1/2} + 0.0216\left[\frac{(P/V)^{0.4}\varrho_c^{0.2}}{\sigma^{0.6}}\right]\left(\frac{V_s}{V_t}\right)^{1/2} \tag{92}$$

When P/V is large, the first term in equation (92) may be neglected. For non-agitated vessels $P/V = 0$ and $H = V_s/V_t$, which corresponds to the condition when the bubbles are freely rising and are evenly distributed over the cross-section of the reactor. Equation (92) correlates well with experimental data when $Re_i^{0.7}(N_iD_i/V_s)^{0.3} < 2 \times 10^4$. For $Re_i^{0.7}(N_iD_i/V_s)^{0.3} > 2 \times 10^4$ the following equation is proposed by Calderbank (Bailey, 1977)

$$H = \frac{a'}{a}\left(\frac{V_s}{V_t}H\right)^{0.5} + 0.015a' \tag{93}$$

where a' is related to a by the following equation

$$\log\frac{2.3a'}{a} = 1.95 \times 10^{-5}Re_i^{0.7}\left(\frac{ND_i}{V_s}\right)^{0.3} \tag{94}$$

where a is the interfacial area per unit volume as given in equation (91) and Re_i is the impeller Reynolds number $= \varrho ND_i^2/\mu_c$.

Richards' (1961) correlation for aerated–agitated vessels containing water is an alternative to Calderbank's correlation for the evaluation of gas holdup, H, which has the following form

$$H = 0.13(P/V)^{0.4}V_s^{1/2} - 0.31 \tag{95}$$

where P is power (HP), V is ungassed liquid volume (m^3) and V_s is superficial gas velocity (m h^{-1}). This equation is valid for $0.02 < H < 0.2$.

Gas holdup correlations for bubble columns are considerably different to those of agitated vessels and are covered in Chakravarty (1973).

Experimental methods have been developed for the determination of the gas–liquid interfacial area (Calderbank, 1967). All of these methods are based on the light-scattering properties of dispersed particles. The details of these methods are beyond the scope of this chapter and are well covered in Calderbank (1967).

2.3   OXYGEN TRANSFER IN GASSED AND AGITATED SYSTEMS

2.3.1   Power Requirement for Agitation

In agitated aerobic bioreactors, the rate of oxygen transfer varies with the power supplied for agitation of the fermentation broth, therefore estimation of the power requirement for effective agitation and oxygen transfer is a prerequisite for the design of aerobic biological reactors.

Only Newtonian fermentation media are considered in this section. By definition, viscosity is constant for Newtonian fluids. Most fermentation media involving bacteria and yeast exhibit

Newtonian fluid behavior. However, fermentation media involving mycelial cells or polymeric compounds (either as a substrate or as a product) exhibit non-Newtonian fluid behavior.

2.3.1.1 Power requirement in non-gassed and agitated systems

The power requirement for non-gassed Newtonian fluids is characterized by a dimensionless power number Po (Rushton, 1950) which by definition is the ratio of the external force to the inertial force exerted by the fluid, that is

$$Po = \frac{Pg_c}{N^3 D_i^5 \varrho} \tag{96}$$

where P is the power supplied to the agitator (for non-gassed systems), N is the rotational speed of the impeller, D_i is the diameter of the impeller, ϱ is the density of the fluid and g_c is the Newton's law conversion factor.

In agitated vessels, the power number is correlated with the impeller Reynolds number (Re_i) which is defined as

$$Re_i = \frac{ND_i^2 \varrho}{\mu} \tag{97}$$

where μ is the viscosity of the fluid. The correlation between Po and Re_i for different types of impellers is depicted in Figure 10. The curves given in Figure 10 are only valid for the geometric ratios specified in this figure. It is also assumed that the agitated fermenters are fully baffled (*i.e.* almost no vortex formation) and therefore the effect of the Froude number (Fr) on power requirement is negligible (*i.e.* Po is only a function of Re_i).

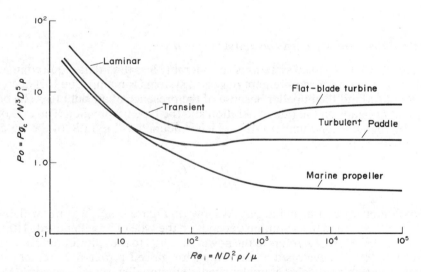

Figure 10 Power number (Po) *versus* Reynolds number (Re) for single impeller agitators (Redrawn from Aiba *et al.*, 1973)

Three different fluid flow regions can be identified in Figure 10: (1) the laminar flow region ($Re_i<10$): $Po = K_i/Re_i$ where K is a constant; (2) the turbulent flow region ($Re_i>10^3$): $Po = K_2 =$ constant; and (3) the transition region ($10<Re_i<10^3$): no analytical expression is available. Figure 10 should be used for this region.

The power number–Reynolds number relationship depicted in Figure 10 is valid only for single impeller systems. The power requirement by multiple impellers may be related to the number of impellers, when the impellers are properly spaced. The proper spacing between impellers is $D_i<H_i<2D_i$ and the number of impellers is $(H_L - 2D_i)/D_i<N_i<(H_L - D_i)/D_i$ where H_i is the spacing between impellers and H_L is the height of the liquid in the vessel. A simple relationship between the power number (Po) and the number of impellers (N_i) was found by Fukuda (1968a,

1968b) and is depicted in Figure 11. As can be seen from Figure 11, $(Po)_2 \approx 2(Po)_1$ and $(Po)_3 \approx 3(Po)_1$ where $(Po)_3$ and $(Po)_2$ are power requirements for three and two impeller systems, respectively.

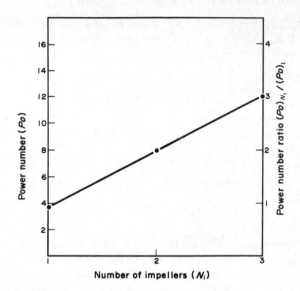

Figure 11 Variation of power number (Po) with the number of impellers (N_i). Liquid volume: 4100 l (Redrawn from Wang *et al.*, 1979)

2.3.1.2   *Power requirement in gassed and agitated systems*

The power requirement in gassed systems is considerably less than that required in non-gassed systems. The decrease in power requirement in gassed systems is mainly due to a decrease in the density of the liquid around the impeller because of the presence of air bubbles. The decrease in power requirement depends on the rate of aeration and the type of impeller. The rate of aeration is characterized by the aeration number (Ae), which is defined as the ratio of the superficial gas velocity to the impeller tip speed

$$Ae = \frac{F_G/D_i^2}{ND_i} = \frac{F_G}{ND_i^3} \tag{98}$$

where F_G is the volumetric flow rate of the gas. As shown in Figure 12, P_G/P varies with the aeration number, *i.e.* $P_G/P = f(Ae)$, and this variation depends on the type of impeller used. The ratio P_G/P varies between 1 and 0.3. Here, P_G refers to the power supplied to the agitator for gassed systems.

Michel and Miller (1962) developed a correlation for gassed power requirement for a large range of operating variables such as impeller diameter, impeller speed and airflow rate. The ranges of physical parameters in their experiments were: $\varrho = 0.8–1.65$ g cm^{-3}, $\mu = 0.9–100$ cP, $\sigma = 27–72$ dynes cm^{-1}, $P_G = 0.005 − 0.1$ HP, $V_L = 3.5–10.5$ l. The empirical correlation developed by Michel and Miller for gassed power absorption is:

$$P_G = K\left(\frac{P^2ND_i^3}{F_a^{0.56}}\right)^{0.45} \tag{99}$$

where K is a constant which is a function of geometry. Although equation (99) was developed for air–water systems, similar functional forms can be used for other liquids. The data obtained by Fukuda *et al.* (1968b) and Michel *et al.* (1962) fit equation (99) reasonably well, therefore this correlation (equation 99) is adequate for predicting P_G over a wide range of fermenter volumes and other physical parameters.

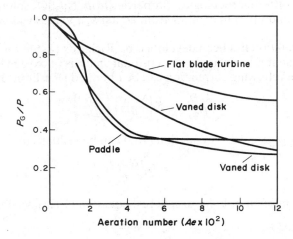

Figure 12 Gassed to ungassed power ratio (P_G/P) *versus* aeration number (Ae) for different types of impellers (Redrawn from Wang *et al.*, 1979)

2.3.2 Oxygen Transfer Coefficient–Power Input Correlations

The volumetric oxygen transfer coefficient, $k_L a$, has been correlated with power input per unit volume of reactor (P_G/V_L) and the superficial air velocity (V_s) for aerated and mechanically agitated reactors. Cooper *et al.* (1944) used the sodium sulfite oxidation method to determine and correlate $k_L a$ with (P_G/V_L) and V_s. Figure 13 is a plot of the data obtained by Cooper *et al.* depicting $k_L a/V_s$ *versus* P_G/V_L. The following correlation has been obtained from this data

$$k_L a = K(P_G/V_L)^{0.95} V_s^{0.67} \tag{100}$$

where K is a constant. The value of K varies with the type of impeller. For a vaned-disk impeller $K = 0.0635$ when metric units are used (*i.e.* $k_L a$, kg mol h^{-1} m^{-3} atm^{-1}; P_G/V_L, HP m^{-3}; and V_s, m h^{-1}). Equation (100) is valid when $P_G/V_L > 0.1$ HP m^{-3}, $H_L \approx D_T$, $V_s < 90$ m h^{-1} for single impellers, and $V_s < 150$ m h^{-1} for two impeller systems.

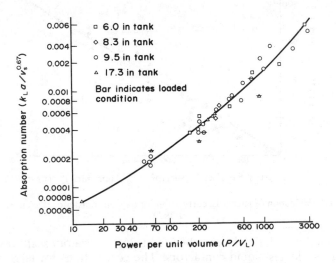

Figure 13 Variation of absorption number (1b mol ft^{-3} h^{-1} atm^{-1} ft$^{-0.67}$ h$^{0.67}$) for oxygen transfer with power input per unit volume (ft lbf min^{-1} ft^{-3}) (Redrawn from Cooper *et al.*, 1944)

For a paddle impeller the correlation is

$$k_L a = 0.038(P_G/V_L)^{0.53} V_s^{0.67} \tag{101}$$

when $P_G/V_L > 0.06$ HP m^{-3}, $V_s < 21$ m h^{-1} and $H_L \approx D_T$ (Aiba, 1973).

Cooper's correlation was obtained by the sodium sulfite technique. Since a Newtonian fermentation broth will have different rheological properties from Na_2SO_3 solution, k_La values obtained from equations (100) and (101) may differ from actual k_La values which would be obtained in a fermentation broth.

An alternative correlation has been developed by Richards (1961) for oxygen transfer in aerated–agitated Newtonian fluids. From the dimensional analysis of oxygen transfer from air bubbles in agitated vessels, the following correlation can be obtained (Rushton, 1951).

$$\frac{k_LD_i}{D_{O_2}} = K_1\left(\frac{ND_i^2\varrho}{\mu}\right)^a\left(\frac{\mu}{\varrho D_{O_2}}\right)^b \tag{102}$$

For constant values of ϱ, μ and D_{O_2}, equation (102) can be reduced to

$$k_L = K_2\left(\frac{ND_i^2\varrho}{\mu}\right)^a\frac{1}{D_i} \tag{102a}$$

With the approximation of air bubbles to spheres, equation (102a) becomes

$$k_L = K_3N^{1/2} \tag{103}$$

since $a = 1/2$ for rigid spheres.

The gas–liquid interfacial area (a) in aerated–agitated vessels is given by equation (91; Calderbank, 1958). Assuming that ϱ, σ and V_t are constants, equation (91) reduces to

$$a = K_4(P_G/V_L)^{0.4}V_s^{0.5} \tag{104}$$

Combining equations (103) and (104) we obtain Richards' correlation

$$k_La = K_5(P_G/V_L)^{0.4}V_s^{0.5}N^{0.5} \tag{105}$$

Equation (105) has been tested using both Cooper's and Richards' data. A plot of these experimental data (k_La *versus* $(P_G/V_L)^{0.4}V_s^{0.5}N^{0.5}$), which is depicted in Figure 14, indicates that Richards' correlation fits the experimental data reasonably well for a large range of experimental variables.

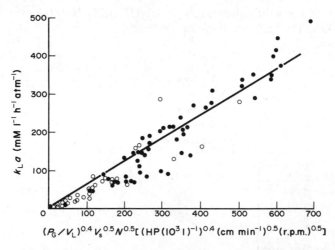

Figure 14 Experimental verification of Richards' correlation for oxygen transfer coefficient (Redrawn from Wang *et al.*, 1979)

The aforementioned correlations were obtained using laboratory scale vessels (6–27 l) and are valid only for small scale gas–liquid contactors. The correlations for large scale fermenters are sparse. Fukuda *et al.* (1968b) have used the sodium sulfite oxidation method to obtain oxygen transfer coefficient correlations for large scale vessels (100–42 000 l). Since multiple impellers are used for large scale fermentation vessels, Fukuda considered N_i as an additional parameter and obtained the following correlation

$$k_La = (\beta + \gamma N_i)(P_G/V_L)^{0.77}V_s^{0.67} \tag{106}$$

where β and γ are constants, N_i is the number of impellers, P_G/V_L is the gassed power per unit volume of liquid (HP per 1000 l) and V_s is the superficial gas velocity (cm min^{-1}); k_La is in mM O$_2$ l^{-1} h^{-1} atm^{-1}. Figure 15 is a plot of the experimental data of Fukuda *et al.* (1968b) correlated with equation (106). The constants in equation (106) are $\beta = 2.0$ and $\gamma = 2.8$ for the experimental data presented in Figure 15.

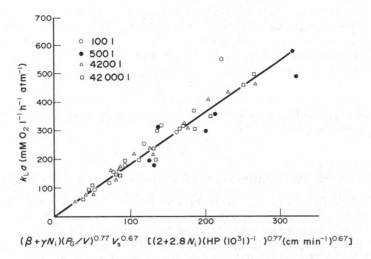

Figure 15 Fukuda's correlation for oxygen transfer coefficient for multiple impeller systems (Redrawn from Wang *et al.*, 1979)

Experimental data obtained in large scale fermenters (100–4200 l) can be fitted to an equation similar to Richards' correlation. In this case, the correlation will have the following form:

$$k_La = (\beta + \gamma N_i)(P_G/V_L)^{0.56}V_s^{0.7}N^{0.4} \qquad (107)$$

As can be seen from equations (100) and (106) the exponent for (P_G/V_L) decreases as the size of the vessel increases. However, the exponent for V_s does not vary with the scale of the vessel.

The oxygen transfer coefficient correlations vary with the geometry of the fermenter vessel. The aforementioned correlations are valid only for mechanically agitated vessels. Hospodka *et al.* (1942) used a 20 000 l Waldhof fermenter for the production of bakers' yeast. The Waldhof fermenter contains a draft tube directly above the impeller which enhances gas–liquid circulation rates and provides better oxygen transfer. The correlation for k_La in a Waldhof fermenter has the following form:

$$k_La = K'(P_G/V_L)^{0.72}V_s^{0.11} \qquad (108)$$

where K' is constant. The exponent on V_s is lower than that for conventional fermenters. This is mainly because of the presence of a draft tube which causes circulation of the gas–liquid mixture in the reactor and therefore the measured gas velocity is not the same as the actual gas flow rate.

2.3.3 Oxygen Transfer Efficiency

The oxygen transfer rate in aerobic bioprocessing can be improved by increasing the power input or the superficial gas velocity, which adds to the cost of aeration. The parameter which should be maximized to yield a cost effective aeration is the amount of oxygen transferred per unit energy input. The oxygen transfer efficiency is defined as kilograms of oxygen transferred to the reaction medium per unit kilowatt-hour energy input. The oxygen transfer efficiency (OTE) provides a direct means of comparing different configurations and scales of aerobic biological reactors. Higher oxygen transfer efficiency results in a more cost effective aeration.

Oxygen transfer efficiencies for various reactor sizes are listed in Table 3. The energy input includes power supplied to the agitator and the power for air compression. The power required for gas compression was determined by assuming 100% efficiency for isentropic compression of air to overcome the hydrostatic head of the fermenter broth. The superficial gas velocity (V_s) and the oxygen absorption rate were kept constant at levels of 0.32 m min^{-1} and 60 mM O$_2$ l^{-1} h^{-1}, respectively, for part of the data presented in Table 3. As can be seen from this table, the OTE decreases with increasing scale of the fermenters. This can be explained by considering the effect of P_G/V_L on the volumetric oxygen transfer coefficient (k_La), that is k_La is directly proportional to $(P_G/V_L)^\alpha$ for constant V_s, *i.e.* $k_La = K(P_G/V_L)^\alpha$. Since the values of K and α for small scale fermenters are larger than those for large scale fermenters, the same percent increase in (P_G/V_L) would result in a larger increase in k_La for small scale fermenters compared to production (large) scale fermenters. This yields a better OTE for small scale equipment.

Another comparison depicted in Table 3 is the effect of the oxygen transfer medium on the OTE. The last two rows in Table 3 indicate that sulfite medium adequately represents the Newtonian fermentation broth involving yeast cells, therefore oxygen transfer efficiencies obtained in sodium sulfite solution are reasonable approximations to OTE values in Newtonian fermentation broths.

The oxygen transfer efficiencies for non-Newtonian fluids are considerably lower than those obtained in Newtonian broths, usually by a factor of 2 to 4 depending on the nature of the non-Newtonian fermentation broth.

2.4　OXYGEN TRANSFER IN NON-NEWTONIAN FERMENTATION BROTH

The rheological properties of the fermentation broth have a profound effect on the performance of biological reactors as they affect the mixing pattern, power requirements and mass transfer rates in the fermentation broth. In some fermentation processes, such as those involving mycelial organisms or polymeric compounds either as substrates or products, the fermentation broth behaves as a non-Newtonian fluid. Since viscous forces are predominant in non-Newtonian fluids, the power required to obtain a certain oxygen transfer rate in those fluids is much higher than that required for Newtonian fluids (Charles, 1978). This section covers the oxygen transfer characteristics of non-Newtonian fermentation broths.

2.4.1　Rheological Characteristics of Non-Newtonian Fluids

For Newtonian fluids shear stress is linearly related to shear rate and the proportionality constant is the fluid viscosity. This relation has the following form:

$$\tau = -\frac{\mu}{g_c}\frac{dv}{dy} \tag{109}$$

where τ is the shear stress (g cm^{-2}), dv/dy is the shear rate (s^{-1}), μ is the viscosity (g cm^{-1} s^{-1}), and g_c is the Newton's law constant (cm s^{-2}). The viscosity of a Newtonian fluid is independent of shear rate and is constant. However, the viscosity of a non-Newtonian fluid is a function of shear rate and is not constant.

Various models have been proposed to describe non-Newtonian fluid behavior. The most widely used model is the power law model. According to this model, shear stress is related to the shear rate in the following manner

$$\tau = -\frac{K}{g_c}\left(\frac{dv}{dy}\right)^n \tag{110}$$

where K is the consistency index of the fluid (g cm^{-1} s^{n-2}) and n is the flow behavior index. Depending on the value of n, two different types of non-Newtonian fluids can be identified. For pseudoplastic fluids, the value of n is between zero and one and for dilatant fluids n is greater than one. Fermentation broths involving mycelial organisms (such as in antibiotic fermentations involving *Streptomyces* or *Penicillium*) or polymeric materials (polysaccharide fermentations) usually exhibit pseudoplastic behavior. There are other non-Newtonian fluids which require a yield stress at zero shear rate. These fluids are known as Bingham fluids. The shear stress–shear rate relationship for Bingham fluids is

$$\tau = \tau_y + \frac{\mu_p}{g_c}\frac{dv}{dy} \tag{111}$$

where τ_y is the yield stress and μ_p is the rigidity coefficient or plastic viscosity. Certain fermen-

Table 3 Comparison of Oxygen Transfer Efficiencies

Type of fermenter broth	Volume of fermenter broth (l)	Oxygen transfer rate (mM O_2 l^{-1} h^{-1})	Power input (HP/1000 l)	Superficial air velocity (m min^{-1})	Compression energy (kW h/kg O_2)	Agitation energy (kW h/kg O_2)	Oxygen transfer efficiency (kg O_2 kW^{-1} h^{-1})
Na$_2$SO$_3$ solution	60	60	0.63	0.32	—	0.23	4.35
Na$_2$SO$_3$ solution	4200	60	2.3	0.32	0.03	0.90	1.075
Na$_2$SO$_3$ solution	42 000	60	2.3	0.32	0.05	0.90	1.05
Yeast fermentation	20 000	22	0.487	0.60	0.05	0.494	1.84
Na$_2$SO$_3$ solution	42 000	22	0.459	0.60	0.083	0.462	1.84

tation broths, such as kanamycin fermentations, exhibit Bingham fluid behavior. Other fermentation broths exhibit yield stress but behave differently from Bingham fluids. For some mycelial fermentations the Casson equation has been found to fit experimental data better than the power law equation for pseudoplastic fluids (Charles, 1978). This equation has the following form:

$$\tau^{1/2} = \tau_y^{1/2} + K_c \left(\frac{dv}{dy}\right)^{1/2} \tag{112}$$

Figure 16 is a plot of shear stress *versus* shear rate for various types of fluids.

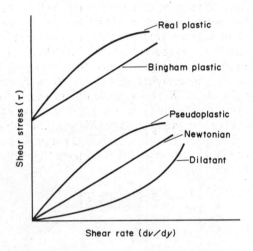

Figure 16　Shear stress *versus* shear rate for Newtonian and non-Newtonian fluids

For power law fluids the shear stress–shear rate relationship is usually expressed in terms of apparent viscosity (μ_a) which is defined as

$$\mu_a = K \left(\frac{dv}{dy}\right)^{n-1} \tag{113}$$

Then, in terms of apparent viscosity, equation (110) can be rewritten as

$$\tau = -\frac{\mu_a}{g_c}\frac{dv}{dy} \tag{114}$$

As can be seen from equation (113), the apparent viscosity is a function of shear rate for non-Newtonian fluids.

2.4.2　Power Requirements for Non-Newtonian Fluids

2.4.2.1　Non-gassed systems

Since the apparent viscosity of non-Newtonian fluids depends on shear rate (or impeller speed), the prediction of power requirement in this case is more complicated than that for Newtonian fluids. As had been done with Newtonian fluids, Metzner and Otto (1957) defined a modified Reynolds number to estimate the power requirement in non-Newtonian fluids. For a pseudoplastic fluid, the modified impeller Reynolds number has the following form:

$$Re_i' = \frac{\varrho D_i^2 N^{2-n}}{0.1K} \left(\frac{n}{6n+2}\right)^n \tag{115}$$

Furthermore, Calderbank and Moo-Young (1959) have shown that the shear rate around an impeller is proportional to the impeller speed (N) for various types of impellers and for pseudoplastic and Bingham fluids. This relationship is:

$$\text{Pseudoplastic or Bingham fluids:} \quad \left(\frac{dv}{dy}\right)_i = \dot{\gamma}_i = 10N \tag{116}$$

$$\text{Dilatant fluids:} \quad \left(\frac{dv}{dy}\right)_i = \dot{\gamma}_i = 12.8N\left(\frac{D_i}{D_T}\right)^{0.5} \tag{117}$$

Using equations (116) and (117), the power requirement for agitation of non-Newtonian fluids can be estimated (Calderbank, 1959). The relationship between the power number (Po) and the modified Reynolds number (Re') has been obtained experimentally by Calderbank and Moo-Young (1959). Figures 17 and 18 are plots of experimental data for Po *versus* Re' for pseudoplastic and dilatant fluids, respectively. As can be seen from these figures the power number is inversely related to the modified Reynolds number in laminar regions (low Re') and Po is almost independent of Re' for turbulent regions (high Re'). One problem in estimating the power requirements for non-Newtonian fluids is the variation of the rheological properties of the fermentation broth during fermentation. As a result of this variation, the power requirement varies with fermentation time. In a batch glucoamylase fermentation using *Endomyces* spp., Taguchi and Miyamoto (1966a) found that both K and n varied significantly during the fermentation which affected the power requirement for the agitation of this pseudoplastic fermentation broth. Figure 19 depicts the variation of the consistency index (K) and flow behavior index (n) with time in glucoamylase fermentation. Variation in the rheological properties of a fermentation broth with time has also been observed by LeDuy *et al.* (1974) during the synthesis of a polysaccharide from sucrose by *Pullularia pullulans*.

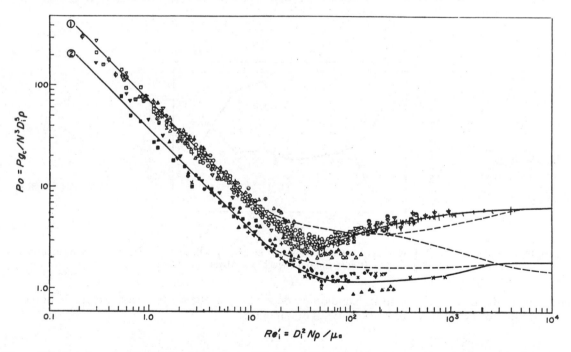

Figure 17 Variation of power number (Po) with modified Reynolds number (Re') for pseudoplastic fluids: (1) 6-bladed turbine, (2) paddle. ●, ○ 4% CMC/water; ■ □ 4.5% CMC/water; ▲, △ perspex/toluene I; ▼, ▽ kaolin/water I; ×, + kaolin/water II; ⊘ perspex/toluene II; ⊡ paint; △ 22% paper powder/paraffin; ▽ 40% clay/water (Reprinted from Calderbank and Moo-Young, 1959)

The power requirement for a non-Newtonian fermentation broth also varies with the dimensions of the fermenter (Taguchi and Miyamoto, 1966). The power number (Po) was correlated with Re', D_i/D_T and W/D_T by Taguchi and Miyamoto (1966) for unaerated, non-Newtonian broths of *Endomyces* as follows:

$$Po = k_1(Re'_i)^a\left(\frac{D_i}{D_T}\right)^b\left(\frac{W}{D_T}\right)^c \tag{118}$$

where W is the impeller width, D_i and D_T are the impeller and tank diameters, respectively, and

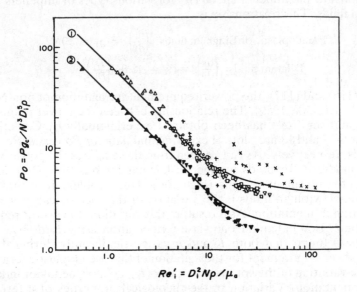

Figure 18 Power number (*Po*)–modified Reynolds number (*Re'*) correlation for dilatant fluids: (1) 6-bladed turbine, (2) paddle. ●, ○ CF/SSI; ■, □ CF/SSII; ▲ △, CF/SSIII; ▼, ▽ CF/SSIV; + GA/BSI; × GA/BSII, ⊕ GA/BSI; ⊗ GA/BSII. CF/SS, cornflour/sugar solution; GA/BS, gum arabic/borax solution. (Reprinted from Calderbank and Moo-Young, 1959)

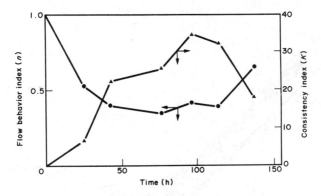

Figure 19 Variation of flow behavior index (*n*) and consistency index (*K*) with time in glucoamylase fermentation by *Endomyces* (Redrawn from Taguchi and Miyamoto, 1966)

the magnitudes of the exponents *a*, *b* and *c* vary depending on the range of Re'_i. Figure 20 depicts the variation of *Po* with Re'_i for various types of impellers.

2.4.2.2 Gassed systems

Taguchi and Miyamoto (1966) investigated the effect of aeration on the power requirement for the agitation of a non-Newtonian fermentation broth and found that the power requirement decreased with aeration. They plotted their data as P_G/P *versus* the aeration number (F_a/ND_i^3) for various Reynolds numbers (Figure 21). At low *Re*, the ratio of P_G/P is not significantly affected by aeration. However, as the Reynolds number increases the ratio of P_G/P decreases significantly with increasing aeration number.

Taguchi and Miyamoto (1966) also found that a modified form of the Michel and Miller correlation (equation 99) can be used to estimate the power requirement in the turbulent regime ($Re'_i > 50$) and over a large range of liquid volumes. Figure 22 is a plot of P_G *versus* $P^2ND_i^3/F_a^{0.56}$ for various values of D_i/D_T obtained by Taguchi *et al.* for glucoamylase fermentation by

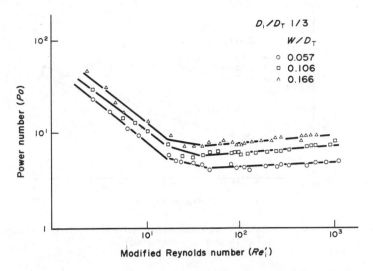

Figure 20 Variation of power number (*Po*) with modified Reynolds number (*Re′*) for gassed non-Newtonian fermentation broth (Redrawn from Taguchi and Miyamota, 1966)

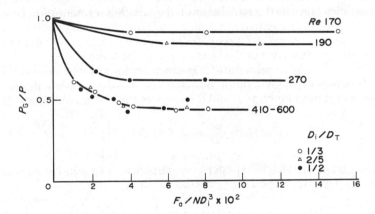

Figure 21 Gassed to ungassed power ratio (P_G/P) *versus* aeration number (*Ae*) for the non-Newtonian broth of glucoamylase fermentation (Redrawn from Taguchi and Miyamoto, 1966)

Endomyces. In the turbulent flow regime, the Michel and Miller correlation adequately fitted their experimental data, that is

$$P_G = K' \left(\frac{P^2 N D_i^3}{F_a^{0.56}} \right)^{0.45} \tag{119}$$

The following equation (smaller exponent) can be used for laminar and transition regimes ($Re_i' < 50$).

$$P_G = K'' \left(\frac{P^2 N D_i^3}{F_a^{0.56}} \right)^{0.27} \tag{120}$$

2.4.3 Oxygen Transfer Correlations for Non-Newtonian Fluids

Oxygen transfer coefficients ($k_L a$) in non-Newtonian fermentation broths are affected by a number of factors characterizing the fluid rheology, flow conditions and the geometry of the system. Among the major factors which influence the $k_L a$ are bubble and cell (mycelia) dimensions, power input, fluid density and viscosity, and geometry of the tank and impeller.

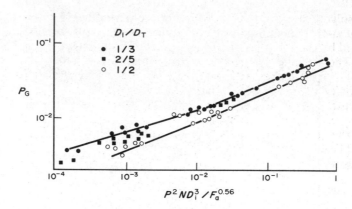

Figure 22 Gassed power (P_G) *versus* $P^2ND_i^3/F_a^{0.56}$ for glucoamylase fermentation by *Endomycopsis* (Redrawn from Taguchi and Miyamoto, 1966)

For a constant power input, the $k_L a$ value for a non-Newtonian fluid is lower than that of a Newtonian fluid. The value of $k_L a$ varies during the course of fermentation as a result of variations in the rheological properties of the non-Newtonian fermentation broth. The value of $k_L a$ has been reported to decrease by about 90% as a result of an increase in the concentration of mycelial *Aspergillus niger* from 0.02 to 2.5%. The inclusion of smaller cell (bacteria or yeast) suspensions or inert particles (such as 0.3 μm alumina) in a mycelial fermentation broth was reported to enhance the oxygen transfer coefficient (Bailey, 1977). This enhancement in the presence of particles is probably due to changes in fluid hydrodynamics near the gas–liquid interface resulting in decreased mass transfer resistance of the fluid film near the bubble.

There are very few studies on oxygen transfer coefficients for non-Newtonian fluids. The correlations given for Newtonian fluids cannot be used for non-Newtonian fluids. However, the general form of the correlation function is still the same as for Newtonian fluids, that is

$$Sh = \frac{k_L D_b}{D_{O_2}} = f(Re, Sc) \tag{121}$$

In studying oxygen transfer rates into paper pulp suspensions (a non-Newtonian fluid) the following correlation has been developed for $k_L a$ (Blakebrough, 1966).

$$\frac{k_L a}{N} = 0.113 \left(\frac{D_T^2 H_L}{WL(D_i - W)}\right)^{1.44} (Nt)^{-1.09} \left(\frac{D_i}{D_T}\right)^{1.02} \tag{122}$$

where H_L is the liquid height, W is the impeller width, t is the characteristic mixing time of the vessel and L is the length of the impeller.

2.5 TRANSPORT BOTTLENECKS IN BIOPROCESSES

The performance of aerobic biological reactors containing high cell densities may be limited by the rate of oxygen transfer, since the rate of oxygen consumption may exceed the rate of oxygen transfer resulting in a very low or zero dissolved oxygen concentration in the liquid medium. In such systems, the productivity will be directly related to the rate of oxygen transfer. The rheological properties of fermentation broths may also have significant influence on the performance of biological reactors. Fermentation broths containing mycelial cells or polymeric materials (such as polysaccharides, either as a substrate or product) generally exhibit non-Newtonian fluid behavior. Moreover, mycelial cells may form pellets through which soluble nutrients have to diffuse to be available for cells growing inside the pellet (Atkinson, 1974). The evolution and removal of carbon dioxide and other fermentation products from fermentation media (especially from mycelial pellets) present another mass transfer problem which could adversely affect the performance of biological reactors. This section covers major transport bottlenecks in biological processes with particular emphasis on mass transfer in microbial pellets.

2.5.1 Critical Oxygen Concentration and Oxygen Uptake

The level of dissolved oxygen concentration in aerobic fermentation media has a profound effect on the rate of microbial metabolism and product formation. The dissolved oxygen concentration is determined by the relative rates of oxygen consumption and oxygen transfer in the fermentation broth. The specific rate of oxygen consumption (Q_{O_2}) by microbial cells [g O_2 consumed (g cells)$^{-1}$ h^{-1}] varies with dissolved oxygen concentration (C_L) in a hyperbolic fashion. If the dissolved oxygen concentration is below a certain level, the Q_{O_2} increases linearly with the DO concentration. However, above the critical value of DO concentration, the Q_{O_2} is constant and is independent of C_L. The critical dissolved oxygen concentration (C_{crit}) is the DO level below which the specific rate of oxygen consumption increases with increasing DO level. When the DO level is above C_{crit}, Q_{O_2} reaches a maximum value. Figure 23 depicts the variation of Q_{O_2} with the dissolved oxygen concentration in a yeast culture.

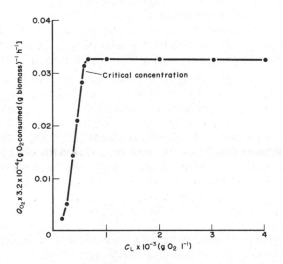

Figure 23 Variation of specific rate of oxygen consumption (Q_{O_2}) with the concentration of dissolved oxygen (C_L) in a yeast culture (Redrawn from Finn, 1967)

In order to eliminate oxygen limitations, the dissolved oxygen concentration at every point in a fermentation broth should be above the critical DO level (*i.e.* $C_L > C_{crit}$). The value of the critical DO level depends on the type and activity (*i.e.* respiration rate) of the cells and the nature of the substrate. Usually, C_{crit} is equal to 5 to 10% of the saturation dissolved oxygen concentration (C^*). At steady state (*i.e.* time invariant DO concentration) the rate of oxygen transfer into the fermentation broth is equal to the rate of oxygen consumption by the microbes, that is

$$k_L a(C^* - C_L) = Q_{O_2} X \tag{123}$$

or

$$C_L = C^* - \frac{Q_{O_2} X}{k_L a} \tag{123a}$$

When the temperature and composition of the fermentation broth and the partial pressure of oxygen in the gas phase are constant, the equilibrium DO level (C^*) is fixed and the only terms which influence the DO level are oxygen demand ($Q_{O_2} X$) and the oxygen transfer coefficient ($k_L a$). The ratio of $Q_{O_2} X / k_L a$ should be low enough to yield $C_L > C_{crit}$ [or $Q_{O_2} X / k_L a < (C^* - C_{crit})$]. For a given cell concentration (X) and oxygen consumption rate (Q_{O_2}), the minimum value of $k_L a$ resulting in $C_L > C_{crit}$ can be calculated using the aforementioned inequality. At very high cell concentrations ($X > 50$ g l^{-1}) this inequality may not be satisfied and the DO level may drop below the critical DO value. In such cases, the oxygen transfer coefficient, $k_L a$, should be increased to a level where the aforementioned inequality is satisfied. This could be done by increasing the power input or the superficial air velocity in agitated fermenters.

The airflow rate and agitation speed can be controlled to maintain the desired dissolved oxygen concentration in the fermentation medium. The feed rate of nutrients (particularly carbon source) can also be controlled to yield the desired DO concentrations.

2.5.2 Oxygen Transfer and Utilization in Microbial Pellets

Microbial pellet formation is usually observed in fermentation processes involving mycelial organisms such as fungi. The environmental conditions inside a microbial pellet are different to those in a well-mixed nutrient solution (Aiba, 1971; Yano, 1961). The transfer of nutrients and product removal constitute the major problems adversely affecting the performance of a fermentation broth containing microbial pellets. In order to be available for the cells growing inside a pellet, the nutrients have to diffuse into the interior of the pellets. Nutrient diffusion and consumption are simultaneous in microbial pellets. When the rate of nutrient utilization is higher than the rate of nutrient diffusion, the reaction is said to be diffusion limited. In such cases, the rate of product formation or microbial growth is determined by the rate of diffusion of the nutrient which has the lowest diffusion rate. In many cases, the rate limiting nutrient is dissolved oxygen since oxygen is a sparingly soluble gas compared to other soluble nutrients.

Since the concentration of dissolved oxygen inside a pellet is lower than that in the bulk liquid and varies with the distance from the pellet–liquid interface, the rate of oxygen consumption is also a function of the position inside the pellet. Consider a spherical mycelial pellet with radius R and assume that the dissolved oxygen concentration at the surface of the pellet is equal to the bulk DO concentration C_O (*i.e.* the liquid phase is well mixed). A material balance for oxygen (or any other limiting nutrient) with respect to a thin spherical shell (with radii r and $r + dr$) of this pellet results in

$$D_{O_2}\left(\frac{d^2C}{dr^2} + \frac{2}{r}\frac{dC}{dr}\right) = Q_{O_2}X \tag{124}$$

where D_{O_2} is the diffusivity of oxygen (assumed constant), C is the concentration of dissolved oxygen, Q_{O_2} is the specific rate of oxygen consumption and X is the cell concentration (cell mass/volume of pellet) inside the pellet. The Q_{O_2} varies with the DO concentration in a hyperbolic fashion when C is lower than C_{crit}, therefore

$$Q_{O_2} = \frac{\mu_{max}C}{Y_o(K_s + C)} \tag{125}$$

where Y_o is the oxygen yield coefficient [g cells (g O_2 consumed)$^{-1}$], μ_{max} and K_s are the maximum specific rate of growth and the saturation constant, respectively.

Combining equations (124) and (125) yields

$$D_{O_2}\left(\frac{d^2C}{dr^2} + \frac{2}{r}\frac{dC}{dr}\right) = \frac{\mu_{max}C}{Y_o(K_s + C)}X = r_{O_2} \tag{126}$$

where r_{O_2} is the total rate of oxygen consumption representing $Q_{O_2}X$ (g O_2 consumed l^{-1} h^{-1}). Equation (126) has to be solved in order to obtain the dissolved oxygen concentration and the oxygen consumption rate at any radial distance inside the pellet. This equation can be rewritten using dimensionless variables, that is

$$\frac{d^2\bar{C}}{d\bar{r}^2} + \frac{2}{\bar{r}}\frac{d\bar{C}}{d\bar{r}} = \frac{r_{O_2}R^2}{D_{O_2}C_o} = \phi^2\frac{\bar{C}}{1 + \alpha\bar{C}} \tag{127}$$

where $\bar{C} = C/C_o$, $\bar{r} = r/R$, $\alpha = \dfrac{C_o}{K_s}$ and $\phi = R\left(\dfrac{\mu_{max}X/Y_oK_s}{D_{O_2}}\right)^{1/2}$ and the dimensionless boundary conditions are

$$\bar{C} = 1 \text{ at } r = R$$

$$\frac{d\bar{C}}{d\bar{r}} = 0 \text{ at } r = 0$$

The parameter ϕ is known as the Thiele modulus and is the square root of the ratio of the first-order reaction rate to the diffusion rate. The saturation parameter, α, is a measure of the deviation from first-order kinetics. For very low values of α, the reaction rate is first order, whereas for large values of α the reaction approaches zero-order kinetics.

It is difficult to evaluate the rate of oxygen consumption inside the pellet (*i.e.* r_{O_2}). However, the observed rate of oxygen consumption (r_o) by the pellet is equal to the diffusion flux of oxygen into the pellet:

$$r_o = \frac{A_p}{V_p}\left(D_{O_2}\frac{dC}{dr}\bigg|_{r=R}\right) \tag{128}$$

where A_p and V_p are the external surface area and volume of a pellet, respectively.

In diffusion limited reaction systems, an effectiveness factor is defined to quantify the magnitude of diffusion limitation. By definition, the effectiveness factor is the ratio of the reaction rate with diffusion limitation (or diffusion flux) to the reaction rate without diffusion limitation (or the rate of reaction in bulk liquid conditions), therefore

$$r_o = \eta r_{O_2}(C_o) \tag{129}$$

where r_o is the observed rate of oxygen consumption, $r_{O_2}(C_o)$ is the rate of oxygen consumption when the DO concentration is C_o and η is the effectiveness factor.

The effectiveness factor for the consumption of oxygen in a microbial pellet, as described by equation (127), in terms of dimensionless parameters is

$$\eta = \frac{3 \mid d\bar{C}/d\bar{r} \mid_{\bar{r}=1}}{\phi^2/(1 + \alpha)} \tag{130}$$

Since $(d\bar{C}/d\bar{r})_{\bar{r}=1}$ is only a function of ϕ and α, then η depends on only ϕ and α, that is

$$\eta = f(\phi, \alpha) \tag{131}$$

In practice, it is difficult to measure the intrinsic kinetic parameters μ_{max} and K_s for microbial pellets. For this reason a modified Thiele modulus is defined on the basis of observed rate of oxygen consumption. The modified modulus is

$$\varphi = \left(\frac{V_p}{A_p}\right)^2 \frac{r_o/C_o}{D_{O_2}} \tag{132}$$

and the effectiveness factor can be expressed in terms of φ and α, that is

$$\eta = g(\varphi, \alpha) \tag{133}$$

The variation in the effectiveness factor with φ and α is given in Figure 24 for first-order ($\alpha \rightarrow$ 0) and zero-order ($\alpha \rightarrow \infty$) reaction kinetics. The curves for saturation (Michaelis–Menten) kinetics will be between the two curves for first-order and zero-order kinetics.

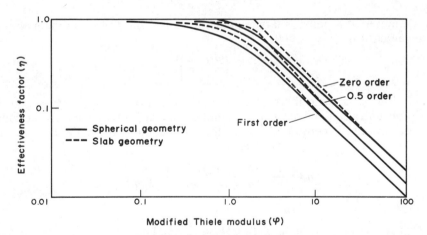

Figure 24 Variation of effectiveness factor (η) with modified Thiele modulus (φ) for substrate utilization in microbial pellets (Redrawn from Bailey and Ollis, 1977)

The solution to equation (127) can only be obtained numerically to determine the oxygen concentration profile within the pellet. However, analytical solutions can be obtained for the first-order and zero-order reaction kinetics. The analysis of these two extreme cases is discussed in the following sections.

2.5.2.1 First-order reaction kinetics

In many practical situations the concentration of dissolved oxygen (or the limiting substrate) within the pellet is low enough for the reaction kinetics to be approximated to a first-order reaction rate (*i.e.* $\alpha \rightarrow 0$). In this case equation (127) can be solved analytically to yield an expression for η

$$\eta = \frac{1}{\phi}\left(\frac{1}{\tan 3\phi} - \frac{1}{3\phi}\right) \tag{134}$$

where

$$\phi = \frac{V_p}{A_p}\left(\frac{k_1}{D_{O_2}}\right)^{1/2} \tag{135}$$

and k_1 is the first-order reaction constant (*i.e.* $k_1 = \mu_{max}X/YK_s$). Equation (134) can be used to plot η *versus* φ (Figure 24) for first-order kinetics using the relationship $\varphi = \eta\phi^2$. As can be seen in Figure 24, for low values of φ ($\varphi < 0.1$) the effectiveness factor is approximately equal to one and reaction is not diffusion limited. However, for large values of φ ($\varphi > 1$), the value of η is smaller than one and the reaction is diffusion limited. Therefore, in order to eliminate diffusion limitations, the radius of the microbial pellet should be kept small enough for the value of the Thiele modulus to be low resulting in $\eta = 1$.

2.5.2.2   *Zero-order reaction kinetics*

If the rate of oxygen consumption inside the microbial pellet is zero order, equation (127) can be solved analytically to yield an oxygen concentration profile of the following form:

$$C = C_o - \frac{k_o}{6D_{O_2}}(R^2 - r^2) \tag{136}$$

where k_o is the first-order rate constant ($k_o = \mu_{max}X/Y$).

In this case, the DO concentration may be zero at a certain radial distance from the center of the pellet. This radial distance, at which $C = 0$ is called the critical radius r_{cr} and is determined by setting $C = 0$ at $r = r_{cr}$ in equation (136), therefore

$$\left(\frac{r_{cr}}{r}\right)^2 = 1 - \frac{6D_{O_2}C_o}{k_oR^2} \tag{137}$$

When the diameter of a pellet is large enough to satisfy the following inequality (138), then there is a real value for r_{cr} (*i.e.* $r_{cr} > 0$).

$$R > \left(\frac{6D_{O_2}C_o}{k_o}\right)^{1/2} \tag{138}$$

If inequality (138) is satisfied there is an interior portion of the pellet (from $r=0$ to $r=r_{cr}$) in which the DO concentration and oxygen consumption rate are zero and anaerobic conditions prevail. In this case, oxygen is consumed only in the outer shell of the pellet and the effectiveness factor is

$$\eta = \frac{k_o\frac{4}{3}\pi(R^3 - r_{cr}^3)}{\frac{4}{3}\pi R^3 k_o} = 1 - \left(\frac{r_{cr}}{R}\right)^3 \tag{139}$$

or

$$\eta = 1 - \left(1 - \frac{6D_{O_2}C_o}{k_oR^2}\right)^{3/2} \tag{139a}$$

If the condition given in equation (138) is not satisfied, then the effectiveness factor is 1 and the rate of substrate utilization within the pellet is uniform.

In general, diffusion limitations in microbial pellets can be eliminated by maintaining a small pellet size in the fermentation medium. The control of pellet size in fungal fermentations is a challenging task which should be given more attention.

2.5.3   Carbon Dioxide Evolution

In many aerobic fermentation systems one of the gaseous fermentation products is carbon dioxide. The removal of carbon dioxide from the fermentation medium is essential in order to avoid the potential adverse effects of a high dissolved CO_2 concentration in the medium. The desorption of CO_2 from the liquid phase follows the same rule as that of gas absorption. The overall rate of CO_2 desorption is given by the following equation

$$n_{CO_2} = K'_L a'(C'_L - C^{*'}) \tag{140}$$

where C'_L is the dissolved CO_2 concentration in the liquid, $C^{*'}$ is the dissolved CO_2 concentration in the liquid which is in equilibrium with the partial pressure of CO_2 in the gas phase, $K'_L a'$ is the overall volumetric CO_2 transfer coefficient and n_{CO_2} is the flux of CO_2.

Correlations for the carbon dioxide desorption coefficient (k'_L) are quite similar to that of the oxygen absorption coefficient and have the same general form, that is

$$Sh = \frac{k'_L D_b}{D_{CO_2}} = f(Re, Sc) \tag{141}$$

or

$$Sh = a' Re^{b'} Sc^{c'} \tag{141a}$$

However, the constants a', b' and c' are different from those for oxygen absorption.

Removal of carbon dioxide from the fermentation medium may be a more pronounced problem in solid substrate fermentation systems and mycelial (*i.e.* fungal) fermentations. In order to achieve effective CO_2 removal in such systems, the particle size of the solid substrates and mycelial pellets should be small and vigorous agitation should be supplied.

In submerged fermentations, the CO_2 desorption is usually enhanced by increasing the CO_2 transfer coefficient (k'_L) and the gas–liquid interfacial area (a'). This is usually achieved by supplying vigorous agitation.

The rate of CO_2 evolution from the fermentation medium can be determined using methods similar to those used in the measurement of the oxygen uptake rate. A carbon dioxide balance between the inlet and off-gas streams gives the rate of CO_2 evolution.

$$\text{Rate of } CO_2 \text{ evolution} = \frac{1}{R_G V_L}\left(\frac{F_i P_i y'_i}{T_i} - \frac{F_o P_o y'_o}{T_o}\right) \tag{142}$$

where the terms are as defined in equation (63a) except that y'_i and y'_o represent mole fractions of CO_2 in the inlet and off-gas streams.

2.6 OTHER FACTORS AFFECTING OXYGEN TRANSFER

The rate of oxygen transfer in a fermentation broth is affected by several physical (temperature, pressure, fluid rheology) and chemical (ionic strength, surfactants, organic substances) factors. These factors may affect either the driving force (ΔC) or the transfer coefficient ($k_L a$), therefore it is possible to enhance the rate of oxygen transfer by controlling the aforementioned factors at their optimal level. This section briefly covers the effect of these factors on oxygen transfer in aerobic fermentation systems.

2.6.1 Physical Factors

2.6.1.1 Temperature

The temperature of the fermentation medium affects both the solubility (*i.e.* equilibrium DO concentration) and the diffusivity (*i.e.* transfer coefficient) of oxygen in the fermentation broth. The solubility of oxygen in water decreases with temperature, but the diffusivity of O_2 in the liquid phase increases (almost linearly) with absolute temperature. The net effect of temperature on the rate of oxygen transfer depends on the range of temperatures considered. At low temperatures (10 °C$<T<$40 °C) the increase in temperature is more likely to increase the rate of oxygen transfer due to an increase in the diffusivity of oxygen. In fact, in a study of oxygen transfer rates in an activated sludge system, O'Connor (Aiba, 1973) reported that the value of $k_L a$ at 30 °C was approximately 15% higher than that at 20 °C. However, at high temperatures (40 °C$<T<$90 °C), the solubility of oxygen drops significantly, which adversely affects the driving force and the rate of oxygen transfer.

2.6.1.2 Pressure

The partial pressure of oxygen in the gas phase mainly affects the solubility of oxygen and therefore the driving force for oxygen transfer. The solubility of O_2 is related to the partial pressure of O_2 in the gas phase by Henry's law

$$P_{O_2} = P_T y_{O_2} = H' C_{O_2}^* \tag{143}$$

where H' is the Henry's law constant, which is a function of temperature, P_T is the total pressure of air and y_{O_2} is the mole fraction of oxygen in the gas phase. An increase in P_{O_2} (at constant temperature) will result in an increase in $C_{O_2}^*$ and this would in turn increase the driving force $(C^* - C_L)$ and the rate of oxygen transfer.

In certain fermentation and wastewater treatment systems, oxygen-enriched air or pure oxygen (instead of air) is used to improve the rate of oxygen transfer. An alternative method is to increase the total pressure of air supplied to the fermenter (*i.e.* the use of highly compressed air) or to operate the system under a constant high pressure head of air in order to improve the rate of oxygen transfer. For better oxygen transfer, high pressure pure oxygen can be used.

2.6.1.3 Rheological properties of the medium

The apparent viscosity (μ_a) of the fermentation medium has a profound effect on the oxygen transfer coefficient ($k_L a$). Fermentation broths containing polysaccharides (either as a substrate or as a product) or mycelial cells exhibit non-Newtonian fluid behavior. The rheological properties of these fermentation broths vary with time as microbial growth continues and products are formed (Roels, 1974). LeDuy *et al.* (1974) have observed the variation of rheological properties of the fermentation medium in extracellular polysaccharide synthesis by *Pullularia pullulans*. Figure 25 depicts the variation of apparent viscosity with fermentation time in the aforementioned polysaccharide fermentation.

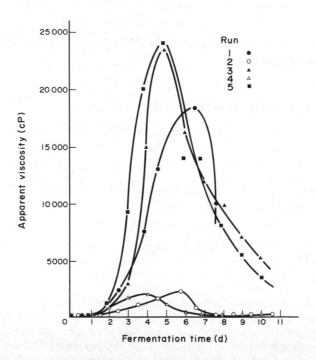

Figure 25 Variation of apparent viscosity (μ_a) of growth medium with time in polysaccharide fermentation by *Pullularia pullulans* (Redrawn from LeDuy *et al.*, 1974)

The morphology of mycelial cells affects the rheological properties of the fermentation broth (Charles, 1978). Disperse filamentous growth tends to cause pseudoplasticity; on the other hand, growth in pellet form tends to cause more Newtonian-like behavior. In most of the reported studies it has been observed that mycelial culture media behave either like Bingham plastic or pseudoplastic fluids. However, there are a few studies indicating Casson plastic behavior. Deindoerfer and Gaden (1955) reported that *Penicillium chrysogenum* culture media show Bingham plastic behavior and apparent viscosity increases with reaction time as the cell concentration increases.

Variation of the culture medium viscosity with fermentation time results in significant variations in the oxygen transfer coefficient (Taguchi, 1968, 1971). Brierley and Steel (1959) measured the $k_L a$ values at different concentrations of *Aspergillus niger* and reported that the increase in apparent viscosity of the culture broth due to an increase in cell concentration during fermentation resulted in a marked decrease in $k_L a$ (Aiba, 1973).

2.6.2 Chemical Factors

2.6.2.1 Surface active agents (surfactants)

Surfactants are easily adsorbed at an air–liquid interface, which results in reduction of the surface tension (σ) and surface free energy. The reduced surface tension results in reduced gas bubble diameter according to equation (90) with a corresponding increase in the gas–liquid interfacial area (a) according to equation (83). Therefore, the inclusion of small concentrations of surfactants in the fermentation medium will increase the interfacial area. However, the mass transfer coefficient, k_L, decreases with the addition of surfactants. The adsorption of surfactant molecules on the gas–liquid interface increases the mass transfer resistance and also reduces the mobility of the interface and hence results in a reduced value for k_L. Usually the increase in interfacial area (a) overcomes the decrease in mass transfer coefficient (k_L) and as a result the volumetric transfer coefficient ($k_L a$) increases with increasing surfactant concentration (Aiba, 1963, 1964, 1973; Bailey, 1977; Mancy, 1963; McKeown, 1963).

Eckenfelder (1961) investigated the effect of surfactants on oxygen transfer from air bubbles into water. Addition of a small amount of sodium lauryl sulfate (NaLSO$_4$) caused a sharp decrease in k_L (about 55% decrease *versus* pure water), but the value of k_L reached a plateau value at high NaLSO$_4$ concentrations. The gas–liquid interfacial area per liquid volume (a) increased slowly with surfactant concentration (between 0 to 75 p.p.m.). As a result, the $k_L a$ value first decreased with NaLSO$_4$ concentration resulting in a minimum at about 10 to 15 p.p.m. and then increased at higher concentrations of surfactant.

Benedek and Heideger (1971) studied the influence of sodium dodecyl sulfate (SDS) on $k_L a$ in a turbine aerator and they observed that the interfacial area (a) increased by a factor of four with the addition of 4 p.p.m. SDS; however the net increase in $k_L a$ was only 15%. Contrary to the results of Eckenfelder, Benedek and Heideger found that $k_L a$ increased continuously with the addition of surfactant. They also found that the addition of 10 p.p.m. Antifoam C caused a 30% decrease, while 0.2 g l^{-1} NaCl resulted in a 140% increase, in the $k_L a$ values.

2.6.2.2 Ionic strength

The volumetric oxygen transfer coefficient ($k_L a$) was found to be strongly dependent on the ionic strength of the liquid medium (Robinson, 1973). Robinson and Wilke (1973) have developed a correlation between $k_L a$ and the ionic strength of the medium for Newtonian fluids, which has the following form:

$$k_L a = \lambda \left(\frac{P_G}{V_L}\right)^n v_s^m \frac{\varrho_L^{0.533} D_{O_2}^{2/3}}{\sigma^{0.6} \mu_L^{1/3}} \tag{144}$$

where λ is the ionic strength function, P_G/V_L is the power input per unit volume of liquid (ft lb min^{-1} ft^{-3}), v_s is the superficial gas velocity (ft s^{-1}), σ is the surface tension (dyne cm^{-1}), D_{O_2} is the diffusivity of oxygen (cm^2 s^{-1}), ϱ_L and μ_L are the density and viscosity of the liquid (g cm^{-3} and g cm^{-1} s^{-1}).

The parameters λ, n and m were empirically correlated with the ionic strength (I) of the solution. By definition, the ionic strength of a salt solution is given by

$$I = \tfrac{1}{2}\Sigma Z_i^2 C_i \tag{145}$$

where Z_i is the charge on species i, and C_i is the concentration of species i. The suggested correlations are:

$$\lambda = 18.9 - \frac{28.7 I_o}{0.276 + I_o} \tag{146}$$

where

$$I_o = I \text{ for } 0 \leqslant I \leqslant 0.40 \text{ g ion l}^{-1}$$
$$I_o = 0.40 \text{ for } I > 0.40 \text{ g ion l}^{-1}$$
$$n = 0.40 + \frac{0.862 I_o}{0.274 + I_o} \text{ for } I_o = I \leqslant 0.4 \tag{147}$$
$$n = 0.9 \text{ for } I > 0.4$$

m increases monotonically from 0.35 to 0.39 for $0 < I < 0.4$

$$m = 0.39 \text{ for } I > 0.4$$

It is thought that the variation of n and m with ionic strength may account in part for the differences in $k_L a$ values reported in previous studies.

2.6.2.3   Organic substances

There are controversial reports on the effect of organic substances on the oxygen transfer coefficient. In general, the influence of an organic compound on $k_L a$ depends on the chemical nature of the compound.

Eckenfelder and Barnhart (1961) studied the effect of n-heptanoic acid concentration on $k_L a$ for oxygen transfer from air bubbles into water. They found that both bubble diameter and surface tension decreased with increasing concentration of n-heptanoic acid between 2 and 100 p.p.m. The overall oxygen transfer coefficient decreased with n-HA concentration for 2 p.p.m.$<n$-HA<20 p.p.m. followed by an increase at higher concentrations of n-HA. This increase is a result of an increase both in k_L and interfacial area (a).

Zieminski *et al.* (1960) investigated the effect of various alcohols, esters and ketones on $k_L a$ in the bubble aeration of water. They reported that the addition of 20 p.p.m. of n-butyl acetate and isopentyl acetate increased the $k_L a$ value by 50% to 100% over that obtained in distilled water. This was the result of a substantial increase in the gas–liquid interfacial area (a), which predominated over the decrease in k_L.

In another study reported by Aiba and Yamada (1961) it was found that the value of $k_L a$ in the presence of 1% peptone was about 40% of the value obtained in water. The addition of 1% peptone resulted in a decrease of about 67% in k_L and a decrease of about 15% in bubble diameter.

2.7   HEAT TRANSFER IN BIOLOGICAL PROCESSES

In order to control the temperature at a desired level, heat is either supplied to or removed from the fermentation broth during the course of fermentation. For this reason, fermentation vessels are usually equipped with either external jackets or internal coils for heating or cooling purposes. It is relatively easy to control the temperature in well-agitated fermenters due to the high degree of turbulence and high heat transfer rates in the bulk fermentation broth (*i.e.* forced convection). However, in static bed microbial reactors, such as trickling filters, temperature control is often difficult and heat transfer takes place by natural convection or by phase change (evaporation–condensation).

2.7.1   Major Heat Transfer Configurations

The two major heat transfer configurations in fermentation vessels are (a) external jackets, and (b) internal coils. Figure 26 depicts these heat transfer configurations. The external jacket system provides sufficient heat transfer area for laboratory and pilot plant scale fermenters. Because of the need for a larger heat transfer area, internal coils are more frequently used in production scale fermenters.

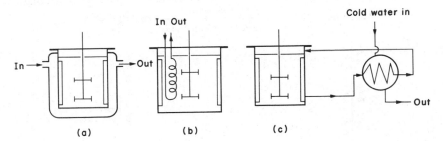

Figure 26   Various heat transfer configurations for biological reactors: (a) external jacket, (b) internal coil and (c) external surface heat exchanger

Although internal coils provide better heat transfer capabilities, the presence of internal coils cause other problems such as microbial film growth on coil surfaces, alteration of mixing patterns and fluid velocities. The heat transfer coefficient between the surface of the coil and the fermentation broth varies with the configuration of the cooling coil, therefore the internal coils may cause problems in the scale-up of biological reactors.

One other heat transfer configuration is the use of external surface-heat exchangers. In this configuration, the fermentation broth is pumped through an external heat exchanger where the heat transfer takes place through the surface of exchanger tubes. With an appropriate design of heat exchanger this system may provide better heat transfer capabilities compared to the aforementioned systems and may eliminate scale-up problems encountered in internal coil configurations.

2.7.2   Heat Balance and Microbial Heat Generation

Microbial growth is usually accompanied by the release of metabolic heat into the fermentation medium. Microbial heat evolution depends on the type and utilization efficiency of the carbon source.

A general heat balance on a fermentation broth is

$$Q_{Gr} + Q_{Ag} = Q_{Acc} + Q_{Exch} + Q_{Evap} + Q_{Sen} \tag{148}$$

where Q_{Gr} is the rate of metabolic heat generation; Q_{Ag} is the rate of heat generation by agitation; Q_{Acc} is the rate of heat accumulation; Q_{Exch} is the rate of heat transfer to the surroundings or exchanger; Q_{Evap} is the rate of heat loss by evaporation; and Q_{Sen} is the rate of sensible enthalpy change of liquid and air streams (exit–entrance).

The value of Q_{Sen} is usually much less than that of Q_{Gr} and may be neglected. The heat loss by evaporation (Q_{Evap}) can be eliminated by using a saturated air stream at the temperature of the fermentation broth. The terms Q_{Exch} and Q_{Ag} can be experimentally determined at different values of agitation speed and air flow rate or Q_{Ag} can be estimated using the power correlations. The heat accumulation (Q_{Acc}) has been measured by monitoring the transient temperature rise of an insulated fermenter (Cooney *et al.*, 1968). Having determined each term in equation (148) the rate of metabolic heat generation (Q_{Gr}) can be determined using this equation. The typical values for various heat terms in equation (148) are as follows (kJ l^{-1} h^{-1}): $Q_{Gr} = 36.16$, $Q_{Ag} = 13.84$, $Q_{Acc} = 47.23$, $Q_{Exch} = 2.72$, $Q_{Evap} = 0.188$, $Q_{Sen} = 0.042$ (Cooney *et al.*, 1968).

One other method of estimating Q_{Gr} is based upon heat of combustion data of the substrate and microbial cells. A simple heat balance on substrate utilization is depicted below.

The heat of combustion of the substrate is equal to the sum of the metabolic heat and the heat of combustion of the cellular material.

$$\frac{\Delta H_S}{Y_S} = \Delta H_C + \frac{1}{Y_H} \tag{149}$$

where ΔH_S is the heat of combustion of the substrate (kJ per g substrate); Y_S is grams of cell mass produced per gram of substrate consumed (g cell per g substrate); ΔH_C is the heat of combustion of cells (kJ per g cells); and $1/Y_H$ is the metabolic heat evolved per gram of cell mass produced (kJ per g cells).

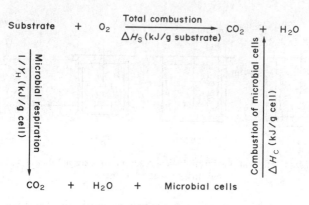

Figure 27

Equation (149) can be rearranged to yield

$$Y_H = \frac{Y_S}{\Delta H_S - Y_S \Delta H_C} \tag{150}$$

If the only oxidant is oxygen and the oxidation products are CO_2 and H_2O, the terms ΔH_S and ΔH_C can be determined from the combustion of cells and substrate. Typical ΔH_C values for bacterial cells are 20 to 25 kJ (g cells)$^{-1}$. The substrate yield coefficient (Y_S) can be determined using the following equation (Pirt, 1975):

$$\frac{1}{Y_S} = \frac{1}{Y_G} + \frac{m}{\mu} \tag{151}$$

where Y_G is the growth yield coefficient; m is the maintenance coefficient; and μ is the specific growth rate.

The rate of metabolic heat evolution in a batch fermentation is (in kJ h^{-1})

$$Q_{Gr} = V_L \mu X \frac{1}{Y_H} \tag{152}$$

The rate of metabolic heat evolution (kJ l^{-1} h^{-1}) has been related to the rate of oxygen uptake (mmol l^{-1} h^{-1}) by Cooney *et al.* (1968) for a number of microorganisms growing on different substrates and the following equation has been obtained:

$$Q_{Gr} \approx 0.12 Q_{O_2} \tag{153}$$

2.7.3　Heat Transfer Coefficient Correlations

Depending on the temperature of the fermentation process the major role of the heat transfer unit (internal coil or external jacket) is either to remove the metabolic heat from the medium or to supply extra heat to the medium to keep the temperature at a desired level.

The rate of heat transfer from the fermentation medium into the cooling fluid is given by the following equation:

$$Q_{Exch} = \bar{h} A (T_f - T_c) \tag{154}$$

where $\bar{h}$ is the overall heat transfer coefficient; A is the heat transfer area; and T_f and T_c are the temperatures of the fermentation broth and cooling fluid, respectively. The major heat transfer resistances involved are: (1) cooling fluid convective heat transfer resistance; (2) wall resistance; and (3) fermentation broth convective heat transfer resistance. The overall heat transfer resistance is equal to the sum of the individual resistances. Therefore, for rectangular geometry

$$\frac{1}{\bar{h}} = \frac{1}{h_c} + \frac{1}{k_w/L} + \frac{1}{h_f} \tag{155}$$

where k_w is the thermal conductivity of the wall; and h_c and h_f are the convective heat transfer coefficients for the cooling fluid and fermentation broth respectively. For heat transfer through internal coils the overall heat transfer resistance is

$$\frac{1}{\bar{h}_o} = \frac{1}{h_o} + \frac{\ln(r_o/r_i)}{k_w/r_o} + \frac{r_o}{h_i r_i} \tag{156}$$

where r_o and r_i are the outside and inside radii of the tubes; and h_o and h_i are the outer and inner convective heat transfer coefficients, respectively.

For steady-state operation, $Q_{Acc} = 0$ and the net rate of heat generation is equal to the rate of heat removal by the cooling fluid, that is

$$Q_{Gr} + Q_{Ag} \approx Q_{Exch} = \bar{h}A(T_f - T_c) \tag{157}$$

The required heat transfer area is then

$$A \approx \frac{Q_{Gr} + Q_{Ag}}{\bar{h}(T_f - T_c)} \tag{157a}$$

The heat generation terms $(Q_{Gr} + Q_{Ag})$ can be estimated as described above. The overall heat transfer coefficient is calculated using equations (155) or (156). The individual heat transfer coefficients (h_o and h_i) are evaluated using literature correlations. Usually, the heat transfer coefficient on the fermentation broth side is smaller than that of the cooling fluid side and the major heat transfer resistance is the liquid film resistance between the solid surface of the cooling coil and the fermentation broth, that is $\bar{h} \approx h_f$ (or h_o).

The general form of the heat transfer correlation for forced convection is:

$$Nu = f(Re, Pr)$$

or

$$\frac{hD}{k_f} = a\left(\frac{D\bar{V}\varrho_f}{\mu_f}\right)^b\left(\frac{Cp\mu_f}{k_f}\right)^c \tag{158}$$

For turbulent flow conditions equation (158) has the following form (Dittus–Boelter correlation):

$$Nu = \frac{hD}{k} = 0.023 Re^{0.8} Pr^b \tag{159}$$

which is valid when $10^4 < Re < 1.2 \times 10^5$; $0.7 < Pr < 120$; $L/D > 60$ (*i.e.* fully developed flow); and $b = 0.4$ for heating and 0.3 for cooling.

If the temperature difference between the wall and the fluid is large, a viscosity correction term is included in the Seider–Tate equation

$$Nu = \frac{hD}{k} = 0.027 Re^{0.8} Pr^{1/3}\left(\frac{\mu_b}{\mu_w}\right)^{0.14} \tag{160}$$

where μ_b and μ_w are fluid viscosity at bulk fluid and wall temperatures.

The Colburn analogy is an alternative equation for the estimation of h:

$$\frac{h}{\varrho\bar{V}Cp}\left(\frac{Cp\mu}{k}\right)^{2/3} = \frac{f}{2} \tag{161}$$

or

$$\frac{h}{\varrho\bar{V}Cp}\left(\frac{Cp\mu}{k}\right)^{2/3} = 0.023\left(\frac{D\bar{V}\varrho}{\mu}\right)^{-0.2} \tag{161a}$$

where f is the friction coefficient which is correlated with Re.

When the fluid density is not uniform, natural convection may be an important heat transfer mechanism and the Grashof number should be considered in this case.

$$Gr = \frac{D^3 g\bar{\varrho}\Delta\varrho}{\mu_f^2} \tag{162}$$

For liquids flowing in horizontal tubes the general form of correlation is

$$Nu = 1.75\left[\frac{D}{L}RePr + 0.04\left(\frac{D}{L}GrPr\right)^{0.75}\right]^{1/3}\left(\frac{\mu_b}{\mu_w}\right)^{0.14} \tag{163}$$

The first term in the brackets is the contribution of forced convection and the second term is that of natural convection. For vertical tubes the coefficient 0.04 is replaced by 0.072 and μ_b/μ_w is 1.

For heat transfer by natural convection from a vertical plane or cylinder the correlation is

$$Nu = \frac{hL}{k} = c(GrPr)^a \tag{164}$$

where $c = 0.13$, $a = 1/3$ for turbulent flow, $3.5 \times 10^7 < GrPr < 10^{12}$; $c = 0.55$, $a = 1/4$ for laminar flow, $10^4 < GrPr < 3.5 \times 10^7$.

The heat transfer coefficient correlations vary depending on the flow conditions, geometry and configuration of the system, rheological properties of the fluid, and the mechanism of heat transfer. Other heat transfer correlations are available in Welty (1976), Foust (1980), Bailey (1977) and Perry (1973).

The major goal in the design of heat exchangers for fermentation vessels is to maximize the heat transfer coefficient and therefore minimize the heat transfer area. In order to achieve this goal, a high degree of turbulence is required in the fermentation medium and better heat transfer configurations need to be developed.

2.8 NOMENCLATURE

Ae	aeration number
A_p	external surface of a support particle or a microbial pellet
a	gas–liquid interfacial area per unit liquid volume
a'	surface area of pores occupied by cells per unit volume of a support particle
C_A	molar concentration of A
C_A^*	liquid phase concentration of A in equilibrium with partial pressure of A in the gas phase
C_{AL}	concentration of A in the liquid phase
C_{Ai}	concentration of A at the gas–liquid interface
C_s	substrate concentration
C_{ss}	substrate concentration on the surface of a porous support
D_{AB}	diffusivity of A in B
D_b	diameter of gas bubbles
$\bar{D}_{bs}$	Sauter mean diameter of gas bubbles
D_{CO_2}	diffusion coefficient of carbon dioxide
D_i	diameter of impeller
D_{O_2}	diffusivity of oxygen
D_s	effective diffusivity of limiting substrate
D_p	diameter of a spherical particle
D_T	tank diameter
d_A	diameter of molecule A
F	Faraday's constant
F_a	volumetric flow rate of air
F_G	volumetric flow rate of gas
G	free energy
Gr	Grashof number
g	gravitational acceleration
g_c	Newton's law constant
H'	Henry's law constant
H	gas holdup
H_L	height of liquid in fermenter
ΔH_C	heat of combustion of cells
ΔH_S	heat of combustion of substrate
$\bar{h}$	average convective heat transfer coefficient
I	ionic strength
i	intensity of current
J_{AZ}	molar flux of A in the Z-direction
K	consistency index of non-Newtonian fluid
K_G	overall mass transfer coefficient based on partial pressure driving force
K_L	overall mass transfer coefficient based on liquid phase driving force
K_{O_2}	saturation constant for specific rate of oxygen consumption
k_c	mass transfer coefficient based on molar concentration

k_G gas film mass transfer coefficient

k_L liquid film mass transfer coefficient

$k_L a$ volumetric oxygen transfer coefficient

k_1 first-order rate constant

k Boltzmann's constant

k_w thermal conductivity of wall

L characteristic length

M_A molecular weight of A

m equilibrium constant between gas and liquid phase concentrations

m_A mass of molecule A

m_s rate of substrate transfer (or utilization)

m_{so} rate of substrate utilization without diffusion limitation

N_A molar flux of A

N rotational speed of impeller

n flow behaviour index of non-Newtonian fluid

n_{CO_2} rate of CO_2 desorption

$n_{O_2,T}$ rate of oxygen transfer

$n_{O_2,c}$ rate of oxygen consumption

$n_{O_2,i}$ molar flow rate of oxygen in inlet gas stream

$n_{O_2,o}$ molar flow rate of oxygen in off-gas stream

P power supplied to the agitator

P_G gassed power requirement

Po power number

p absolute pressure

p_{Ai} partial pressure of A at the gas–liquid interface

p_{AG} partial pressure of A in the gas phase

p_A^* partial pressure of A in equilibrium with the liquid phase concentration of A

Q_{Acc} rate of heat accumulation

Q_{Ag} rate of heat generation by agitation

Q_{Evap} rate of heat loss by evaporation

Q_{Exch} rate of heat transfer to surroundings or exchanger

Q_{Gr} rate of metabolic heat generation

Q_{Sen} rate of sensible enthalpy change

Q_{O_2} specific rate of oxygen consumption

$Q_{O_2,max}$ maximum specific rate of oxygen consumption

R radius of spherical support particle or mycelial pellet

Re_i impeller Reynolds number

R_G gas constant

R_{eq} equivalent radius of non-spherical particles

r radial distance from the center of a sphere

r_{AB} molecular separation at collision

r_{cr} critical radius

Sc Schmidt number

Sh Sherwood number

T absolute temperature

$T_{b,A}$ normal boiling point of liquid A

t time

V linear fluid velocity

V_p volume of an inert support particle or a microbial pellet

V_L liquid volume

v_{rms} root mean square velocity of gas bubbles

v_s superficial gas velocity

v_t terminal velocity of gas bubbles

v_A molal volume of liquid A at normal boiling point

W impeller width

X biomass concentration in liquid phase

Y_H g biomass produced per unit metabolic heat evolved

Y_{O_2} oxygen yield coefficient

Y_S g biomass produced per g substrate consumed

y mole fraction of oxygen in air

Z	number of charges on an ion
α	dimensionless saturation parameter
ε_{AB}	energy of molecular attraction
μ	viscosity
μ_a	apparent viscosity
μ	specific rate of microbial growth
μ_{max}	maximum specific rate of microbial growth
η	effectiveness factor
ϱ	density
σ	surface tension
ψ	electrical potential
ϕ	Thiele modulus
φ	modified Thiele modulus
φ'	association factor for solvent
τ	shear stress
τ_y	yield stress

2.9   REFERENCES

Aiba, S. and K. Kobayashi (1971). Comments on oxygen transfer within a mold pellet. *Biotechnol. Bioeng.*, **13**, 583.
Aiba, S. and K. Toda (1963). The effect of surface active agents on oxygen absorption in bubble aeration. I. *J. Gen. Appl. Microbiol.*, **7**, 100.
Aiba, S. and K. Toda (1964). Effect of surface active agents on oxygen absorption in bubble aeration. II. *J. Gen. Appl. Microbiol.*, **10**, 157.
Aiba, S. and T. Yamada (1961). Oxygen absorption in bubble aeration. Part 1. *J. Gen. Appl. Microbiol.*, **7**, 100.
Aiba, S., M. Hara and J. Someya (1963b). Mass transfer in fermentation: Oxygen absorption in gluconic acid fermentation. *J. Ferment. Technol.*, **41**, 74.
Aiba, S., A. E. Humphrey and N. F. Millis (1973). *Biochemical Engineering*, 2nd edn. Academic, New York.
Arnold, B. H. and R. Steel (1958). In *Biochemical Engineering*, ed. R. Steel. Macmillan, New York.
Astarita, G. (1967). *Mass Transfer with Chemical Reaction*. Elsevier, Amsterdam.
Atkinson, B. (1974). *Biochemical Reactors*. Pion, London.
Bailey, J. E. and D. F. Ollis (1977). *Biochemical Engineering Fundamentals*. McGraw-Hill, New York.
Baird, M. H. I (1960). Ph.D. Thesis, University of Cambridge.
Benedek, A. and W. J. Heideger (1971). Effect of additives on mass transfer in turbine aeration. *Biotechnol. Bioeng.*, **13**, 663.
Bird, R. B., W. E. Stewart and E. N. Lightfoot (1960). *Transport Phenomena*. Wiley, New York.
Bittar, E. D. (ed.) (1975). *Membranes and Ion Transport*, vols. 1–3. Wiley, New York.
Blakebrough, N. (ed.) (1967). *Biochemical and Biological Engineering Science*, vol. 1. Academic, New York.
Blakebrough, N. and K. Sambamurthy (1966). Mass transfer and mixing rates in fermentation vessels. *Biotechnol. Bioeng.*, **8**, 25.
Boos, W. (1974). Bacterial transport. *Annu. Rev. Biochem.*, **43**, 123–146.
Brierley, M. R. and R. Steel (1959). Agitation–aeration in submerged fermentation. Part 2. Effect of solid disperse phase on oxygen absorption in a fermenter. *Appl. Microbiol.*, **7**, 57.
Bull, D. N. and L. L. Kempe (1971). Influence of surface active agents on oxygen absorption to the free interface in a stirred fermenter. *Biotechnol. Bioeng.*, **13**, 529.
Calderbank, P. H. (1958). Physical rate processes in industrial fermentation. Part 1. The interfacial area in gas–liquid contacting with mechanical agitation. *Trans. Inst. Chem. Eng.*, **36**, 443.
Calderbank, P. H. (1967). Mass transfer in fermentation equipment. In *Biochemical and Biological Engineering Science*, ed. N. Blakebrough, vol. 1, pp. 102–180. Academic, New York.
Calderbank, P. H. and M. B. Moo-Young (1959). The prediction of power consumption in the agitation of non-Newtonian fluids. *Trans. Inst. Chem. Eng.*, **37**, 26–33.
Calderbank, P. H. and M. B. Moo-Young (1961). The power characteristics of agitators for the mixing of Newtonian and non-Newtonian fluids. *Trans. Inst. Chem. Eng.*, **39**, 337.
Calderbank, P. H. and J. Rennie (1962). The physical properties of foams and froths. *Trans. Inst. Chem. Eng.*, **40**, 191.
Calderbank, P. H., D. S. L. Johnson and J. Loudon (1970). Mechanics and mass transfer of single bubbles in free rise through some Newtonian and non-Newtonian liquids. *Chem. Eng. Sci.*, **25**, 235.
Chakravarty, M., S. Begum, H. D. Singh, J. N. Baruah and M. S. Iyengar (1973). Gas hold-up distribution in gas-lift column. *Biotechnol. Bioeng. Symp. Ser.*, **4**, 363.
Charles, M. (1978). Technical aspects of the rheological properties of microbial cultures. In *Advances in Biochemical Engineering*, ed. A. Fiechter, vol. 8, pp. 1–62. Springer-Verlag, New York.
Cooney, C. L., D. I. C. Wang and R. I. Mateles (1968). Measurement of heat evolution and correlation with oxygen consumption during microbial growth. *Biotechnol. Bioeng.*, **11**, 269.
Cooper, C. M., G. A. Fernstrom and S. A. Miller (1944). Performance of agitated gas–liquid contactors. *Ind. Eng. Chem.*, **36**, 504–509.
Coppock, P. D. and G. T. Meiklejohn (1951). The behavior of gas bubbles in relation to mass transfer. *Trans. Inst. Chem. Eng.*, **29**, 75.

Danckwerts, P. V. (1970). *Gas–Liquid Reactors*. McGraw-Hill, New York.

Darby, R. T. and D. R. Goddard (1950). Studies of the respiration of the mycelium of the fungus *Myrotechium verucara*. *Am. J. Bot.*, **37**, 379.

Deindoerfer, F. H. and E. L. Gaden, Jr. (1955). Effects of liquid physical properties on oxygen transfer in penicillin fermentation. *Appl. Microbiol.*, **3**, 253.

Eckenfelder, W. W., Jr. (1959). Absorption of oxygen from air bubbles in water. *J. Sanitary Eng. Div.*, *Proc. A.S.C.E.*, **85**, No. SA4, 89–99.

Eckenfelder, W. W., Jr. and E. L. Barnhart (1961). The effect of organic substances on the transfer of oxygen from air bubbles in water. *AIChE J.*, **7**, 631.

Endo, K. and Y. Oyama (1958). Size of droplet disintegrated in liquid–liquid contacting mixer. *Inst. Phys. Chem. Res.*, *Sci. Pap. (Jpn.)*, **52**, 131 (*Sci. Pap. Phys. Chem. Res. (Tokyo) Chem. Abstr.*).

Finn, R. K. (1954). Agitation–aeration in the laboratory and in industry. *Bacteriol. Rev.*, **18**, 254–274.

Finn, R. K. (1967). Agitation and aeration. In *Biochemical and Biological Engineering Science*, ed. N. Blakebrough, vol. 1, pp. 69–99. Academic, New York.

Foust, A. S., L. A. Wenzel, C. W. Clump, L. Maus and L. B. Andersen (1980). *Principles of Unit Operations*, Wiley, New York.

Fukuda, H., Y. Sumino and T. Kansaki (1968a). Scale-up of fermenters. I. Modified equations for volumetric oxygen transfer coefficient. *J. Ferment. Technol.*, **46**, 829–837.

Fukuda, H., Y. Sumino and T. Kansaki (1968b). Scale-up of fermenters. II. Modified equations for power requirement. *J. Ferment. Technol.*, **46**, 838–845.

Hammerton, D. (1953). Ph.D. Thesis, University of Birmingham.

Higbie, R. (1935). Rate of absorption of a pure gas into a still liquid during short period of exposure. *Trans. Am. Inst. Chem. Eng.*, **31**, 365–389.

Hirschfelder, J. O., R. B. Bird and E. L. Spotz (1949). Viscosity and other physical properties of gases and gas mixtures. *Trans. ASME*, **71**, 921.

Hixon, A. W. and E. L. Gaden, Jr. (1950). Oxygen transfer in submerged fermentation. *Ind. Eng. Chem.*, **42**, 1792–1801.

Hospodka, J., Z. Caslavsky, K. Beren and F. Stross (1964). The polarographic determination of oxygen uptake and transfer rate in aerobic steady-state yeast cultivation on laboratory plant scale. In *Continuous Cultivation of Microorganisms*, ed. I. Malek, K. Beran and J. Hospodka, vol. 2, pp. 353–367.

Johnson, M. J., J. Borkowski and C. Engblom (1964). Steam sterilizable probes for dissolved oxygen measurement. *Biotechnol. Bioeng.*, **6**, 457.

Kaback, H. R. (1972). Transport across isolated bacterial membranes. *Biochim. Biophys. Acta*, **265**, 367–416.

Kafarov, V. V. and B. M. Babanov (1959). Interfacial surface of mutually insoluble liquids produced by mechanical mixing with an impeller. *Zh. Prikl. Khim.*, **32**, 789.

Kinzer, G. D. and R. Gunn (1951). The evaporation temperature and thermal relaxation-time of freely falling water-drops. *J. Meteorol.*, **8**, 71.

Kotyk, A. and K. Janacek (1975). *Cell Membrane Transport: Principles and Techniques*, 2nd edn. Plenum, New York.

LeDuy, A., A. Marsan and B. Coupal (1974). A study of rheological properties of a non-Newtonian fermentation broth. *Biotechnol. Bioeng.*, **16**, 61–76.

Lehninger, A. L. (1975). *Biochemistry*, 2nd edn., pp. 779–805. Worth, New York.

Mancy, K. H. and D. A. Okun (1963). Effect of surface active agents on the rate of oxygen transfer. In *Advances in Biological Waste Treatment*, ed. W. W. Eckenfelder, Jr. and J. McCabe, p. 111. Pergamon, Oxford.

Mateles, R. I. (1971). Calculation of the oxygen required for cell production. *Biotechnol. Bioeng.*, **13**, 581–582.

McDowell, R. V. and I. E. Myers (1955). Preprint 49, *Bubble and Drop Symposium*, AIChE meeting, Detroit, MI.

McKeown, J. J. and D. A. Okun (1963). Effects of surface active agents on oxygen bubble characteristics. In *Advances in Biological Waste Treatment*, ed. W. W. Eckenfelder, Jr. and J. McCabe, p. 113. Pergamon, Oxford.

Metzner, A. B. and R. E. Otto (1957). Agitation of non-Newtonian fluids. *AIChE J.*, **3**, 3–10.

Michel, B. J. and S. A. Miller (1962). Power requirements of gas–liquid agitated systems. *AIChE J.*, **8**, 262–266.

Moo-Young, M. and H. W. Blanch (1981). Design of biochemical reactors: Mass transfer criteria for simple and complex systems. *Adv. Biochem. Eng.*, **19**, 1–69.

Oyama, Y. and K. Endoh (1955). Power characteristics of gas–liquid contacting mixers. *Chem. Eng. (Jpn.)*, **19**, 2–11.

Perry, R. H. and C. H. Chilton (1973). *Chemical Engineers' Handbook*, 5th edn. McGraw-Hill, New York.

Pirt, J. S. (1975). *The Principles of Microbe and Cell Cultivation*. Wiley, New York.

Richards, J. W. (1961). Studies in aeration and agitation. *Prog. Ind. Microbiol.*, **3**, 143–172.

Richards, J. W. (1964). Power input to fermenters and similar vessels. *Br. Chem. Eng.*, **8**, 158–163.

Robinson, C. W. and C. R. Wilke (1973). Oxygen absorption in stirred tanks: A correlation for ionic strength. *Biotechnol. Bioeng.*, **15**, 755.

Roels, J. A., J. van den Berg and R. M. Voncken (1974). The rheology of mycelial broths. *Biotechnol. Bioeng.*, **16**, 181.

Rose, W. D. and M. R. J. Wyllie (1950). Specific surface areas and porosities from photomicrographs. *Bull. Am. Assoc. Petrol. Geol.*, **34**, 1748.

Rushton, J. H. (1951). The use of pilot-plant mixing data. *Chem. Eng. Prog.*, **47**, 485.

Rushton, J. H., E. W. Costich and H. J. Everett (1950). Power characteristics of mixing impellers. Part 2. *Chem. Eng. Prog.*, **46**, 467–476.

Sheng-Li, Pang (1953). Ph.D. Thesis, University of Washington.

Sherwood, T. K., R. L. Pigford and C. R. Wilke (1975). *Mass Transfer*. McGraw-Hill, New York.

Steinberger, R. L. and R. E. Treybal (1960). Mass transfer from a solid soluble sphere to a flowing liquid stream. *AIChE J.*, **6**, 227.

Stryer, L. (1981). *Biochemistry*, 2nd edn. Freeman, San Francisco, CA.

Taguchi, H. (1971). The nature of fermentation fluids. *Adv. Biochem. Eng.*, **1**, 1.

Taguchi, H. and A. E. Humphrey (1966). Dynamic measurement of volumetric oxygen transfer coefficient in fermentation systems. *J. Ferment. Technol.*, **44**, 881–889.

Taguchi, H. and S. Miyamota (1966). Power requirement in non-Newtonian fermentation broth. *Biotechnol. Bioeng.*, **8**, 43–54.

Taguchi, H., T. Imanaka, S. Teramoto, M. Takatsa and M. Sato (1968). Scale-up of glucoamylase fermentation by *Endomyces* sp. *J. Ferment. Technol.*, **46**, 823–828.

Treybal, R. E. (1980). *Mass Transfer Operations*. McGraw-Hill, New York.

Umbreit, W. W., R. H. Burris and J. F. Stauffer (1972). *Manometric and Biochemical Techniques*, 5th edn., p. 13. Burgess, St. Paul, MN.

van Krevelen, D. W. and P. J. Hoftijzer (1950). Studies of gas-bubble formation: Calculation of interfacial area in bubble contactors. *Chem. Eng. Prog.*, **46**, 29.

Vardar, F. and M. D. Lilly (1982). The measurement of oxygen transfer coefficients in fermenters by frequency response techniques. *Biotechnol. Bioeng.*, **24**, 1711–1719.

Wang, D. I. C., C. L. Cooney, A. L. Demain, P. Dunnil, A. E. Humphrey and M. D. Lilly (1979). *Fermentation and Enzyme Technology*. Wiley, New York.

Welty, J. R., C. E. Wicks and R. E. Wilson (1976). *Fundamentals of Momentum, Heat and Mass Transfer*. Wiley, New York.

Wernan, W. C. and C. R. Wilke (1973). New method for evaluation of dissolved oxygen probe response for $k_L a$ determination. *Biotechnol. Bioeng.*, **15**, 571.

Whitman, W. G. (1923). The two-film theory of gas absorption. *Chem. Methods. Eng.*, **29**, 147.

Wilhelm, R. H., W. A. Donohue, D. J. Valesano and G. A. Brown (1966). Gas absorption in a stirred vessel: Locale of transfer action. *Biotechnol. Bioeng.*, **8**, 55–69.

Wilke, C. R. (1949). Estimation of liquid diffusion coefficients. *Chem. Eng. Prog.*, **45**, 218.

Wilke, C. R. and P. Chang (1955). Correlation of diffusion coefficients in dilute solutions. *AIChE J.*, **1**, 264.

Wilke, C. R. and C. Y. Lee (1955). Estimation of diffusion coefficients for gases and vapors. *Ind. Eng. Chem.*, **47**, 1253.

Wise, W. S. (1951). The measurement of the aeration of culture media. *J. Gen. Microbiol.*, **5**, 167–177.

Wise, D. L., D. I. C. Wang and R. I. Mateles (1969). Increased oxygen mass transfer rates from single bubbles in microbial systems at low Reynolds numbers. *Biotechnol. Bioeng.*, **11**, 647.

Yano, T., T. Kodama and K. Yamada (1961). Fundamental studies on the aerobic fermentation. Part III. Oxygen transfer within a mold pellet. *Agric. Biol. Chem.*, **25**, 580.

Yoshida, F., A. Ikeda, S. Imakawa and Y. Miura (1960). Oxygen absorption rates in stirred gas–liquid contactors. *Ind. Eng. Chem.*, **52**, 435.

Zieminski, S. A., C. C. Goodwin and R. L. Hill (1960). The effect of some organic substances on oxygen absorption in bubble aeration. *Tappi*, **43**, 1029.

3

Fermenter Design and Scale-up

M. CHARLES
Lehigh University, Bethlehem, PA, USA

3.1 INTRODUCTION

The fermenter designer's primary objective is so obvious that it should be stated frequently: a fermenter should be designed so that it will operate reliably and be controllable enough to fit into a general operating plan structured to achieve overall plant optimization over a reasonable range of processing fluctuations. To achieve this goal we should (but all too often we do not) consider the following simultaneously): fermenter productivity and yield, fermenter operability and reliability, product purification, the relationship between purification and fermentation, water management, energy requirements, waste treatment, *etc*. We must keep in mind that fermentation is only part of the overall process, and that 'optimizing' this one stage may not always be beneficial. For example, it is not always true that maximum fermentation productivity will yield lowest overall costs; indeed, it is quite possible that over some range of productivity, increased productivity will result in higher overall costs. Such situations can occur when increased productivity results in lower yield or in broth properties which make the recovery process more difficult.

The designer must determine first the general nature of the fermenter to be used. Unfortunately, in all too many cases the work done in R & D together with existing 'conventional wisdom' will have narrowed the choice rather markedly, often without very good reason. In most cases, a more or less 'standard' stirred tank fermenter (STF) will be chosen. The designer's primary job usually becomes to scale up the 'conventional' fermenter and to design in features such that control will be possible over reasonable ranges of the important process variables (*e.g.* oxygen concentration, temperature, pH) and that the operation will be reliable and contamination-free. To achieve these ends he must provide for, at least, adequate heat and oxygen transfer; aseptic and sterilization procedures which are as close as possible to being 'idiot proof'; reliable foam control; good spatial definition of environmental conditions; simple, rapid and thorough cleaning systems; responsive, reliable and appropriate monitoring and control systems; appropriate materials of construction and reliable fabrication methods; and the possibility that the fermenter will be used for more than one product. He must achieve all this at the lowest possible cost.

At first sight this does not appear to be a difficult job, particularly in light of the fact that most fermenters are more or less 'standard' STFs. However, the fact is that it is not a simple matter, and there are very few who can do it well. In this chapter we will highlight the state of the art concerning the basic operating characteristics of fermenters and will try to point out where the gaps in our knowledge create real problems for the designer. We will devote considerable space to the discussion of newer designs in the hope that we can stimulate serious thinking concerning practical applications.

In a review of this type, authors tend to be influenced markedly by information and emphases in the open literature. For example, the reader should recognize from the outset that in the past oxygen transfer has received more emphasis than any other design factor. We cannot help but consider that here, but we will attempt to emphasize other important factors and to point out important inadequacies.

Finally, we hope the reader will appreciate that the complexity of fermentations always will cause significant design problems and uncertainties. Improving our design theories and correlations will be important, but only in the sense of providing more valuable guides. They will prove most valuable when used in conjunction with modern automatic control and scale translation studies involving systematic, coordinated, carefully planned and well-documented studies at pilot and production scales.

3.1.1   Rheology

Despite the fact that for a given fermenter, heat transfer, mass transfer and bulk flow behavior all are governed largely by the rheology of the fermentation broth, little attention has been given to it. Furthermore, there appears to be considerable confusion and misinformation, some of which has found its way into the literature. The interested reader is directed to the reviews of Blanch and Bhavaraju (1976), Charles (1976) and Pace (1980) for a more extensive discussion.

3.1.1.1   *Viscosity*

Viscosity, μ, for a fluid in *simple shear flow* is by *definition*

$$\mu = \tau/\dot{\gamma} \tag{1}$$

where τ = shear stress and $\dot{\gamma}$ = shear rate. In somewhat simplified terms, a fluid is in simple shear flow when it has only one velocity component. For a more complete discussion see Coleman and Noll (1966).

To measure viscosity (as defined) we must use a device which insures simple shear flow (sometimes called 'viscometric flow'). A variety of reliable (as well as unreliable) viscometers providing various types of viscometric flows are available readily; however, one must exercise considerable caution in operating them, and in interpreting the data they provide. Readers unfamiliar with such instruments would do well to consult Charles (1976), Van Wazer *et al.* (1963), Skelland (1967) and Middleman (1968).

There are several commercial and experimental devices which purport to measure viscosity but which do not provide viscometric flow; obviously, such devices do not satisfy the definition and cannot be relied upon to give true viscosity data. This creates no significant problems if the data are used for purposes of evaluating or controlling a fixed process, for quality control, *etc*. However, problems can arise if the information is used for purposes of scale-up, comparison of fermenters displaying different fluid mechanical behaviors, calculation of heat and mass transfer characteristics, *etc*.

3.1.1.2 Newtonian fluids

By definition, a Newtonian fluid is one for which the viscosity is independent of shear rate (*i.e.* the relationship given by equation (1) is strictly linear for a specific fluid at a single temperature). Water, solutions of low molecular weight compounds and suspensions of relatively non-deformable sphere-like particles are typical Newtonian fluids. In general, the viscosity of a Newtonian fluid will vary with temperature, composition, pH and ionic strength.

3.1.1.3 Non-Newtonian fluids

Fluids for which viscosity is a function of shear rate (equation 1 is not linear) are by definition non-Newtonian. Those of primary importance in fermentation are pseudoplastic fluids and fluids exhibiting a yield stress. The viscosity of a pseudoplastic fluid decreases with increasing viscosity. Several mathematical models are available to predict such behavior (Charles, 1976), but usually the simple power law model suffices for design purposes:

$$\mu = K\dot{\gamma}^{n-1} \tag{2}$$

Here *n* is the flow behavior index and *K* is the consistency index. For fluids having a yield stress

$$\tau - \tau_{\mathrm{y}} = \mu_{\mathrm{p}}\dot{\gamma} \tag{3}$$

Here τ_{y} is the yield stress and μ_{p} is the plastic viscosity or rigidity. In general, μ_{p} is a function of shear rate. Note that the fluid will not flow until the applied shear stress exceeds the yield stress.

Non-Newtonian viscosity can have very important practical effects on bulk flow and on heat and mass transfer. For example, in a stirred tank the shear rate is highest near the impeller and decreases rather sharply with distance from it (Norwood, 1960). If the fluid is pseudoplastic, viscosity will be relatively low near the impeller but quite high at any position not close to it. Therefore, mixing and transport phenomena will be good only in the immediate region of the impeller. Similar and even more pronounced problems can arise if the fluid has a high yield stress. These problems have important implications concerning equipment design (see sections dealing with design of specific types of fermenters).

3.1.1.4 Unicellular fermentation broths

Yeast and bacterial broths usually exhibit concentration-dependent, relatively low Newtonian viscosities. Surprisingly, there are very few published data. In most cases, viscosities during fermentation do not get high enough to cause any significant difficulty. However, viscosities can become quite high during product recovery; considerable caution is advisable in that viscosity can rise *very* quickly with concentration at concentration around 85 g packed cells per 100 g suspension (Modeer, 1974).

There are some cases in which unicellular broths exhibit high viscosity, non-Newtonian behavior: these generally involve the production of an extracellular polysaccharide such as xanthan, which exhibits very high viscosity and pronounced pseudoplasticity (Charles, 1976). High concentrations of protein also can cause some rheological problems, but this has not received any significant attention in the literature.

3.1.1.5 Mycelial fermentation broths

Mycelial broths tend to be highly viscous and non-Newtonian. In broad terms, the rheological characteristics are determined by the mycelial concentration and morphology. The characteristics are quite complex and have not been studied extensively. With the exception of work done by Metz *et al.* (1979) and Roels (1974) the very important effects of morphology have generally been neglected.

What little published information exists indicates that most broths containing disperse, filamentous growth tend to be pseudoplastic and to have significant yield stress. Unfortunately, there is a great deal of disagreement and ambiguity in the literature, as well as a considerable amount of data which must be regarded as suspect. The primary reasons for this are a lack of understanding of the fundamental rheological concepts and experimental difficulty in measuring the viscosity of mycelial broths (Charles, 1976). The designer must exercise great caution when relying on published data.

3.1.2 Oxygen Transfer

The rate of oxygen transfer, OTR, from the bulk gas phase to the bulk liquid phase is governed by the rate equation

$$\text{OTR} = k_L a(C^* - C) \tag{4}$$

where $k_L a$ is the mass transfer coefficient based on the liquid phase (s^{-1}), C is the actual dissolved oxygen concentration in the bulk liquid phase and C^* is the dissolved oxygen concentration in equilibrium with the existing oxygen partial pressure in the gas phase.

At atmospheric pressure, C^* is only about 0.25 mM; therefore, the maximum transfer rate is 0.25 $k_L a$ (mmol l^{-1} h^{-1}). Typical $k_L a$ values range from 0.02 s^{-1} to 0.25 s^{-1} giving maximum transfer rates (at atmospheric pressure, with air) ranging from 18 mmol l^{-1} h^{-1} to 225 mmol l^{-1} h^{-1}. Such rates will support growth rates up to only 7–8 g l^{-1} h^{-1}. Higher rates require unreasonably high power inputs or the use of higher pressures and/or pure oxygen.

Generally, high power inputs are required to obtain high $k_L a$ values: the more viscous and non-Newtonian the fluid becomes, the higher the power input. For example, a power input of about 1.3 kW m^{-3} is required often to obtain oxygen transfer rates of only 20–40 mmol l^{-1} h^{-1} in viscous fungal broths in standard stirred tank fermenters. Other devices are more efficient than the stirred tank, but even these require relatively high power inputs (see Section 3.3). Obviously, this is viewed currently as a very important factor and is receiving a great deal of attention. Unfortunately, practical progress is not being made rapidly. Clearly, attempts to minimize energy requirements should be encouraged, but one should not lose sight of the overall goal: it is not necessarily the case that complete energy optimization of the fermenter will be consistent with plant optimization in terms of energy or profit.

The theory of oxygen transfer has received a great deal of attention and need not be reviewed here. Suffice it to say that the theory, as it now stands, is at best a useful guide in practical design. This is understandable in light of the complex nature of fermentation. For extensive discussions of oxygen transfer theory see the reviews of Blanch (1976), Moo-Young (1981), Schügerl (1981) and Brauer (1979).

Many empirical oxygen transfer correlations have been presented in the literature and have been reviewed extensively. They have some limited use in practical design, but almost all have some very serious shortcomings. For further discussion see subsequent sections dealing with specific fermenters and the reviews cited in the previous paragraph.

3.1.3 Power Input Requirements

The bulk of the power input in aerobic fermentations is required to ensure adequate oxygen transfer. In practice, power inputs to STFs range from about 0.5 to about 2.5 HP per 100 gal; agitation power accounts for 85–95% of the total, and compression accounts for 5–15%. The mechanical energy input can account for as much as 25% of the fermenter heat load thereby contributing significantly to the ever-present heat transfer problem (see Section 3.1.5).

The need for very high power inputs imposes considerable constraints on fermenter design and on the maximum size of STFs. For example, it is impractical to transmit *via* a single shaft to an STF more than about 1000 HP. In theory, multiple shafts could be used, but this would create mechanical design and asepsis problems. Such constraints have led to a great deal of research to reduce power requirements. They also have led to more favorable consideration of non-STF fermenters such as the airlift. The ICI 3000 m^3 (working volume) pressure cycle fermenter (PCF) is a prime example (Winkler, 1983; see Figure 1). The power input to the PCF is approximately 2.5 MW (Andrew, 1981). Power input requirements for specific fermenter types will be discussed in subsequent sections. The reader also is referred to recent reviews (Moo-Young, 1981; Brauer, 1979; Schügerl, 1981).

3.1.4 Spatial Variations

Spatial variations of environmental conditions, compositions, *etc.*, exist in any fermenter. In some cases they are designed in (as in loop fermenters); in others they exist as imperfections (as in imperfectly mixed STFs). Variations exist even in what we usually consider to be a perfectly mixed tank. Speaking strictly, complete homogeneity cannot be achieved in any stirred tank. For example, shear rate, shear stress, local Reynolds number (N_{Re}) and local turbulence all can vary greatly with distance from the impeller.

Clearly, our usual definition of homogeneity is a functional one based on just a few factors. The extent to which this is adequate depends on the particular case: if shear stress appears to have no significant effect on the fermentation, the fact that it varies spatially is not important provided that we always do the fermentation in a given vessel (or one just like it). The problems caused by ignoring spatial variations of *seemingly* unimportant variables generally arise when we scale up or change the fermenter geometry, operating conditions, *etc.* (See Section 3.3 for a more detailed discussion.)

Our ability to understand, to characterize and to control spatial variations determines, in large measure, the success of our efforts to design fermenters and to control fermentations. We usually base control actions on point measurements of individual sensors. We infer some 'average' value of the sensed variable, but unless we have reasonably good information (empirical or calculated) concerning spatial variations (which generally vary with time) the point measurements can be meaningless and misleading. For example, large spatial variations of dissolved oxygen concentrations are not uncommon in STFs (Maxon, 1959; Phillips, 1961; Steel, 1966).

Similarly, if we wish to do strictly rational design wherein we design in or design out spatial variations we must be able to predict the variations. In general, the complex rheology and fluid mechanics preclude meaningful predictions based on theoretical calculations; in most cases we must resort to empirical correlations such as mixing time correlations for STFs. Usually the importance of these considerations increases with increasing the viscosity and rheological complexity of the culture broth. Details for specific fermenter types will be discussed in subsequent sections.

3.1.5 Heat Transfer

Providing for adequate heat transfer is often a major problem we encounter in designing large fermenters for aerobic operation. Metabolic activities can generate as much as (in vigorous, aerobic fermentations) 100–200 BTU gal^{-1} h^{-1} of thermal energy, while mechanical energy inputs of 0.5 and 2.5 HP per 100 gal (mechanical agitation plus gas expansion) can generate an additional 10–60 BTU gal^{-1} h^{-1}. The overall heat transfer coefficient for a typical production fermenter (most are 'standard' STFs) is usually between 50 and 150 BTU ft^{-2} h^{-1} $°F^{-1}$ (the low value is typical of viscous, non-Newtonian broths while the high value is typical of relatively low viscosity, unicellular broths). The cooling water inlet temperature can range from 13 to 21 °C. Therefore, for a 2.5 °C approach at the cooling water outlet, the heat transfer area required for a fermentation conducted at 28 °C in a 50 000 gal fermenter can range from about 1000 ft^2 to about 32 000 ft^2; the corresponding cooling water flow rate ranges from 500 to about 3800 g.p.m. The available jacket area is approximately 1600 ft^2. The problem is clear: obviously a jacket alone usually will not supply adequate heat transfer surface unless we make compromises, *e.g.* less agitation/aeration and slower growth, which will result in lower productivity.

Usually, most designers turn to internal cooling coils, external heat exchange loops or mechanical refrigeration (when all else fails) to minimize the heat transfer problem without significantly decreasing production rates. Internal coils overcome the problem of the decrease in jacket area per unit volume of fermenter with increasing fermenter volume; internal coil area is essentially scale independent. In *some* cases coils also have higher heat transfer coefficients than jackets. However, they can occupy a considerable volume, interfere with flow patterns and create cleaning and sterilization problems. Also, pinhole leaks often are a serious source of contamination. It also is worth noting that in many cases coil heat transfer coefficients are decreased greatly by heavy overgrowths of microorganisms. The trade off between jacketing and coils depends on the specific fermentation (*e.g.* viscosity characteristics) and on the details of vessel and agitation design. Unfortunately, very little heat transfer data have been published for real fermentation broths in STFs or any other type of fermenter. Indeed, very little has been published in general concerning fermentation heat transfer.

Refrigerated coolant will decrease the heat transfer area required but is quite costly. For example, the heat load for a 50 000 gal fermenter can reach 6×10^6 BTU h^{-1} and the total refrigeration cost can be up to \$60 per hour. Clearly, this depends on the particular fermentation and can vary significantly during a given fermentation. The point is that the high cost coupled with significant maintenance problems requires considerable justification dependent on the economic factors of the overall process.

External loops remove the constraints imposed by the fermenter design and uncouple mass transfer, *etc.*, from heat transfer. A further advantage is that external loops can provide much larger heat transfer coefficients than can jackets or internal coils. One of the major disadvantages is that pumping can be inefficient and/or expensive; very few pumps can handle air-laden broths efficiently at the flows and pressures required for external circulation. Those that can tend to be very expensive (*e.g.* the IZ pump of Vogelbusch). However, it is important to note that the often-raised argument, that in the case of STFs all the pumping power is wasted, is quite incorrect and misleading. Clearly, the only power 'wasted' is that required to overcome frictional losses in the loop; the remainder can serve to agitate the broth in the fermenter. A well-designed re-entry configuration will insure effective use of the kinetic energy of the returning broth thereby decreasing the agitator power requirement. This idea carried further leads to 'agitatorless' recirculation fermenters such as the deep jet and the jet loop. Finally, the potentials for contamination are obvious but the problems can be handled readily.

Clearly, thermophilic organisms can decrease heat transfer problems. Unfortunately many of the thermophilic strains are not as productive as their mesophilic relatives. Nevertheless, this alone should not be a deterrent to their use; one must consider the *overall* economic effects including the cost of heat transfer. Heat transfer is extremely important in large-scale fermentation and will become more important as we move toward increasing microbial concentrations and the use of faster organisms provided to us by the new biotechnology. Probably more than any other mechanical factor, heat transfer will limit vessel size and fermenter productivity. It warrants far greater study than it has received.

3.1.6 Asepsis

This obviously important aspect of large-scale fermenter design seldom receives adequate attention when design is discussed in the literature. The implication (intended or not) appears to be that design for asepsis is essentially a mechanical consideration and has little to do with heat and mass transfer, *etc.* This attitude is quite prevalent among academic researchers and is found often in industrial R & D groups. This is indeed unfortunate, for while researches or developers should not have to worry about the details of designing steam-locks, aseptic seals, *etc.*, they should keep in mind the problems of large-scale asepsis with a view toward heading off at early stages problems which might require unfavorable design compromises at the production scale.

Large-scale aseptic design *per se* is not discussed to any significant extent in the literature. There are a few references (Lundell, 1976; Solomons, 1969) which might prove valuable to the novice, but experience and word-of-mouth are still the best current sources.

3.1.7 Foam

Foam is a major problem during many fermentations: significant losses are incurred as a result of foam-outs. The severity of the problem depends on virtually all of the fermentation parameters ranging from medium composition to mechanical design and operation of the fermenters. For further discussion of this important practical problem the reader is referred to Solomons (1969).

3.1.8 Fermenter Utility

In a large segment of the fermentation industry, a given fermenter often is used to produce at least two products. The designer should attempt to anticipate such use, and to design into the fermenter enough flexibility to deal with several anticipated products. Obviously, this will require some compromises at the design stage, but these generally allow one to approach overall optimization more closely than compromises which must be made after the fermenter is built.

3.1.9 Construction

Unfortunately, we do not have enough space here to discuss the details of materials of construction, fabrication, *etc*. In general, decisions concerning these aspects will depend strongly on whether or not the vessel must be designed for aseptic operation. If it must be, then one is best advised to deal with engineering firms who have proved their abilities to design and construct fermenters (and fermentation plants). Conventional chemical engineering design and construction firms, no matter how good they are, usually think of a fermenter as being just another reactor, and overlook small but critical details, particularly with regard to asepsis. One must remember that asepsis can be designed in relatively easily by fermenter-design professionals; but the errors in an improperly designed system are very difficult and sometimes impossible to correct once the fermenter has been constructed.

3.1.10 Scale Translation

In practice, it usually is difficult to translate (up or down) reliably a fermentation from one scale to another, particularly an aerobic fermentation involving viscous, non-Newtonian broths. Indeed, this is a problem which seldom is addressed in the literature directly, meaningfully or with much accurate quantitative rationality. Also, much of the information and philosophy expressed in the literature can be quite misleading.

The behavior of a given population of microorganisms should be scale invariant so long as *all* aspects of the environment also are scale invariant and the size of the initial population is adjusted for scale. Therefore, we would be able to translate easily from one scale to another if we could provide precisely the same environment at both scales. Alternatively, if we knew the responses of the organisms to all the important variables, and if we could predict the temporal and spatial histories of the environments, we would be able to scale up accurately and to exercise precise control at any scale.

In practice, neither approach is possible; therefore, highly empirical methods based on the concepts of similitude and dimensionless groups have been developed. These methods are based generally on the fact that through dimensional analysis important variables such as power input, heat and mass transfer coefficients, *etc*., can be related to system geometry, agitation speed, physical properties, *etc*., by means of dimensionless groups. For example, power input to an STF is correlated by means of

$$N_P = a_1 N_{Re}{}^{b_1} f(\text{geometry}) \tag{5}$$

where a_1 and b_1 are constants, and N_P and N_{Re} are defined as follows:

$$N_P = \frac{P g_c}{N^3 D_i{}^5 \varrho} = \text{power number} \tag{6}$$

$$N_{Re} = \frac{N D_i{}^2 \varrho}{\mu} = \text{Reynolds number} \tag{7}$$

where P is power input, N is impeller speed, D_i is impeller diameter, ϱ is broth density and μ is broth viscosity. Note that implicit in the formulation of equation (5) is the assumption that we have taken into account all of the important variables, but we have not. For example, equation (5) is applicable only when gravitational and surface forces are not important. See Johnstone and Thring (1957) for a more extensive discussion.

In most cases, published correlations do not specifically include terms for the effects of geometry [f (geometry) in equation 5]. Therefore, one must expect that at least the constants of such correlations will be scale dependent if geometric similarity is not maintained. (Geometric similarity requires that the ratios of significant dimensions remain scale invariant. For further discussion see Johnstone and Thring (1957).) Furthermore, most published correlations do not take into account variables that may become important outside the range(s) employed in the small-scale experiments used to develop the correlation and as the scale is increased.

In general, if we are to apply the empirical methods successfully, we must be able to identify the factors (*e.g.* oxygen transfer, shear) which are most important and which should be scale invariant (see subsequent sections for details concerning particular types of fermenters). Obviously, we cannot maintain constant all of the factors as we scale up; we are forced to assume that those we do not hold constant will not change enough to affect the large-scale fermentation deleteriously. Furthermore, it often is difficult for us to apply the empirical methods reliably because we do not have adequate correlations (Oosterhuis, 1981) and because very often reported successes and failures (reported less often) are difficult to explain.

These difficulties underscore the importance of the general concept of 'scale translation' discussed by Aiba (1973). In practice this involves the coordinated use of production and pilot plants in a series of systematic scale-down and scale-up experiments: operating conditions in the pilot plant are adjusted until the pilot fermentation behaves the same way as the production fermentation. Once this has been done to the point wherein the scale effects of operating variables and physical characteristics are understood reasonably well, the pilot vessel can be used most effectively to optimize the production fermentation and to provide accurate predictions of production-scale changes that will result from changes in culture, *etc.* It also is apparent that if the information is accumulated intelligently, it will be most useful in future attempts to scale up other fermentations. This approach should be encouraged in the private sector.

3.2   TYPES OF FERMENTERS

The 'standard' submerged culture stirred tank fermenter is used almost universally in the fermentation industry. Some of the reasons for this are quite sound (*e.g.* controllability); others are based on historical conventional wisdom, which has been questioned quite seriously and appropriately. The questions have led to the development and intensive study of new types of fermenters. Many of the new designs are reputed to be superior to the STF in the following ways: better heat and mass transfer; more reliable; easier to scale up; able to cope better with viscous, non-Newtonian broths. Also, some (*e.g.* airlift) can be built much larger than STFs because of the means by which power is transmitted to the fermentation broth. However, with very few exceptions (*e.g.* ICI's pressure cycle fermenter, Vogelbusch's deep jet fermenter), there has been little large-scale application of these new designs. Nevertheless, there is good reason to believe that this will change in the near future.

In the remainder of this section we will introduce some of these fermenters. They will be discussed in greater detail in subsequent sections.

3.2.1   Defined-flow Fermenters

In a defined-flow fermenter (DFF), the bulk liquid flow (at least) is forced to follow a defined path. The claims of the developers of such fermenters include all of those cited in Section 3.2 plus the claim of better overall control.

The major categories of DFFs are distinguished readily on the basis of applied motive power: air driven or direct mechanical drive.

3.2.1.1   Air-driven DFFs

Air-driven DFFs all are essentially modifications of the airlift fermenter. The ICI pressure cycle fermenter (PCF) illustrated in Figure 1 is an example of a modified airlift which has been operated on a remarkable scale (approximately 3000 m^3 broth volume) (Winkler, 1983). The PCF appears to operate quite well so long as the broth viscosity is relatively low. Also, it is apparent that organisms grown in a PCF are subjected to a wide range of conditions. The importance of the effects of these characteristics will depend on the particular fermentation. Further discussion will be presented in sections dealing specifically with DFFs.

The bubble column fermenter is similar to the airlift but has no provisions for direct control of the bulk liquid flow.

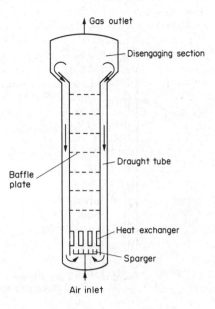

Figure 1 Pressure cycle fermenter (ICI)

3.2.1.2 Mechanically driven DFFs

One major class of mechanically driven DFF comprises modifications of the STF: the classical example is the Waldhof draught-tube fermenter which has long been used on a commercial scale. Others include the completely filled reactor developed at ETH Zürich (Figure 2) and the double-impeller system developed by LeGrys and Solomons (Figure 3).

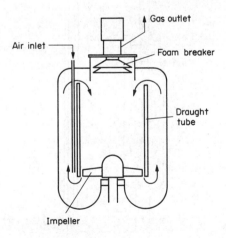

Figure 2 Completely filled reactor

A second major class of DFF comprises the so-called 'loop' fermenters illustrated in Figure 4. Design details vary, but in all cases mechanical energy is delivered by a pump located in a loop external to the main body of the fermenter. There has been a great deal of interest in such designs because they appear to have the following important advantages: less power required per kg of oxygen transferred; independent control of heat and mass transfer with the opportunity for much better heat transfer than can be achieved in a design without an external loop; the promise of more reliable scale-up; the practical possibility to build much larger fermenters. For example, the Vogelbusch deep jet aeration fermenter has been scaled up to 2000 m³ (for yeast production) for

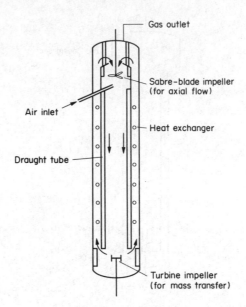

Figure 3   Double impeller design of LeGrys and Solomons

which Vogelbusch reported a power requirement of only 0.5 kWh per kg O_2 (compared to 1.0 or greater for a comparable STF). The various loop reactors are discussed further in sections dealing with recirculation fermenters.

Finally, external loops have been used on industrial STFs primarily for purposes of improving heat transfer.

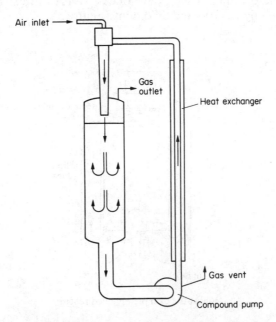

Figure 4   Deep jet aeration fermenter (Vogelbusch)

3.2.1.3   *Immobilized-cell fermenters*

In an immobilized-cell fermenter (ICF) cells are retained by physical attachment to a solid which does not leave the fermenter, control of hydrodynamic conditions which obviate cell migration from the reactor even if the culture fluid leaves, or a barrier such as a membrane. Usually, the primary advantages of ICFs are that they use higher cell concentrations, make continuous processing more feasible, and decrease yield and time losses associated with generating new cells. In addition there are some cases wherein there are other benefits such as improved mass transfer.

A great deal of research and development is being done on ICFs. It is clear that fairly extensive commercial application will come soon. See Volume 2 Chapter 13 for a detailed discussion of this important subject.

3.3 SCALE-UP OF STIRRED TANK FERMENTERS

3.3.1 General Design

3.3.1.1 The standard STF

The 'standard' STF is illustrated in Figure 5. This overall design is used widely, but there are important variations which can strongly affect oxygen transfer, power input, mixing quality, *etc.* The primary variations are in geometric ratios and in types and numbers of impellers used. Generally, the design decisions are made on the bases of conventional 'wisdom', professional advice (*e.g.* from equipment manufacturers), personal experience and hearsay.

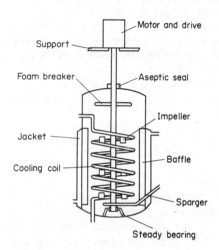

Figure 5 'Standard' stirred-tank fermenter

Geometric ratios discussed most often are D_i/D_t, H/D_t, D_s/D_i, D_b/D_r [D_i = impeller diameter, D_t = tank diameter, H = tank height, D_s = distance from impeller to tank bottom, D_b = baffle width] and ratios of various impeller dimensions (*e.g.* blade length: impeller diameter). While there is considerable literature dealing with these factors, there is very little applicable directly to fermenters. There have been very few carefully done, systematic studies involving aerated, solid/liquid systems having rheological and other (*e.g.* surface tension) properties similar to those of 'typical' fermentation broths.

The most widely used and studied STF impeller is the flat-blade turbine. Some of the many others that have been studied, although not used to any great extent in aerobic fermentations, include:

(1) The marine propeller which gives better axial flow but poorer oxygen transfer than the turbine.

(2) The 'vibromixer' which has been said to provide good oxygen transfer and relatively low shear.

(3) The multirod impeller which was used by Steel and Maxon (1966b) to agitate a very viscous streptomycete broth in a 15 000 liter fermenter. They reported that compared to the turbine, the multirod impeller gave better shear distribution, lower power requirements and relatively small power change during the fermentation.

(4) The gas inducing agitator (Joshi, 1977; Topiwala, 1974) which comprises a hollow impeller attached to a hollow shaft; the impeller has at least one opening exposed to the liquid. As the

impeller rotates, the pressure at the opening(s) decreases thereby inducing gas flow down the shaft and into the liquid. Mass transfer/power input characteristics appear to be comparable to those of turbines for low viscosity liquids. The rate of gas induction can be quite adequate for fermentation but it appears to depend very strongly on the physical properties of the liquid. This may limit the usefulness of gas inducing agitators. However, it is best to reserve judgement because their characteristics have not yet been studied thoroughly nor has there been much done to optimize the designs.

3.3.1.2 Modified STFs

In most cases modified STFs are examples of mechanically driven, defined-flow fermenters (see Section 3.2.1.2).

(1) The Waldhof fermenter does not appear to have any significant advantage over the standard STF other than its ability to cope well with foam (Tsao, 1966).

(2) The LeGrys–Solomons (1977) design uses two independently driven impellers (see Figure 3), a turbine at the bottom of the vessel to disperse gas and promote oxygen transfer, and a sabre blade impeller in a draft tube to induce axial flow. Each impeller is designed and operated independently to optimize its function. The designers report that the induced liquid flow velocity is greater than gas bubble terminal velocities; therefore, the sparger is placed at the top of the vessel thereby saving compression costs. The draft tube also serves as a heat exchanger. Additional heat transfer is achieved by coils in the annular region.

(3) The completely filled reactor (CFR) (see Figure 2) has been reported (Karrer, 1978) to provide greater, more efficient oxygen transfer than the conventional STF. Karrer's results show that $k_L a$ can be as high as 0.4 s^{-1} at a power input of only $2 \times 10^3 \text{ W m}^{-3}$; at the same power input, the STF usually will have a $k_L a$ value of less than 0.2 s^{-1}. Karrer also has reported that CFRs have much shorter mixing times than STFs, even for fairly viscous fluids. Other advantages cited include elimination of head space and higher gas hold-ups than is typical for standard STFs.

3.3.2 Design Correlations

3.3.2.1 Agitation power

Power characteristics of aerated, agitated fermenters have been studied, discussed and reviewed extensively (Moo-Young, 1981; Schügerl, 1981; Brauer, 1979; Banks, 1977; Blanch, 1976; Wang, 1979). In spite of this we still do not have reliable correlations useful for full-scale design. What systematic studies have been done generally have produced correlations which are rather limited by the (usually) relatively small fermenters used, the small ranges of geometric factors studied, and the fact that little or no attention has been paid to rheological and other physical properties. Furthermore, fermentation broths have been used in only a very few studies, and all too often these have not been systematic enough to yield useful correlations.

This is not to say that the correlations are without value. They do provide some guidance; but all too often they given us results not reliable enough for rational design. In some cases two correlations will predict very divergent results over some ranges of design variables and very similar results over others. Often, even when the results predicted by the correlations agree closely, they do not agree with practical experience. The problem is particularly bad for calculation of oxygen transfer rate in general and for power input to non-Newtonian fluids. Calculations of power input to Newtonian fluids do not *appear* usually to cause problems; however, the apparent agreement among results predicted by several correlations can be puzzling. A few examples should help to illustrate these points.

Two of the most frequently cited power correlations for aerated, Newtonian fluids are those of Oyama and Endoh (1955) and of Michel and Miller (1962). The Oyama–Endoh correlation is of the form

$$P_G/P_0 = f(N_a) \tag{8}$$

where P_G is the gassed power, P_0 is the non-gassed power and N_a is the aeration number defined as

$$N_a = \frac{Q}{ND_i^3} \tag{9}$$

The correlation is available in graphical form for various types of impellers. In general, P_G/P_0 is less than 1.

The Michel–Miller correlation is given by

$$P_G = a_1\left[\frac{P_0^2 N D_i^3}{Q^{0.56}}\right]^{a_2} \tag{10}$$

In the original study, fixed values of a_1 and a_2 appeared to be valid for a wide range of fluid viscosities (water to glycerol).

Consider now a 100 000 liter (working volume) fermenter designed and operated as follows: $D_t = 4.0$ m, $D_i = 1.60$ m, $N = 120$ r.p.m., $Q = 20\,000$ SLPM, $\mu = 1$ cP. The non-gassed power input is 707 HP as calculated by using the standard N_{Po}/N_{Re} correlation for turbine impellers (Metzner, 1957; Wang, 1979). Power input for the gassed fermenter is calculated *via* the Michel–Miller correlation as 895 HP and as 548 HP *via* the Oyama–Endoh correlation, a difference of 62%. The small difference is not surprising because (i) the correlations were developed using different operating conditions and fermenter designs; (ii) neither correlation is based on data at the 100 000 liter size; (iii) neither correlation contains any geometric factors (this is true of most such correlations) and should not be expected to be scale or geometry invariant.

For aerated, non-Newtonian fluids there are more obvious and significant problems than there are for Newtonian fluids: the effects of rheological properties are much more pronounced and there are very few correlations. The work of Taguchi and Miyamoto (1966) is one of the few based on experiments employing real non-Newtonian fermentation broths. The authors correlated P_G/P_0 as a function of aeration number but found that for a given culture broth they obtained a family of curves with N_{Re} as a parameter. They found also that all the curves shifted as the fermentation progressed. However, Taguchi (1968) found that for the same *Endomyces* broth a Michel–Miller-type correlation (*not* identical with that for Newtonian fluids) fitted his data reasonably well over a wide range of conditions and scales. There are no other correlations to speak of for fermentation broths.

Now consider a 100 000 liter fermenter designed and operated as follows: $D_t = 4.0$ m; $D_i = 1.6$ m; $N = 120$ r.p.m.; $Q = 20\,000$ SLPM. Assume that the broth is a pseudoplastic fluid having a flow index (n) of 0.5 and consistency (K) of 20 dynes cm^{-2} $s^{-0.5}$ (these values are in the ranges cited by Taguchi, 1966). Using the Michel–Miller-type correlation developed by Taguchi (1968) for non-Newtonian broths, we get 112 HP for the power input. This is considerably lower than is required normally in practice under similar conditions.

When we turn to Taguchi's P_G/P_0 *vs* N_a curves, we find they are unusable because our N_{Re} is well beyond any value he studied, and each set of curves is only valid for a specific value of what is called 'apparent viscosity', a term which Taguchi did not define. One other point is well worth noting: the Michel–Miller correlation for Newtonian fluids differs considerably from that for non-Newtonian fluids as developed by Taguchi (1968). For large systems, the results predicted by the two correlations can differ by a factor of as much as two to three. One must wonder how 'non-Newtonian' a broth must be before a non-Newtonian correlation can be used. One also must wonder how different the correlation would be for other non-Newtonian broths having rheological properties different from those of the *Endomyces* culture broths studied by Taguchi. Unfortunately, there is nothing in the literature which can help us to resolve these questions for practical fermentation broths.

3.3.2.2 Oxygen transfer

Oxygen transfer also has received a great deal of attention, but here too we have no reliable design correlations for mass transfer coefficients. The reasons are primarily the same as those cited (at the beginning of Section 3.3.2.1) for the lack of reliable power correlations. For an extensive compilation and description of the numerous published correlations for the oxygen transfer coefficient, $k_L a$, the reader is referred to recent reviews of the subject (Shügerl, 1981; Moo-Young, 1981; Wang, 1979); here we will consider just a few for purposes of comparison.

Correlations cited frequently for Newtonian fluids are given in Table 1. Comparisons of the results predicted by these correlations are presented in Table 2. The large discrepancies are not surprising given the fundamental shortcomings of the correlations (again, see comments at the beginning of Section 3.3.2.1).

There are only a few published correlations for oxygen transfer to non-Newtonian fluids. We

Table 1 Oxygen Transfer Correlations for Newtonian Fluids[a]

Correlation	Comments	Ref.
(A) $K_v = K_1(P/V)^{0.95}V_s^{0.67}$	No consideration of geometric factors or rheology	Cooper (1944)
(B) $K_v = K_1(P/V)^{\alpha}$	Stated to be independent of H/D ratio over range of 1 to 3; number of impellers from one to three; minor geometric changes. Only one gas linear velocity used	Oldshue (1966)
(C) $K_v = K_1(P/V)^{0.4}V_s^{0.5}N^{0.5}$	Same comments as for Cooper correlation	Richards (1961)
(D) $K_v = K_1(2.0 + 2.8N_i) \times (P/V)^{0.77}V_s^{0.67}$	Number of impellers has a pronounced effect; differs considerably from the other correlations cited	Fukuda (1968)

[a] All correlations used as presented in Wang (1979).

Table 2 Oxygen Transfer Rates Predicted by Various Correlations for Newtonian Fluids

P/V (kW m^{-3})	V_s (cm min^{-1})	N (min^{-1})	N_i	Oxygen transfer rate[a,b] (mmol l^{-1} h^{-1}) A	B	C	D
4.0	180	60	1	1200	572	26	206
		180	1	1200	572	46	206
		60	2	1200	572	26	327
		180	2	1200	572	46	327
		60	3	1200	572	26	448
		180	3	1200	572	46	448
2.65	180	60	1	816	432	21	151
		180	1	816	432	32	151
		60	2	816	432	21	239
		180	2	816	432	32	239
		60	3	816	432	21	327
		180	3	816	432	32	327
1.32	180	60	1	422	266	16	88
		180	1	422	266	29	88
		60	2	422	266	16	140
		180	2	422	266	29	140
		60	3	422	266	16	192
		180	3	422	266	29	192

[a] Oxygen transfer rate based on a log-mean average oxygen partial pressure of 0.26 atm.
[b] The letters A–D refer to the correlations cited in Table 1.

consider here only those of Taguchi (1968) and of Loucaides and McManamey (1973). Taguchi's correlation is

$$k_La = K_1(P/V)^{0.33}V_s^{0.56} \tag{11}$$

It is special in that it is based on experiments done over a range of 20–30 000 l with real *Endomyces* sp. fermentation broths.

The Loucaides–McManamey correlation is a bit more involved. For (P/V) below a critical value, $(P/V)_b$,

$$k_La/B = (k_La/B)_0 + C_8(P/V)V_s^{0.3} \tag{12}$$

where B is the Henry's law constant for oxygen and

$$(k_La/B)_0 = C_7\frac{D_t}{H}ND_iD_t^{-0.5} + C_{7b} \tag{13}$$

C_7, C_{7b} and C_8 are empirical constants: $C_7 = 2.4 \times 10^{-4}$ m$^{-0.5}$, $C_{7b} = 2.1 \times 10^{-4}$ s^{-1} and $C_8 = 3.7 \times 10^{-6}$ W^{-1} m$^{2.7}$ s$^{-0.7}$. $(P/V)_b$ is given by

$$(P/V)_b = 200\left[\frac{(ND_i)^2 - 0.185}{D_t^{0.45}H}\right] \tag{14}$$

Above the critical point

$$k_L a/B = [(k_L a/B)_0 + C_8(P/V)_b V_s^{0.3}](P/V)^{0.5}(P/V)_b^{-0.5} + (k_L a/B)_0 \qquad (15)$$

The correlation is based on experiments done at bench scale using paper pulp slurries to simulate fungal fermentation broths.

In Table 3 we present calculated oxygen transfer rates predicted by both correlations for wide ranges of conditions. The results are highly variable: in some cases, the agreement between the predicted values is quite good; in others it is quite poor. In general, the results predicted by both correlations do not agree well with oxygen transfer rates for fungal fermentations under practical operating conditions. Again, this is not particularly surprising, for essentially the same reasons as cited previously concerning the general weaknesses of most published correlations. In this regard, it is important to note particularly that *Endomyces* culture broths and paper pulp slurries do not have the same rheological characteristics and that neither correlation considers them explicitly.

Table 3 Oxygen Transfer Rates Predicted by Correlations for Non-Newtonian Fluids

						Oxygen transfer rate[a] (mmol l^{-1} h^{-1})	
D_r	H	V_s	N_i	D_i	P/V		Loucaides and
(m)	(m)	(cm s^{-1})		(m)	(kW m^{-3})	Taguchi[b]	McMananey[c]
5.0	5.0	25	1	1.34	1.09	15	56
		101	1	1.34	0.763	28	60
		202	1	1.34	0.638	40	63
4.0	8.0	40	1	1.06	0.261	12	27
		160	1	1.06	0.182	23	30
		241	1	1.06	0.164	28	31
		40	2	1.06	0.522	15	30
		80	2	1.06	0.437	20	31
		160	2	1.06	0.365	29	33
3.5	10.5	53	2	0.93	0.237	13	21
		105	2	0.93	0.198	18	23
		210	2	0.93	0.166	26	23
		53	3	0.93	0.355	15	24
		105	3	0.93	0.298	20	24
		210	3	0.93	0.249	30	24

[a] Transfer rate based on an agitation speed of 150 r.p.m. and a log-mean average oxygen partial pressure of 0.26 atm.
[b] Taguchi (1968) as presented in Wang (1979).
[c] Loucaides and McMananey (1973).

The importance of the effects of rheological characteristics on $k_L a$ for non-Newtonian broths requires special attention. Ranade and Ulbrecht (1978) found that for pseudoplastic solutions (polyacrylamide) $k_L a$ decreased by a factor of three when the flow index decreased from 0.975 to 0.80 and the consistency increased from 2.33 N S^{n}m^{-2} to 5.45 N S^{n}m^{-2}. Jarai (1969) reported that during streptomycete fermentations, $k_L a$ changed very markedly with changing non-Newtonian characteristics. Similar observations were made by Tuffile and Pinho (1970). Unfortunately, none of the experiments were done in a way that would yield a useful design correlation; nevertheless, their work did make clear that for such correlations to be useful practically, they must contain rheological properties. This has yet to be done satisfactorily.

3.3.2.3 Bulk mixing

Good bulk mixing is essential for good control and good heat and mass transfer in STFs. The quality of bulk mixing can be expressed in terms of the mixing time, θ_m, which is defined usually as the time required by a homogeneous, agitated liquid to reach a specified degree of homogeneity after a tracer pulse (*e.g.* acid or base, concentrated salt, heated fluid) has been added to it.

Mixing time is a function of fermenter geometry, impeller design, operating conditions and fluid rheological properties. There are several correlations available for Newtonian fluids (Novak, 1975; Nagata, 1975; Hoogendorn, 1967; Norwood, 1960; Fox, 1956; Godleski, 1962);

however, they can give widely divergent estimates of mixing time. Furthermore, they were developed for unsparged systems containing no cell mass; therefore, they may not be accurately applicable to aerated fermenters (see below). However, this may not be a serious problem because Newtonian broths having relatively low viscosities usually can be mixed adequately with relatively little difficulty.

Non-Newtonian broths, on the other hand, usually are very difficult to mix well in conventional STFs and this causes many significant problems (Phillips, 1961; Maxon, 1959; Steel, 1966a, 1966b), particularly for those conducting large-scale fermentations. Nevertheless, very little has been reported concerning the effects on mixing time of reactor design, operating conditions and rheological characteristics.

In one of the few systematic studies done to determine the effects on mixing time of rheological characteristics and operating conditions, Moo-Young (1972) developed a correlation between $N\theta_m$ and N_{Re} where N is the impeller speed and θ_m is the mixing time. The correlation shows that below $N_{Re} = 10^3$, θ_m for non-Newtonian fluids is very much greater than for Newtonian fluids (at the same N_{Re}). Unfortunately, the work was done with a homogeneous polymer solution and the effects of aeration were not determined.

Blakebrough and Sambamurthy (1966) found that mixing time for a simulated non-Newtonian mycelial broth (1.6% paper pulp — rheological properties not reported) can be correlated with the momentum factor, M_f:

$$M_f = ND_i \times \text{NWL}(D_i - W) \tag{16}$$

The correlation does not explicitly include the rheological properties; therefore, it is difficult to predict its applicability to fluids other than the paper pulp slurry studied. However, the authors did show that θ_m for aerated systems was greater than for non-aerated systems, the difference increasing with decreasing N_{Re}.

Wang and Fewkes (1976) reported that θ_m for *Streptomyces niveus* broths can be correlated by the method of Norwood and Mentzner (1960):

$$N_M' = \frac{\theta_m(ND_i^2)^{2/3}g^{1/6}D_i^{1/2}}{y^{0.5}D_t^{3/2}} = f(N_{Re}') \tag{17}$$

where

$$N_{Re}' = \frac{D_i^2 N^{2-n}\varrho}{0.1K}\left(\frac{n}{6n+2}\right)^n \tag{18}$$

They obtained good qualitative agreement with the original Norwood–Metzner correlation, but they found considerable quantitative differences which might be due to aeration effects. Unfortunately the authors did not specify whether θ_m was determined under aerated or non-aerated conditions, nor did they mention the extent of gas hold-up. See Charles' (1976) review for further discussion.

No mixing-time correlations have been reported for aerated, non-Newtonian culture fluids containing microbial polysaccharides.

It is obvious that impeller design can affect the bulk mixing characteristics of a fermentation broth considerably. However, only Steel and Maxon (1966b) have reported a serious study. They showed that their multiple rod impeller was superior to turbine impellers in providing homogeneous conditions in fungal broths.

Other investigators (Nagata, 1972; Loucaides, 1973; Moo-Young, 1972) have shown that turbine impellers are not well suited to mixing viscous, non-Newtonian fluids (see Leamy, 1973) for a different opinion). The reason for this is that near the impeller the shear rate is high and hence the viscosity is low resulting in good mixing and mass transfer. However, even at relatively small distances from the impeller, the shear rate is low, the viscosity is high, and mixing and mass transfer are poor. The result is poor bulk mixing and only a small region of good mass transfer.

The viscosity gradients in non-Newtonian broths also tend to cause most of the inlet air to flow toward the impeller which results in (1) lower gas residence time (channeling), (2) increased bubble size, (3) further deterioration of bulk mixing, and (4) poor distribution of oxygen.

Conditions can be improved by larger turbines and increased turbine speed, but both increase power consumption considerably. Therefore, impellers (Charles, 1976) which have been found best suited for mixing other viscous, non-Newtonian fluids (*e.g.* polymers) should be studied for application to fermentation. It also is clear that much more attention must be paid to rheological characteristics in future studies of mixing and mass transfer in non-Newtonian fluids.

3.3.2.4 Scale-up methods

Scaling up an STF involves a considerable number of compromises; in general, we cannot provide at large scale exactly the same environment as exists in our small-scale vessels. For example, if we choose to maintain geometric similarity then, at least in theory, we may assume that the general forms of correlations for heat and mass transfer, power input, *etc.*, should be essentially scale invariant. Thus, if we wish to maintain a constant $k_L a$:

$$(k_L a)_1 = \alpha (N_{Re})_1^{a_1} (V_s)_1^{a_2} \tag{19}$$

$$(k_L a)_2 = \alpha (N_{Re})_2^{a_1} (V_s)_2^{a_2} \tag{20}$$

and

$$(N_{Re})_2 = (N_{Re})_1 \left[\frac{(V_s)_1}{(V_s)_2} \right]^{a_1/a_2} \tag{21}$$

Fixing the linear velocity $(V_s)_2$ then fixes $(N_{Re})_2$. If the fluid properties are assumed to be constant, then all other important variables are fixed. For example:

$$h_2 = f[(N_{Re})_2 (V_s)_2] \tag{22}$$

$$N_2 = \frac{(N_{Re})_2 \eta}{(D_i)_2^2 \varrho} \tag{23}$$

$$(P/V)_2 = f[N_2, (N_{Re})_2, Q_2] \tag{24}$$

$$\dot{\gamma}_2 = \dot{\gamma}_1 (N_2/N_1) \tag{25}$$

Obviously, other factors (*e.g.* N_{Re}, P/V) could be used as bases for scale-up. But the important point is that insisting on geometric similarity imposes significant constraints. One is constrained to decide from small-scale experiments which is the most important variable to hold constant. One must recognize also that the approach as presented is a bit simplistic and naive. For example:

1. Consider the previous example for maintaining constant $k_L a$, but this time assume the fluid is non-Newtonian. From equation (23)

$$(N_{Re})_2 = \frac{(D_i)_2^2 N_2 \varrho}{\mu_2} = \left[\frac{(V_s)_1}{(V_s)_2} \right]^{a_1/a_2} \frac{(D_i)_1^2 N_1 \varrho}{\mu_1} \tag{26}$$

Therefore

$$N_2 = N_1 \left[\frac{(D_i)_1}{(D_i)_2} \right]^2 \frac{\mu_2}{\mu_1} \left[\frac{(V_s)_1}{(V_s)_2} \right]^{a_1/a_2} \tag{27}$$

and

$$\frac{\mu_2}{\mu_1} = \frac{K_2}{K_1} \left(\frac{N_2}{N_1} \right)^{n-1} \tag{28}$$

Therefore

$$N_2 = N_1 \left(\frac{(D_i)_1}{(D_i)_2} \right)^{2/2-n} \left[\frac{(V_s)_1}{(V_s)_2} \right]^{a_1/a_2(2-n)} \tag{29}$$

This assumes that at corresponding points during the fermentation, K and n will be the same at both scales; however, they can vary during the fermentation. Obviously, even if the assumption is valid, which usually is quite unlikely, the example illustrates some of the problems that may arise even for this relatively simple case. The reality is less kind.

2. Many of the available correlations are scale dependent. This is most apparent for non-Newtonian fluids.

3. Physical properties (*e.g.* surface tension, viscosity) do change during a fermentation and generally the pattern of changes varies with scale.

4. Gas linear velocity at large scale can be easily more than an order of magnitude greater than that at small scale, even if the VVM in the large vessel is smaller than that in the small vessel. This can cause very large differences in factors such as gas hold-up, foaming, *etc.*

5. It is almost impossible to maintain scale independent the effects of fermenter internals.

6. The effects of surface aeration change markedly with scale. A correlation is available (Fuchs, 1971) but it has very limited applicability.

7. The *reliable* correlations needed are unavailable.

Obviously, one is under no obligation to maintain geometric similarity; indeed, in many cases it is impractical or imprudent to do so. Also, as noted by Oldshue (1966), departure from geometric similarity generally gives greater flexibility to the designer. For example, non-geometric scale-up allows us to maintain constant $k_L a$ and shear (at the impeller only), *etc*. (Shear is considered by many as being quite important, particularly for many fungal fermentations (Wang, 1979).) However, we should *expect* the usual correlations for $k_L a$, *etc*., to be scale dependent if we abandon geometric similarity. Therefore, it will be more difficult to specify important design parameters (*e.g.* N, D_i) reliably. Furthermore, one still is faced with the problems associated with conditions changing during the fermentation (at all scales).

Given the preceding, it is not surprising that the usual scale-up methods comprise a great deal of art and empiricism. One of the more frequently cited and seemingly more popular methods is based on maintenance of constant $k_L a$ along with, in some instances, constant shear (at the impeller). Evidence of successful application of this approach has been published as well as *some* evidence of failure (Oosterhuis, 1981; Bartholomew, 1960). Clearly, to apply this approach practically, one must: (1) have reasonably reliable methods to determine the power input required to reach the desired $k_L a$ and shear; (2) be able to measure $k_L a$ reliably; (3) be able to predict the effects of such factors as increased pressure, surface aeration, *etc*.; (4) be able to predict how spatial distributions of important variables will change with scale and if any such changes will be detrimental.

Obviously, the most reliable way to get the required information is to do experiments using existing fermenters. But this is expensive and can cause serious scheduling problems. It also tends to discourage innovative designs.

Alternatively, one could rely on company production records for similar fermentations. Such records can be very valuable; but all too often the data are not collected at all or are not collected usefully.

In far too many cases reliable information simply is not available and cannot be obtained readily. Therefore, too many scale-ups are based on often baseless conventional wisdom, hearsay and wishful thinking. As a result the track record is mixed for existing scale-up methods. Furthermore, it is difficult to explain successes or failures rationally.

Historically, this seems a most appropriate time to develop reliable scale-up methods. To this end, fermentation technologists must use highly instrumented scale-translation facilities (see Section 3.1.10) and computers much more extensively. Computers should be used at both ends of the translation for control, data acquisition, on-line calculations (*e.g.* material balances, $k_L a$, RQ), generation of correlations, data filing and report generation.

Such systematic, dedicated programs will provide the basis for scale-up methods as reliable as those used routinely in the chemical process industry. Certainly, there still will be considerable uncertainty, but we should be able to design accurately enough such that automatic control systems will be able to drive fermentation plants to operate optimally. In this regard, control system design always should be made an integral part of the overall design process from the beginning. This sounds too obvious to warrant saying, but unfortunately there are far too many examples that violate this basic principle.

Finally, far more attention must be devoted to heat transfer. This was discussed in Section 3.1.5 but is so important that it bears repetition.

3.4 REFERENCES

Aiba, S., A. E. Humphrey and N. Millis (1973). *Biochemical Engineering*, 2nd edn., chap. 8. Academic, New York.
Banks, G. T. (1977). In *Topics in Enzyme and Fermentation Biotechnology*, ed. A. Wiseman, vol. 1, chap. 4. Wiley, New York.
Bartholomew, W. H. (1960). In *Advances in Applied Microbiology*, ed. W. W. Umbreit, vol. 2, pp. 289–300. Academic, New York.
Blakebrough, N. and K. Sambamurthy (1964). *J. Appl. Chem.*, **14**, 413–422.
Blakebrough, N. and K. Sambamurthy (1966). *Biotechnol. Bioeng.*, **8**, 25–42.
Blanch, H. W. and S. M. Bhavaraju (1976). *Biotechnol. Bioeng.*, **18**, 745–790.
Brauer, H. (1979). In *Advances in Biochemical Engineering*, ed. A. Fiechter, vol. 13, pp. 87–119. Springer-Verlag, New York.
Charles, M. (1976). In *Advances in Biochemical Engineering*, ed. A. Fiechter, vol. 8. Springer-Verlag, New York.
Coleman, B. D., H. Markowitz and W. Noll (1966). *Viscometric Flows of Non-Newtonian Fluids*. Springer-Verlag, New York.
Cooper, C. M., G. A. Fernstrom and S. A Miller (1944). *Ind. Eng. Chem.*, **36**, 504–509.
Ettler, P., J. Paca and V. Cechner (1976). Abstracts of papers presented at the 5th International Fermentation Symposium, Berlin, Section 3.08, p. 46.
Fox, E. and V. E. Gex (1956). *AIChE J.*, **2**, 539.

Fuchs, R., D. Y. Ryu and A. E. Humphrey (1971). *Ind. Eng. Chem., Prod. Res. Dev.*, **10**, 190.

Fukuda, H., Y. Sumino and T. Kenzaki (1968). *J. Ferment. Technol.*, **46**, 829–837, 838–845.

Godleski, E. S. and J. C. Smith (1962). *AIChE J.*, **8**, 617.

Hoogendoorn, C. J. and A. P. Den Hartog (1967). *Chem. Eng. Sci.*, **22**, 1689.

Jarai, M., E. Gyori and J. Tombor (1969). *Biotechnol. Bioeng.*, **11**, 605–622.

Joshi, J. B. and M. M. Sharma (1977). *Can. J. Chem. Eng.*, **55**, 683–695.

Johnstone, R. E. and Thring (1957). *Pilot Plants, Models, and Scale-up Methods in Chemical Engineering*. McGraw-Hill, New York.

Karrer, D. (1978). *Der Total Gefüllte Bioreaktor*, Dissertation no. 6254, ETH Zürich.

Kaszab, I., I. Hogge, S. Komocsi and J. Szilagyi (1981). *Process Biochem.*, **16**, February/March, 38.

Leamy, G. H. (1973). *Chem. Eng. (New York)*, **278**, October 15, 115.

Loucaides, R. and W. J. McManamey (1973). *Chem. Eng. Sci.*, **29**, 2165–2178.

Lundell, R. and P. Laiho (1976). *Process Biochem.*, **11**, April, 13–17.

Maxon, W. D. (1959). *Biotechnol. Bioeng.*, **1**, 311.

Metz, B., N. W. F. Kossen and J. C. Van Suijdam (1979). In *Advances in Biochemical Engineering*, ed. A. Fiechter, vol. 1, pp. 103–155. Springer-Verlag, New York.

Metzner, A. B. (1956). In *Advances in Chemical Engineering*, ed. T. B. Drew, vol. 1, pp. 77–153. Academic, New York.

Michel, J. B. and S. A. Miller (1962). *AIChE J.*, **8**, 262–266.

Middleman, S. (1968). *Flow of High Polymers*, chap. 2. Interscience, New York.

Miura, Y. (1976). In *Advances in Biochemical Engineering*, ed. A. Fiechter, vol. 4, pp. 3–40. Springer-Verlag, New York.

Modeer, B. (1974). *Process Biochem.*, **9**, 23.

Moo-Young, M., K. Tichar and F. A. L. Dullien (1972). *AIChE J.*, **18**, 178–182.

Moo-Young, M. and H. W. Blanch (1981). In *Advances in Biochemical Engineering*, ed. A. Fiechter, vol. 19, pp. 1–69. Springer-Verlag, New York.

Nagata, S., M. Nishikawa, T. Katsube and T. Takaish (1972). *Int. J. Chem. Eng.*, **12**, 175.

Nagata, S. (1975). *Mixing Principles and Applications*, pp. 204–206. Wiley, New York.

Norwood, K. W. and A. B. Metzner (1960). *AIChE J.*, **6**, 432.

Novak, V. and F. Riegler (1975). *Chem. Eng. J.*, **9**, 63.

Oldshue, J. Y. (1966). *Biotechnol. Bioeng.*, **8**, 3–24.

Oosterhuis, N. M. G. and M. W. F. Kossen (1981). *Biotechnol. Lett.*, **3**, 645–650.

Oyama, Y. and K. Endoh (1955). *Chem. Eng. (Jpn.)*, **19**, 2–11.

Paca, J., P. Ettler and V. Gregr (1976). *J. Appl. Chem. Biotechnol.*, **26**, 309–317.

Paca, J., P. Ettler and V. Gregr (1978). *J. Ferment. Technol.*, **56**, 144–151.

Pace, G. W. (1980). In *Fungal Biotechnology*, cd. J. E. Smith, pp. 95–110. Academic, New York.

Phillips, D. H. and M. J. Johnson (1961). *Biotechnol. Bioeng.*, **3**, 277.

Ranade, V. R. and J. J. Ulbrecht (1978). *AIChE J.*, **24**, 796–803.

Richards, J. W. (1961). *Prog. Ind. Microbiol.*, **3**, 143–172.

Roels, J. A., J. Van den Berg and R. M. Voncken (1974). *Biotechnol. Bioeng.*, **16**, 181–208.

Schügerl, K. (1981). In *Advances in Biochemical Engineering*, ed. A. Fiechter, vol. 19, pp. 71–174. Springer-Verlag, New York.

Skelland, A. H. P. (1967). *Non-Newtonian Flow and Heat Transfer*, chap. 2. Wiley, New York.

Solomons, G. L. (1969). *Materials and Methods in Fermentation*, chap. 4. Academic, New York.

Solomons, G. L. (1980). In *Fungal Biotechnology*, ed. J. E. Smith, pp. 53–80. Academic, New York.

Steel, R. and W. D. Maxon (1966a). *Biotechnol. Bioeng.*, **8**, 97–108.

Steel, R. and W. D. Maxon (1966b). *Biotechnol. Bioeng.*, **8**, 109–115.

Taguchi, H. and S. Miyamoto (1966). *Biotechnol. Bioeng.*, **8**, 43–54.

Taguchi, H., T. Imanaka, S. Teramoto, M. Takatsaka and M. Sato (1968). *J. Ferment. Technol.*, **46**, 823–828.

Topiwala, H. H. and G. Hamer (1974). *Trans. Inst. Chem. Eng.*, **52**, 113–120.

Tsao, G. T. and W. D. Cramer (1978). *AIChE Symp. Ser.*, **67** (108), 158–163.

Tuffile, C. M. and F. Pinho (1970). *Biotechnol. Bioeng.*, **12**, 849–871.

Van Wazer, J. R., J. W. Lyons, K. Y. Lian and R. E. Colwell (1963). *Viscosity and Flow Measurement*. Interscience, New York.

Wang, D. I. C. and R. C. J. Fewkes (1976). Paper presented at the 32nd Meeting of the Society of Industrial Microbiology, Jekyll Island, GA.

Wang, D. I. C., C. L. Cooney, A. L. Demain, P. Dunill, A. E. Humphrey and M. D. Lilly (1979). *Fermentation and Enzyme Technology*, chaps. 9 and 10. Wiley, New York.

Wilkenson, W. L. (1960). *Non-Newtonian Fluids*. Pergamon, New York.

Winkler, M. A. (1983). In *Principles of Biotechnology*, ed. A. Wiseman, chap. 4. Surrey University Press, London.

4

Imperfectly Mixed Bioreactor Systems

A. MOSER
Technical University Graz, Austria

4.1 INTRODUCTION

The mixed-flow-type reactor (stirred tanks, STR) is the main reactor type that has been used in biotechnology, both in the past and at present. However, in bioprocessing practice several systems are known which exhibit plug-flow-type behaviour (PFR). In plug-flow reactors no mixing of fluid elements occurs longitudinally along the flow path, which is one of the main differences between STRs and PFRs. Therefore they can be classified generally as imperfectly mixed systems. The composition of the inflowing medium changes along the axis according to a certain concentration gradient. The system is comparable to a multistage process with an infinite number of stirred vessels between which a continuous transition arises. In other words the system is a spatially variant batch culture but is temporarily invariant in the steady state.

Practical implementation of PFRs, in particular tubular reactors for bioprocessing, was thought to be difficult if not impossible in the past, especially for aerobic processing where backmixing was thought to be significant. However, theoretical considerations according to chemical reaction engineering laws suggested several potential advantages for PFRs (Grieves *et al.*, 1964). They are essentially simple devices with a well-defined flow and without dead zones, and thus can be scaled up with greater confidence. In several bioprocesses, including enzymes, single cell protein (SCP), antibiotics, polysaccharide and waste processing, and in the case of complex kinetics, *e.g.* in the declining and stationary growth phase or with inhibition or catabolite repression, the use of PFRs may give a more favourable yield (maximum productivity) with a simultaneous high degree of conversion which generally cannot be obtained in STRs. Other disadvantages of STRs include the build-up of solid deposits on vessel walls (filamentous growth, solid substrates, polymeric metabolites), cell damage by high shear forces in mechanically stirred tanks and foaming.

In addition to bioreactors with well-defined concentration gradients like PFRs and recycle reactors (RRs), the phenomenon of imperfect mixedness also exists in STRs due to local deviations, especially in technical scale production when it causes problems in scale-up and inefficiency in performance.

Both types of imperfectly mixed bioreactor system are reviewed in this chapter. After discussing the advantages of PFRs for bioprocessing compared with STRs, the construction and engineering characteristics of some newly designed tubular reactors are described, with particular reference to biofloc, biofilm and special bioprocessing. Finally a research methodology which incorporates a systematic treatment of the interactions between metabolism and bioreactor operations in general and imperfect mixedness in particular is outlined.

4.2 COMPARISON OF GRADIENTLESS REACTORS AND REACTORS WITH CONCENTRATION GRADIENTS

4.2.1 Basic Reactor Flow Behaviour: Macro- and Micro-mixing

The flow behaviour of the two extreme cases of continuous reactor operation is illustrated in Figure 1 (after Zwietering, 1959). It is interesting to note that macromixing, identical with the residence time distribution RTD, is normally directly connected with micromixing, including the extreme situations of maximum mixedness (mm flow) in ideal CSTRs and total segregation (ts flow) in ideal CPFRs (see Wen and Fan, 1975). The consequences which arise from these differences in basic reactor behaviour can be easily understood by considering the fundamental law of conservation. Written in one-dimensional form for one phase (*e.g.* liquid phase), the mass balance equation is generally given by

$$\frac{\partial c_i}{\partial t} = - v_z \frac{\partial c_i}{\partial z} + D_{\text{eff}} \frac{\partial^2 c_i}{\partial z^2} + k_{\text{Tr}}(c_i^* - c_i) \pm r_i \tag{1}$$

to account for convection, axial dispersion, interfacial transport and reaction.

From this general balance equation the definition of reaction rate, with respect to the rate of formation ($+r_i$) or consumption ($-r_i$), in the case of stationary operation of CPFRs with $v_z = F/A$ can be derived as

$$\pm r_i = v_z \frac{dc_i}{dz} \tag{2}$$

In the case of discontinuous operation of STRs (DCSTR) representing a non-stationary operational mode the rate is given by

$$\pm r_i = \frac{dc_i}{dt} \tag{3}$$

while in the case of CSTRs at steady state

$$\pm r_i = \frac{F}{V}(c_{\text{ex}} - c_{\text{in}}) \tag{4}$$

with

$$t_r = V/F = 1/D \tag{5}$$

The evaluation of kinetic parameters from CPFRs and DCSTRs therefore needs a graphical, numerical or analytical differentiation of data, as CFPRs and DCSTRs are integral reactors. Integral evaluation is often complicated, even when using computer solutions, because it usually involves the integration of kinetic functions (Moser, 1981).

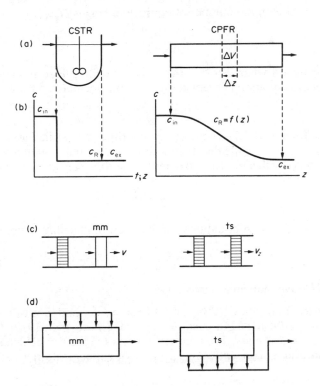

Figure 1   (a) Continuous plug-flow reactor (CPFR) *versus* continuous stirred tank reactor (CSTR); (b) comparison of concentration profiles; (c) visualization of flow behaviour: maximum mixedness (mm) and total segregation (ts) according to Weinstein and Adler (1967); (d) realization of flow behaviour according to Zwietering (1959)

The design of bioreactor operations with optimal conversion and/or optimal productivity has also to take into account the flow behaviour in PFRs (see Figure 1). Figure 2 shows schematically the *RTD* of ideal CSTRs and CPFRs. In CPFRs all fluid elements remaining in the reactor with the same t_r contribute to production, while in CSTRs only fluid elements with $t > t_r$ contribute. This general situation occurs in several bioprocess operations, *e.g.* in sterilization and drying processes (time of sterilization $t_{st} = t_r$) and microbial productions with a maturation time ($t_M = t_r$).

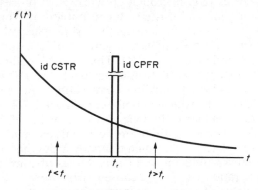

Figure 2   Residence time distribution (RTD) of ideal CSTR and CPFR according to pulse method, $f(t)$, and its consequences for bioprocessing; only cells with a maturation time, which is necessary for microbial production, greater than mean residence time t_r contribute to production

The solution to equation (1) can be given with the aid of the well-known $f(t)$ functions (Levenspiel, 1979).

Using the dispersion model with the Bodenstein number *Bo* as the model parameter, we obtain:

$$f(t/t_r) = \sqrt{\frac{Bo}{4\pi t/t_r}} \exp\left[-\frac{(1 - t/t_r)^2}{4t/t_r}Bo\right] \tag{6}$$

or the tank-in-series model with the number of equivalent stages N_{eq}:

$$f(t/t_r) = \frac{N_{eq}^{N_{eq}}(t/t_r)^{N_{eq}-1}}{(N_{eq}-1)!} e^{-N_{eq}t/t_r} \tag{7}$$

True plug-flow is achieved when $Bo \geqslant 7$ or $N_{eq} \geqslant 5$, while mixed-flow behaviour is realized when $N_{eq} = 1$. In this case equation (7) reduces to the non-normalized form

$$f(t) = De^{-Dt} \tag{8}$$

which is the *RTD* pulse function of the chemostat. Experimental verification of the $f(t)$ functions is achieved with the pulse method, while the step method results in $F(t)$ values which are the integral values of the $f(t)$ functions. Both model approaches to macromixing are related by the equation

$$N_{eq} = 1 + \frac{1}{2}\sqrt{Bo^2 + 1} \tag{9}$$

which is valid for $Bo \geqslant 10$. These equations can be used in calculating conversion in non-ideal reactors.

4.2.2 Dependence of Conversion on Macromixing

As a better comparison between CSTRs and CPFRs the dependence of conversion on macromixing can be demonstrated using a graphical procedure (Levenspiel, 1979). The starting point is again the law of conversion of mass. From equations (2)–(5) we can calculate the holding times required to convert material at concentration c_{in} into exit concentration c_{ex} at steady-state:

For CPFR (or DCSTR): $$t_{DCST} = t_{PF} = \int_{c_{in}}^{c_{ex}} \frac{dc_i}{r_i} \tag{10}$$

For CSTR: $$t_{ST} = \frac{c_{ex} - c_{in}}{r_i} \tag{11}$$

The holding time t_{PF} can be compared with t_{ST} from a graphical plot of $1/r_i$ *vs.* c_i, as shown in Figure 3. While t_{PF} is the area under the curve from c_{in} to c_{ex}, t_{ST} is the area of the rectangle with sides of length $c_{ex} - c_{in}$ and $1/r_i$ according to equations (10) and (11).

This procedure may be applied to the variety of bioprocesses, which can be classified into two basic types of reaction: normal catalytic processes with ordinary kinetics (*e.g.* enzyme technology) and autocatalytic processes (*e.g.* microbial growth). Real processes are a combination of both basic types (*e.g.* fermentation and waste water technology). While for ordinary kinetics r_i decreases with increasing c_i, r_i increases for autocatalytic processes. Without using mathematical functions, this behaviour is shown in the curves in Figure 3 for the evaluation of conversion in the case of CSTRs, CPFRs and recycle reactors (RR). Thus, for Monod-type kinetics, the CPFR is superior to the CSTR in all cases of enzyme technology while the CSTR is to be preferred in the case of fermentation processes as long as only low conversion (high exit concentration; point I in Figure 3b) is needed. For high conversions a combination of a CSTR with a CPFR is generally optimum, even when this statement strongly depends on the value of the saturation constant K_S (Topiwala, 1973). The choice of reactor, therefore, can only be made when the value of K_S is determined with a high degree of accuracy, which normally is not the case (Moser, 1981).

The design of recycle reactors results in even better values of conversion at the same holding time when the proper recycle ratio r ($= F_r/F$) is chosen. This procedure is schematically depicted in Figure 3c. The optimum value of r can be found by choosing the reciprocal of the rate at $c_{i,1}$ equal to the average reciprocal rate in the reactor (by making the areas indicated in Figure 3c equal). Quantitatively:

$$-\frac{1}{r_i}\bigg|_{c_{i,1}} = \frac{\int_{c_{i,1}}^{c_{i,ex}}(dc_i/-r_i)}{c_{i,ex} - c_{i,1}} \tag{12a}$$

with

$$c_{i,1} = \frac{c_{i,in} + rc_{i,ex}}{r + 1} \tag{12b}$$

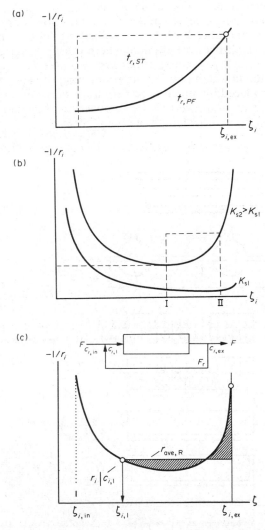

Figure 3 Graphical design of continuous reactor operation on the basis of coupling balances and kinetics: plot of reciprocal rate of consumption or formation $(1/r_i)$ *versus* conversion ζ_i, enabling the graphical estimation of mean residence time t_r in CSTRs and CPFRs (see text). (a) Bioprocess with ordinary kinetics; (b) bioprocess with partly autocatalytic and partly ordinary kinetics (fermentation technology); (c) design of recycle reactor operation for fermentation processes with an optimum value of recycle ratio r in order to minimize reactor size (Levenspiel, 1979)

4.3 DESIGN AND CHARACTERIZATION OF CONTINUOUS PLUG-FLOW REACTORS

Owing to the expected difficulties in realizing plug-flow in aerobic systems, a cascade of CSTRs with $N_{eq} \geqslant 5$ can be used to approximate plug-flow behaviour (Mason and Piret, 1950; Powell and Lowe, 1964). In addition, the flow behaviour of CPFRs can also be realized in tower or bubble column reactors (Štěrbáček and Šáchová, 1973; Greenshields and Smith, 1974; see also Volume 2, Chapter 5) and also directly in tubular reactors with microbial flocs and films.

4.3.1 Tubular Reactors Involving Bioflocs in Fermentation and Waste Water Technology

Horizontal vessels are generally well known in chemical engineering. Rotary drum reactors, rotating cylinders, pipeline contactors and rotary kilns play an important role in industry in processes such as mixing, heat transfer, polymerization and gas–liquid mass transfer, *etc.* (Helmrich and Schügerl, 1979; Moritz and Reichert, 1981; Collin *et al.*, 1979; Joshi and Sharma, 1976; Zuiderweg and Bruinzeel, 1971; Shah and Sharma, 1975).

The paucity of suitable reactors for bioprocessing was surmounted by the development of several types of tubular reactor construction which are summarized in Figure 4. By analogy with the natural flow in rivers or in long open channels (oxidation ponds or lagoons in waste water technology) simple coils or spiral tubes have been applied, *e.g.* in beer brewing (Denshchikov, 1962; Cook, 1968) in order to realize a wort gradient where yeast passes through, resulting in high conversion and high product quality. Inclined tubes are also used (Hall and Howard, 1968). A more advanced system of pipe flow suitable for the production of yeast was described by Olsen (1958). Air or oxygen is introduced by injectors or venturi orifices near the entrance of each section and the CO_2, together with unutilized O_2, is removed with deaeration pumps. Turbulent flow is achieved in the pipe system with a length of about 250 m.

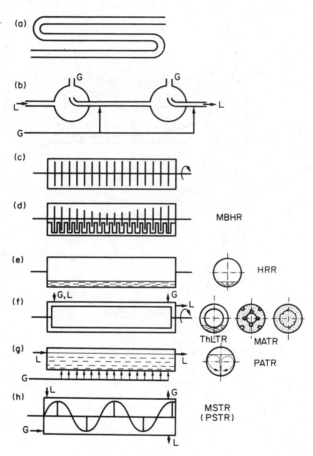

Figure 4   Horizontal tubular bioreactors for bioprocessing: (a) ordinary pipe; (b) with degasifiers; (c) biodisc; (d) multiple blade; (e) horizontal rotary reactor, HRR; (f) thin-layer tubular reactor, ThLTR, and mechanically agitated and aerated tubular reactor, MATR; (g) pneumatically aerated and agitated tubular reactor, PATR; (h) mechanically or pneumatically scraped tubular reactor. Further information and references are given in the text

A quite different design of horizontal vessel with biofilms (the biodisc) was developed for the treatment of waste water Borchardt, 1971). The device consists of a series of closely spaced discs anchored to a shaft supported just above the surface of the liquid. Thus a unit area of biological slime is alternately submerged to absorb food and then raised out of the liquid to oxidize the absorbed components. A comparable design, called the multiple blade horizontal reactor (MBHR), was presented by Means *et al.* (1962) for the handling of mycelial fermentations without the formation of a biofilm. The tube consists of nine cylindrical compartments, joined end to end, where each part is sealed off from its neighbouring compartment by a separating plate having an overflow hole in the upper half. The combined splashing of the fermenter walls and the shearing action between the agitator blades and comb-shaped baffle plates installed vertically at the bottom of each compartment in the fermenter increased mixing and suppressed the formation of mycelial deposits. Similar horizontal cylindrical chambers with several rotating discs are known in the literature for different technologies (Herrick *et al.*, 1935; Heden *et al.*, 1955; hydrocarbon fermentation: Kyowa Hakko Kogyo Ltd., 1968; Mako, 1971; continuous cultivation of microbes: Société de Bactogène, 1954; gas–liquid contacting, paddle-wheel reactor; Zlokarnik, 1975). Another design shown in Figure 4e is the horizontal rotary reactor (HRR) with an unbaffled rotating drum, originally constructed for the accurate measurement of oxygen transfer rates (Phillips *et al.*, 1961; Phillips, 1969) and subsequently used for carrying out several fermentations (Ghose and Mukhopadhyay, 1976; Mukhopadhyay and Ghose, 1976; Mukhopadhyay and Malik, 1980). The HRR was constructed by Bioengineering/Switzerland and is also called the 'rotaschon-reactor'.

Furthermore, a type of annular reactor, the thin-layer tubular reactor (ThLTR; Figure 4f), was designed by Gorbach (1969), and used for the verification of Danckwerts' renewal theory of mass transfer (Moser, 1973a). It has also found application in biotechnology in which less foaming was observed (Moser, 1973b). In order to increase the oxygen transfer rate in this reactor type, the mechanically agitated and aerated tubular reactor (MATR; Figure 4f) was further developed and characterized by A. Moser (1973b, 1973c, 1977a) for yeast production. Thin-layer reactors including HRR, TLR and MATR were summarized in a recent article (Moser, 1982a).

A pneumatically aerated tubular reactor (PATR; Figure 4g) has been designed for biological waste water treatment (Moser, 1973b; Moser, F., 1977a, 1977b; Moser, F. *et al.*, 1977; Wolfbauer *et al.*, 1979; Moser, 1982c). The criteria for the choice of continuous reactors with long residence times have been reviewed by Langensiepen (1980). A special feature of the PATR, apart from the absence of any mechanical devices is the fact that air or O_2 is introduced over the entire length of the reactor. Consequently oxygen supply can be adapted to oxygen consumption, thereby avoiding oxygen limitation.

Last but not least, a series of scraped tubular reactors agitated either by mechanical means (MSPFR) or pneumatically with gas jets (PSPFR) was designed by Moo-Young and coworkers (1979). One of these designs is schematically shown in Figure 4h. Wall growth is minimized by using rotating internal coils, a moving belt of internal discs, or helical ribbons and orifices directly in the tube. These scrapers partially segregate the liquid into moving compartments, where cross-flow aeration, effected by orifices at the bottom is realized as in the case of the PATR. These devices have been used for the production of several materials, including lipase by yeast and cellulase by fungi. Since the production of both lipase and cellulase is subject to catabolite repression, better performance may be expected in CPFRs.

One basic problem encountered with the use of tubular bioreactors is the maintenance of constant cell population inside the bioreactor because of wash-out. This problem can be solved by partial recirculation of the effluent. Such a recycle reactor is schematically shown in Figure 5a. The influence of recycle ratio r $(= F_r/F)$ on the plug-flow behaviour has been intensively studied in the case of the MATRR. By analogy with chemical reactors it was shown that the Bodenstein number decreases when r increases so that plug-flow behaviour is negligible when $r > 1$ (see Section 4.5 and Figure 12c).

A tubular loop reactor (TLR) has been designed and constructed, and was found to behave in a similar manner to the conventional STR when operating in batch (Russell *et al.*, 1974). As shown in Figure 5b, a gas phase separator chamber or a cyclone is installed in the loop in order to disengage the gas. Correlations of the type used in stirred tank systems are used to describe the influence of power and aeration rate on the mass transfer coefficient. Yeast cultures grown on hydrocarbon and glucose substrates show growth characteristics similar to those obtained in conventional stirred tanks. Finally a design procedure for commercial scale continuous TLRs has been presented by taking into consideration the fermentation kinetics, the flow behaviour (two-phase flow) (see equation 1) and O_2 transfer correlations (Ziegler *et al.*, 1977). The TLR approximates plug-flow behaviour for aerobic systems with better utilization of the available air during a single cycle at a liquid velocity of 2–3 m s^{-1} in a 22 m long tube. It is evident that this recycle reactor must be treated as a CSTR with respect to substrates in the liquid phase when high values of recycle ratio are used (Gillespie and Carberry, 1966).

A somewhat similar principle to the recycle reactor is the basis of the design of the cycle tube cyclone reactor (CTCR; Figure 5c), developed by Ringpfeil and coworkers (Liepe *et al.*, 1978; Ringpfeil, 1980a, 1980b). In this fermenter, oxygen-rich gas is contacted with the liquid broth and the gas–liquid dispersion is pumped from the top to the bottom of a column, such that no significant rise in the number of gas bubbles occurs. After about 70% of the oxygen has been transferred with the aid of perforated plates in the column at varying distances, the broth is degassed in a cyclone and is then pumped into the cycle again. In contrast to conventional STRs, the inflowing gas with maximum oxygen partial pressure is not introduced into the zone of maximum mixedness but rather into the zone of minimum agitation. In this tube reactor,

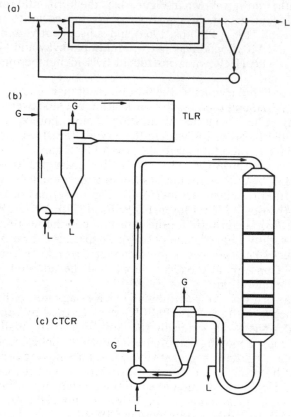

Figure 5   Tubular recycle reactors: (a) the ThLTR with external recycle; (b) the tubular loop reactor, TLR; (c) the cycle tube cyclone reactor, CTCR. References are given in the text

agitation increases as the oxygen in the gas bubbles is depleted, by the use of decreasing baffle spacing as the gas–liquid dispersion passes down the column, thus maintaining constant mechanical action and driving force throughout the reactor.

The design of CTCRs makes it possible to achieve optimum performance in the case where cell growth is performed at both the maximum specific growth rate and under conditions of maximum cell concentration with minimum energy consumption. It is essential to consider the utilization of substrate and oxygen simultaneously with oxygen depletion of the gas bubbles. As depletion is formally a first-order process where reaction velocity is controlled by the instantaneous oxygen concentration driving force, the required flow characteristics for a bioreactor would be plug-flow of the gas/liquid mixture without backmixing. Such a requirement seems to be contradictory to the requirement for complete mixedness for efficient cell growth. Fortunately, the time scale for the O_2 transfer process is measured in seconds whilst that for cell doubling is measured in tens of minutes. Thus, optimal bioreactor design must realize the required plug-flow within seconds for O_2 transfer during one cycle, and complete mixing within tens of minutes to avoid substrate limitation (Figure 3b). This general claim is fulfilled in quite different ways, especially in the PATR and in the CTCR or TLR. While plug-flow behaviour is realized in the PATR and MATR for the liquid phase, where agitation and aeration increase radial mixing but significantly decrease longitudinal mixing, in the CTCR and the TLR plug-flow is realized for the gas phase at low conversion levels.

A detailed characterization of these bioreactor systems to enable a thorough evaluation of their potential is still lacking; because of the paucity of information on their design. Scattered quantitative data are given for oxygen transfer rates ($k_L a$), Bodenstein number, power consumption, degree of hinterland (*Hl*) and reaction kinetics. Some comparable data are compiled in a few reviews (Moser A., 1973b, 1977, 1982, 1983a). With regard to the oxygen transfer rate, the following correlations have been reported in the literature:

For HRR: $$k_L = 0.12 \times 10^{-3} v + 1.65 \times 10^{-3} \tag{13}$$

For ThLTR: $Sh_l = 0.14Re^{0.5}Sc^{0.5}$ (14)

For MATR: $Sh_L = 0.1Re + 1.5$ (15a)

               $k_L a = c(P/V)^{0.5 \text{ to } 1.2}$ (15b)

For TLR: $k_L a = c(P/V)^{0.4 \text{ to } 0.9}(v_{S,G})^{0.4 \text{ to } 0.2}$ (16)

Correlations for the extent of dispersion have been reported only in the case of the PATR (scale-up operation):

$$D_L/d_h^{1.6} = 1.78(F_G/V_L)^{0.32}$$ (17)

4.3.2 Tubular Reactors Involving Biofilms

The significance of microbial films in fermenters has been pointed out from detailed studies by Atkinson and coworkers (Atkinson, 1973, 1974; Atkinson and Fowler, 1974; Atkinson and Knights, 1975). Industrial fermentations using biofilms include biological waste water treatment (trickling filter, fluidized beds, rotating disc; see Figure 4c), the 'quick vinegar process', animal tissue culture and bacterial leaching. Biofilm reactors have the advantage over biofloc reactors in that they can be operated at higher dilution rates. If the objective of the process is the reduction in effluent concentration of the limiting substrate, as in the case of waste water technology, then a piston-flow unit permits more efficient operation at the same residence time, as already discussed (see Figure 3b). The question as to whether a CPFR is superior to a CSTR is widely discussed in the literature, especially in the case of biological waste water treatment (Tucek and Chudoba, 1969; Toerber *et al.*, 1974; Turian *et al.*, 1975; Moser, F., 1977a; Wolfbauer *et al.*, 1978; Water Research Centre, 1981; Heckershoff and Wiesmann, 1981). The discrepancies which have been observed can only be explained by incorporating the biosorption phenomenon of the activated sludge with the relevant kinetic model (Moser, 1982c). Within recent years it has been realized that in order to approach a basic understanding, the different phenomena have to be singled out for specific investigation. According to the integrating strategy as a basis for a general methodology (Moser, 1982b), special bioreactors under idealized conditions are required with the means for adequate control of significant phenomena like film thickness, interfacial area and mass transfer coefficients. These so-called model bioreactors are therefore a prerequisite for a systematic process design (Moser, 1983b) with little resemblance to bioreactors in practice. This situation is especially predominant in the case of biofilm reactor operation. In spite of many decades of practical experience, and the large number of design formulae available, there are still no general correlations for the systematic design of biofilm reactors due to their complexity. The main difficulty may be attributed to the interaction between bioreactions and transfer processes, especially coupling between external and internal transport limitation (*e.g.* Fujie *et al.*, 1977; Watanabe *et al.*, 1980). Only a few biofilm reactors which enable the proper estimation of process kinetics are known. Atkinson (1974) used the sloping plane biological film fermenter to derive the general biological rate equation which incorporates internal transport limitation. Difficulties were posed by non-aseptic operation and external transport limitation under laminar flow conditions. The thickness of the biofilm was controlled by mechanical means. Tomlinson and Snaddon (1966) used a horizontal rotary reactor with the catalyst (cells) attached to the inner surface of an empty tube, typical of immobilized enzyme reactors and heterogeneous catalysis. The effects of external transport disappear with the use of completely mixed microbial film fermenters (Atkinson and Fowler, 1973) or with the use of the adsorbed cells on the outer wall of an internal tube, where liquid–solid mass transfer can be increased by increasing the rotational speed of the internal tube (La Motta, 1976). This type of reactor system has been used in biofilm processing by Kornegay and Andrews (1969) with a vertical reactor and by Gorbach (1969) with a horizontal reactor (MATFR; *cf.* Figures 4e and 4f). The biofilm thickness is hydrodynamically controlled in this case. The special properties of the MATFR are that the interfacial area is given by the geometry of the system and the mass transfer coefficient obeys Danckwerts' surface renewal theory (Moser, A., 1973a, 1977a, 1977b). A microprobe technique was proposed for the direct measurement of concentration profiles in the biofilm (Bungay *et al.*, 1969; Whalen *et al.*, 1969), facilitating model verification.

4.3.3 Tubular Reactors for Special Bioprocesses

4.3.3.1 Classical bioprocessing

The principle of plug-flow behaviour plays an important role in sterilization technology (see Figure 2) and also in enzyme technology (see Figure 3a). For applications involving substrate and/or product inhibition, CPFRs are usually superior to CSTRs (Lilly and Dunnill, 1972). Open tubular reactors, which pose several kinetic problems when used with immobilized enzymes, have become popular because of their use in gas segmented continuous flow analyzers (Bowers, 1982).

4.3.3.2   Solid substrate processing

The horizontal packed-bed tubular fermenter also seems to be promising in the case of sugar cane fermentation, in the Ex-Ferm process, where excellent sugar consumptions and ethanol yields were obtained (Rolz, 1980). In order to increase ethanol production in this process, equal weights of cane and water are required. Efficient mixing of the high solids loading is achieved by using a rotating fermenter with a horizontal perforated drum (Er-el *et al.*, 1981). Analogous problems arise when handling solid substrates (Lindenfelser and Ciegler, 1975; Hesseltine, 1977; Cannel and Moo-Young, 1980a, 1980b). Standard cement mixers, horizontal rotating drums with baffles, trays, windrows and towers are used in this special situation of handling media with a high solids content.

Similarly, plug-flow reactor systems are advantageously used as biogas plants, inclined tube settlers, vertical and horizontal tubes (*e.g.* Braun, 1982).

4.3.3.3   Shear-sensitive tissue cell cultivation

The rotating drum also represents the basis for carrying out monolayer cultures of animal cells for vaccine production ('Gyrogen' with tubes, Chemap/Switzerland, Girard *et al.*, 1980). Here surface-attached growth in roller bottles is copied and transferred to technical scale so that finally cell culture and virus multiplication could be carried out *in situ* in the same apparatus.

The handling of fragile tissue cells requires the use of special reactors such as tubular reactors. On the basis of cost of nutrient medium which dominates the economics of this bioprocess, a vertical CPFR with indirect oxygen supply was used to overcome operational problems usually encountered in conventional reactors (Pollard and Khosrovi, 1978).

4.3.3.4   Photobioreactors

Finally, another type of bioprocess illustrates the favourable application of tubular devices. Of the various means for exploiting solar energy, photosynthetic biomass production emerges as the only likely technique over large areas. The development of a suitable reactor was emphasized by Pirt (1981), who used a tubular loop reactor with a certain extent of light and dark zones. The need for an adequate supply of light forces layouts of the plants that differ essentially from those for the production of heterotrophic microorganisms. Under axenic conditions the sterile production of large quantities of microalgae and other phototrophic organisms using a 110 l tube plant was also described by Jüttner (1982). A systematic design of photoreactors on the basis of light intensity distribution in the reactor was described by Märkl and Vortmeyer (1975). See also Pirt *et al.*, (1983).

4.3.4   Comparison Between Tubular and Tower Reactors

Sparged vessels usually take the form of stirred tanks or vertical towers. For operation on a very large scale, however, these designs have their drawbacks: superficial gas rates become unallowably high and a high power input is required for compression of the feed gas to overcome the high static liquid head in tall towers. The necessity of developing unconventional bioreactors was recently stressed by Schügerl (1980) who pointed out some of the major disadvantages of stirred tanks (problems with bearings and sealings, high power consumption, low heat transfer and high costs for cooling, cell damage and low substrate conversion). Some of these problems can be easily solved by using horizontal vessels in the form of tubular reactors. On the basis of oxygen transfer efficiency η_O [kg O_2 transferred per kW power consumed] for aeration and agitation a comparison between tubular and tower reactors is depicted in Figure 6 (Moser, F., 1980). Values are calculated for different concentrations of substrate in the inflow (S_0). It can be concluded that only for low values of S_0 (*e.g.* 200 mg BOD_5 l^{-1}) is the total energy needed in tubes lower than in towers, when their height is more than a minimum of 15 m (h_{min}). With increasing influent concentrations the tower reactor is superior to the tubular reactor with respect to η_0. It is to be noted that the maximum value of η_0 in the case of tubular reactors is 2.2 and is independent of the height h, while in tower reactors $\eta_0 = f(h)$, varying from 2.3 ($h = 6$ m) to 3 ($h = 25$ m). However, a comparison should also include other factors of importance, *e.g.* the extent of axial dispersion. As already mentioned (Shiotani and Yamane, 1981) the CO_2 gas evolved during fermentation causes several undesirable problems in vertical towers as it increases the dead space and hydrostatic pressure. Horizontal devices should be preferred to vertical reactors because with the latter, nutrients entering the bottom are displaced upwards by CO_2 evolution, resulting in a reduction in the effective residence time of the feed solution. This often leads to channelling with a consequent decrease in plug-flow behaviour. A simple solution in this case is achieved by employing cross-aeration in the horizontal PATR (see Figure 4g). See also Rolz (1980).

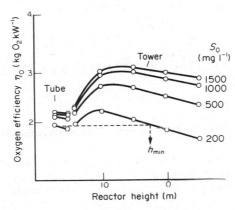

Figure 6 Comparison between horizontal (tubes) and vertical reactors (towers) on the basis of oxygen efficiency η_O as a function of reactor height for different values of influent concentration S_0 (Moser, 1980)

4.4 IMPERFECTLY MIXED STIRRED TANK REACTORS

4.4.1 Experimental Evidence

The performance of a continuous flow bioreactor is customarily expressed in terms of the steady state concentration of cells and substrates as a function of dilution rate D or the mean residence time t_r. Deviations from the predicted behaviour based on a model of the CSTR were first observed by Herbert *et al.* (1956). Obviously the use of an ideal reactor is an oversimplified description of the real bioreactor behaviour, just as it is for chemical reactors (van de Vusse, 1962). Other deviations which arise as a consequence of imperfect mixedness in the liquid phase have also been reported. These deviations can affect the cell yield by endogenous metabolism at low values of D (Hansford and Humphrey, 1966), productivity of cell mass especially with respect to hydrocarbon fermentation (Einsele, 1972), the glucose effect on yeasts (Fiechter, 1974, 1978), cell yields in batch cultures of yeast (Moser, A., 1977), cellulase productivity (Mukhopadhyay and Malik, 1980), antibiotic fermentation (Bylinkina *et al.*, 1973) and mycelial growth (Einsele and Karrer, 1980). Furthermore, some interesting results have been reported in the literature with regard to the comparison between CSTRs and CPFRs. Grieves *et al.* (1964) emphasized that a piston-flow unit is able to yield a greater maximum production rate of cell mass than a CSTR when applying a recycle. This was experimentally verified in the case of beer brewing in a tubular reactor with recycle which was reported as 'mixed plug-flow' (Moser, A., 1973, 1977). Similar results of an optimal recycle rate, which depends on Bodenstein (*Bo*) and Damköhler numbers (*Da*), are reported for bubble columns (Schügerl, 1977) and enzyme reactors (Wandrey and Flaschel, 1979). These findings are in agreement with general results for recycle reactors (Levenspiel, 1979; see Figure 3c). To understand these deviations a more structured approach to bioreactor operation must be followed by considering the importance of micromixing.

4.4.2 Characterization of Micromixing

There are various ways of expressing the extent of mixing in bioreactor systems by analogy with chemical reactors (Zlokarnik, 1967; Blakebrough, 1972; Käppel, 1976; Einsele, 1976; Oldshue *et al.*, 1978; Kipke, 1984). The parameter normally used for quantification and correlation purposes is the mixing time (terminal mixing time) t_m which is related to the degree of mixing m ($m = 1 - \Delta c/c_\infty$, with c_∞ = average concentration after homogeneous mixing and Δc = fluctuation concentration). The value of t_m, however, does not provide a sufficient criterion for micromixing and is not at all significant, as it does not include liquid–solid and solid–cell mass transfer (Einsele *et al.*, 1978). Bryant (1977) used the concept of circulation time distribution (CTD) for this purpose. The fluid is circulated regularly from a region with total segregated (ts) plug-flow through a region of maximum mixedness (mm) (*e.g.* Mukataka *et al.*, 1981). This approach seems promising for modelling (see Section 4.5). The degree of segregation J given by Danckwerts (1958) and modified by Zwietering (1959) has also been used to characterize the degree of mixing. Using this approach the simultaneous effect of macro- and micro-mixing on growth processes was calculated by Fan *et al.* (1970, 1971). In this case $J = 1$ for total segregation and $J = 0$ for maximum mixedness. Steiner (1981) extended this approach by combining segregated and mixed flows with mixing time curves.

4.5 MATHEMATICAL MODELLING OF THE EFFECT OF MICRO- AND MACRO-MIXING ON BIOPROCESSES

4.5.1 Modelling Imperfect Mixedness

Figure 7 summarizes the flow profiles observed in conventional stirred tanks (a) and the approaches used for modelling (b).

A two-region mixing model was successfully applied by Sinclair and Brown (1970) to explain the experimental deviations observed in the CSTR by Herbert *et al.* (1956). Basically, this model is identical with the so-called two-environment model of micromixing, first introduced by Ng and Rippin (1964) and then successfully applied to bioprocessing in the more empirical reversal of the original arrangement of mm region and ts environment. The 'reversed two-environment' model (Tsai *et al.*, 1971) contains an entry region with a mm flow pattern and an exit environment with a ts flow pattern. Similar attempts have been carried out by Toda and Dunn (1982) by simulating several combinations of backmix plug-flow units, representing the flow behaviour of mm–ts sequences (see Figure 7b). Brown *et al.* (1979) adapted a multiloop recirculation model previously proposed by van de Vusse (1964), shown in Figure 8, to characterize the flow behaviour of the CSTR.

The result of such analyses of the influence of imperfect mixedness on bioprocessing is shown in Figure 9 (Toda and Dunn, 1982). The steady-state behaviour of the cell concentration in this X–D diagram is shown as a function of the culture recycle rate r. It is interesting to note that similar plots of the effect of internal flow rates have been obtained by Sinclair and Brown (1970) and Brown *et al.* (1979), even though somewhat different model approaches were used. Recently Bajpaj and Reuss (1982) applied the two-environment model to bioprocessing by coupling micromixing and microbial kinetics in the case of growth of *Saccharomyces cerevisiae*. The agreement with experimental observations was found to be good using the CTD model, which permits the application to other reactor operation modes such as batch and fed-batch cultures, where previously mentioned models in the case of continuous reactor operation cannot be used.

4.5.2 Modelling the Influence of Macro- and Micro-mixing on Bioprocessing

The work of Toda and Dunn (1982) shows that the combination of the backmix and plug-flow fermenters provides a better performance than one single CSTR, with regard to the continuous production of substances which depend on the maturity of growing cells. However, the productivity of cell mass, of products which follow the kinetic model of Luedeking and Piret (1959) and repressible substances which follow the kinetic model of van Dedem and Moo-Young (1973) is still greater in the CSTR. These results, which are shown in Figure 10, are extensions of the basic concepts of Bischoff (1966) as shown in Figure 3b.

The combined effects of macro- and micro-mixing on growth processes have also been simulated by the concept of segregation (J) (Tsai *et al.*, 1969; Fan *et al.*, 1970, 1971) involving several

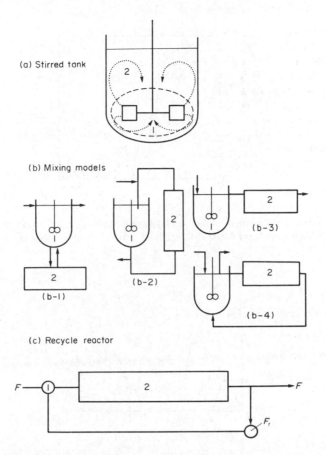

Figure 7 Schematic illustration of imperfect mixedness: (a) experimental observation of zones with mm (1) and zones with ts (2) in the case of stirred tank reactors; (b) representation of mixing models: (b-1) two-region mixing model, (b-2) two-environmental mixing model, (b-3) reversed-two-environmental mixing model, (b-4) combined backmix–plug-flow configuration; (c) experimental verification of micromixing in recycle reactor operation with a mixing point 1 (mm) and a plug-flow reactor with non-mixing zone 2. Further details in the text

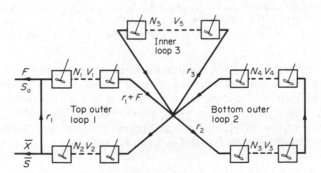

Figure 8 Multi-loop recirculation mixing model according to van de Vusse (1962) for continuous culture systems. N_i = number of mixing modules in series, V_i = volume of modules, r_i = feed rate to one mixing module ($m^3\,h^{-1}$), F = feed rate to complete system (Brown *et al.*, 1979)

combinations of CSTRs and CPFRs. They found that in a combination of the CSTR and CPFR the exit cell concentration exceeded that from other systems when t_r is not large. This is in agreement with the optimized results of Bischoff (1966). Figure 11 illustrates a typical plot of the comparison between the exit concentrations of CSTRs and CPFRs, showing not only the influence of segregation (compare $CSTR_{mm}$ and $CSTR_{ts}$) but also the effect of macromixing (compare $CSTR_{mm}$ and

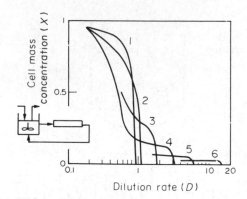

Figure 9 Cell mass concentration *versus* dilution rate in a backmix–plug-flow configuration according to model (b-4) in Figure 7 at varying recycle ratio *r*: (1) 0.9; (2) 0.5; (3) 0.2; (4) 0.1; (5) 0.05; (6) 0.02 (Toda and Dunn, 1982)

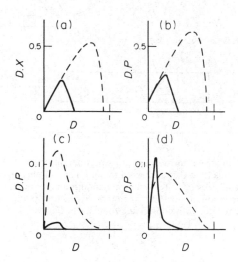

Figure 10 Productivity of fermentation product in backmix-plug-flow configuration in comparison to a single CSTR. Solid lines indicate the combined configuration, broken lines the single-stage CSTR. (a) Productivity of cell mass; (b) productivity of Luedeking–Piret-type product formation; (c) productivity of repressible product; (d) productivity of maturation time-dependent product (Toda and Dunn, 1982)

CPFR). The highest conversion is obtained in the CPFR at higher values of t_r as in the case of the $CSTR_{mm}$, where high productivity is achieved at lower values of t_r but at lower conversions.

Furthermore, the influence of micromixing was shown to be important in the case of a small number of CSTRs in series. With an increasing number of CSTRs in the cascade, thus approximating plug-flow behaviour, micromixing effects become insignificant. Nevertheless, experimental verification of these findings, which is still lacking, is essential.

4.5.3 The Recycle Reactor as a Model Bioreactor for Imperfect Mixing Studies

A generalized recycle model, shown in Figure 7c, was introduced by Gillespie and Carberry (1966) and Rippin (1967) and was subsequently used to describe various levels of micromixing in chemical reactors and bioreactors with a fixed residence time distribution (RTD) (Dohan and Weinstein, 1973; Moser and Steiner, 1974, 1975). The model is based on the assumption, which was experimentally confirmed in the MATR with recycle (see Figure 4f), that micromixing predominantly occurs instantaneously and completely at the point 1 between the flow rate of inlet fluid F and recycled fluid F_r, due to the action of a recirculation pump. Both the RTD (macromixing)

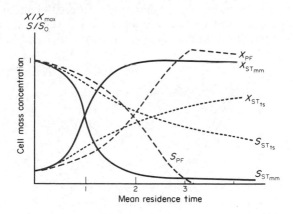

Figure 11 Comparison of the exit concentrations from the CPFR and the CSTR including different states of micromixing (maximum mixedness mm and total segregation ts) in a diagram of cell mass and substrate concentration *versus* mean residence time (Tsai *et al.*, 1969)

and the flow segregation characteristics of such a loop reactor are dependent upon the recycle ratio r ($= F_r/F$). Thus, this ratio provides a relatively simple but potentially valuable means of influencing the mixing behaviour in reactors.

The function of the total RTD in the system, $f_{tot}(t)$, is the sum of the internal RTDs, $f_{int}(t)$, given by equation (6).

$$f_{tot}(t/t_r) = \sum_{i=1}^{i_{end}} \left(\frac{1}{r+1}\right)\left(\frac{r}{r+1}\right)^{i-1} f_{int}(t/t_r) \tag{18}$$

By increasing the recycle ratio the resulting $f_{tot}(t/t_r)$ will be broader than the original internal RTD, $f_{int}(t/t_r)$. Quantification is usually achieved with the aid of concepts from probability theory using the first and second moment of the distribution function (t_r; and the variance σ^2, for the internal and total reactor system; see Dohan and Weinstein, 1973; Moser and Steiner, 1975).

The basic idea of the investigations with a recycle reactor is that by changing r and $f_{int}(t)$ at the same time, but in such a way that $f_{tot}(t)$ remains the same (macromixing effect is constant) or changes very little, the effects of both micro- and macro-mixing can be separated. This is not possible in normal reactor configurations, where micromixing is coupled to macromixing. In this case the degree of segregation can be expressed as a function of the recycle ratio r (Dohan and Weinstein, 1973)

$$J = 1 - \frac{\dfrac{r}{r+1}\left(\sigma^2_{tot} + 1 - \dfrac{r}{r+1}\right)}{\sigma^2_{int}} \tag{19}$$

The relationship between J and r is shown in Figure 12a. The curves end at the minimum value of J corresponding to the maximum value of r possible for each value of σ^2_{tot}. The curve for $\sigma^2_{tot} = 1$, representing the single CSTR, is the envelope for all possible states of mixing. Experimental verification in the case of the MATR with recycle revealed that both σ^2_{int} and σ^2_{tot} are strongly dependent on r. Figure 12b presents some typical values of the change of the variance of the internal RTD with increasing r (Moser and Steiner, 1974), while Figure 12c shows the result of the decrease in macromixing from plug-flow to maximum mixedness of the CSTR behaviour with an increase in r (Moser, A., 1977b). Here the variance of the internal RTD, σ^2_{int}, is varied to account for the experimental findings shown in Figure 12b. Application of the concepts of micro- and macro-mixing to bioprocessing showed that for optimum reactor operation a high degree of micromixing (large σ^2_{int}) is required (Dohan and Weinstein, 1973). However, a large σ^2_{int} is not favourable to conversion in those cases in which an intermediate degree of mixing resulted in maximum conversion, which have been demonstrated in Figures 3a–3c and 10. Optimum conversion clearly depends on the coupling between the two components of mixing. It has been shown that the generalized recycle model is suitable as a model of micromixing only for a constant RTD of the whole system with suitable modifications (Dudukovic, 1977). The most suitable model for micromixing is thought to be the two-environment model (*e.g.* Weinstein and Adler, 1967) with some of the modifications discussed earlier, *e.g.* consecutive segregated maximum mixedness model, parallel maximum mixedness model, *etc.* (see Section 4.5.1 and Figure 7).

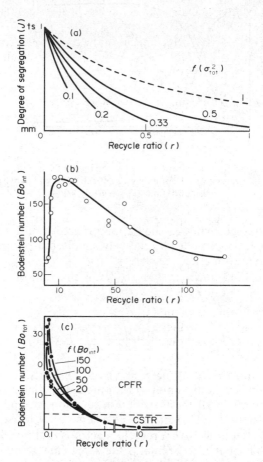

Figure 12   Quantification of micromixing: (a) relation between degree of segregation J (according to Danckwerts, 1958) and recycle ratio r, in a recycle reactor (*cf.* Figure 7c) with the variance σ_{tot}^2 (proportional to Bo_{tot} as a measure of RTD) as parameter (Dohan and Weinstein, 1973); (b) relation between the Bodenstein number of the internal RTD (Bo_{int}, proportional to σ_{int}^2) and recycle ratio r, verified in a recycle reactor according to Figure 7c (Moser and Steiner, 1974); (c) computer simulation of the coupling between micro- and macro-mixing shown in a plot of Bo_{tot} (proportional to σ_{tot}^2) *versus* recycle ratio r with Bo_{int} ($\sim \sigma_{int}^2$) as parameter. The broken line indicates the region of transition from CPFR to CSTR behaviour (Moser A., 1977?)

Nevertheless, a recycle reactor represents a simple device for experimental verification of mixing models as emphasized before. Recycle reactors can be operated in both batch and continuous mode for this purpose. Additionally they usually have only a small cycle time distribution (CTD) which is very favourable compared with normal stirred tanks.

4.6   SUMMARY AND CONCLUSIONS

Imperfectly mixed bioreactor systems include stirred tanks and reactors with concentration gradients, *e.g.* plug-flow and recycle reactors. Tubular reactors, commonly employed in chemical processing, are seldom applied in bioprocessing apart from their use in waste water treatment (simple oxidation ponds), enzyme technology (liquid–solid phase reactors) and sterilization (liquid phase reactors). The main reason for this fact is the problems in realizing high OTR simultaneously with plug-flow behaviour in the case of gas–liquid processing. Some bioreactors such as draft tube reactors with large height-to-diameter ratios have often been incorrectly referred to as tubular reactors because they exhibit some plug-flow characteristics with respect to gas–liquid mass transfer. However, with regard to biological reactions in the liquid phase, they are essentially well-mixed reactors. Similar objections exist in the case of most of the recycle reactors, which are being increasingly applied in industry. Therefore, it is important to distinguish between plug-flow in the gas phase and in the liquid phase and to classify bioreactor systems according to this criterion. Depending on the purpose of the design, plug-flow in the gas phase is advantageous for complete utilization of oxygen in the air (*e.g.* in the CTCR) while plug-flow in the liquid phase is desirable for high substrate conversion (*e.g.* in the PATR).

Tubular reactors offer some potential advantages over conventional stirred tank reactors:

(1) High productivity and optimum conversion can be achieved simultaneously.

(2) Mixing within tubular devices is more uniform, eliminating dead spaces and resulting in a more reliable scale-up of the reactors.

(3) Surface area to volume ratio is significantly higher resulting in facilitated transfer processes. As a consequence, mass transfer is achieved with comparably less power consumption in the case of horizontal devices, and heat transfer to the liquid is easily achieved. This fact becomes crucial in some bioprocessing, *e.g.* with solid substrates, photoreactions (maximum exposure to light), shear-sensitive tissues, *etc.* See, for example, Bauer and Moser (1985).

(4) Since bioprocessing occurs with gradients of concentration and/or temperature over the length of the tubes, adaptation of technical manipulations to biological demands can be carried out (*e.g.* temperature programming as a function of axial distance, substrate dosing, CO_2 removal, *etc.* in the case of inhibition and/or repression kinetics). This property is also advantageous for fundamental investigations of biological processes.

(5) Horizontal or vertical devices can be used (tubular or tower reactors), each having several advantages. With horizontal configurations plug-flow is not significantly disturbed by the CO_2 evolved and hydrostatic pressure effects are negligible.

(6) Practical advantages exist due to the closed system of tubes (no aerosol formation in the case of waste treatment; easy operation under pressure and/or with pure O_2). The technical arrangement is easily handled due to the fact that basic elements are highly developed in practice (pipes, pumps, standard fittings). Additionally, wall growth seems to be controlled more easily due to the use of high liquid velocities. A promising new method prevents surface growth by using an antimicrobial coating (Hüttinger and Müller, 1980; Ratnam *et al.*, 1982).

The problem of ensuring sufficient oxygen supply in the case of aerobic bioprocesses has generally been solved in bench, pilot and large scale operations. However, there is a lack of comparable data concerning the physical characteristics (*OTR, P/V, Bo, J, Hl, etc.*) of the various systems. For better comparison the use of a biological test system as a reference (see Lehmann *et al.*, 1982; Moser, 1983c) is emphasized. Furthermore, the implementation of recent approaches in the analysis and design of new bioreactors using regime analysis (Moser, 1984b), on the basis of the macroscopic principle using the formal kinetic concept as part of an integrating strategy (Moser, 1982b), is strongly recommended (Moser, 1983a, 1983b, 1983c, 1984c). Using this approach together with the concept of structured modelling of reactors and cells, correlations between key variables with greater accuracy and of biological significance should be extracted in the near future. As an example, the work of scale-up of bioreactors using a scale-down approach is cited (Oosterhuis, 1984).

As a part of the systematic process design, a model bioreactor must be used, presenting idealized conditions which enable the separation of biological and physical parameters. Such an ideal bioreactor has been used by Moser (1983b) in connection with bioflocs and biofilms.

Finally, the general modelling of imperfect mixedness occurring in stirred tank and tubular reactors can be readily verified in experiments using a recycle reactor. Effects of both the macro- and micro-mixing are directly dependent on the recycle ratio. As a consequence, the author recommends the application of recycle reactors for the experimental determination of the influence of the degree of mixing on bioprocesses. Combinations of stirred tank and plug-flow models are usually adequate to characterize the mixing behaviour of bioreactor systems.

Furthermore, more scale-conscious and thus reliable empirical correlations should be elaborated from technical scale bioreactors for better engineering calculations (*e.g.* Stenberg and Andersson, 1984a, 1984b, 1984c) in respect of mixing time, power consumption, gas hold-up and mass transfer.

As a general conclusion it may be stated that tubular reactors represent a valuable contribution to our understanding of optimal conversions in large-scale bioreactors. Apart from the aspects of commercial realization, tubular bioreactor systems represent an effective method for determining bioprocess kinetics, as well as evaluation of the physiological state of microbial cells.

(This article is directly connected with a research grant of FFWF/Austria, Project No. 4496.)

4.7 NOMENCLATURE

General

A	m^2	Area
Bo	—	Bodenstein number; $Bo = vL/D_L$
c_i	$kg\ m^{-3}$	Concentration of component i
c_i^*	$kg\ m^{-3}$	Saturation value of concentration
D	$m^2\ s^{-1}$	Diffusion coefficient
D	h^{-1}	Dilution rate
Da	—	Damköhler number ($Da = \mu_{max}t_r$)
D_L	$m^2\ s^{-1}$	Dispersion coefficient in liquid phase
D_{eff}	$m^2\ s^{-1}$	Effective dispersion coefficient
d_h	m	Hydraulic diameter
F	$m^3\ h^{-1}$	Flow rate
F_r	$m^3\ h^{-1}$	Flow rate in recirculation
$f(t)$	—	Mathematical RTD function
Hl	—	Degree of hinterland of a gas/liquid reactor; $Hl = V_L/A_{G/L}\delta_L$
h	m	Height
J	—	Degree of segregation
i	—	Number of recirculations
k_r	$(h^{-1})n = 1$	Reaction rate constant
k_{Tr}	h^{-1}	Transport rate constant
$k_L a$	h^{-1}	Volumetric mass transfer coefficient for OTR
K_s	$kg\ m^{-3}$	Saturation coefficient in Monod kinetics
L	m	Length
m	—	Degree of mixing
N_{eq}	—	Number of equivalent stages
P	$kg\ m^{-3}$	Product concentration
P/V	$kJ\ m^{-3}$	Specific power consumption
r	—	Recycle ratio
r_i	$kg\ m^{-3}\ h^{-1}$	Rate of consumption or formation of component i
Re	—	Reynolds number; $Re = vL/v$
S	$kg\ m^{-3}$	Substrate concentration
Sh	—	Sherwood number; $Sh = k_L L/D$
t	h	Time
t_r	h	Mean residence time
t_m	h	Mixing time
t_M	h	Maturation time according to Brown and Vass (1973) (time needed for 'ripening' of a product)
t_{st}	h	Sterilization time
V	m^3	Volume
v	$m\ s^{-1}$	Velocity
$v_{S.G}$	$m\ s^{-1}$	Superficial gas velocity
z	—	Coordinate

Greek Letters

δ_L	m	Thickness of liquid film
η_O	—	Oxygen efficiency
μ_{max}	h^{-1}	Maximum specific growth rate
v	$m^2\ s^{-1}$	Viscosity
σ^2	—	Variance of RTD function
ζ	—	Fractional conversion

Indices

end	End value
ex	Exit

G	Gas
in	Inlet
int	Internal
i_{end}	End number of recirculations
L	Liquid
max	Maximum
min	Minimum
mm	Maximum mixedness
0	Zero
O	Oxygen
PF	Plug flow reactor
R	Reactor
ST	Stirred tank reactor
tot	Total
ts	Total segregation
s	Substrate or solid
∞	Infinite value
z	Coordinate

4.8 REFERENCES

Atkinson, B. (1973). The role of microbial films in fermentation. *Pure Appl. Chem.*, **36**, 279–304.

Atkinson, B. (1974). *Biochemical Reactors*. Pion, London.

Atkinson, B. and H. W. Fowler (1974). The significance of microbial film in fermenters. *Adv. Biochem. Eng.*, **3**, 221–277.

Atkinson, B. and A. J. Knights (1975). Microbial film fermenters: their present and future application. *Biotechnol. Bioeng.*, **17**, 1245–1267.

Atkinson, B. *et al.* (1980). Process intensification using cell support systems. *Process Biochem.*, April/May, 24–32.

Bajpaj, R. K. and M. Reuss (1982). Coupling of mixing and microbial kinetics for evaluating the performance of bioreactors. *Can. J. Chem. Eng.*, **60**, 384–391.

Bauer, A. and A. Moser (1985). Mixing and mass transfer in the horizontal stirred tank. In *Proceedings of the 5th European Conference on Mixing*, Würzburg/FRG. BHRA Press.

Bischoff, K. B. (1966). Optimal continuous fermentation reactor design. *Can. J. Chem. Eng.*, **44**, 281–284.

Blakeborough, N. (1972). Mixing effects in biological systems. *Chem. Eng.*, **17**, 58–63.

Borchardt, J. A. (1971). Biological waste treatment using rotating discs. *Biotechnol. Bioeng. Symp.*, **2**, 131–140.

Bowers, H. D. (1982). Immobilized enzymes in analytical chemistry. *Trends Anal. Chem.*, **1**, 191–198.

Braun, R. (1982). *Biogas-Methangärung organ. Abfallstoffe*. Springer-Verlag, Vienna.

Brown, D. E. and R. C. Vass (1973). *Biotechnol. Bioeng.*, **15**, 321–330.

Brown, D. E. *et al.* (1979). Application of an aerated mixing model to a continuous culture system. *Biotechnol. Lett.*, **1**, 159–164.

Bryant, J. (1977). Characterization of mixing in fermenters. *Adv. Biochem. Eng.*, **5**, 101–123.

Bungay, H. R. *et al.* (1969). Microprobe techniques for determining diffusivities and respiration rates in microbial slime systems. *Biotechnol. Bioeng.*, **11**, 675–772.

Bylinkina, E. S. *et al.* (1973). Studies of agitation intensity in antibiotic fermentation broth using isotropic tracers. *Biotechnol. Bioeng. Symp.*, **4**, 331–342.

Cannel, E. and M. Moo-Young (1980a). Solid state fermentation systems. *Process Biochem.*, June/July, 2–7.

Cannel, E. and M. Moo-Young (1980b). Solid state fermentation systems. *Process Biochem.*, August/September, 24–28.

Collin, G. *et al.* (1979). Pyrolitische Rohstoff-Rückgewinnung aus unterschiedlichen Sonderabfällen in einem Drehtrommelreaktor. *Chem.-Ing.-Tech.*, **51**, 220–224.

Cook, A. H. (1968). Der Einfluss von kontinuierlichen Brauverfahren auf die Bierqualität. *Mitteil. Vesuchanstalt Gärungsgewerbe Wien*, **22**, 33–38.

Danckwerts, P. V. (1958). The effect of incomplete mixing on homogeneous reactions. *Chem. Eng. Sci.*, **8**, 93–102.

Denshchikov, M. (1962). Development of apparatus for continuous beer fermentation. In *Proc. Symp. Cont. Culture Microorg.*, 2nd, pp. 221–233. Czechoslovak Academy of Sciences, Prague.

Dohan, L. A. and H. Weinstein (1973). Generalized recycle reactor model for micromixing. *Ind. Eng. Chem. Fundam.*, **12** (1), 64–69.

Dudukovic, M. P. (1977). On the use of the generalized recycle model to interpret micromixing in chemical reactors. *Ind. Eng. Chem. Fundam.*, **16** (3), 385–388.

Einsele, A. (1972). *Die Funktion des Stofftransportes in Bioreaktoren beim microbiellen Kohlenwasserstoffabbau*. Ph.D. Thesis, ETH Zürich.

Einsele, A. (1976). Characterisierung von Bioreaktoren durch Mischzeiten. *Chem. Rundschau*, **29**, 53–55.

Einsele, A. and D. Karrer (1980). Design and characterization of a completely filled stirred reactor. *Eur. J. Appl. Microbiol. Biotechnol.*, **9**, 83–91.

Einsele, A. *et al.* (1978). Mixing times and glucose uptake measured with a fluorometer. *Biotechnol. Bioeng.*, **20**, 1487–1492.

Er-el, Z. *et al.* (1981). Ethanol production from sugar cane segments in a high solids drum fermenter. *Biotechnol. Lett.*, **3**, 385–390.

Fan, L. T. *et al.* (1970). Effect of mixing on the washout and steady-state performance of continuous cultures. *Biotechnol. Bioeng.*, **12**, 1019–1068.

Fan, L. T. *et al.* (1971). Simultaneous effect of macromixing and micromixing on growth processes. *AIChE J.*, **17**, 689–696.

Fiechter, A. (1974). Regulatory aspects of yeast metabolism and their consequences for cell mass production. In *Proc. Int. Symp. Yeasts, Vienna, 4th*, ed. H. Klaushofer and U. B. Slaytr, part 2, pp. 17–38. Univ. Bodensultur, Vienna.

Fiechter, A. (1978). Specification of bioreactors. Requirements in view of the reaction and scale of operation. In *Proc. Eur. Congr. Biotechnol., 1st.* (Dechema Monograph No. 82), pp. 17–36.

Fujie, K. *et al.* (1977). Mass transfer in the liquid phase with tubular waste-water treatment contactor. *J. Ferment. Technol.*, **55**, 532–543.

Ghose, T. K. and S. N. Mukhopadhyay (1976). Kinetic studies of gluconic acid fermentation in horizontal rotary fermenter by *Pseudomonas ovalis. J. Ferment. Technol.*, **54**, 738–750.

Gillespie, B. M. and J. J. Carberry (1966). Reactor yield at intermediate mixing levels — an extension of van de Vusse analysis. *Chem. Eng. Sci.*, **21**, 472–475.

Girard, A. C. *et al.* (1980). Monolayer studies of animal cells with the gyrogen equipped with tubes. *Biotechnol. Bioeng.*, **22**, 477–493.

Gorbach, G. (1969). Die kontinuierliche Dünnschichtfermentation. *Monatsschrift Brauerei*, **22**, 49–52.

Greenshields, R. N. and E. L. Smith (1974). The tubular reactor in fermentation. *Process Biochem.*, April, 1–28.

Grieves, R. B. *et al.* (1964). Piston-flow reactor model for continuous industrial fermentations. *J. Appl. Chem.*, **14**, 478–487.

Hall, R. D. and G. Howard (1968). Inclined fermenter tube for beer brewing. *US Pat.* 3 407 069.

Hamer, G. (1982). Recycle in fermentation processes. *Biotechnol. Bioeng.*, **24**, 511–531.

Hansford, G. S. and A. E. Humphrey (1966). The effect of equipment scale and degree of mixing on continuous fermentation yield at low dilution rates. *Biotechnol. Bioeng.*, **8**, 85–96.

Heckershoff, H. and U. Wiesmann (1981). Der Belebtschlammreaktor im ein- und vier-stufigen Betrieb. *Chem.-Ing.-Tech.*, **53**, 268–270.

Heden, G. C. *et al.* (1955). An improved method for the cultivation of microorganisms by the continuous technique. *Acta Path. Microb. Scand.*, **37**, 42–49.

Helmrich, H. and K. Schügerl (1979). Drehrohrreaktoren in der Chemietechnik. *Chem.-Ing.-Tech.*, **51**, 771–778.

Herbert, D. *et al.* (1956). The continuous culture of bacteria. A theoretical and experimental study. *J. Gen. Microbiol.*, **14**, 601–623.

Herrick, N. T. *et al.* (1935). Apparatus for the application of submerged mould fermentation under pressure. *Ind. Eng. Chem.*, **27**, 681–689.

Hesseltine, C. W. (1977). Solid-state fermentation. *Process Biochem.*, July/August, 24–27; November, 29–32.

Hüttinger, K. J. and H. Müller (1980). Antimikrobiell wirkende Aufpropfschichten-Aufpropfreaktionen, Wirkung, Anwendungen. *Chem.-Ing.-Tech.*, **52**, 77.

Joshi, J. B. and M. M. Sharma (1976). Mass transfer characteristics of horizontal agitated contactors. *Can. J. Chem. Eng.*, **54**, 560–565.

Jüttner, F. (1982). Mass cultivation of microalgae and phototrophic bacteria under sterile conditions. *Process Biochem.*, March/April, 2–7.

Käppel, M. (1976). Development and application of a method for measuring the mixture quality of miscible liquids. *Int. Chem. Eng.*, **19**, 196–215, 431–445, 571–590.

Kipke, K. (1984). Mixing time. In *Process Variables in Biotechnology*, pp. 120–126. Dechema, Frankfurt.

Kornegay, B. H. and J. F. Andrews (1969). Application of the continuous culture theory to the trickling filter process. In *Proc. Ind. Waste Conf., 24th*, pp. 1398–1409.

Kyowa Hakko Kogyo Co. Ltd. (1968). Fermenting hydrocarbons using a horizontal stirring tank. *Jpn. Pat.* 72 38 185.

La Motta, E. J. (1976). External mass transfer in a biological film reactor. *Biotechnol. Bioeng.*, **18**, 1359–1370.

Langensiepen, H. W. (1980). The choice of continuous phase laboratory reactors with long residence time. *Chem. Eng. Sci.*, **35**, 492–498.

Lehmann, J. *et al.* (1982). Reference tests for fermentations. In *Arbeitsmethoden für die Biotechnologie.* Dechema, Frankfurt.

Lehmberg, J. *et al.* (1977). Transverse mixing and heat transfer in horizontal rotary drum reactors. *Powder Technol.*, **18**, 149–163.

Levenspiel, O. (1979). *Chemical Reactor Omnibook.* OSU Book Stores, Cornvallis/Oregon.

Liepe, F. *et al.* (1978). The present state and the perspective of development of industrial high performance fermenters for aerobic fermentations. In *Prepr. Eur. Congr. Biotechnol., 1st, Interlaken*, part 1, p. 78.

Lilly, M. D. and P. Dunnill (1972). Engineering aspects of enzyme reactors. *Biotechnol. Bioeng. Symp.*, **3**, 221–228.

Lindenfelser, L. A. and A. Ciegler (1975). Solid-substrate fermenter for ochratoxin A production. *Appl. Microbiol.*, **29**, 323–327.

Luedeking, R. and E. L. Piret (1959). *Biotechnol. Bioeng.*, **1**, 393–412.

Mako, P. F. (1971). Improved fermentation apparatus for use in hydrocarbon fermentation processes. *US Pat.* 3 594 277.

Mason, D. R. and E. L. Piret (1950). Continuous flow stirred tank reactor systems. Development of transient equations. *Ind. Eng. Chem.*, **42**, 817–829.

Märkl, H. and D. Vortmeyer (1975). Berechnung der Verteilung von Bestrahlungsstärbe und Adsorption in einem diffus von aussen beleuchteten zylindrischen Photoreaktor. *Chem.-Ing.-Tech.*, **47**, MS 172.

Means, C. W. *et al.* (1962). Design and operation of a pilot-plant fermenter for the continuous propagation of filamentous microorganisms. *Biotechnol. Bioeng.*, **4**, 5–16.

Moo-Young, M. and H. W. Blanch (1981). Design of biochemical reactors. Mass transfer criteria for simple and complex systems. *Adv. Biochem. Eng.*, **19**, 2–69.

Moo-Young, M. *et al.* (1979). Design of scraped tubular fermenters. *Biotechnol. Bioeng.*, **21**, 593–607.

Moritz, H. U. and K. H. Reichert (1981). Zur kontinuierlichen Perlpolymerisation im gerührten Rohrreaktor. *Chem.-Ing.-Tech.*, **53**, 386.

Moser, A. (1973a). Untersuchungen des Gas/Flüssigkeitsstofftransportes im Dünnschichtfermenter. *Chem.-Ing.-Tech.*, **45**, 1313–1317.

Moser, A. (1973b). Über die reaktionstechnischen Möglichkeiten der Verwendung eines Rohrreaktors zur Durchführung kontinuierlichen Fermentationen. In *Proc. Symp. Tech. Microbiol. 3rd*, ed. H. Dellweg, pp. 61–66. Verlag-Versuchs-und Lehranstalt für Spiritusfabrikation und Fementationstechnologie im Insitut für Gärungsgwerbe und Biotechnologie, Berlin.

Moser, A. (1973c). Tubular fermenter for aerobic processes. I: Physical characteristics. *Biotechnol. Bioeng. Symp.*, **4**, 399–411.

Moser, A. (1977a). Dünnschichtreaktoren in der Biotechnologie. *Chem.-Ing.-Tech.*, **49**, 612–625.

Moser, A. (1977b). *Fundamentals of Systematic Process Development in Biotechnology*. Thesis of habilitation, TU Graz.

Moser, A. (1981). *Bioprozesstechnik*. Springer, Vienna.

Moser, A. (1982a). Bioreactors with thin-layer characteristics. *Biotechnol. Lett.*, **4**, 281–286.

Moser, A. (1982b). Integrating strategy—a basis for biotechnological methodology. *Biotechnol. Lett.* **4**, 73–78.

Moser, A. (1982c). Kinetics applied in process design for biological waste water treatment. In *Proc. International Waste Treatment and Utilization*, 2nd, ed. M. Moo-Young *et al.*, vol. 2, pp. 177–192. Pergamon, Oxford.

Moser, A. (1983a). Tubular bioprocessing—a review. In *Proc. Can. Chem. Eng. Conf.*, *33rd, Toronto*, vol. 2, pp. 417–423. Can. Soc. Chem. Eng., Toronto.

Moser, A. (1983b). Multi-purpose bioreactor for bioprocess kinetic analysis. In *Proc. Adv. Ferm. Conf., London* (*Suppl. Process Biochem.*, 202), pp. 202–211.

Moser, A. (1983c). Kinetics of batch fermentations. In *Proc. Biotech. 83*, pp. 961–972.

Moser, A. (1984a). Kinetic strategy for bioprocess design. In *Biotechnology—A Comprehensive Treatise*, ed. H. Rehm and G. Reed, vol. 2, chap. 14. Verlag Chemie, Weinheim.

Moser, A. (1984b). *Acta Biotechnol.*, **4**, 3–9.

Moser, A. (1985). In *Bioprocess Engineering*, in press. Springer, New York.

Moser, A. and W. Steiner (1974). Verweilzeitverteilungsverhalten eines Rohrreaktors mit Rückführung. *Chem.-Ing.-Tech.*, **46**, 695.

Moser, A. and W. Steiner (1975). Verweilzeitverteilung und Mischverhalten eines Rohrreaktors mit Rückführung. *Chem.-Ing.-Tech.*, **47**, 211.

Moser, F. (1977a). Ein Rohrreaktor zur Abwasserreinigung. *Verfahrenstechnik*, **11**, 670–673.

Moser, F. (1977b). Abwasserreinigung in Röhren. Paper presented at 6th Reaction Engineering Meeting, TU Graz, May.

Moser, F. (1980). Verfahrenstechnische Aspecte der biologische Abwasserreinigung. In *Grundlagen der Abwassererreinigung*, ed. F. Moser, vol. 2, pp. 431–419. Oldenburg, Munich.

Moser, F. *et al.* (1977). A novel aeration basin for biological waste water treatment. *Prog. Water Technol.*, **8**, 235–237.

Mukataka, S. *et al.* (1981). Circulation time and degree of fluid exchange between upper and lower circulation regions in a stirred vessel with a dual impeller. *J. Ferment. Technol.*, **59**, 303–307.

Mukhopadhyay, S. N. (1978). Oxygen transfer in a laboratory horizontal fermenter under varying rotation speed. *J. Ferment. Technol.*, **56**, 558–560.

Mukhopadhyay, S. N. and T. K. Ghose (1976). A simple dynamic method of $k_L a$ determination in a laboratory fermenter. *J. Ferment. Technol.*, **54**, 406–419.

Mukhopadhyay, S. N. and R. K. Malik (1980). A comparative study on the production of cellulase of *Trichoderma* spp. in STR and HTR. In *Proc. Symp. Bioconv. Biochem. Eng.*, *2nd*, ed. T. K. Ghose, vol. 2, pp. 295–305. IIT, New Delhi.

Ng, D. Y. G. and D. W. T. Rippin (1964). The effect of incomplete mixing on conversion in homogeneous reactions. In *Proc. Eur. Symp. Chem. React. Eng.*, *3rd*, pp. 161–166. Pergamon, Oxford.

Oldshue, J. Y. *et al.* (1978). Fluid mixing variables in the optimization of fermentation production. *Process Biochem.*, November, 16–19.

Olsen, A. J. C. (1958). Fermentation vessel and process. *Br. Pat.* 827 404.

Oosterhuis, N. M. G. (1984). Scale-up of bioreactors. Ph.D. Thesis, TU Delft.

Phillips, K. L. (1969). Reactor systems for processes with extreme gas–liquid transfer requirements. In *Fermentation Advances*, ed. D. Perlman, pp. 465–490. Academic, New York.

Phillips, K. L. *et al.* (1961). Oxygen transfer in fermentations. *Ind. Eng. Chem.*, **53**, 749–754.

Pirt, S. J. (1981). Microbial photosynthesis in the harnessing of solar energy. In *Proc. Eur. Congr. Biotechnol.*, *2nd*, p. 54. Soc. Chem. Ind., London.

Pirt, S. J. *et al.* (1983). A tubular bioreactor for photosynthetic production of biomass from carbon dioxide: design and performance. *J. Chem. Technol. Biotechnol.*, **33**, 35–39.

Pollard, R. and B. Khosrovi (1978). Reactor design for fermentation of fragile tissue cells. *Process Biochem.*, May, 31–37.

Powell, O. and J. R. Lowe (1964). Theory of multi-stage continuous cultures. In *Continuous Culture of Microorganisms*, ed. I. Malik *et al.*, pp. 45–57. Czech Academy of Sciences, Prague.

Ratnam, D. A. *et al.* (1982). Effects of attachment of bacteria to chemostat walls in a microbial predator–prey relationship. *Biotechnol. Bioeng.*, **24**, 2675–2694.

Ringpfeil, M. (1980a). Biotechnologie auf dem Weg zur Wissenschaft. Paper presented at the 2nd Socialist Countries Symposium on Biotechnology.

Ringpfeil, M. (1980b). Fermenter design for SCP production. Paper presented at the 6th International Fermentation Symposium, London, Ontario.

Rippin, D. W. T. (1967). The recycle reactor as a model of incomplete mixing. *Ind. Eng. Chem. Fundam.*, **6**, 488.

Rolz, C. (1980). A new technology to ferment sugar cane directly: the Ex-ferm process. *Process Biochem.*, August/September, 2–6.

Russell, T. W. F. *et al.* (1974). The tubular loop batch fermenter: basic concepts. *Biotechnol. Bioeng.*, **16**, 1261–1272.

Schügerl, K. (1977). Apparative Aspekte von Blasensäulen-Bioreaktoren. *Chem.-Ing.-Tech.*, **49**, 605–611.

Schügerl, K. (1980). Neue Bioreaktoren für aerobe Bioprozesse. *Chem.-Ing.-Tech.*, **52**, 951–965.

Shah, A. K. and M. M. Sharma (1975). Mass transfer in gas–liquid (horizontal) pipeline contactors. *Can. J. Chem. Eng.*, **53**, 572–576.

Shiotani, T. and T. Yamane (1981). A horizontal packed-bed bioreactor to reduce CO_2-gas holdup in the continuous production of ethanol by immobilized yeast cells. *Eur. J. Appl. Microbiol. Biotechnol.*, **13**, 96–101.

Sinclair, C. G. and D. E. Brown (1970). Effect of incomplete mixing on the analysis of the static behaviour of continuous culture. *Biotechnol. Bioeng.*, **12**, 1001–1017.

Sittig, W. and H. Heine (1977). Erfahrungen mit grosstechnisch eingesetzten Bioreaktoren. *Chem.-Ing.-Tech.*, **49**, 595–604.

Société de Bactogène (1954). Improvements in methods and apparatus for the cultivation of microorganisms. *Br. Pat.* 704 872.

Steiner, W. (1981). Charakterisierung von Bioreaktoren. In *Proc. Rotenburger Symp. Ferment.*, *2nd*, ed. R. M. Lafferty, pp. 56–68. Springer Verlag, Vienna.

Stenberg, O. and B. Andersson (1984a). *Chem. Eng. Res. Des.*, in press.

Stenberg, O. and B. Andersson (1984b). *Chem. Eng. Res. Des.*, in press.

Stenberg, O. and B. Andersson (1984c). *Chem. Eng. Sci.*, in press.

Štěrbáček, Z. and M. Šáchová (1973). Non-ideal flow phenomena in tubular fermentation systems—fundamentals and influence on equipment design. *Pure Appl. Chem.*, **36**, 365–376.

Toda, K. and I. J. Dunn (1982). Continuous culture in combined backmix–plugflow–tubular loop fermenter configurations. *Biotechnol. Bioeng.*, **24**, 651–668.

Toerber, E. D. *et al.* (1974). Comparison of completely mixed and plug flow biological systems. *J. Water Pollut. Control Fed.*, **46**, 1995–2014.

Tomlinson, T. G. and D. M. Snaddon (1966). Biological oxidation of sewage by films of microorganisms. *Air Water Pollut.*, **10**, 865–872.

Topiwala, H. H. (1973). The application of kinetics to bioreactor design. *Biotechnol. Bioeng. Symp.*, **4**, 681–690.

Tripathi, G. *et al.* (1971). Mass transfer at rotating cylinders. *Indian J. Technol.*, **9**, 237–241.

Tsai, B. I. *et al.* (1969). The effect of micromixing on growth processes. *Biotechnol. Bioeng.*, **11**, 181–205.

Tsai, B. I. *et al.* (1971). The reversed two-environmental model of micromixing and growth processes. *J. Appl. Chem. Biotechnol.*, **21**, 307–312.

Tucek, F. and J. Chudoba (1969). Purification efficiency in aeration tanks with complete mixing and piston flow. *Water Res.*, **3**, 559–570.

Turian, R. M. *et al.* (1975). The dispersed flow model for a biological reactor as applied to the activated sludge process. *Can. J. Chem. Eng.*, **53**, 431–437.

Ueda, K. (1958). Studies on continuous fermentation, part VII: continuous fermentation of molasses with a combined system of agitated vessel and flow pipe. *J. Agric. Chem. Soc. Jpn.*, **32**, 26–31.

Van Dedem, G. and M. Moo-Young (1973). *Biotechnol. Bioeng.*, **15**, 419–439.

van de Vusse, J. G. (1964). Multi-loop recirculation model of mixing. *Chem. Eng. Sci.*, **19**, 994–997.

Wandrey, C. and E. Flaschel (1979). Process development and economic aspects in enzyme engineering: acylase L-methionine system. *Adv. Biochem. Eng.*, **12**, 148–218.

Watanabe, T. *et al.* (1980). Nitrification kinetics in a rotating biological disk reactor. *Prog. Water Technol.*, **12**, 233–251.

Water Research Centre (1981). *Proceedings of a Conference on Activated Sludge Bulking Prevention and Cure*, *Cambridge*. Water Research Centre.

Weinstein, H. and R. J. Adler (1967). Micromixing effects in continuous chemical reactors. *Chem. Eng. Sci.*, **22**, 65–75.

Wen, C. Y. and L. T. Fan (1975). *Models for Flow Systems and Chemical Reactors*. Dekker, New York.

Whalen, W. J. *et al.* (1969). Microelectrode determination of oxygen profile in microbial slime systems. *Environ. Sci. Technol.*, **3**, 1297–1308.

Wolfbauer, O. *et al.* (1978). Reaction engineering models of biological waste water treatment and the kinetics of the activated sludge process. *Chem. Eng. Sci.*, **33**, 953–960.

Wolfbauer, O. *et al.* (1979). Ein Gas/Flüssigkeits-Rohrreaktor mit geringer Rückvermischung. *Chem.-Ing.-Tech.*, **51**, 535.

Ziegler, H. *et al.* (1977). The tubular loop fermenter: oxygen transfer, growth kinetics and design. *Biotechnol. Bioeng.*, **19**, 507–525.

Ziegler, H. *et al.* (1980). Oxygen transfer and mycelial growth in a tubular loop fermentor. *Biotechnol. Bioeng.*, **22**, 1613, 1635.

Zlokarnik, M. (1967). Eignung von Rührerin zum Homogenisierien von Flüssigkeitsgemischen. *Chem.-Ing.-Tech.*, **39**, 539–548.

Zlokarnik, M. (1975). Der Schaufelradreaktor—ein spezieller Reaktortyp für Reaktionen im System gasformig/flüssig. *Verfahrenstechnik*, **9**, 442–445.

Zuiderweg, F. J. and C. Bruinzeel (1971). Liquid mixing and oscillators in sparged horizontal cylindrical vessels. In *Proc. Eur. Symp. Ser. Chem. React. Eng.*, *4th*, pp. 183–189. Pergamon, Oxford.

Zwietering, T. N. (1959). The degree of mixing in continuous flow systems. *Chem. Eng. Sci.*, **11**, 1–16.

5

Nonmechanically Agitated Bioreactor Systems

K. SCHÜGERL
Technischen Universität Hannover, Federal Republic of Germany

5.1 INTRODUCTION

In the early stages of biotechnology very different reactor types were used. However, in recent decades the stirred tank reactor has succeeded in becoming the standard bioreactor.

Since, generally speaking, the product-specific, fixed and variable costs decrease with increasing production capacity of a plant, considerable efforts are made in the industry to build large single line units. This holds especially true for the manufacture of mass products. This trend requires very large bioreactors which are difficult to construct and to handle, *e.g.* considerable construction problems arise with mechanically stirred reactors in the volume range of 1000 m³ and above. The power requirement is extremely high when keeping the specific power input (P/V) constant. This leads to high costs for energy and cooling water and problems with heat removal. Furthermore, the quick and uniform distribution of energy, dissolved oxygen, carbon source (substrate) and other nutrients as well as the pH correction and antifoam agents in the large reactor volume cause considerable difficulties. The traditional stirred tank reactors did not meet these demands.

Besides the economical and technical reasons, biological factors (high shear stress) rule out the use of stirred tanks for certain applications.

Accordingly, many new reactors have been developed, and well-known types have been improved, offering technical, economical and sometimes biological advantages over the standard stirred tank in special production processes or, for example, in waste water treatment. Large-scale production of single cell proteins and waste water treatment require large reactor volumes for reasons of economy. For example, ICI installed a single cell protein (SCP) unit in Billingham having a total reactor volume of about 2600 m^3, an effective reactor volume of 2100 m^3 and a total power input of 7 MW; and in Leverkusen Bayer AG installed a waste water treatment plant having a total volume of about 20 000 m^3 (reactor height 30 m, diameter 26 m).

The operators of industrial plants publish little key information on the performance of such plants. Therefore, the main information sources are publications on laboratory and pilot plant investigations. Often only results with nonbiological model systems are published. Even in biological systems the reactors are often not comparable, because the biological test systems or the cultivation conditions are different.

For these reasons only incomplete characterization and comparisons are possible.

5.2 CLASSIFICATION OF NONMECHANICALLY AGITATED BIOREACTOR SYSTEMS

Because of the diversity of nonmechanically agitated bioreactors, it is difficult to classify them according to a unified system satisfactory to all aspects of biotechnology. Therefore, the following will consider only the technological aspects; the bioreactors are classified first according to the method of introducing energy: (i) power input by liquid kinetic energy; (ii) power input by gas compression.

The construction details and the operation modes are considered within these two main classes. Only reactors for aerobic processes are examined.

5.3 POWER INPUT BY LIQUID KINETIC ENERGY

The liquid is accelerated by a liquid pump with an efficiency η given by

$$\eta = \eta_h \eta_m \tag{1}$$

where η_h is hydraulic efficiency, including clearance losses and edge-side friction, and η_m is mechanical efficiency due to friction losses in the bearings and packing gland.

In the case of centrifugal pumps $\eta_h \simeq 0.62$ to 0.92, $\eta_m \simeq 0.88$ to 0.98 and $\eta \simeq 0.55$ to 0.90 and for reciprocating pumps $\eta_h \simeq 0.85$ to 0.98, $\eta_m \simeq 0.88$ to 0.97 and $\eta \simeq 0.75$ to 0.95. The total loss varies with the rate of discharge F_L (m^3 s^{-1}) and the stirring speed N (s^{-1}). With increasing discharge and speed of rotation η increases.

The energy fed into the liquid is utilized very differently, depending on the gas-dispersing mechanism. The main aim is the homogenization and dispersion of the phases. Usually the gas phase is dispersed into the liquid phase. For this purpose different types of nozzle are used. These various nozzles and gas inlet devices, described in the following paragraphs, are illustrated in Figures 1 to 3, and will be referred to throughout this chapter as 'nozzle A', 'nozzle F1', 'gas inlet G', *etc.*

Nozzle A: two-phase free jet nozzle

Nozzle B: two-phase nozzle: central nozzle for liquid and ring nozzle for gas feed

Nozzle C: two-phase nozzle with mixing chamber: central nozzle for liquid and mixing chamber for gas feed

Nozzle D: radial flow nozzle with gas dispersion and bubble distribution

Nozzle E: vertical flow nozzle

Nozzle F: plunging jet nozzle with different length of the jet. If the jet is shorter than its critical length, the jet hits the liquid surface (nozzle F1). If the jet is longer than its critical length, the decomposed jet (droplets) hits the liquid surface (nozzle F2). If the nozzle

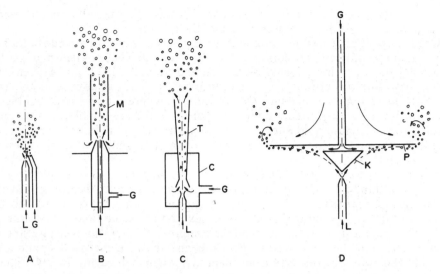

Figure 1 Different types of nozzles for gas dispersion used in the reactor bottom

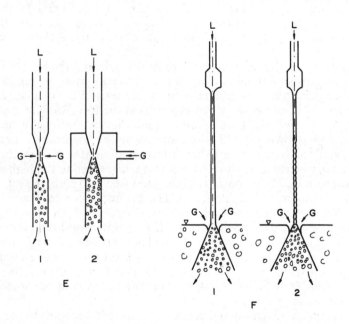

Figure 2 Different types of nozzles for gas dispersion used at the top of reactor

is positioned at the liquid surface, the jet length is zero (nozzle F3) (Kürten and Zehner, 1980). (F3 is not shown in Figure 2.)

A, B, C and D are used at the bottom of the reactor; E and F are mounted into the reactor head.

In nozzles B and C the liquid is pumped into the central nozzle and the gas is sucked out (ejector nozzle), or the compressed gas is expanded (injector nozzle) into the ring nozzle or into the mixing chamber. Recently nozzle B has been operated in the alternative way: the liquid is pumped into the ring nozzle and the gas is sucked out (ejector) or the compressed gas is expanded (injector) into the central nozzle. Here these nozzles will be called nozzle B$^+$ (see top of Figure 7, p. 109). Such nozzles were recommended by Goto *et al.* (1979, 1981) as well as by Räbiger and Vogelpohl (1983). Goto *et al.* used compressed air and placed the nozzle at the bottom of the reactor. Räbiger and Vogelpohl operated these nozzles as ejector as well as injector nozzles and fastened them at the top of the reactor such that nozzle B^+ is immersed into the liquid (Figure 7).

Actually the injector nozzles are hybrid systems, because they use the kinetic energy of the liquid as well as compressed air for power input into the reactor.

When using the two-phase free jet (nozzle A), the gas is dispersed by means of a high speed liquid jet. The same phenomenon occurs in nozzles B, C and D. The mechanism of gas dispersion in ejector nozzles has been investigated by Kürten and Maurer (1977) and Hallensleben *et al.* (1977). A fast jet of liquid is produced by the central nozzle. The gas flows from the annular gas nozzle into the momentum exchange tube (M in Figure 1), where it forms large bubbles. The fast jet of liquid, which is pulsating, breaks up into fine drops in the gas bubble; these drops impinge with high velocity on the gas/liquid interface destroying it. The small bubbles thus produced are finely dispersed in the momentum exchange tube by the high turbulent dynamic pressure, due to the high Reynolds number.

When using an injector nozzle, a mixing chamber (C in Figure 1) is employed. The gas is sucked into the liquid jet due to the underpressure caused by the jet. According to Witte (1962), a shock forms in the mixing tube (T in Figure 1). In the region of this mixing shock the flow changes from a jet flow to a homogeneous bubble flow. At the same time the pressure increases considerably. The mixing shock is thicker than the compression shock in the transition of a super-sonic flow into a subsonic flow. The mixing shock is connected with energy dissipation. The prob-able reason for the different dispersion mechanisms, when comparing the two-phase nozzle having momentum exchange tubes with those having mixing chambers, is the fact that in the first type the liquid is being sucked in from the surrounding area. This results in a smaller gas fraction, and even with large driving-jet velocities of 40 m s^{-1}, no larger zone with supersonic velocity of gas/liquid mixture can form. In contrast, supersonic velocity is obtained with two-phase nozzles having mixture chambers. When using a radial flow nozzle (G in Figure 1), the high velocity jet impinges on the tip of the cone (K in Figure 1) and disperses the gas which is compressed through the holes near the cone tip, into the liquid. The flat plate (P in Figure 1) distributes the bubbles to a larger area.

If nozzle B operates the gas input through the central nozzle and the liquid input across the ring nozzle (nozzle B^+), gas dispersion takes place at the interface of the gas/liquid jet, and the redis-persion of the bubbles occurs at the interface of the liquid jet and the two-phase flow. This nozzle can be used at the bottom (Goto *et al.*, 1979) as well as at the top (Räbiger and Vogelpohl, 1983) of the reactor. Nozzles E and F (Figure 2) are also installed at the reactor head. In the vertical flow nozzle (E) the down-flowing liquid is accelerated by the constriction in the nozzle. At the smallest cross-section of the nozzle, where the static pressure is at its lowest, the gas is sucked into the liquid through a narrow slit or perforated ring and is dispersed in the high velocity liquid flow (Vollmüller and Walburg, 1972). A downward-directed free jet (plunging jet, F in Figure 2) can also be used for gas entrainment and dispersion. The gas dispersion mechanism for plunging jets was investigated by Van de Sande and Smith (1976), Burgess and Molloy (1973), McKeogh and Ervine (1981) and many others (see Schügerl, 1982b). A downward-directed jet of liquid is pro-duced by a nozzle. This jet strikes the surface of the liquid, breaks through it and penetrates into the body of the liquid (F in Figure 2). The identity of the jet is maintained until it is surrounded by an envelope of the entrained gas. The breakup of this gas envelope results in the formation of small, tightly packed bubbles which move downward and laterally in the liquid in such a manner that the bubble swarm forms a cone.

At lower velocities the jet is laminar and breaks up at a critical length. In the laminar region the critical length is proportional to the velocity of the jet. If the jet is shorter than its critical length, a smooth jet strikes the liquid surface (F1 in Figure 2). If it is larger than the critical length, the jet decomposes into discrete drops which entrain air upon impact with the pool (F2 in Figure 2). As the jet velocity is increased, a transitional region is reached in which the critical length decreases with increasing jet velocity. When the velocity is increased still further, a turbulent region is reached. Here the critical length is again proportional to the jet velocity. The internal turbulence of the jet plays a major role in the degree of air entrainment, jet surface roughness and breakup length of the jet (McKeogh and Ervine, 1981). When the velocity is increased even more, a region is attained in which the critical length of the jet approaches a constant value as a result of air resistance. For the same discharge rate it is better to select a small nozzle diameter, in order to increase the liquid velocity and with it the energy input and gas entrainment.

A special construction of the plunging jet loop reactor was recommended by Kürten and Zehner (1980). The nozzle is lowered to the liquid surface area. Thus, the reactor is operated with zero jet length (immersed jet loop reactor). Especially in gas–liquid concurrent down-flow reactors the gas is fed into the liquid through a short connecting pipe mounted to the wall of the conical diverging channel (Herbrechtsmeier *et al.*, 1981) (G in Figure 3). The gas phase is dispersed by the high dynamic turbulence pressure maintained by the high Reynolds number.

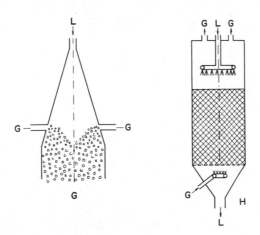

Figure 3 Gas (G) and liquid (H) dispersion devices at the top of reactor

In packed bed (trickle bed) reactors with a counterflow of the phases the liquid is dispersed and distributed in the gas phase (H in Figure 3). This reactor can also be operated in a concurrent mode (Shah, 1979). In multistage tower reactors with countercurrent flow of the phases the power input is ensured by the kinetic energy of the liquid as well as by gas compression (Voigt and Schügerl, 1979). In a multistage reactor with concurrent gas and liquid phases the nozzle B^+ is used to disperse the gas in the liquid phase (Goto *et al.*, 1979). Gas dispersal mechanisms for various bioreactors, in which the power input is maintained by kinetic energy of the liquid due to an external liquid pump, are summarized in Table 1.

Table 1 Bioreactors with Power Input by Kinetic Energy of Liquid due to External Liquid Pump

Reactor	See Figure	Gas dispersion[a]
Nozzle tower	4a	Nozzle A, B, C or D
Nozzle tower loop	4b	Nozzle A, B, C or D
Nozzle loop	5b	Nozzle A, B or C
Plunging loop	5a	Nozzle F
Immersed jet loop		Nozzle F (plunging jet loop with zero jet length)
Plunging jet channel	5c	Nozzle E
Trickle bed	5d	Liquid dispersion device H
Multistage tower loop	5e	Perforated plate and gas compression
Multistage jet loop		Nozzle B^+ (concurrent gas–liquid flow)
Tubular loop	5f	Tubular liquid flow at high *Re* numbers
Tower downflow	5g	Device G
Nozzle downflow loop	7	Nozzle B^+

[a] See Figures 1–3 for details of nozzles A–H.   [b] Additive power input by gas compression is possible.

5.3.1 Nozzle Tower Reactor (Figure 4a)

This is actually a bubble column with a two-phase nozzle as aerator. A bubble column with an ejector nozzle B and an injector nozzle C was investigated by Schügerl *et al.* (1976). Both of them are very efficient aerators, much better than perforated or porous plates.

The relative gas hold-up H can be calculated by equation (2) for synthetic media with low viscosity.

$$H = 0.91 Fr^{1.19} \tag{2}$$

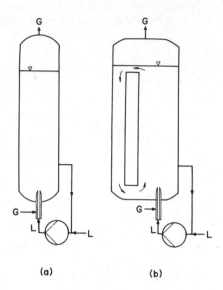

Figure 4 Nozzle tower (a) and nozzle tower loop (b) reactors

The Froude number *Fr* is defined as

$$Fr = \frac{V_s}{\sqrt{gd_S}}$$

where V_s = superficial gas velocity (m s^{-1}), d_S = Sauter bubble diameter (m) and g = acceleration of gravity (m s^{-2}). Equation (1) is valid for $0.05 < H < 0.60$ and $0.02 < Fr < 1.0$.
 The geometric specific interfacial area of bubbles is given by equation (3)

$$a = \frac{6H}{d_S} \tag{3}$$

By means of the two-phase nozzles a specific interfacial area a is obtained for different substrates: $a \leqslant 2000$ m^{-1} for methanol solutions and $a \leqslant 5000$ m^{-1} for ethanol solutions.
 Because of the extremely high oxygen transfer rates, it was not possible to determine sufficiently reliable volumetric mass transfer coefficients, $k_L a$.
 A specific MPC nozzle (Mitsubishi Precision Co Ltd), which is similar to nozzle D, was used by Fujie (1980) for his investigations. For gas bubbles smaller than $d_S = 2$ mm, the following relationship was found:

$$Sh_s = 0.5 Re_s \tag{4}$$

where the Sherwood and Reynolds numbers are defined as follows:

$$Sh_s = \frac{d_S k_L}{D_v}$$

$$Re_s = \frac{d_S V_r}{v}$$

where D_v = molecular diffusivity (m^2 s^{-1}), V_r = relative bubble (slip) velocity (m^2 s^{-1}) and v = kinematic viscosity of medium (m^2 s^{-1}).
 A special type of nozzle C (slit nozzle) is used in large tower reactors (26 m in diameter, 30 m in height) for waste water treatment (Zlokarnik, 1979). Four Bayer Tower Biology® units are operated for a daily effluent throughput of 90 000 m^3 having a BOD content of 95 tons. In these units a network of four-slit nozzles is used to aerate the waste water. The two-phase jets impinge on the bottom due to the 45° inclination of the nozzles with regard to the horizontal bottom surface. Thus the bubbles are well distributed in the large cross-section of the column (Bayer Tower Biology Brochure).
 In a quadrangle reactor, horizontal nozzles B, each with a twisting element and a momentum exchange tube, are used as in the BASF waste water treatment process (Krupp Brochure). A pool

of 15 000 m³ activated sludge is aerated by 56 nozzles with an oxygen transfer rate capacity of 30 000 kg O_2 per day.

5.3.2 Nozzle Tower Loop Reactor (Figure 4b)

This is an air lift tower loop reactor with a two-phase nozzle as aerator. At low liquid recirculation rates it behaves as a nozzle tower reactor (Schügerl, 1976). At high liquid recirculation velocities gas holdup and specific interfacial area diminish. No correlations have been published yet for these reactor types.

5.3.3 Nozzle Loop Reactor (Figure 5b)

Two types of nozzle loop reactors were investigated fairly thoroughly. Blenke (1979) gave a detailed review on loop reactors with nozzles A and B, and Müller and Sell (1983) reported on measurements with nozzle D in a so-called 'Biohigh-Reactor' for waste water treatment.

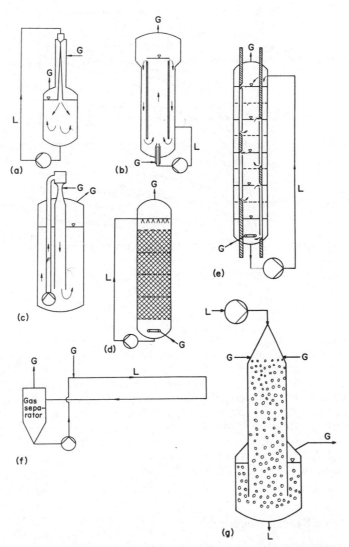

Figure 5 Different types of reactors with power input by kinetic energy of liquid due to external liquid pump: (a) plunging jet loop reactor, (b) nozzle loop reactor, (c) plunging jet channel reactor, (d) trickle bed reactor, (e) multistage tower loop reactor, (f) tubular loop reactor, (g) tower downflow reactor

In Figure 6 the specific interfacial area is shown as a function of the superficial gas velocity for different power inputs to a sulfite solution due to the variation of kinetic energy of a liquid jet according to Hirner and Blenke (1977). They recommended the following relationships:

For relative gas holdup:

$$H = 2.57 \times 10^{-4} V_s Re_N^{1.4} \tag{5}$$

where $Re_N = V_N D_N / \nu$, the nozzle Reynolds number, D_N = the nozzle diameter and V_N = liquid velocity in the nozzle. Equation (5) holds true for $3 \times 10^4 < Re_N < 10^5$, $0.004 < V_s < 0.04$ m s^{-1}, and $0.02 < H < 0.40$.

For the specific interfacial area:

$$a = \frac{A}{V_R} = 5.35 \times 10^3 V_s^{0.4} \left(\frac{P_N}{V_R} \right)^{0.66} \tag{6}$$

where a is the specific gas/liquid interfacial area with regard to the two-phase volume V_R (m^3), A = gas/liquid interfacial area (m^2) and P_N = power input due to liquid jet (kW). Equation (6) is valid for $0.05 < P_N/V_R < 1$ kW m^{-3}, $0 < V_s < 0.041$ m s^{-1} and $50 < a < 1000$ m^{-1}. Since in the sulfite system $k_L = 4.6 \times 10^{-4}$ m s^{-1}, which does not depend on V_s, equation (6) also holds true for $0.023 < k_L a < 0.46$ s^{-1}.

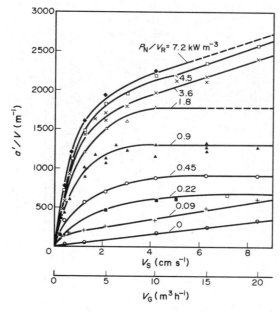

Figure 6   Specific interfacial area as a function of the aeration rate and specific power input as a parameter in a nozzle loop reactor (Blenke, 1979)

A 2 m^3 loop reactor with nozzle B was used for the cultivation of *Candida lypolytica* on *n*-paraffin substrate (Seipenbusch and Blenke (1980).

Adler and Fiechter (1983) cultivated *Trichosporon cutaneum* and *Candida tropicalis* in a 2 m^3 loop reactor with nozzle B and achieved a maximum oxygen transfer rate (Q_{O_2}) of 2.1 kg m^{-3} h^{-1} at a specific power input (P/V_L) of 1 kW m^{-3}, *i.e.* an efficiency (E_{O_2}) of 2.1 kg kWh^{-1}.

In the 22 m high Biohigh Reactor® (44 m in upper diameter and 17 400 m^3 volume) developed by Hoechst, nozzles D with a draft tube were used to aerate an activated sludge. In four factories of Hoechst AG seven large Biohigh Reactors are expected to be constructed within the next few years with a total volume of 130 000 m^3 (Zlokarnik, 1982).

5.3.4   Plunging Jet Loop Reactor (Figure 5a)

Plunging jet loop reactors are used in different constructions. The main variables are the distance between nozzle F and the liquid pool surface, the geometry of the pool and the angle of jet inclination from the vertical position. In all of them three different regions of mass transfer can be distinguished: (1) mass transfer to the turbulent free liquid jet passing through a gaseous space, (2) mass transfer to the free liquid surface of the pool and (3) mass transfer between bubble dispersion and the pool liquid. The relative contributions of regions (1) and (2) are low; thus they are usually neglected. The investigations of mixing processes in the pool indicate that the mixing time is much less than the mean residence time of the liquid in the pool; thus the pool can be considered well mixed. Bin and Smith (1982) recommended the following simple relationship for the oxygen transfer into the medium:

$$k_L a = 9 \times 10^{-2} P_N \tag{7a}$$

$$\text{or} \qquad Q_{O_2} = 3 P_N \tag{7b}$$

where P_N is the power input due to the liquid jet (kW) and Q_{O_2} the oxygen transfer into the liquid (kg h^{-1}). Equation (7) holds true in the ranges $10^{-6} < k_L a < 2 \times 10^{-2}$ m^3 s^{-1} and $2 \times 10^{-5} < P_N < 0.75$ kW.

The ratio $Q_{O_2} : P_N$ is called oxygen efficiency.

$$E_{O_2} = \frac{Q_{O_2} V_L}{P_N} = 3 \text{ kg kWh}^{-1}$$

According to Van de Donk (1981), for small nozzles with a diameter D_N in the range 3 to 8 mm the following is valid:

$$E_{O_2} = 2.5 P_N^{-0.21} \tag{8a}$$

With large conical nozzles ($D_N = 30$ to 100 mm):

$$E_{O_2} = 3.6 P_N^{-0.2} \tag{8b}$$

Equation (8) holds true for $0.016 < P_N < 1$ kW.

The jet length has only a slight effect on the oxygen transfer. From the results of Van de Sande (1974), De Wijs (1976) and Van de Donk (1981) it can be concluded that oxygen transfer depends on the geometric ratio of jet length L_j to the nozzle diameter D_N, raised to the power of about 0.2 in the range of $8 < L_j/D_N < 300$. The angle of jet inclination affects the oxygen transfer to some extent. The highest oxygen transfer is obtained for vertical jets.

A 20 m^3 plunging jet reactor was used to produce single cell protein from whey (Moebus and Teuber, 1979). However, the main application of this reactor is in waste water treatment.

The manufacturer of the Meijel sewage treatment plant of DSM at Geleen claims that the oxygen transfer amounts to 54 kg O$_2$ h^{-1} at a high jet velocity and 21 kg O$_2$ h^{-1} at a low jet velocity in a pool of 2250 m^3 with an oxygen transfer efficiency E_{O_2} of 2 kg O$_2$ kWh^{-1}. According to Böhnke (1970) a value of E_{O_2} of 3.3 kg O$_2$ kWh^{-1} was found with waste water with a jet angle of 70°. With a pig slurry the transfer efficiency was found to be rather high ($E_{O_2} = 8$ kg O$_2$ kWh^{-1}) (Sneath, 1978).

5.3.5 Plunging Jet Channel Reactor (Figure 5c)

In plunging jet channel reactors the potential energy of the recirculated liquid is converted into kinetic energy. In order to make this conversion complete the liquid is accelerated in a slit-shaped nozzle (E).

At the narrowest position, where the velocity is highest and the static pressure lowest, the air is sucked through slits or holes into the nozzle and dispersed (Müllner, 1973). The gas bubbles formed in the nozzle are carried by the down-flowing liquid in the channel to the bottom of the vessel. At the end of the channel they move upwards in the vessel and leave the liquid. The specific oxygen transfer rate depends on the height of the liquid layer. With increasing height the specific oxygen transfer with regard to the power input increases. In pilot plant operation 4 kg O$_2$ kWh^{-1} was attained if the efficiency of the liquid pump is not considered. A similar reactor ('U-tube aerator') was described by Mitchell and Lev (1970). They also presented economics calculations for waste water treatment with this reactor.

5.3.6 Trickle Bed Reactor (Figure 5d)

The liquid is distributed at the head of the reactor and then trickles down the packings on which the microorganisms are fixed as a microbial film (*e.g.* Fujie *et al.*, 1977). The air is introduced at the bottom of the tower and flows countercurrent to the liquid. This type of reactor is popular in aerobic waste water treatment. Some papers have considered mass transfer into the microbial film (*e.g.* Fujie *et al.*, 1979). Several companies offer such reactors (*e.g.* Uhde brochure).

In anaerobic waste water treatment the liquid percolates the packings without any air flow (*e.g.* Young and Dahab, 1982). The layout of such reactors has been considered only with regard to their chemical application (Shah, 1979).

Somewhat unusual is the use of wire gauze packings in countercurrent absorption columns (Niranjan *et al.*, 1982). According to these authors the values of effective interfacial area offered by multifilament wire gauze packings are much higher than those of conventional packings and these values are independent of the superficial liquid and gas velocities.

5.3.7   Multistage Tower Loop Reactor (Figure 5e)

In multistage countercurrent tower reactors, perforated trays or sieve-trays are used as stage separators and the liquid level is controlled by the overflow pipes. The relative gas holdup and the volumetric mass transfer coefficients were determined in model media of low and high viscosity (Voigt *et al.*, 1979, 1980; Hecht *et al.*, 1980). Cell mass productivity of a yeast cultivation was estimated as a function of the geometric parameters (Voigt and Schügerl, 1981).

Similar reactor types with different contacting devices of up to 50 m^3 volume were investigated in detail by Viesturs *et al.* (1980, 1981). Oxygen transfer rates of 8 to 10 $\text{kg m}^{-3} \text{ h}^{-1} \text{ bar}^{-1}$ and specific oxygen transfer rates of 1.6 $\text{kg O}_2 \text{ kWh}^{-1}$ at 2 bar were measured in these reactors.

5.3.8   Tubular Loop Reactor (Figure 5f)

The air is introduced into a liquid flow at rates of up to $3-4 \text{ m s}^{-1}$ across a porous filter separated from the liquid phase in the cyclone and removed from the system at this point. The liquid flows as a single phase through the pump and flow meter.

Candida tropicalis on glucose and hexadecane (Ziegler *et al.*, 1977) as well as different mycelial organisms (Ziegler *et al.*, 1980) were cultivated in this reactor. The performance of these reactors corresponds to that of the stirred tank reactors with regard to the specific oxygen transfer rate. However, the energy dissipation rate is uniform in this reactor in contrast to stirred tank reactors. Specific energy dissipation rates in this reactor of up to 8 kW m^{-3} do not harm the mycelial cultures. According to Liepe *et al.* (1978), in this type of reactor very high cell mass concentrations (up to 165 g (dry) mass 1^{-1}) and high oxygen transfer rates (up to 60 $\text{kg m}^{-3} \text{ h}^{-1}$) were achieved. However, the power input necessary for this was not published by these authors.

5.3.9   Tower Downflow Reactor (Figures 5g and 7)

Two types of tower downflow reactor are recommended: one with the gas inlet G (Figure 5g) and the other with the nozzle B (as in Figure 1 but with liquid and gas inlets exchanged; see Figure 7). Both of them are recommended for waste water treatment. However, that shown in Figure 5g has up to now only been used for chemical treatment (Friedel *et al.*, 1980; Herbrechtsmeier *et al.*, 1981, 1982). The reactor in Figure 7 is also used for biological treatment (Räbiger and Vogelpohl, 1983). According to Herbrechtsmeier *et al.* (1982), in four-stage tower downflow reactors a specific gas–liquid interfacial area of 1000 m^{-1} can be attained with 1 kW m^{-3} specific power input. The main advantage of these reactors is that oxygen in the air can be completely consumed.

5.3.10   Liquid Fluidized Bed Reactor

The number of publications on fluidized bed bioreactors has increased considerably in recent years. There are three main areas of application of biofluidization: (1) enzymes immobilized on a solid matrix; (2) biofluidization of pure cultures of whole cells immobilized on a solid matrix; (3) application of biofluidization to a wide variety of waste water treatment processes (Baker *et al.*, 1981).

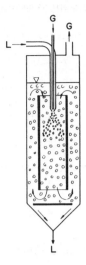

Figure 7 Nozzle downflow loop reactor with nozzle B[+]

Halwachs *et al.* (1977) used a multistage fluidized bed with immobilized chymotrypsin to separate D- and L-phenylalanine. Margaritis and Wallace (1982) applied a fluidized bed of immobilized cells of *Zymomonas mobilis* to produce ethanol. The fluidized bed consisted of two draft tubes. The liquid was introduced into the annular space between the two coaxial draft tubes, and was recirculated within the inner draft tube as well as in the annular space between the wall and the outer draft tube.

Oak Ridge National Laboratory used a tapered fluidized bed which has been applied to a wide variety of waste water treatment processes (Scott and Hancher, 1976). Fluidized bed towers of a height to diameter ratio of about 30 were also used for biological treatment of nitrate waste water (Walker *et al.*, 1981). Furthermore, an anaerobic biological fluidized bed was used to generate methane from industrial wastes (Hickley and Owens, 1981).

The use of motionless mixers in fluidized beds can improve their performance considerably (Flaschel *et al.*, 1982).

5.4 POWER INPUT BY GAS COMPRESSION

The compressor efficiency depends on its type; for example, for rotary compressors $\eta = 80$ to 90%, for piston compressors $\eta = 74$ to 90%, and for turbocompressors $\eta = 75$ to 80%. The compressed gas is dispersed into the liquid through a sparger. However, under conditions of industrial equipment the primary bubbles formed are dispersed by turbulence. Therefore the final degree of gas dispersion in these reactors is also controlled by the energy dissipation rate. As long as the medium suppresses bubble coalescence, the primary bubble size is much smaller than the dynamic equilibrium bubble size d_e, and if the tower reactor is small, the aerator considerably influences the two-phase properties (Schügerl *et al.*, 1977). However, in large industrial reactors the two-phase property is mainly influenced by the mean energy dissipation rate in the reactor.

5.4.1 Single-stage Tower (Bubble Column) Reactor (Figure 8a)

These reactors are usually equipped with a perforated or porous plate or a perforated ring (a so-called nozzle) as an aerator at the bottom of the tower. The gas is dispersed by the aerator and the bubbles rise in the liquid due to their buoyancy forces. The liquid flows upward, *i.e.* a concurrent gas liquid flow is maintained.

Equation (1) can be used to calculate the relative gas holdup H. The volumetric mass transfer coefficient is only a function of the superficial gas velocity V_s and the Sauter bubble diameter d_S (Schügerl *et al.*, 1977):

$$k_La = 0.0023(V_s/d_S)^{1.58} \tag{9}$$

Equation (9) holds true for $0.01 < k_L a < 0.8$ s^{-1} and $3 < (V_s/d_S) < 43$ s^{-1} in synthetic media consisting of nutrient salts and alcohols.

The influence of the gas velocity on H is conventionally expressed by

$$H \propto V_s^n \tag{10}$$

At low velocities (bubbly flow) n varies from 0.7 to 1.2, while at high velocities (turbulent or heterogeneous flow) it varies from 0.4 to 0.7.

The volumetric mass transfer coefficient is also a function of V_s (Deckwer, 1980):

$$k_L a = b V_s^m \tag{11}$$

where the exponent $m = 0.8$ for water and electrolyte solutions and b is a function of the aerator type and the medium properties. Equation (11) holds true for $0.01 < V_s < 25$ cm s^{-1} and $0.003 < k_L a < 0.1$ s^{-1}.

According to Shah and Deckwer (1981) the coefficient of longitudinal dispersion in tower reactors can be calculated by

$$Pe_L = 2.83 Fr^{0.34} \tag{12}$$

where $Pe_L = (V_s D_t)/D_L$ (Peclet number of liquid), $Fr = V_s^2/D_t g$ (Froude number), D_t = tower diameter (m) and D_L = longitudinal dispersion coefficient (m s^{-1}).

Single-stage towers are versatile bioreactors. They are used for the production of bakers' yeast, for waste water treatment (Verhaagen, 1978), beer and vinegar production (Smith and Greenshields, 1974), fungus cultivation (Morris *et al.*, 1973) and the production of diphtheria toxine (Nikolajewski *et al.*, 1982). When using flocculating cells, high cell mass concentration and metabolite productivity can be attained.

5.4.2   Single-stage Air Lift Tower Loop Reactors

In recent years, these types of bioreactors have become very popular. When aerating a tower reactor, liquid is carried upwards with the bubbles. Therefore, if the mean liquid velocity is lower than the bubble velocity, a fraction of the liquid has to move downwards. In small towers this occurs at the wall. This countercurrent liquid flow can be better controlled by splitting the reactor into a riser and a downcomer section. Four different constructions have become popular: tower reactors with a coaxial cylindrical draft tube (Figure 8b), tower reactors with a cylindrical outer loop (Figure 8c), tower reactors with a split cylinder (Figure 8d) and tower reactors with an inner loop channel (Figure 9).

5.4.2.1   *Single-stage air lift tower loop reactor with a draft tube* (Figure 8b)

The largest unit of 40 m^3 total volume (25 m^3 liquid volume), 1.6 m tower diameter (1.3 m draft tube diameter), 2.2 m head diameter and 16 m aerated liquid height has been operated for years by Hoechst AG, Höchst, for the production of SCP (*Methylomonas clara*) on methanol substrate with a capacity of 1000 tons per year.

A smaller unit of 2300 l aerated liquid volume, 57.5 cm tower diameter, 34.5 cm draft tube diameter and 5.32 m aerated liquid height was used to produce SCP (*Candida lipolytica*) on *n*-paraffin by a consortium of the companies Hoechst, Uhde and Gelsenberg. This unit is being investigated now by Adler and Fiechter (1983) using *T. cutaneum* and *C. tropicalis*. The 2300 l unit is equipped with a two-phase nozzle. In this latter reactor an OTR of 2.1 kg m^{-3} h^{-1} at a specific power input P/V_L of 1 kW m^{-3}, *i.e.* a specific oxygen transfer rate of 2.1 kg O$_2$ kWh^{-1} was attained (Adler and Fiechter, 1983). In the 25 m^3 unit an OTR of 6 kg m^{-3} h^{-1} and a specific oxygen transfer rate of 2 kg O$_2$ kWh^{-1} was achieved (Uhde Brochure).

Euzen *et al.* (1978) investigated two units with volumes of 1900 l and 3300 l and attained specific oxygen transfer rates of 2 kg O$_2$ kWh^{-1}. According to these authors, the specific interfacial area a (m^{-1}) is a simple function of the specific power input P/V (kW m^{-3}).

$$a = K(P/V_L)^n \tag{13}$$

where $0.6 < n < 0.7$ and K is a function of the geometry of the system and the operating con-

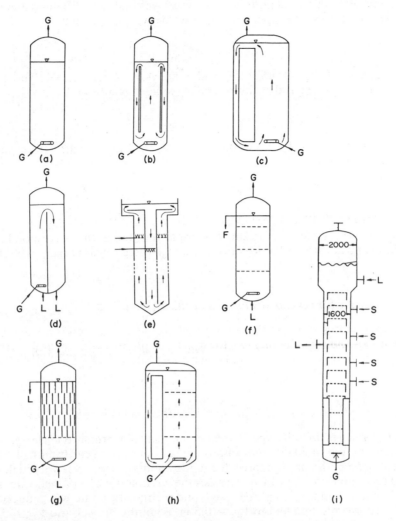

Figure 8 Different types of tower reactors with power input by gas compression: (a) single-stage tower reactor, (b) single-stage air lift tower reactor with draft tube, (c) single-stage air lift reactor with outer loop, (d) single-stage air lift tower loop reactor with split cylinder, (e) downstream air tower loop (deep shaft) reactor, (f) tower reactor with stage separating plates, (g) tower reactor with static mixer, (h) multistage reactor with outer loop, (i) multistage reactor with draft tube

ditions. Gulf Oil also used a tower loop reactor to produce SCP based on *n*-paraffins (Cooper *et al.*, 1975). Gas oil was used as a substrate by British Petroleum to produce SCP in a tower loop reactor (Lainé and du Chaffaut, 1975).

Hoechst recommends tower loop reactors for waste water treatment (Müller *et al.*, 1978). In a semicommercial unit (5 m in diameter, 22 m in aerated liquid height) specific oxygen transfer rates of 3 to 3.5 kg O_2 kWh^{-1} were attained (Uhde Hoechst Brochure).

Tower loop reactors are sometimes used for the cultivation of plant cells (Townsley *et al.*, 1983).

5.4.2.2 *Single-stage air lift reactor with an outer loop* (Figure 8c)

A 1000-ton-per-year SCP pilot plant 'pressure cycle fermenter' was operated by ICI in Billingham (Gow *et al.*, 1975). However, no construction and operational data were published on this unit. Detailed investigations were published only on bench-scale towers.

A 2.75 m high tower 15 cm in diameter with a downcomer 2.5 cm in diameter and 50 l of medium volume was investigated by Schügerl (1982a) for the production of yeasts and bacteria. Another 10 m high tower (8.5 m in aerated liquid height), 10 cm in diameter with a downcomer 5 cm in diameter, and 80 l of aerated liquid volume was investigated by Weiland and Onken (1981). According to Weiland (1978) the superficial liquid velocity V_{Ls} is given by

$$V_{Ls} = 1.15 V_s^{0.41} \tag{14}$$

which is valid for $0.002 \leqslant V_{Ls} \leqslant 0.5$ m s^{-1} and $0.005 \leqslant V_s \leqslant 0.15$ m s^{-1}. The gas holdup was evaluated with a porous plate gas distributor and for ionic strengths > 0.1 as

$$H = 55.8 V_s^{0.725} \tag{15}$$

which is valid for $0.01 \leqslant H \leqslant 0.12$ and $0.005 \leqslant V_s \leqslant 0.14$ m s^{-1} and with a perforated plate (regardless of ionic strength) as

$$H = 62.3 V_s^{0.803} \tag{16}$$

in the range $0.008 \leqslant H \leqslant 0.12$ and $0.005 \leqslant V_s \leqslant 0.16$ m s^{-1}.

The volumetric mass transfer coefficients $k_L a$ are lower than in the corresponding bubble column. The medium and aerator influence on $k_L a$ is less in these reactors than in bubble columns.

5.4.2.3   *Single-stage air lift reactor with a split cylinder* (Figure 8d)

Only laboratory investigations on these reactors have been reported (Orazem *et al.*, 1979). A special construction is the thin channel reactor which is equipped with several vertical plates, the channels between them serving as risers and downcomers alternately (Gasner, 1974).

5.4.2.4   *Single-stage air lift tower with an inner loop channel* (Figure 9)

The best-known reactor of this type is the commercial ICI pressure cycle fermenter in Billingham for the production of 50 000 tons of protein (Pruteen) per year (Smith, 1980). At an aerated liquid height of 55 m the driving force due to the bubble buoyancy is very high. This causes a very high liquid recirculation rate which increases the gas holdup to very high values. To improve the separation of the gas and liquid phases the liquid recirculation rate was reduced by the insertion of 19 baffles (perforated plates) to $V_{Ls} = 0.5$ m s^{-1} in the upriser and $V_{Ls} = 3$ to 4 m s^{-1} in the downcomer. The gas holdup was found to be still very high: $H = 0.52$ in the upriser and 0.48 in the downcomer.

The main problems of operating reactors with bacteria of high growth rates in a large medium volume (2.1×10^6 l) are high power inputs, oxygen and substrate transfer into the medium, and their uniform distribution as well as the transfer of CO_2 out of the medium in order to keep the CO_2 concentration low. Oxygen is mainly transferred into the medium at the bottom (4.5 to 5 bar) while CO_2 is removed at the top (2 to 2.5 bar). The dissolved gases are dispersed along the length of the riser.

The uniform distribution of the substrate (methanol) in the 2100 m^3 medium caused considerable difficulties. With nonuniform distribution of the substrate the yield coefficient was found to drop from 0.65 to 0.46. In order to attain a high yield coefficient and productivity the longitudinal substrate profile is fitted to the longitudinal dissolved oxygen profile by injecting the substrate at 5000 to 8000 different positions along the column. The proper operation of this cross-flow reactor is maintained by on-line computer control.

According to the information of John Brown Ltd (Anonymous, 1983) a new reactor with 5600 m^3 volume is being constructed with a capacity of 100 000 tons of Pruteen a year. In this reactor, a better energy efficiency will be attained by injecting the substrate into the medium through 20 000 nozzles distributed along the reactor.

In order to keep the cost of downstream processing low, the reactor is operated at high cell mass concentrations. Hence the specific oxygen demand is high: 10 kg O_2 m^{-3} h^{-1}. This can only be attained at a high specific power input of about 6.6 kW m^{-3} at which the efficiency of the oxygen transfer rate is fairly low (1.5 kg O_2 kWh^{-1}). The oxygen efficiency is also low (50%) (Walker and Wilkinson, 1979). At half this concentration the specific oxygen demand is lower at 3 kg O_2 m^{-3} h^{-1}. This can be attained at a moderate specific power input P/V_L of 1.5 kW m^{-3}.

Hence the efficiency of the oxygen transfer rate is higher (2 kg O_2 kWh^{-1}) (Hines, 1978). The latter is a fairly conventional value.

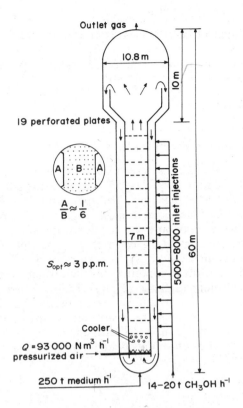

Figure 9 ICI pressure cycle fermenter

5.4.3 Downstream Air Lift Tower Loop (Deep Shaft) Reactor (Figure 8e)

The same concept which was used by ICI in the pressure cycle fermenter for production of Pruteen was also utilized in the deep shaft pressure cycle fermenter for waste water treatment. However, since the oxygen demand for waste water treatment is usually less than that for the cultivation of *Methylophilus methylotrophus* (Pruteen production), the power demand is also lower. A value of 3 kg O_2 m^{-3} h^{-1} at a P/V_L of 1 kW m^{-3} is usually satisfactory. At a lower specific power input the efficiency of the oxygen transfer rate is higher (3 kg O_2 kWh^{-1}) (Hines, 1978). ICI has since improved the performance of the deep shaft reactor, since a year later a much higher transfer rate was reported, *i.e.* 6 kg O_2 kWh^{-1} at 90% oxygen utilization (Walker and Wilkinson, 1979). In contrast to the protein production unit, most of the air is introduced into the downflow tube and no baffles are permitted because of potential blockage to flow. The depth is also greater, normally in the range of 50–150 m. Air introduced in this way is under a lower pressure than if it were introduced at the bottom. Liquid circulation is established by injection of compressed air through a sparger at a relatively shallow depth on the upflow side. The air is then gradually shifted to the downflow side (Hines, 1978). For the oxygen transfer rate the following equation is approximately valid (Hines, 1978):

$$Q_{O_2} = 900 \frac{V_s}{D_L} \, (\text{kg } O_2 \, \text{m}^{-3} \, \text{h}^{-1}) \tag{17}$$

where D_L is the depth of the shaft (m). A V_s value of 0.15 m s^{-1} is about the energy optimum with regard to the drive mechanics. Hence in a shaft at the energy optimum and with $D_L = 135$ m a Q_{O_2} value of 1 kg O_2 m^{-3} h^{-1} can be achieved. Q_{O_2} can be increased up to 3 kg O_2 m^{-3} h^{-1} by decreasing D_L and increasing V_s.

5.4.4 Tower Reactors with Different Built-ins

To improve the performance of tower reactors different constructions have been recommended:

multistage tower with perforated plates as stage-separating trays (Figure 8f);
multistage loop combination (Figure 8h and 8i);
static mixer (Figure 8g).

5.4.4.1 *Tower reactors with stage-separating plates* (Figure 8f)

When using sieve or perforated plates in tower reactors with concurrent gas and liquid phases, gas layers are formed above the aerated liquid layers in each of the stages. Hence, a separation of gas and liquid phases occurs on each stage and a redispersion of the gas phase is necessary. An improvement in the performance of these reactors with regard to the oxygen transfer rate in comparison with the single-stage tower may only be expected if the stage-separating plates and their distance are well adjusted with respect to the two-phase properties. A comparison of the volumetric mass transfer coefficients measured in such reactors with different types and numbers of plates indicates that depending on these conditions a single-stage, a six-stage or a twelve-stage tower may exhibit the best performance (Schügerl *et al.*, 1977). When a highly efficient aerator is used at the bottom and the trays are inefficient with regard to the gas dispersion, the single-stage tower can exhibit a better performance than the multistage. In the multistage tower, the distribution of the residence time is more uniform than in the single-stage tower (Schügerl *et al.*, 1977). Such towers have been used for the continuous cultivation of *E. coli* (Kitai *et al.*, 1969) and for waste water treatment (Hsu *et al.*, 1975).

5.4.4.2 *Tower reactors with multistage loop combination*

There are a number of different ways to combine a loop reactor with a multistage one:

tower reactor with an outer loop and trays (Figure 8h);
tower reactor with an inner loop and trays (Figure 8i);
tower reactor with a loop channel and trays (Figure 9).

The pilot plant pressure cycle fermenter of ICI actually belongs to the first type, the pilot plant tower fermenter of the Mitsubishi Gas Chemical Co. for SCP production with a capacity of 500 tons SCP per year (Kuraishi *et al.*, 1979) to the second type and the commercial ICI pressure cycle fermenter to the third type.

According to the literature these trays serve to create turbulence in these reactors. This is, of course, misleading since in these reactors high turbulence intensity is attained without the use of baffles. They disperse the large turbulent eddies to smaller ones, redisperse the gas bubbles and reduce the liquid recirculation rate to improve the separation of the gas and liquid phases at the tower head.

Since the ICI units have been discussed already, only the Mitsubishi unit is considered here. It has an overall height of 11.5 m (8.5 m tower, 3 m head), is 1.6 m in tower diameter, 2.0 m in head diameter and has a volume of 20 m^3. The substrate is injected through four nozzles along the tower to obtain a higher productivity by adjusting the longitudinal substrate concentration profile to the dissolved oxygen concentration profile (Kuraishi *et al.*, 1979) (see also Section 5.4.2.4).

5.4.4.3 *Tower reactors with static mixers*

Two types of reactor have been evaluated:

tower reactor with a static mixer (Figure 8g);
tower loop reactor with a static mixer (Figure 8h but instead of trays static mixers are used).

Tower reactors with static mixers were investigated by Maclean *et al.* (1977) and Wang and Fan

(1978). The intensity of longitudinal dispersion is considerably reduced and the volumetric mass transfer coefficient is increased by the use of static mixers. They recommend equation (18):

$$k_L a = m V_{Ls}^\beta H \tag{18}$$

The constants m and β were tabulated for different types of static mixers and tower configurations by Wang and Fan (1978). Tower loop reactors with static mixers were investigated by Gebauer (1983). He cultivated *E. coli* in a tower loop reactor with and without static mixers. In the presence of static mixers Q_{O_2} is considerably higher than in their absence.

5.4.5 Gas Fluidized Bed Reactor

Gas fluidized beds have been used for the cultivation of the yeast *S. cerevisiae* (Moebus *et al.*, 1982; Mishra *et al.*, 1982) and for ethanol production by *S. cerevisiae* (Moebus and Teuber, 1982). The cells were agglomerated and stirred during fluidization to avoid the formation of large lumps. A stationary state was maintained by careful operation. However, the cell and ethanol productivities were significantly lower than those of submerged cultivations and fermentations.

5.5 COMPARISON OF DIFFERENT REACTORS. SUMMARY

In Table 2, different bioreactors are compared with regard to their oxygen transfer efficiencies. For comparison, the efficiencies of some stirred tank reactors are also given. Because E_{O_2} depends on the specific power input, the height of the liquid level and the medium properties, this comparison can only be considered a qualitative one, especially because some values are probably erroneous. In spite of this, one can recognize that the efficiency of oxygen transfer of nonmechanically agitated bioreactor systems is generally higher than that of stirred tanks.

Table 2 Comparison of Different Bioreactors with Regard to the Efficiency of Oxygen Transfer E_{O_2}

Bioreactor	E_{O_2} (kg O_2 kWh^{-1})	Ref.
Stirred tank with		
tubular impeller	1.4	Zlokarnik (1980)
turbine stirrer	2.0–2.5	Zlokarnik (1980)
propeller	0.8–1.1	Zlokarnik (1980)
Nozzle tower loop	2.1	Adler and Fiechter (1983)
Plunging jet loop	0.88	Zlokarnik (1980)
	3.0	Bin and Smith (1982)
	2.0	Meijel Sewage Plant
	3.3	Böhnke (1970)
Plunging jet channel	4.0	Müllner (1973)
Multistage tower loop	1.6	Viesturs *et al.* (1980)
Single-stage air lift tower loop		
(ICI pressure cycle fermenter)	1.5 at 6.6 kW m^{-3}	Walker and Wilkinson (1979) (SCP production)
	2.0 at 1.5 kW m^{-3}	Hines (1978)
Single-stage air lift tower loop	2.0	Uhde Brochure (SCP production)
	2.0	Euzen *et al.* (1978)
	3–3.5	Uhde Hoechst Brochure (waste water treatment)
Deep shaft	3 at 1 kW m^{-3}	Hines (1978)
Tower loop with perforated plate	3.39	Zlokarnik (1980)
Static mixer (kenics)	3.3	Zlokarnik (1980)
Tower loop with slit jet	3.6	Zlokarnik (1980)
Tower loop with injector nozzle	3.6	Zlokarnik (1980)
Tower loop with porous plate	4.0	Zlokarnik (1980)

Since nonmechanically agitated bioreactors were developed fairly recently, they have, in general, not been optimized yet. Therefore, we can expect improvement in their efficiency in the years to come.

5.6 SYMBOLS

A	interfacial area	m^2
a	specific interfacial area	m^{-1}
D_v	molecular diffusivity	$m^2\,s^{-1}$
D_N	nozzle diameter	m
D_T	tower diameter	m
D_L^*	longitudinal dispersion coefficient	$m^2\,s^{-1}$
D_L	depth of the shaft	m
d_S	Sauter bubble diameter	m
E_{O_2}	efficiency of oxygen transfer	$kg\,kWh^{-1}$
F_L	rate of discharge	$m^3\,s^{-1}$
Fr	Froude number (equation 2)	—
g	acceleration of gravity	$m\,s^{-2}$
H	gas holdup	—
k_L	mass transfer coefficient	$m\,s^{-1}$
L_j	jet length	m
N	stirring speed	s^{-1}
P	power input	kW
P_N	power input due to liquid jet	kW
Pe	Peclet number (equation 12)	
Q_{O_2}	oxygen transfer rate	$kg\,m^{-3}\,h^{-1}$
Re_N	Reynolds number (equation 5)	—
Re_s	Reynolds number (equation 4)	—
Sh_s	Sherwood number (equation 4)	—
V_G	gas flow rate	$m^3\,h^{-1}$
V_L	liquid volume	m^3
V_{Ls}	superficial liquid velocity	$m\,s^{-1}$
V_N	liquid velocity in nozzle	$m\,s^{-1}$
V_R	aerated liquid volume	m^3
V_r	relative bubble (slip) velocity	$m\,s^{-1}$
V_s	superficial gas velocity	$m\,s^{-1}$
ν	kinematic viscosity of liquid	$m^2\,s^{-1}$

5.7 REFERENCES

Adler, I. and A. Fiechter (1983). Characterisierung von Bioreaktoren mit biologischen Testsystemen. *Chem.-Ing.-Tech.*, **55**, 322–323.

Anonymous (1983). *Biotechnology News Watch*, **3** (4), 5.

Baker, C. G. J., A. Margaritis and M. A. Bergougnou (1981). Fluidization principles and applications to biotechnology. In *Advances in Biotechnology*, ed. M. Moo-Young, vol. 1, pp. 635–641. Pergamon, Toronto.

Bayer Tower Biology® Brochure. E. 589–777/68619, Bayer AG.

Bin, A. K. and J. M. Smith (1982). Mass transfer in a plunging liquid jet absorber. *Chem. Eng. Commun.*, **15**, 367–383.

Blenke, H. (1979). Loop reactors. In *Advances in Biochemical Engineering*, ed. T. K. Ghose, A. Fiechter and N. Blakebrough, vol. 13, pp. 121–214. Springer, Berlin.

Böhnke, B. (1970). Extendable sewage plant in package type construction for rapidly expandable communities up to 25,000 inhabitants. In *Adv. Water Poll. Res.*, *Proc. Int. Conf.*, *5th*, ed. S. H. Jenkins, paper II-9, TD 420.A27. Pergamon, Oxford.

Burgess, J. M. and N. A. Molloy (1973). Gas absorption in the plunging liquid jet reactor. *Chem. Eng. Sci.*, **28**, 183–190.

Cooper, P. G., R. S. Silver and J. P. Boyle (1975). Semi-commercial studies of petroprotein process based on *n*-paraffins. In *Single-cell Protein II*, ed. S. R. Tannenbaum and D. I. C. Wang, pp. 454–466. MIT Press, Cambridge.

Deckwer, W. D. (1980). Physical transport phenomena in bubble column bioreactors. In *Advances in Biotechnology*, ed. M. Moo-Young, vol. 1, pp. 465–476. Pergamon, Toronto.

De Wijs, J. J. (1976). International Report of TN-FT, Technische Hogeschool, Delft.

Euzen, J. P., P. Trambouze and H. van Landeghem (1978). Comparison of the performances of various fermenters and selection criteria. In *Chemical Reaction Engineering*, *Houston*, *5th International Symposium* (*ACS Symp. Ser.*, *no. 65*), pp. 153–162.

Flaschel, E., P. F. Fauquex and A. Renken (1982). Zum Verhalten von Flüssigkeits-Wirbelschichten mit Einbauten. *Chem.-Ing.-Tech.*, **54**, 54–56.

Friedel, L., P. Herbrechtsmeier and R. Steiner (1980). Mean gas holdup in downflow bubble columns. *Ger. Chem. Eng.* (*Engl. Transl.*), **3**, 342–346.

Fujie, K., T. Furuya and H. Kubota (1979). Effectiveness factor of microbial film in aerobic reaction. *J. Ferment. Technol.*, **57**, 99–106.

Fujie, K., N. Ishihara and H. Kubota (1980). Mass transfer coefficient of dispersed fine bubbles in electrolytic solutions. *J. Ferment. Technol.*, **58**, 477–484.

Fujie, K., T. Sekizawa and H. Kubota (1977). Mass transfer in the liquid phase with tubular waste-water treatment contactor. *J. Ferment. Technol.*, **55**, 532–543.

Gasner, L. L. (1974). Development and application of the thin channel rectangular air lift mass transfer reactor to fermentation and waste-water treatment systems. *Biotechnol. Bioeng.*, **16**, 1179–1195.

Gebauer, A. (1983). Ongoing Ph.D. thesis, University of Hannover.

Goto, S., R. Okamoto and T. Inui (1981). Flow characteristic simulation for tower fermenter with two-fluid nozzle. *J. Ferment. Technol.*, **59**, 73–76.

Goto, S. *et al.* (1979). Application of tower type fermenter with two-fluid nozzle to biomass production from methanol. *J. Ferment. Technol.*, **57**, 341–348.

Gow, J. S. *et al.* (1975). SCP production from methanol: bacteria. In *Single-cell Protein II*, ed. S. R. Tannenbaum and D. I. C. Wang, pp. 370–384, MIT Press, Cambridge.

Hallensleben, J. *et al.* (1977). Blasenbildung und -verhalten im dynamischen Bereich. *Chem.-Ing.-Tech.*, **49**, 663.

Halwachs, W., C. Wandrey and K. Schügerl (1977). Application of immobilized chymotrypsin in a multistage fluidized bed reactor. *Biotechnol. Bioeng.*, **19**, 1667–1677.

Hecht, V., J. Voigt and K. Schügerl (1980). Absorption of oxygen in countercurrent multistage bubble columns III. Viscoelastic liquids. *Chem. Eng. Sci.*, **35**, 1325–1330.

Herbrechtsmeier, P. and H. Schäfer (1982). Development of a cascade downflow reactor as a high performance equipment for physical absorption and desorption processes. *Ger. Chem. Eng. (Engl. Transl.)*, **5**, 369–374.

Herbrechtsmeier, P., H. Schäfer and R. Steiner (1981). Gas absorption in downflow bubble columns for the ozone–water system. *Ger. Chem. Eng. (Engl. Transl.)*, **4**, 258–264.

Hickley, R. F. and R. W. Owens (1981). Methane generation from high-strength industrial wastes with the anaerobic biological fluidized bed. *Biotechnol. Bioeng. Symp.*, **11**, 399–413.

Hines, D. A. (1978). The large scale pressure cycle fermenter configuration. *Prepr. Eur. Congr. Biotechnol.*, *1st, 1978*, 55–64.

Hirner, W. and H. Blenke (1977). Gasgehalt und Phasengrenzfläche in Schlaufen- und Strahlreaktoren. *Verfahrenstechnik*, **11**, 297–303.

Hsu, K. H., L. E. Erickson and L. T. Fan (1975). Oxygen transfer to mixed cultures in tower system. *Biotechnol. Bioeng.*, **17**, 499–514.

Jet Nozzle Aerating System BASF, Brochure of Krupp Industry and Steel Construction, Duisburg-Rheinhausen.

Kitai, A., H. Tone and A. Ozaki (1969). Performance of perforated plate column as a multistage continuous fermenter. *Biotechnol. Bioeng.*, **11**, 911–926.

Kuraishi, M. *et al.* (1979). SCP-process development with methanol as a substrate. *Proc. Int. Congr. Microbiol.*, *12th, 1979*, pp. 111–124. Dechema, Frankfurt.

Kürten, H. and B. Maurer (1977). Gasdispergierung im turbulenten Scherfeld. In *Particle Technology*, ed. H. Brauer and O. Molerus, pp. H47–H69. GVC, Nürnberg.

Kürten, H. and P. Zehner (1980). Der Gasumlaufreaktor — Ein Strahlschlaufenreaktor zur Eigenbegasung. Vortrag auf der GVC Jahrestagung 1980 in Strassburg.

Lainé, B. M. and J. du Chaffaut (1975). Gas-oil as a substrate for single-cell protein production. In *Single-cell Protein II*, ed. S. R. Tannenbaum and D. I. C. Wang, pp. 424–437, MIT Press, Cambridge.

Liepe, F. *et al.* (1978). The present state and perspective development of industrial high performance fermenters for aerobic fermentations. *Prepr. Eur. Congr. Biotechnol.*, *1st, 1978*, part 1, 78.

McKeogh, E. J. and D. A. Ervine (1981). Air entrainment rate and diffusion pattern of plunging liquid jets. *Chem. Eng. Sci.*, **36**, 1161–1172.

Maclean, G. T. *et al.* (1977). Oxygen transfer and axial dispersion in an aeration tower containing static mixers. *Biotechnol. Bioeng.*, **19**, 493–505.

Margaritis, A. and J. B. Wallace (1982). The use of immobilized cells of *Zymomonas mobilis* in a novel fluidized bioreactor to produce ethanol. *Biotechnol. Bioeng. Symp.*, **12**, 147–159.

Meijel Sewage Treatment Plant. DSM Brochure.

Mishra, I. M., S. A. El-Temtamy and K. Schügerl (1982). Growth of *Saccharomyces cerevisiae* in gaseous fluidized beds. *Eur. J. Appl. Microbiol. Biotechnol.*, **16**, 197–203.

Mitchell, R. and A. D. Lev (1970). Economic comparison of U-tube aeration with other methods for aerating waste-water. *Chem. Eng. Progr. Symp. Ser.*, **67** (107), 558–565.

Moebus, O. and M. Teuber (1979). Herstellung von Single Cell Protein mit *Saccharomyces cerevisiae* in einer Tauchstrahlbegasungsanlage. *Kieler Milchwirtschaftliche Forschungsberichte*, **31**, 297–361.

Moebus, O. and M. Teuber (1982). Production of ethanol by solid particles of *Saccharomyces cerevisiae* in a fluidized bed. *Eur. J. Appl. Microbiol. Biotechnol.*, **15**, 194–197.

Moebus, O., M. Teuber and R. Reuter (1982). Verfahren zur Herstellung von Äthanol durch Fermentation von Kohlenhydraten, *Ger. Pat. 3 105 581*.

Müller, G. *et al.* (1978). Abwasserreinigung in einem Bio-Hochreaktor. *Chem.-Tech.*, **7**, 257–260.

Müller, H. G. and G. Sell (1984). Der Radialstrombegaser — ein leistungsfähiger Belüfter für biologische Abwasser-Reinigungsanlagen. *Chem. Ing. Tech.*, **56**, 399–401.

Müllner, J. (Waagner Biro GmbH) (1973). Verfahren und Einrichtung zur Behandlung von Flüssigkeit und Trüben. *Austrian Pat. 319 864*.

Morris, G. G., R. N. Greenshields and E. L. Smith (1973). Aeration in tower fermenters containing microorganisms. *Biotechnol. Bioeng. Symp.*, **4**, 535–545.

Nikolajewski, H. E. *et al.* (1982). Production of *Corynebacterium diphtheriae* toxin in a bubble column fermenter. *J. Biol. Standardization*, **10**, 109–114.

Niranjan, K., S. B. Sawant, J. B. Joshi and V. G. Pangar (1982). Counter current absorption using wire gauze packings. *Chem. Eng. Sci.*, **37**, 367–374.

Orazem, M. E., L. T. Fan and L. E. Erickson (1979). Bubble flow in the downflow section of an airlift tower. *Biotechnol. Bioeng.*, **21**, 1579–1606.

Prescott, S. C. and C. G. Dunn (1982). In *Industrial Microbiology*, ed. G. Reed, 4th edn., pp. 609–611. Avi Publishing, Westport, Conn.

Räbiger, N. and A. Vogelpohl (1983). Der Kompaktreaktor, ein neuentwickelter Schlaufenreaktor mit hoher Stoffaustauschleistung. *Chem.-Ing.-Tech.*, **55**, 486–487.

Schügerl, K. (1976). Unpublished results.

Schügerl, K. (1982a). Characterization and performance of single and multistage tower reactors with an outer loop for cell mass production. In *Advances in Biochemical Engineering*, ed. A. Fiechter, vol. 22, pp. 93–224, Springer, Berlin.

Schügerl, K. (1982b). New bioreactors for aerobic processes. *Int. Chem. Eng.*, **22**, 591–610.

Schügerl, K., J. Lücke and U. Oels (1977). Bubble Column Bioreactors. Tower bioreactors without mechanical agitation. In *Advances in Biochemical Engineering*, ed. T. K. Ghose, A. Fiechter and N. Blakebrough, vol. 7, pp. 1–84. Springer, Berlin.

Scott, C. D. and C. W. Hancher (1976). Use of tapered fluidized bed as a continuous bioreactor. *Biotechnol. Bioeng.*, **18**, 1393–1403.

Seipenbusch, R. and H. Blenke (1980). The loop reactor for cultivating yeast on *n*-paraffin substrate. In *Advances in Biochemical Engineering*, ed. A. Fiechter, vol. 15, pp. 1–40, Springer, Berlin.

Shah, Y. T. (1979). Gas Liquid Solid Reactor Design. McGraw-Hill, London.

Shah, Y. T. and W. D. Deckwer (1981). In *Scale-up in the Chemical Process Industries*, ed. R. Kabel and A. Bisio, Wiley, New York.

Smith, E. L. and R. N. Greenshields (1974). Tower fermentation systems and their application to aerobic processes. *The Chemical Engineer*, January, 28–34.

Smith, S. R. L. (1980). Single cell protein. *Philos. Trans. R. Soc. London, Ser. B*, **290**, 341–354.

Sneath, R. W. (1978). The performance of a plunging jet aerator and aerobic treatment of pig slurry. *Water Pollut. Control*, **77**, 408.

Townsley, P. M. *et al.* (1983). The recycling air lift transfer fermenter for plant cells. *Biotechnol. Lett.*, **5**, 13–18.

Uhde Brochure, Lo II 19150079.

Uhde-Hoechst Brochure, 'Bio-High Reactor'.

Uhde, Konstoff-Tropfkörper-System 'Hydropak'. Brochure B & K VI. 195400078.

Van de Donk, J. A. C. (1981). Water aeration with plunging jets. Ph.D. thesis. Technische Hogeschool Delft.

Van de Sande, E. (1974). Air entrainment by plunging water jets. Ph.D. thesis. Technische Hogeschool Delft.

Van de Sande, E. and J. M. Smith (1976). Jet break-up and air entrainment by low velocity turbulent water jets. *Chem. Eng. Sci.*, **31**, 219–224.

Verhaagen, J. (1978). *LAWPR Specialized Conference on Aeration, Amsterdam, 19–22 Sept.*, Preprints, 150–157.

Viesturs, U. E. *et al.* (1980). Investigations of fermenters with various contacting devices. *Biotechnol. Bioeng.*, **22**, 799–820.

Viesturs, U. E. *et al.* (1981). Fermenters with the power introduced by aerating gas. *Biotechnol. Bioeng.*, **23**, 1171–1191.

Voigt, J. and K. Schügerl (1979). Absorption of oxygen in countercurrent multistage bubble columns I. Aqueous solutions with low viscosity. *Chem. Eng. Sci.*, **34**, 1221–1229.

Voigt, J. and K. Schügerl (1981). Comparison of single- and three-stage tower loop reactors. *Eur. J. Appl. Microbiol. Biotechnol.*, **11**, 97–105.

Voigt, J., V. Hecht and K. Schügerl (1980). Absorption of oxygen in countercurrent multistage bubble columns II. Aqueous solutions with high viscosity. *Chem. Eng. Sci.*, **35**, 1317–1323.

Vollmüller, H. and R. Walburg (1972). Bubble size in gassing with venturi nozzles. Joint Meeting of the VTG and ICE on 'Bubble and Foam', Nürnberg. *VDI-Ber.*, No. 182, 23–29.

Walker, J. and G. W. Wilkinson (1979). The treatment of industrial effluents using the deep shaft process. *Ann. N. Y. Acad. Sci.*, **326**, 181–191.

Walker, J. F., Jr. *et al.* (1981). In *Biotechnol. Bioeng. Symp.*, **11**, 415–427.

Wang, K. B. and L. T. Fan (1978). Mass transfer in bubble columns packed with motionless mixers. *Chem. Eng. Sci.*, **33**, 945–952.

Weiland, P. (1978). Untersuchung eines Airliftreaktors mit äusserem Umlauf im Hinblick auf seine Anwendung als Bioreaktor. Dissertation Universität Dortmund.

Weiland, P. and U. Onken (1981). Fluid dynamics and mass transfer in an airlift fermenter with external loop. *Ger. Chem. Eng. (Engl. Transl.)*, **4**, 42–50.

Witte, J. H. (1962). Dissertation, Technische Hogeschool Delft.

Young, J. C. and M. F. Dahab (1982). Operational characterisitics of anaerobic packed-bed reactors. *Biotechnol. Bioeng. Symp.*, **12**, 303–316.

Ziegler, H. *et al.* (1977). The tubular loop fermenter: oxygen transfer, growth kinetics and design. *Biotechnol. Bioeng.*, **19**, 507–525.

Ziegler, H., I. J. Dunn and J. R. Bourne (1980). Oxygen transfer and mycelial growth in a tubular loop fermenter. *Biotechnol. Bioeng.*, **22**, 1613–1635.

Zlokarnik, M. (1979). Sorption characteristics of slot injectors and their dependency on the coalescence behavior of the system. *Chem. Eng. Sci.*, **34**, 1265–1271.

Zlokarnik, M. (1980). Eignung und Leistungsfähigkeit von Volumenbelüftern für biologische Abwasserreinigungsanlagen. *Korrespondenz Abwasser*, **27** (3), 194–209.

Zlokarnik, M. (1982). Bioengineering of aerobic wastewater purification — development and trends. In *Proc. Austrian–Italian–Yugoslav Chem. Eng. Conf., 3rd, Graz, 1982*, vol. 2, pp. 563–511.

6

Dynamic Modelling of Fermentation Systems

M. L. SHULER
Cornell University, Ithaca, NY, USA

6.1 INTRODUCTION

A living cell is an immensely complex self-regulated chemical reactor which responds to environmental stimuli (such as changes in nutrient levels, temperature and pH) by altering its internal composition and biosynthetic capabilities. Such changes are not instantaneous but reflect finite time lags in the various biochemical pathways in the cell. Mathematical models that aspire to reflect the basic nature of living organisms must recognize the dynamic nature of such organisms.

Such models are built to fulfill at least one of the following objectives: (1) discrimination among possible mechanisms for the control of cellular processes, (2) bioreactor design and optimization, and (3) process control. The requirements placed on the model building process will differ with respect to the ultimate objective of the model builder.

Typically a model which seeks to be useful in mechanism discrimination at the subcellular level must be very general (and hence complex) and contain a low level of empiricism. Such models must accurately reflect the basic biochemistry of the cell. A high level of detail will invariably require large numbers of parameters; it must be realized that this does not reflect on the validity of a model. A 100-parameter model with no adjustable parameters may be intrinsically more valid than a two-parameter model where both parameters must be adjusted. In models where various subprocesses are self-regulated, and also regulated by the products of other subprocesses, the overall system response may be more dependent on model structure than on the values of the kinetic parameters associated with any individual subsystem.

Such complex models must be closely tied to experimental data to retain validity. Experiments are required for independent parameter estimation and to provide an information base for the formulation of hypotheses about a subcellular control system or pathway. Model predictions incorporating the various hypotheses must be tested against experimental results. Comparison to predictions about the dynamic behavior of the system offers a more stringent test of validity than does comparison to steady-state experiments. The process is simply that experimental evidence suggests models which lead to testable predictions and to further experiments which lead to refinements in the model resulting in new hypotheses and experiments, *etc.*

The other extreme is the formulation of models solely for process control. In this case the model builder is restricted to variables which can be readily determined on-line. Since the number of variables which can be reliably measured on-line is small (particularly in commercial systems), the model builder will use much simpler models than those intended for mechanism discrimination. Such models generally contain a moderate level of empiricism, particularly when explicit measurements of the product are impossible and productivity must be correlated with other more easily measured parameters. Consequently, models intended for process control will be valid for a relatively narrow operating range of abiotic conditions and will have a minimal number of parameters of which a large fraction may have to be obtained using curve-fitting procedures. The ultimate extreme would be the so-called 'black box' models.

Intermediate between these extremes are models intended to develop a more basic understanding of bioreactor performance or for the actual optimization of a process. Generality is important if a broad range of reactor conditions and types are to be explored and consequently the level of empiricism which can be tolerated is low. Since emphasis is on productivity, the level of biochemical detail required will be tied directly to the nature of the product. Models used for optimization will be mature models already subjected to substantial experimental verification; the results of the optimization undoubtedly require experimental validation but such an experimental program would be less extensive than for either mechanism discrimination or control.

6.2 DEFINITIONS AND IMPLICATIONS

6.2.1 Balanced Growth

Campbell (1957) was probably the first to introduce the term 'balanced growth'. He wrote: '. . . it will be convenient to say that growth is balanced over a time interval if, during that interval, every extensive property of the growing system increases by the same factor'. His definition was based on the behavior of a large population of cells.

Barford and coworkers (1982) have sought to broaden this definition to include the growth of individual cells and sustained oscillations by the culture as a whole. In this chapter we will accept the extension of the definition to an individual cell but not to the case of sustained oscillation in a whole culture since the average concentration of cellular components per unit cell weight would be time dependent. Balanced growth for an individual cell requires that each division cell be an exact replica of the previous cycle.

If a culture is in balanced growth, each individual cell need not be in balanced growth (see the data of Powell, 1958) but on the average a 'typical' cell within the culture will fulfill the definition of balanced growth.

6.2.2 Model Characteristics

Tsuchiya and coworkers (1966) in a pioneering review article suggested a conceptual framework for classifying models of microbial cultures. This framework has been retained, although the terminology has been modified through the years. Harder and Roels (1982) offer a well written summary of distinctions among models.

In this chapter we will concentrate on models which are deterministic rather than probabilistic. A deterministic model allows the exact prediction of future behavior based on specifying the current state vector (essentially values for all variables in the model). Deterministic models become increasingly valid as the number of individual members in the population increase. Generally a total population greater than 10 000 is sufficient to treat the system as deterministic. Special consideration must be given to synchronized or to synchronous cultures where 'all' cells initially divide at the same time and cell number increases in a stepwise fashion. After a few generations asynchrony develops as the distribution of cell division times broadens. Such behavior is deterministic in that with a large cell number the future time course is predictable. However, the development of asynchrony depends upon a random or probabilistic event within a population. Models seeking to simulate such behavior must include some mechanism to recognize such randomness.

Models are generally 'structured' or 'unstructured'. An unstructured model assumes that only a single variable such as cell number or dry weight is sufficient to describe the biosphere; in essence only the quantity of biomass is important. A structured model allows the division of the biosphere into two or more components. A model which is chemically structured divides the biosphere into chemical components. These components may be real and measurable such as DNA, RNA, protein, *etc.* Alternatively, chemical structure may be imparted with less well-defined components such as 'synthetic component', 'structural component', or similar terms. A model may be non-chemically structured by recognizing that in a pure culture the biosphere consists of cells of different cell sizes and ages and the biosynthetic capabilities of a cell depend on age or size. With a mixed-culture a non-chemically structured model would recognize the existence of different species and would consider the interactions among species. Often the term 'structured model' implies only chemical structure. In this chapter an effort is made to recognize explicitly the two possible forms of structure. In a structured model both quality and quantity of the biosphere are important.

Another distinction arises due to the nature of a microbial culture; it consists of many distinct cells. A 'segregated' or 'corpuscular' model is one that explicitly recognizes that a population consists of individuals each of whom may have distinct properties. A 'non-segregated' or 'distributive' or 'continuum' model does not explicitly recognize the existence of individuals but rather the cell mass is viewed as a lumped biomass which interacts as a whole with its environment.

As long as the properties of interest can be adequately represented by averages, the non-segregated approach is satisfactory. However, if properties with moments higher than first-order are important, then the lack of recognition of the existence of individual cells can be important. For example, suppose that 10% of the total population is responsible for 90% of the product formation. Shifts in the distribution of cell types in the population could be important. With the use of genetically engineered organisms it will be quite possible for a population to contain a wide variety of cell types differing in gene dosages (Imanaka and Aiba, 1981). For such cultures some recognition of segregation in the model will be important.

The mathematical requirements for the non-segregated and segregated models to give identical results have been described (Harder and Roels, 1982; Ramkrishna, 1979). Essentially, the continuum approach can be derived from the segregated approach if: (1) the rate function of a sequence of enzymatic reactions, R, can be factorized out of the probability–density function, and (2) the properties of the cell are statistically independent. Under these conditions it can be demonstrated that the correct formulation of chemically structured, non-segregated models requires the use of intrinsic concentrations (*e.g.* mass of component i per unit mass of total bio-material) for all biotic components. Abiotic components can be expressed as extrinsic concentrations (*i.e.* component mass per unit of reactor volume). Fredrickson (1976) was the first to articulate this requirement based on physical considerations.

The simplest type of model is unstructured and non-segregated; the Monod equation is an example of such a model. Fredrickson and coworkers (1971) have shown that only structured models can possibly predict the response of a microbial culture in unbalanced growth. Thus the Monod equation can only work under balanced growth conditions. Generally, exponential growth in batch culture and steady-state growth in a single-stage chemostat are considered the only common balanced growth situations. Probably neither exactly fulfills Campbell's definition of balanced growth. Barford and coworkers (1982) cite examples of exponential growth in batch culture which are not balanced. For a truly unstructured model to apply to steady-state chemostat growth, cell composition would have to be the same at all dilution rates; experimental measurements have shown that cell composition varies with dilution rate.

For any transient response structured models must be used. The rest of the chapter will be devoted to models which contain sufficient structure (chemical and/or non-chemical) to be useful in predicting the dynamic response of fermentation systems.

6.3 MODELS OF CELLS IN SUBMERGED CULTURE

6.3.1 Chemically Structured Non-segregated Models

Two of the first chemically structured models proposed were those by Williams (1967) and by Ramkrishna *et al.* (1967). Both were two-component models. Williams (1967) lumped the cell into a synthetic component (primarily RNA) and a structural–genetic component (primarily DNA and protein). Ramkrishna *et al.* (1967) divided the cell into a G-mass (RNA and DNA) and D-mass (proteins). As pointed out by Fredrickson (1976), both models are invalid since intrinsic concentrations were not used in formulating the kinetic expressions. Nonetheless these papers are of importance because they marked a conceptual breakthrough in the modelling of microbial dynamics.

The proper formulation of a chemically structured model requires that the reaction kinetics be formulated in terms of intrinsic concentrations, that is the amount of the component per unit cell volume or per unit cell mass. Clearly the enzymes within a cell respond to intracellular concentrations and not to the extrinsic concentrations (amount of intracellular component per unit reactor volume). For the nutrient concentration it is allowable to use extrinsic concentrations. However, the use of an extrinsic concentration assumes that the intracellular concentration of nutrient is, at any instant, proportional to external nutrient concentration. If a nutrient is stored within a cell and can be released to provide an internal nutrient source, then the assumption of instantaneous equilibrium will not be valid. In such cases a component to represent the intracellular nutrient concentration should be included.

According to Fredrickson (1976) the general formulation of a mass balance in a non-flow reactor is:

$$\mathrm{d}/\mathrm{d}t(m\hat{V}C_j) = m\hat{V}\sum_i r_{ij} \tag{1}$$

where m = total biomass in the system at time t; $\hat{V}$ = volume of biomaterial per unit of biomass (essentially the reciprocal of the cellular dry weight density); C_j = mass of the jth component of biomaterial per unit volume biomaterial at time t (essentially the intrinsic concentration of j); and r_{ij} = rate per unit volume of biomaterial at which the jth component of biomaterial appears (or disappears) because of the *ith* process. Normally $\hat{V}$ will be a constant. Equation (1) can be rearranged to yield:

$$\mathrm{d}C_j/\mathrm{d}t = \sum_i r_{ij} - \mu C_j \tag{2}$$

where:

$$\mu = 1/m\,\mathrm{d}m/\mathrm{d}t \tag{3}$$

The term 'μC_j' represents the dilution of intracellular components brought about by growth. It is this term which has been neglected in many structured models.

The above equations could also be written in terms of mass concentrations. That is X_j (mass of jth component per unit mass of biomaterial) instead of C_j. Clearly $X_j = C_j\hat{V}$. Other formulations of the same concept can be used.

As an example, consider the correct formulation of Williams' model (1967). In the original model the following components were defined: D = concentration (mass per unit volume of reactor) of structural and genetic components; R = concentration (mass per unit volume of reactor) of synthetic components; A = concentration (mass per unit volume of reactor) of limiting nutrient; and M = total biomass = $R+D$. All of the above are extrinsic concentrations. Williams assumed that component A was extracted from the medium and used to produce component R. Component D was assumed to be formed from R. The following equations were then postulated for R and D formation:

$$\mathrm{d}R/\mathrm{d}t = k_1AM - k_2RD \tag{4}$$

$$\mathrm{d}D/\mathrm{d}t = k_2RD \tag{5}$$

The above equations are conceptually incorrect since they are based on extrinsic concentrations of intracellular components.

The correct expression for equation (4) can be written:

$$\underset{\substack{\text{Rate of}\\\text{change of R}\\\text{concentration}}}{\mathrm{d}R/\mathrm{d}t} = \underset{\substack{\text{Biomass}\\\text{concentration}}}{M} \quad [\underset{\substack{\text{Rate of}\\\text{nutrient}\\\text{uptake per}\\\text{unit mass}\\\text{of biomaterial}}}{k_1AM/M} - \underset{\substack{\text{Rate of conversion}\\\text{of R into D per}\\\text{unit biomaterial}}}{k_2(R/M)(D/M)}] \tag{6}$$

or

$$\mathrm{d}R/\mathrm{d}t = k_1AM - k_2\frac{RD}{M} \tag{7}$$

which is clearly different from equation (4). Equation (1) could also have been used directly to derive equation (8):

$$d(R/M)/dt = k_1 A(R/M + D/M) - k_2 \frac{RD}{MM} - \mu R/M \tag{8}$$

Equation (8) reduces directly to (7) when the differentiation on the left hand side of (8) is carried out, and the definition for μ ($\mu = 1/M \, dM/dt$) is substituted into equation (8).

The correct expression for the D component is:

$$dD/dt = M[k_2(R/M)(D/M)] \tag{9}$$

which is different from equation (5).

The above discussion says nothing about the reasonableness of the kinetic expressions chosen by Williams (1967), but only the correct conversion of those kinetic concepts into mathematical terms. A potential point of disagreement could be, for example, whether biomolecular kinetics or saturation kinetics would be preferable.

Harder and Roels (1982) have reviewed many of the two-component and three-component models that have evolved from the pioneering efforts of Williams (1967) and Ramkrishna *et al.* (1967). Of particular interest is their three-component model: K is RNA, G is protein and R is the remainder of the biomass consisting primarily of carbohydrates and precursor molecules. The model allows for turnover in the K and G components. The rates of ATP consumption by the various components can be matched to ATP generation through substrate catabolism. The model was applied to an activated sludge and found to fit the results of continuous culture experiments and to show fair agreement to some preliminary transient experiments.

The three-component model is attractive since it reflects a reasonable fidelity to basic biochemical facts, no cellular component is excluded, and the various components are defined in terms of components which can be directly measured. If an investigator were interested in the formation of a specific product, an additional component (or more) would probably be required. A three- or four-component model has sufficient structure to mimic much of the general behavior of a real culture. With the appropriate definitions of components at least some of the parameters in the model can be evaluated independently. The mathematical complexity of such models is relatively modest; solutions can be obtained with minimal demands on the capabilities of modern computers. This type of model would be well suited to exploring possible bioreactor options for suspended cell cultures.

Another conceptual approach to structured non-segregated models is the 'cybernetic' perspective proposed by Ramkrishna (1983). Basically this approach contends that the cell's internal machinery has the facility to set effectively an optimal policy in using available resources. Such an approach replaces the need for accurate kinetic expressions with an 'optimal' decision-making unit. This decision-making apparatus can then control the rates of synthesis or activity of key enzyme systems. Ramkrishna (1983) has discussed the application of this approach to diauxic growth.

Examples of other types of model potentially useful for designing bioreactors for the formation of non-growth associated products include those by Van Dedem and Moo-Young (1973, 1975). Van Dedem and Moo-Young (1973) considered a model for the prediction of extracellular enzyme synthesis and included sufficient structure to account for the genetic control of enzyme synthesis. The formulation is mechanistically sound and offers an interesting general model. Van Dedem and Moo-Young (1975) extended the same conceptual framework to making predictions with respect to diauxic growth. The model made reasonable qualitative estimates of both batch growth and transient behavior in the chemostat. Many model parameters were estimated, at least roughly, from the literature but not all could be unequivocally justified. Model predictions were not compared quantitatively with experimental data. Also, the model is incorrectly formulated since intrinsic concentrations were not used.

6.3.2 Models of Mixed Cultures

The above models have recognized structure within a population by presuming that the response of the system was independent of the distribution of types of organism present. For pure cultures such an assumption is valid as long as cellular differentiation is unimportant. In some cases waste treatment systems containing a wide variety of species have been modelled by first lumping the biomass into a single mass and then dividing that mass into several subcomponents.

Such models have proved useful but are clearly limited since the interactions among members of the population may be essential to maintain the stability of the system.

Another form of structured but non-segregated models is obtained by dividing the total biomass in a mixed culture into components based on species rather than on compositional variables such as RNA, DNA, *etc*. Such an approach makes good sense if each species performs a unique function within the population. It is not necessary that each species component be further divided into compositional categories (although such an approach could be very valuable); the division of the culture into species is sufficient to give the culture 'structure'.

A large number of models for mixed cultures exist. An excellent overview can be obtained from articles by Fredrickson (1977, 1983), Kuenen (1983), and Bazin *et al*. (1983). The classification scheme offered by Fredrickson (1983) is given in Figure 1. Such interactions can give rise to a wide variety of dynamic responses. Even when the manipulated parameters in a chemostat (*e.g.* temperature, feed concentration, flow rate, *etc*.) are held constant, the system may exhibit sustained oscillations. Small perturbations in flow rate or substrate levels in the feed can cause very strong transient responses and possible destabilization of the system. Some examples of the dynamic behavior that can be encountered in steady-flow systems are given in Figure 2.

Mixed cultures are of importance in many natural food fermentations, waste treatment and

Effect of presence of B on growth rate of A	Effect of presence of A on growth rate of B	Qualifying remarks	Name of interaction
−	−	Negative effects caused by removal of resources	COMPETITION
−	0		COMPETITION
−	−	Negative effects caused by production of toxins or inhibitors	ANTAGONISM
−	0		AMENSALISM
−	+	Negative effects caused by production of lytic agents; positive effects caused by solubilization of biomass	EOCRINOLYSIS
+	0	Positive effect caused by production by B (host) of a stimulus for growth of A (commensal) or by removal by B of an inhibitor for growth of A	COMMENSALISM
+	+	See remarks for commensalism. Also presence of both populations not necessary for growth of both	PROTO-COOPERATION
+	+	See remarks for commensalism. Also presence of both populations is necessary for growth of either	MUTUALISM
−	+	B feeds on A	FEEDING (includes predation and suspension-feeding)
−	+	The parasite (B) penetrates the body of its host (A) and therein converts the host's biomaterial or activities into its own	PARASITISM
+	+ (or perhaps 0)	A and B are in physical contact; interaction highly specific	SYMBIOSIS
−	−	Competition for space	CROWDING

Figure 1 Scheme of classification of binary population interactions. The roles of A and B may be reversed. Top part of figure is for indirect interactions, while the bottom is for direct interactions. (From Fredrickson, 1983, with the permission of the American Chemical Society, Washington, DC)

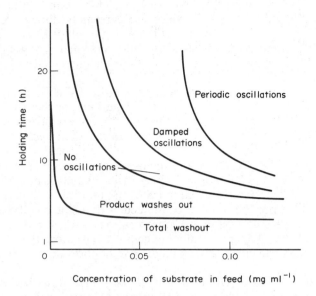

Figure 2 Stability regions of a model for the predator–prey interactions of *Dictyostelium discoideum* and *Escherichia coli* in continuous culture. The model was based on saturation kinetics. Five types of steady-state behavior can be predicted as a function of combinations of holding time and concentration of the limiting nutrient (glucose in this case). The prediction of sustained periodic oscillations was confirmed experimentally with good correlation between the model predictions and experimental data. (From Tsuchiya *et al.*, 1972, with the permission of the American Society for Microbiology)

natural ecosystems. They also represent the case where models of dynamic behavior are essential. Models of populations in which the behavior of each component species is modelled by a chemically structured model have not been accomplished. Such a model would have a much greater potential of truly representing the wide variety of dynamic responses that can be obtained with mixed cultures.

6.3.3 Segregated and Chemically Unstructured Models

Models which are termed segregated but chemically unstructured are based on the presumption that a single variable such as cell age or cell size can completely describe the physiological state of a cell. Thus any cell of say the same size must have the same composition and biosynthetic capabilities. The population model has 'structure' in the sense that the biosynthetic capabilities and composition of the population can be altered as there is a shift in the controlling variable such as size. Such models have the potential to predict transient responses.

Ramkrishna (1979) has summarized a number of aspects of formulating segregated models and reviewed some important aspects of previous studies, however, these models have generally had less impact on biotechnologists working with bioreactors than structured non-segregated models have had.

Shu's (1961) model for product formation is a possible exception. It makes use of an age density function, and product formation is tied to cell age. It is a versatile model and can reproduce the transient profiles typical in a wide variety of fermentations. However, it is difficult to evaluate all the necessary parameters from basic biochemical principles and the model, in practice, has a high degree of empiricism. Such a model may be useful in bioreactor design but not in mechanism discrimination.

As Bailey (1980) has pointed out, the development of segregated models with a significant level of chemical structure has been impeded by the difficulties in obtaining experimental data for model building and verification. The rapid measurement of DNA, protein and RNA (and potentially others) of a single cell can be accomplished with the appropriate fluorescent stains and flow cytometry. The availability of such measurements will undoubtedly act as an impetus to the development of segregated models which allow cells to contain chemical structure.

6.3.4 Population Models Based on Single-cell Models—Segregated and Chemically Structured Models

Models derived from the population point of view, which contain chemical structure as well as recognize segregation, result in equations which are extremely difficult to solve. Shuler *et al.* (1979) described a complex model for the growth of a single cell of *E. coli.* It was suggested that population models containing chemical structure and recognizing segregation might be constructed from a finite-representation technique using each single-cell model to represent some subfraction of the total population.

Nishimura and Bailey (1980) in an important paper starting from the perspective of a single cell of *E. coli* have constructed a model giving analytical solutions for the distributions of cell mass, DNA content, chromosome configuration and total cell numbers. The model requires that the growth rate be specified so that it responds implicitly rather than explicitly to changes in nutrients. The model makes very good predictions of the transient response of such a culture to a shift-up in growth rates. Bailey (1983) has reviewed the use of this general approach to the eukaryotes *Schizosaccharomyces pombe* and *Saccharomyces cerevisiae* as well as bacteria.

Shuler and Domach (1983) have reviewed much of the literature concerning the development of models of single cells. Since the number of molecules in a single cell is small, the use of the normal types of kinetic expressions based on concentrations is not strictly allowable. However, if the model cell is to be typical of a large number of cells (at least more than 100), then such kinetic expressions are acceptable. Such an understanding is implicit in almost all of the single-cell models developed.

Shuler and Domach (1983) and Domach *et al.* (1984) have described a complex single-cell model for *E. coli* (see Figure 3). Almost all of the model parameters were estimated from data in the literature. Four parameters associated with cross-wall formation could be evaluated only after the model was run at one growth rate where glucose was rate-limiting. Although the model is complex, it contains only four parameters adjusted within predetermined limits. Such a model provides an ideal framework for the quantitative testing of the plausibility of biological mechanisms. Shuler and Domach (1983) use the model as a basis for testing mechanisms for the control of initiation of chromosome synthesis in *E. coli.*

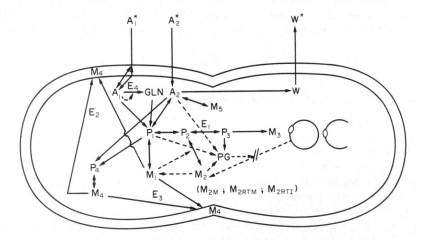

Figure 3 An idealized sketch of the model *E. coli* B/rA growing in a glucose–ammonium salts medium with glucose or ammonia as the limiting nutrient. At the time shown the cell has just completed a round of DNA replication and initiated cross-wall formation and a new round of DNA replication. Solid lines indicate the flow of material, while dashed lines indicate flow of information. The symbols are: A_1, ammonium ion; A_2, glucose (and associated compounds in the cell); W, waste products (CO_2, H_2O and acetate) formed from energy metabolism during aerobic growth; P_1, amino acids; P_2, ribonucleotides; P_3, deoxyribonucleotides; P_4, cell envelope precursors; M_1, protein (both cytoplasmic and envelope); M_{2RTM}, immature 'stable' RNA; M_{2RTI}, mature 'stable' RNA (r-RNA and t-RNA—assume 85% r-RNA throughout); M_{2M}, messenger RNA; M_3, DNA; M_4, non-protein part of cell envelope (assume 16.7% peptidoglycan, 47.6% lipid and 35.7% polysaccharide); M_5, glycogen; PG, ppGpp; E_1, enzymes involved in the conversion of P_2 to P_3; E_2, E_3, molecules involved in directing cross-wall formation and cell envelope synthesis—the approach used in the prototype model was used here but more recent experimental support is available; GLN, glutamine; E_4, glutamine synthetase; * indicates that the material is present in the external environment. (From Shuler and Domach, 1983, with the permission of the American Chemical Society, Washington, DC)

A population model can be constructed from the single-cell model without the addition of any adjustable parameters (Domach, 1983; Shuler and Domach, 1983; Domach and Shuler, 1984b), However, a cause for asynchrony must be specified and included in the model; in this case a random variation in the quantity of enzyme responsible for cross-wall formation was chosen (Domach and Shuler, 1984a). Domach and Shuler (1984b) have described the use of such a model for the prediction of the dynamic response of a population of *E. coli* in a single-stage chemostat to a shift in dilution rate. A comparison of experiment to model predictions is given in Figures 4 and 5. Recalling that no adjustable parameters were utilized in developing the population model, the correlation of prediction with experiment is quite remarkable. Thus it appears possible to predict the dynamic response of a large fermenter based solely on basic biochemistry without recourse to empirical expressions. However, such models, while mathematically straightforward, are quite tedious to develop and require substantial computer time.

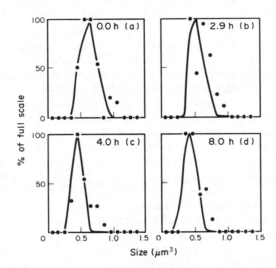

Figure 4 Shift in predicted (●) and observed (—) size distributions due to a flow perturbation in a chemostat. The organism modelled was *Escherichia coli* B/rA at 37 °C. Predictions were made using the single-cell model depicted in Figure 3 as a base for a population model using a finite representation scheme. 225 model cells were included in the population scheme; a smoother predicted size distribution would have been obtained if more model cells had been used. Nonetheless, the model accurately predicts the time-dependent shift in cell size and gives a reasonable approximation to the breadth and skew of the size distribution. The initial steady-state distribution is shown in (a) just prior to the decrease in flow. The initial dilution rate of 0.91 h^{-1} was changed to 0.65 h^{-1} at time $t = 0$. (From Domach and Shuler, 1984b)

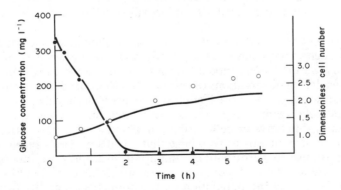

Figure 5 Prediction of transient changes in substrate concentration and dimensionless cell number. The model prediction is denoted by a solid line, while the observed values of substrate concentration and dimensionless cell number are given by ● and ○ respectively. The data are for the experiment described in Figure 4. The model is a population model based on an ensemble of single-cell models as described in Figure 3

6.3.5 Models with Time Delays

Computationally simple models that predict dynamic behavior are particularly desirable for process control. Rather than explicitly introducing a complex kinetic network the effects of cell adaptation can be included through the use of time delays. Nominally unstructured models modified by inclusion of time delays are potentially promising candidates for making predictions of dynamic behavior.

The underlying rationale for such models can be found in the concept of relaxation times (Harder and Roels, 1982). The concept originated as a means of realistically describing complex thermodynamic systems. A relaxation time characterizes the rate of adaptation of an internal process to changes in the external or abiotic conditions. The system, the biotic phase, is then described in terms of relaxation times and externally observable variables. The smaller the relaxation time the more quickly the internal mechanism adapts to changes in input.

A typical cell is characterized by a large number of processes with widely varying relaxation times, *e.g.* allosteric controls with relaxation times of about 1 s (range 10^{-4} to 10^2 s) to evolutionary changes with times of 10^6 s or larger. Not all of these internal processes are usually of importance to the prediction of the behavior of interest. If the rate of change of a variable in the abiotic environment is slow compared to the rate of adaptation of an internal mechanism to that change, then the dynamics of that internal mechanism may be neglected; it will always be at a quasi steady-state with respect to the external variable. In the above example the relaxation time of the internal process is much smaller than a characteristic time associated with the external system. On the other hand, if the relaxation time of the internal process is much larger than the external relaxation time, then that internal process can generally be ignored from a short-term viewpoint such as for process control. For example, the 'normal' dynamic response of a population in continuous culture to perturbations in flow are dissipated in two or three residence times but such changes may have long-term effects in the selection of a subpopulation of cells. Such a selection might not become apparent for many more cell generations.

Consequently, the dynamic behavior of a system could be satisfactorily estimated by only considering those internal processes that have the relaxation times of external changes. These internal processes with smaller relaxation times can be considered to be in a quasi steady-state while those processes with larger relaxation times can be ignored. For process control where the major fluctuations in the abiotic environment can be anticipated, a model recognizing a small number of relaxation times may be quite adequate.

The application of the use of transfer functions to biological systems is an example of the concept of relaxation times. An important example is the model suggested by Young and Bungay (1973). With this model they were able to predict the transient response of a chemostat to perturbations in flow or substrate feed concentrations; parameters predicted were biomass, substrate, protein, RNA and cell number. Results that might be expected are given in Figure 6. The values of the time constants could be estimated from experiments using a 'black box' approach. The application of such techniques in the field of process control is well known (Coughanowr and Kopel, 1965). The essential limitations to this approach are: (1) that only predetermined external variables are changing and at a rate consistent with the experiments to evaluate the time constants, and (2) the transfer approach assumes a linearized system. Since biological systems are highly non-linear, the transfer function will be valid only for relatively small perturbations.

Time delays can also be included in models based on physiological reasons rather than 'black-box' models. Many such models have been recently reviewed by MacDonald (1982). An early example is the discussion of a linear model with discrete delay which was invoked by Finn and Wilson (1953) in considering observations of sustained oscillations of a yeast population in a chemostat. Others have also suggested more complicated expressions making use of not only discrete delays but distributed delays employing a memory function. Such delays may act to approximate the complicated relationship between cell numbers and biomass in a population, or to include the effects of inertial nutrient pools, or to recognize that a cell's previous physiological history will affect its dynamic response to perturbations. Important examples are models as suggested by Powell (1969) which use a memory function to assess the influence of the history of the nutrient concentrations experienced by the population on the population's ability to respond to perturbations. Harder and Roels (1982) have described the use of Powell's model for predicting the specific rate of product formation to specific growth rate.

Models with time delay can usefully simulate a variety of responses. The evaluation of parameters from 'black-box' experiments can provide workable models pertinent to the control of real systems in terms of variables which can be readily measured. While such models are potentially attractive for process control and do have a conceptual justification, they are limited to situations where the potential perturbations are known. The accuracy of the predictions depends on the size of the perturbation of the external variable. Such models are not particularly useful in

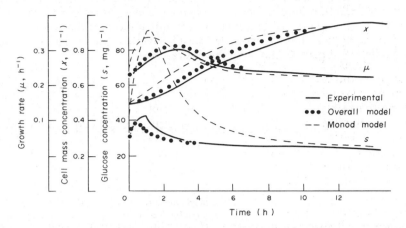

Figure 6 Comparison of predictions from a model derived from a system-analysis perspective, predictions from a Monod model, and experiment. The experimental system was a chemostat for a glucose-limited culture of *Saccharomyces cerevisiae* operating at a dilution rate of 0.20 h^{-1}. In this particular experiment the system was perturbed with a stepwise increase in feed glucose concentration from 1.0 and 2.0 g l^{-1}. x is biomass concentration, μ is growth rate and s is substrate concentration. (From Young and Bungay, 1973, with the permission of *Biotechnol. Bioeng.* and Wiley, New York)

discriminating among hypotheses of how cells function or in predicting the performance of a large variety of bioreactor types employing the given cells.

6.4 MODELS OF CELLS IN SURFACE CULTURES

Although most commercial fermentation processes in the West make use of submerged cultures, surface cultures offer potential advantages (Hesseltine, 1972). In Japan the growth of molds on solid particles (*i.e.* the koji process) is important as a source of enzymes and as a first step in the production of sake. Solid substrate fermentation has been practised successfully on a large scale. Certain mold products, *e.g.* mold spores to be used as insecticides (Miller *et al.*, 1983), require a high level of cellular differentiation and can be best obtained with solid substrate differentiation.

Such a process would, at least at the microscopic level, be always a dynamic one and presents some unique modelling challenges. A colony is always changing, and the system is much more heterogeneous than in submerged culture; spatial considerations cannot be neglected. The difficulties of these challenges coupled with the current low level of commercial activity with solid substrate fermentations has resulted in little real progress in this area. Prosser (1982) has reviewed a number of the suggested models for mold growth on solid substrates. Most models examine a specialized aspect of colony growth such as changes in macromolecular composition during vegetative growth, hyphal tip shape and extension, and growth of individual hyphae. Models for colony formation on solid media exist and are adequate to predict rate of extension of colony and branching patterns. Such models fail to address three important points: cellular differentiation, product formation associated with cellular differentiation, and interaction with nutrients in the solid media.

The rational design of solid substrate fermenters requires dynamic models. Such models would ideally include explicit recognition of the abiotic environment (gas phase and solid medium) and consider differentiation and product formation, and the interaction of colonies through the competition for nutrients or excretion of metabolic byproducts. Mathematically, ordinary differential equations describing mold growth need to be matched to partial differential equations for nutrient (or extracellular byproduct) profiles within the solid media. The macroscopic reactor model would be constructed from the models at the microscopic level.

6.5 SUMMARY

Transient responses of cell populations invariably result in unbalanced growth. Only models that contain structure have the inherent capability of accurately modelling population dynamics.

Structure, in the broadest sense, means that the modeller recognizes that both the quantity and quality of the cell population determines the dynamic behavior of the population. Models vary greatly in complexity and degree of empiricism. The objective that the model is to fulfill determines selection of the model, thus the modeller must be aware of the range of model types and be able to pick a modelling approach matching the desired goal.

ACKNOWLEDGMENTS

The author's work in this area has been supported in part by NSF Grant CPE-7921259.

6.6 REFERENCES

Bailey, J. E. (1980). Biochemical reaction engineering and biochemical reactors. *Chem. Eng. Sci.*, **35**, 1854–1886.
Bailey, J. E. (1983). Single-cell metabolic model determination by analysis of microbial populations. In *Foundations of Biochemical Engineering, ACS Symposium Series, No. 207*, ed. H. W. Blanch, E. T. Papoutsakis and G. Stephanopoulos, chap. 6, pp. 135–157. American Chemical Society, Washington, DC.
Barford, J. P., N. B. Pamment and R. J. Hall (1982). Lag phases and transients. In *Microbial Population Dynamics*, ed. M. J. Bazin, chap. 3, pp. 55–89. CRC Press, Boca Raton, FL.
Bazin, M. J., C. Curds, A. Dauppe, B. A. Owen and P. T. Saunders (1983). Microbial predation dynamics. In *Foundations of Biochemical Engineering, ACS Symposium Series, No. 207*, ed. H. W. Blanch, E. T. Papoutsakis and G. Stephanopoulos, chap. 11, pp. 253–264. American Chemical Society, Washington, DC.
Campbell, A. (1957). Synchronization of cell division. *Bacteriol. Rev.*, **21**, 263–272.
Coughanowr, D. R. and L. B. Koppel (1965). *Process Systems Analysis and Control*. McGraw-Hill, New York.
Domach, M. M. (1983). Refinement and Use of a Structured Model of a Single Cell of Escherichia coli for the Description of Ammonia-Limited Growth and Asynchronous Population Dynamics. Ph.D. Thesis, Cornell University, Ithaca, NY.
Domach, M. M. and M. L. Shuler (1984a). Testing of a potential mechanism for *E. coli* temporal cycle imprecision with a structured model. *J. Theor. Biol.*, **106**, 577–585.
Domach, M. M. and M. L. Shuler (1984b). A finite representation model for an asynchronous culture of *E. coli*. *Biotechnol. Bioeng.*, **26**, 877–884.
Domach, M. M., S. K. Leung, R. E. Cahn, G. G. Cocks and M. L. Shuler. (1984). Computer model for glucose-limited growth of a single cell of *Escherichia coli* B/rA. *Biotechnol. Bieng.*, **26**, 203–216.
Finn, R. K. and R. E. Wilson (1953). Population dynamics for a continuous propagator for microorganisms. *J. Agric. Food Chem.*, **2**, 66–69.
Fredrickson, A. G. (1976). Formulation of structured growth models. *Biotechnol. Bioeng.*, **18**, 1481–1486.
Fredrickson, A. G. (1977). Behavior of mixed cultures of microorganisms. *Annu. Rev. Microbiol.*, **31**, 63–87.
Fredrickson, A. G. (1983). Interactions of microbial populations in mixed culture situations. In *Foundations of Biochemical Engineering, ACS Symposium Series, No. 207*, ed. H. W. Blanch, E. T. Papoutsakis and G. Stephanopoulos, chap. 9, pp. 201–227. American Chemical Society, Washington, DC.
Fredrickson, A. G., D. Ramkrishna and H. M. Tsuchiya (1971). The necessity of including structure in mathematical models of unbalanced microbial growth. *Chem. Eng. Prog. Symp. Ser.*, *No. 108*, **67**, 53–59.
Harder, A. and J. A. Roels (1982). Application of simple structured models in bioengineering. In *Advances in Biochemical Engineering*, ed. A. Friechter, vol. 21, pp. 55–107. Springer-Verlag, New York.
Hesseltine, C. W. (1972). Solid substrate fermentations. *Biotechnol. Bioeng.*, **14**, 517–532.
Imanaka, T. and S. Aiba (1981). A perspective on the application of genetic engineering: stability of recombinant plasmid. *Ann. N. Y. Acad. Sci.*, **369**, 1–14.
Kuenen, J. G. (1983). The role of specialists and generalists in microbial population interactions. In *Foundations of Biochemical Engineering, ACS Symposium Series, No. 207*, ed. H. W. Blanch, E. T. Papoutsakis and G. Stephanopoulos, chap. 10, pp. 229–251. American Chemical Society, Washington, DC.
MacDonald, N. (1982). Time delays in chemostat models. In *Microbial Population Dynamics*, ed. M. J. Bazin, chap. 2, pp. 33–53. CRC Press, Boca Raton, FL.
Miller, L. K., A. J. Lingg and L. A. Bulla, Jr. (1983). Bacterial, virial, and fungal insecticides. *Science*, **219**, 715–721.
Nishimura, Y. and J. E. Bailey (1980). On the dynamics of Cooper–Helmstetter–Donachie procaryote populations. *Math. Biosci.*, **51**, 305–328.
Powell, E. O. (1958). An outline of the pattern of bacterial generation times. *J. Gen. Microbiol.*, **18**, 382–417.
Powell, E. O. (1969). Transient changes in the growth rate of microorganisms. In *Continuous Cultivation of Microorganisms*, ed. I. Malek, K. Bevan, Z. Fencl, V. Munk, J. Ricica and H. Smrckova, pp. 275. Academic, New York.
Prosser, J. I. (1982). Growth of fungi. In *Microbial Population Dynamics*, ed. M. J. Bazin, chap. 5, pp. 125–166. CRC Press, Boca Raton, FL.
Ramkrishna, D. (1979). Statistical models of cell populations. In *Advances in Biochemical Engineering*, ed. T. K. Ghose, A. Fiechter and N. Blakebrough, vol. 11, pp. 1–47. Springer-Verlag, New York.
Ramkrishna, D. (1983). A cybernetic perspective of microbial growth. In *Foundations of Biochemical Engineering, ACS Symposium Series, No. 207*, ed. H. W. Blanch, E. T. Papoutsakis and G. Stephanopoulos, chap. 7, pp. 161–178. American Chemical Society, Washington, DC.
Ramkrishna, D., A. G. Fredrickson and H. M. Tsuchiya (1967). Dynamics of microbial propagation: models considering inhibitors and variable cell composition. *Biotechnol. Bioeng.*, **9**, 129–170.
Shu, P. (1961). Mathematical models for the product accumulation in microbial processes. *J. Biochem. Microbiol. Technol. Eng.*, **3**, 95–109.
Shuler, M. L. and M. M. Domach (1983). Mathematical models of the growth of individual cells. Tools for testing bioche-

mical mechanisms. In *Foundations in Biochemical Engineering*, *ACS Symposium Series*, *No. 207*, ed. H. W. Blanch, E. T. Papoutsakis and G. Stephanopoulos, chap. 5, pp. 93–133. American Chemical Society, Washington, DC.

Tsuchiya, H. M., A. G. Fredrickson and R. Aris (1966). Dynamics of microbial cell populations. *Adv. Chem. Eng.*, **6**, 125–206.

Tsuchiya, H. M., J. F. Drake, J. L. Jost and A. G. Fredrickson (1972). Predator–prey interactions of *Dictyostelium discoideum* and *Escherichia coli* in continuous culture. *J. Bacteriol.*, **110**, 1147–1153.

Van Dedem, G. and M. Moo-Young (1973). Cell growth and extracellular enzyme synthesis in fermentations. *Biotechnol. Bioeng.*, **15**, 419–439.

Van Dedem, G. and M. Moo-Young (1975). A model for diauxic growth. *Biotechnol. Bioeng.*, **17**, 1301–1312.

Williams, F. M. (1967). A model of cell growth dynamics. *J. Theor. Biol.*, **15**, 190–207.

Young, T. B., III and H. R. Bungay (1973). Dynamic analysis of a microbial process: a systems engineering approach. *Biotechnol. Bioeng.*, **15**, 377–393.

7

Instrumentation for Monitoring and Controlling Bioreactors

W. B. ARMIGER
BioChem Technology Inc., Malvern, PA, USA

7.1 INTRODUCTION

The development of new and more sophisticated systems for monitoring fermentations is making a major contribution toward a better understanding of microbial processes. The increasing utilization of these new techniques allows for better environmental control and monitoring of metabolic processes, resulting in improvements in productivity.

Traditionally, mutation and strain selection have been utilized to breed specific metabolic controls into or out of microorganisms. This has been necessary since techniques for monitoring and controlling cellular metabolism on-line have not been available. However, the recent development of on-line sensors and data analysis systems is leading toward the eventual on-line regulation and control of microbial systems.

Instrumentation systems can be grouped into three basic categories: in-line, on-line and off-line. Off-line systems involve the removal of discrete samples at periodic intervals for subsequent treatment and analysis. These techniques involve standard wet chemical analysis and the use of automated laboratory instrumentation systems. However, since these techniques typically involve turn around times which are too long (hours or days) for use in any kind of on-line monitoring or feedback control scheme, they will not be discussed in this chapter. On-line systems involve techniques utilizing the continuous sampling of a process stream with the subsequent conditioning and analysis such that the effective 'response time' is within the range for meaningful process control decisions. Examples of these types of systems include sampling of gas streams for on-line analysis by a mass spectrometer or gas chromatograph, continuous dialysis of culture broth for analysis by a high pressure liquid chromatograph, or continuous diffusion through a porous teflon membrane to detect volatile components in the culture broth. Finally, in-line systems are those for which probes and sensors are in direct contact with the broth to give a continuous, non-invasive, rapid response signal. These include such sensors as temperature, pH, dissolved oxygen and pressure. The on-line sensors are typically the ones most utilized for control purposes.

In general, to evaluate and compare the performance of various monitoring techniques, consideration should be given to the response time, gain, sensitivity, accuracy and stability of the instrument. In addition, systems which are non-invasive and capable of sterile operation are essential for an in-line system.

The fundamental purpose of the sensor or transducer is the translation of a physical or chemical event into an electrical signal. Examples are a thermocouple which converts a temperature difference into an electric potential and a pH electrode which converts hydrogen ion activity into an electric potential. The precise relationship between the physical/chemical activity and the electric signal is often very complicated and as a result is often treated as a 'black box' in which the input event is related to the output signal by a transfer function. The ability to describe mathematically the transfer process function is essential for stable process control operation. The concepts of response time and gain are used to characterize a transfer function by subjecting the system to a step change and evaluating the response. This is illustrated in Figure 1 (Fleischaker *et al.*, 1981) showing a step change in x from 0 to 1 and the subsequent electrical response from e_0 to e_1, where e_0 and e_1 are the steady state values. The response time (t_r) is the time required for the output signal to reach 90% of the final steady state value, *i.e.* $0.9(e_1-e_0)$. The gain is defined as the output (e) divided by the input (x) and may or may not be a linear function over the response range. Thus, a plot of the gain *versus* input describes the calibration curve for the transducer. The term sensitivity is defined as the change in output divided by the change in input ($\Delta e/\Delta x$). Of course, the acceptable values for response time and gain will depend upon the particular measurement, for example a microorganism with a doubling time of 2 h would require an instrumentation system with a response time of 2.5 min or less to be able to resolve a change in cell mass to within 5%.

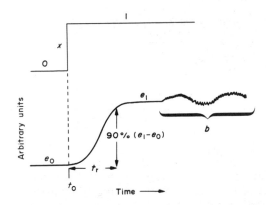

Figure 1   Typical transducer output response to a step change in input (Fleischaker, 1981)

The accuracy of the transducer is determined by the signal to noise ratio (S/N). Referring to Figure 1, the signal is e_1-e_0 and the noise is defined as the root mean square (RMS) of the fluctuations about the time average of signal e_1. The high frequency S/N component consists of the fluctuations with frequencies greater than $1/t_r$ and the low frequency S/N component consists of those fluctuations with frequencies less than $1/t_r$. The overall accuracy of the measurement is affected more by the high frequency S/N ratio. As a rule of thumb, an S/N ratio of 10 or greater is desirable.

The stability of the signal is determined by the signal drift which is related to the low frequency portion of the signal noise. The drift will dictate how often the instrument must be recalibrated or compensated by base-line changes. Obviously, systems which cannot be recalibrated or compensated on-line can pose serious limitations. For example, it is not possible to correct base-line drift of a dissolved oxygen probe over a 40 h fermentation and this can cause reliability problems. On the other hand, base-line changes in pH readings due to temperature fluctuations can be compensated for electronically.

In general, sensors for fermenters can be divided into two major groups, *i.e.* those monitoring the physical environment and those measuring chemical species. The systems for measuring chemical species can involve materials in the bulk environment of the fermentation broth or substances in the intracellular environment. Table 1 is a listing of the types of measurements in each category.

Table 1 Types of Sensors used in Fermenters[a]

Systems involved in controlling the physical environment	*Systems involved in controlling the chemical environment*
Temperature	pH
Pressure	Redox
Agitator shaft power	Dissolved oxygen
Foam	Dissolved carbon dioxide
Flow rate (gas and liquid)	Oxygen in exit gas
Turbidity	Carbon dioxide in exit gas
Viscosity	Precursor level and feed rate
	Sugar (carbon) level and feed rate
	Protein (nitrogen) level and feed rate
	Mineral ion level
	Mg^{2+}, K^+, Ca^{2+}, Na^+, Fe^{3+}, SO_4^{2-}, PO_4^{3-}
	RNA
	DNA
	NAD, NADH
	ATP, ADP, AMP

[a] Aiba *et al.*, 1973.

Sensors can also be grouped into monitoring systems which have been described (Aiba *et al.*, 1973) as 'gateway' sensors (Table 2) since they allow for the calculation of important process information based upon several parameters. Gateway sensors have the greatest application in instrumentation systems which have a microprocessor or some other means of on-line computational ability.

Table 2 Gateway Sensors[a]

Sensor	*Information*
pH	Acid product formation
Dissolved oxygen	Oxygen transfer rate
Oxygen in exit gas	Oxygen uptake rate
Gas flow rate	
Carbon dioxide in exit gas	Carbon dioxide evolution rate
Gas flow rate	
Oxygen uptake rate	Respiratory quotient
Carbon dioxide evolution rate	
Sugar level and feed rate	Yield and cell density
Oxygen uptake rate	

[a] Aiba *et al.*, 1973.

The subject of on-line data analysis is covered in the chapter by Zabriskie (Volume 2, Chapter 10). This chapter will be limited to describing instrumentation systems and techniques for measuring the basic process variable.

7.2 IN-LINE MEASUREMENTS

In-line measurements are made by direct reading probes in a non-destructive manner within the physical environment of the fermenter. Since these sensors are directly in contact with the process, they are not subject to errors caused by loss of aeration or poor mixing in a recirculation loop, by longer response times due to sampling procedures, or by inadequate sampling techniques.

7.2.1 Extracellular Environment

7.2.1.1 Temperature

Since heat can be transferred freely through the cell wall and directly influence intracellular metabolism, temperature measurement and control are often critical process variables. Typical sensors used for fermentation applications include platinum resistance sensors (RTD), thermistors, thermocouples and filled bulbs. The advantages and best applications for each type of sensor have been described previously (Tannen and Nyiri, 1979).

In general, biological systems are not very efficient in substrate conversion, for example *Escherichia coli* utilizes only about 50% of the available free energy for the conversion of glucose to cell mass, with the remaining 50% being liberated as heat (Wang *et al.*, 1976). Thus, the measurement of temperature at appropriate points to obtain a process heat balance is another important use of temperature sensors.

7.2.1.2 Pressure

The measurement of static head pressures in a biorector is important for a variety of reasons. The static head pressure obviously influences the partial pressure conditions which will affect the solubility of gases such as oxygen and carbon dioxide, as well as other volatile compounds such as alcohols. In addition, a positive head pressure in a fermenter will assist in maintaining sterile conditions. The measurement of internal pressure is also critical during vessel sterilization to assure that proper conditions for sterilization have been achieved. Finally, the measurement of pressure in the main supply line of steam, water and air is necessary to monitor the proper functioning of the system.

In each of these applications it is possible to use pressure gauges as indicators or pressure transducers. Some of the considerations in the selection of a device are as follows: (1) at non-sterile points where the measurement is not needed in a process calculation, Bourdon gauges are typically used along with a pressure switch if an alarm function is desired; (2) at sterile operating points, a diaphragm is used between the aseptic environment and the gauge; and (3) for fermenter head pressure measurements utilized for on-line process calculations, a pressure transducer with appropriate seals to withstand sterilization conditions is required. One such device is a linear variable differential transformer (LVDT) protected by a diaphragm. Typical specifications for such a device are a range of 3–15 to 0–3000 p.s.i., linearity of ±0.25%, stability of 0.01%/ °F (from 10 °F to 140 °F), and a response time of 100 ms.

7.2.1.3 Agitator speed and power consumption

Agitator speed is measured directly by a tachometer on the agitator shaft. The electric signal is used to control the agitator speed to ensure adequate mixing and maintain the required level of oxygen transfer. The monitoring and control of agitator speed is relatively simple to obtain, however, accurate measurements of the actual power delivered to the fluid is a more difficult task. Traditionally, power consumption has been measured by a Hall effect wattmeter attached to the agitator motor. Even though the Hall effect wattmeter has an accuracy of ±1% full-scale, the validity of the data is obscured by the fact that the measurement represents the total motor power uptake and indicates the actual power delivered to the liquid by the impellers. The design of the agitator system (shaft, bearings, seals, gears, pulleys, belt, *etc.*) will significantly contribute to the loss of power between the motor and the impeller.

Dynamometers and strain gauges have been used in pilot plant fermenters to determine directly the power input to the fluid (Harmes, 1972). The detector consisted of four strain gauges attached directly to the internal portion of the shaft of a 270 l fermenter. ϕ (defined as the power at the shaft divided by the power measured at the motor armature) increased with increasing motor speed. For this agitator configuration at maximum r.p.m., the power lost between the motor and the impeller was about 30% of the total energy supplied.

A compromise solution between using a wattmeter and strain gauges is to use an in-line torque measuring external to the vessel. With this system, losses due to the bearings and the seals will still be a factor in the measurement. However, these losses can be determined under a no-load condition and then subtracted from the measurement. Even though this technique is subject to some error, it is the best practical solution in many situations.

7.2.1.4 Foam detection

Foam detection and control can be accomplished by mechanical foam breakers or the use of chemical antifoams. Mechanical foam breakers are centrifugal devices and can be overwhelmed in a heavy foam situation. However, they do offer the advantage of not having to add chemicals to the media formulation.

Chemical antifoam systems require a sensor which is usually nothing more than a teflon coated electrode. The foam completes the circuit between the tip of the probe and the wall of the vessel activating a solenoid valve controlled by a timer.

7.2.1.5 Liquid and gas flow rates

Flow rates are an important process variable since they are used as a basis for calculating material balances around a fermenter. Rotameters are an economical means of measuring flow when an electric signal is not needed. Rotameters are available in which the position of the float in the meter is converted to an electric signal, but these systems are not generally accurate enough for material balance calculations. Flow meters with electrical transducers providing outputs for data analysis and control include magnetic flow rate sensing elements, thermal mass flow rate sensing elements, laminar flow rate sensing devices and turbine flow meters. Selection of the device will depend upon the magnitude of the flow rate and the nature of the fluid. Standard literature sources (Liptak, 1972) are available for determining selection criteria.

7.2.1.6 Volume

The measurement of the working volume of a fermenter is especially important for fed-batch and continuous processes. Three types of systems are generally used for measuring volume on-line: (1) liquid level sensors; (2) weight measuring devices; and (3) differential pressure sensors. For small-scale and pilot-plant systems, weight measurement devices such as load cells are the most reliable and accurate means for determining volume. For large-scale systems, differential pressure between the top and bottom of the tank is used to indicate liquid height and hence volume.

7.2.2 Chemical Environment

7.2.2.1 pH

Steam sterilizable combination pH probes have been available for a long time. They were one of the first truly direct reading sensors. The current probes can be repeatedly sterilized (over 100 times at 121 °C) and some models are even designed to be removed under pressure, recalibrated, resterilized and placed back into operation in the process. Table 3 lists several sources of sterilizable in-place pH electrodes.

Table 3 Steam Sterilizable pH Sensors[a]

Manufacturer	Type	Model	Temperature range (°C)	Pressure range
Activion Glass, Ltd.	Combination	11–801	10–140	N/A
		11–902	10–140	
Ingold	Combination	761/764 series	−10–130	6 bar
Leeds & Northup	Combination	117494	20–130	3.5 bar
		117495		

[a] Tannen and Nyiri, 1979.

7.2.2.2 Dissolved oxygen

Dissolved oxygen probes have also been available for a long time and are one of the key sensors in following the progress of a fermentation. Table 4 lists some of the commercially available dissolved oxygen probes for sterile and non-sterile fermentations. It has been shown (Siegell and Gaden, 1962; Taylor *et al.*, 1971) that the dissolved oxygen probe measures the oxygen partial pressure and not the dissolved oxygen concentration. Thus, accurate dissolved oxygen readings should be corrected for changes in vessel pressure. Dissolved oxygen probes have been used to determine the volumetric oxygen transfer coefficient ($k_L a$) and the oxygen uptake rate (QX) (Bandyopadhyay *et al.*, 1967; Taguchi and Humphrey, 1966). These variables can be related by the following equation (Aiba *et al.*, 1973):

$$\frac{d\tilde{C}}{dt} = k_L a(C^* - \tilde{C}) - Q_{O_2}X \tag{1}$$

where C^* = concentration of dissolved oxygen which is in equilibrium with pO_2 in the bulk gas phase (mM); C = the actual dissolved oxygen concentration in the bulk of the liquid (mM); $k_L a$ = the volumetric oxygen transfer coefficient (h^{-1}); Q = specific rate of oxygen uptake ($mM\ t^{-1}$); X = cell mass concentration.

Table 4 Commercially Available Dissolved Oxygen Probes[a]

Manufacturer	Type of sensor	DO measurement range (p.p.m.)	Temperature compensation	Pressure compensation	Response time (s)[c]
Beckman Instruments	Polarographic	0–9.99	Yes	No	30
Instrumentation Laboratory, Inc.[b]	Polarographic	0–20	Yes	Yes	20
New Brunswick Scientific Co., Inc.[b]	Galvanic	0–8	No	Yes (vented)	45
Transidine General Corp.	Polarographic	0–1999	No	No	—
Yellow Springs Instruments[b]	Galvanic	0–8	No	Yes (vented)	60–120

[a] Tannen and Nyiri, 1979. [b] Steam sterilizable (minimum 121 °C). [c] 90% of reading at 20 °C.

7.2.2.3 Dissolved carbon dioxide

Carbon dioxide is one of the end products of the complete oxidation of carbon substrates utilized by microorganisms. The effect of dissolved carbon dioxide on cellular metabolism has been reported (Nyiri and Lengyel, 1965; Shibai *et al.*, 1973). However, work on studying the effects of dissolved carbon dioxide on cell growth and product formation has been limited due to the lack of steam sterilizable CO_2 probes. Recently, two sensors for dissolved CO_2 have been described (Puhar *et al.*, 1980; Shoda and Ishikawa, 1981). Both probes are based upon measuring pH in the following equilibrium relationship:

$$CO_2 + H_2O \rightleftharpoons H_2CO_3 \rightleftharpoons H^+ + HCO_3^- \tag{2}$$

At equilibrium

$$K = \frac{[H^+][HCO_3^-]}{[CO_2][H_2O]} \tag{3}$$

At constant temperature, the value of K is constant. Since in the fermentation broth the concentration of H_2O and HCO_3^- are large compared to H^+ and CO_2, the concentration of CO_2 is proportional to H^+ concentration.

Figure 2 shows the construction of the probe described by Puhar *et al.* (1980). The long-term stability of the electrode is primarily dependent upon the pH probe and has been reported to be stable over a period of weeks. The response of the probe after repeated sterilizations (120 °C for 20 min) is shown in Table 5.

7.2.2.4 Redox probe

Steam sterilizable redox probes based upon a combined platinum–reference electrode system are available. However, the interpretation of the data is difficult. Lengyel and Nyiri (1965) report that under aerobic conditions, the oxidation–reduction potential is essentially a measurement of

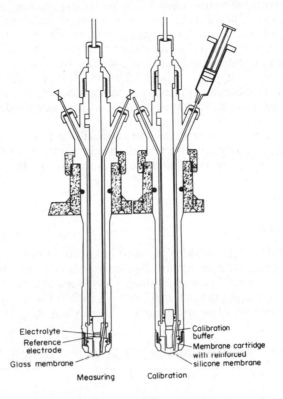

Figure 2 Basic design of the pCO$_2$ electrode (Puhar *et al.*, 1980)

Table 5 Influence of Sterilization of Dissolved CO$_2$ Probe[a]

Number of sterilizations	ΔmV of gas mixture[b] after sterilization	ΔmV of buffer A after sterilization	Slope s (mV)
0	—	—	54.2
1	8.7	8.8	54.0
2	26.5	26.5	54.1
3	34.7	31.3	53.9
4	29.7	29.4	53.1
5	24.0	23.1	53.8
6	25.2	24.4	54.9

[a] Puhar *et al.*, 1980.   [b] 27.5 mbar pCO$_2$ after sterilization.

the dissolved oxygen concentration. Furthermore, the usefulness of the data is questionable because the probe is measuring the oxidation–reduction potential in the bulk environment which is probably quite different from the intracellular environment. For an anaerobic process, the probe is useful for monitoring a highly reduced environment. However, this situation is also subject to the fundamental limitation of monitoring the extracellular environment. Another disadvantage of the redox signal is that it is a function of pH (Ishizaki *et al.*, 1974). Ishizaki *et al.* (1974) observed a 33 mV decrease in redox potential for each increase of 1 pH unit. In spite of these reported difficulties, some useful correlations of redox data have been reported for specific fermentation processes (Akashi *et al.*, 1978; Dahod, 1982; Shibai *et al.*, 1974). Dahod (1982) has compared the dissolved oxygen reading with the redox potential for a penicillin fermentation. Since oxygen is the terminal electron acceptor, the dissolved oxygen probe measures the oxidation potential of the terminal reaction while the redox probe can measure the oxidation potential of all the fermentation products excreted into the broth. According to Dahod (1982), the concentration of intermediate metabolites on an oxidative pathway is controlled by the availability of oxygen for the terminal reaction. It is argued that there will be more reduced species in the extracellular environment and a lower redox potential if oxygen is lacking than if sufficient oxygen is present. Thus, in a pelleted fermentation, the dissolved oxygen in the bulk liquid can be high while the dissolved oxygen in the center of the pellet is quite low. A dissolved oxygen probe showing a high reading can indicate sufficient oxygen transfer when in fact this is not the case. A redox probe could potentially be useful in correctly interpreting this situation.

Redox potential has been correlated with product formation and shifts in metabolism. Akashi *et al.* (1978) observed that optimal production of L-leucine by *B. lactofermentum* occurred when the redox potential was between −200 mV and −220 mV. Shibai *et al.* (1974) found that *B. subtilis* produced different metabolites as its primary product at different redox levels. Lactic acid was the major product at −200 mV, 2,3-butyleneglycol was the major product at −195 mV, and acetone the major product at −160 mV.

In summary, even though the intracellular and extracellular redox potential can vary independently and the redox probe is only sensitive to extracellular redox potential, the measurement can provide useful correlations for product formation particularly in the regions below the useful range of a dissolved oxygen probe.

7.2.2.5 Specific ion probes

Specific ions such as NH_4^+, Mg^{2+}, Na^+, Ca^{2+} and PO_4^{3-} are important in the regulation of cellular metabolism. Table 6 is a list of commercially available specific ion electrodes that could be of interest for a biological process. However, these electrodes are not steam sterilizable and in general do not have an operating range for pH that is suitable for many fermentation processes. As a result, there have been very few applications of these electrodes in fermentation processes.

Table 6 Commercially Available Ion Sensors for Biological Processes[a, b]

Ion species	pH range for direct measurement	Concentration range (mol)
Ammonium	11–13	$1–10^{-6}$
Calcium	6–8	$1–10^{-5}$
Carbon dioxide	5	$10^{-2}–10^{-4}$
Chloride	2–11	$1–8 \times 10^{-6}$
Magnesium	—	10^{-4c}
Nitrate	3–10	$10^0–6 \times 10^{-6}$
Nitrate	0–2	$10^{-2}–5 \times 10^{-7}$
Potassium	3–10	$1–10^{-5}$
Sodium	9–10	Saturation to 10^{-6}
Phosphate	—	5×10^{-3c}
Sulfate	—	5×10^{-3c}

[a] Based on Orion Research Institute Literature, 8th edn. May, 1977. [b] Tannen and Nyiri, 1979. [c] Minimum.

7.2.3 Intracellular NADH

All living cells contain nicotinamide adenine dinucleotide (NADH) and nicotinamide adenine dinucleotide phosphate (NADPH). These molecules in their reduced form are important redox carriers in microbial metabolism. The concentration of NAD(P)H present in each cell is an index of its energy inventory and metabolic activity.

NAD(P)H fluoresces at 460 nm when excited by 340 nm light. Since this characteristic light emission is relatively unique, this principle provides a basis on which to monitor intracellular NAD(P)H concentration by optical means. The culture fluorometer schematically represented in Figure 3 applies this principle to a fermentation culture by the determination of the fluorescence of a suspended cell population in the vicinity of the sensor.

The applications of culture fluorescence have been summarized by Zabriskie (1979) and Beyeler *et al.* (1981). This technique has been employed to follow NADH oscillations which occur in yeast in response to abrupt changes in the cellular environment (Chance *et al.*, 1964). Harrison and Chance (1970) used the technique to monitor the accumulation of NADH during the transition from aerobic to anaerobic growth in yeast. It has been used to monitor growth and utilization of the primary carbon source in fermentation broths (Zabriskie, 1976; Zabriskie and Humphrey, 1978a). Reports by Watteeuw *et al.* (1979) and Einsele *et al.* (1979) have shown that

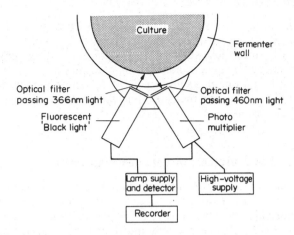

Figure 3 Schematic of fermenter fluorometer assembly (Zabriskie, 1979)

culture fluorescence is useful in measuring intracellular NADH profiles in response to step changes in the extracellular substrate concentration. By monitoring these step changes, it has been possible to detect shifts in metabolic pathways resulting from changes in substrate concentration. Ristroph *et al.* (1977) used the ability of culture fluorescence to measure shifts in metabolism as a means to control the feeding of ethanol to a fed-batch *Candida utilis* fermentation. Einsele *et al.* (1978) have reported the use of culture fluorescence to monitor step changes in quinine concentration as a means of determining bulk mixing times for a reactor vessel.

All of the studies on culture fluorescence have utilized fluorometer systems that were especially built for laboratory applications because instrumentation has not been commercially available. Recently, a commercial unit has become available in the form of a compact sterilizable probe (Figure 4). This unit is a practical means of overseeing a fermentation process in real time. The sensor excites the process liquid with a low intensity, short wavelength (340 nm), UV lamp. The process liquid in turn emits light energy in the visible spectrum (460 nm). This energy is detected and converted to a format suitable for process monitoring and control. The control electronics nests in a standard 19 in. relay rack or bench enclosure. The probe is compatible with industry standard 25 mm diameter INGOLD™ fittings for easy insertion in pilot plant or production vessels.

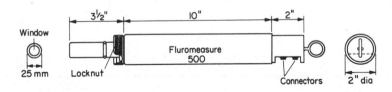

Figure 4 Fluorometer probe (Courtesy of BioChem Technology, Inc., Malvern, PA, USA)

From the previously reported studies, it is evident that culture fluorescence measurements may provide very useful information for monitoring and controlling bioreactor processes. One of the major advantages of this technique is that it is responsive to intracellular as opposed to extracellular events. The measurement has further advantages in that it has a very rapid response (about 250 ms), is continuous and does not require the collection of discrete samples. Furthermore, it has been demonstrated with a variety of microorganisms (bacteria, yeast and fungi) on a number of substrates such as defined media, complex media and media with suspended solids.

In conclusion, culture fluorescence is a relatively simple technique which is capable of generating on-line very valuable biological information about a process. Further work is necessary to demonstrate its utility for a variety of industrial processes, but the availability of a commercial instrument now makes this practical. It should become one of the important measurements for process control of bioreactors.

7.3 ON-LINE MEASUREMENT

On-line measurement systems are utilized in situations where the sensor is not able to withstand *in situ* sterilization conditions or when the sample requires some form of conditioning prior to using a detection system. Since these sensor systems are not in direct contact with the process, they are subject to errors caused by lack of aeration or poor mixing in a recirculation loop, longer response time due to sampling procedures and inadequate sampling techniques. In spite of these potential problems, a number of on-line measurement systems have been developed and are described in the following sections.

7.3.1 Off-gas Analysis and Measurement of Volatile Compounds

The measurement of gases in the inlet and exhaust streams is a useful means of monitoring and controlling fermentations, for example cell mass and growth rates can be calculated from oxygen uptake rates (Zabriskie and Humphrey, 1978b) or carbon dioxide evolution rates (Mou and Cooney, 1976; Mou, 1979). The respiratory quotient (RQ), the ratio of the CO_2 evolution rate to the oxygen uptake rate, has been used to control the glucose additions to yeast fermentations (Nagai *et al.*, 1976; Spruytenburg *et al.*, 1979; Wang, 1977). The reader is referred to Volume 2, Chapter 8 for a more detailed discussion. In addition to carbon dioxide and oxygen, other gases and volatile components of the fermentation broth such as ammonia, methane, butanol, ethanol, water vapor, nitrogen, hydrogen, acetaldehyde and acetone can be useful to monitor in the off-gas.

There are several kinds of systems which have been used for monitoring off-gases. These include paramagnetic analyzers for carbon dioxide and ammonia, gas chromatographs and mass spectrometers (Tables 7–10).

The paramagnetic oxygen analyzer is limited by its sensitivity to changes in flow rate (Swartz

Table 7   Carbon Dioxide Gas Analyzers[a]

Manufacturer	Model	Range (%CO_2)	Response time (s)[b]	Accuracy	Drift (zero, span per 24 h)
Anarad, Inc. Cavitron Corp.	AR500	0–30	5	± 1%	± 1%
Beckman Instruments	865	0–5	0.5	± 1% FS	± 1%
Infrared Instruments	IR703	0–1	5	± 1% C	1 °C
	IR702	0–30			
Mine Safety (M.S.A.)	200	0–30	5	± 1%	± 1%

[a] Tannen and Nyiri, 1979.   [b] 90% for step change.

Table 8   Oxygen Gas Analyzers[a]

Manufacturer	Model	Range (%O_2)	Response time (s)[b]	Accuracy (%FS)	Sample temperature range (°F)	Measurement method
Beckman Instruments	755	0–1 to 0–100	7	± 2	−20–120	Paramagnetic
Mine Safety Appliances	802	0–0.5 to 0–100	80	± 2	32–118	Paramagnetic
Taylor Instrument (Sybron)	OA 184	0–25 to 0–100	7	± 1	14–122	Paramagnetic (both inlet and outlet)

[a] Tannen and Nyiri, 1979.   [b] 90% for step change.

Table 9   Process Gas Chromatograph

Manufacturer	Model	Detectors
Bendix	0020	Thermal conductivity Flame ionization Flame photometric

Table 10 Process Mass Spectrometer

Manufacturer	Model	Gases	Range (%)	Response time (s)[a]	Accuracy (%FS)
Perkin–Elmer	1200	Mass numbers 2–135	0–0.1 to 0–100	1	1%

[a] 90% for step change.

and Cooney, 1978). It is also limited by its accuracy (1–2% FS) which can lead to significant errors in calculating oxygen uptake rates when the change in oxygen concentration across the fermenter is less than 1%. The IR analyzers do not suffer from these limitations, however, the response time associated with both instruments (including the sampling system) restricts their use to single fermentation vessels.

Gas chromatographs (GC) have been utilized for the off-line determination of a variety of volatile compounds such as methane, acetaldehyde and ethanol (Abbott *et al.*, 1973; George and Gaudy, 1973). Nyiri *et al.* (1976) reported utilizing off-gas analysis to monitor the effects of varying the carbon/nitrogen feed ratio on the biosynthesis of ethanol and biomass during a baker's yeast fermentation.

Quadruple mass spectrometers (MS) are available for the continuous monitoring of gases with mass numbers ranging from 2–132. The major advantages of the mass spectrometer are its rapid response time (about 1 s) and its sensitivity. The rapid response time makes it practical to utilize one MS to monitor a large number of fermenters. Thus, in many plant and pilot plant applications, an MS can be the most cost effective method and accurate means for off-gas analysis. A number of applications of mass spectrometers have been reported (Buckland, 1981; Reuss *et al.*, 1975; Weaver *et al.*, 1976).

Volatile compounds in fermentation broths can be measured using the gas analyzers described above with a porous teflon tubing/inert gas carrier system. Phillips and Johnson (1961) originally developed the porous teflon tubing sampling technique for measuring dissolved oxygen. Dairaku and Yamane (1979) and Yamane *et al.* (1981) have reported the use of a similar sampling system to monitor ethanol for controlling a yeast fermentation. A similar technique using a porous membrane probe (Figure 5) has been used to monitor CO_2, NH_4, methanol, ethanol, acetone and butanol (Heinzle *et al.*, 1981).

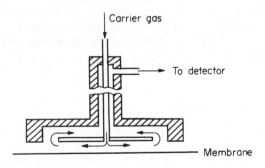

Figure 5 Probe for porous teflon membrane film (Heinzle *et al.*, 1981)

7.3.2 Flow Microfluorometry

Laser flow microfluorometry (FMF) has recently been applied to the study of population dynamics in fermentation processes (Agar and Bailey, 1981; Fazel-Madijlessi and Bailey, 1979; Fazel-Madijlessi *et al.*, 1980). The major advantage of this technique is its ability to measure characteristics of individual cells rather than the average of an entire population. Therefore, in principle, this technique can be used to determine physiological characteristics of subpopulations within pure cultures. Such detailed analysis of population dynamics could provide new insights into microbial kinetics useful for process control.

Figure 6 shows a schematic of an FMF assembly. The unit operates by passing a suspension of cells one at a time through a detector region. By controlling the cells with various dyes prior to injection in the system, different cellular components can be monitored by detecting fluorescence and light scattering. An advantage of this technique is that several detector systems can be used simultaneously. Thus, cells can be characterized for such parameters as size, DNA content, RNA content, protein content and specific enzymes. Systems are currently available which will analyze cell populations at rates of up to 3×10^4 cells per second.

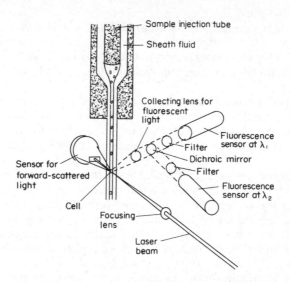

Figure 6   Schematic of assembly for cytofluorometry (Fleischaker *et al.*, 1981)

Several accounts of FMF for studying single components in bacteria (Bailey *et al.*, 1977; Paau *et al.*, 1977) and yeast (Gilbert *et al.*, 1978; Hutter *et al.*, 1975; Hutter and Eipel, 1978; Slater *et al.*, 1977) have been reported. Simultaneous measurements of protein and double-stranded nucleic acid in *Bacillus subtilis* (Bailey *et al.*, 1978) and protein and DNA in *Saccharomyces cerevisiae* and *Candida utilis* (Hutter and Eipel, 1979) illustrate morphological and biochemical changes during batch culture. These simultaneous measurements provide a detailed account of physiological and morphological changes in a batch culture. Since the yield and productivity of many important industrial processes are related to cell morphology and the physiological state of the organism (*i.e.* secondary metabolites), this measurement technique could become an important tool for controlling the growth of a cell population.

7.3.3 Dialysis and Continuous Filtration

The measurement of chemical species directly in the fermentation is limited by the availability of sterilizable probes. The ability to extract a cell-free continuous sample would allow many analytical instruments and non-sterilizable probes to be used. Several techniques for accomplishing this task have been described. Zabriskie and Humphrey (1978c) reported the use of continuous dialysis *via* a modified fermenter baffle (Figure 7) to monitor glucose on-line. A similar device in the form of a probe (Figure 8) has been described by Gardiner and Briggs (1974). Chotani and Constantinides (1982) also reported monitoring glucose on-line using a continuous filtration cell for generating samples. Pungor *et al.* (1981) utilized a continuously recirculating sample through a membrane system (Figure 9) under vacuum to pull off volatile components for analysis by a mass spectrometer.

7.3.4 Multiple Internal Reflection Spectrometry

Multiple internal reflection (MIR) spectrometry has recently become available as an on-line process technique. It is based upon the principle that as a beam of radiation passes through a transparent medium, it is reflected internally from a surface with a small portion of energy lost as the beam projects beyond the reflecting surface. If air is in contact with the surface of the medium, very little energy is lost in transmission. However, if an absorbing liquid is in contact

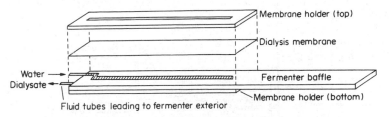

Figure 7   Details of fermenter continuous dialyzer design (Zabriskie and Humphrey, 1978c)

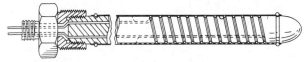

Figure 8   Sampling device for removing dialyzable fluids from a liquid for analysis (Gardiner and Briggs, 1974)

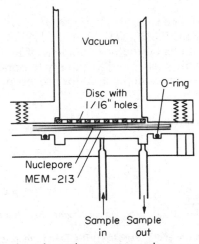

Figure 9   Sampling through membrane system under vacuum (Pungor *et al.*, 1981)

with this surface, energy will be absorbed at the wavelength at which the substance normally absorbed energy. In a liquid sample, the depth of penetration is usually of the order of 10 μm so that very little energy is absorbed in a single reflection. However, if the optical system is designed for a series of reflections (usually about 10), absorption can be detected. Instruments are available which utilize a cylindrical internal reflection cell. This geometry has the advantage of being easily sealed by use of an O-ring and eliminates restrictions to the smooth flow of sample through the chamber so that relatively dense, high-viscosity materials can be analyzed. Furthermore, since particles have no effect on the internal reflection measurements. it is not necessary to remove them prior to sampling. A schematic of a flow-through chamber and optics are shown in Figure 10. The system has been utilized to monitor a number of aqueous streams: dissolved sugar and carbon dioxide in beverages; sugar, alcohol, and CO_2 in beer; starch/fructose ratios in enzyme conversion reactions; sugar and citric acid in citrus juice streams; and water, carbohydrate, fat, or protein concentrations in liquid and semiliquid food processing streams. Similar applications are certainly possible in monitoring fermentation reactions and other bioprocess blending and mixing operations.

7.3.5   Other Off-line Measurement Systems

A number of other off-line measurement systems are under development. Some will always require an off-line sampling technique while some systems (*i.e.* enzyme probes) could be utilized in-line if a sterilizable probe could be developed. There are several reviews of enzyme electrodes and enzyme based sensors (Barker and Somers, 1978; Danielsson *et al.*, 1979, 1981). Specific electrodes for glucose (Danielsson, 1979), catechol (Neujahr, 1979), phenol (Neujahr and Kjellen, 1979; Kjellen and Neujahr, 1980), and alcohols (Hikuma *et al.*, 1979) have been described. The measurements of cell mass have been demonstrated in continuous flow systems based

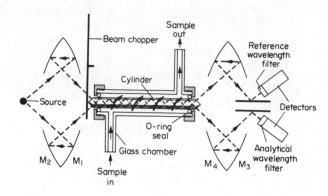

Figure 10   Schematic of assembly for multiple internal reflection spectrometry (Wilks, 1982)

upon turbidity (Lee and Lim, 1980) and viscosity (Perley *et al.*, 1979) and upon a 'filtration probe' technique (Nestaas and Wang, 1981a, 1981b; Nestaas *et al.*, 1981).

In summary, the primary limitations on the development of new sensors for monitoring bioreactors are availability of sterilizable probes and the ability to obtain a representative sample of the culture. Many analytical techniques could be used for process monitoring and control if these two problems could be solved.

7.4   REFERENCES

Abbott, B. J., A. I. Laskin and C. J. McCoy (1973). Growth of acinetobacter–calcoaceticus on ethanol. *Appl. Microbiol.*, **25**, 787–792.

Agar, D. W. and J. E. Bailey (1981). Continuous cultivation of fission yeast: classical and flow microfluorometry observations. *Biotechnol. Bioeng*, **23**, 2217–2229.

Aiba, S., A. E. Humphrey and N. F. Millis (1973). *Biochemical Engineering*, 2nd edn., pp. 331–374. Academic, New York.

Akashi, K., S. Ikeda, H. Shibai, K. Kobayashi and Y. Hirose (1978). Determination of redox potential levels critical for cell respiration and suitable for L-leucine production. *Biotechnol. Bioeng.*, **20**, 27–41.

Bailey, J. E., J. Fazel-Madjilessi, D. N. McQuitty, L. Y. Lee, J. C. Allred and J. A. Oro (1977). Characterization of bacterial growth by flow microfluorometry. *Science*, **198**, 1175–1176.

Bailey, J. E., J. Fazel-Madjilessi, D. N. McQuitty, L. Y. Lee, J. C. Allred and J. A. Oro (1978). Measurement of structured microbial population dynamics by flow microfluorometry. *AIChE J*, **24**, 570–577.

Bandyopadhyay, B., A. E. Humphrey and H. Taguchi (1967). Dynamic measurement of the volumetric oxygen transfer coefficient in fermentation systems. *Biotechnol. Bioeng.*, **8**, 533–544.

Barker, S. A. and P. J. Somers (1978). Enzyme electrodes and enzyme based sensors. In *Topics in Enzyme and Fermentation Biotechnology*, ed. A. Wiseman, vol. 2, chap. 3, pp. 120–151. Wiley, New York.

Beyeler, W., A. Einsele and A. Fiechter (1981). On-line measurements of culture fluorescence: method and application. *Eur. J. Appl. Microbiol. Biotechnol.*, **13**, 10–14.

Buckland, B. C. (1981). Analysis of fermentation exhaust gas using a mass spectrometer. Presented at *Third International Conference on Computer Applications in Fermentation Technology*, Manchester, England.

Chance, B., R. W. Estabrook and A. Ghosh (1964). Damped sinusoidal oscillations of cytoplasmic reduced pyridine nucleotide in yeast cells. *PNAS*, **51**, 1244–1251.

Chotani, G. and A. Constantinides (1982). On-line glucose analyzer for fermentation applications. *Biotechnol. Bioeng.*, **24**, 2743–2745.

Dahod, S. K. (1982). Redox potential as a better substitute for dissolved oxygen in fermentation process control. *Biotechnol. Bioeng.*, **24**, 2123–2125.

Dairaku, K. and T. Yamane (1979). Use of the porous teflon tubing method to measure gaseous or volatile substances dissolved in fermentation liquids. *Biotechnol. Bioeng.*, **21**, 1671–1676.

Danielsson, B., B. Mattiasson, R. Karlsson and F. Winqvist (1979). Use of an enzyme thermistor in continuous measurements and enzyme reactor control. *Biotechnol. Bioeng.*, **21**, 1749–1766.

Danielsson, B., C. F. Mandenius, F. Winquist, B. Mattiasson and K. Mosbach (1981). Fermenter monitoring and control by enzyme thermistors. In *Advances in Biotechnology*, ed. M. Moo-Young, vol. I, chap. 71, pp. 445–451. Pergamon, Oxford.

Einsele, A., D. L. Ristroph and A. E. Humphrey (1978). Mixing times and glucose uptake measured with a fluorometer. *Biotechnol. Bioeng.*, **20**, 1487–1492.

Einsele, A., D. L. Ristroph and A. E. Humphrey (1979). Substrate uptake mechanisms for yeast cells — A new approach utilizing a fluorometer. *Eur. J. Appl. Microbiol. Biotechnol.*, **6**, 335–339.

Fazel-Madjilessi, J. and J. E. Bailey (1979). Analysis of fermentation processes using flow microfluorometry: single-parameter observations of batch microbial growth. *Biotechnol. Bioeng.*, **21**, 1995–2010.

Fazel-Madjilessi, J., J. E. Bailey and D. N. McQuitty (1980). Flow microfluorometry measurements of multicomponent cell composition during batch bacterial growth. *Biotechnol. Bioeng.*, **22**, 457–462.

Fleischaker, R. J., J. W. Weaver and A. J. Sinskey (1981). Instrumentation for process control in cell culture. In *Advances in Applied Microbiology*, ed. D. Perlman and A. I. Laskin, vol. 27, pp. 137–167. Academic, New York.

Gardiner, W. and K. Briggs (1974). *US Pat.* 3 830 106.

George, T. K. and A. F. Gaudy (1973). Transient response of continuously cultured heterogeneous populations to changes in temperature. *Appl. Microbiol.*, **26**, 796–803.

Gilbert, M. F., D. N. McQuitty and J. E. Bailey (1978). Flow microfluorometry study of diauxic batch growth of *Saccharomyces cerevisiae*. *Appl. Environ. Microbiol.*, **36**, 615–617.

Harmes, C. S., III (1972). Design criteria of a fully computerized fermentation system. *Ind. Microbiol.*, **13**, 146–156.

Harrison, D. E. F. and B. Chance (1970). Fluorimetric technique for monitoring changes in the level of reduced nicotinamide nucleotides in continuous cultures of microorganisms. *Appl. Microbiol.*, **19**, 446–450.

Heinzle, E., O. Bolzern, I. J. Dunn and J. R. Bourne (1981). A porous membrane–carrier gas measurement system for dissolved gases and volatiles in fermentation systems. In *Advances in Biotechnology*, ed. M. Moo-Young, vol. I, chap. 70, pp. 439–444. Pergamon, Oxford.

Hikuma, M., T. Kubo, T. Yasuda, I. Karube and S. Suzuki (1979). Microbial electrode sensor for alcohols. *Biotechnol. Bioeng.*, **21**, 1845–1853.

Hutter, K. J. and H. E. Eipel (1978). DNA determination of yeast by flow cytometry. *FEMS Microbiol. Lett.*, **3**, 35–38.

Hutter, K. J. and H. E. Eipel (1979). *Eur. J. Appl. Microbiol. Biotechnol.*, **6**, 223.

Hutter, K. J., C. Boose, H. Oldiges and C. C. Eneis (1975). Investigations about the synthesis of DNA, RNA, and proteins of selected populations of microorganisms by cytophotometry and pulse cytophotometry. Part III. Determination of intracellular substances of partially synchronized aerobe and respiratory deficient yeast cells. *Chem. Mikrobiol. Technol. Lebensm.*, **4**, 101–104.

Ishizaki, A., H. Shibai and Y. Hirose (1974). Basic aspects of electrode potential change in submerged fermentation. *Agric. Biol. Chem.*, **38**, 2399–2406.

Kjellen, K. G. and H. Y. Neujahr (1980). Enzyme electrode for phenol. *Biotechnol. Bioeng.*, **22**, 299–310.

Lee, C. and H. Lim (1980). New device for continuously monitoring the optical density of concentrated microbial cultures. *Biotechnol. Bioeng.*, **22**, 639–642.

Lengyel, Z. L. and L. Nyiri (1965). An automatic aeration control system for biosynthetic processes. *Biotechnol. Bioeng.*, **7**, 91–100.

Liptak, B. G. (ed.) (1972). *Instrument Engineers Handbook*, vol. 1, 2, and suppl. 1. Chilton Book Company, Philadelphia, PA.

Mou, D. G. (1979). Toward optimal penicillin fermentation by monitoring and controlling growth through computer-aided mass balancing. Ph.D. Thesis, Massachusetts Institute of Technology, Cambridge, MA.

Mou, D. G. and C. L. Cooney (1976). Applications of dynamic calorimetry for monitoring fermentation processes. *Biotechnol. Bioeng.*, **18**, 1371–1392.

Nagai, S., Y. Niechizawa and T. Yamagata (1976). *Fifth International Fermentation Symposium*, Abstract, p. 30. Verlag Versuchs- und Lehranstalt für Spiritusfabrikation und Fermentationstechnologie, Berlin.

Nestaas, E. and D. I. C. Wang (1981a). A new sensor, the 'filtration probe', for quantitative characterization of the penicillin fermentation. I. Mycelial morphology and culture activity. *Biotechnol. Bioeng.*, **23**, 2803–2813.

Nestaas, E. and D. I. C. Wang (1981b). A new sensor, the 'filtration probe', to monitor and control antibiotic fermentations. In *Advances in Biotechnology*, ed. M. Moo-Young, vol. I, chap. 69, pp. 433–438. Pergamon, Oxford.

Nestaas, E., D. I. C. Wang, H. Suzuki and L. B. Evans (1981). A new sensor, the 'filtration probe', for quantitative characterization of the penicillin fermentation. II. The monitor of mycelial growth. *Biotechnol. Bioeng.*, **23**, 2815–2824.

Neujahr, H. Y. (1980). Enzyme probe for catechol. *Biotechnol. Bioeng.*, **22**, 913–918.

Neujahr, H. Y. and K. G. Kjellen (1979). Bioprobe electrode for phenol. *Biotechnol. Bioeng.*, **21**, 671–678.

Nyiri, L. K. and Z. L. Lengyel (1965). Studies on automatically aerated biosynthetic processes. I. The effect of agitation and CO_2 on penicillin formation in automatically aerated liquid cultures. *Biotechnol. Bioeng.*, **7**, 343–354.

Nyiri, L. K., G. M. Toth, C. S. Krishnaswami and D. V. Parmenter (1976). On-line analysis and control of fermentation processes. In *Gesellschaft fur Biotechnologische Forschung mbH.*, ed. R. P. Jefferis, III, pp. 37–46. Verlag Chemie.

Paau, A. S., J. R. Cowles and J. A. Oro (1977). Flow microfluorometric analysis of *Escherichia coli*, *Rhizobium meliloti*, and *Rhizobium japonicum* at different stages of the growth cycle. *Can. J. Microbiol.*, **23**, 1165–1169.

Perley, C. P., J. R. Swartz and C. L. Cooney (1979). Measurement of cell mass concentration with a continuous-flow viscometer. *Biotechnol. Bioeng.*, **21**, 519–523.

Phillips, D. H. and M. J. Johnson (1961). Measurement of dissolved oxygen in fermentations. *J. Biochem. Microbiol. Technol. Eng.*, **3**, 261–275.

Puhar, E., A. Einsele, H. Buhler and W. Ingold (1980). Steam-sterilizable pCO_2 electrode. *Biotechnol. Bioeng.*, **22**, 2411–2416.

Pungor, E., Jr., E. Schaefer, J. C. Weaver and C. L. Cooney (1981). Direct monitoring of a fermentation in a computer–mass spectrometer–fermentor system. In *Advances in Biotechnology*, ed. M. Moo-Young, vol. I, chap. 63, pp. 393–398. Pergamon, Oxford.

Reuss, M., H. Piehl and F. Wagner (1975). Application of mass spectrometry to measurement of dissolved gases and volatile substances in fermentation. *Eur. J. Appl. Microbiol.*, **1**, 323–325.

Ristroph, D. L., C. M. Watteeuw, W. B. Armiger and A. E. Humphrey (1977). Experience in the use of culture fluorescence for monitoring fermentations. *J. Ferment. Technol.*, **55**, 599–608.

Schimz, K., B. Rutten and M. Tretter (1981). Determination of adenosine nucleotides with luciferin/luciferase from crude firefly lantern extract on a bioluminescence analyzer. In *Advances in Biotechnology*, ed. M. Moo-Young, vol. I, chap. 73, pp. 457–462. Pergamon, Oxford.

Siegell, S. D. and E. L. Gaden, Jr. (1962). Automatic control of dissolved oxygen levels in fermentation. *Biotechnol. Bioeng.*, **4**, 345–356.

Shibai, H., A. Ishizaki, H. Mizuno and Y. Hirose (1973). Effects of oxygen and carbon dioxide on inosine fermentation. *Agric. Biol. Chem.*, **37**, 91–97.

Shibai, H., A. Ishizaki, K. Kobayashi and Y. Hirose (1974). Simultaneous measurement of dissolved oxygen and oxidation–reduction potentials in the aerobic culture. *Agric. Biol. Chem.*, **38**, 2407–2411.

Shoda, M. and Y. Ishikawa (1981). Carbon dioxide sensor for fermentation systems. *Biotechnol. Bioeng.*, **23**, 461–466.

Slater, M. L., S. O. Sharrow and J. J. Gart (1977). Cell cycle of *Saccharomyces cerevisiae* in populations growing at different rates. *Proc. Natl. Acad. Sci. USA*, **74**, 3850–3854.

Spruytenburg, R., I. J. Dunn and J. R. Bourne (1979). Computer control of glucose feed to a continuous aerobic culture of *Saccharomyces cerevisiae* using the respiratory quotient. In *Biotechnology and Bioengineering Symposium No. 9*, ed. W. B. Armiger, pp. 359–368. Wiley, New York.

Swartz, J. R. and C. L. Cooney (1978). Instrumentation in computer-aided fermentation. *Process Biochem.*, **13** (2), 3–7.

Taguchi, H. and A. E. Humphrey (1966). Dynamic measurement of the volumetric oxygen coefficient in fermentation systems. *J. Soc. Ferment. Technol.*, **44**, 881.

Tannen, L. P. and L. K. Nyiri (1979). Instrumentation of fermentation systems. In *Microbial Technology*, ed. D. Perlman and H. Peppler, 2nd edn., vol. 2, pp. 331–374. Academic, New York.

Taylor, G. W., J. P. Kondig, S. C. Nagle and K. Higuchi (1971). Growth and metabolism of L cells in a chemically defined medium in a controlled environment culture system part 1—effects of oxygen tension on L cell cultures. *Appl. Microbiol.*, **21**, 928–933.

Wang, H. (1977). Computer control of yeast fermentations. Ph.D. Thesis, Massachusetts Institute of Technology, Cambridge, MA.

Wang, H. Y., D. G. Mou and J. P. Swartz (1976). Thermodynamic evaluation of microbial growth. *Biotechnol. Bioeng.*, **18**, 1811–1814.

Watteeuw, C. M., W. B. Armiger, D. L. Ristroph and A. E. Humphrey (1979). Production of single cell protein from ethanol by fed-batch process. *Biotechnol. Bioeng.*, **21**, 1221–1237.

Weaver, J. C., M. K. Mason, J. A. Jarrel and J. W. Peteson (1976). Biochemical assay by immobilized enzymes and a mass spectrometer. *Biochim. Biophys. Acta*, **438**, 296–303.

Wilks, P. A., Jr. (1982). Sampling method makes on-stream IR analysis work. *Ind. Res. Dev.*, September, 132–135.

Yamane, T., M. Matsuda and E. Sada (1981). Application of a porous teflon tubing method to automatic fed-batch culture of microorganisms. II. Automatic constant-value control of fed substrate (ethanol) concentration in semibatch culture of yeast. *Biotechnol. Bioeng.*, **23**, 2509–2524.

Zabriskie, D. W. (1976). Real-time estimation of aerobic batch fermentation biomass concentration by component balancing and culture fluorescence. Ph.D. Thesis, University of Pennsylvania, Philadelphia, PA.

Zabriskie, D. W. (1979). Use of culture fluorescence for monitoring of fermentation systems. In *Biotechnology and Bioengineering Symposium No. 9*, ed. W. B. Armiger, pp. 117–123. Wiley, New York.

Zabriskie, D. W. and A. E. Humphrey (1978a). Estimation of fermentation biomass concentration by measuring culture fluorescence. *Appl. Environ. Microbiol.*, **35**, 337–343.

Zabriskie, D. W. and A. E. Humphrey (1978b). Real-time estimation of aerobic batch fermentation biomass concentration by component balancing. *AIChEJ*, **24**, 138–146.

Zabriskie, D. W. and A. E. Humphrey (1978c). Continuous dialysis for the on-line analysis of diffusible components in fermentation broth. *Biotechnol. Bioeng.*, **20**, 1295–1301.

8

Instrumentation for Fermentation Process Control

D. N. BULL
Satori Corporation, Upper Montclair, NJ, USA

8.1 INTRODUCTION

Fermentation is the central process of modern biotechnology when practiced on an industrial scale. Successful application of fermentation technology requires large amounts of information. During the course of a fermentation the status of the process is constantly evolving, especially in batch fermentations (which still account for the bulk of all industrial fermentations performed) in which the status is changing from moment to moment. The concentrations of substrates and nutrients are changing due to biological activity and also frequently because these compounds are fed to the fermenter. Biomass is increasing in bulk and, indeed, the composition of the biomass is changing physically, morphologically and biochemically. Various products of biological activity are accumulating. It is to be hoped that desired products are formed primarily, but undesirable products are also present to some degree.

The physical and chemical status of the fermenter is subject to manipulation; such factors as agitation and aeration rates, temperature, pH and so on may be changing, purposefully or otherwise.

In general, the more information we have about a particular fermentation the better. It is true, of course, that in many cases our understanding of the underlying process is so poor that large amounts of information can be confusing. However, diligent and judicious use of such information can only help to increase our understanding in the long run. We will accept, for present purposes, the premise that we would like to have as much information about a given process as is feasible and economically justifiable.

There are two fundamental questions which must be addressed in this context. First, what kind of information can be acquired? Second, what will we do with the information?

Information is generally in the form of numerical data. These data can be acquired by means of measurement devices, sensors, electrodes, analyzers and so on. In some cases, measurements can be made *in situ* and in real time. In other cases, measurements must be made in the laboratory on samples taken from the fermenter. Either way, such measurements constitute estimates of the true status of the variables involved. The variables being estimated are known as state variables, while the estimates from the measurement devices are output variables.

Such output variables can be used, along with other information about the process, to gain what are known as process state variable estimates. In many cases, the state variable estimate is the output variable itself. The agitation rate is an example of such a variable. Other state variables cannot be measured directly and other information must be used to make guesses about the magnitude of such variables. For example, a process model, along with stochastic estimating techniques (that is techniques for treatment of random, noisy information), can be used to make indirect estimates of process state variables.

So the information we can gather consists of numerical data which are estimates, either direct or indirect, of the state variables of a fermentation.

The most important single purpose to which these data are applied is that of process control. A fermentation can be directed to a desirable goal, that is a particular status at a particular point, by using the state estimate data to apply a control strategy. The result of that control strategy is a series of manipulations of control variables. For example, the armature voltage applied to the agitator motor of a fermenter is manipulated to control the agitation rate.

Another application of the state variable estimates is in process modelling and development of a fermentation strategy. Data collected from a fermentation can be analyzed on-line or off-line and used to make decisions about how to run future fermentations. Thus we can develop a data base with respect to a given fermentation or class of fermentations and use that data base to devise an optimum strategy of operation.

Both real-time process control and off-line data base management are tasks which are best performed with the assistance of a computer. The actual measurement of the primary variables is done by local analyzers and measuring devices. Usually control of these variables is also done by local controllers. Frequently these local controllers are actually small, dedicated microprocessors (Bull and Masucci, 1982). The strategy of use of these dedicated controllers can be directed by a supervisory computer, usually a minicomputer. Strategies for computer control of fermentations are the subject of the next chapter in this volume. Methods of accumulation of data bases, as well as their implementation in the overall strategy of operation of fermentation plants, will also not be discussed in this chapter. This chapter will cover the methods used for the measurement of output variables, and application of the resulting data to feedback control loops.

8.2　ESTIMATION OF FERMENTATION OUTPUT VARIABLES

The classic monograph on instrumentation for fermentation is that of Solomons (1969). This work actually was a review of the various materials and devices used in the actual conduct of fermentations, and it is still useful for that purpose. Numerous other reviews and progress reports have appeared since, most recently those of Tannen and Nyiri (1979) and Mor (1982). Table 1 is a list of some of the principal variables which can be measured directly in a fermenter. Table 2 is a selection of some of the more important quantities which can be estimated off-line. Table 3 gives some of the variables which can be calculated in a relatively straightforward way using the results of measurements such as those listed in Tables 1 and 2. These tables are not exhaustive but are representative. Estimates of still other quantities are possible using such techniques as digital filtering and prediction algorithms and applying them according to the theory of stochastic processes. More information about these techniques will be presented below.

It is not possible to give a comprehensive catalog of the myriad devices and sensors which are applied to fermentation processes. The review of Tannen and Nyiri (1979) is a good starting point for those unfamiliar with such equipment. Rather we will examine the criteria which must be satisfied for successful measurement in a biological environment. Then we will review a few of the devices which are most frequently employed and mention some of the new techniques which seem to be coming into common usage.

Table 1   A Selection of Variables which can Usually be Measured 'Directly'

Elapsed fermentation time
Vessel temperature
Temperature, pressure and flow rate of cooling water
Agitation rate
Gross (occasionally net) mechanical power input
Vessel pressure
Inlet air flow rate
Other inlet gas flow rate(s)
Inlet gas temperature(s)
Inlet gas pressure(s)
Exhaust gas composition
Exhaust gas temperature
Exhaust gas flow rate
Dissolved oxygen
pH
Amounts and/or rates of acid and base addition
Redox potential
Amounts and/or rates of addition of redox control reagents
Amounts and/or rates of nutrient addition
Amounts and/or rates of substrate addition
Liquid level, volume, or mass of vessel contents
Foam level or detection

Additional variables which can sometimes be measured usefully

   Turbidity (optical density) of culture
   Fluorescence
   Dissolved carbon dioxide
   Various ions (especially with specific ion electrodes)
   Sugars or other chemical compounds by electrodes or by on-line wet chemistry analyzers
   Product concentrations
   By-product concentrations
   Apparent broth viscosity

Table 2   Selected Useful Variables which can be Measured
Off-line

Sugar or other substrate concentrations
Protein concentrations
Nitrogen levels
Other nutrients (*e.g.* phosphorus, ammonia, DNA, RNA, *etc.*)
Enzyme activities
ATP, NADH, *etc.*
Product concentrations
By-product and intermediate product concentrations
Biomass (in clear media)
Suspended solids volume (by centrifugation)
Dry weight of unfiltered broth
Dry weight of culture filtrate
Viable cell count or 'active center' count (by plating)
Total cell count (sometimes)
Viscosity of broth

8.2.1   Real-time Measurements

Figure 1 is a schematic representation of a typical measurement system for a fermentation. To make a real-time measurement *in situ* at least the following components must be present.

Table 3 Selected Quantities Which are Commonly
Calculated from Measured Variables

Oxygen uptake rate
Oxygen transfer coefficient
Carbon dioxide evolution rate
Respiratory quotient
Substrate utilization rates and concentrations
Nutrient utilization rates and concentrations
Product evolution rates and concentrations
By-product evolution rates and concentrations
Various yields, conversions and selectivities
Biomass concentration
Biomass growth rate
Specific growth rate
Various yield coefficients
Maintenance coefficients
Balances based on First Law of Thermodynamics, *e.g.*
    Heat balance
    Carbon balance
    Nitrogen balance
    Oxygen balance
Power input
Apparent broth viscosity
Process identification and control parameters, *e.g.*
    Kinetic rate constants
    Dynamic process response times
    Process model parameters
    Controller set points and gains
    Optimal process paths

(1) A primary sensing device or detector. There are three main classes: sensors which penetrate into the interior of the fermentation vessel, sensors which operate on samples which are continuously drawn from the fermenter, and sensors which do not come in contact with the fermentation fluid. Examples of the first class are pH and redox electrodes, dissolved oxygen electrodes, pressure transducers, level sensors, temperature sensors, and so on. Such devices, in addition to the usual requirement of being suitable for the primary sensing tasks, must also be designed for aseptic operation. Thus an adequate barrier between the interior and the exterior of the fermenter must be provided to prevent contamination of the fermentation. A number of schemes are feasible to accomplish this, but must modern, well designed devices employ 'O' rings of a suitable elastomer. Occasionally double 'O' rings are used and in extreme cases steam may be passed into a channel between the 'O' rings. Such schemes, however, are not intended to prevent contamination but rather are used to prevent escape of material from the fermenter into the ambient space. Schmidli (1982) and Schmidli and Swartz (1982) have provided excellent reviews of the state of the art with regard to aseptic operation of fermentation systems. An example of the second class of sensor, which operates on a continuous sample, might be a gas analyzer for analysis of the effluent gas from the fermenter. The third class of sensor is typified by devices which measure agitator shaft rotation rates, or temperature sensors which are used to detect the jacket cooling water temperature.

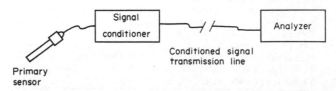

Figure 1 Elements of a measurement system. On rare occasions the primary sensor is connected directly to the analyzer

(2) A device for conditioning the signal from the sensor or transducer; this is generally an electronic amplifier. The signal produced by the primary sensor is usually a current or a voltage. Occasionally pneumatic signals are still used and there are certain instances, such as in measurement of agitation rates, where a digital signal is produced. Primary measuring devices often have

very high impedance, consequently the current or voltage produced is extremely small in magnitude. Moreover, such small signals may be noisy or otherwise subject to external disturbances. Careful design of the sensing system is necessary to eliminate ground loops and to minimize common mode disturbances (Williams, 1980). In principle if the signal from the transducer is of a reasonable quality, then that signal can be transmitted directly to the analyzer. In practice it is preferable to condition all signals in such a way as to ensure high accuracy, low noise, and good rejection of common mode voltages. Thus signal conditioning amplifiers should be located reasonably close to the transducer, and conditioned, high level signals transmitted to the analyzers. Noise suppression is also very important. Local considerations will determine whether such noise suppression should be done before or after signal conditioning and transmission. Most signals of interest to fermentation engineers can be treated by means of simple low-pass passive filters, although other filters including active filters are sometimes desirable. When a computer is used for signal analysis digital filters can be used (Oppenheim and Schafer, 1975; Franklin and Powell, 1980).

(3) A means of transmitting the conditioned signal to the analyzer. Voltage signals can be used over short distances. Cable capacitance and other factors usually lead to unacceptable signal deterioration if the cable length is more than about 100 m. Current signals are preferred over longer distances, but producing such signals is somewhat more complex. A third method is to frequency encode the signal. This requires voltage or current to frequency converters at the measuring point and frequency to voltage or current converters at the receiving point. A prudent practice is to employ optical coupling of the transmitted signal to the analyzer, thus further ensuring a high quality noise free signal at the analyzer.

(4) Finally there must be an analyzer proper which converts the received signal to a form which is recognizable to the operator. Thus, for example, a pH electrode produces a voltage of about 0 V at pH 7 and a voltage change of about 59 mV for a change of one pH unit. The analyzer must convert this voltage signal into a form which can be recognized in pH units and it must compensate for differences in response among different electrodes, and for other factors such as temperature changes.

These then are the minimal requirements for on-line measurement of fermentation variables. Under certain conditions, as for example when there are many similar measurements at a remote location, multiplexing of signals may be desirable. In general there are two methods of multiplexing in common use. One method is to pass each signal through a signal conditioning amplifier and then multiplex with solid state relays. High level signals are then transmitted sequentially to a single analyzer. Problems with relay noise are avoided by this method, but there is the added expense of an amplifier for each signal. An alternative method is to multiplex the low level signals and use only a single conditioning amplifier. A drawback to this technique is that many low level signals are unacceptably degraded in the switching process. Also high impedance sensors such as pH electrodes are characterized by long settling times after switching, so multiplexing rates are rather low. In any case, multiplexing is a technique which is most frequently used in conjunction with computers. Data acquisition by computer is now cost effective and any user should seriously consider such methods when planning the instrumentation of a fermentation plant.

8.3 BASIC VARIABLES FOR FERMENTATION MONITORING

There are a small number of variables which taken together constitute what might be considered a basic set for monitoring a fermentation. Additional variables can also be measured according to the needs of a particular situation. Such a minimal set is itself variable according to the type of fermenter used. For the most common application, the conventional aerobic submerged culture stirred tank bioreactor, the minimal set is the temperature, the agitation rate, the aeration rate, the dissolved oxygen activity, the pressure in the head space of the tank, and the pH. Some practitioners might also include the oxygen and carbon dioxide partial pressures in the exhaust gas. Other variables can be, and often are, measured, but this group is probably the minimum which can be used for meaningful monitoring of a fermentation. In special cases the minimal group can be reduced in scope, but the set is widely accepted in modern practice. In certain cases there is a tendency for foaming to occur and a foam detector is then also required. All of the variables in the basic group are also subject to control by various means. Basic control concepts will be surveyed briefly below. In this section we will be concerned only with measurement systems.

8.3.1 Temperature

The temperature of the culture fluid in a fermentation tank can be measured in several ways. Thermocouples are the cheapest and simplest devices, but they do not have good resolution, and a cold junction (or cold junction compensation) must be provided. Thermistors are also inexpensive and are highly sensitive, but they are also highly non-linear. Miniature transistorized integrated circuits are available which have many of the features of thermistors but which display better linearity. The device with the best linearity is the platinum resistance temperature detector or RTD. A current must be applied to the RTD in order to detect changes in resistance with temperature. The principal difficulty in the use of the RTD is that the resistance of the lead wires to the RTD can be significant compared to the resistance of the RTD itself. Changes in ambient temperature can contribute significant errors to the temperature measurement due to changes in the resistance of the lead wires, which are part of the measurement bridge. Proper design of the bridge circuit, including use of compensating loops in the lead wires will obviate these problems. RTD devices, although more costly than some other sensors, are probably best for critical applications.

8.3.2 Agitation

There are also several ways to measure the rate of agitator shaft rotation. It should be noted that most large fermenter agitators are single speed or two speed devices. However, monitoring of that rate is still desirable. Modern frequency controllers for control of the speed of AC motors now make variable speed drives practical even for relatively large fermenters. Virtually all mechanically agitated smaller fermenters have variable speed drives. The agitation rate is usually measured either by means of magnetic or optical sensors or by some sort of tachometer–generator. The magnetic and optical sensors detect the rate of rotation of a disc mounted to the shaft of either the agitator or the drive motor. The disc is slotted either mechanically or optically, and the sensors produce an electrical pulse each time a slot passes the detector. A frequency counter can then be used to determine the rotation rate. Tachometer–generators are small generators mounted on the agitator or drive motor. The output voltage of the generator is a linear function of the rotation rate.

8.3.3. Aeration

The aeration rate can be measured by means of traditional devices such as rotameters, but for on-line data acquisition, devices which produce an electrical signal are required. For air or gas flow rates less than about $400 \, l \, min^{-1}$ thermal mass flowmeters are most frequently employed. These instruments have the advantage of being independent of temperature and pressure over a fairly wide range of operating conditions. Since they are based on the principle of the hot wire anemometer, the response time is relatively slow. For larger gas flow rates, measurement of the pressure drop across some sort of restriction such as an orifice plate or a nozzle is usually preferred. The flow rate through these devices is usually proportional to the square root of the pressure drop, so the useful range is only about ten to one.

8.3.4 Dissolved Oxygen

Many schemes have been used over the years to measure dissolved oxygen in a fermentation fluid, but some adaptation of the Clark type electrode is still almost universal. There are two principal designs: a polarographic electrode for which a polarization voltage must be applied and the galvanic electrode in which reduction of oxygen occurs at the cathode spontaneously. A membrane which is permeable to oxygen but not to most other species which would interfere with the reaction separates the fermentation fluid from the electrode. Oxygen diffuses across the membrane and a current is produced between the cathode and the anode. The current is proportional to the activity of oxygen in the fermentation broth, not to concentration. Polarographic electrodes tend to produce smaller currents than galvanic electrodes, but often have faster response times. Many technical difficulties exist with these electrodes and it is beyond the scope of this chapter to discuss them. Lee and Tsao (1979) have reviewed the use of such electrodes for fermentation measurements.

8.3.5 Pressure

Pressure can be measured by means of Bourdon tube gauges, but transducers are used for data acquisition and monitoring purposes. A number of such devices are suitable for fermentation purposes. Chief among the requirements are resistance to temperatures encountered during the sterilization process, sensitivity to relatively small pressure changes, and a clean, thread-free design for aseptic operation. For small fermenters physical size of the transducer is also important.

8.3.6 pH

Measurement of pH in a fermenter is somewhat problematical. Conventional glass membrane combination electrodes are available from several manufacturers, which are capable of withstanding repeated steam sterilization. The chief problems with pH electrodes in fermentation applications are fouling of the liquid junction due to the nature of the fermentation fluid and the difficulties encountered in calibrating the electrodes during the course of the fermentation. Special electrode holders are available which allow the electrode to be removed aseptically from the fermenter, recalibrated and then reinserted aseptically back into the fermenter. The impedance of glass membrane pH electrodes is quite high so circuits, amplifiers, cables and so on must be carefully designed to avoid ground loops.

8.3.7 Exhaust Gases

Oxygen is a gas which exhibits paramagnetism and this property is exploited to measure the concentration of oxygen in the effluent gas from the fermenter. Interferences are usually minor since few other gases are appreciably paramagnetic. This technique requires that a continuous sample stream be diverted from the exhaust gas through the analyzer. The sample stream must be filtered and dried before entering the analyzer. It is also possible to insert an oxygen electrode essentially identical to a dissolved oxygen electrode directly into the exhaust gas stream. Such installations, while conceptually simple, require frequent maintenance. Carbon dioxide in the exhaust gas can be measured by passing a sample stream through a non-dispersive IR analyzer. A more elegant (and more expensive) method of measurement of both oxygen and carbon dioxide, as well as other gases and vapors in the exhaust stream, is by means of a mass spectrometer. Pungor *et al.* (1980, 1981) describe the use of a computer coupled mass spectrometer for continuous on-line monitoring of effluent gas. Another configuration was reported by Heinzle *et al.* (1981). Such systems can be multiplexed to analyze the effluent from several fermenters sequentially (Bull, 1978; Buckland and Fastert, 1982). With careful design such systems can analyze gas streams at the rate of one fermenter every 4 or 5 s or faster.

8.3.8 Other Variables

Measurement of the variables enumerated so far constitutes a basic system for simple monitoring of an aerobic fermentation. Many other variables can be measured as outlined by Tannen and Nyiri (1979) and by Mor (1982). The redox potential is sometimes measured. The technique is essentially the same as that used for pH. In an aerobic fermentation the redox potential is intimately related to the dissolved oxygen level. The amounts and rates of addition of various nutrients and substrates are also frequently important. Such measurements are normally made by determining the change of weight of the vessel holding the nutrient solution by means of a load cell, or by detecting the change in liquid level of the nutrient by a device such as a capacitance probe, or by measuring changes in the hydraulic head of the nutrient solution by means of a differential pressure cell.

It is generally desirable to know the level of mechanical power input to the fluid in the fermenter. Many schemes have been used in the past, but for the most part only the gross power applied by the agitator motor is measured. Strain gauges mounted on the impeller shaft require special shaft construction, and the electrical connections used cause problems with contamination. Torsion dynamometers can be used but they are expensive and bulky. The simplest procedure is to measure the power delivered by the agitator motor by means of a transducer such as a Hall effect wattmeter.

Electrodes are now available which can be used to estimate the level of dissolved carbon dioxide in the fermenter (Puhar *et al.*, 1980; Shoda and Ishikawa, 1981). These electrodes consist of a pH electrode immersed in a bicarbonate buffer solution in an electrode holder. A membrane which is permeable to carbon dioxide provides a barrier between the bicarbonate solution and the fermentation broth. Carbon dioxide permeates the membrane and alters the pH of the electrode buffer. At equilibrium, the pH of the bicarbonate buffer is a measure of the activity of carbon dioxide in the fermentation broth. As is the case with dissolved oxygen electrodes, the carbon dioxide electrode measures activity rather than concentration. The solubility must be known to determine the concentration.

Most recent work to extend the range of useful measurements which can be made on-line in a fermenter have dealt in some way with measurements of the chemical environment in the fermenter (Bull, 1983). Lack of specific sensors is currently one of the most important limitations in the development of effective fermentation control systems (Spriet and vanSteenkiste, 1978; Blachere *et al.*, 1978). Recent reports have centered on three areas: electrodes with immobilized enzymes, electrodes with immobilized organisms and on-line analysis of sample streams by means of automatic analyzers (Enfors and Molin, 1978; Sakato *et al.*, 1981; Rechnitz, 1981). The enzyme thermistor is an interesting device for measurement of sugars and other substrates in fermenters. Danielsson and Mosbach (1982) have reviewed the status of these devices. Roughly 40 other enzyme electrodes have been described in the literature (Mor, 1982; Bull, 1983). Electrodes which make use of immobilized microbes are the subject of much work. The subject was reviewed by Suzuki and Karube (1981). Bull (1983) gives a few more recent references. In principle any substance which can be measured by assay of a grab sample can be analyzed by means of a continuous sample stream connected to a wet chemistry analyzer (Mor, 1982). Liquid chromatography and ion exchange chromatography are very promising in this respect (Swartz, 1982).

Other methods of sensing conditions in a fermenter are also under development. Beyeler *et al.* (1981) described the use of culture fluorescence, while Kell (1980) and Biryukov *et al.* (1981) reported on ion selective electrodes for on-line fermentation measurements.

To date, very few methods, including those mentioned here, can supply information on the environment inside cells. Real-time measurements are not yet possible, although on-line determinations can be made in certain special cases. Thus solvent extraction, cell disruption, or other such methods can be used in conjunction with automatic wet chemistry analyzers to make on-line analyses of cellular components in a sample stream.

8.4 STATE ESTIMATION BY INDIRECT METHODS

We have discussed a number of variables which can be measured by means of various methods of chemical or biochemical analysis or by several physical measurements. The number of state variables which can be directly measured in such ways is still quite small compared to the ensemble of desirable measurements. A methodology is emerging to deduce the values of state variables indirectly from other state variables by use of models. In fact, such indirectly measured variables have even been used to control fermentations (Mattiasson *et al.*, 1982). The technique is well developed for the chemical process industries (Ray, 1981), but is not yet in general use in the fermentation industry. Yamashita *et al.* (1969) first proposed the technique for estimation of biomass by means of material balances. The technique was limited by noise in the process from run to run. Stephanopoulos and San (1981) applied filtering methods to compensate for stochastic components of the measurement signals. Others have used different filters (Moilanen, 1980; Howell and Jones, 1982). Detailed explanation of such methods is beyond the scope of the present work. Ray (1981) gives a particularly clear introduction to the subject. It is necessary to build up a model of the system under investigation using a data base accumulated as the fermentation is repeated. From the data base an estimate of the intensity of the random noise components can be achieved. These correlations can then be used to derive estimates of process variables in real time using a dedicated computer.

8.5 FERMENTATION PROCESS CONTROL

8.5.1 A Feedback Control Loop

The data which are gathered by means of instrumentation such as that under discussion can be used in a more immediate fashion to control some of the state variables. This is done in most cases by means of the concept of feedback control. A simple scheme for feedback control is shown schematically in Figure 2. It should be recognized that the monitoring system of Figure 1 is contained in the part of the figure enclosed in dotted lines. The process represented by the central box in the figure is evolving in some fashion and the control system is designed to force that evolution to follow some specific path. The process itself is subject to random disturbances. In the terminology of modern process control, the variable $x(t)$ is one of many state variables and is used to gain a value for the output variable $y(t)$. The value of the output variable is 'fed back' to a control system. The controller compares the value of the output variable to a desired value, the set point, and computes an error. The error is then inserted into an algorithm and a control action is decided upon according to that algorithm. The control action is represented as a control variable, $u(t)$. Such a scheme constitutes a simple feedback control loop. The great majority of all schemes used to control fermentation processes are based on this principle. A few control strategies are based on 'feedforward' control, whereby a process model is used to predict what will happen based on the measurement of disturbances to the process, and control action initiated as a result. This requires that all disturbances be known precisely, and that the process model be complete and highly accurate. Such information is virtually never achievable, so even feedforward control is normally combined with feedback control action.

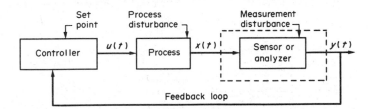

Figure 2 A simple feedback process control loop. The block within the dotted lines is equivalent to the system of Figure 1: $x(t)$=process state, $y(t)$=measured quantity, $u(t)$=control signal

8.5.2 Final Control Elements

The control variable $u(t)$ of Figure 2 is applied to the process by means of a final control element. In the case of the agitator the controller signal directs the alteration of the armature voltage and current. In the case of pH control the final control element is frequently a pump for acid or base. Perhaps the most frequently used final control element is a pneumatically operated valve. Such valves are used for regulation of air flow, vessel pressure, heating and cooling water flow for temperature control, and for acid and base addition, nutrient addition, substrate addition and antifoam addition. Control valves of this type are opened or closed by modulating the pressure of instrument air supplied to a membrane on the valve. Movement of the membrane causes movement of the valve stem. Special precautions are necessary to ensure aseptic operation of such control elements (Schmidli, 1982; Schmidli and Swartz, 1982). Piping configurations must be carefully designed for similar reasons (Bull *et al.*, 1983).

8.5.3 Basic Control System Concepts

Although the basic idea of feedback control is quite simple, the application can be very complex. The algorithm executed by the controller can range from very simple on/off control action to intricate strategies which take into account interactions among many of the variables. Often there are control loops nested within other control loops, such as in cascade control, and so on. A convenient survey of the principal control modes encountered in process control was presented by Rinard (1982). Thorough treatment of the theory and application of modern process control is the subject of numerous fine texts such as those of Douglas (1972a, 1972b) and Ray (1981). We will review here some of the basic principles as applied to fermentations (Bull and Masucci, 1982).

8.5.3.1   On/off control

In certain cases the control action can be applied manually. Control of pH is one example. An operator observes the pH of the process, and adds either acid or base to bring about the desired value. Dissolved oxygen can be controlled roughly by manual adjustment of the agitation rate, the pressure, or some other appropriate variable. Generally such control is not very effective, and automatic control is preferred.

The simplest automatic control is known as on/off control. Figure 3 is a schematic representation of a basic fermenter temperature control system. Such a system can be operated using on/off control with fairly good results. The temperature controller senses the temperature in the fermenter by means of a sensor, as outlined in Section 8.2.1. If the temperature is too low, the steam or hot water valve is opened completely. If the temperature is too high, the cooling water supply valve is opened completely. The control valves are either completely open or completely closed, hence the term on/off control. This type of control system is suitable when the heating or cooling loads are relatively constant. Most batch fermentations, on the other hand, are exothermic and autocatalytic, that is they produce heat and the amount of heat increases as the fermentation progresses. If the cooling water flow rate is adjusted to give good regulation when the heat load is low, then the fermenter will tend to run too hot when the heat load is high. If adjusted for good regulation at high heat loads, the temperature will oscillate at low heat loads, or there will be an offset, depending on the system design.

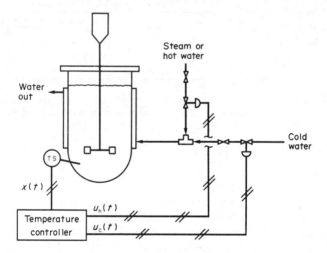

Figure 3   A simple temperature control system. TS=temperature sensor, $\#$ = pneumatic or electric control line, $x(t)$=process state, $u_h(t)$=heating control signal, $u_c(t)$=cooling control signal. The example is highly simplified compared to an actual control system

8.5.3.2   PID control

These problems can be overcome if the control valves are modulated so that the flow rates are automatically adjusted to compensate for changes in load. Classical controllers employ some sort of PID control algorithm to make the error signal (the difference between the desired temperature and the actual temperature) as small as possible. Three modes of control are used. First, the opening of the valve can be made a (usually) linear function of the error. The control signal is proportional to the magnitude of the error:

$$u(t) = K[y(t) - S] = Ke \qquad (1)$$

Here S is the set point, that is the desired value, of the output variable $y(t)$. K is the controller gain, and *e* is the error. In older terminology $1/K$ in per cent is known as the proportional band. Unfortunately, unless the gain is itself a function of the error, pure proportional action will result in offset, that is it will not correct for changes in load. In principle, the amount of offset can be made arbitrarily small by adjusting the gain, but if the gain is too large the system will be unstable and the controlled variable will oscillate unacceptably.

A second control action, integral control, is added to the corrective action to reduce offset. The additional control action is proportional to the integral of the error with respect to time. Such an action will reduce the amount of offset automatically:

$$u(t) = Ke + I\!\int e\,d\theta \tag{2}$$

$$= Ke + K/\tau\!\int e\,d\theta \tag{3}$$

In classical terminology τ is the integral time, and I is the reset rate. Integral control action can be used to eliminate or reduce offset without the need for excessive gain. However, the lower gain will tend to increase the response time of the system. If the response becomes too slow then a third control action proportional to the rate of change of the error can be added. Such derivative control is not very often needed in fermentation control systems, but it is occasionally useful. The term PID control stems from the names of the terms in the control algorithm. Some control strategies depend almost entirely on integral control, for example DC motor control.

8.5.3.3 *Complex PID control*

PID control is the underlying concept of most modern process control schemes. The actual algorithm employed may be quite complex, but can usually be decomposed into a series of PID terms, with various modifiers to take account of special features of the process dynamics. For example, factors such as 'Smith predictors' can be added to the PID algorithm to compensate for time delays in the control system. Interested readers should consult the text of Ray (1981) for further details.

Even in digital process control schemes, the usual practice for fermentation systems is simply to employ a digitized version of the PID algorithm. The exception is when the sampling rate of the output variable by the digital controller is of the same order of magnitude as the changes in the output variable or less. This is not a normal event in a well designed system since fermentation rates are slow compared to the sampling rates possible with modern microprocessors. When such a problem is encountered, then special digital control algorithms, usually based on the *z* transform, and derived from the theory of complex variables, are used (Franklin and Powell, 1981).

8.5.3.4 *Cascaded feedback control*

Feedback control in its various forms can be used for the control of many variables in fermentation processing. Cascade control is particularly important in several situations. Control of dissolved oxygen is a useful example. There are numerous ways to control the dissolved oxygen activity in a fermentation broth, including use of pH, temperature, nutrient addition rates and so on. The most commonly used schemes manipulate the agitation rate, the air or gas flow rate, the total fermenter pressure, or the partial pressure of oxygen in the inlet gas to control the dissolved oxygen. Combinations of these control variables are also used, either simultaneously or sequentially. In general the oxygen uptake rate can be expressed as

$$q_{O_2} + dC/dt = k_L a\Delta C \tag{4}$$

In many cases dC/dt is small compared to q_{O_2} and can be regarded as zero. Oxygen uptake is proportional to the interfacial area between the gas bubbles and the fermentation liquid, and to the driving force, which is a concentration difference. The interfacial area is affected primarily by the gas flow rate and the agitation rate. The constant of proportionality or oxygen transfer coefficient k_L is also influenced, but to a relatively limited extent. The driving force can be altered by changing the oxygen partial pressure and the vessel pressure. If the biological rate of oxygen utilization is large then the dissolved oxygen level can fall to zero, and the rate of oxygen uptake can become the rate limiting step for the fermentation, hence the need for dissolved oxygen control. This can be efficiently accomplished through cascade control.

An example of cascade control is shown in Figure 4. Dissolved oxygen is detected by a sensor in the fermenter which produces a current proportional to the dissolved oxygen activity. The dissolved oxygen controller, known as the primary controller, computes a control action $u(t)$ by means of a PID algorithm. Now, instead of using that control signal to manipulate the agitation rate, for example, directly, the control signal is received as a set point by a secondary, or slave controller. The slave controller computes another control output, using a second PID algorithm, which is finally used to control, say, the agitation rate. The PI control algorithm for the primary controller might be:

$$u_1(t) = K_1[y_1(t) - S_1] + K_1/\tau_1 \int (y_1 - S_1)\,d\theta \tag{5}$$

The PI control algorithm for the secondary (agitation) controller is then:

$$u_2(t) = K_2[y_2(t) - u_1(t)] + K_2/\tau_2 \int (y_2 - u_1)\,d\theta \tag{6}$$

The measured dissolved oxygen concentration is $y_1(t)$, the dissolved oxygen set point is S_1. The output of the primary controller $u_1(t)$ becomes the set point of the secondary controller in equation (6). The actual control signal for the agitation rate is $u_2(t)$. K_1 and K_2 and τ_1 and τ_2 are the gains and integral times for the primary and secondary controllers, respectively. The overall control action can be expressed in a single equation by substituting for $u_1(t)$ in equation (6):

$$\begin{aligned}
u_2(t) = {} & K_2 y_2(t) - K_1 K_2[y_1(t) - S_1] + K_2/\tau_2 \int y_2\,d\theta \\
& + K_1 K_2 (\tau_1 + \tau_2)/(\tau_1 \tau_2)\,(y_1 - S_1)\,d\theta \\
& - K_1 K_2/\tau_1 \tau_2 \iint (y_1 - S_1)\,d\phi d\theta
\end{aligned} \tag{7}$$

Such a cascaded control system has long been recognized as resulting in a faster response and better control over a wider range of conditions (Douglas, 1972b). When microprocessors or other computing devices are used, then it is a straightforward matter to set up a hierarchy of control, for example first the agitation rate might be varied as the secondary controlled variable. When some predetermined maximum agitation rate is reached, then the air flow could be varied, then the pressure. With the help of the computer quite general protocols or profiles of control action can be set up. Microprocessor controllers are available commercially which can accomplish such control action easily.

8.5.4 Supervisory Computer Control

Decisions about the optimum control action to pursue can be reached by the use of supervisory computers, although direct digital control by a large computer is still sometimes practiced. In Figure 5 a schematic representation is shown of how such supervisory control might be carried out. The function of the computer from a control aspect is to instruct the individual controllers as to the 'best' way to accomplish the process objective. Detailed consideration of these matters is not appropriate here, but in general the computer can estimate values for the state variables, as in Section 8.4, identify the process dynamics by model building from the raw data, and optimize the set points for the controllers according to the process model as applied to a process objective. This is in addition to the conventional computer applications such as report generation, graphics, and so on.

8.6 FERMENTATION PROCESS DYNAMICS

The last subject to be treated here is that of fermentation process dynamics. The subject is quite complex and, moreover, is not well developed at all in the field of fermentation. Guy (1981) presented a series of articles which constitute a good practical introduction to chemical process dynamics. The material can be extended to fermentation processes and a literature is just beginning to develop in this regard. Ray (1981) deals with the subject in a more academic way, but still in a manner useful to a practicing engineer. For present purposes, an attempt will be made to show how control system design is affected by considerations of fermentation process dynamics.

Methods of achieving good fermentation process regulation have been outlined, but these methods, to be effective, must be related to an understanding, however empirical or qualitative, of the process to be controlled. Ideally the understanding is expressed in a process model. Equation (3) or equivalent expressions for the controller output do not predict what the result will be

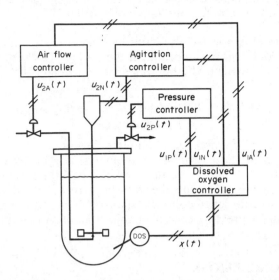

Figure 4 A cascaded dissolved oxygen control system. Three cascades are shown, each of which is independent of the others. DOS=dissolved oxygen sensor, $x(t)$=process state, $u_{1P}(t)$, $u_{1N}(t)$, $u_{1A}(t)$=primary control signals for pressure, agitation and aeration, $u_{2P}(t)$, $u_{2N}(t)$, $u_{2A}(t)$=secondary (or slave) control signals for pressure, agitation and aeration, respectively.

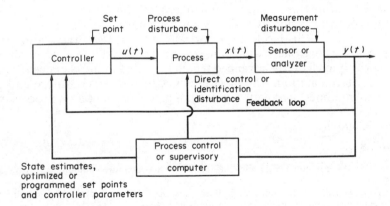

Figure 5 One type of computerized process control system. Notice that the computer can control the process directly (DDC or Direct Digital Control), it can intervene in the operation of the feedback control loop (Supervisory Control), or it can allow feedback control to proceed undisturbed (Data Acquisition Mode). If the controllers themselves are comput-ing devices (*e.g.* microprocessor based controllers) then the operation is sometimes referred to as direct digital control, even without a process control computer

of the controller action. For example, in classical control theory, a model of a single control loop such as that shown in Figure 2 might be as follows:

$$y(s) = g(s)u(s) + g_d(s)d(s) \tag{8}$$

where $y(s)$ is the Laplace transform of the output variable $y(t)$, and $u(s)$ and $d(s)$ are the corre-sponding Laplace transforms of the control action and the process disturbances, respectively. The relationship is embodied in the transfer functions $g(s)$ and $g_d(s)$. Such a model can also be expressed in a time domain model. For the simplest linear case:

$$dx(t)/dt = Ax(t) + Bu(t) + Dd(t) \qquad x(0) = x_0 \tag{9}$$

$$y(t) = Cx(t) \tag{10}$$

The use of ordinary differential equations implies that we are dealing with 'lumped parameter' models, that is the status of any variable is independent of position in the reactor. For a stirred tank reactor this is usually correct for control purposes. For a plug flow fermenter this would not be true, and a 'distributed parameter' model, with partial differential equations would be required.

In reality the linear model is not usually descriptive and it is much more useful to employ generalized non-linear models. More important, if the model is extended to a system of equations, then the interactions between the variables can be dealt with. Such a model can be expressed as

$$dx(t)/dt = f(x(t),u(t),d(t)) \qquad x(0) = x_0 \qquad (11)$$
$$y(t) = h(x(t),u(t)) \qquad (12)$$

where vector quantities are now employed. Complete coverage of this subject is not possible here. However, most fermentation processes are highly non-linear (witness the venerable example of the Monod equation), and so are the responses of the associated control systems. For example, the response of a pressure controller will be highly dependent on the air flow rate through the fermenter. A well designed control system would account for this sort of interaction by providing for changes in the pressure controller gain as a function of the air flow rate.

8.7 SUMMARY AND CONCLUSIONS

Complete coverage of the field of instrumentation for fermentation process control is not possible in this short format. Entire subjects have not even been mentioned, for example automatic sterilization of fermenters and other uses for automatic sequencing of valves or other control action. What has been attempted is to outline the basic elements of a good measurement and control system, with special reference to the unique aspects of fermentations, and to provide an introduction to the advanced control concepts which will be employed in the future as fermentation process control matures into a modern engineering technology.

The present status of this technology is somewhat behind the status of automatic control in the chemical process industries. This has been historically true for the fermentation industry. New, advanced microprocessor controllers which are designed especially for fermentation control are now available. In principle all of the advanced techniques used in the chemical process industries are now available for fermentation. The most pressing need now is for more and better sensors and on-line analytical methods, and better understanding of fermentation process dynamics.

8.8 REFERENCES

Beyeler, W., A. Einsele and A. Fiechter (1981). On-line measurements of culture fluorescence. *Eur. J. Appl. Microbiol. Biotechnol.*, **13**, 10–14.

Biryukov, V. V., S. B. Itsygin, V. M. Kantere, V. F. Bashutkin, A. L. Zharov and P. V. Choban (1981). Use of ion selective electrodes in culture fluids of antibiotic producers. *Khim.-Farm. Zh.*, **15**, 71–75.

Blachere, H. T., P. Peringer and A. Cheruy (1978). New developments in instrumentation and control in fermentation. *Dechema Monograph*, vol. 82, pp. 65–87. Verlag Chemie, Weinheim.

Buckland, R. C. and H. Fastert (1982). Analysis of fermentation exhaust gas using a mass spectrometer. In *Computer Applications in Fermentation Technology*, pp. 119–126. Society of Chemistry in Industry, London.

Bull, D. N. (1978). Rapid, low-cost data acquisition from multiple fermentors, using a desk top computer. Presented at ASM Annual Meeting, Las Vegas, NV.

Bull, D. N. (1983). Automation and optimization of fermentation processes. In *Annual Reports on Fermentation Processes*, ed. G. T. Tsao, vol. 6, pp. 359–375. Academic, New York.

Bull, D. N. and C. Masucci (1982). Current trends in fermentation process control. *J. Paren. Sci. Technol.*, **36** (2), 64–71.

Bull, D. N., R. W. Thomas and T. E. Stinnett (1983). Bioreactors for submerged culture. In *Advances in Biotechnological Processes*, ed. A. Mizrahi and A. L. van Wezel, vol. 1, pp. 1–30. Alan R. Liss, New York.

Danielsson, B. and K. Mosbach (1982). The prospects for enzyme-coupled probes in fermentation. In *Computer Applications in Fermentation Technology*, pp. 137–145. Society of Chemistry in Industry, London.

Douglas, J. M. (1972a). *Process Dynamics and Control*, vol. 1. Prentice–Hall, Englewood Cliffs, NJ.

Douglas, J. M. (1972b). *Process Dynamics and Control*, vol. 2. Prentice–Hall, Englewood Cliffs, NJ.

Enfors, S.-O. and N. Molin (1978). Enzyme electrodes for fermentation control. *Process Biochem.*, **13** (2), 9–11.

Franklin, G. F. and J. D. Powell (1980). *Digital Control of Dynamic Systems*. Addison–Wesley, Reading, MA.

Guy, J. L. (1981). Fundamentals of chemical process dynamics. *Chem. Eng. (London)*, **88** (13), 74–80, *et seq.*

Heinzle, E., O. Bolzern, I. J. Dunn and J. R. Bourne (1981). A porous membrane-carrier gas measurement system for dissolved gases and volatiles in fermentation systems. In *Advances in Biotechnology*, ed. M. Moo-Young, C. W. Robinson and C. Vezina, vol. I, pp. 439–444. Pergamon, Toronto.

Howell, J. A. and M. G. Jones (1982). Problems in on-line parameter estimation for a structured model. In *Computer Applications in Fermentation Technology*, pp. 57–65. Society of Chemistry in Industry, London.

Kell, D. B. (1980), The role of ion-selective electrodes in improving fermentation yields. *Process Biochem.*, **15** (1), 18–23.

Lee, Y. H. and G. T. Tsao (1979). Dissolved oxygen electrodes. In *Advances in Biochemical Engineering*, ed. T. K. Ghose, A. Fiechter and N. Blakebrough, vol. 13, pp. 35–86. Springer-Verlag, Berlin.

Mattiasson, B., C.-F. Mandenius, J.-P. Axelsson, B. Danielsson and P. Hagander (1982). Computer control of fermentation using biosensors. Presented at Biochemical Engineering III, Engineering Foundation Conference, Santa Barbara, CA.

Moilanen, U. (1980). State estimation of a fermentation process. In *Proceedings of the 19th IEEE Conference on Decisions Control*, pp. 396–398. IEEE, Piscataway, NJ.

Mor, J.-R. (1982). A review of instrumental analysis in fermentation technology. In *Computer Applications in Fermentation Technology*, pp. 109–118. Society of Chemistry in Industry, London.

Oppenheim, A. V. and R. W. Schafer (1975). *Digital Signal Processing*. Prentice–Hall, Englewood Cliffs, NJ.

Puhar, E., A. Einsele, H. Buehler and W. Ingold (1980). Steam-sterilizable pCO_2 electrode. *Biotechnol. Bioeng.*, **22**, 2411–2416.

Pungor, E., Jr., C. R. Perley, C. L. Cooney and J. C. Weaver (1980). Continuous monitoring of fermentation outlet gas using a computer coupled MS. *Biotechnol. Lett.*, **2**, 409–414.

Pungor, E., Jr., E. Schaefer, J. C. Weaver and C. L. Cooney (1981). Direct monitoring of a fermentation in a computer–mass spectrometer–fermentor system. In *Advances in Biotechnology*, ed. M. Moo-Young, C. W. Robinson and C. Vezina, vol. I, pp. 393–398. Pergamon, Toronto.

Ray, W. H. (1981). *Advanced Process Control*. McGraw-Hill, New York.

Rechnitz, G. A. (1981). Bioselective membrane electrode probes. *Science*, **214**, 287–291.

Rinard, I. H. (1982). A road map to control-system design. *Chem. Eng. (London)*, **89** (24), 46–58.

Sakato, K., H. Tanaka and H. Samejima (1981). Electrochemical measurements of cell populations. *Ann. N. Y. Acad. Sci.*, **369**, 321–334.

Schmidli, B. L. (1982). Asepsis in fermentation. Lecture notes, SIM Workshop on Fermentation Technology, St. Paul, MN.

Schmidli, B. L. and R. W. Swartz (1982). Design considerations for aseptic pilot fermentation. 184th National Meeting of the ACS, Kansas City, MO.

Shoda, M. and Y. Ishikawa (1981). Carbon dioxide sensor for fermentation systems. *Biotechnol. Bioeng.*, **23**, 461–466.

Solomons, G. L. (1969). *Materials and Methods in Fermentation*. Academic, London.

Spriet, J. A. and G. C. vanSteenkiste (1978). New approach towards measurement and identification for control of fermentation systems. *IMACS Symposium on Simulation of Control Systems*, pp. 245–248. North-Holland, Amsterdam.

Stephanopolous, G. and K.-Y. San (1981). State estimation for computer control of biochemical reactors. In *Advances in Biotechnology*, ed. M. Moo-Young, C. W. Robinson and C. Vezina, vol. I, pp. 399–406. Pergamon, Toronto.

Suzuki, S. and I. Karube (1981). Microbial sensors for fermentation control. In *Advances in Biotechnology*, ed. M. Moo-Young, C. Vezina and K. Singh, vol. III, pp. 355–360. Pergamon, Toronto.

Swartz, R. W. (1982). Analytical methods in fermentation. Lecture notes, SIM Workshop on Fermentation Technology, St. Paul, MN.

Tannen, L. P. and L. K. Nyiri (1979). Instrumentation of fermentation systems. In *Microbial Technology*, ed. H. J. Peppler and D. Perlman, 2nd edn., vol. II, pp. 331–374. Academic, New York.

Williams, T. J. (1980). *Digital Process Control and Mathematical Modeling of Industrial Systems*. Purdue Laboratory for Applied Industrial Control, W. Lafayette, IN.

Yamashita, S., H. Hoshi and T. Inagaki (1969). Automatic control and optimization of fermentation processes: glutamic acid. In *Fermentation Advances*, ed. D. Perlman, pp. 441–465. Academic, New York.

9
Systems for Fermentation Process Control

M. J. ROLF and H. C. LIM
Purdue University, West Lafayette, IN, USA

9.1 INTRODUCTION

Control is an important aspect in the overall system design of fermentation processes. In recent years, more emphasis has been placed in this area partly due to the increasing economic pressure to improve productivity and yield. The fermentation industry has lagged somewhat behind the other process industries in implementing advanced control technology. The same types of control technologies that are used in those industries can also be applied in the fermentation industry. However, there are three particularly difficult problems in controlling fermentation processes and these are due to the fact that a living system is involved. First, there is a lack of appropriate instrumentation, although much work is being done to improve this situation. Next, fermentation processes involve a large number of complex and highly interacting biochemical reactions and transport phenomena. These are difficult to understand, let alone control. Finally, the microorganisms themselves have a complex regulatory system within the cells and the external control system can only manipulate the extracellular environment in hopes of affecting the intracellular mechanisms. Many of the typical and also most difficult control problems found in the other process industries can be found in one particular fermentation process.

9.2 CONTROL TECHNIQUES

The types of control techniques that are significant for fermentation processes are listed in Table 1. This is simply a list of important control techniques which are currently being used or developed for fermentation processes.

Manual control is the simplest type of control as it requires a human operator to manipulate the control element (*e.g.* aeration valve, agitation motor, feed pump, *etc.*). Although it is a very poor regulatory technique, manual control is still important in start-up and shut-down procedures and for backing up classical automatic and computer controllers.

Sequencing operations (*e.g.* sterilization, cleaning, filling, emptying, sampling, *etc.*) have often been handled using mechanical devices such as cam controllers and timers. There has been a

Table 1 Types of Control for Fermentation Processes

Manual Control	Computer control
Classical automatic control	DDC (replacement of classical controllers)
Sequence	SPC (setpoint control)
Analog (PID)	Supervisory
On/off	Feedforward
	Modern

trend to replace these mechanical devices with electronic timer devices, and, more recently, with programmable logic systems and computers. In many cases, significant economic gains can result from improved sequencing. Due to the discrete nature of sequence control, computerization is an excellent method for achieving this. With a computer, the time base for the sequence is more accurate and does not need to be calibrated. The logic, as well as the time base, can also be readily changed within the computer.

Classical control techniques are covered fairly extensively in standard references (Coughanowr and Koppel, 1965; Douglas, 1972; Shinskey, 1979) and will not be defined here. However, a few qualifying remarks are in order. Most controllers use negative feedback (see Figure 1a) in which a measured process output (controlled variable) is subtracted from a desired value (setpoint) to generate an error signal, ε. The controller recognizes the error signal and manipulates a process input (control element) to reduce the error. PID type controllers, which utilize proportional, integral and derivative action or combinations thereof, are the most common type of controller. The manipulated variable, m, which is generated by a PID controller, is given by the following equation:

$$m = K_c\left(\varepsilon + \frac{1}{\tau_I}\int_0^t \varepsilon \mathrm{d}\tau + \tau_D\frac{\mathrm{d}\varepsilon}{\mathrm{d}t}\right) \tag{1}$$

where K_c, τ_I and τ_D are the proportional, integral and derivative constants, respectively. A PID type controller is usually satisfactory for about 80% of all control applications.

A special case of a proportional only controller is the on/off controller. Here, the controller gain, K_c, is infinite, but control action is restricted by the limits of the control element. On/off controllers are usually used in conjunction with on/off control elements (*e.g.* solenoid valves, fixed speed pumps, constant load heaters, *etc.*). This type of control is satisfactory when the variations in the controlled variable are small and change slowly with time. In many fermentation processes, the temperature and pH loops have these characteristics and thus on/off control is adequate, even preferred, due to its simplicity.

Although most recent applications of PID controllers utilize analog electronics or digital computers, there are many pneumatic controllers in operation. This is due to two factors. In the past, pneumatic instrumentation was the mainstay of the process industries. Due to their ruggedness and reliability, many existing installations still have pneumatic controllers. Secondly, pneumatic valves have many desirable characteristics and are still specified for some new installations.

9.2.1 Computer Control

The use of computers for the control of fermentation processes has increased significantly in the last decade due to the realization that the computer can be a vital instrument for process control and optimization. The decrease in cost and improved reliability in recent years have made the use of computers even more attractive. The advent of microprocessors has had an enormous impact in the area of real-time computer applications to fermentation processes. For a fermentation plant, the primary objective of computer control is to produce products as economically as possible. The computer is generally used to provide quality control, save operator time, furnish automatic documentation, and decrease per loop control costs. Computer control has found its way not only in plants, where the economic advantages are more obvious, but also in pilot plants and research laboratories. In the later cases, the computer provides fast and efficient data acquisition, the ability to monitor and control experimental conditions, and flexibility in the operation of the system.

One aspect in which the capabilities of the computer can truly be realized is in the implementation of advanced control and optimization strategies. It was not until the 1970s that much of the

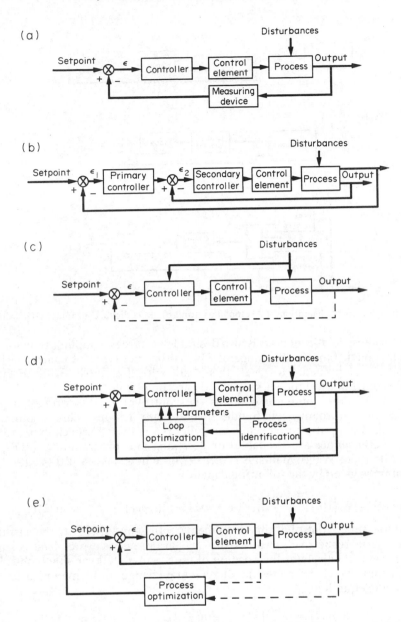

Figure 1 Control techniques: (a) feedback control; (b) cascade control; (c) feedforward control; (d) adaptive control; (e) optimization.

modern control theory developed in the 1950s and 1960s could be utilized practically. This change was sparked primarily by tremendous economic and technological advances in computer hardware, especially with the advent of microcomputers.

Computer control of fermentation has been an extremely active area as evidenced by numerous reviews in the literature (Armiger and Humphrey, 1979; Bull, 1983, Dobry and Jost, 1977; Hampel, 1979; Hatch, 1982; Jefferis, 1975; Rolf and Lim, 1982; Weigand, 1978; Zabriskie, 1979).

Setting the control configuration of the low-level loops is one of the major decisions to be made in computer control. The low-level control loops are those which feed back directly upon one of the basic process measurements, for example when the manipulated variable of acid or base addition is used to control the pH. The decision is whether to use a classical hardware controller with the computer being used to provide the setpoint (SPC; setpoint control), or to replace the hardware controllers with DDC (direct digital control) where the computer is used to implement low-level control algorithms as well (see Figure 2a and 2b).

With SPC the computer is involved only in providing the setpoints for the low-level loops, therefore in the event of computer failure, the low-level loops remain operational. Less reliance

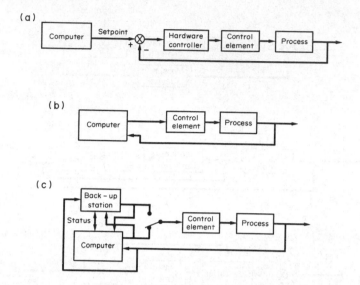

Figure 2 Computer control configurations: (a) setpoint control; (b) direct digital control; (c) DDC with back-up

on the computer for these low-level tasks is also desirable as it frees computer time for other tasks which are more involved. One disadvantage of SPC is that the types of control techniques for hardware controllers are limited. However, this is not critical as most low-level loops can be handled by simple controller techniques such as PID and on/off.

In the case of DDC, the computer contains the controller in the form of a program or subroutine. Often, one subroutine will handle several control loops. This is sometimes called table-driven DDC as the inputs, outputs and controller parameters for each loop are stored in a software table. Several guides are available for designing and programming DDC algorithms (Bristol, 1977; Goff, 1966; Webb; Williams *et al.*, 1973). When classical PID control is approximated, the algorithm is given by the following equation:

$$m[nT] = K_c\left\{\varepsilon[nT] + \frac{T}{\tau_I}\sum_{i=0}^{n}\varepsilon[(n - i)T] + \frac{\tau_D}{T}(\varepsilon[nT] - \varepsilon[(n - 1)T])\right\} \tag{2}$$

where T is the sample time (controller time increment). $m[nT]$ and $\varepsilon[nT]$ are the current manipulated variable and error, respectively, while $\varepsilon[(n - i)T]$ are previous errors. This is known as the position algorithm since it calculates the position of the manipulated variable (control element). The velocity algorithm, which calculates the incremental change in the manipulated variable, is given by the following equation:

$$\Delta m[nT] = K_c\left\{\varepsilon[nT] - \varepsilon[(n - 1)T] + \frac{T}{\tau_I}\varepsilon[nT] + \right.$$
$$\left. \frac{\tau_D}{T}(\varepsilon[nT] - 2\varepsilon[(n - 1)T] + \varepsilon[(n - 2)T])\right\} \tag{3}$$

The velocity algorithm has the advantage that it can be modified to prevent integral overshoot and subsequent oscillation. Additionally, it is convenient when the control element is an incremental or integrating type.

The advantages offered by DDC are listed in Table 2. Due to these advantages, there is a general trend to utilize DDC for fermentation processes. If DDC is used, those low-level control loops which are critical to the operation of the process must have some sort of back-up in case of computer failure (see Figure 2c). Another disadvantage of using DDC for the low-level control loops is that the computer must be available at all times and can not be diverted to some other function while the DDC programs are running. The evolution of microcomputers and hierarchical computer systems has been a tremendous help in this area. Low-level control can be carried out in microcomputers while higher computers accomplish other tasks more efficiently. Also, in multicomputer systems, the burden of computer failure is somewhat relieved as failure of one computer will not cripple the entire process or plant.

The determination of the setpoints of the low-level control loops, whether they involve hardware or DDC controllers, can be done in three ways by the computer: supervisory control,

Table 2 Potential Advantages of Direct Digital Control

Flexibility and versatility
  Control techniques are not limited
  Control algorithms are modified readily
  Controller parameters are adjusted readily
  One control routine for several loops

Reduced hardware requirements
  No need to design, construct, calibrate, or service analog controllers

Improved response in some cases
  Better regulation
  Smoother start-up and shut-down
  Bumpless transfer in switching with manual systems

Automatic documentation of control actions, output responses and manipulated variables

---

cascade control and advanced (modern) control. With supervisory control, the setpoints are determined by computer–operator interaction, predefined values or profiles, or scheduling considerations. As an example consider the bioreactor temperature control loop. In many situations this loop can be handled by an on/off control technique in which the controller has three states: cooling, off and heating. To eliminate oscillations that can occur with this type of double-sided on/off control, the supervisory computer program can decide whether to control with cooling or heating at any particular point in time. It would be convenient to design the low-level controller such that a positive-valued setpoint would cause control by heating and a negative-valued setpoint would cause control by cooling. Consider also a case in which it would be desirable to have the bioreactor temperature change according to a predefined profile. The supervisory program could then determine the magnitude of the setpoint as well as the sign.

Cascade control is used when the feedback loop is less directly associated with the primary variable which is to be controlled. In this case, a secondary (inner) loop is added to regulate disturbances closer to the source (see Figure 1b). Here, the primary (outer) controller manipulates the setpoint of the secondary controller. Typically, dissolved oxygen concentration is controlled in this manner. Figure 3 depicts a cascaded dissolved oxygen controller which is implemented on a hierarchical computer system. The low-level loops involving agitation speed and airflow rate are the secondary controllers and are contained in a microcomputer. A minicomputer contains the primary dissolved oxygen controller which determines the setpoints for the agitation speed and airflow rate controllers. Other secondary loops which can be used for dissolved oxygen control are bioreactor pressure and oxygen concentration in the inlet airflow stream. Several variables have also been cascade controlled using the medium feed rate as the secondary loop. These primary variables include cell, substrate and product concentrations and the respiratory quotient (RQ).

One type of control technique which is particularly amenable to computer control is feed-

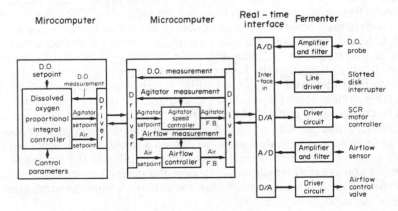

Figure 3 Hierarchical configuration for dissolved oxygen control (Reproduced from Rolf *et al.*, 1982 by permission of Wiley)

forward control (see Figure 1c). With feedforward control, disturbances to the process are measured and the controller takes action to counteract the effect of the disturbances before they reach the process output. This is a formidable task as the disturbances must not only be identified and measured, but their effect on the process output must be known. However, there is one area in which this type of control is applicable to fermentation processes: control loops involving medium feed rates. Feedforward control can be combined with the cascade controllers described previously by measuring the substrate (or other nutrient) concentration in the medium feed. Any disturbances can then be accounted for when controlling the cell concentration, *etc.* A different type of feedforward control, called ratio control, can also be used where the flow rates of two separate feed streams must be held at a desired ratio.

9.2.2 Modern Control

While classical control deals primarily with a process having one input and one output, modern control can handle multiple inputs and outputs. Modern control encompasses a wide variety of techniques (Athans and Falb, 1966; Bell and Griffin, 1969; Bryson and Ho, 1975; Citron, 1969; Ray, 1981) in which the control scheme is optimized in a certain sense. The optimization may be in the response of a particular loop (or set of loops) or it may be in the overall performance of the process. In addition to controller design, modern control encompasses other techniques which are integral parts of controller synthesis. These include overall system optimization, process identification and estimation of variables which are difficult or impossible to measure.

A mathematical model of the process is required before the system can be optimized. A model will describe mathematically how the outputs respond to changes in the input. Thus, if the desired outputs are known, it is possible to determine the necessary inputs to achieve this. The so-called open-loop policy, or control function then results. When the inputs can be expressed in terms of the process outputs instead of a time function, the so-called closed-loop policy or control law results. Since the model relates the inputs with the outputs, knowledge of the inputs and outputs allows one to learn about the process and is the basis of process identification. Models are also used in estimating the state of the process if the state is unmeasurable or corrupted by noise. Dynamic models can be used to simulate the actual process on a computer. This is useful in testing the feasibility and also debugging the control system design.

Multivariable control can be achieved in two ways. One approach is to use optimal control theory which considers the multiplicity of inputs and outputs. The other approach is to eliminate the interaction among individual loops by introducing a decoupler which transforms interacting multiple loops into a number or non-interacting single loops. Then any of the well known techniques for single input–single output systems can be applied. Compensation for dead-time (time delay) in loops can be incorporated into these techniques, but is frequently handled with special dead-time compensators (DTC) (Ray, 1981). Multivariable control is typically used to control the cell and substrate concentrations in a bioreactor by manipulating the medium feed rate and feed substrate concentration. Cell concentrations of two species in a mixed culture can also be controlled in this way (Wilder *et al.*, 1980).

When there are variations in the process parameters or the process itself, an adaptive control system must be used to automatically compensate for the variations. This is usually done by identifying the process dynamics on-line and then adjusting the controller by some specified criteria (see Figure 1d). Consider as an example a fermentation involving a filamentous microorganism (*e.g.* penicillium molds, *etc.*). Due to changes in the morphology of the culture over the course of the fermentation, drastic changes in the heat and mass transfer characteristics of the culture can occur. A temperature or dissolved oxygen controller designed for one phase of the fermentation may be inadequate during another phase. Adding adaptation to these loops, in the form of changing the setpoint or changing the controller parameters, may be necessary in this case.

With optimal control, the control system is designed by maximizing a specific analytical performance index. The problem can be stated as follows: given a mathematical model of the process, a set of constraints on the process inputs, and a performance index of the control system, find the control function (or control law) which will maximize the performance index. Since the control function is a function of time, not a constant value, variational calculus must be used to determine the time profiles of the inputs.

The overall objective of the control system is to achieve optimality for the process as a whole. This is done by determining the optimal setpoints for the individual control loops, thus maximizing a performance index for the entire process (see Figure 1e). At this point it is convenient to discuss separately steady state optimization and dynamic optimization.

Continuous bioreactors are subject to steady state optimization. The objective is to find the optimum constant (steady state) values of the control variables, such as temperature, pH, dissolved oxygen concentration and medium feed rate, which will maximize a given performance index. An appropriate performance index is the profit that can be realized. Since the operating cost is usually fixed, the optimal steady state conditions may require the maximization of the difference in the price associated with the product (P) and the cost of the substrate (S):

$$IP = \alpha FP - \beta FS_i \qquad (4)$$

where F is the medium feed rate, S_i is the feed substrate concentration, and α and β are the unit prices associated with the product and substrate, respectively. If the product is cell mass X, then:

$$IP = \alpha FX - \beta FS_i \qquad (5)$$

When S_i is fixed at an upper limit, as is the usual case, the maximization of IP is equivalent to maximizing the productivity, FP (or FX). Therefore, the productivity is often maximized.

If a reliable model is available, the optimization can be done off-line using ordinary calculus and/or a search technique. When no reliable model is available, or when the process changes, the optimization must be carried out on-line using a multivariable optimum seeking technique. Here, the control variables are incremented and the new steady state that is attained is assessed. Based on the observed response, the next control variable increments are chosen and this procedure is repeated until no further improvement in the productivity can be achieved. This is a time-consuming procedure as one has to wait for the complete steady state response after each change in the control variables. Therefore, it is desirable to develop a technique based on short-time transient responses in order for the optimization to proceed more rapidly.

Batch and fed-batch bioreactors are operated dynamically and therefore are subject to dynamic optimization. The objective is now to find the best time profiles, rather than constant values, for the control variables which will maximize a performance index. Batch and fed-batch bioreactors can be operated in a cyclic mode by repeating the operation batch after batch, therefore an additional control variable is available that is not in steady state optimization, the initial volume. Thus, for a batch bioreactor, the temperature (T), pH and dissolved oxygen concentration (DO) can be programmed as functions of time for each cycle along with selecting the best initial volume V_0. For fed-batch bioreactors, there are additionally the inlet and outlet medium feed rates which may be programmed as functions of time.

An appropriate performance index for dynamic optimization is:

$$IP = \int_0^{t_f} (\alpha FP - \beta F_i S_i) \mathrm{d}t / t_f \qquad (6)$$

where t_f is the cycle time. This performance index is maximized by selecting V_0, $F(t)$, $F_i(t)$, $T(t)$, pH(t) and DO(t). For a more detailed description of dynamic optimization of fermentation processes, readers are referred to Rolf and Lim (1982).

One particular problem with fermentation processes is that sensors are not available for direct on-line measurement of many of the variables needed for control purposes. Several estimation schemes are available for fermentation processes, as reviewed by Rolf and Lim (1982). One method relies on a material balance around the bioreactor for a component that can be readily measured. This is coupled with a known yield factor to estimate the desired variable. Another method makes use of elemental balances through an overall stoichiometric equation. This method removes reliance on yield factors provided that the elemental composition of all components is known. Other methods utilize rigorous mathematical models to estimate the desired variables. All of these methods fall under a general class of estimators for deterministic systems called state observers.

In certain situations, there is considerable process noise or the measurements may be corrupted by noise as well as being limited in numbers. These situations require special attention and involve the general topic of state estimation and filtering theory. This topic has been well developed for process control and it involves finding the best estimate of the state such that the variance of the estimation error is minimized. For such a problem, a number of estimation schemes have been proposed, among which the most widely accepted is the extended Kalman filter (Jazwinski, 1970).

9.3 COMPUTER CONTROL SYSTEMS

Computers can now be used to carry out most fermentation control functions, that is replacing classical controllers and implementing modern control, therefore the configuration of a computer control system is considered.

9.3.1 Functional Components

System requirements inevitably vary with the particular application as each fermentation process may have its own uniqueness and complexity, therefore any system configuration is process dependent and must be planned carefully. However, there are some basic functional components which are common to most computer control systems for fermentation processes. These are listed in Table 3 and are also described in detail by Rolf and Lim (1982). A few remarks about current trends in these areas are in order.

Table 3 Functional Components of a
Computer Control System

Instrumentation
Computer–Instrumentation Interface
Computer System
Computer hardware
System software, data bases
Displays, operator–machine interface
Data logging, documentation
Back-up, error detection
Low-level control
Advanced control

Efforts are being made to develop more sterilizable sensors and hence improve the controllability of the fermentation process. Methods to remove constituents automatically from the bioreactor are also being developed so that conventional instrumentation can be used.

When designing interfacing equipment for existing facilities, special consideration may be required for pneumatic and electromechanical devices. Solid state technology is generally available for most types of equipment and is preferred for new systems due to lower cost and higher reliability.

Rapid and interactive communication between the operator and the computer is essential in order to facilitate control and operational decisions. Graphics terminals, especially those with color, are being used to deal with this problem very effectively.

Due to the advancement of computer hardware, especially the development of microcomputers, it is now practically realizable to implement most computer system structures that can be devised.

9.3.2 Hardware Structures

The number of applications in which hierarchical and distributed computer systems are used for fermentation processes has increased recently. This is partly due to the availability of low cost microcomputers for use in these multicomputer systems. Another reason is that the control functions have a hierarchical structure starting from low-level control, to supervisory control, to optimization and adaptation, and finally to management and production decisions. Also, a fermentation facility is itself distributed in many ways. This distribution can occur in both type and number of variables, control techniques, unit operations, plant functions, personnel functions and even physical locations. It seems only natural that multicomputer systems would evolve to separate some of the hierarchical and distributed levels stated above. Multicomputer systems are not necessary for every application in computer-controlled fermentation, but the technology available today makes them justifiable and economical in many cases.

There are a number of pertinent multicomputer structures (Eades, 1982; Martinovic, 1983; Williams and Kompass, 1976) and some of the common types are shown in Figure 4. Main computers are usually found in the form of a minicomputer. An application of a two-level hierarchical system was shown in Figure 3. The radial and multidrop structures are essentially extensions of the hierarchical structure since the substation computers, usually microcomputers, cannot communicate with each other. There is presently a trend in configuring systems with ring and distributed structures. In these structures, each computer can communicate with any other computer, thus allowing greater flexibility and reducing the consequences of failures. Minicomputers are often added to the ring and distributed structures as they are still needed to perform much of the data analysis, advanced control and optimization. It has been suggested that each substation of the distributed structure becomes distributed itself (Martinovic, 1983). This can be achieved by adding an independent data bus, much like the data hiway, in each of the individual substations. This provides a totally distributed system which is completely flexible.

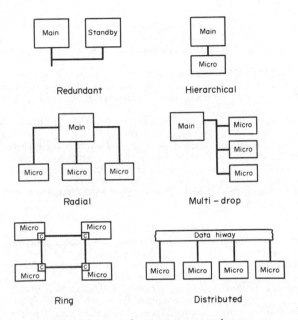

Figure 4 Structures for computer control systems

The use of computers in a stand-alone configuration is also quite popular for fermentation processes. This generally occurs in laboratories, pilot plants and plants in which the control system is less developed. In some cases, an inexpensive microcomputer may be the only way in which computer control can be integrated into a fermentation process. Other applications of microcomputers is in the evolution of intelligent interfaces and instrument systems.

The so-called programmable controllers are extremely popular pieces of industrial equipment. These are usually microprocessor-based controllers which provide low-level control and a limited amount of supervisory control. There may also be some provisions for communicating with other programmable controllers or a supervisory computer. The specific functions available are dependent on the particular vendor. With these programmable controllers, there is the possibility of both stand-alone operation and incorporation into a multicomputer control system.

9.4 CONCLUSIONS

Computer technology has had a significant impact in fermentation, especially in recent years. Integration of computers into fermentation systems is generally justified by the computer's capability for process monitoring, data acquisition, data storage, low-level control and error detection. However, the capabilities of the computer can and should be further exploited with the implementation of advanced control and optimization strategies. This area can provide significant improvement in product yield and productivity.

It was beyond the scope of this chapter to describe all of the details of each control technique. However, the intent was to provide the necessary information in regard to the types of control techniques available and their significance to fermentation processes. Specific examples of the application of some of these techniques can be found in the reviews mentioned earlier. Finally, it must be stressed that control considerations should be taken into account early in the design stage for a fermentation process. This will not only assure controllability of the fermentation process, but also optimal overall performance of the entire system.

ACKNOWLEDGEMENTS

This work was supported in part by a grant from the National Science Foundation, CPE 7918902.

9.5 REFERENCES

Armiger, W. B. and A. E. Humphrey (1979). Computer applications in fermentation technology. In *Microbial Technology*, ed. H. J Peppler, vol. 2, pp. 375–401. Academic, New York.

Athans, M. and P. L. Falb (1966). *Optimal Control, An Introduction to the Theory and Its Applications*. McGraw-Hill, New York.

Bell, D. and W. J. Griffin (eds.) (1969). *Modern Control Theory and Computing*. McGraw-Hill, London.

Bristol, E. H. (1977). Designing and programming control algorithms for DDC systems. *Control Eng.*, **24** (1), 24–26.

Bryson, A. E. and Y. Ho (1975). *Applied Optimal Control, Optimization, Estimation, and Control*. Hemisphere, Washington, DC.

Bull, D. N. (1983). Automation and optimization of fermentation processes. In *Annual Reports on Fermentation Processes*, ed. G. T. Tsao, vol. 6, pp. 359–375. Academic, New York.

Citron, S. J. (1969). *Elements of Optimal Control*. Holt, Rinehart and Winston, New York.

Coughanowr, D. R. and L. B. Koppel (1965). *Process Systems Analysis and Control*. McGraw-Hill, New York.

Dobry, D. D. and J. L. Jost (1977). Computer applications to fermentation operations. In *Annual Reports on Fermentation Processes*, ed. D. Perlman, vol. 1, pp. 95–114. Academic, New York.

Douglas, J. M. (1972). *Process Dynamics and Control*. Prentice-Hall, Englewood Cliffs, NJ.

Eades, C. G. (ed.) (1982). *Trends in On-line Computer Control Systems*. IEE Conference Publication Number 208, London.

Goff, K. W. (1966). A systematic approach to DDC design. *ISA Journal*, December, 44–54.

Hampel, W. A. (1979). Application of microcomputers in the study of microbial processes. In *Advances in Biochemical Engineering*, ed. A. Fiechter, vol. 13, pp. 1–33. Springer-Verlag, New York.

Hatch, R. T. (1982). Computer applications for analysis and control of fermentation. In *Annual Reports on Fermentation Processes*, ed. G. T. Tsao, vol. 5, pp. 291–311. Academic, New York.

Jazwinski, A. H. (1970). *Stochastic Processes and Filtering Theory*. Academic, New York.

Jefferis, R. P. (1975). Computer in the fermentation pilot plant. *Process. Biochem.*, **10** (3), 15–18.

Martinovic, A. (1983). Architectures of distributed digital control systems. *Chem. Eng. Prog.*, **79** (2), 67–72.

Meskanen, A., R. Lundell and P. Laiho (1976). Engineering of fermentation plants. Part 3. Automation. *Process. Biochem.*, **11** (5), 31–36.

Ray, W. H. (1981). *Advanced Process Control*. McGraw-Hill, New York.

Rolf, M. J. and H. C. Lim (1982). Computer control of fermentation processes. *Enzym. Microb. Technol.*, **4**, 370–380.

Rolf, M. J., P. J. Hennigan, R. D. Mohler, W. A. Weigand and H. C. Lim (1982). *Biotechnol. Bioeng.*, **24**, 1191–1210.

Shinskey, F. G. (1979). *Process-Control Systems, Application, Design, Adjustment*. McGraw-Hill, New York.

Webb, J. C. DDC Hardware and Software, General Electric Company.

Weigand, W. A. (1978). Computer applications to fermentation processes. In *Annual Reports on Fermentation Processes*, ed. D. Perlman, vol. 2, pp. 43–72. Academic, New York.

Wilder, C. T., T. W. Cadman and R. T. Hatch (1980). Feedback control of a competitive mixed culture. *Biotechnol. Bioeng.*, **22**, 89–106.

Williams, T. J., F. J. Mowle, W. A. Weigand, G. V. Reklaitis, R. E. Goodson and H. C. Lim (1973). *Digital Computer Applications to Process Control*. Purdue University, West Lafayette, IN.

Williams, T. S. and J. Kompass (eds.) (1976). *Hierarchical and Distributed Computer Control*. Third Annual Advanced Control Conference. Control Engineering, Chicago.

Zabriskie, D. W. (1979). Real-time applications of computers in fermentation processes. *Ann. N. Y. Acad. Sci.*, **326**, 223–239.

10
Data Analysis

D. W. ZABRISKIE
BioChem Technology Inc., Malvern, PA, USA

10.1 INTRODUCTION

This chapter will be concerned with computerized data analysis for on-line process applications. The terms 'on-line' or 'in real time' will be used synonymously to indicate that the time needed to measure, acquire and analyze the data is so small with respect to process response times that it may be neglected.

Typical on-line data functions are listed in Table 1. The first task is to acquire the measured data from the interface located between the sensor instrumentation and the on-line computer. Measured data can be verified to check for instrumentation malfunctions and appropriate countermeasures can be implemented. Often sensor signals are noisy so that it can be beneficial to filter certain measurements before further processing by the data analysis software. Some measurements need to be converted to units with the most significance to the operator using conversion factors and calibration curves. Indirect variables derived from measurements such as the O_2 uptake rate and respiratory quotient can be calculated next. Rate data are obtained by differentiating extensive properties and the cumulative totals are obtained by numerically integrating rate data. The software can generate messages to alarm the operator that a variable falls outside allowable limits, to suggest means to correct the alarm condition, to remind the operator to complete a task, or to warn the operator of conditions which could result in a future alarm. The results of on-line data analysis can be displayed in tabular and graphical formats. Data reduction is generally employed to prevent the operator from being overwhelmed with information by displaying data summaries containing only those results of the most significance on a routine basis.

Table 1 On-line Data Analysis Tasks

Acquisition of measurements
Verification of measured data
Filtering
Units conversion
Calculation of indirect measurements
Differentiation
Integration
Calculation of estimated variables
Generation of system, alarm, warning and diagnostic messages
Data reduction
Tabulation of results
Graphical presentation of results
Process simulation
Supply of inputs to process control algorithms
Storage of results

The other results are obtainable on demand. All the results need to be accessible by the process control algorithms and are written to a mass storage device, usually a tape or disk.

Related data analysis activities include process modeling, process simulation and control. Process modeling is generally an off-line activity and will not be covered in detail here. The reader is referred to Chapter 6 in Volume 2 which covers this subject. Process optimization and control are discussed in Chapter 8.

A variety of data analysis programs have been described in the literature (Bowski *et al.*, 1981; Bravard *et al.*, 1979; Dobry and Jost, 1977; Jefferis, 1975; Meiners and Rapmundt, 1983; Nyiri, 1972a, 1972b; Nyiri *et al.*, 1975; Rolf *et al.*, 1982). A typical tabulation of results from an on-line data analysis system is shown in Table 2. It summarizes the data generated by a computer-coupled fermentation system employing a program developed by BioChem Technology, Inc., Malvern, PA, called LADAC, an acronym for log, analyze, display and control. This report pertained to a batch fermentation and was updated at five minute intervals. Only variables of most interest to the operator are shown. Other variables such as raw sensor electrical signals in mV or mA, cumulative totals, *etc.* are stored on the computer disk and may be retrieved upon request. The report contains directly measured quantities such as temperature, pressure and pH, and indirect measurements such as the O_2 uptake rate, CO_2 evolution rate and respiratory quotient. Certain variables such as the biomass concentration and growth rate are estimated quantities. Each variable is numbered to facilitate on-line data correlation. Any variable can be tabulated or plotted against another by using these identification numbers in a brief series of commands. All of the results are stored on a computer disk to be used for off-line analysis or archival purposes.

There are many motivations for installing on-line data analysis capabilities. The ability to display all the pertinent on-line information using units and a format meaningful to the operator obviously facilitates the process monitoring function. The generation of on-line process warnings, alarms and diagnostic messages, and the representation of the process data graphically are important aspects of this function. On-line data analysis programs supply information needed to control the process. The mode of control may be a simple empirical procedure implemented by the operator, such as initiating the feed of a product precursor after the growth rate, estimated by the on-line data analysis program, falls below a specific value. In other situations, automated heuristic procedures may be involved, such as regulating the flow rate of carbohydrate to a fed-batch fermenter to maximize product yield by maintaining the respiratory quotient, calculated by the on-line data analysis program, at a fixed value. In the future, on-line analysis results may be used as inputs to multivariable optimal control strategies or adaptive learning networks. On-line data analysis can have important contributions for process scale-up. For example, it may be desirable to test a newly developed process in pilot plant equipment operated to simulate heat and mass transfer limitations that will be encountered in an existing full-scale plant. On-line determination of the heat and mass transfer coefficients would be valuable in this application. The maintenance of a process data base is an important capability. In the pilot plant, the data base is essential for off-line analysis typical of process research and development. In a manufacturing plant, the data base can be a convenient and cost-effective way of documenting Good Manufacturing Practice to satisfy US Food and Drug Administration (FDA) and related agency requirements when compounds for clinical applications are being produced.

On-line data analysis has only become practical since the 1970s with the introduction of digital computing systems, real-time operating software and appropriate interfacing hardware with suf-

Table 2

Run: F09904	Date/Time	21/ 9:43:30
Date: 1/20/83	Elapsed time	13:43:30
Specification file: F09904.A1		13:725 h
Output file: F09904.B1	Record number:	55

1	Temperature	30.8	°C
2	Pressure	0.0	p.s.i.g.
3	Liquid volume	240.0	l
4	Aeration rate	340	S.l.p.m.
5		1.42	VVM
6	Agitation Rate	297	r.p.m.
7	Dissolved oxygen	33	% sat
8		2.57	p.p.m.
9	pH	5.0	
10	Culture fluorescence	19	CNTS
11		2.20	MV
12	O_2 in off-gas	16.41	%
13	CO_2 in off-gas	4.24	%
14	N_2 in off-gas	75.76	%
15	H_2O in off-gas	2.36	%
16	Other in off-gas	1.23	%
17	Oxygen uptake rate	165.2	$mmol\ l^{-1} h^{-1}$
18	CO_2 evolution rate	163.1	$mmol\ l^{-1} h^{-1}$
19	Respiratory quotient	0.99	
20	$k_L a$	1511.7	h^{-1}
21	Biomass	24.7	$g\ l^{-1}$
22	Growth rate	8.17	$g\ l^{-1} h^{-1}$
23	Specific growth rate	0.33	h^{-1}
24	Temperature setpoint	30.0	°C
25	pH setpoint	5.0	
26	Agitation rate setpoint	50	r.p.m.
27	Pumping rate setpoint	*****	$ml\ h^{-1}$
28	Dissolved oxygen setpoint	5	% sat

ficient capabilities at reasonable costs. Prior to this period, central control rooms were generally used in large, modern facilities to display the measured variables from sensors located throughout the plant using a variety of meters, chart recorders and other output devices. The utility of this data was limited, however. Since analog recording of all measured variables was generally impractical, some data were recorded manually at varying frequency. Displayed data often needed to be manually converted to units of significance to the operator. All indirect measurement and estimated variables had to be computed off-line. These factors together with the layout of the control room made the correlation of process variables difficult. Anyone who has tried to graphically correlate data from one chart recorder with that recorded on another, both recorders using different chart speeds and scaling factors, can appreciate the problems. Data storage and retrieval were exceptionally difficult by today's standards. The application of the on-line computer has virtually eliminated all of these problems.

The discussion to follow will consider three types of process variables. The term 'direct measurement' will be used to describe data derived from sensors expressed in appropriate units. The term 'gateway measurements' has been used with the same meaning (Humphrey, 1971). Sometimes trivial calculations are employed to change units using a conversion constant or a calibration curve. Data derived from a temperature sensor expressed in mA, °C or °F constitute an example of a measured variable. The term 'indirect measurement' will be used to describe combinations of measured variables defined by a formula common to all fermentations which are especially useful for data interpretation. Indirect measurements are not directly measurable using a sensor. The oxygen uptake rate is an example of an indirect variable since it depends on an equation involving the measured variables of aeration rate, culture volume and concentrations of CO_2, O_2, N_2 and H_2O in the sparging and effluent gases. The term 'estimated variables' will refer to quantities which cannot be determined directly using a sensor or employing a simple formula. They generally require more complicated numerical procedures and involve mathematical models of the process. Table 3 lists typical direct measurements, indirect measurements and estimated variables.

Table 3　Results on On-line Data Analysis

Direct measurements	Indirect measurements	Estimated variables
Fermenter	Mass transfer rates:	Metabolic heat generation rate
Temperature	O_2 uptake rate	Holdup
Pressure	CO_2 evolution rate	Biomass concentration
Liquid weight	Evaporation rate	Growth rate
Foam level	Component feed rate(s)	Substrate concentration(s)
Gas sparging rate(s)	Component production rate(s)	Substrate consumption rate(s)
Sparging gas temperature(s)	Total	Precursor concentration(s)
Sparging gas pressure(s)	Cumulative mass addition/	
	removal:	Precursor consumption rate(s)
Agitation rate	O_2, CO_2, H_2O, feed	Product concentration(s)
	component(s), product	
	component(s)	
Agitator power consumption		
rate		Product production rate(s)
Flow rate in jacket	Total	Yields
Jacket inlet temperature	Component ratios:	Efficiencies
Jacket outlet temperature	Respiratory quotient	Apparent viscosity
Gas analyzers:	Feed ratios for C, N, P, K	Impeller Reynolds number
O_2, CO_2, N_2, H_2O, NH_3,	Mass transfer coefficient	Aeration number
ethanol, other(s)	Heat transfer rate	Power number
Liquid analyzers:	Cumulative heat transferred	
O_2, CO_2, NH_3, pH, redox	Heat transfer coefficient	
potential, turbidity, density,	Culture volume	
culture fluorescence, other(s)	Agitator tip speed	
Supply tank(s)	Agitation torque	
Temperature		
Pressure		
Liquid weight		
Feed rate		
Feed pressure		
Feed temperature		

10.2 DIRECT MEASUREMENTS

Data analysis associated with direct measurements is straightforward. Generally a uniform set of units is established for each fermentation installation. In many cases, the output from the sensing instruments will need to be converted to comply with the designated unit system. This can simply involve multiplying the measurement by a conversion constant, for example, to change a measurement in ft^3 min^{-1} to l min^{-1}. Other instruments require a calibration curve for conversion. This may be accomplished on-line using a correlating equation which expresses the measurement as a function of the signal value. The correlation is established off-line using calibration data. An example is the measurement of gaseous CO_2 using an infrared analyzer where a calibration curve is required to relate the meter reading to % CO_2 (v/v).

Procedures which have the effect of reducing the noisy component of a measured signal are referred to as filters. Filters may be implemented by the electronics associated with the sensor, or by an algorithm called a digital filter that is executed after the measurement has been acquired by the computer. Generally filter efficiency, data acquisition frequency and measurement response time are closely related. The selection of a filter must carefully consider these factors with the objectives for data analysis and process control systems. Digital filters have not been widely applied to the on-line analysis of fermentation processes.

Digital filters in their simplest form provide a time averaged signal as output. Wang *et al.* (1977) used averaging intervals of up to 15 minutes in their work with computer-aided bakers' yeast fermentations. Jefferis *et al.* (1976) have employed a least squares recursive filter in their work to estimate cell density based on optical absorbance at 600 nm. More sophisticated filters have been applied when a kinetic model of the process is available. The Kalman filter is one example which will be discussed in Section 10.4 below.

10.3 INDIRECT MEASUREMENTS

10.3.1 Mass Transfer Rates

The amounts of specific chemical components that are added to or removed from the fermenter are important indirect measurements. They are calculated by making a mass balance around the fermenter for the component of interest. This may be expressed mathematically as:

$$\text{Transferred} = \text{inputs} - \text{outputs} \tag{1}$$

$$W = \sum_{j=1}^{n} F_j C_j - \sum_{j=1}^{m} F_j C_j \tag{2}$$

where F is the volumetric flow, C is the concentration, n is the number of input streams, and m is the number of output streams. It is beneficial to know W as a rate (*e.g.* mol h^{-1}) and as a cumulative total for the process cycle (*e.g.* mol). Usually a mathematical conversion from one representation to the other is necessary. When volumetric flow rates are being measured, the mass transfer rate data must be numerically integrated to obtain a cumulative total. When the volume of a supply tank is being measured, the mass transferred must be numerically differentiated to obtain the rate. Sometimes mass transfer rates are reported on a per unit of liquid volume basis (*e.g.* mol $1^{-1} h^{-1}$).

Consider for example a fed-batch fermentation where gaseous ammonia is sparged to control pH and corn steep liquor (CSL) is being added to the fermenter. The rate of elemental nitrogen addition would be given by:

$$W = F_1 C_{g,1} + F_2 C_{g,2} + F_3 C_{g,3} - F_4 C_{g,4} \tag{3}$$

Here elemental nitrogen enters with the ammonia gas (stream 1), air (stream 2) and CSL (stream 3), and is removed in the off-gas (stream 4) as molecular nitrogen and a small amount of ammonia. Similar balances are necessary for water, methanol, ethanol and other components present in appreciable amounts in gas and liquid phases.

Often in the case of gases, only the sparging gas and off-gas streams are significant so that equation (2) can be simplified to:

$$W = F_{in} C_{g,\,in} - F_{out} C_{g,\,out} \tag{4}$$

where 'in' refers to the sparging gas and 'out' refers to the off-gas. In the case for O_2, W is referred to as the O_2 transfer rate (OTR). The rate at which O_2 is consumed by the organism is referred to as the oxygen uptake rate (OUR). Although these quantities are not identical, usually the changes in O_2 concentration in the culture liquid are sufficiently slow that OUR may be equated with the OTR. A similar relationship exists between the CO_2 transfer rate and the CO_2 evolution rate (CER) when the changes in dissolved CO_2 are sufficiently slow. The CER is the rate at which CO_2 is evolved by the organism. When equation (4) is applied to water, the result is referred to as the evaporation rate.

Procedures for the measurement of gas transfer rates have been described by Fiechter and Von Meyenburg (1968), Nyiri *et al.* (1975) and Flickinger *et al.* (1980). It is assumed that F_{in}, $C_{g,\,in}$ and $C_{g,\,out}$ are directly measurable quantities. In general, F_{in} is not equal to F_{out}. F_{out} may be estimated using a gas component which is an inert in the system. In this case W in equation (4) is equal to zero so that equation (4) can be rearranged to solve for F_{out} as a function of F_{in}, $C_{g,\,in}$ and $C_{g,\,out}$. Nitrogen or argon is a typical choice for the inert gas component.

10.3.2 Component Ratios

The respiratory quotient (RQ) is a dimensionless quantity defined as:

$$RQ = \frac{CER}{OUR} \tag{5}$$

and is a measure of the number of moles of CO_2 evolved per mole of O_2 consumed. The RQ is an especially significant variable to the fermentation technologist in aerobic cultures. Complete oxidation of a specific carbon source to CO_2 and H_2O yields a value of RQ characteristic of the compound as shown in Table 4. Deviations from the value for complete oxidation of a specific

Table 4 RQ Values for Complete Oxidation of
Selected Carbon Sources

Carbon source	Formula	n	RQ
Carbohydrates	$C_nH_{2n}O_n$	n	1.00
Alkanes	C_nH_{2n+2}	n	$2n/3n + 1$
methane	CH_4	1	0.50
decane	$C_{10}H_{22}$	10	0.65
Alcohols	$C_nH_{2n+2}O$	n	0.67
methanol	CH_4O	1	0.67
ethanol	C_2H_6O	2	0.67
Carboxylic acids	$C_nH_{2n+2}O_2$	n	$2n/3n - 1$
acetate	$C_2H_6O_2$	2	0.80
propionate	$C_3H_8O_2$	3	0.75
Lactate	$C_3H_6O_3$	—	1.00
Glycerol	$C_3H_8O_3$	—	0.75
Citrate	$C_6H_8O_7$	—	1.33
Formate	CH_2O_2	—	0.50
Glyoxalate	$C_2H_3O_3$	—	2.00
Oxalate	$C_2H_2O_4$	—	4.00

substrate indicate that some of the substrate is being diverted to biochemical pathways which produce other products. Some microorganisms, for example, can metabolize ethanol to CO_2 and H_2O (RQ = 0.67) or to acetate and H_2O (RQ = 0). An RQ less than 0.67 can indicate that both pathways are operating simultaneously. RQ values can vary from 0.67 to 5 or more during the growth of bakers' yeast on glucose. High values of RQ indicate the expression of the Crabtree effect where glucose is fermented to ethanol (RQ = ∞) under aerobic conditions. Low values of RQ indicate that accumulated ethanol is being completely oxidized to CO_2 and H_2O (RQ = 0.67). The transition between growth and acid formation phases during the production of citric acid by *Aspergillus niger* is accompanied by a decrease in RQ (Clark and Lentz, 1961).

Another value of potential significance is the cumulative respiratory quotient (CRQ), though this figure has not been reported. The CRQ may be defined as:

$$CRQ = \frac{\int CERdt}{\int OURdt} \tag{6}$$

It is an overall indicator of the extent of process deviation from complete respiration.

Other ratios which are sometimes monitored are the cumulative amounts of carbon, nitrogen, phosphorus and potassium added to a fed-batch fermenter. These may be compared with the ratios established for the products. Differences may indicate nutrient deficiencies, nutrient excesses or the accumulation of unexpected by-products. The carbon to nitrogen ratio (C/N) has been used to control the aerobic production of *Candida utilis* by Nyiri *et al.* (1977).

10.3.3 Mass Transfer Coefficient

The establishment of mass transfer rates allows the calculation of the overall mass transfer coefficients between gas and liquid phases. Usually the diffusive resistance within the bubbles can be neglected in fermenters. In small fermenters, the mass transfer rate is given by:

$$W = k_La(C^* - C) \tag{7}$$

where C is the bulk concentration in the liquid phase, C^* is the concentration in the liquid in equilibrium with the gas at the interface and k_La is the overall mass transfer coefficient in the liquid phase. The concentration C^* is not measurable but may be computed from C_g using a gas/liquid equilibrium relationship such as Henry's Law expressed as:

$$C^* = KC_g \tag{8}$$

Care must be used in supplying the correct value of K since it is strongly dependent on temperature and, to a lesser extent, the concentrations of other species in the aqueous phase. In small fermenters, liquid and gas phases are mixed so that $C_{g, out}$ measured in the off-gas can be used to calculate C^*.

In large fermenters, the liquid phase may still be regarded as being perfectly mixed, but the gas phase will vary between $C_{g, in}$ and $C_{g, out}$ depending on position within the fermenter. In this case the mass transfer rate may be expressed by:

$$W = k_L a(C^* - C)_{\text{log mean}} \tag{9}$$

where

$$(C^* - C)_{\text{log mean}} = \frac{(C^*_{in} - C) - (C^*_{out} - C)}{\ln\left(\dfrac{C^*_{in} - C}{C^*_{out} - C}\right)} \tag{10}$$

Again, equation (8) is used to compute the appropriate values of C^*_{in} and C^*_{out} using measured values of $C_{g, in}$ and $C_{g, out}$ respectively.

Conceptually, a value of $k_L a$ can be computed for each component where appropriate mass transfer rate and concentration data are available. However, the hydrodynamics which are the principle determinant of mass transfer coefficients are common to all components being exchanged between bubbles and the culture liquid in a fermentation. Usually the mass transfer coefficient for aerobic processes is calculated on the basis of the OUR because of its importance in scaling-up fermenters.

Another procedure called the dynamic method may be used to compute OUR and $k_L a$ using only dissolved oxygen (DO) measurements (Mukhopadhyay and Ghose, 1976; Ruchti *et al.*, 1981). This involves the monitoring of DO after the sparger in the fermenter has been turned off temporarily. The rate of DO decline is related to the OUR during this period. After this rate is established, the sparger is turned on again. The response of the DO measurement during its recovery to its original value is related to OUR and $k_L a$. Although a computer-coupled fermentation system can be designed to implement this procedure automatically on a routine basis, the results are not as reliable as the approach described earlier. The dynamic method is sensitive to DO probe dynamics, mass transfer from the fermenter head space and a variety of other artifacts. For these reasons, it is not used in most computer-coupled fermentation systems.

10.3.4 Energy Balance Around the Fermenter

During most fermentations, heat generated by metabolic reactions, mechanical energy from agitation and other sources must be removed by cooling the fermenter. The metabolic component is obtained by calculating an energy balance around the fermenter.

$$\text{Energy accumulated} = \text{inputs} - \text{outputs} \tag{11}$$

Assuming the fermenter is adiabatic, and that changes in potential energy, flow kinetic energy and heat of solution are negligible, equation (11) becomes:

$$WC_p\frac{dT}{dt} = P + Q_M + \sum_{j=1}^{n} F_j \Delta H_j^0 - \sum_{j=1}^{m} F_j \Delta H_j^0 \tag{12}$$

where C_p is the liquid heat capacity, T is temperature, t is time, P is agitation power, Q_M is the metabolic heat, ΔH^0 is the enthalpy per unit mass related to a reference condition, n is the number of input streams and m is the number of output streams. In this context, the stream inputs and outputs include not only feed and harvest lines, but also the flow of cooling fluid to and from the fermenter heat exchanger (*e.g.* jackets, coils or external heat exchangers). All of the variables in equation (12) are measurable except Q_M which may be calculated by rearranging equation (11). When the fermentation temperature is controlled at a constant value, the derivative in equation (12) becomes zero and the calculations are simplified. Generally Q_M is expressed as a rate, but sometimes the cumulative energy transferred is calculated. These values can also be reported on a per unit of culture volume basis. A discussion of an experimental system in which these calculations were performed on-line is presented in Swartz and Cooney (1978).

10.3.5 Heat Transfer Coefficient

Like the mass transfer coefficient, the heat transfer coefficient (U) has special engineering significance in scaling up bioreactors. It is defined by an equation analogous to equation (7) for the mass transfer coefficient:

$$Q_C = UA(T_C - T)_{\text{log mean}} \tag{13}$$

where

$$(T_C - T)_{\text{log mean}} = \frac{(T_{\text{in}} - T) - (T_{\text{out}} - T)}{\ln\left(\dfrac{T_{\text{in}} - T}{T_{\text{out}} - T}\right)} \tag{14}$$

Here A is the surface area used to cool the fermenter, T_{in} is the coolant inlet temperature, T_{out} is the coolant outlet temperature and T is the process temperature to be maintained. Q_C is computed from an enthalpy balance around the cooling system. Assuming adiabatic conditions apply, this is given by:

$$Q_C = F_{\text{in}}\Delta H_{\text{in}}^0 - F_{\text{out}}\Delta H_{\text{out}}^0 \tag{15}$$

Pursuing the analogy with mass transfer further, there is a dynamic method of determining the heat transfer rate and heat transfer coefficient using temperature measurements of the culture. After cooling water flow has been stopped temporarily, the heat generation rate is related to the rate of temperature increase during this period. After cooling water flow has been restored, the response of the temperature sensor is related to the heat generation rate and the heat transfer coefficient. Again this procedure can be implemented and the results analyzed automatically using a computer-coupled fermenter. The accuracy of the technique is limited, however, by various complications including the sensible heat effects of the fermenter parts, changes in agitation power dissipation, and heat losses from the fermenter which interfere with the dynamics of the measurement cycle. Furthermore, even small temperature excursions such as those required by this procedure can be detrimental to certain microorganisms. Like the dynamics method for mass transfer, this technique is not widely used in computer-coupled fermentation systems.

10.3.6 Other Indirect Measurements

There are a variety of other indirect measurements that are useful for on-line monitoring. Culture liquid volume is usually a calculated number based on measurements of culture liquid mass and measurements or estimates of liquid density. Agitator tip speed is calculated from the measured agitation rate in revolutions per minute (r.p.m.) and the impeller diameter. This speed can be an important variable to monitor in fermenters with variable speed agitators containing organisms that are sensitive to shear such as certain filamentous fungi, streptomyces, protoplasts and tissue culture cells.

It is generally impractical to measure directly power consumed for agitation in small fermenters since a disproportionate amount of energy is dissipated by the agitator seal. Instead agitator torque measurements are made using strain gauges mounted in the agitator shaft within the fermenter. The net power for agitation can be computed from:

$$P = N\tau \tag{16}$$

where τ is the torque and N is the angular velocity of the agitator. In large scale fermenters, the power consumption may be measured directly in which case equation (16) may be used to compute the torque on the agitator shaft.

10.4 ESTIMATED VARIABLES

Estimated variables differ from direct and indirect measurements in that a process model is required for their determination. The model is a simplified description of the process based on a set of assumptions. The assumptions are selected to retain the major features of the process while disregarding the insignificant details. Models are limited in the range over which they can accurately describe the process because of the simplifying assumptions. Models are generally expressed as equations which may describe reaction stoichiometry, reaction rates (kinetics) and hydrodynamics. Models may be based on theoretical considerations, experimental data or a combination of both.

10.4.1 Applications of Macroscopic Material Balances

By the law of conservation of matter, the rate of mass transferred to and from the fermenter, given by equation (2), must be balanced by the amount accumulated and reacted within the fermenter. Usually the accumulation in the gas phase is negligible so that this law may be expressed mathematically as:

$$W = V\frac{dC}{dt} + V\sum_{k=1}^{p} R_k \tag{17}$$

where p is the number of reactions involving the component of interest, and R is the rate at which the component is reacted in the liquid.

Consider the aerobic process for producing bakers' yeast from molasses in which ethanol accumulates as a by-product. The overall reaction can be expressed as:

$$aC_6H_{12}O_6 + bO_2 + cNH_3 \rightarrow dC_6H_{10}NO_3 + eC_2H_6O + fCO_2 + gH_2O \tag{18}$$

Based on this equation, a balance for each element may be constructed as:

$$\text{C:} \quad 6a = 6d + 2e + f \tag{19}$$

$$\text{H:} \quad 12a + 3c = 10d + 6e + 2g \tag{20}$$

$$\text{O:} \quad 6a + 2b = 3d + e + 2f + g \tag{21}$$

$$\text{N:} \quad c = d \tag{22}$$

In mathematical terms, equations (19)–(22) form a system of four simultaneous equations in seven unknown variables (a–g). The system is said to have three degrees of freedom, meaning that three additional independent constraints are required before the system can be solved. These constraints may be other independent equations in the seven variables, specified values of selected variables established, for example, by measurements, or a combination of both.

Cooney *et al.* (1977) solved this system on-line by measuring the OUR, CER and the ammonia transfer rate (ATR). Application of equation (17) for each of these measurements gives:

$$\text{OUR} = \frac{dC_{O_2}}{dt} + b \tag{23}$$

$$\text{CER} = \frac{dC_{CO_2}}{dt} + f \tag{24}$$

$$\text{ATR} = \frac{dC_{NH_3}}{dt} + c \tag{25}$$

The derivatives in equations (23)–(25) were insignificant under these experimental conditions and could be neglected. Therefore, b, f and c in equations (19)–(22) can be substituted using equations (23)–(25). Equations (19)–(22) can now be solved for a, d, e and g to give on-line estimates of the sugar reaction rate, the growth rate, the ethanol production rate and water production rate. The rates can be integrated numerically on-line to give accumulated consumption of sugar and production of yeast, ethanol and water. Process yields are also available on an accumulative basis. For example, the yield of yeast from carbohydrate would be computed from:

$$Y_{x/s} = \frac{\int d\,dt}{\int a\,dt} \tag{26}$$

It is sometimes useful to compute yield on an instantaneous basis, for example, when a batch is to be harvested when the instantaneous yield decreases below a specified value. The instantaneous yield of yeast from carbohydrate would be calculated from:

$$Y'_{x/s} = d/a \tag{27}$$

Although material balancing is an extremely useful procedure, it also has some limitations. The overall reaction and empirical formulae for products and reactants must be available. In complex fermentations, however, the reaction may not be completely defined, such as when unknown by-products accumulate. It may not be possible to use a single reaction to describe the process for its entire duration. Empirical formulae can change during the fermentation, especially for products such as biomass (Atkinson and Mavituna, 1983). When two or more products with similar empirical formulae are formed, only the amount of the combined products may be determined. This could occur, for example, during the production of an enzyme product in the presence of other protein by-products. Usually, the fermenter must be equipped with rather extensive instrumentation in order to measure a sufficient number of variables in the component balance to effect closure of the system of equations. This becomes more difficult as the overall reaction becomes more complex owing to the limited number of sensors available to make the required measurement on-line.

Much attention has been given recently to a more generalized material balancing approach where the empirical formulae are not completely specified for the overall reaction (Erickson *et al.*, 1978; Heijnen and Roels, 1981; Roels, 1980). This, of course, increases the number of unknown variables in the problem, but these can be offset by other constraints. For example, the overall equation must be balanced with respect to electronegativity. This can be accomplished by calculating the number of electrons available for transfer to O_2 for each component in the reaction, sometimes referred to as the degree of reductance. Since the number of available electrons from products and reactants must be equal, this becomes another constraint. Certain empirical regularities which may provide other useful constraints have been identified such as the degree of reductance for biomass which is generally 4.291 g available electron equivalents per quantity of biomass containing 1 gram atom of carbon, and the mass fraction of carbon in biomass is approximately 0.462 (Minkevich and Eroshin, 1973). Other constraints may be obtained by incorporating kinetic models in the material balance such as models for growth or product formation. In other cases it has been found beneficial to divide the process into phases, and perform a separate balance for each phase (Mou and Cooney, 1983a, 1983b).

10.4.2　Application of Process Models

Process models may be used with material balances in on-line data analysis applications to estimate values of the process state variables. Models can reduce the degrees of freedom associated with system component balances, or stated another way, they can reduce the number of measured variables needed to solve the system. In the example discussed in the previous section, a model using one equation expressed in the seven unknown variables could have been used to increase the number of independent equations to five. This would have reduced the degrees of freedom to two, making the measurement of the ATR unnecessary and enabling the system to be solved using only gas exchange rate measurements.

In other cases, models can be used to simplify component balances. Jefferis *et al.* (1972) used this approach to estimate yeast concentration and growth rate from OUR measurements in a fermentation similar to that described in Section 10.4.1. The model proposed originally by Marr *et al.* (1962) related the rate of microbial consumption of O_2 to biomass concentration and growth rate as:

$$R = MX + 1/Y_{x/O_2}^{max}\frac{dX}{dt} \tag{28}$$

where M is the maintenance coefficient, X is the biomass concentration and Y_{x/O_2}^{max} is the maximum yield of biomass that could be obtained if M were zero. Under the experimental conditions employed, the derivative in equation (17) could be neglected so that the substitution of equation (28) with equation (17) results in:

$$\text{OUR} = MX + 1/Y_{x/O_2}^{max}\frac{dX}{dt} \tag{29}$$

Equation (29) may be integrated on-line together with OUR data to provide estimates for X and the growth rate. This may be compared with the three measured variables that were required to obtain these estimates in the example discussed in Section 10.4.1. The accuracy of the estimates provided by equation (29) was improved by Zabriskie and Humphrey (1978) by accounting for changes in the yield of metabolic energy from O_2 which depend upon the balance of three catabolic reactions that occur during the process. These changes were modeled by introducing a metabolic correction function whose value depended on the value of RQ.

10.4.3 Other Estimated Variables

There are a variety of other variables which can be estimated with the aid of process models. Reaction conversion efficiencies can be reported as alternatives to yields and may be calculated as:

$$E_{p/s} = \frac{Y_{p/s}}{Y_{p/s}^{\max}} \tag{30}$$

where $Y_{p/s}^{\max}$ is the maximum yield calculated on the basis of the theoretical reaction stoichiometry. In those situations where the heat transfer rate is not measured, the metabolic heat can be estimated. Minkevich and Eroshin (1973) and Cooney *et al.* (1969) have noted that the heat of reaction is usually proportional to the OUR for aerobic fermentations.

Hydrodynamic quantities of engineering significance can also be estimated. The aeration number (N_A) is a dimensionless number representing the ratio of superficial gas velocity to the impeller tip velocity and is defined by:

$$N_A = \frac{F}{ND_i^3} \tag{31}$$

The aeration number and power measurements may be used with empirical correlations such as those described by Ohyama and Endoh (1955) to estimate the power required to agitate the fermenter in the absence of sparging at the same agitation rate. The result can be used to calculate another dimensionless number called the power number (N_P) defined by:

$$N_P = \frac{P_0}{N^3 D_i^5 \varrho} \tag{32}$$

where P_0 is the power without aeration and ϱ is the liquid density. The power number represents the ratio of external forces to inertial forces in an agitated tank. It may be used with other empirical correlations such as the one described by Rushton *et al.* (1950) to obtain a value of the impeller Reynold's number (N_{Re}). This is another dimensionless group which represents the ratio of inertial forces to viscous forces. It is defined by:

$$N_{Re} = \frac{D_i^2 N \varrho}{\mu} \tag{33}$$

where μ is the apparent viscosity. Mixing is said to be laminar when $N_{Re} < 10$, turbulent when $N_{Re} > 10\ 000$ and transient for intermediate values. Once the value of N_{Re} has been established and knowing D_i, N and ϱ, the apparent viscosity may be calculated from equation (33). The portion of the volume occupied by the gas bubbles in the culture dispersion is called the gas holdup and may be estimated using P_0 and aeration rate data with correlations such as those proposed by Calderbank (1958) or Richards (1961).

10.4.4 Application of Modern Estimation Theory

A system is said to be 'deterministic' when the model represents the process exactly, model parameters are known exactly, and all measurements are exact, *i.e.* there are no uncertainties associated with the process. So far, our estimated variables have been calculated assuming a deterministic system. In practice, however, neither the parameters nor the measurements are known exactly. Real systems are said to be 'stochastic' since process uncertainties such as random occurrences and measurement errors are inevitable.

Estimation procedures are generally concerned with establishing values of selected parameters or dependent variables in a process model. The term state variables is often used synonymously with dependent variables. Procedures for state and/or parameter estimation in stochastic systems are referred to as filters since they may be thought of as extracting a pure signal from a noisy signal. The simple filters discussed in Section 10.2 were concerned only with the smoothing of a single measurement. Other filters differ in that they estimate unmeasured states and imprecisely known parameters.

Although a number of filter algorithms have been described, the Kalman–Bucy filter has been used most widely with fermentation processes (Stephanopoulos, 1981; Svrcek *et al.*, 1974). A description of this technique is beyond the scope of this chapter but the reader is referred to Jazwinski (1970) for further details. The function of the filter may be illustrated, however, considering an example discussed by Wells (1971). The process is modeled using a system of four differential equations involving four state variables x_1, x_2, x_3, and x_4. The task is to estimate values of the state variables given noisy measurements of only three of the state variables indicated as z_1, z_2 and z_3. Figure 1 shows the noisy measured values of z_1, z_2 and z_3 as a function of time. Figure 2 compares the results of the filter $\hat{x}_1$, $\hat{x}_2$, $\hat{x}_3$ and $\hat{x}_4$ with the ideal responses x_1, x_2, x_3 and x_4. The filter provided excellent estimates of the state variables in spite of the noisy measurements.

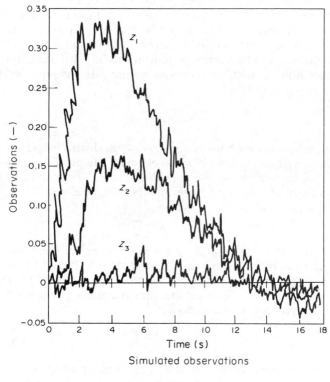

Simulated observations

Figure 1

In practice these techniques are difficult to implement and are not yet widely used with on-line data analysis programs by the fermentation industry. Their effectiveness has been demonstrated in the chemical processing industry, but their importance in on-line fermentation processes has not been assessed thoroughly.

10.5   SOFTWARE CONSIDERATIONS

The most useful on-line data analysis programs are those which do not require programming skills to use and are designed to operate reliably. It is not often recognized for programs with these objectives that the data analysis calculations only constitute a minor part of the overall package. The majority of the software is concerned with input/output (I/O) operations between the system and the user. Making this interface 'user friendly' is especially challenging for the system engineer. The abilities to use the program without extensive training, to correct mistakes readily, to modify data from the terminal, to display selected results and to be free from format restrictions are important aspects of user friendliness. The overall framework of the LADAC program developed by BioChem Technology is shown in Figure 3 and will be used to illustrate some of the considerations behind developing a reliable, user-orientated program.

An off-line program is run by the user to specify the various system options values of parameters through a computer terminal. The user can supply this information by completing a ques-

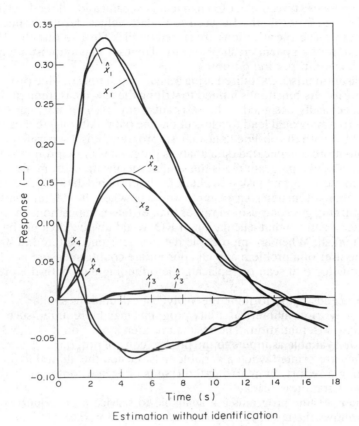

Estimation without identification

Figure 2

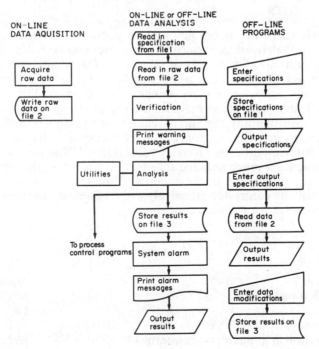

Figure 3

tion and answer sequence, or can elect to change any specification directly using a specification identification code. The latter method is used to update values during on-line operation of the data analysis program. The specifications are stored on disk file 1 so that these data need not be reentered in the event of a system crash or restart. The user may request a tabulation of all the current specifications for inspection purposes.

The on-line measurements are acquired using a small and simple on-line program which has the highest system priority. Its function is limited to writing the raw measurement data on temporary disk file 2. It is specifically designed to be resistant from crashing since an interruption in the operation of this program would lead to a loss of on-line data. Although conceptually its function could be incorporated with the on-line data analysis program, it is a separate program in this case since the larger and more complicated data analysis program is more prone to crash.

The on-line data analysis program reads the system specifications from file 1 and the raw measurement data from file 2. Its first task is to check for improbable data which could cause a system crash or other system malfunction. An example would be when the off-gas analyzer was unexpectedlt measuring sparging gas composition rather than off-gas composition. In this case, the OUR would have a value of zero so that calculation of RQ would involve dividing by zero which could lead to a program crash. When an improbable data value is encountered, a warning message is printed to alert the user of a problem and an appropriate countermeasure is taken, such as skipping the calculations for that scan or replacing the offending data with the previous acceptable value.

The program proceeds by performing the analytical calculations similar to those described in Sections 10.2.–10.4. Various numerical utility programs may be required such as procedures for data integration and differentiation. The results are then stored on disk file 3 for archival purposes and are made available as inputs to the process control programs. The results are checked and alarm messages are printed when a variable falls outside the allowable limits. The program concludes by issuing a report summarizing the results of greatest interest to the operator using standard tabular and graphical formats.

A family of other off-line programs are available to service user requests during or after an experiment. One allows the user to change any data value stored on file 3 manually. Because of the interactions among variables, a single change may require updates of other variables dependent on the changed variable. The data analysis program can be rerun for the appropriate time points to implement the change consistently. Other programs enable the user to tabulate or plot data from file 3 according to specifications entered through the terminal.

10.6   FUTURE PROSPECTS

On-line data analysis capability is becoming more prevalent in modern fermentation pilot plants. Process information derived from calculations, sometimes complex and tedious, is now as available to the operator as are directly measured results. This ability to access and manipulate large quantities of process data efficiently will certainly lead to more effective ways to monitor and control fermentation processes. For the moment, however, on-line data analysis concentrates on obtaining a rather macroscopic understanding of the process, usually by helping to define the environment surrounding the microorganism or to estimate quantities which are not measurable. The deficiency in on-line sensors is well recognized as being an obstacle to this industry so that suitable instrumentation will probably be developed. The role of on-line data analysis may change then to ascertain more detailed information regarding the intracellular metabolic processes responsible for production of the product of interest. Only when these intercellular events can be monitored and controlled effectively will on-line data analysis and process control reach their full potential in fermentation technology.

10.7   NOMENCLATURE

Symbol	Description	Dimensions*
ATR	Ammonia transfer rate	$M\,t^{-1}, M\,t^{-1}\,L^{-3}$
A	Area	L^2
C	Concentration in liquid phase	$M\,L^{-3}$
C^*	Concentration in liquid phase at interface	$M\,L^{-3}$
C_g	Concentration in gas phase	$M\,L^{3}$

C_p	Heat capacity at constant pressure	$E\,M^{-1}\,T^{-1}$
CER	CO_2 evolution rate	$M\,t^{-1}, M\,t^{-1}\,L^{-3}$
CRQ	Cumulative respiratory quotient	—
D_i	Impeller diameter	L
E	Reaction efficiency	—
F_j	Volumetric flow in stream j	$L^3, L\,t^{-3}$
F_{in}	Volumetric flow of inlet stream	$L^3, L\,t^{-3}$
F_{out}	Volumetric flow of outlet stream	$L^3, L\,t^{-3}$
ΔH^0	Enthalpy per unit mass related to reference conditions	$E\,M^{-1}$
k	Number of reactions	—
K	Gas/liquid equilibrium constant	—
$k_L a$	Gas/liquid mass transfer coefficient	t^{-1}
m	Number of outlet streams	—
M	Maintenance coefficient	t^{-1}
n	Number of inlet streams	—
N	Agitation rate	t^{-1}
N_A	Aeration number	—
N_P	Power number	—
N_{Re}	Reynolds number	—
OTR	O_2 transfer rate	$M\,t^{-1}, M\,t^{-1}\,L^{-3}$
OUR	O_2 uptake rate	$M\,t^{-1}, M\,t^{-1}\,L^{-3}$
P	Power used to agitate a dispersion	$E\,t^{-1}, E\,t^{-1}\,L^{-3}$
P_0	Power used to agitate an ungassed liquid	$E\,t^{-1}, E\,t^{-1}\,L^{-3}$
ϱ	Density	$M\,L^{-3}$
Q	Heat	$E, E\,t^{-1}, E\,t^{-1}\,L^{-3}$
Q_c	Cooling	$E, E\,t^{-1}, E\,t^{-1}\,L^{-3}$
Q_M	Metabolic heat	$E, E\,t^{-1}, E\,t^{-1}\,L^{-3}$
R	Reaction rate	$M\,t^{-1}, M\,t^{-1}\,L^{-3}$
RQ	Respiratory quotient	—
t	Time	t
T	Temperature	T
T_C	Temperature of coolant	T
τ	Torque	E
U	Overall heat transfer coefficient	$E^{-1}\,t^{-1}\,T^{-1}$
μ	Viscosity	$M\,L^{-1}\,t^{-1}$
V	Volume	L^3
W	Mass	$M, M\,t^{-1}, M\,t^{-1}\,L^{-3}$
x	State variables	—
$\hat{x}$	Estimated value of state variable	—
X	Biomass concentration	$M\,t^{-1}$
Y	Yield	—
$Y_{x/s}$	Cumulative yield of biomass from substrate	—
$Y'_{x/s}$	Instantaneous yield of biomass from substrate	—
Y_{x/O_2}^{max}	Maximum yield of biomass from O_2	—
$Y_{p/s}^{max}$	Maximum yield of product from substrate	—
z	Measured value of state variable	—

*Dimension designations:

M	mass (moles)
L	length
t	time
T	temperature
L^3	volume
E	energy (*i.e.* $M\,L^2\,t^{-2}$)

10.8 REFERENCES

Atkinson, B. and F. Mavituna (1983). *Biochemical Engineering and Biotechnology Handbook*, chaps. 6 and 9. Nature Press, New York.

Bowski, L., C. Perley and J. M. West (1981). A minicomputer system for analyzing and reporting pilot plant fermentor data. Paper presented at 182nd American Chemical Society Meeting, New York.

Bravard, J. P., M. Cordonnier, J. P. Kernevez and J. M. Lebeault (1979). On-line identification of parameters in a fermentation process. *Biotechnol. Bioeng.*, **21**, 1239–1249.

Calderbank, P. H. (1958). Physical rate processes in industrial fermentation. Part 1. The interfacial area in gas-liquid contacting with mechanical agitation. *Trans. Inst. Chem. Eng.*, **34**, 443.

Clark, D. S. and C. P. Lentz (1961). Submerged citric acid fermentation of sugar beet molasses. Effect of pressure and recirculation of oxygen. *Can. J. Microbiol.*, **7**, 477.

Cooney, C. L., D. I. C. Wang and R. I. Mateles (1969). Measurement of heat evolution and correlation with oxygen consumption during microbial growth. *Biotechnol. Bioeng.*, **11**, 269.

Cooney, C. L., H. Y. Wang and D. I. C. Wang (1977). Computer-aided material balancing for prediction of fermentation parameters. *Biotechnol. Bioeng.*, **19**, 55–67.

Dobry, D. D. and J. L. Jost (1977). Computer applications to fermentation operations. In *Annual Reports on Fermentation Processes*, ed. D. D. Perlman, vol. 1, pp. 95–114. Academic, New York.

Erickson, L. E., I. G. Minkevich and V. K. Eroshin (1978). Application of mass and energy balance regularities in fermentation. *Biotechnol. Bioeng.*, **20**, 1595–1621.

Fiechter, A. and K. Von Meyenburg (1968). Automatic analysis of gas exchange in microbial systems. *Biotechnol. Bioeng.*, **10**, 535–549.

Flickinger, M. C., N. B. Jansen and E. H. Forrest (1980). Mobile self contained off-gas analysis units for fermentation pilot plants. *Biotechnol. Bioeng.*, **22**, 1273–1276.

Heijnen, J. J. and J. A. Roels (1981). A macroscopic model describing yield and maintenance relationships in aerobic fermentation processes. *Biotechnol. Bioeng.*, **23**, 739–763.

Humphrey, A. E. (1971). Proc. of LABEX symposium on computer control of fermentation process. London.

Jazwinski, A. H. (1970). *Stochastic Processes and Filtering Theory*. Academic, New York.

Jefferis, III, R. P. (1975). Computer in the fermentation pilot plant. *Process Biochem.*, April, 15–28.

Jefferis, III, R. P., A. E. Humphrey and L. K. Nyiri (1972). Indirect measurement of microbial density and growth rate by oxygen balancing. Paper presented at 164th American Chemical Society Annual Meeting, New York.

Jefferis, III, R. P., H. Winter and H. Vogelmann (1976). Digital filtering for automatic analysis of cell density and productivity. In *Gesellschaft für Biotechnologische Forschung mbH*, ed. R. P. Jefferis, III, pp. 141–152. Verlag Chemie, Weinheim.

Marr, A. G., E. H. Nilson and D. J. Clark (1962). The maintenance requirement of *Eschrichia coli. Ann. N.Y. Acad. Sci.*, **102**, 536.

Meiners, M. and W. Rapmundt (1983). Some practical aspects of computer applications in a fermentor hall. *Biotechnol. Bioeng.*, **25**, 809–844.

Minkevich, I. G. and V. K. Eroshin (1973). Theoretical calculation of mass balance during the cultivation of microorganisms. *Biotechnol. Bioeng. Symp.*, **4**, 21–25.

Mou, D. and C. L. Cooney (1983a). Growth monitoring and control through computer-aided on-line mass balancing in a fed-batch penicillin fermentation. *Biotechnol. Bioeng.*, **25**, 225–255.

Mou, D. and C. L. Cooney (1983b). Growth monitoring and control in complex medium: a case study employing fed-batch penicillin fermentation and computer-aided on-line mass balancing. *Biotechnol. Bioeng.*, **25**, 257–269.

Mukhopadhyay, S. N. and T. K. Ghose (1976). A simple dynamic method of $k_L a$ determination in laboratory fermenter. *J. Ferment. Technol.*, **54**, 406–419.

Nyiri, L. K. (1972a). A philosophy of data acquisition, analysis, and computer control of fermentation processes. *Developments in Industrial Microbiology.*, **13**, 136–145.

Nyiri, L. K. (1972b). Application of computers in biochemical engineering. In *Advance in Biochemical Engineering*, ed. T. K. Ghose, A. Fiechter, and N. Blakebrough, vol. 2, chap. 1, pp. 49–95. Springer-Verlag, New York.

Nyiri, L. K., G. M. Toth and M. Charles (1975). On-line measurement of gas-exchange conditions in fermentation processes. *Biotechnol. Bioeng.*, **17**, 1663–1678.

Nyiri, L. K., G. M. Toth, C. S. Krishnaswami and D. V. Parmenter (1977). On-line analysis and control of fermentation processes. In *Gesellschaft für Biotechnologische Forschung mbH*, ed. R. P. Jefferis, III, pp. 37–46. Verlag Chemie, Weinheim.

Ohyama, Y. and K. Endoh (1955). Power characteristics of gas–liquid contacting mixers. *Chem. Eng. (Jpn.)*, **19**, 2.

Richards, J. W. (1961). Studies in aeration and agitation. *Prog. Ind. Microbiol.*, **3**, 143.

Roels, J. A. (1980). Application of macroscopic principles to microbial metabolism. *Biotechnol. Bioeng.*, **22**, 2457–2514.

Rolf, M. J., P. J. Hennigan, R. D. Mohler, W. A. Weigand and H. C. Lim (1982). Development of a direct digital-controlled fermentor using a microminicomputer hierarchical system. *Biotechnol. Bioeng.*, **24**, 1191–1210.

Ruchti, G., I. J. Dunn and J. R. Bourne (1981). Comparison of dynamic oxygen electrode methods for the measurement of $K_L a$. *Biotechnol. Bioeng.*, **23**, 277–290.

Rushton, J. H., E. W. Costrich, and H. J. Everett (1950). Power characteristics of mixing impellers. Part 2. *Chem. Eng. Prog.*, **46**, 467.

Stephanopoulos, G. N. (1981). State estimation for computer control of biochemical reactors. In *Proc. Ferment. Symp., 6th, London, Ontario*, ed. M. Moo-Young, C. W. Robinson and C. J. Vezino, vol. 1, pp. 399–406. Pergamon, Toronto.

Svrcek, W. Y., R. F. Elliot and J. E. Zajic (1974). The extended Kalman filter applied to a continuous culture model. *Biotechnol. Bioeng.*, **16**, 827–846.

Swartz, J. R. and C. L. Cooney (1978). Instrumentation in computer-aided fermentation. *Process Biochem.*, February, 3.

Wang, H. Y., C. L. Cooney and D. I. C. Wang (1977). Computer-aided baker's yeast fermentations. *Biotechnol. Bioeng.*, **19**, 69–86.

Wells, C. H. (1971). Application of modern estimation and identification techniques to chemical processes. *AIChE J.*, **17**, 966–973.

Zabriskie, D. W. and A. E. Humphrey (1978). Real-time estimation of aerobic batch fermentation biomass concentration by component balancing. *AIChE J.*, **24**, 138–146.

11

Immobilization Techniques— Enzymes

R. A. MESSING
Horseheads, NY, USA

11.1 INTRODUCTION

The objective of immobilization is the economic application of enzyme systems. Other benefits such as ease of control and uniformity of conversions may be derived from immobilization techniques. The greatest return from immobilization is achieved with expensive enzymes since there is a definitive cost associated with immobilizing processes. It is clearly apparent, then, that a very inexpensive enzyme should not be employed as an immobilized derivative in a process unless another advantage such as avoidance of contamination or immune response is attained. The key to quantifying the economics of an immobilized enzyme is the determination of the cost of the individual components of the immobilized system *versus* the value of the system's performance.

There are five markedly different techniques of immobilizing enzymes: (1) adsorption of the enzyme on a carrier surface; (2) covalent coupling of an enzyme to a carrier surface; (3) crosslinking between enzyme molecules (copolymerization); (4) entrapment of an enzyme in a matrix; and (5) encapsulation or confinement of an enzyme solution in a membrane structure. Although each immobilization technique is unique, they are by no means pure processes. Adsorption on a carrier surface, either intentionally or unintentionally, will involve crosslinking between enzyme molecules to some extent. Covalent coupling to a carrier surface usually involves adsorption of enzymes on that surface and crosslinking between enzyme molecules. Entrapment of an enzyme in a matrix may involve adsorption, covalent coupling to the surface or crosslinking between the enzyme molecules. Encapsulation or confinement of an enzyme solution within a membrane structure minimizes contamination by these mixed immobilization effects, however, some adsorption on the membrane surface and some crosslinking of the molecules do occur.

After intensive investigation over a quarter of a century, it has become clearly apparent that there is no universal immobilization technique. Each immobilization process has both advantages and disadvantages. The choice of technique, therefore, should be dictated by the specific conditions of the application which would selectively employ the positive attributes of the specified immobilization.

11.1.1   Definition of a Carrier

For the purpose of this discussion, a carrier is defined as the support material utilized to immobilize the enzyme. This support may be a matrix system, a membrane or a solid surface.

11.1.2   Carrier Durability under Application Conditions

When a carrier is employed for the immobilization of an enzyme, the durability of that carrier during use is second in importance only to the enzyme activity. If the carrier, whether it be a matrix or a membrane, is not stable under the pH, ionic strengths and solvent conditions of the application, the carrier will be disrupted or dissolved and the enzyme will be released into solution. This does not necessarily mean that the carrier must be durable under all pH, ionic strengths and solvent conditions, but only those of the specific application. In addition to the chemical environment a most important consideration is that of the physical environment of the application. A rigid structure, such as an inorganic material, may be readily abraded in a continuous stirred reactor. This will not only lead to dissolution of the enzyme but also to formation of non-uniform particles which may disrupt the reactor performance. Under these conditions it would be far more advantageous to choose an elastic type carrier for this application.

11.1.3   Adverse Effects of Carriers

There are reactions that are exhibited by carrier surfaces that do not involve the enzyme–substrate complex, however, these reactions may have an adverse effect on the total system in which the immobilized enzyme is employed. Immobilized enzyme systems utilized to interface with blood and tissue, either in the form of an extracorporeal shunt or as an insertion in the blood system or tissues, for the treatment of enzyme deficiencies or leukemia represent a typical problem with respect to the choice of a carrier for this type of application. Glass and many ceramics initiate or enhance the blood clotting reaction. If these materials were to be employed as a carrier for the enzyme, the clotting reaction would ultimately lead to the restriction of the blood flow and the death of the patient.

Another adverse effect of one class of carrier is that of an induced immune response. Protein supports, with perhaps the exception of pure collagen, evoke the immune response. The utilization of this type of carrier for interfacing with the circulatory system and tissues might lead to anaphylactic shock and ultimately death.

11.2   ENZYME IMMOBILIZATIONS

11.2.1   Adsorption

Although operationally adsorption is the most economical and simple process to perform, the forces involved are the most complex. Contrary to the term 'inert' frequently applied to characterize carriers or supports, these materials are quite active. As a matter of fact, the author is not aware of any material that does not have some surface contribution. It is the surface activity of the support acting in concert with a functional moiety or characteristic group on the surface of the enzyme protein that is responsible for the bonding or immobilization of the enzyme. The bonds that exist between the enzyme protein and the carrier depend upon the nature of the carrier and the nature of the enzyme protein surface. These bonds may be ionic, hydrogen, coordinate covalent, covalent, hydrophobic or a combination of any of the aforementioned bonds.

An enzyme may be immobilized by bonding to either the external or internal surface of a carrier. If the enzyme is immobilized externally, the carrier particle size must be very small in order to achieve an appreciable surface for bonding. These particles may have diameters which range from 500 Å to about 1 mm. The advantage of immobilizing the enzyme on an external surface is that no pore diffusion limitations are encountered. The disadvantages of external adsorption are: (1) a relatively low surface area for bonding; (2) the enzyme is more subject to physical abrasion, inhibitory effects and the turbulence of the bulk solution; (3) the enzyme is more exposed to microbial attack; and (4) smaller particles result in high pressure drops in a continuous packed-bed reactor and are difficult to retain in fluidized beds or continuous-stirred reactors.

Although immobilization of an enzyme to the internal surface of a porous carrier has one marked disadvantage, many advantages may be harvested from this approach. The major disadvantage of internal pore immobilization is that of pore diffusion limitations. If the pore diameter is optimized with respect to either the enzyme molecule or the substrate, whichever is the larger, then internal immobilization offers several more magnitudes of surface area for the immobilization of the enzyme than can be achieved with very small particles by external immobilization. Unlike the situation resulting from external immobilization, the enzyme immobilized on an internal surface is protected from abrasion, inhibitory bulk solutions and microbial attack. Generally speaking, a more stable and active enzyme system may be achieved with internal pore immobilization.

11.2.1.1 *Forces frequently involved in internal surface immobilization*

Although the type of bond established between the carrier and the enzyme may vary with the support material and the surface properties of the enzyme, the basic forces underlying the immobilization of enzymes on internal pore surfaces can be assessed from the adsorption studies on porous glass (Messing, 1969). These studies indicated that a force was required to attract the enzyme to the internal surface. In this case, the dissociated silanol negative charge of the glass attracted the positively charged amine groups on the surface of the protein. An additional process involves the diffusion of the protein to the internal surface. Finally, the bonding process involved multiple hydrogen bonds and ionic amine–silanol bonds. It is clearly apparent, then, that the three forces involved in internal surface immobilization are: an attractive surface force, diffusion and the formation of multiple hydrogen bonds.

11.2.1.2 *Optimization of pore diameters for internal surface immobilization*

In a given porous material having a narrow distribution of pore diameters, as the pore diameter decreases the surface area increases. In a preliminary investigation (Messing, 1970) the effective bonding surface appeared to be a function of the pore diameter and the molecular weight of the protein. This relationship was reassessed and refined in a subsequent more comprehensive study (Messing, 1974a). The objective of this study was to utilize fully the internal surface of porous carriers for attaching the enzyme and yet maintain as high a surface area as possible. To attain this objective, the smallest pore diameter with a very narrow pore distribution is required that entry of the limiting molecule (the enzyme when it is larger than the substrate) is just possible. Studies of glucose oxidase and catalase with relation to pore diameter indicated that the optimum pore diameter for immobilizing the highest quantity of active enzyme was related to the major axis of the unit cell dimension of the limiting molecule rather than the molecular weight of that molecule. The highest activity of an enzyme system attained by internal surface immobilization is achieved when the pore diameter is chosen such that it is twice the major axis of the unit cell (spin diameter) of the largest molecule in the system whether it be the largest enzyme or the substrate.

11.2.1.3 *Modification and preconditioning of carrier materials before immobilization*

The carriers generally used for immobilization are very active materials. A major concern should be that these materials are thoroughly cleaned and fully activated with respect to their surfaces prior to the exposure of the enzyme solution. Carrier materials will adsorb volatile organic substances from the air, as well as microbes. These contaminants will mask the functional groups on the surface and thus prevent the immobilization of the enzyme. The cleaning processes for the specific carrier will depend upon the basic durability characteristics of that carrier. A carrier that is very stable in an acid environment may be cleaned with an acid. On the other hand, a carrier that is stable in a basic environment may be cleaned with a base. A variety of organic solvents such as alcohol, acetone or carbon tetrachloride may be employed for the removal of fine organic films. The durability of the carrier in these solvents should be ascertained before the treatment.

A simple procedure for cleaning the surface of inorganic carriers involves the exposure of the carrier to temperatures in excess of 450 °C in the presence of sufficient air or oxygen to volatilize the carbonaceous material. A simple furnace elevated to 500 °C for 1.5 h with an ample supply of air is adequate to thoroughly clean the internal surfaces of small quantities of carrier.

Perhaps the most frequent and more costly omission in the immobilization process is that of preconditioning the carrier surface for the enzyme. Each enzyme has its own stability characteristics with respect to pH, ionic strengths, activators and cofactors. A frequent error is that an enzyme solution is applied directly to the surface of a dry carrier without prior exposure to, for example, the preferred pH or salt conditions. Due to this omission, a dramatic loss in the total enzyme activity is encountered. The losses may be attributed either to the very adverse pH environment contributed by the non-preconditioned surface or the transfer of a stabilizing salt or ion from the enzyme environment to the carrier. Prior to the immobilization process, the carrier should be treated for a sufficient length of time (greater than 1 h at room temperature) with a sufficient volume (at least 10 volumes of solution for each volume of carrier) of the buffer–activator solution which offers the greatest stability to the particular enzyme that will be immobilized.

Carriers appear to selectively bind such cofactors as metal ions. The carrier, therefore, is in competition with the enzyme requiring metal ions for activation. These metal ions are frequently removed from the functional site of the enzyme and are subsequently bound to the carrier. If this occurs, then the enzyme will become inactive. It has been noted that in many instances if the carrier is exposed to a solution of the appropriate metal ions required for this enzyme, the resultant preparation will be fully active. Pretreatment of the carrier with a concentration of metal ions approximately 10 times higher than that required for activation of the enzyme, in a volume at least twice the volume of the carrier, for approximately one hour at 37 °C, has generally been effective for saturating the carrier requirements for the metal ion. This pretreatment will generally avoid competition by the carrier for enzyme metal ions.

11.2.1.4 *Methods for immobilization by adsorption*

There are four procedures that have been used for the immobilization of enzymes by adsorption. They are the static process, the dynamic batch process, the reactor loading process and the electrodeposition process.

(i) The static process

This technique is the most inefficient of the four processes and requires the most time. The enzyme is immobilized on the carrier by simply allowing the solution containing the enzyme to contact the carrier without agitation or stirring. Generally, the loading of the enzyme on the carrier surface is not uniform and is rather low. The carrier must be either exposed to high concentrations of the enzyme or for long periods of time at lower concentrations (many days) in order to achieve a modestly active immobilized enzyme preparation.

(ii) The dynamic batch process

This is the procedure most frequently employed for laboratory preparations of immobilized enzymes. The carrier is placed into the enzyme solution and the carrier and enzyme solution are either mixed by stirring or agitated continuously on a shaker. This process is rather effective and normally results in relatively uniform high loading if adequate concentrations of enzymes are employed.

When using this process precautions should be taken to ensure that agitation is not so vigorous that the carrier will be abraded and disrupted, however, it needs to be sufficiently vigorous to allow a low density carrier surface to be adequately exposed to the enzyme solution.

(iii) The reactor loading process

This process is most frequently employed for the commercial preparation of immobilized enzymes. The carrier is placed into the reactor that will be subsequently employed for processing. The enzyme solution is then transferred to the reactor and the carrier is loaded in a dynamic environment by either circulating the enzyme or by agitation of the enzyme solution and carrier.

This process may be utilized with continuously stirred reactors, packed-bed reactors or fluidized-bed reactors.

(iv) The electrodeposition process

Electrodeposition is a process in which the carrier is placed proximal to one of the electrodes in an enzyme bath, the current is turned on, the enzyme migrates to the carrier and is deposited upon the surface. It should be ascertained prior to the employment of a specific carrier whether that particular carrier is durable in an electric field. In an electric field, ions will be removed from the surface of the carrier. If these ions are involved in the stability or the activity of the enzyme, then either the ions should be added to the bath solution or replaced in the immobilized preparation as soon as possible.

11.2.1.5 Additional phenomena exhibited by carrier immobilized systems

(i) Non-specific adsorption or accumulations

A carrier contains surface active groups that will not only attract the enzyme which is to be immobilized, but will also subsequently attract contaminating ions or macromolecules. Further, the enzyme immobilized on the carrier surface contains surface active groups that will attract these same contaminants. If a sufficient number of large contaminating molecules are adsorbed on the active site of the enzyme, the activity of that preparation will be diminished. Metal ions that are inhibitory to the enzyme may be adsorbed and will decrease the activity of the preparation. Various methods of overcoming these effects have been devised, including removal of the metal ions by chelating agents and the removal of the macromolecules by detergent rinses or enzyme treatment.

An example of a carrier surface accumulation that blocked the entry of the substrate to the enzyme which was immobilized on an internal pore surface is that of immobilized papain (Figure 1). The substrate employed for this study was casein, which upon acidification or mild hydrolysis, tends to clot or form insoluble particles. Insoluble particles may have been induced either by the relatively acid surface of the titania carrier or by the hydrolysis of casein accumulated on the carrier particles. As a result the mouth of the pore became inaccessible to the unhydrolyzed casein and thus did not have access to the active site of the enzyme. The original delivery of the substrate was in a downward plug-flow pattern. After the activity declined over an interval of 20 days, the bed was washed by fluidization with an upward flow of water and the insoluble particles were removed. The activity of the column was partially restored, which indicated that the substrate gained access to the enzyme. The substrate was then delivered by downward flow and an activity decay rate similar to that observed previously was noted with the simultaneous appearance of the loosely bound insoluble particles. The bed was then washed again by fluidization and the activity was again restored.

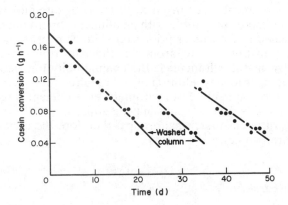

Figure 1 Performance of papain immobilized by adsorption in a continuous plug-flow reactor with a casein substrate at 37 °C (From Messing, 1975a with the permission of Academic Press, New York)

(ii) Reversible activity loss

It has been noted repeatedly that the activity of an immobilized enzyme can increase by 20–45% of the initial activity (Figure 2; Messing, 1974a). A study of glucose isomerase immobilized on controlled-pore alumina (Figure 2) in a packed-bed under continuous operation indicated that it takes approximately four to five days for this immobilized preparation to demonstrate its full activity. Initially, some of the enzyme-active sites may be bound to the carrier or to an adjacent enzyme molecule. Since the affinity of the active site is greater for the substrate (glucose) than for either the carrier or the adjacent enzyme molecule, those sites become available as the glucose diffuses to the vicinity of the bound sites in the depth of the pore.

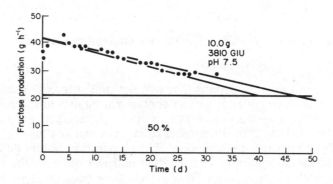

Figure 2 Column performance of glucose isomerase immobilized by adsorption in controlled-pore alumina (From Messing, 1975b with the permission of Elsevier, New York)

11.2.2 Covalent Coupling to a Carrier Surface

Covalent attachment to surfaces generally offers the advantage of an immobilized enzyme system that can be utilized under a broad spectrum of pH conditions, ionic strengths and uncontrolled variable conditions. The disadvantage of this type of attachment is that generally it requires a multistep process to immobilize the enzyme. These steps may include the attachment of a coupling agent following by an activation process or the attachment of a functional group, and finally the attachment of the enzyme.

Covalent attachment may be directed to a specific group (amine, hydroxyl, tyrosyl, *etc.*) on the surface of the enzyme. Thus, the active site of the enzyme can be avoided by judiciously choosing a group on the surface of the enzyme protein that is not involved in the site.

Frequently, covalent coupling is preferred to other processes where the enzyme may contain polymeric units or prosthetic groups. There appears to be less tendency to disrupt the complex nature of these enzymes since specific bonds can be formed with the functional group.

Enzyme composites may be formed with covalent techniques that are appropriately tailored for specific substrates or environments. For instance, if the substrate molecule is hydrophobic and aliphatic, then it would be more compatible with an aliphatic environment and a long chain coupling agent may be attached to the carrier prior to the attachment of the enzyme. On the other hand, if the substrate is hydrophic and aromatic, then an aromatic coupling agent may be attached to the surface so that it will increase the compatibility for the enzyme–substrate reaction. The hydrophobic composite may additionally be required for enzyme reactions occurring in partially organic solvent environments. If a hydrophilic environment is required, then a coupling agent containing a number of hydroxyl groups may be utilized.

Each carrier type has different functional groups, therefore the attachment to these carriers will be discussed as separate topics.

11.2.2.1 Carbohydrate carriers

The attachment to carbohydrate carriers is generally through a hydroxyl group. Axen and coworkers (1967) and Porath *et al.* (1967) developed the cyanogen bromide technique for coupling enzymes to carbohydrates. This technique involves the attachment of the cyano group to two adjacent hydroxyls on the surface according to the scheme shown in Figure 3.

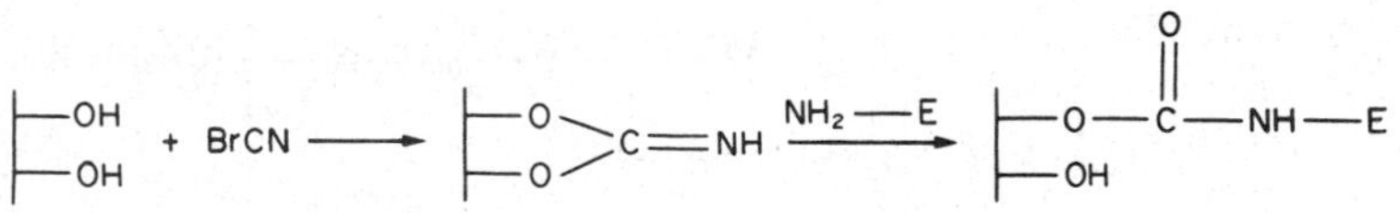

Figure 3 Cyanogen bromide technique for coupling enzymes to carbohydrate carriers

One of the earliest procedures employed for attachment to carbohydrates was that of azide coupling (Figure 4; Mitz and Summaria, 1961).

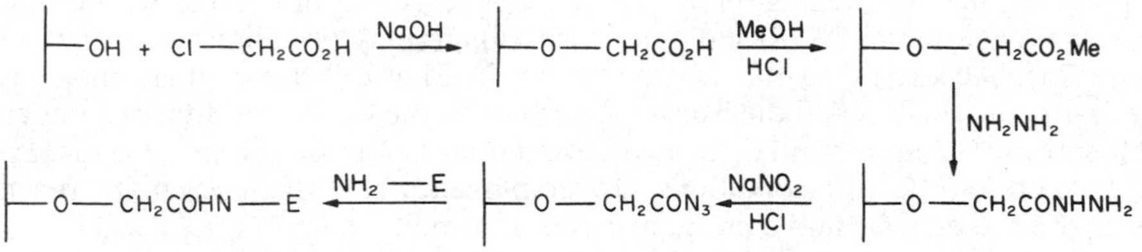

Figure 4 Azide coupling of enzymes to carbohydrate carriers

The carbodiimide procedure developed by Weliky and Weetall (1965) results in attachment through the carboxyl group which may be formed on cellulose, in a manner similar to the first reaction under azide coupling. The reaction scheme then follows the path shown in Figure 5.

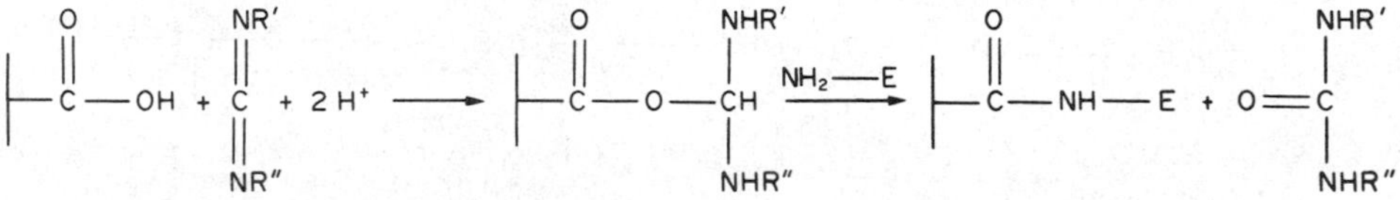

Figure 5 Carbodiimide technique for coupling enzymes to carbohydrate carriers

Other techniques that have been utilized to attach directly to hydroxyl groups of carbohydrates are the cyanuric chloride developed by Kay and Crook (1967) and the transition metal-activation developed by Barker and coworkers (1971).

11.2.2.2 Protein and amine bearing carriers

A variety of multifunctional reagents have been utilized for the covalent attachment of enzymes to protein- and amine-bearing carriers. Primarily difunctional reagents have been employed. In order to bind selectively to tyrosyl groups, diazobenzidine and its derivatives have been utilized. Glutaraldehyde forms a Schiff base with amine groups. Both aliphatic and aromatic isocyanates under alkaline conditions will form substituted urea bonds with amine groups. Similarly, isothiocyanates will form substituted thiourea bonds with amines.

11.2.2.3 Inorganic carriers

Perhaps the most frequently used technique for the covalent attachment to inorganic surfaces is that of silane coupling (Messing and Weetall, 1970; Weetall, 1969). This technique involves attaching a silane through an inorganic functional group to the inorganic surface leaving the organic moiety, such as an amine, available for covalent attachment of the enzyme. The most frequently used silane is γ-aminopropyltriethoxysilane and this scheme is as shown in Figure 6. The terminal amine may be functionalized by bifunctional coupling agents (as discussed in Section 11.2.2.2) or by protein- and amine-bearing carriers.

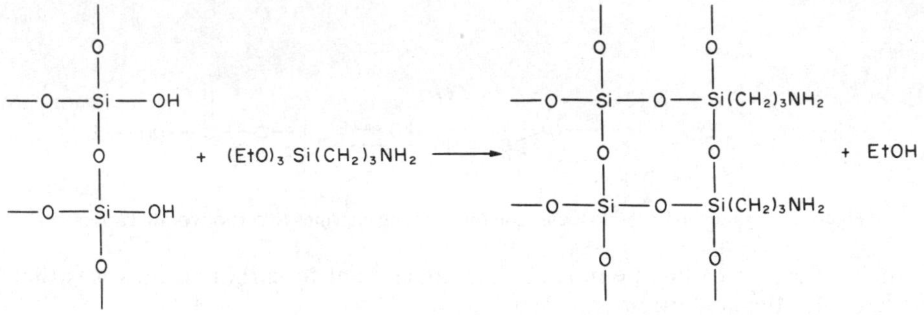

Figure 6   Silane coupling of enzymes to inorganic carriers

Another process for immobilizing on inorganic supports is that of polymeric isocyanate bonding (Messing and Yaverbaum, 1978). Polymeric isocyanate may be represented by the structure below (Figure 7a). According to the scheme shown in Figure 7 a carbamate bond is formed between the inorganic surface and the isocyanate group. If the enzyme is attached under alkaline conditions, a substituted urea bond is formed between an amine on the protein surface and the isocyanate. If moderately acid conditions are employed, then the isocyanate reacts with a hydroxyl group on the enzyme and a urethane bond is formed.

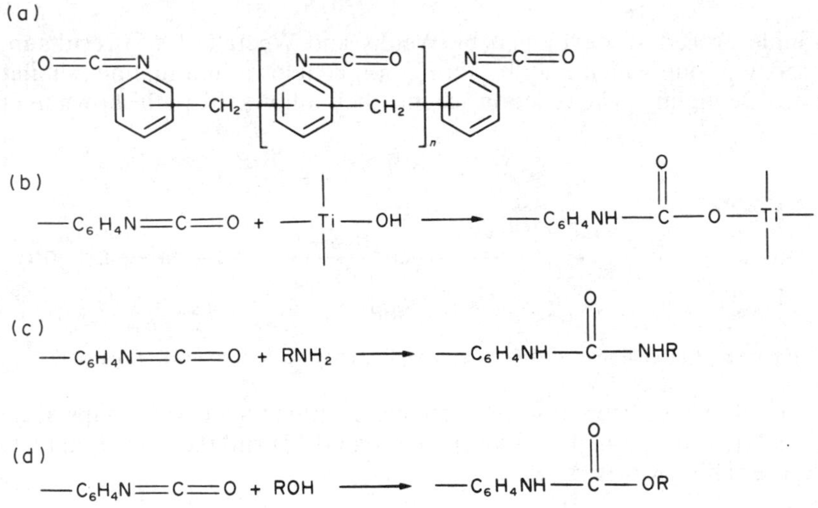

Figure 7   Polymeric isocyanate bonding of enzymes to inorganic carriers: (a) polymeric isocyanate; (b) carbamate bond formation; (c) substituted urea bond formation; (d) urethane bond formation

Other techniques that have been employed for covalent attachment to inorganic surfaces are the tin(II) bridge (Messing, 1974b) and an aromatic amine coupling process that employed *o*-dianisidine (Messing and coworkers, 1976). Although an ionic amine bond was formed between the coupling agent and the silica surface, it was further stabilized by two hydrogen bonds formed between the methoxy groups and the silica surface.

11.2.3   Crosslinking Between Enzyme Molecules or Copolymerization

The purpose of crosslinking between enzyme molecules or copolymerizing enzyme molecules with other monomers is essentially to obtain an insoluble preparation that can be readily handled or manipulated in a continuous reactor. These immobilized preparations are generally rather difficult to handle because they are gummy materials or gels. In particulate form they present multiple problems of high pressure drops in a packed-bed reactor and poor fluidizing properties in a fluidized-bed reactor. If, however, these preparations can be made in a membrane configuration, then they can be utilized in a spiral-wound reactor module similar to that described by Vieth and Venkatasubramanian (1976). The spiral-wound module is used as a packed-bed reactor.

The enzyme activity of immobilized preparations prepared by crosslinking with multifunctional reagents will depend upon the degree of crosslinking. If high concentrations of crosslinking agents are employed, more bonds will be formed which will generally result in lower activities. The additional bonds formed at high concentrations not only block the entrance of the substrate to the active sites but also may crosslink the active site itself. The effect of the quantity of the crosslinking agent, diazobenzidine, on papain was demonstrated by Silman and coworkers (1966). The effect of glutaraldehyde crosslinking on carboxypeptidase A was demonstrated by Quiocho and Richards (1964, 1966).

The preparation of an immobilized trypsin prepared by copolymerization with ethylene and maleic anhydride was described by Bar-Eli and Katchalski (1960). Levin and coworkers (1964) described trypsin preparations formed by copolymerizing the enzymes with peptides. All of these techniques have been utilized to study the charge effect in immobilized enzymes. They have not proven to be commercially significant.

11.2.4 Entrapment in a Matrix

Entrapment may be viewed as the physical confinement of an enzyme in a matrix network. It is distinguished from crosslinking of enzymes and copolymerization of enzymes in that the enzyme is not chemically part of the matrix.

This immobilization technique has been extensively explored since the early work of Bernfeld and Wan (1963). The most frequently employed matrix for entrapment has been polyacrylamide, but starch, collagen and silicone rubber have also been used. Bauman and coworkers (1965) immobilized cholinesterase in starch. Vieth and Venkatasubramanian (1976) have thoroughly evaluated the entrapment of enzymes in collagen matrices. Pennington and coworkers (1968) pioneered the entrapment of enzymes in silicone rubber. Of course, the initial studies of Bernfeld an Wan (1963) were performed with polyacrylamide. An early study of an entrapped enzyme in polyacrylamide is that of glucose oxidase for employment in an enzyme electrode (Updike and Hicks, 1967).

One of the major concerns with respect to entrapped enzymes is that of leaching. The enzyme may migrate out of the pore if the pore is too large. In many cases, this leaching may be overcome by simply crosslinking the enzyme after entrapment with a bifunctional reagent such as glutaraldehyde.

Another concern that can be expressed with respect to entrapment is that of pore diffusion limitations. If the substrate is a rather large molecule, such as a protein, it may be restricted from entry into the pore.

Although the entrapment of enzymes has been widely used for sensing applications, it has not achieved a great deal of success in industrial processes.

11.2.5 Encapsulation or Confinement in a Membrane Structure

Encapsulation or microencapulation can be distinguished from entrapment by the fact that a solution of the enzyme is separated from the bulk solution by a membrane in encapsulation, while entrapment involves a lattice-immobilized enzyme. Encapsulation offers the opportunity to immobilize larger quantities of enzyme per unit volume of immobilized preparation than any other procedure. Perhaps the biggest disadvantage of this immobilization technique is that only relatively small substrate molecules can be utilized with the intact membranes. The pioneer of this technique was Chang (1964). Although this procedure probably will not be widely adopted for industrial applications, the potential for therapeutic treatment of various diseases is most intriguing and promising.

Chang (1977) has described the encapsulation of enzymes in a variety of membranes. Two of the membranes that he used were cellulose nitrate and nylon. The scheme that was employed for microencapsulation by Chang is shown in Figure 8.

The potential of the above approach was demonstrated when a combination of urease and an ammonia adsorbent was encapsulated and administered to the gastro-intestinal tract. Blood urea levels were significantly reduced utilizing this procedure (Chang and Loa, 1970).

Another interesting technique for encapsulating enzymes is that which employs red blood cells

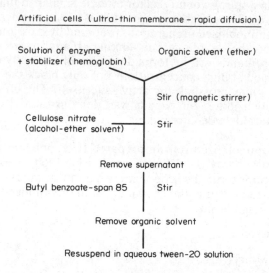

Figure 8   Enzyme microencapsulation (Chang, 1977)

as the encapsulating membrane (Ihler and coworkers, 1973). The cells are swollen in hypotonic solution. During this swelling process, minute holes, sufficiently large to allow the entrance of the enzyme, appear in the membrane. After the cells are loaded with enzyme solution, they are exposed to isotonic solutions and the holes are resealed. If erythrocytes were to be harvested from a patient to be treated for an enzyme deficiency and loaded with the appropriate enzyme, there should be no foreign body response upon injection of this encapsulated enzyme.

The most promising of the delivery systems is liposome encapsulation (Gregoriadias, 1977). Frequently the terminology utilized for this immobilization is liposome entrapment, however, this procedure is truly an encapsulation. The liposome is a lipid vesicle consisting of one or more concentric phospholipid bilayers alternating with aqueous compartments within which enzymes and other water soluble solutions of materials may be encapsulated. The formation of liposomes occurs upon the confrontation of water-insoluble polar lipids with water in the presence of an appropriate energy form which can disrupt the lipid layer (sonication).

One of the interesting aspects of the liposome technique is that it may be employed either as a delivery system with the subsequent liberation of the enzyme, or the enzyme may remain encapsulated in which case the small substrate molecules penetrate the phospholipid layers, react with the enzyme in the aqueous phase and the products are eliminated through the bilayer. The first technique of delivery and liberation is possible because the phospholipid bilayer, with the appropriate charge, appears to become preferentially associated with liver and spleen cells. The liposomes penetrate the membrane of these cells and subsequently fuse with the cell lysosomes at which time the enzyme is liberated. Thus, the liver and the spleen appear to be appropriate target organs for the application of this technique. In addition, liposomes can be tailored by the degree of saturation and modification of surface charge, such that in the future liposomes may be tailored for other target tissues. It would appear for the present that enzyme deficiencies or malfunctions in liver and spleen cells related to carbohydrate hydrolysis may benefit from this approach.

The preparation of liposomes is detailed in Figure 9.

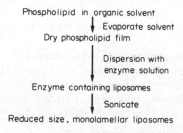

Figure 9   Liposome formation (Gregoriadis, 1977)

May and Li (1972) described a procedure called liquid-membrane encapsulation. This membrane is formed by making an emulsion to two immiscible phases and dispersing the emulsion in a third phase. The emulsion can be either oil in water or water in oil. The enzyme solution is contained within the water soluble phase. This particular encapsulation has not found very wide acceptance in enzyme application.

11.3 CONCLUSIONS

The most important conclusion that can be drawn from studies performed to date with immobilized enzyme systems is that the immobilization technique should be chosen according to the specific application.

11.4 REFERENCES

Axen, R., J. Porath and S. Ernback (1967). Chemical coupling of peptides and proteins to polysaccharides by means of cyanogen halides. *Nature (London)*, **214**, 1302.

Bar-Eli, A. and E. Katchalski (1960). A water-insoluble trypsin derivative and its use as a trypsin column. *Nature (London)*, **188**, 856.

Barker, S. A., A. N. Emery and J. M. Novais (1971). Enzyme reactors for industry. *Process Biochem.*, **6** (10), 11.

Bauman, E. K., L. H. Goodson, G. G. Guilbalt and D. N. Kramer (1965). Preparation of immobilized cholinesterase for use in analytical chemistry. *Anal. Chem.*, **37**, 1378.

Bernfeld, P. and J. Wan (1963). Antigens and enzymes made insoluble by entrapping them into lattices of synthetic polymers. *Science*, **142**, 678.

Chang, T. M. S. (1964). Semipermeable microcapsules. *Science*, **146**, 524.

Chang, T. M. S. (1977). Encapsulation of enzymes, cell contents, cells, vaccines, antigens, antiserum, cofactors, hormones, and proteins. In *Biomedical Applications of Immobilized Enzymes and Proteins*, ed. T. M. S. Chang, vol. I, pp. 69–89. Plenum, New York.

Chang, T. M. S. and S. K. Loa (1970). Urea removal of urease and ammonia adsorbents in the intestine. *Physiologist*, **13**, 70.

Gregoriadis, G. (1977). Liposomes as carriers of enzymes and proteins in medicine. In *Biomedical Applications of Immobilized Enzymes and Proteins*, ed. T. M. S. Chang, vol. I, pp. 191–218. Plenum, New York.

Ihler, G., R. Glew and F. Schnure (1973). Enzyme loading of erythrocytes. *Proc. Natl. Acad. Sci. USA*, **70**, 2663.

Kay, G. and E. M. Crook (1967). Coupling of enzymes to cellulose using chloro-*s*-triazines. *Nature (London)*, **216**, 514.

Levin, Y., M. Pecht, L. Goldstein and E. Katchalski (1964). A water-insoluble polyanionic derivative of trypsin. I. Preparation and properties, *Biochemistry*, **3**, 1905.

May, S. W. and N. N. Li (1972). The immobilization of urease using liquid–surfactant membranes. *Biochem. Biophys. Res. Commun.*, **47**, 1179.

Messing, R. A. (1969). Molecular inclusions. Adsorption of macromolecules on porous glass membranes. *J. Am. Chem. Soc.*, **91**, 2370.

Messing, R. A. (1970). Relationship of pore size and surface area to quantity of stabilized enzyme bound to glass. *Enzymol.*, **39**, 12.

Messing, R. A. (1974a). Simultaneously immobilized glucose oxidase and catalase in controlled-pore titania. *Biotechnol. Bioeng.*, **16**, 897.

Messing, R. A. (1974b). A stannous bridge for coupling urease to controlled-pore titania. *Biotechnol. Bioeng.*, **16**, 1419.

Messing, R. A. (1975a). *Immobilized Enzymes for Industrial Reactors*, p. 81. Academic, New York.

Messing, R. A. (1975b). Adsorption of proteins on glass surfaces and pertinent parameters for the immobilization of enzymes in the pores of inorganic carriers. *J. Non-Cryst. Solid*, **19**, 277.

Messing, R. A. and H. H. Weetall (1970). Chemically coupled enzymes. *US Pat.* 3 519 538.

Messing, R. A. and S. Yaverbaum (1978). Proteins bonded *via* polyisocyanates to inorganic support materials. *US Pat.* 4 071 409.

Messing, R. A., L. F. Bialousz and R. E. Lindner (1976). Aromatic amine coupling of *Aspergillus niger* lactase to controlled-pore silica with *o*-dianisidine. *J. Solid-Phase Biochem.*, **1**, 151.

Mitz, M. A. and L. J. Summaria (1961). Synthesis of biologically active cellulose derivatives of enzymes. *Nature (London)*, **189**, 576.

Pennington, S. N., H. D. Brown, A. B. Patel and S. K. Chattopadhyay (1968). Silastic entrapment of glucose oxidase–peroxidase and acetylcholinesterase. *J. Biomed. Mater. Res.*, **2**, 443.

Porath, J., R. Axen and S. Ernback (1967). Chemical coupling of proteins to agarose. *Nature (London)*, **215**, 1491.

Quiocho, F. A. and F. M. Richards (1964). Intermolecular cross linking of a protein in the crystalline state: carboxypeptidase-A. *Proc. Natl. Acad. Sci. USA*, **52**, 833.

Quiocho, F. A. and F. M. Richards (1966). The enzymic behavior of carboxypeptidase-A in the solid state. *Biochemistry*, **5**, 4062.

Silman, I. H., M. Albu-Weissenberg and E. Katchalski (1966). Some water-insoluble papain derivatives. *Biopolymers*, **4**, 441.

Updike, S. J. and G. P. Hicks (1967). The enzyme electrode. *Nature (London)*, **214**, 986.

Vieth, W. R. and Venkatasubramanian (1976). Collagen-immobilized enzyme systems. In *Methods in Enzymology*, ed. K. Mosbach, vol. 44, pp. 243–263. Academic, New York.

Weliky N. and H. H. Weetall (1965). The chemistry and use of cellulose derivatives for the study of biological systems. *Immunochemistry*, **2**, 293.

Weetall, H. H. (1969). Trypsin and papain covalently coupled to porous glass: preparation and characterization. *Science*, **166**, 615.

12

Immobilization Techniques—Cells

J. KLEIN and K.-D. VORLOP
Technischen Universität Braunschweig, Federal Republic of Germany

12.1 INTRODUCTION

Immobilization of whole cells is now a well-established method in the field of enzyme technology, both scientifically and industrially. The scientific status is documented by an ever-increasing number of publications; the industrial breakthrough is demonstrated by the successful performance of several industrial plants (Chibata, 1979).

This development mainly occurred during the past decade, while the first landmark papers, which bridged the gap between some old, but rather accidentally developed technologies (*i.e.* vinegar production) and the art of rational design of an immobilized cell system, appeared around 20 years ago (Hattori and Furusaka, 1959, 1960; Mosbach and Mosbach, 1966).

The really important point was the observation that enzymes are active and stable for long periods of time, if kept within the cellular domain together with all other cell constituents, whether the cells are dead or viable, but in a resting state. The development of immobilized cell systems happened on the following lines.

First of all, single-enzyme catalyzed reactions were considered, where the preservation of cell viability was not a prerequisite or where cell viability was destroyed deliberately by heat treatment or chemical treatment to improve the catalyst stability. It should be mentioned however that cell viability could also be advantageous in these groups of applications, because reactivation or enhancement of activity by cell growth becomes possible. All these immobilized cell systems have to compete technically and economically with immobilized enzyme systems.

The next step in heterogeneous biocatalysis was the controlled immobilization of viable cells, which were however kept in the resting state and were able to catalyze multienzyme and cofactor dependent reactions. It is obvious that this group of reactions really is the domain of immobilized whole cells, where no practical alternative exists on the immobilized enzyme level.

The final step was the immobilization of living and growing immobilized cells, where (limited) growth is a necessary prerequisite for the bioconversion. This development was not only important for the expansion of the field of microbial cell immobilization, but even more was the key point to include plant cells and mammalian cells as well.

Comparing the task of cell immobilization on the one hand and that of enzyme immobilization on the other, the size of the cells makes it much easier to develop rather simple immobilization techniques such as entrapment. With increasingly high demands from cell physiology in connection with viable and growing cells the task is more difficult, but in most cases solutions could be presented. In addition to the principal strategies of (a) biocatalyst aggregation (crosslinking flocculation), (b) attachment of the biocatalyst to a pretreated carrier (adsorption, covalent bonding), and (c) entrapment (porous network, microencapsulation), the concentration and the spatial distribution of the biocatalyst species can be influenced by cell growth within a carrier (*e.g.* cell growth from immobilized spores). Further, by gradual change in reaction conditions the physiology and activity of the immobilized species can be adapted to new levels in continuous processes.

The overall composition of an immobilized cell system is chemically less well defined than an immobilized enzyme preparation. Careful attention has to be given to the loss of cell matter or reaction by-products into the reaction medium and their action as toxins.

It is the objective of this chapter to present in a comprehensive fashion the status of the methodological development of whole cell immobilization. Selected applications will be mentioned in discussion of the feasibility and flexibility of the immobilization methods.

For a number of years the field of whole cell immobilization has been reviewed extensively (Abbott, 1976, 1977; Berger, 1981, 1982; Brodelius, 1978; Bucke and Wiseman, 1981; Chibata and Tosa, 1977, 1981; Chibata *et al.*, 1983; Durand and Navarro, 1978; Fukui and Tanaka, 1982b; Jack and Zajic, 1977a; Klein and Wagner, 1977, 1978, 1983; Kolot, 1981; Mosbach, 1982; Vandamme, 1976; Vishnoi, 1980), again demonstrating the extent of interest in this research area.

12.2 IMMOBILIZATION PROCEDURES

12.2.1 Adsorption

Adsorption of cells on to a preformed carrier is a classical method. For example, the process for vinegar production using woodchips as carriers for *Acetobacter* adsorption has been in existence since 1823. More than 100 years had to pass by before the method of cell immobilization was revitalized, again using an adsorption technique (Hattori and Furusaka, 1959, 1960). In retrospect, this is not surprising because the adsorption technique is very simple and very mild.

A preformed carrier of very variable structure is mixed with the cell suspension and the cells adhere to the surface in a more or less complete way. Since the carrier is inert and no additional chemicals are involved in the process, the immobilization is carried out under the same physiological conditions in which the cells are kept in suspension. It is therefore not surprising that viable cells (Hattori and Furusaka, 1959, 1960) and living cells for ethanol production (Navarro and Durand, 1977) or in steroid conversion (Atrat and Groh, 1981) could be successfully immobilized.

As can be seen from Table 1, carriers of different origin have been used to solve various immobilization problems. Since the carrier is preformed and carrier preparation does not interfere with the immobilization step, a large freedom in the choice of a suitable carrier exists. This is demonstrated by the fact that both inorganic and organic materials can be applied, and that within each group, selection from a wide range of materials, from very simple (and inexpensive) natural products such as sand or woodchips to highly sophisticated structures such as controlled pore glass or Concanavalin A-Sepharose, is possible.

Keeping these advantages of simplicity, flexibility and physiology in mind, the question has to

Table 1   Immobilization of Whole Cells by Adsorption

Support material	Cells	Reaction	Ref.
(a) Adsorption			
Chitosan (crosslinked)	Animal cells	Cell growth	Nilsson and Mosbach (1980)
Gelatin (crosslinked)	Animal cells	Cell growth	Nilsson and Mosbach (1980)
Gelatin	Lactobacilli	Lactose/lactic acid	Compere and Griffith (1976)
PVC, bricks	*Saccharomyces carlsbergensis*	Beer	Corrieu *et al.* (1976)
Porous glass	*Saccharomyces carlsbergensis*	Glucose/ethanol	Navarro and Durand (1977)
Anthracite	*Pseudomonas* sp.	Phenol degradation	Scott and Hancher (1977)
Cotton fibers	*Zymomonas mobilis*	Glucose/ethanol	Anon. (1981)
Vermiculite	*Zymomonas mobilis*	Glucose/ethanol	Bland *et al.* (1982)
Pectate (crosslinked)	*Saccharomyces cerevisiae*	Glucose/ethanol	Vijayalakshmi *et al.* (1979)
Cordierite, Duralite Noire®	Mixed culture	Waste water treatment	Messing (1982)
(b) Ionic adsorption			
Dowex I. E. resin	*E. coli*	Succinic acid	Hattori and Furusaka (1959)
DEAE-cellulose	*Nocardia erythropolis*	Steroid conversion	Atrat and Groh (1981)
(c) Bioadsorption			
Concanavalin A-Sepharose	Erythrocytes		Mattiasson and Borrebaeck (1978)

be raised as to why the overall use of this technique is still limited. These limiting factors are adhesion capacity and adhesive strength.

To obtain a high cell loading capacity the inner surface of a porous particulate material has to be accessible for adsorption as well as the outer. Therefore, the pore diameter has to be small enough to give a high surface area, but on the other hand large enough to allow for cell penetration. In the case of bacteria and yeast, a ratio of pore to cell diameter of 4 to 5 has been reported (Messing *et al.*, 1979). With respect to accessibility on the one hand and specific surface area on the other, an optimum pore size can be determined (Messing *et al.*, 1979; Navarro and Durand, 1977). Adsorption capacities are usually reported on a mg g^{-1} basis, which makes it difficult to compare carriers of different chemical nature. These values range from 2 to 248 mg g^{-1}, where in the first case porous silica and in the second case wood chips have been used (Durand and Navarro, 1978). Recent studies on immobilized yeast in ethanol fermentation confirm the finding that wood chips is the material which provides highest adsorption capacity (Moo-Young *et al.*, 1980). On a cell number basis, values of 10^9 cells g^{-1} for *E. coli* and 10^7–10^8 cells g^{-1} for yeast have been determined on porous glass carriers (Messing *et al.*, 1979). This should be compared to 6 × 10^9 cells ml^{-1} for yeast cells entrapped in *K*-carrageenan (Chibata, 1979) showing the much higher loading capacity of the latter method.

The second point of concern is adhesive strength. Interaction forces generally will not be very strong, with the exception of those species which have developed special mechanisms for surface-anchored growth. A detailed analysis of the thermodynamics of adhesion as related to the surface energy of carriers and cells has been given (Gerson and Zajic, 1979). These problems have found substantial interest in soil microbiology and in processes like bacterial leaching, although experimental data in the area of biocatalysis are scarce (Ash, 1979). Some information on the influence of pH (Navarro and Durand, 1977; Marcipar *et al.*, 1979), ionic strength (Vijayalakshmi *et al.*, 1979) and support composition (Marcipar *et al.*, 1979) is available. Not only van der Waals forces but also electrostatic forces have to be considered, which seems obvious in the case of adsorption of negatively charged cells on positively charged anion exchange resins (Seyan and Kirwan, 1979). However, the same cells adhere to negatively charged inorganic carriers such as glass or ceramic, and therefore partial covalent bonding has to be postulated to account for such observations. The experimental data available so far are not at all conclusive in a mechanistic sense and more experimental data on surface properties of cells and carriers are required to develop a better understanding. Furthermore, specific chelating agents may be advantageous in developing stronger surface to cell interaction (Vijayalakshmi *et al.*, 1979; Kolot, 1981; Kennedy *et al.*, 1980).

In an overall evaluation of the method of cell adsorption, the advantages of simplicity and negligible changes in cell physiology are overshadowed to some extent by the disadvantages of limited cell loading and limited adhesion stability compared to cell entrapment. Thus, application in any reactor with higher shear fields caused by high fluid flow rates or intensive mixing is prohibited due to cell wash-out. The main fields of technical application are those cases where inexpensive carriers and lower fluid velocities in fixed bed reactors are required, as in large volume anaerobic waste water treatment. Another area of interest will be the use of polymeric microcarriers for the cultivation of animal and mammalian cells which require surface anchorage for cell growth (Nilsson and Mosbach, 1980).

12.2.2 Covalent Bonding

Covalent bonding is another way of attaching cells to the surface of a preformed carrier. The considerations of loading capacity as a function of pore size distribution and surface area are therefore identical to those discussed in the case of adsorption.

The difference lies in the way the cells are coupled to the surface, *viz.* covalent bonds as opposed to secondary valence bonding. Covalent bonding is by far the most important and widely applied technique in the field of enzyme immobilization. It is decidedly puzzling why this technique is not much more than a curiosity in the field of cell immobilization.

In this respect, one has to realize that the expected advantage of higher bonding stability has to be paid for by two disadvantages when compared to adsorption. On the one hand, the number of possible carriers due to the specific requirements for functionalized surfaces is limited, and on the other, chemical coupling agents or reactive groups are introduced giving rise to nonphysiological conditions and possible toxic effects on enzyme or cell activity.

Table 2 gives a summary of some of the pertinent papers devoted to this method. Special emphasis should be given to the work of Kennedy and his group (1976, 1978, 1980) in relation to their efforts for developing a series of inorganic carriers which simultaneously provide functionalities for covalent bonding. The key factor is the formation of insoluble oxides of transition metals like titanium or zirconium starting from their soluble hydroxides, and simultaneous bonding of the cells, even in a living state (Kennedy *et al.*, 1980; Kennedy, 1978).

Table 2   Immobilization of Whole Cells by Covalent Bonding

Reactants	Cells	Reaction	Ref.
CM cellulose + carbodiimide	*B. subtilis*	L-Histidine/urocanic acid	Jack and Jajic (1977b)
Glycidylmethacrylate polymer + glutaraldehyde	*Aspergillus niger*	Glucose/gluconic acid	Nelson (1976)
Carriers with aldehyde, oxirane and amine groups	*E. coli*	Penicillin G/6-APA	Žůrkobá *et al.* (1978)
Carriers with epoxy and halocarbonyl groups	*Proteus rettgeri*	Penicillin G/6-APA	Nelson (1976)
Zr(IV) oxide	*E. coli*	Oxygen uptake	Kennedy *et al.* (1976)
Ti(IV) oxide, *etc.*	*Acetobacter* sp.	Wort/vinegar	Kennedy (1978)
Cellulose + cyanuric chloride	*Saccharomyces cerevisiae*	Glucose/ethanol	Gainer *et al.* (1980)

12.2.3 Cell to Cell Crosslinking

The simplest way of achieving cell aggregation in the form of larger particles with higher cell density is by flocculation. However, few microorganisms are available with sufficiently high flocculating activity, although many studies are under way to improve this situation, *e.g.* in anaerobic waste water treatment (Lettinga *et al.*, 1980). A more generally applicable way is to enhance flocculation and stabilize the cell aggregates by the addition of crosslinking chemicals.

One should distinguish between low molecular weight bi- or multi-functional reagents such as glutaraldehyde which give rise to chemical crosslinks, and higher molecular weight, usually ionic, polymers which assist in producing a physical crosslink.

However, in the first case one has to realize that the coupling agent reacts not only with the cell surface, but also to an appreciable extent with the proteins in the interior of the cells. The cells therefore will generally lose their viability and thus their ability to catalyze multienzyme reactions, limiting this technique to single enzyme reactions (Chibata *et al.*, 1974). It is obvious that high cell densities and, thus, possibly high activities are special features of these preparations.

By addition of polyelectrolytes like chitosan (Tsumara and Kasumi, 1976) flocculation can be enhanced not only with respect to the degree and velocity of cell attraction but also with respect to the stability of the floccules. Application of two oppositely charged polymers has been reported as well (van Keulen *et al.*, 1981). Further steps of treatment may be extrusion drying, granulation and finally additional chemical crosslinking with glutaraldehyde, to obtain biocatalysts with high stability and in some cases very high specific activity, *e.g.* in the glucose to fructose isomerization reaction. As can be seen from Table 3, this method of immobilization has not been widely used in spite of its simplicity. This may possibly be attributed to the denaturing effect of the immobilization procedure.

Table 3   Immobilization of Whole Cells by Cell to Cell Crosslinking

Reactants	Cells	Reaction	Ref.
Diazotized diamines	*Streptomyces* sp.	Glucose/fructose	Lartigue and Weetall (1976)
Glutaraldehyde	*E. coli*	Fumaric acid/L-aspartic acid	Chibata *et al.* (1974)
Flocculation by polyelectrolytes	*Aspergillus niger*	Invert sugar/gluconic acid	Lee and Long (1974)
Flocculation by chitosan	*Lactobacillus brevis*	Glucose/fructose	Tsumura and Kasumi (1976)

12.2.4   Microencapsulation

Microcapsules are defined as usually spherical particles, where a liquid or a suspension is enclosed by a dense, but semipermeable polymeric film. In this case a suspension of cells is separated from the bulk solution by a membrane. Transport of nutrients across the membrane is the major limiting factor to the application of this technique.

Typical film-forming encapsulation materials such as cellulose derivatives have been evaluated (Fukushima *et al.*, 1976; Miyoshi *et al.*, 1977). Liquid membranes have been studied as well (Mohan and Li, 1975), although capsule break up will put severe limitations on any large scale use of this latter technique. Microencapsulation has recently found special interest in connection with the immobilization of mammalian or human cell lines. In this approach, the cells are first entrapped in an alginate bead and then treated with a polycation solution like polylysine. An insoluble polyelectrolyte complex is then formed with the required membrane characteristics. Finally, the noncomplexed alginate is redissolved from the microbead leaving the cells in suspension in the microcapsule. Immobilized pancreas cells were grown in culture and survived for six weeks, while nonencapsulated control cells only survived about 24 days (Anon., 1979; Lim and Sun, 1980).

Another attractive application has been the successful production of monoclonal antibodies from microencapsulated hybridoma cells (Anon., 1982). This latter application may demonstrate the special strength and potential of this method in the field of medicine and health care, and possibly also in the development of artificial organs. Table 4 gives an overview of typical contributions to this method's development.

Table 4   Immobilization of Whole Cells by Microencapsulation

Materials	Cells	Reaction	Ref.
Cellulose acetate	*Comamonas* sp.	Production of 7-ACA	Fukushima *et al.* (1976)
Ethylcellulose	*Streptomyces* sp.	Glucose/fructose	Miyoshi *et al.* (1977)
Polyester	*Streptomyces* sp.	Glucose/fructose	Ghose and Chand (1978)
Ice particles coated with cellulose triacetate	*Saccharomyces cerevisiae*	Glucose/ethanol	Mitsubishi Rayon Co. (1981)
Eudragit®	*Saccharomyces cerevisiae*	Glucose/ethanol	Eng (1980)
Liquid membranes	*Micrococcus denitrificans*	Reduction of nitrate and nitrite	Mohan and Li (1975)
Alginate-polylysine	Pancreas cells	—	Lim and Sun (1980)
Alginate-polylysine	Hybridoma cells	Monoclonal antibodies	Anon. (1982)

12.2.5   Entrapment in Polymeric Networks

Physical entrapment in porous polymeric carriers is by far the most widely used technique for whole cell immobilization. The size of the immobilized species makes it rather simple to prepare porous networks which combine complete cell retention with high enough a porosity for substrate and product transport. It is clear, however, that the network has to be formed in the presence of the cells to be entrapped and that the network-forming reactions therefore have to be adapted to the physiological requirements of the stability of the enzymes or cells.

The reaction conditions are characterized by the solvent (aqueous or nonaqueous), temperature, possible addition of catalysts and, last but not least, by the chemical composition, functionality and size of the network precursor molecules (monomer, oligomer, polymer). van der Waals or other secondary valence bonding, ionic bonding and covalent bonding can all be distinguished in the crosslinking reactions. These determine not only the toxicity and rate of crosslinking but also the stability of the polymeric carrier under operational conditions. Details of the various mechanisms of network formation are discussed below.

12.2.5.1 Gelation of polymers

Gelation is a temperature-controlled phase transition in polymer–solvent systems, where a homogeneous polymer solution is transformed to a homogeneous gel without change in composition. Usually gelation occurs by lowering the temperature (*e.g.* gelatin), but systems are known where the reverse effect is observed (*e.g.* Curdlan). Systems of the latter type can be of interest in connection with the application of thermophilic organisms. The cells are mixed with the polymer solution at an appropriate concentration and are completely immobilized by entrapment in the solidified gel (Vorlop, 1982).

Besides gelatin, mainly agar or agarose has been used (Toda and Shoda, 1975). Small particles of regular size have been prepared using specific forms in which the solution is solidified by cooling (Brodelius and Nilsson, 1980). A much more practical procedure, however, is the use of a water-immiscible oil phase (paraffin oil, soy bean oil, butyl phthalate) for the dispersion of the solution and the conversion of this emulsion into a solid/liquid dispersion by cooling the whole system. In this way, small spherical particles in a controlled diameter range can be obtained, depending on the degree of mixing in the liquid/liquid system (Nilsson *et al.*, 1983).

The very simple method of gelation suffers from the fact that rather soft, mechanically unstable gels are obtained. Therefore, a technical application of this method is very unlikely. The situation may be different, however, in the application of plant, animal and mammalian cells, where the very mild immobilization conditions (if the temperature of gelation is acceptable) are of higher importance than the mechanical properties (Nilsson *et al.*, 1983). The reversibility of gelation may be of interest if the cells or intracellular products have to be recovered.

Systems where gelation of a polymer solution is necessarily followed by a covalent crosslinking reaction are covered in Section 12.2.5.4.

12.2.5.2 Precipitation of polymers

In this category a polymer solution is used as a dispersion medium for the cells. In this case, however, coagulation by phase separation of the polymer is induced by changing physicochemical parameters other than temperature, such as the solvent or pH. Since the polymeric precipitate usually has to be stable in aqueous media, the starting solution of the polymer has to be prepared with an organic solvent or a water/solvent mixture. Water-miscible solvents include DMF, THF, ethanol and acetone. Typical examples of applications of the method are given in Table 5.

Table 5 Entrapment of Whole Cells by Polymer Precipitation

Polymers	Cells	Reaction	Ref.
Cellulose triacetate	*E. coli*	Penicillin G/6-APA	Dinelli (1972)
Cellulose	*Actinoplanes missouriensis*	Glucose/fructose	Linko *et al.* (1980)
Polystyrene	*Candida tropicalis*	Phenol degradation	Hackel *et al.* (1975)
Eudragit®	*E. coli*	L-Serine + indole/L-tryptophan	Klein *et al.* (1979a)

The drawback of this procedure is the intensive contact of the cells with the nonphysiological solvent, possibly limiting the application of the method to single enzyme reactions with dead cells.

12.2.5.3 *Ionotropic gelation of polymers*

If a water-soluble polyelectrolyte is mixed with the appropriate, usually multivalent, counterions, solidification by polysalt formation occurs. In connection with the formation of highly water-swollen structures of controlled morphology, the term 'ionotropic gelation' was introduced. The most well-known example is the Ca alginate gel, which is obtained by gelation of an Na alginate solution in a $CaCl_2$ bath.

Table 6 summarizes those polymer–counterion systems which have been successfully applied to whole cell immobilization. One has to distinguish between polyanions and polycations since they determine the stability regions of the respective polysalts: polyanion salts become unstable at higher pH values, while polycation salts are redissolved at lower pH. Another important aspect for biochemical transformation is the stability against phosphate buffer: alginate gels are completely redissolved, while chitosan gels require the presence of phosphate to stabilize the network.

Table 6 Ionic Polymers and Related Counterions in the Preparation of Ionotropic Gels for the Entrapment of Whole Cells

	Poly-ions	*Counterions*
CO_2^-	Alginate	Ca^{2+}, Al^{3+}, Zn^{2+}, Co^{2+}, Ba^{2+}, Fe^{2+}, Fe^{3+} ...
	Pectin	Ca^{2+}, Al^{3+}, Zn^{2+}, Co^{2+} ... and Mg^{2+}
	Carboxymethylcellulose	Ca^{2+}, Al^{3+} ...
	Carboxy-guar-gum	Ca^{2+}, Al^{3+} ...
	Copolystyrene-maleic acid	Al^{3+} ...
PO_3^{2-}	Phospho-guar-gum	Ca^{2+}, Al^{3+} ...
SO_3^{2-}	Carrageenan	K^+, Ca^{2+} ...
	Furcellaran	K^+, Ca^{2+} ...
	Cellulose sulfate	K^+
NH_3^+	Chitosan	Polyphosphates
		$[Fe(CN)_6]^{4-}$, $[Fe(CN)_6]^{3-}$
		Polyaldehydocarbonic acid
		Poly-1-hydroxy-1-sulfonate-propene-2

A large variety of polymers with different structure and functional groups exists with regard to polyanions. Alginate and *K*-carrageenan are the most important and most widely used products. A more detailed discussion of these immobilization procedures will be given in Sections 12.3.1 and 12.3.2. Chitosan as a polycation has shown very promising properties and will quite probably find more successful applications in the near future.

Since the entrapment of a controlled amount of cells is very simple and generally non-toxic, various cells could be immobilized with a complete preservation of viability. Examples of ionotropic gels used in the field of whole cell immobilization are given in Table 7.

Table 7 Application of Ionotropic Gels for the Entrapment of Whole Cells

Matrix	*Cells*	*Reaction*	*Ref.*
Al alginate	*Candida tropicalis*	Phenol degradation	Hackel *et al.* (1975)
Ca alginate	*Saccharomyces cerevisiae*	Glucose/ethanol	Kierstan and Bucke (1977)
Ca, Al carboxymethylcellulose	*Candida tropicalis*	Phenol degradation	Klein *et al.* (1979b)
Al copolystyrene maleic acid	*Candida tropicalis*	Phenol degradation	Klein *et al.* (1979c)
Mg pectinate	Fungi	Glucose/fructose	Vorlop (1982)
K-Carrageenan	*E. coli*	Fumaric acid/L-aspartic acid	Takata *et al.* (1977)
Chitosan polyphosphate	*E. coli*	L-Serine + indole/L-tryptophan	Vorlop and Klein (1981)
Chitosan–$K_4[Fe(CN)_6]$	*Pseudomonas* 3ab	Methanol + glycine/L-serine	Behrendt (1981)
Chitosan alginate	*Saccharomyces cerevisiae*	Glucose/ethanol	Vorlop and Klein (1983)
Carrageenan–polyethylenimine	*Aspergillus niger*	Glucose/citric acid	Miles Laboratories, Inc. (1982)

Ionotropic gelation is a reversible procedure. This can be advantageous in connection with the recovery of the entrapped species or in connection with multistep immobilization procedures, *e.g.* in microcapsule formation (Lim and Sun, 1980) or epoxy bead preparation (Klein and Eng,

1979). Attempts have been made, on the other hand, to stabilize the ionic gel, *e.g.* by covalent crosslinking (Birnbaum *et al.*, 1981).

12.2.5.4 Polycondensation

In polymer technology, polycondensation reactions are used to produce covalent networks of high mechanical and chemical stability. Typically, oligomeric, liquid precursors which are at least bifunctional are cured with a multifunctional component of appropriate cofunctionality. Functional groups which are often used include hydroxy, amino, epoxy and isocyanate groups in various combinations. The problem to overcome was how to adapt the rather severe reaction conditions (high temperatures, very basic or very acidic pH values) to cell physiology. While the presence of reactive functional groups like isocyanate will always cause some concern, the problem of development of successful immobilization procedures, including those for living cells, has been solved. Table 8 summarizes those systems which have been reported in the literature.

Table 8   Polymer Networks from Polycondensation for the Entrapment of Whole Cells

Matrix	Cells	Reaction	Ref.
Epoxy	*E. coli*	Penicillin G/6-APA	Klein and Eng (1979a)
Polyurethane	*Arthrobacter simplex, etc.*	Steroid conversion	Tanaka *et al.* (1979)
Polyurethane	*Pleurotus ostreatus*	Penicillin V/6-APA	Klein and Kluge (1981)
Kaurit® (urea–formaldehyde)	*E. coli*	Penicillin G/6-APA	Vorlop (1982)
Silica gel	*Saccharomyces cerevisiae*	Glucose/ethanol	Tschang *et al.* (1980)
Xerogel (PVA–tetraethoxysilane)	*Saccharomyces cerevisiae*	Glucose/ethanol	Yamada *et al.* (1980)
Silicone rubber	*Saccharomyces cerevisiae*	Glucose/ethanol	Kressdorf (1982)
PEI–glutaraldehyde	*Bovista plumbea*	Penicillin V/6-APA	Gestrelius (1980)
PEI–polyacrolein	*Arthrobacter simplex*	Cortisol/prednisolone	Klein and Manecke (1982)
Activated PVA crosslinked *via* disulfide bonds	*Arthrobacter simplex*	Cortisol/prednisolone	Klein and Manecke (1982)
Collagen and glutaraldehyde	*Streptomyces phaeochromogenes*	Glucose/fructose	Vieth *et al.* (1973)
Gelatin and glutaraldehyde	*Actinoplanes missourienses*	Glucose/fructose	Roels and van Tilberg (1979)
Albumin and glutaraldehyde	*E. coli*	β-Galactosidase activity	Petre *et al.* (1978)

The first reported successful immobilization was performed with the epoxy system (Klein and Eng, 1979a). While the direct entrapment of cells in epoxy/amino systems showed only limited success, the combination of the alginate procedure of ionotropic gelation with the polycondensation reaction was the key step in this development. The epoxy precursor and the curing component together with the cells were dispersed in a sodium alginate solution and calcium alginate particles were prepared in the usual way (Section 12.3.2). Within a time period of about 24 h, including partial drying, the curing reaction was completed forming an interpenetrating epoxy–alginate network. Finally the alginate was redissolved by dispersing the particles in a phosphate buffer solution. In this way the alginate gel served to stabilize the catalyst particle during the rather long epoxy-curing reaction period at the low temperature (20–40 °C) and a pH of about 8–9, and to introduce porosity to an otherwise nonporous polymeric material.

Careful mixing of the various catalyst components has been found to reduce the toxic action of the functional groups to the cells, thus allowing for the immobilization of yeast cells in the living state (Klein and Kressdorf, 1982).

The second and, at the present, more flexible and applicable system is the polyurethane carrier obtained from isocyanate groups containing precursors (Fukushima *et al.*, 1978; Klein and Kluge, 1981).

In this procedure, immobilization is effected in a one-step operation by mixing the prepolymer with the cells and a controlled amount of water. Depending on the ratio of isocyanate groups to water, the reaction product, CO_2, gives rise to either a foam structure of the polyurethane product or a homogeneous gel when the rate of CO_2 formation is very low. When using this polymer system special care has to be given to the chemical matrix structure with regard to a more hydrophilic or hydrophobic nature. In this way the reactant partition between bulk phase and catalyst phase can be influenced. This has been shown to be a very positive factor in the conversion of hydrophobic substrates (Fukui and Tanaka, 1982a). For a more detailed discussion see Section 12.3.3.

A special group of polycondensation networks is obtained by chemical crosslinking of functional polymers with low molecular weight reagents. A typical example is the crosslinking of a gelatine bead, obtained by gelation, with glutaraldehyde (Bachmann *et al.*, 1981). The most extensive studies with this group of carriers have been performed with the collagen/glutaraldehyde system in the preparation of membrane structures with immobilized cells (Vieth *et al.*, 1973; Constantinides, 1980). For additional systems see Table 8.

12.2.5.5 Polymerization

Covalent polymeric networks can be very conveniently prepared by crosslinking copolymers of vinyl groups containing monomers, oligomers or polymers.

The classical system is the free radical initiated copolymerization of acrylamide and bisacrylamide, which was the first successful attempt in whole cell immobilization by entrapment (Mosbach and Mosbach, 1966). The same system was used in the first successful industrial application of an immobilized cell process (Chibata *et al.*, 1974). While this technology has been replaced by the ϰ-carrageenan method (Sato *et al.*, 1979), it is still a convenient method for initial laboratory studies (Kim and Ryu, 1980; Kokubu *et al.*, 1981; Koshcheenko *et al.*, 1981).

As mentioned above, the polymerization reaction can proceed from monomeric, oligomeric or polymeric precursors. Discussing the common monomer systems first, besides acrylamide (AAm) the monomers methacrylamide and hydroxyethylmethacrylate have been used (Klein and Schara, 1980; Kumakura *et al.*, 1978, 1979). Methacrylamide seemed to be advantageous with respect to mechanical product stability, but caused more toxic problems. Water-insoluble monomers have been used as well (Klein *et al.*, 1979b; Yoshii *et al.*, 1981). Crosslinking in most cases is achieved by copolymerization with a divinyl compound, such as methylenebis(acrylamide). Initiation is effected by water-soluble free radical initiators, like $(NH_4)_2S_2O_8$ in combination with activators. Decrease of polymerization temperature down to 4 °C has proved to be advantageous for attaining higher activity (Klein and Schara, 1980; Koshcheenko *et al.*, 1981).

Another approach to initiation has been the application of γ-radiation in the polymerization of mono-, di- or tri-methacrylate monomers. An advantage of this method may be the fact that very low temperatures (−24 °C) can be used (Kumakura *et al.*, 1978, 1979; Yoshii *et al.*, 1981). One has to realize that not only the toxicity of the monomers but also the heat evolution from the polymerization reaction are equally important factors in enzyme deactivation and related activity losses.

Polymerization technology usually is very simple, if the polymerizing solution is mixed with the cell suspension and bulk polymerization is applied. The flat, polymeric block is then pressed through a sieve plate to obtain smaller but irregularly shaped particles (Chibata *et al.*, 1974). A more elegant method is suspension polymerization, which requires the selection of an appropriate suspension agent and surfactant to stabilize the dispersion. While in earlier studies hydrocarbons have been used, dibutyl phthalate proved to be a much better, nontoxic suspending liquid (Klein *et al.*, 1979b; Klein and Schara, 1980). A special advantage of the suspension method is the much better temperature control as compared to the bulk method. Therefore, activity yields in PAAm close to 100%, including viable cells, could be obtained (Schara, 1977).

A second type of polymerization starts from oligomeric precursors. Extensive studies on such systems have been performed by the group of Fukui (Omata *et al.*, 1979).

By coupling methacrylate and various groups to polyethylene glycol oligomers of controlled chain length, photocrosslinkable resins with well-defined network densities, have been prepared. By incorporation of propylene glycol spacers, polymeric carriers with more hydrophobic character have been obtained (Omata *et al.*, 1979).

A third possibility for the preparation of polymerization networks is the use of high polymers with pendant vinyl groups. Such a polymer can be prepared by esterification of poly(vinyl alcohol) with acrylic acid and crosslinking by photoinitiation (Manecke and Beier, 1983; Klein and Manecke, 1982).

Another example is the coupling of glycidyl methacrylate to chitosan (Behrens, 1983).

For a summary of polymeric carriers from polymerization reactions see Table 9. There are many more contributions in the literature which show various applications without contributing to the development of the method of immobilization itself. Generally the polymeric structure is mechanically unstable, the loading capacity of cells is limited (10% on wet weight basis for bacteria), and catalyst shaping is laborious.

Table 9 Polymer Networks from Polymerization for the Entrapment of Whole Cells

Matrix	Cells	Reaction	Ref.
Monomers			
Acrylamide	*Umbilica pustulata*	Production of orcinol	Mosbach and Mosbach (1966)
Methacrylamide	*Candida tropicalis*	Phenol degradation	Klein and Schara (1980)
2-Hydroxyethyl methacrylate	*Streptomyces*		
	phaerochromogenes	Glucose/fructose	Kumakura *et al.* (1978)
Oligomers/polymers			
Maleic polybutadiene (PBM-2000)	*Nocardia rhodocrous*	Steroid transformation	Omata *et al.* (1979)
Activated poly(propylene glycol)			
(ENT 4000, ENTP 2000, ENTB 1000)	*Nocardia rhodocrous*	Steroid transformation	Omata *et al.* (1979)
PVA esterified with acrylic acid	*Arthrobacter simplex*	Cortisol/prednisolone	Manecke and Beier (1983)

12.3 SELECTED PROCEDURES

Immobilization by entrapment in a polyacrylamide network is still the most extensively used method; however, the application is practically restricted to laboratory studies. With regard to technical applications in food- and health-related areas, a definite trend towards the use of natural polymers can be observed. Another area of increasing interest is defined by the application of nonaqueous media for the conversion of hydrophobic compounds.

The intention of this section is to give some more detailed information about those methods which recently have found particularly wide application. This may be helpful in any study where a fast and effective immobilization procedure is needed, without having to go through the tedious task of selecting one of the numerous methods referred to in Section 12.2.

12.3.1 The *K*-Carrageenan Method

The *K*-carrageenan method was introduced by Chibata and his group in 1977 (Yamamoto *et al.*, 1977; Tosa *et al.*, 1979). *K*-Carrageenan is a naturally occurring polysaccharide isolated from seaweed and widely used as a food additive, *i.e.* it is readily available and nontoxic. The polymer is composed of β-D-galactose sulfate and 3,6-anhydro-α-D-galactose units.

K-Carrageenan gels most readily and strongly under a variety of conditions. Cooling the solution from 40 °C to 10 °C is the most simple way, but the gels obtained are rather soft. Improved mechanical stability can be obtained by using gel-inducing agents, such as various metal ions (*e.g.* K^+, Rb^+, Cs^+, Ca^{2+}, Al^{3+}), aliphatic amines (*e.g.* hexamethylenediamine), aromatic diamines (*e.g.* *p*-phenylenediamine), amino acids and their derivatives (*e.g.* histamine, DL-histidine hydrazide) or water-miscible organic solvents (*e.g.* ethanol, acetone). The examples indicated are the most important in each case. By comparing different microorganisms and enzymatic reactions, gelation with K^+ showed the best performance.

In a typical procedure *K*-carrageenan is dissolved in physiological saline previously warmed to 70–80 °C at a concentration between 2% and 5% (*e.g.* 3.4% w/v) and this solution is kept at 40 °C. Simultaneously, the cell suspension is warmed up to 40–50 °C. After mixing, the carrageenan/cell suspension is dropped from a sampler pipette into a cold 0.1 M KCl solution for gel formation (Tosa *et al.*, 1979). Except for the heating step, which may cause some irreversible enzyme deactivation, this method is very mild, such that most of the immobilized cells are in the living state. This has been demonstrated by the growth of yeast cells up to a cell density of 5.4×10^9 cells ml^{-1} gel, starting from 3.5×10^6 cells ml^{-1} gel after immobilization (Chibata, 1979).

In many other applications, the living state of the cells is not required, as in aspartase-catalyzed L-aspartic acid production with immobilized *E. coli* cells. In this case a post-treatment of the gel particles described above with different hardening and crosslinking agents was used, to improve the operational stability of the biocatalyst considerably. Such procedures include consecutive treatment with hexamethylenediamine (0.02 to 0.12 M) and glutaraldehyde (0.006 to 0.6 M) solutions (Tosa *et al.*, 1979) or contact with a polyethylenimine solution (Tosa *et al.*, 1982). In this way catalytic half-life times of 680 days (L-aspartic acid production) and 243 days (L-malic acid production) were reached.

The *K*-carrageenan method has found industrial use (Chibata, 1979; Chibata and Tosa, 1981)

and numerous other studies have been reported in the literature, especially for the immobilization of living and growing cells. A selection of typical applications is given in Table 10.

Table 10 Applications of *K*-Carrageenan for the Immobilization of Whole Cells

Cell	Product	Ref.
Saccharomyces cerevisiae	Ethanol	Wada *et al.* (1979)
Zymomonas mobilis	Ethanol	Grote *et al.* (1980)
Enterobacter aerogenes	2,3-Butanediol	Chua *et al.* (1980)
Serratia marcescens	L-Isoleucin	Wada *et al.* (1980b)
E. coli	L-Aspartic acid	Sato *et al.* (1979)
Pseudomonas dacuntiae + E. coli	L-Alanine	Sato *et al.* (1982)
Brevibacterium fuscum	12-Ketochenodeoxycolic acid	Sawada *et al.* (1981)
Bacillus amyloliquefaciens	α-Amylase	Shinmyo *et al.* (1982)
Trichoderma reesei	Cellulase	Frein *et al.* (1982)

12.3.2 The Alginate Method

The application of alginate for the purpose of whole cell immobilization was first reported in 1975 (Hackel *et al.*, 1975). In these experiments, Al counterions were used, while Kierstan and Bucke (1977) in a subsequent paper introduced the now widely used Ca alginate combination. As with *x*-carrageenan the alginates are produced from seaweed and they are readily available as food additives. Alginates are heteropolymer carboxylic acids, coupled by $1 \rightarrow 4$-glycosidic bonds of β-D-mamuronic (M) and α-L-guluronic acid (G) units. The uronic acids are arranged in a blockwise fashion along the chain. The same macromolecule contains the two homopolymeric blocks, $(M)_n$ and $(G)_n$, together with the blocks of the alternating sequence $(MG)_n$ (McDowell, 1977). Alkali and magnesium alginates are soluble in water, whereas alginic acids and the salts of polyvalent metal cations are insoluble. By simple dropping of a sodium alginate solution into a $CaCl_2$ solution, rigid near-spherical gels are formed by ionotropic gelation, in which the interaction of alginates and the strength and texture of the resulting gels are dominated by the association of $(G)_n$ sequences, while the $(M)_n$ and $(MG)_n$ blocks play only a subordinate part in the gel network. But independent of the type of uronic acid, both of them hold Ca ions strongly enough so that Ca alginate gels themselves have no tendency to redissolve (McDowell, 1977). However, in the presence of any monovalent ion like Na^+, a minimum amount of Ca^{2+} has to be present, to prevent swelling of the gel. This minimum Na^+/Ca^{2+} ratio may vary from 5:1 (M-type) to 20:1 (G-type). Furthermore, different molecular weight samples are available, which determine the solution viscosity at a given polymer concentration. Typical values may be found between 1% and 8% of Na alginate.

Ca alginate beads can be prepared in a broad range of particle sizes. The typical biocatalyst particles will be in the size range of 0.5 to 3.5 mm. The particle size is primarily determined by the surface tension of the alginate solution and not by the inner diameter of the extrusion die. The most effective way to prepare smaller particles is to use a controlled blow-off stream of compressed air concentric to the extrusion die. If special care is given to the construction of the coaxial droplet formation device, very small particles in the size range of between 10 and 120 μm can be obtained.

A special property of the Ca alginate system is the fact that the particle size can be reduced by partial or complete drying, since the resulting shrunken particles reswell to only a limited extent if re-equilibrated with aqueous media (Klein and Wagner, 1978).

The precipitation bath used for the solidification of the Na alginate solution droplets usually contains a 2% $CaCl_2$ solution, although this may be reduced to 0.05% if required. The application of counterions other than Ca^{2+} has been demonstrated recently by using Al^{3+} in a pilot scale study (Fukushima and Hanai, 1982). The temperature of gelation can be chosen between 0 and 80 °C, depending on the nature of the cells. Biomass loading in the direct entrapment step can be varied up to 30% wet weight. By applying the drying procedure and subsequent reduction of the particle size, the cell loading can be increased to values of 100% wet weight and even higher (Klein and Wagner, 1979; Vorlop *et al.*, 1980).

Furthermore, the problem of scale-up of the process of biocatalyst preparation should be mentioned. If the single sampler pipette for droplet formation is substituted by a plate with 42 outlets, each with an inner diameter of 0.4 mm, combined with compressed air tubings, about 5 kg alginate solution can be processed in one hour (Klein and Vorlop, 1983).

Some typical applications of the alginate method are given in Table 11.

Table 11 Applications of Alginates for the Immobilization of Whole Cells

Cell	Product, etc.	Ref.
Candida tropicalis	Phenol degradation	Hackel *et al.* (1975)
Saccharomyces cerevisiae	Ethanol	Kierstan and Bucke (1977)
Morinda citrifolia, etc.	Anthrachinone, *etc.*	Brodelius *et al.* (1979)
Acetobacter sp.	Gluconic acid	Vorlop *et al.* (1980)
Methanosarcina barkeri	Methane	Scherer *et al.* (1981)
Vicia fabe (protoplasts)	—	Scheurich *et al.* (1980)
Red blood cells	—	Pilwat *et al.* (1980)
Zymomonas mobilis	Ethanol	Margaritis *et al.* (1981)
Photobacterium phosphoreum	Light	Makiguchi *et al.* (1980)
Lactobacillus delbrueckii	Lactic acid	Strenross *et al.* (1982)

12.3.3 Polyurethane Biocatalysts

Polyurethanes have many advantages which make them attractive as materials for the entrapment of microbial cells. They are chemically inert, cheap, completely elastic and mechanically stable with respect to abrasion.

Polyfunctional monomers or oligomers (prepolymers) with isocyanate end groups can react with compounds containing hydroxy or amino groups to form a polyurethane network. For the immobilization of cells, prepolymers with a low NCO content are preferred, because of their lower toxicity. These prepolymers are usually synthesized from polyetherdiol and toluene diisocyanate. Polyurethanes with different hydrophilicity or hydrophobicity can be obtained by using different polyetherdiols (Fukushima *et al.*, 1978). Commercially available prepolymers can also be used for cell entrapment (Klein and Kluge, 1981).

The immobilization process is very simple. The prepolymers are mixed with a water suspension of cells and the mixture quickly and readily crosslinks. *In situ* formation of CO_2 gives an open macroporous structure (PU foam). The foam expansion can be reduced if the polyurethane formation is performed under elevated pressure. Foam formation can be totally suppressed if gelation is performed at low temperatures. In this case the originating CO_2 dissolves completely and the PU becomes a solid gel (Klein and Kluge, 1981). Gels can be used either as a film or cut into small pieces, whereas polyurethane foams, with their open-cell structure which allows flow through the pores, can be used in fixed bed reactors. Spherical foam or gel particles have been made by suspending the liquid prepolymer–cell mixture in a stirred paraffin (Klein and Kluge, 1981).

In the field of polyurethane biocatalysts, Fukui and coworkers have done some fundamental work (Tanaka *et al.*, 1979; Fukui and Tanaka, 1982a). They have also studied the immobilization of cells in gels of differing hydrophobicity and the resulting effects on the bioconversion of lipophilic compounds such as steroids and terpenoids. For these bioconversions the activity of cells entrapped in hydrophobic gels usually is much higher than that of cells entrapped in hydrophilic gels. These reactions have been carried out in organic solvents (Fukui *et al.*, 1980).

Table 12 Applications of Prepolymers of Variable Hydrophobicity for the Immobilization of Whole Cells

Prepolymer	Cells	Reaction	Ref.
PU 6, PU 9	*Arthrobacter rhodocrous*	Steroid transformation	Tanaka *et al.* (1979)
PU 3, PU 6	*Nocardia rhodocrous*	Conversions of testosterone	Fukui *et al.* (1980)
PU 6	*Alcaligenes eutrophus*	Hydrogenase activity	Egerer *et al.* (1982)
PU 3, PU 6	*Rhodotorula minuta*	Menthyl succinate/menthol	Omata *et al.* (1981)
Hypol® 3000	*E. coli* ATCC 11105	Penicillin G/6-APA	Klein and Kluge (1981)
	Catharantus roseus	Isocitrate dehydrogenase activity	Felix and Mosbach (1982)
Desmodur T80	*E. coli* ATCC 11105	Penicillin G/6-APA	Klein and Kluge (1981)

The polyurethane biocatalysts seem to be applicable to the bioconversion of various water-soluble, water-insoluble or strongly hydrophobic compounds. Different applications of this technique are given in Table 12.

12.4 PHYSICAL CHARACTERIZATION

The subject of biocatalyst characterization has been covered in detail in a monograph by Buchholz (1979a), in which both immobilized enzymes and immobilized cells have been considered. In this section some brief comments will be given on those methods which are especially important for immobilized cell systems; for experimental details and typical examples the reader is referred to the above-mentioned monograph.

A first aspect of characterization is the mechanical stability under certain operating conditions, such as in the packed bed or stirred tank reactor. While in a laboratory test reaction a soft poly(acrylamide) or gelatin gel may well be sufficient, in a large scale industrial application minimum stability requirements have to be met.

With respect to any packed bed application, the compressibility of a catalyst under pressure, caused by the weight of the packing and by the frictional force from fluid flow, has to be determined.

As a first and more fundamental step in this direction, a test procedure has been developed (Klein and Washausen, 1979b; Klein *et al.*, 1980) which gives quantitative information on the deformation behavior of a single catalyst particle. The data derived from this experiment include (a) the critical force for bead destruction, (b) an elasticity coefficient (compression per unit force), and (c) the ratio of elastic (reversible) and plastic (irreversible) deformation.

The technique can also be used to study the compression behavior of a packed bed. In another study a commercial rheometer has been used to determine the maximum load prior to gel disruption as a function of various conditions of gel formation (Takata *et al.*, 1980). The second step is the direct measurement of the pressure drop along the column (Klein and Kluge, 1979, 1981; Bungard *et al.*, 1979; Cheetham *et al.* 1979).

As has been shown experimentally, there is a quantitative correlation between the single particle behavior and the compression behavior of a packed bed under the action of identical forces (Klein and Kluge, 1979). In connection with the measurement of pressure drop as a function of flow rate, the change of the height of catalyst packing has to be measured as well. Again, emphasis has to be given to reversible and irreversible compression and to hysteresis effects. The viscoelastic behavior of Ca alginate columns has also been discussed (Cheetham, 1979). Another stability test is especially related to the method of cell adsorption, where the critical fluid velocity (shear rate) for the transition from cell retention to cell washout is determined (Kolot, 1981).

With the increased importance of gas-forming reactions like CO_2 production in ethanol fermentations not only compression stress but also the tensile stress is of importance, since such a stress may finally lead to catalyst particle disruption. A method to evaluate such tensile stress situations experimentally has been described (Krouwel and Kossen, 1980; Krouwel, 1982).

With regard to the use of immobilized cells in a well-stirred suspension, surface abrasion is of interest, and a corresponding experimental device has been developed which makes use of the increasing turbidity of the suspension liquid as a measure of abrasion rate (Klein and Eng, 1979b; Klein *et al.*, 1980). In this way it becomes possible to compare different carrier materials and to establish rational criteria for the limiting values of cell loading for a given carrier and cell type (Klein and Kressdorf, 1982).

Any application of an immobilized cell system requires the transport of substrates and products and, with the exception of convectively driven membrane configurations, diffusion processes have to take place. As discussed in the following section, knowledge of diffusion coefficients is a prerequisite for the quantitative correlation of efficiency data. A method for the experimental determination of diffusion coefficients in porous polymer networks has been described (Klein and Washausen, 1979a; Klein and Manecke, 1982). With respect to the hindrance to diffusion, not only the matrix composition (*e.g.* polymer concentration) but also the cell concentration is of importance. The effective diffusivity, D_{eff}, can be correlated with the diffusivity in free solution, D_0, by equation (1).

$$D_{eff} = D_0 \exp[-a(V_P + bV_B)] \tag{1}$$

where V_P is the volume fraction of the polymer matrix, V_B is the volume fraction of the immobilized biomass, a is a parameter which depends on the size of the diffusing molecule, and b is given by the dry weight content of the immobilized cells.

Another parameter which gives more indirect information about the hindrance to mass transfer is the porosity itself. Typical methods well known in catalysis research, such as mercury porosimetry, are difficult to apply. Some information about the maximum pore size of a swollen polymer network can be obtained from size exclusion chromatography with polymer samples of known molecular weight or molecular size (Klein *et al.*, 1983). However, surface porosity and bulk phase porosity have to be carefully distinguished.

Finally, scanning electron microscopy can be an important tool to gain some insight into the structure of an immobilized cell system. The problem to overcome is the possibility of strong structural changes during the course of sample preparation. However, structural differences with respect to matrix porosity and spatial distribution of the immobilized cells can well be observed (Chibata, 1979; Wada *et al.*, 1979, 1980a; Siess and Dinés, 1981).

In summary, the various methods and their combination will be helpful in designing an immobilized cell system for optimal performance. In cases where high cell concentrations in the system are required, one has to be aware of the fact that not only the nature of the matrix but also the structure and the concentration of the cells will control the property of the system significantly.

12.5　ACTIVITY, EFFICIENCY AND STABILITY

12.5.1　Activity

It is the objective of any immobilization procedure to retain the activity of the immobilized cells to a maximal extent. The term 'activity' is then related to the specific function, *i.e.* typically a biocatalytic reaction, and therefore will be linked to very different cell functions depending on the level of complexity of the biochemical process from monoenzyme reactions to living and growing whole cell reactions. A given immobilization procedure which may be very successful in the first case may be very poor in the other if cell viability cannot be preserved.

The activity of an immobilized cell preparation has two important aspects: (a) the relative activity [comparing the activity of the immobilized cells with the same number of free cells (r_1)] and (b) the absolute specific activity [the rate of reaction based on unit weight or unit volume of the whole catalyst (r_2)]. The first is a measure of enzyme or cell deactivation caused by the immobilization procedure, while the second includes the method-dependent possibility of immobilizing more or less cells per unit catalyst volume.

In trying to establish a more general relation between activity yield and immobilization procedure, our interest has to focus on the three main groups: adsorption, crosslinking and entrapment. Immobilization by covalent bonding or microencapsulation is only of minor importance.

For the immobilization of nonliving cells (*i.e.* typically for monoenzyme reactions), the methods of crosslinking and entrapment have a high priority, since high activity levels in crosslinking are based more on very high cell densities, while in the case of entrapment high values of r_1 can be obtained. Within the group of entrapment procedures the methods of ionotropic gelation (*e.g.* in carrageenan or alginates) or polymerization are to be recommended.

For the immobilization of living cells the methods of adsorption and entrapment have to be considered. Adsorption, as a very mild method, will be very advantageous with regard to high values of r_1, but the limits of r_2 are caused by limits in cell loading. Relatively high cell loadings can be obtained with entrapment techniques, but depending on the particular procedure, different limits on r_1 have to be expected. Very high r_1 values, close to 100%, can be obtained with those methods where natural high polymers are used to prepare the entrapment network. This applies to the methods of gelation (gelatine, agar) and ionotropic gelation (carrageenan, alginate) and has been demonstrated for microbial cells as well as plant and mammalian cells. As a second priority the methods of polymerization of monomers or oligomers can be mentioned, if the reaction temperature and reaction time can be properly controlled.

Network formation by polycondensation always includes the presence of chemically reactive functional groups and will always result in some loss of activity. However, living cells can be successfully immobilized if this activity loss is acceptable or can be compensated for by subsequent cell growth in the matrix.

12.5.2　Efficiency

The efficiency, η, of an immobilized cell preparation is defined as the ratio of the actual reaction rate, r_2, to the maximum possible reaction rate, r_m (*i.e.* if all active immobilized cells participate fully in the reaction). Loss of efficiency is caused by transport limitation of a rate-controlling substrate or cosubstrate (*e.g.* oxygen) resulting from boundary layer diffusion or intraparticle pore diffusion.

Boundary layer resistance depends on the hydrodynamic conditions in the macroenvironment surrounding a catalyst particle (Buchholz, 1979b), and hence the resulting equations for the actual reaction rates do not differ, whether immobilized enzymes or whole cells are considered. The hydrodynamic conditions are controlled by the relative velocity of the particle in the bulk solution, the particle diameter and solvent properties, such as viscosity or diffusivity. In a dimensionless form the Sherwood number, Sh, is a measure of the rate constant for mass transfer, k_s. It can be shown that $Sh = 2$ refers to the state of maximum boundary layer thickness (no convection). On the other hand, the boundary layer resistance becomes negligible if $Sh \geqslant 10$. The Sherwood number can be calculated on the basis of the Reynolds number, Re, and the Schmidt number, Sc, if the parameters a, m and n in equation (2) are known.

$$Sh = aRe^m Sc^n \tag{2}$$

It is well established that in a fixed bed Re — and thus Sh — is a function of the fluid velocity as in equation (3) (Karabelas *et al.*, 1971).

$$Sh \sim Re \sim u^{1/3} \tag{3}$$

In a stirred vessel the relationship

$$Sh \sim Re \sim n^{3/4} \tag{4}$$

holds, where n is the number of rotations per unit time (Aiba *et al.*, 1973).

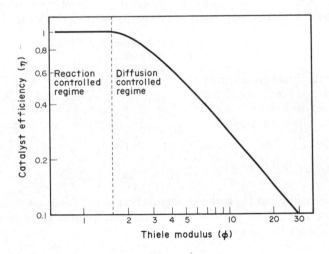

Figure 1 Schematic representation of the relation between catalyst efficiency, η, and Thiele modulus, ϕ

When these equations for the rate of mass transfer are coupled in the proper way with the rate equation for substrate consumption by the catalyst, the following relations are obtained:

$$[S]_b/r_{br} = \text{constant}/u^{1/3} + 1/k_1' \text{ (fixed bed)} \tag{5}$$

$$[S]_b/r_{br} = \text{constant}/n^{3/4} + 1/k_1' \text{ (stirred vessel)} \tag{6}$$

where $[S]_b$ is the bulk phase substrate concentration, r_{br} is the actual reaction rate, and k_1' is a first order rate constant (*e.g.* assuming Michaelis–Menten kinetics for $[S]_b \ll K_M$).

Various studies of immobilized enzyme and cell preparations in fixed bed and stirred tank reactors have indeed found a linear relationship between $[S]_b/r_{br}$ and $u^{-1/3}$ or $n^{-3/4}$ (Buchholz, 1979b; Klein *et al.*, 1984).

Intraparticle pore diffusion depends on the microenvironment (*i.e.* the matrix composition and structure), on the size and number of immobilized cells, and on the size of the diffusing molecules. The various parameters have been combined in equation (1) for the calculation of the effective diffusivity, D_{eff}. As with boundary layer resistance, it is not the absolute rate of diffusion, but the relative mass transfer rate compared to the rate of reaction which is of importance. Here again, dimensionless groups are well suited to give a quantitative description of the interaction between pore diffusion and reaction.

It is well established in heterogeneous reaction kinetics that the efficiency, η, of a catalyst depends in a very typical manner on the 'Thiele modulus', ϕ, as shown schematically in Figure 1. In this function

$$\eta = f(\phi) \tag{7}$$

where the catalyst efficiency, η, is defined as

$$\eta = \frac{\text{rate of reaction of immobilized cells}}{\text{rate of reaction of free cells}} \tag{8}$$

The efficiency is based on the same number of catalytically active cells. Practically, η, can be obtained either by comparing the reaction rate (r) of immobilized cells in particles of radius R to the reaction rate of very small particles $(R \rightarrow 0)$, or, if the concentration of immobilized active cells is known, by comparing the specific reaction rates (r_s).

$$r_s = -\frac{1}{X_{\text{act}}} \frac{d[S]}{dt} \tag{9}$$

where X_{act} is the active biomass:

$$\eta = \frac{r_{\text{immob.cells, }R}}{r_{\text{immob. cells, }R \rightarrow 0}} \tag{10a}$$

$$\eta = \frac{r_{s,\text{immob.cells}}}{r_{s,\text{ free cells}}} \tag{10b}$$

The Thiele modulus, ϕ, on the other hand, is a dimensionless expression for the relative rate of reaction and diffusion:

$$\phi \approx \frac{\text{rate of reaction}}{\text{rate of diffusion}} \tag{11}$$

Depending on the type of rate equation and on the geometry of the catalyst, different expressions for ϕ can be found in the literature (Boersma *et al.*, 1979; Klein *et al.*, 1979c; Klein and Schara, 1981; Klein and Manecke, 1982; Klein and Vorlop, 1983; Roels and van Tilberg, 1979).

For a simple irreversible reaction of order n and a spherical catalyst particle, the following expression holds:

$$\phi = \frac{R}{3} \sqrt{\left(\frac{(n + 1)}{2} \frac{k_r X_{\text{act}}[S]^{n-1}}{D_{\text{eff}}} \right)} \tag{12}$$

Here, k_r is the rate constant and $[S]$ is the concentration of the rate-controlling substrate. Reactions which require oxygen as a cosubstrate can often be formulated to be zero order (for low K_M values). Equation (12) then becomes

$$\phi = \frac{R}{3} \sqrt{\left(\frac{K_r X_{\text{act}}}{2 D_{\text{eff}}[O_2]} \right)} \tag{13}$$

This equation has been used successfully in the quantitative description of oxidation reactions, such as degradation of phenol (Klein *et al.*, 1979c) and production of gluconic acid (Klein and Vorlop, 1983).

In the case of Michaelis–Menten kinetics, equation (11) becomes (Bischoff, 1965)

$$\phi = \frac{R}{3} \sqrt{\left(\frac{r_{\text{max}}}{2 K_M D_{\text{eff}}} \right) \left(\frac{[S]}{K_M + [S]} \right) \left[\frac{[S]}{K_M} - \ln \left(1 + \frac{[S]}{K_M} \right) \right]^{-1/2}} \tag{14}$$

The cleavage reaction of penicillin G towards 6APA could be quantitatively evaluated on this basis (Klein and Vorlop, 1983).

A careful inspection reveals the fact that the rate expression incorporated in equation (14) is of a simplified form. Product and substrate inhibitions, and pH gradients within the biocatalysts were not considered in the derivation of equation (14). Using numerical calculation procedures an exact solution of this problem is possible (Klein and Vorlop, 1984).

It becomes obvious from equation (12) that the Thiele modulus and thus the catalytic efficiency of immobilized cell preparations depend on the particle size R, the specific activity of the cells, the loading of immobilized active cells X_{act}, the effective diffusivity D_{eff}, and the substrate concentration $[S]$. Assuming, that the most active cells are used at an optimal substrate concentration

[S], the parameters R, X_{act} and D_{eff} can be controlled by the immobilization technique used. While it is generally advantageous to prepare small particles, due to the interrelation of X_{act} and D_{eff} (see equation 1), no general recommendation can be given with regard to the optimum choice of X_{act}. Variation of the particle size is usually the most practical way to estimate the relevance of diffusion limitation in any given application.

The Thiele modulus as given in equations (13) and (14) holds for a given substrate concentration [S] (= constant). In a practical reaction system this applies to a steady state situation in a stirred tank reactor. In any reaction system, where [S] is a function of time (batch reactor) or position (plug flow reactor) an operational effectiveness factor, η_0, has to be calculated by integration over the change of concentration and thus ϕ under consideration.

While efficiency values as high as possible should always be obtained, a very high efficiency, however, is not necessarily a positive factor in itself. If the cell mass to be immobilized is not extremely expensive, the absolute activity, r_2, which determines the conversion rate per unit reactor volume, will be practically and economically more important. Therefore, the optimum values of the particle size and cell loading can be determined to define the optimum catalyst composition (Klein and Vorlop, 1983) at an efficiency level well below 100%. Limits in efficiency are also related to the minimum values of particle size. While the decrease in particle size, especially below 1 mm, is the most important step to increase the efficiency, particles with diameters smaller than 0.2 mm will be difficult to handle in any large scale operation.

Finally, it should be mentioned that problems with regard to efficiency should be expected as it becomes possible to achieve very high specific activities as a result of recent advances in genetic engineering (Klein and Wagner, 1980). While the values of D_{eff} may be expected to vary within a narrow range, the values of r_m might be increased by a factor of 10 or even 100. As a result the appreciable gain in absolute activity, r_2, can be expected to overcome the possible drawback of decreasing efficiency values.

12.5.3 Stability

Stability is the third parameter used to characterize the kinetic performance of an immobilized cell system. In this context, a careful distinction has to be made between storage stability on the one hand and operational stability on the other. The first is related to storage conditions, where no substrate is present and where physical parameters such as temperature are different from reaction conditions. Such storage situations are of interest if an immobilized cell has to be used only periodically, *e.g.* depending on the availability of substrate or the necessity of analytical application. Ultimately, the immobilization could be used as a procedure for improved storage itself, if reversible immobilization procedures are applied to free the cells (Pilwat *et al.*, 1980).

Stability values reported in the literature are usually related to reaction conditions and are more precisely referred to as operational stabilities. It is one of the important features of whole cell immobilization that operational stabilities of the fixed cells are considerably higher compared with those in free suspension. We have to refer to the various reviews cited in the introduction for detailed figures. Operational stabilities usually are expressed as values of $t_{1/2}$, which is the time at which the activity loss of 50%, compared to the starting activity, is observed. Alternatively, the corresponding number of batches can be used.

In a more generalized fashion, typical operational stability levels can be observed depending on the type of immobilized cell preparation. The highest values of $t_{1/2} \geqslant 600$ days have been reported in cases of immobilized denatured cells, stabilized by chemical post-treatment (Tosa *et al.*, 1979). Such an approach is of course limited to monoenzyme reactions, but on the other hand, very high stability levels are a prerequisite to successfully compete with immobilized enzymes.

A second and lower stability level is found in the case of immobilized resting cells for multi-enzyme reactions. In the case of phenol degradation with immobilized *Candida tropicalis* cells, a half-life of between 20 and 40 days was observed, whereas free cells could only be used for 20 hours (Klein *et al.*, 1978). This example clearly indicates that the polymer entrapment led to a drastic increase in operational stability.

Different stability criteria exist in the case of immobilized living and growing cells. Owing to the continuous reproduction of biomass the enzyme stability itself may be high but the increase of biomass can cause problems with regard to the mechanical stability of the matrix which ultimately may be destroyed. The method of immobilization and the mechanical properties of the matrix will therefore be significant factors affecting the operational stability.

As with catalyst efficiency, a high value of the operational stability may not be an advantage in itself. From an economic viewpoint, the reactor productivity, which relates the amount of product to the unit amount of catalyst is of primary importance. At a very high activity level, therefore, an operational stability of a few days may well be acceptable.

Furthermore, the close interdependence of activity, efficiency and operational stability has to be kept in mind. It can be shown theoretically and experimentally that an increasing dominance of transport limitation will result in increasing values of $t_{1/2}$. In general, an immobilized cell biocatalyst with a large particle diameter (with low efficiency, η) has a longer half-life than a small biocatalyst (with high efficiency) (Kobayashi *et al.*, 1980; Klein and Vorlop, 1983). In such a case the improved but artificial operational stability cannot give higher productivities and the necessity to extend the period of operation is disadvantageous. On the basis of the strong coupling of activity and efficiency, as discussed above, the operational stability is coupled to the activity level as well. Any advanced design of an immobilized cell preparation and the related process will have to include an optimization procedure with respect to the interrelation of activity, efficiency and stability.

12.6 CONCLUDING REMARKS

In summarizing this chapter, it can be stated that the methodology of whole cell immobolization has proceeded to a level comparable with that of enzyme immobilization.

Practically all types of whole cells have been successfully immobilized at considerable levels of retainment of enzyme functions and complex whole cell functions linked to cell viability.

A large variety of methods for immobilization is now available, including different strategies and different materials. With respect to the materials, the whole range from natural products to synthetic products, from simple and inexpensive to sophisticated and expensive materials has been shown to be applicable. With respect to biological aspects, the trend from single enzyme function to complex functions of plant and mammalian cells can be observed. With regard to the procedure, a trend towards mild and simple entrapment techniques in natural polymer matrices can be envisaged.

In spite of its high scientific potential, only a few processes have found industrial application due to the fact that major problems remain to be solved. The key factor in the advancement of immobilized cell technology will not be the immobilization procedure alone, but the development of cell lines which combine technically important reaction paths with technically acceptable activity and stability levels. Genetic engineering will be an important contribution to future achievements in this respect, as will cell physiological studies on the mechanisms of deactivation and possibly stabilization. Apart from the requirements of high activity combined with high efficiency, which are always important, the development of immobilization procedures will have to contribute to improved partition and transport from the bulk phase to the immobilized cell phase. The improvement of transport rate and transport selectivity may be controlled by particle size, specific interactions or hydrophobic/hydrophilic balance, and may include not only aqueous, but also nonaqueous systems.

While the focus of today's application is mainly on bioconversion in the food and pharmaceutical area on the one hand, and enzyme-based analytical chemistry on the other, new areas such as synthesis of organic intermediates and energy production will possibly be of importance in the future.

As in the past, cooperation of scientists from different disciplines such as biology, chemistry, physics and engineering will be required to solve the future problems.

12.7 REFERENCES

Abbott, B. J. (1976). Preparation of pharmaceutical compounds by immobilized enzymes and cells. *Adv. Appl. Microbiol.*, **20**, 203–257.
Abbott, B. J. (1977). Immobilized cells. *Annu. Rep. Ferment. Processes*, **1**, 203–233.
Aiba, S., A. E. Humphrey and N. F. Millis (1973). In *Biochemical Engineering*, p. 173. Academic Press, New York.
Anon. (1979). Live-cell inplants: hope for diabetics. *Chem. Week*, April 25, 53.
Anon. (1981). Production methods in industrial microbiology. *Sci. Am.*, **9**, 181–196.
Anon. (1982). Method may boost monoclonal antibody output. *Chem. Eng. News*, January 11, 22–23.
Ash, D. G. (1979). Adhesion of microorganisms in fermentation processes. *Spec. Publ. Soc. Gen. Microbiol.*, **2**, 57–86.
Atrat, P. and H. Groh (1981). Steroid transformation with immobilized microorganisms (Part VI). *Z. Allg. Mikrobiol.*, **21**, 3–6.

Bachmann, S., L. Gebicka and Z. Gasyna (1981). Immobilization of glucose isomerase on radiation modified gelatin gel. *Starch*, **33**, 63–66.

Behrendt, U. (1981). *Versuche zur Optimierung der L-Serin-Bildung mit dem methylotrophen Bakterium Pseudomonas 3ab*. Ph.D. Dissertation, TU Braunschweig.

Behrens, W. (1983). *Darstellung kovalent vernetzter Trägermaterialien für die Immobilisierung von Mikroorganismen*. Diploma Thesis, TU Braunschweig.

Berger, R. (1981). Immobilisierung mikrobieller Zellen und deren Nutzung zur Substratwandlung – eine Literaturstudie. *Acta Biotechnol.*, **1**, 73–102.

Berger, R. (1982). Immobilisierung mikrobieller Zellen und deren Nutzung zur Substratwandlung – eine Literaturstudie. *Acta Biotechnol.* **1**, 343–358.

Birnbaum, S., R. Pendleton, P. O. Larsson and K. Mosbach (1981). Covalent stabilization of alginate gel for the entrapment of living whole cells. *Biotechnol. Lett.*, **3**, 393–400.

Bischoff, K. B. (1965). Effectiveness factors for general reaction rate forms. *AIChE J.*, **11**, 351–355.

Bland, R. R., H. C. Chen, W. J. Jewell, W. D. Bellamy and R. R. Zall (1982). Continuous high rate production of ethanol by *Zymomonas mobilis* in an attached film expanded bed fermentor. *Biotechnol. Lett.*, **4**, 323–328.

Boersma, J. G., K. Vallenga, H. G. J. DeWilt and G. E. H. Joosten (1979). Mass-transfer effects on the rate of isomerization of D-glucose into D-fructose, catalyzed by whole-cell immobilized glucose isomerase. *Biotechnol. Bioeng.*, **21**, (10), 1711–1724.

Brodelius, P. (1978). Industrial applications of immobilized biocatalysts. *Adv. Biochem. Eng.*, **10**, 75–129.

Brodelius, P., B. Deus, K. Mosbach and M. H. Zenk (1979). Immobilized plant cells for production and transformation of natural products. *FEBS Lett.*, **103**, 93–97.

Brodelius P. and K. Nilsson (1980). Entrapment of plant cells in different matrices. *FEBS Lett.*, **122**, 312–316.

Buchholz, K. (ed.) (1979a). *Characterization of Immobilized Biocatalysts* (DECHEMA Monograph No. 84). Verlag Chemie, Weinheim.

Buchholz, K. (1979b). In *Characterization of Immobilized Biocatalysts* (DECHEMA Monograph No. 84), ed. K. Buchholz, pp. 212–223. Verlag Chemie, Weinheim.

Bucke, C. and A. Wiseman (1981). Immobilized enzymes and cells. *Chem. Ind.*, **4**, 234–240.

Bungard, S. J., R. Reagan, P. J. Rodgers and K. R. Wyncoll (1979). The use of whole cell immobilization for the production of glucose isomerase. *ACS Symp. Ser.*, **106**, 139–146.

Cheetham, P. S. J. (1979). Physical studies on the mechanical stability of columns of calcium alginate gel pellets containing entrapped microbial cells. *Enzyme Microb. Technol.*, **1**, 183–188.

Cheetham, P. S. J., K. W. Blunt and C. Bucke (1979). Physical studies on cell immobilization using calcium alginate gels. *Biotechnol. Bioeng.*, **21**, 2155–2168.

Chibata, I., T. Tosa and T. Sato (1974). Immobilized aspartase-containing microbial cells: preparation and enzymatic properties. *Appl. Microbiol.*, **27**, 878–885.

Chibata, I. and T. Tosa (1977). Transformation of organic compounds by immobilized microbial cells. *Adv. Appl. Microbiol.*, **22**, 1–27.

Chibata, I. (1979). Development of enzyme engineering—application of immobilized cell system. *Kem.-Kemi*, **12**, 705–714.

Chibata, I. and T. Tosa (1981). Use of immobilized cells. *Annu. Rev. Biophys. Bioeng.*, **10**, 197–216.

Chibata, I., T. Tosa and M. Fujimura (1983). Immobilized living microbial cells. *Annu. Rev. Ferment. Proc.*, **6**, 1–22.

Chua, I. W., A. Erarslan, S. Kinoshita and H. Taguchi (1980). 2,3-Butanediol production by immobilized *Enterobacter aerogenes* AM 1133 with *x*-carrageenan. *J. Ferment. Technol.*, **58**, 123–127.

Compere, A. L. and W. L. Griffith (1976). Fermentation of waste materials to produce industrial intermediates. *Dev. Ind. Microbiol.*, **17**, 247–252.

Constantinides, A. (1980). Steroid transformation of high substrate concentrations using immobilized *Corynebacterium simplex* cells. *Biotechnol. Bioeng.*, **22**, 119–136.

Corrieu, G., A. Blachere, A. Ramirez, I. M. Navarro, G. Durand, B. Duteurtre and M. Moll (1976). An immobilized yeast fermentation pilot plant used for production of beer. In *Proc. Int. Ferment. Symp., 5th, Berlin, 1976*, p. 294.

Dinelli, D. (1972). Fiber-entrapped enzymes. *Process Biochem.*, August, 9–12.

Durand, G. and I. M. Navarro (1978). Immobilized microbial cells. *Process Biochem.*, September, 14–23.

Egerer, P., H. Simon, A. Tanaka and S. Fukui (1982). Immobilization and stability of NAD-dependent hydrogenase from *Alcaligenes eutrophus* and of whole cells. *Biotechnol. Lett.*, **4**, 489–494.

Eng, H. (1980). *Polymerfixierung von Mikroorganismen zur Biokatalyse: Epoxid-Matrix*. Ph.D. Dissertation, TU Braunschweig.

Felix, H. R. and K. C. Mosnzymes in permeabilized and immobilized cells. *Biotechnol. Lett.*, **4**, 181–186.

Frein, E. M., B. S. Montenecourt and D. E. Eveleigh (1982). Cellulose production by *Trichoderma reesei* immobilized on K-carrageenan. *Biotechnol. Lett.*, **4**, 287–292.

Fukui, S., S. A. Ahmed, T. Omata and A. Tanaka (1980). Bioconversion of lipophilic compounds in non-aqueous solvents. Effect of gel hydrophobicity on diverse conversions of estosterone by gel-entrapped *Nocardia rhodocrous* cells. *Eur. J. Appl. Microbiol. Biotechnol.*, **10**, 189–301.

Fukui, S. and A. Tanaka (1982a). Bioconversion of lipophilic or water insoluble compounds by immobilized biocatalysts in organic solvent systems. *Enzyme Eng.*, **6**, 191–200.

Fukui, S. and A. Tanaka (1982b). Immobilized microbial cells. *Annu. Rev. Microbiol.*, **36**, 145–172.

Fukushima, M., T. Fujii, K. Matsumota and M. Morishita (1976). *Jpn. Pat.* 76 70 884.

Fukushima, S., T. Nagai, K. Fujita, A. Tanaka and S. Fukui (1978). Hydrophilic urethane prepolymers: convenient materials for enzyme entrapment. *Biotechnol. Bioeng.*, **20**, 1465–1469.

Fukushima, S. and S. Hanai (1982). Pilot operation for continuous alcohol fermentation of molasses in an immobilized bioreactor. *Enzyme Eng.*, **6**, 347–348.

Gainer, J. L., D. J. Kirwan, J. A. Foster and E. Seyhan (1980). Use of adsorbed and covalently bound microbes in reactors. *Biotechnol. Bioeng. Symp.*, **10**, 35–42.

Gerson, D. F. and J. E. Zajic (1979). The biophysics of cellular adhesion. *ACS Symp. Ser.*, **106**, 29–57.

Gestrelius, S. (1980). Immobilized penicillin acylase for production of 6-APA from penicillin V. *Enzyme Eng.*, **5**, 439–442.

Ghose, T. K. and S. Chand (1978). Kinetic and mass transfer studies on the isomerization of cellulose hydrolyzate using immobilized streptomyces cells. *J. Ferment. Technol.*, **56**, 315–322.

Grote, W., K. J. Lee and P. L. Rogers (1980). Continuous ethanol production by immobilized cells of *Zymomonas mobilis. Biotechnol. Lett.*, **2**, 481–486.

Hackel, U., J. Klein, R. Megnet and F. Wagner (1975). Immobilization of microbial cells in polymeric matrices. *Eur. J. Appl. Microbiol.*, **1**, 291–293.

Hattori, T. and J. Furusaka (1959). Chemical activities of *Escherichia coli* cells adsorbed on a resin. *Biochim. Biophys. Acta*, **31**, 581–582.

Hattori, T. and J. Furusaka (1960). Chemical activities of *Escherichia coli* adsorbed on a resin. *J. Biochem. (Tokyo)*, **48**, 831–837.

Jack, T. R. and J. G. Zajik (1977a). The immobilization of whole cells. *Adv. Biochem. Eng.*, **5**, 126–145.

Jack, T. R. and J. E. Zajic (1977b). The enzymatic conversion of L-histidine to urocanic acid by whole cells of *Micrococcus luteus* immobilized on carbodiimide activated carboxymethylcellulose. *Biotechnol. Bioeng.*, **19**, 631–648.

Karabelas, A. J., T. H. Wegner and T. J. Hanratty (1971). Use of asymptotic relations to correlate mass transfer data in packed beds. *Chem. Eng. Sci.*, **26**, 1581–1589.

Kennedy, J. F., S. A. Barker and J. D. Humphreys (1976). Microbial cells living immobilized on metal hydroxides. *Nature (London)*, **261**, 242–244.

Kennedy, J. F. (1978). Microbial cells immobilized and living on solid supports and their application to fermentation processes. *Enzyme Eng.*, **4**, 323–328.

Kennedy, J. F., J. D. Humphreys, S. A. Barker and R. N. Greenshields (1980). Application of living immobilized cells to the acceleration of the continuous conversions of ethanol (wort) to acetic acid (vinegar)–hydrous titanium(IV) oxide-immobilized acetobacter species. *Enzyme Microb. Technol.*, **2**, 209–216.

Kierstan, M. and C. Bucke (1977). The immobilization of microbial cells, subcellular organelles, and enzymes in calcium alginate gels. *Biotechnol. Bioeng.*, **19**, 387–397.

Kim, Y. S. and D. D. Y. Ryu (1980). Ampicillin biosynthesis by immobilized enzyme. *Arch. Pharmacol. Res.*, **3**, 7–12.

Klein, J., U. Hackel, P. Schara, P. Washausen and F. Wagner (1976). Polymer entrapment of microbial cells: phenol degradation by *Candida tropicalis. Int. Ferment. Symp., 5th, Berlin, 1976*, p. 295.

Klein, J. U. Hackel, P. Schara, P. Washausen, F. Wagner and C. K. A. Martin (1978). Polymer entrapment of microbial cells: preparation and reactivity of catalytic systems. *Enzyme Eng.*, **4**, 339–341.

Klein, J. and F. Wagner (1978). In *Immobilized Whole Cells* (DECHEMA Monograph, No. 82), pp. 142–164. Verlag Chemie, Weinheim.

Klein, J. and H. Eng (1979a). Immobilization of microbial cells in epoxy carrier systems. *Biotechnol. Lett.*, **1**, 171–176.

Klein, J. and H. Eng (1979b). In *Characterization of Immobilized Biocatalysts* (DECHEMA Monograph No. 84), ed. K. Buchholz, pp. 292–299. Verlag Chemie, Weinheim.

Klein, J., H. Eng, M. Kluge and K.-D. Vorlop (1979a). New polymer networks for the entrapment of whole microbial cells. *IUPAC Proceedings Macro Mainz*, **3**, 1547–1549.

Klein, J., U. Hackel, P. Schara and H. Eng (1979b). Polymernetzwerke zum Einschluss von Mikroorganismen. *Angew. Makromol. Chem.*, **76/77**, 329–350.

Klein, J., U. Hackel and F. Wagner (1979c). Phenol degradation by *Candida tropicalis* whole cells entrapped in polymeric ionic networks. *ACS Symp. Ser.*, **106**, 101–118.

Klein, J. and M. Kluge (1979). Packed bed flow resistance. In *Characterization of Immobilized Biocatalysts* (DECHEMA Monograph No. 84), ed. K. Buchholz, pp. 285–291. Verlag Chemie, Weinheim.

Klein, J. and F. Wagner (1979). Immobilized whole cells. In *Characterization of Immobilized Biocatalysts* (DECHEMA Monograph No. 84), ed. K. Buchholz, pp. 265–335. Verlag Chemie, Weinheim.

Klein, J. and P. Washausen (1979a). Diffusion. In *Characterization of Immobilized Biocatalysts* (DECHEMA Monograph No. 84), ed. K. Buchholz, pp. 300–302. Verlag Chemie, Weinheim.

Klein, J. and P. Washausen (1979b). Pressure stability compression experiments. In *Characterization of Immobilized Biocatalysts* (DECHEMA Monograph No. 84), ed. K. Buchholz, pp. 277–284. Verlag Chemie, Weinheim.

Klein, J. and P. Schara (1980). Entrapment of living microbial cells in covalent polymeric networks: I. Preparation and properties of different networks. *J. Solid Phase Biochem.*, **5**, 61–78.

Klein, J. and F. Wagner (1980). Immobilization of whole microbial cells for the production of 6-APA. *Enzyme Eng.*, **5**, 335–345.

Klein, J., P. Washausen, M. Kluge and H. Eng (1980). Physical characterization of biocatalyst particles obtained from polymer entrapment of whole cells. *Enzyme Eng.*, **5**, 359–362.

Klein, J., H. Eng and F. Wagner (1981). Estimation of catalytic efficiency of immobilized cell biocatalysts. *Commun. Eur. Congr. Biotechnol., 2nd, Eastbourne*, 118.

Klein, J. and M. Kluge (1981). Immobilization of microbial cells in polyurethane matrices. *Biotechnol. Lett.*, **3**, 65–70.

Klein, J. and P. Schara (1981). Entrapment of living microbial cells in covalent polymeric networks: II. A quantitative study on the kinetics of oxidative phenol degradation by entrapped *Candida tropicalis* cells. *Appl. Biochem. Biotechnol.*, **6**, 91–117.

Klein, J. and G. Manecke (1982). New developments in the preparation and characterization of polymer-bound biocatalysts. *Enzyme Eng.*, **6**, 181–189.

Klein, J. and B. Kressdorf (1982). Immobilization of living whole cells in an epoxy matrix. *Biotechnol. Lett.*, **4**, 375–380.

Klein, J. and K. D. Vorlop (1983). Immobilized cells — catalyst preparation and reaction performance. *ACS Symp. Ser.*, **207/12**, 377–392.

Klein, J., J. Stock and K. D. Vorlop (1983). Pore size and properties of spherical Ca-alginate biocatalysts. *Eur. J. Appl. Microbiol. Biotechnol.*, **18**, 86–91.

Klein, J. and F. Wagner (1983). Methods for the immobilization of microbial cells. *Appl. Biochem. Bioeng.*, **4**, 11–51.

Klein, J., K.-D. Vorlop and F. Wagner (1984). The implication of reaction kinetics and mass transfer on the design of bio-catalytic processes with immobilized cells. *Enzyme Eng.*, **7**, 437–449.

Klein, J. and K.-D. Vorlop (1984). Kinetic aspects of process development for penicillin G. Cleavage with immobilized cell biocatalysts. *Ger. Chem. Eng.*, **7**, 233–240.

Kobayashi, T., K. Katagiri, K. Ohmiya and S. Shimizu (1980). Effect of mass transfer on operational stability of immobilized enzyme. *J. Ferment. Technol.*, **58**, 23–31.

Kokubu, T., K. Karube and S. Suzuki (1981). Protease production by immobilized mycelia of *Streptomyces fradiae*. *Biotechnol. Bioeng.*, **23**, 29–39.

Kolot, F. B. (1981). Microbial carriers — strategy for selection. *Process Biochem.*, August/September, 2–9 (part I); October/November, 30–33, 46 (part II).

Koshcheenko, K. A., G. V. Sukhodolshaya, V. S. Tyurin and G. K. S. Skryabin (1981). Physiological, biochemical and morphological changes in immobilized cells during repeated periodical hydrocortisone transformations. *Eur. J. Appl. Microbiol. Biotechnol.*, **12**, 161–169.

Kressdorf, B. (1982). Unpublished results.

Krouwel, P. G. and N. W. F. Kossen (1980). Gas production by immobilized microorganisms: theoretical approach. *Biotechnol. Bioeng.*, **22**, 681–687.

Krouwel, P. G. (1982). *Immobilized cells for continuous solvent production; IBE and ethanol fermentation*. Thesis Dr. Tech. Sci., Delft University Press.

Kumakura, M., M. Yoshida and I. Kaetsu (1978). Immobilization of microbial cells by radiation-induced polymerization of glass-forming monomers and immobilization of *Streptomyces phaechromogenes* cells by polymerization of various hydrophobic monomers. *J. Solid Phase Biochem.*, **3**, 175–183.

Kumakura, M., M. Yoshida and I. Kaetsu (1979). Immobilization of *Streptomyces phaechromogenes* by radiation-induced polymerization of glass-forming monomers. *Biotechnol. Bioeng.*, **21**, 679–688.

Lartique, D. J. and H. Weetall (1976). *US Pat.* 3 939 041.

Lee, G. K. and M. E. Long (1974). *US Pat.* 3 821 086.

Lettinga, G., A. F. M. van Velsen, S. W. Hobma, W. de Zeeuw and A. Klapwijk (1980). Use of the upflow sludge blanket (USB) reactor concept for biological waste water treatment, especially for anaerobic treatment. *Biotechnol. Bioeng.*, **22**, 699–734.

Lim, F. and A. M. Sun (1980). Microencapsulated islets as bioartificial endocrine pancreas. *Science*, **210**, 908–910.

Linko, P., K. Poutanen, L. Weckström and Y. Y. Linko (1980). Preparation and kinetic behavior of immobilized whole cell biocatalysts. *Biochemie*, **62**, 387–394.

Makiguchi, N., M. Arita and Y. Asai (1980). Immobilization of a luminous bacterium and light intensity of luminous materials. *J. Ferment. Technol.*, **58**, 17–21.

Manecke, G. and W. Beier (1983). Immobilisierung von *Corynebacterium simplex* durch Photovernetzung von vinyliertem Polyvinylalkohol zur mikrobiologischen Steroidumwandlung. *Angew. Makromol. Chem.*, **113**, 179–202.

Marcipar, A., N. Cochet, L. Brackenridge and J. M. Lebeault (1979). Immobilization of yeasts on ceramic supports. *Biotechnol. Lett.*, **1**, 65–70.

Margaritis, A., P. K. Bajpai and J. B. Wallace (1981). High ethanol productivities using small Ca-alginate beads of immobilized cells of *Zymomonas mobilis*. *Biotechnol. Lett.*, **3**, 613–618.

Mattiasson, B. and C. Borrebaeck (1978). An analytical flow system based on reversible immobilization of enzymes and whole cells utilizing specific lectin–glucoprotein interactions. *FEBS Lett.*, **85**, 119–123.

McDowell, R. H. (1977). *Properties of alginates*. Alginate Industries.

Messing, R. A., R. A. Oppermann and F. B. Kolot (1979). Pore dimensions for accumulating biomass. *ACS Symp. Ser.*, **106**, 13–28.

Messing, R. A. (1982). High-rate, continuous waste processor for the production of high BTU gas using immobilized microbes. *Enzyme Eng.*, **6**, 173–180.

Miles Laboratories, Inc. (1982). *US Pat.* 4 347 320.

Mitsubishi Rayon Co. Ltd. (1981). *Jpn. Pat.* 81 21 591.

Miyoshi, T., Y. Ishimatsu and S. Kimura (1977). *Jpn. Pat.* 77 120 185.

Mohan, R. R. and N. N. Li (1975). Nitrate and nitrite reduction by liquid membrane-encapsulated whole cells. *Biotechnol. Bioeng.*, **17**, 1137–1156.

Moo-Young, M., J. Lamptey and C. W. Robinson (1980). Immobilization of yeast cells on various supports for ethanol production. *Biotechnol. Lett.*, **2**, 541–548.

Mosbach, K. and R. Mosbach (1966). Entrapment of enzymes and microorganisms in synthetic cross-linked polymers and their application in column techniques. *Acta Chem. Scand.*, **20**, 2807–2810.

Mosbach, K. (1982). Use of immobilized cells with special emphasis on the formation of products formed by multistep enzyme systems and coenzymes. *J. Chem. Tech. Biotechnol.*, **32**, 179–188.

Navarro, J. M. and G. Durand (1977). Modification of yeast metabolism by immobilization onto porous glass. *Eur. J. Appl. Microbiol.*, **4**, 243–254.

Nelson, R. P. (1976). *US Pat.* 3 957 580.

Nilsson, K. and K. Mosbach (1980). Preparation of immobilized animal cells. *FEBS Lett.*, **118**, 145–150.

Nilsson, K., S. Birnbau, S. Flygare, L. Linse, U. Schröder, U. Jeppsson, P. O. Larsson, K. Mosbach and P. Brodelius (1983). A general method for the immobilization of cells with preserved viability. *Eur. J. Appl. Microbiol. Biotechnol.*, **17**, 319–326.

Omata, T., A. Tanaka, T. Yamane and S. Fukui (1979). Immobilization of microbial cells and enzymes with hydrophobic photocrosslinkable resin prepolymers. *Eur. J. Appl. Microbiol. Biotechnol.*, **6**, 207–215.

Omata, T., N. Iwamoto, T. Kimura, A. Tanaka and S. Fukui (1981). Stereoselective hydrolysis of dimethyl succinate by gel-entrapped *Rhodotorula winuta* var. *texensis* cells in organic solvent. *Eur. J. Appl. Microbiol. Biotechnol.*, **11**, 199–204.

Petre, D., C. Noel and D. Thomas (1978). A new method for cell immobilization. *Biotechnol. Bioeng.*, **20**, 127–134.

Pilwat, G., P. Washausen, J. Klein and U. Zimmermann (1980). Immobilization of human red blood cells. *Z. Naturforsch.*, *Teil C*, **35**, 352–356.

Roels, J. A. and R. van Tilberg (1979). Temperature dependence of the stability and the activity of immobilized glucose isomerase. *ACS Symp. Ser.*, **106**, 147–172.

Sato, T., Y. Nishida, T. Tosa and I. Chibata (1979). Immobilization of *Escherichia coli* cells containing aspartase activity with *K*-carregeenan. Enzymic properties and application for L-aspartic acid production. *Biochim. Biophys. Acta*, **570**, 179–186.

Sato, T., S. Takamatsu, K. Yamamoto, I. Umemura, T. Tosa and I. Chibata (1982). Production of L-alanine from ammonium fumarate using two types of immobilized microbial cells. *Enzyme Eng.*, **6**, 271–272.

Sawada, H., S. Kinoshita, T. Yoshida and H. Taguchi (1981). Continuous production of 12-ketochenodeoxycholic acid in a column reactor containing immobilized living cells of *Brevibacterium fuscum*. *J. Ferment. Technol.*, **59**, 111–114.

Scott, C. D. and C. W. Hancher (1977). Use of a tapered fluidized bed as a continuous bioreactor. *Biotechnol. Bioeng.*, **18**, 1393–1403.

Seyan, E. and D. Kirwan (1979). Nitrogenase activity of immobilized *Azetobacter vinelandii*. *Biotechnol. Bioeng.*, **21**, 271–281.

Shinmyo, A., H. Kimura and H. Okada (1982). Physiology of α-amylase production by immobilized *Bacillus amyloliquefaciens*. *Eur. J. Appl. Microbiol. Biotechnol.*, **14**, 7–12.

Siess, M. H. and S. Dinés (1981). Behavior of *Saccharomyces cerevisiae* cells entrapped in polyacrylamide gel and performing alcoholic fermentation. *Eur. J. Appl. Microbiol. Biotechnol.*, **12**, 10–15.

Stenross, S. L., Y. Y. Linko, P. Linko, M. Harju and M. Heikonen (1982). Lactic acid fermentation with immobilized *Lactobacillus* sp. *Enzym. Eng.*, **6**, 299–301.

Schara, P. (1977). Einschluss von Mikroorganismen in hydrophile, kovalente Polymernetzwerke—zum Aufbau und zur Reaktivität von Biokatalysatoren für Mehrphasenreaktionen. *Ph.D. Dissertation, TU Braunschweig*.

Scherer, P., M. Kluge, J. Klein and H. Sahm (1981). Immobilization of the methanogenic bacterium *Methanosarcina barkeri*. *Biotechnol. Bioeng.*, **23**, 1057–1065.

Scheurich, P. H. Schnabl, U. Zimmermann and J. Klein (1980). Immobilization and mechanical support of individual protoplasts. *Biochim. Biophys. Acta*, **598**, 645–651.

Takata, I., T. Tosa and I. Chibata (1977). Screening of matrix suitable for immobilization of microbial cells. *J. Solid Phase Biochem.*, **2**, 225–236.

Takata, I., K. Yamamoto, T. Tosa and I. Chibata (1980). Immobilization of *Brevibacterium flavum* with carregeenan and its application for continuous production of L-malic acid. *Enzyme Microb. Technol.*, **2**, 30–36.

Tanaka, A., I. N. Jin, S. Kawamoto and S. Fukui (1979). Entrapment of microbial cells and organelles with hydrophilic urethane prepolymers. *Eur. J. Appl. Microbiol. Biotechnol.*, **7**, 351–354.

Toda, K. and M. Shoda (1975). Sucrose inversion by immobilized yeast cells in a complete mixing reactor. *Biotechnol. Bioeng.*, **17**, 481–497.

Tosa, T., T. Sato, T. Mori, K. Yamamoto, I. Takata, Y. Nishida and I. Chibata (1979). Immobilization of enzymes and microbial cells using carrageenan as matrix. *Biotechnol. Bioeng.*, **21**, 1697–1709.

Tosa, T., I. Takata and I. Chibata (1982). Stabilization of fumarase activity of *Brevibacterium flavum* cells by immobilization with *K*-carrageenan and polyethyleneimine. *Enzyme Eng.*, **6**, 237–238.

Tschang, C. J., H. Klefenz and W. Zahn (1980). *Ger. Offen.* 2 911 776.

Tsumura, N. and T. Kasumi (1976). Immobilization of glucose isomerase in microbial cell. *Int. Ferment. Symp.*, *5th, Berlin, 1976*, 291.

Vandamme, E. J. (1976). Immobilized microbial cells as catalysts. *Chem. Ind.*, 1070–1072.

van Keulen, M., K. Vellenga, G. H. E. Joosten (1981). Kinetics of the isomerization of D-glucose isomerase containing arthrobacter cells in immobilized and non-immobilized form. *Biotechnol. Bioeng.*, **23**, 1437–1448.

Vieth, W. R., S. S. Wang and R. Saini (1973). Immobilization of whole cells in a membraneous form. *Biotechnol. Bioeng.*, **15**, 565–569.

Vijayalakshmi, M., A. Marcipar, E. Segard and G. B. Broun (1979). Matrix-bound transition metal for continuous fermentation tower packing. *Ann. N. Y. Acad. Sci.*, **326**, 249–254.

Vishnoi, P. S. (1980). Biorganic transformations through immobilized microbial cells. *J. Sci. Ind. Res.*, **39**, 578–588.

Vorlop, K.-D., J. Klein and F. Wagner (1980). Immobilization of microbial cells by polymer precipitation. *6th. Intern. Ferment. Sympos., London (Canada) 1980*, Poster, Abstract p. 122.

Vorlop, K.-D. and J. Klein (1981). Formation of spherical chitosan biocatalysts by ionotropic gelation. *Biotechnol. Lett.*, **3**, 9–14.

Vorlop, K.-D. (1982). Unpublished results.

Vorlop, K.-D. and J. Klein (1983). New developments in the field of cell immobilization — formation of biocatalysts by-ionotropic gelation. In *Enzyme Technology*, ed. R. M. Lafferty. pp. 219–235. Springer International, New York.

Wada, M., J. Kato and I. Chibata (1979). A new immobilization of microbial cells. Immobilized growing cells using carrageenan gel and their properties. *Eur. J. Appl. Microbiol. Biotechnol.*, **8**, 241–247

Wada, M., J. Kato and I. Chibata (1980a). Continuous production of ethanol using immobilized growing yeast cells. *Eur. J. Appl. Microbiol. Biotechnol.*, **10**, 275–287.

Wada, M., T. Uchida, J. Kato and I. Chibata (1980b). Continuous production of L-isoleucine using immobilized growing *Serratia marcescens* cells. *Biotechnol. Bioeng.*, **22**, 1175–1188.

Yamada, H., S. Shimizy, Y. Tani and T. Hino (1980). Synthesis of coenzymes by immobilized cell system. *Enzyme Eng.*, **5**, 405–411,

Yamamoto, K., T. Tosa, K. Yamashita and I. Chibata (1977). Kinetics and decay of fumarase activity of immobilized *Brevibacterium ammoniagenes* cells for continuous production of L-malic acid. *Biotechnol. Bioeng.*, **19**, 1101–1114.

Yoshii, F., T. Fujimura and I. Kaetsu (1981). Stabilization of chloroplast by radiation-induced immobilization with various glass-forming monomers. *Biotechnol. Bioeng.*, **23**, 833–841.

Žůrkobá, E., K. Bouchal, M. Slaviček, V. Vojtišek, R. Zeman, F. Švec, J. Drobnik and J. Kálal (1978). Immobilization of bacterial cells of *Escherichia coli* on polymer carriers with reactive groups. *18th Microsymposium on Macromolecules, Prague*, Paper M15/1–4.

13
Combined Immobilized Enzyme/Cell Systems

M. CHARLES and J. PHILLIPS
Lehigh University, Bethlehem, PA, USA

13.1 COMBINED SYSTEMS

What little research and development there has been on the simultaneous use of immobilized cells and immobilized enzymes has been limited primarily to the systems illustrated in Figure 1. The objective in the vast majority of the studies has been to produce ethanol more efficiently.

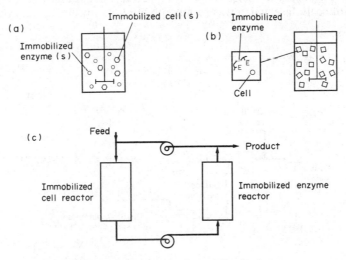

Figure 1 Immobilized enzyme/immobilized cell reactors

13.1.1 Dual Catalysts for the Conversion of Xylose to Ethanol

Chiang *et al.* (1981) used *Schizzosaccharomyces pombe* together with isomerase immobilized in whole cells to produce ethanol from xylose: the objective was to develop a process for the conversion of hemicellulose hydrolyzates to ethanol. Using 45 g l^{-1} of yeast and 50 g l^{-1} of enzyme preparation in a single reactor at pH 6 and 30 °C, Chiang obtained approximately 35 g l^{-1} of ethanol in 45 h from 120 g l^{-1} of xylose. The yield was approximately 63% of theoretical and the productivity was about 0.8 g l^{-1} h^{-1}. The operating conditions were a compromise: the best conditions for the isomerase are 70 °C and pH 7.5; the yeast functions best at 30 °C, pH 4–5.

Jeffries (1981) used a two reactor system (Figure 1c) to produce ethanol from xylose. The immobilized enzyme reactor used xylose isomerase immobilized in whole cells, while the fermenter employed *Candida tropicalis* (ATCC 1369) immobilized in *K*-carrageenin particles. Jeffries obtained approximately 41 g l^{-1} of ethanol after 144 h; the productivity was approximately 1.2 g l^{-1} h^{-1}. The author did not seem to think the process had much promise, however, he did not take advantage of the opportunity offered by the two reactors to provide different operating conditions for each of the immobilized catalysts (their optimum operating conditions differ considerably).

Hagerdahl and Mosbach (1979) used bakers' yeast coimmobilized with β-glucosidase to convert cellobiose to ethanol. The β-glucosidase was bound covalently to sodium alginate and then the complex and yeast were coimmobilized in 2 mm (average) calcium alginate particles (see Figure 1b). The β-glucosidase hydrolyzed cellobiose to glucose which was consumed by the yeast (the yeast cannot use cellobiose) thereby keeping the glucose concentration too low to inhibit the enzyme significantly. The bifunctional catalyst was used in a continuous reactor to convert 50 g l^{-1} of cellobiose to 22 g l^{-1} of ethanol (80% yield). The catalyst remained in stable operation for at least 4 weeks.

Kierstan *et al.* (1982) immobilized β-glucosidase (from *Talaromyces emersonii*) on Sepharose 4B and then coimmobilized the IME and yeast in calcium alginate. In a batch reactor, the catalyst produced 13 g l^{-1} of ethanol in 23 h.

In another example of this approach, Lee *et al.* (1983) coimmobilized in calcium alginate free *Zymomonas mobilis* and β-glucosidase immobilized on concanavalin A Sepharose. The catalyst contained 16.3 g l^{-1} of β-glucosidase/Sepharose and 40 g l^{-1} of *Z. mobilis*. A 200 ml fermenter containing 75 ml of the dual catalyst, and operating continuously, gave 10 g l^{-1} of ethanol at a productivity of 16 g l^{-1} h^{-1} (the dilution rate was 1.78 h^{-1}); the yield was approximately 50% of theoretical. At a lower dilution of 0.2 h^{-1} the ethanol concentration was 16.1 g l^{-1}.

Hahn-Hagerdahl *et al.* (1981) used yeast coimmobilized with isomerase as part of a system designed to convert cellulose to ethanol. The system, illustrated in Figure 2, also employed an ultrafiltration reactor wherein cellulose was hydrolyzed to a sugar mixture (primarily glucose and cellobiose) by soluble cellulase which was retained by the membrane and recycled continuously.

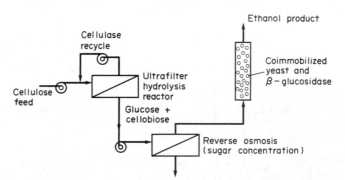

Figure 2 Enzymatic hydrolysis of cellulose to glucose: the use of membrane reactors and of coimmobilized yeast and β-glucosidase

The coimmobilized enzyme/cell catalyst comprised a calcium alginate gel-bound mixture of *Saccharomyces cerevisiae* and *Aspergillus niger* β-glucosidase bound covalently to sodium alginate. A continuous reactor operating at a dilution rate of 0.2 h^{-1} yielded almost 6% ethanol when fed a stream containing 5% glucose and 5% cellobiose; the productivity was 12 g l^{-1} h^{-1} and the yield was essentially quantitative.

13.1.2 Conversion of Cellulose to Ethanol

Strictly speaking, processes which have been developed for the conversion of cellulose to ethanol do not make use simultaneously of immobilized cells and immobilized enzymes; nevertheless, there are enough similarities to warrant their inclusion in this section.

The simultaneous saccharification and fermentation process (SSF), developed by Emert and Katzen (1980), Becker *et al.* (1980) and Blotcamp *et al.* (1978), employed soluble *Trichoderma reesei* (QM9414) cellulase together with *Saccharomyces cerevisiae* (ATCC 4132) in the same vessel to effect a single step conversion of cellulose to ethanol. Becker *et al.* (1980) reported that at pilot scale 60 g l^{-1} of Solka Floc was converted to 18 g l^{-1} of ethanol in 24 h; the productivity was 0.75 g l^{-1} h^{-1} and the yield was approximately 53% of the theoretical yield. The process was conducted at 40 °C which was a compromise between the optimum temperatures for the yeast and the enzyme. For the same mode of operation, 70 g l^{-1} of pulp mill waste was converted to 22 g l^{-1} of ethanol; the yield and productivity were comparable to those obtained with Solka Floc. Becker *et al.* (1980) reported further that the pilot plant results were poorer than the corresponding bench-scale results. Apparently, there were design and operating problems which caused contamination and control problems in the pilot plant.

According to Emert and Katzen (1981), the SSF process gave yields which were 25–40% greater than those reached in conventional two-step processes. They contended that the primary reason for the improvement is the fact that the yeast removed hydrolysis products (*e.g.* glucose) which inhibited the cellulase enzyme complex. Emert and Katzen reported also that a 50 million gallon per year ethanol plant could be built for $112 million (1981 dollars); the projected operating cost was $1.428 per gallon. Obviously, the apparent difficulties encountered with the pilot plant operation must lead one to view these figures as preliminary estimates only.

Finally, it is worth noting that Blotkamp *et al.* (1978) reported that ethanol at concentrations up to 5% did not inhibit reducing sugar production in simple saccharification. This seems to conflict directly with a previous publication by the same group (Takagai *et al.*, 1977) wherein they reported that ethanol at a concentration of only 3% decreased the cellulolytic activity by 50%.

Hahn-Hagerdahl *et al.* (1981) used a novel extractive technique to effect one-step conversion of cellulose to ethanol. They exploited the fact that the polymer pair Dextran T-40/Carbowax PEG (and other pairs of water soluble polymers) formed two phases in aqueous systems. When this two phase system was agitated, cellulose and yeast tended to remain in the dispersed phase along with the cellulase enzyme (Figure 3). The retention of cellulase activity in the dispersed phase was due in part to adsorption of the enzyme onto cellulose and in part to the greater solubility of β-glucosidase in the dispersed phase (this probably is not true for all polymer pairs). The combined action of the β-glucosidase and yeast eliminated product inhibition of the cellulase. Ethanol was extracted continuously into the continuous phase and was removed easily by simple decantation.

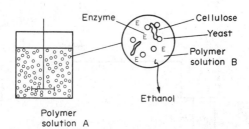

Figure 3 Direct fermentation of cellulose to ethanol: the use of a two-phase water-soluble polymers system

13.1.3 Simultaneous Saccharification and Fermentation of Starch

Lee *et al.* (1983) effected the conversion to ethanol of liquefied corn starch by employing soluble amyloglucosidase (AMG) and *Zymomonas mobilis* in a continuous fermenter/hollow fiber ultrafiltration (UF) recycle system (see Figure 4). The molecular weight cut-off of the UF membrane was 5000, which is well below the 100 000 molecular weight of AMG, therefore all the AMG added initially remained in the system. The process temperature was limited by the use of *Z. mobilis* to 35 °C, however, AMG functions best at approximately 60 °C, therefore an enzyme dose much higher than normal was required to prevent overall rate limitation by the hydrolysis reaction.

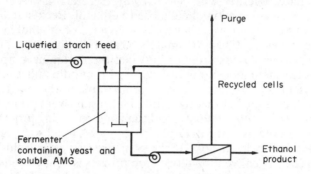

Figure 4   Continuous ethanol fermentation employing ultrafiltration to retain yeast and AMG

In a typical bench-scale experiment employing 7.5% (w/v) AMG (Novo 150L), 130 g l^{-1} (glucose equivalent) of liquefied starch was converted to approximately 65–60 g l^{-1} of ethanol over a range of dilution rates from 0.25 h^{-1} to 1.0 h^{-1}. The maximum productivity was 60 g l^{-1} h^{-1}; the cell mass concentration was approximately constant at 25 g l^{-1} over the entire dilution rate range. At the maximum productivity, the reducing sugar concentration in the product stream was approximately 5 g l^{-1}; the yield was approximately 90% of the theoretical yield.

Lee used a very high AMG concentration (between 1 and 2 orders of magnitude greater than used normally in batch saccharification) to ensure complete saccharification. However, he claimed that if there is no significant loss in AMG activity during a 50 h continuous fermentation, the relative AMG concentration based on starch processed will be only 0.15%. The claim, which appears sound on a theoretical basis, does not seem to have been verified experimentally.

Charles (1983) developed a simultaneous saccharification/fermentation (S/F) system wherein the fermentation temperature was 30 °C and the saccharification temperature was 60 °C. Two variations of the process are illustrated in Figure 5 (flow diagram of a 100 l pilot plant piped for both versions): one used soluble AMG, the other used immobilized (IME) AMG. In each case, the fermentation broth was circulated constantly, through a hollow-fiber, microporous, cross-flow filter (0.4 μm pores) until hydrolysis was complete. In the soluble enzyme version, dextrins and AMG passed through the filter wall and then were reacted at 60 °C; in the IME version, only dextrins passed through the filter. Cells, which were retained by the filter, were recycled to the fermenter.

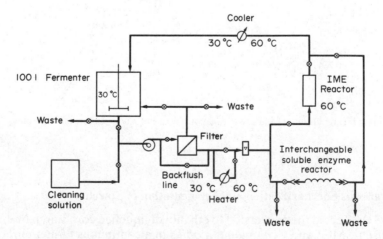

Figure 5   S/F pilot plant

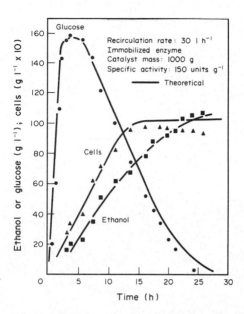

Figure 6 Fermentation history: pilot-plant (100 l) production of ethanol from liquefied starch *via* S/F (immobilized enzyme version)

Both versions of the process were tested at the 100 l scale. The results showed that the IME version (see Figure 6 for typical 100 l fermentation history) can be operated more economically than current commercial S/F processes wherein AMG is added directly to the batch fermenter operating at 30 °C. The soluble enzyme version of Charles' process did not appear to be economically viable.

Ueda *et al.* (1980) combined liquefaction, saccharification and fermentation into a single-step process to produce ethanol from raw cassava starch. In this process a koji, produced by *Aspergillus* spp. grown on wheat bran, was mixed directly with homogenized cassava roots, yeast suspension and water. The mixture was fermented at 30 °C (pH 3.5 initially) for approximately 5 days to give an alcohol yield of up to 99.6% of theoretical. Information concerning the final ethanol concentration was not given. Subsequent developments and modifications (unpublished) appear to have improved the process to the point where it has been considered for commercial application.

13.2 REFERENCES

Becker, D. K., P. J. Blotkamp and G. H. Emert (1980). *ACS Div. Fuel Chem.*, **25**, 297–308.
Blotkamp, P. J., M. Takagai, M. S. Pemberton and G. H. Emert (1978). *AIChE Symp. Ser.*, **74** (No. 181), 85–90.
Charles, M. (1983). Final Report, Dept. of Energy Contract.
Chiang, L.-C., H.-Y. Hsiao, P. P. Ueng, L.-F. Chen and G. T. Tsao (1981). In *Third Symposium on Biotechnology in Energy Production and Conservation*, ed. C. D. Scott, pp. 263–274. Wiley, New York.
Emert, G. H. and R. Katzen (1980). *Chem. Technol.*, **10**, 610–614.
Emert, G. H. and R. Katzen (1981). *Biomass as a Non-Fossil Fuel Source*, chap. 11. ACS, Washington, DC.
Hagerdahl, B. and K. Mosbach (1979). Paper presented at the International Congress on Engineering and Food, Food Processing Engineering, August 27–31, Helsinki, Finland.
Hahn-Hagerdahl, B., E. Andersson, M. Lopez-Leiva and B. Mattiasson (1981). In *Third Symposium on Biotechnology in Energy Production and Conservation*, ed. C. D. Scott, pp. 651–661. Wiley, New York.
Jeffries, T. W. (1981). In *Third Symposium on Biotechnology in Energy Production and Conservation*, ed. C. D. Scott, pp. 315–324. Wiley, New York.
Kierstan, M., A. McHalc and M. P. Coughlan (1982). *Biotechnol. Bioeng.*, **24**, 1461–1463.
Lee, J. H., R. J. Pagan and P. L. Rogers (1983). *Biotechnol. Bioeng.*, **25**, 659–669.
Takagai, M., S. Abe, S. Suzuki, G. H. Emert and N. Yata (1977). In *Proceedings of the Bioconversion Symposium*, pp. 551–571. ITT, Delhi, India.
Ueda, S., C. T. Zenin, D. A. Monteiro and Y. K. Park (1981). *Biotechnol. Bioeng.*, **23**, 291–299.

UPSTREAM AND DOWNSTREAM PROCESSING

14
Introduction

C. L. COONEY
Massachusetts Institute of Technology, Cambridge, MA, USA

The commercialization of new products and processes made possible through biotechnology requires a coordinated coupling of unit operations in order to develop efficient processes. The usual objective of a process is to convert relatively inexpensive raw materials into more valuable products or services. While much of the excitement in biotechnology today has been catalyzed by discoveries and developments in fundamental molecular biology and molecular genetics, it is the role of the process engineer to translate these discoveries and observations into usable processes. In this sense, biotechnology represents a synthesis achieved by close cooperation between scientists and technologists. This synthesis is easily visualized by the schematic diagram of a typical biochemical process illustrated in Figure 1. The central point of this process is the bioreactor. In this unit operation, value is added through biocatalysis, fermentation or cell culture to make valuable products and services. The bioreactor, however, does not exist in isolation. Its efficient and successful operation depends upon adequate upstream processing. Furthermore, recovery of the final product requires a series of steps referred to, collectively, as downstream processing. The objective of Volume 1, Section 2 of *Comprehensive Biotechnology* is to address the upstream and downstream processing unit operations needed in biochemical process development. Section 2 is intended to be complementary to Section 1, which focuses on bioreactor design, operation and control.

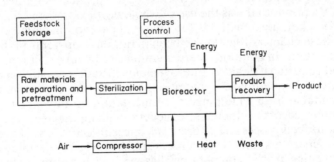

Figure 1 Typical flow sheet for a biochemical process

In the first part of this volume, consideration is given to the problems of gas, liquid and solid handling prior to the use of these materials for a fermentation or bioconversion. The chapter on solids and liquids handling illustrates the diversity of unit operations available for handling key materials used in biochemical processes. It is interesting to note that one of the limitations in scaling down a biochemical process to permit process development in the laboratory is the ability to scale down the solids and liquid handling processes. Many of these operations are performed much more easily on a larger scale. This problem is exacerbated by the fact that many laboratory researchers neglect to consider the problems in scale up of materials handling, focusing solely on the issue of the bioreactor. The chapter on gas compression, while brief, describes the fundamentals of this critical unit operation. While more than 85% of the world's total fermentation capacity is anaerobic and thus does not require gas compression, the remaining 15 to 20%, which is aerobic, generates products of far greater value than anaerobic processes. With few exceptions, most of the processes described in Volume 3 of *Comprehensive Biotechnology* are aerobic processes, thus underlining the importance of efficient gas compression. This operation is one of the most energy intensive and must be coordinated with the design and operation of the bioreactor itself.

Chapters 17 and 18 deal with two of the most critical operations required for successful operation of the biochemical process. These are air and media sterilization. Air sterilization is dominated by the use of filtration as a mechanism for total organism and particle removal. Given the large volumes of air required in aerobic processes, it is clear that efficient and economic removal of organisms is critical to the process design. Reflecting on this unit operation over the past three decades, it has undergone a major change moving away from the use of fibrous depth filters to the implementation of cartridge filters. Today, it is safe to say that all new fermentation facilities utilize cartridge air filtration and many older facilities have been retrofitted. In the area of media sterilization, thermal treatment of the medium to effect destruction of all viable microorganisms is the preferred method. In some situations, such as product sterilization and sterilization of complex organic media used in cell culture, thermal treatment is inappropriate because of extensive nutrient or product degradation. The importance of media filtration is likely to increase in parallel with the increased use of animal and plant cell tissue culture. For most fermentation media, thermal treatment will suffice. The primary change occurring in the past decade has been increased use of continuous sterilization over batch sterilization. This offers the advantages of increased energy efficiency, less nutrient destruction, better use of fermentation vessels and, most importantly, it facilitates scale up by minimizing overheating of fermentation media.

In a single chapter on heat management and fermentation processes this important problem is brought into perspective. One of the major limitations in scale up of large bioreactors is removal of heat. This problem occurs because of the typical low temperature of bioreactor operation relative to cooling water temperatures. Efficient removal of heat becomes increasingly important to high productivity processes.

It is clear that many of the important products emerging through new biotechnology are intracellular proteins. As a consequence, disruption of cells becomes an important unit operation to release these intracellular products prior to their isolation and purification. While a number of mechanisms for cell disruption are available, the use of high pressure homogenizers and high speed ball mills dominate this operation. These techniques are both energy- and capital-intensive. There is also relatively little experience in scaling these operations up to process very large volumetric throughput of cells. Clearly, there are opportunities for new developments.

After fermentation, it is almost always necessary to effect a solid–liquid separation to separate cell and broth. This unit operation is required whether the product is intracellular or extracellular. One of the few exceptions to this is the use of whole broth extraction. This topic is dealt with later in the book. The alternative methods for solving separation include centrifugation and filtration. As seen by these chapters, there have been major developments in the use of membrane filtration for cell separation. In addition, there continues to be an evolution of improvements in the use of high-speed centrifugal devices. Improvements include not only superior materials permitting higher speeds and more efficient operation, but also a variety of designs for centrifugal separators that are suitable for a wide range of fermentation broth characteristics and cell concentrations. In addition, recent developments in genetic engineering necessitate the increased use of contained devices for cell processing. Recent advances in centrifuges have led to sterilizable and highly contained separators suitable for processing these organisms under the most stringent conditions. It is interesting that a 'competition' has evolved between centrifugal separation of cells and the use of membrane filtration for cell harvesting. Two chapters describing tangential-flow and hollow-fiber filters illustrate the merits of membrane filtration. In recent years, an improved understanding of the theory and an extended data base have greatly enhanced interest in using this technique for cell harvesting.

Membrane filtration has become important not only for concentration of cells but, perhaps even more so, for concentration of protein solutions. Two chapters in this volume deal with the use of ultrafiltration. This unit operation has become one of the most important energy-efficient methods for protein concentration. New developments in membranes, as well as improved process configurations, continue to make this an attractive tool in biochemical process synthesis.

Liquid–liquid extraction is clearly one of the major workhorses in biochemical process development. It has been used extensively for the recovery of antibiotics and other low molecular weight organic materials produced by fermentation or bioconversion. Its attractiveness is based on the energy efficiency of this unit operation, the selectivity of solvent extraction and the scalability of the operation. Although a number of mechanical devices are available, high speed centrifugal liquid contactors are perhaps the most widely used. It is most exciting to witness the evolution of liquid–liquid extraction fo biopolymers. Using the principle of immiscibility of two polymer solutions, it is possible to employ biphasic aqueous systems to effect selective partitioning of proteins between the two phases. Dextran and polyethylene glycol or polyethylene glycol and phosphate buffer represent the two most common biphasic systems examined to date. Commercial utilization of this technology has only just begun. It is expected, however, that this unit operation will become a powerful tool in the development of large scale processes for recovery of high purity proteins.

One of the major and most versatile unit operations of downstream processing is chromatography. Chromatography exists in a variety of configurations and is based on separation by charge, hydrophobicity, size and molecular recognition. Ion exchange chromatography is a widely used and scalable technology. It is important in the recovery of antibiotics, as well as proteins. The selectivity of ion exchange permits purification, and the tight binding of products to an ion exchange column permits concentration. As a consequence, this operation is frequently used early in the separation process. Molecular sieve chromatography has been particularly important in protein recovery, the reason being that many proteins differ in molecular size. In combination with other chromatographic and extraction procedures, it becomes a powerful component of downstream processing. Perhaps the ultimate in chromatographic separations are those based on molecular recognition. The concept of affinity chromatography has been popularized over the past 20 years and, in fact, commercialized in a number of recent new processes. There is great expectation that the development of immuno-adsorption chromatography and other specific ligands will accelerate the use of this technique. Hydrophobic chromatography permits separation of molecules on the basis of hydrophobicity and provides the process engineer with yet another tool to separate molecules on a variety of physical properties. One of the very exciting developments in chromatographic separations has been high performance liquid chromatography (HPLC). The technique has begun to evolve from an analytical laboratory tool to a large scale unit operation. Developments in packing chemistry and packing methods to permit uniform flow through larger columns has allowed the design and operation of large HPLC systems. It is expected that, with an increased understanding of the flow mechanics, improved support materials and developments of concepts such as multipoint injection, this technology will become increasingly important in new processes for a wide variety of biochemical products.

There are several other technologies that have proved to be important or are expected to be important in future processes. These include the use of distillation; this has been the prime method of separation of volatile compounds such as ethanol, acetone, butanol, isopropanol, *etc.* During the past decade, major changes in distillation processes have evolved as a consequence of increased prices and as illustrated here, energy efficient designs have greatly improved this unit operation and its applicability to fermentation processes. One of the most exciting new techniques for extraction is supercritical fluid extraction. Taking advantage of the physical properties of supercritical fluids, namely to control solute solubility through changes in pressure and temperature and the high molecular diffusivities in these solvents, they have become very interesting candidates for use in biochemical processes. There is at the present time, however, a dearth of experimental data which would allow one to estimate *a priori* how supercritical solvents will apply to dilute aqueous salt solutions or complex organic mixtures. While it is a common feeling that this technology is most applicable to non-polar, low molecular weight compounds, there is clearly room for innovation in the use of this technique to solve other separation problems. One last unit operation that is increasingly important in both low and high molecular separations is electrodialysis. This technique is an interesting approach to purification and concentration of low molecular weight salts and is becoming an important technology for desalting of high molecular weight products, proteins in particular. Improvements in process configuration and membrane design make this technique attractive in a variety of applications.

In closing, it is hoped that the experimentalist and design engineer will find this volume of *Comprehensive Biotechnology* a useful and important compilation of key unit operations in biochemical process design. To the experimentalist, it describes those technologies that, for the most part, are quite scalable and usable in large scale biochemical processes. To the design engineer, it provides a reference source and, in many cases, the starting point to enter the literature to seek design equations and data bases required for process synthesis. In the coming years, there no doubt will evolve a number of important unit operations to permit improved separation of biochemical products. Biotechnology is an exciting field. It has become important to manufacturing technology. As a consequence, there are many problems to be solved in the successful synthesis of new processes for the biochemical process industry.

15

Solids and Liquids Handling

C. E. DUNLAP
A. E. Staley Manufacturing Company, Decatur, IL, USA

15.1 INTRODUCTION

Materials handling in the bioprocessing industry is essentially similar to the handling of solids, liquids and mixtures of the two in conventional chemical process industry applications. There are, however, several unique problems and situations which arise in biological or biochemical processes which are usually not encountered in other processes, and their proper solution must be found to insure successful operation.

Fermentation and some enzymatic processes require varying degrees of sterility, grain processes require dust control, and protein coagulation requires low shear pumping. These and other unique bioprocessing problems occur in the handling of process flow streams. This chapter introduces the basic technology of liquid and solid mixing and transport, and includes specific information on bioprocesses where appropriate. The chapter provides brief coverage of the theory of the transport processes and the design and availability of equipment, but provides references for further pursuit of these topics.

15.2 TRANSPORT AND HANDLING OF LIQUIDS

15.2.1 Properties and Rheology of Liquids

Most fluids encountered in the biochemical process industries are conventional, nonbiological liquids which obey the standard rules of Newtonian fluid flow. However, in many bioprocesses there is some need to transport and handle liquids which have special needs due to their sensitivity to shear or thermal factors, the need for sterility, or non-Newtonian rheology. Knowledge of the same fluid properties is needed to design proper handling systems for any fluid.

15.2.1.1 Physical and transport properties

Only a few properties of a liquid need to be determined to allow the description of its flow characteristics. However, for proper design of pumps and piping systems it is necessary to determine these properties and to insure that the information includes the conditions at which the system will operate.

Density (ϱ; kg m^{-3}). The density of a fluid is its mass per unit volume. The density of most liquids can be considered to be independent of pressure over the range found in bioprocesses. However, density is a small function of temperature, generally showing an inverse dependence. Density is sometimes expressed in terms of specific gravity which is the density of a liquid at a defined temperature relative to the density of water at some reference conditions, often either 60 °F (15.56 °C) or 4 °C.

Density is sometimes expressed as degrees Baumé, degrees A.P.I., degrees Twaddell or brix. These units are usually unique to certain industries and conversion factors are available such as those in Perry *et al.* (1963).

Viscosity (μ; Pa s, poise). Viscosity is the transport property defined as the proportionality constant between the shear stress and the rate of shear in a fluid. The expression is

$$\tau = \mu \frac{dv}{dt} \tag{1}$$

where τ = shear stress, μ = viscosity and dv/dt = rate of shear. Fluids for which the viscosity is not a function of the rate of shear are referred to as Newtonian fluids. Most thin, homogeneous liquids are Newtonian, but many liquids found in bioprocesses do not follow this behavior and are referred to as non-Newtonian.

Several typical non-Newtonian behavior patterns are shown in Figure 1. Pseudoplastic fluids show decreased viscosity at higher rates of shear. This is the most common type of non-Newtonian behavior and is thought to be caused by alignment of large solute molecules on a microscopic scale, thus permitting easier flow.

Dilatant fluids thicken with increased shear; this condition is usually found in slurries. It is thought to be caused by physical interference between particles in adjacent fluid planes when their respective motion increases. A third type of non-Newtonian behavior is that of a Bingham plastic fluid which shows no flow until some minimum shear stress is reached. After flow is started it may resemble Newtonian behavior. Some pastes and gels are Bingham plastic. A fluid which

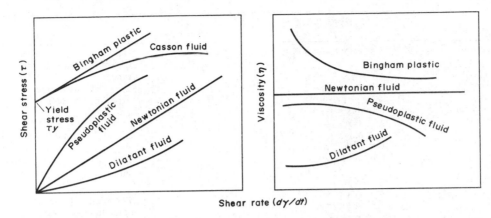

Figure 1   Rheologic behavior of typical Newtonian and non-Newtonian fluids

follows the Casson curve has the stress limit like a Bingham plastic, but resembles a pseudoplastic once the flow starts.

Comprehensive treatment of the rheology of fluids is given by Van Wazer *et al.* (1963) and Sherman (1970) among others. Data for the viscosity of many liquids can be found in Washburn (1926).

15.2.1.2   Rheology of typical bioprocess liquids

The rheology of many liquids important in bioprocessing can be very complex. Doughs, fermentation broths, protein and starch solutions, and fibrous slurries all depart significantly from Newtonian behavior, and one is advised to seek pertinent literature or experimental data for the system of interest before the design of transfer systems.

Fermentation broths. The rheology of fermentation broths has been reviewed by Blanch and Bhavaraju (1976), Deindoerfer and West (1960) and Roels *et al.* (1974).

The rheology of dilute broths and cultures of yeast or nonchain-forming bacteria is usually Newtonian; however, most mycelial broths and cultures in which extracellular polymers are formed follow Bingham plastic or pseudoplastic rheology. Summaries of the available experimental data on the rheology of fermentation broths have been prepared by Blanch and Bhavaraju (1976) and Charles (1978) (See Table 1).

Industrial gums. Gums are used for their viscosity enhancing properties in foods and industrial processes. The basic rheology of gum solutions has been studied by Kabre *et al.* (1964), and rheology is usually pseudoplastic. Whistler and BeMiller (1959) cover basic viscosity data of a number of important industrial gums. Chang and Ollis (1982) describe the rheology of a pseudoplastic fermentation broth containing xanthan gum.

Starch solutions. The viscosity and rheological behavior of starch pastes and solutions has been reviewed by Myers and Knauss (1965). Starch rheology is complicated by the dependence of the solution viscosity on concentration, temperature, shear rate, time of shear and type of starch. Since knowledge of starch rheology is essential in the starch processing industries, numerous test methods have been developed to measure rheology of solutions under a given set of test conditions. Instruments such as the Brabender Visco-Amylo-Graph and the Corn Industries Viscometer are used for these measurements.

15.2.1.3   Special properties of biological liquids

In addition to density and rheological properties, one often must know the sensitivity of bioprocessing liquids to temperature and shear in terms of the functionality or activity of any solute molecules. The denaturation of proteins, pasting of starches, inactivation of enzymes and destruction of microbial cells are all possible consequences of excess shear or temperature. These properties should be determined on the actual process stream of interest before handling systems are designed in order to prevent product damage when the system is installed.

Table 1 Summary of Rheologic Behavior of Several Fermentation Broths (Blanch and Bhavaraju, 1976)

Organism	Constitutive equation	Effect of cell concentration	Comments
P. chrysogenum	$\sqrt{\tau} = \sqrt{\tau_0} + K_c\sqrt{\gamma}$	$\tau \propto X^2$	Based on fiber-like morphology; Casson equation fits entire shear range
P. chrysogenum	$\tau = \tau_0 + K\gamma$		$\tau_0 = \alpha X^2$ also agrees with experimental data
Kanamycin broth	$\tau = \tau_0 + K\gamma$		
Endomyces sp.	$\tau = K\gamma^N$		$0.3 \leqslant N \leqslant 0.9$
P. chrysogenum	$\tau = \tau_0 + K\gamma$ or $\tau = K\gamma^N$		$2 \times 10^{-3} \leqslant K \leqslant 35 \times 10^{-3}$ N m^{-1}s^{-1}
S. grisens	$\tau = \tau_0 + K\gamma$		τ_0 and plastic viscosity varied sinusoidally with fermentation age
S. aurofaciens	$\tau = K\gamma^N$		
Digested sewage sludge	$\tau = AX^{a_1}\gamma^{a_2 + a_3\ln X}$ or $\tau = Ae^{a_1 x}\gamma^{a_2 + a_3 X}$ or $\tau = \tau_0 + K\gamma$		The first model gave better agreement with data than the second
P. pullulans	$\tau = K\gamma^N$		n and K are functions of broth age and also of polysaccharide concentration
Xanthomonas campestris	$\tau = K\gamma^N + \tau_0$		

15.2.2 Flow Measurement and Control

The knowledge of liquid volumetric and mass flow rates is essential in the operation and control of chemical processes. A large number of different devices have been developed which measure flow rates of fluids using principles based on several different phenomena. All of the devices are used in the bioprocessing industry. Each device has strengths and weaknesses for various applications, but at least one type can usually be found for any application which may arise.

15.2.2.1 Theory of flow measurement

Fluids flowing through channels develop velocity profiles which vary in shape according to the channel geometry, flow rate and fluid properties. Since most flow systems in the bioprocess industries use cylindrical pipes for liquid flows, we will discuss only that case, and not weirs, flumes, *etc*. Velocity profiles at a given axial point in round pipes are regular in the cylindrical and axial directions and change with the radial distance from the center of the pipe. Maximum velocity occurs at the center of straight pipes. Some flowmeters such as the pitot tube measure the velocity at a point in the flow field, and others such as orifice meters and rotometers measure some averaged velocity of the liquid. Knowledge of the average velocity is usually most useful when one is attempting to derive mass flow information from flow velocity measurement. The theory of fluid flow is developed in depth in a number of texts, one of which is by Bird *et al.* (1960).

15.2.2.2 Velocity measuring devices

Pitot and Annubar tubes. The simple pitot tube is composed of a tube bent to place one measuring port normal to the fluid flow and a second port parallel to the flow direction. The differential pressure (ΔP) is generated between the two ports due to the difference between the 'stagnation pressure' and the line pressure. This pressure difference is proportional to the liquid velocity. ΔP can be measured by using any of a number of devices, from a simple U-tube manometer to a dP cell connected to a computational device, and correlated with the liquid's velocity.

Since the simple pitot tube only measures fluid velocity at one point in the tube, it is not often used to assist in measuring mass flow rates. Of more use is the Annubar-type device which has several strategically placed ports to measure some average stagnation pressure. Thus, an average flow rate can be measured.

Both of these devices incorporate a stagnant column of liquid which transmits pressure between the pressure ports and the ΔP measuring device. If solids are present in the liquid, the pressure ports are subject to fouling and solids can settle out in the stagnant liquid inside the tubes. Neither device is widely used in the bioprocessing industry.

Flow restriction devices. Several widely used flowmeters use the principle of measuring the ΔP resulting from an induced velocity change to correlate with fluid flow rate. Figure 2 shows four of these pressure impulse generating devices. The orifice meter, the simplest and most widely used of these flowmeters, illustrates the principle of their operation as follows. Fluid flowing at point A has a pressure P_a measured at pressure tap a; the actual orifice causes the fluid to flow through a restriction resulting in a higher velocity at point B; this pressure is measured at pressure tap b. The mean fluid velocity can then be calculated as:

$$u_0 = \frac{C_0}{\sqrt{1-\beta^4}} \sqrt{\frac{2g(P_a-P_b)}{\varrho}} \tag{2}$$

where u_0 = velocity, C_0 = discharge or orifice coefficient which is a constant for each orifice arrangement, β = ratio between the orifice and pipe diameters, P_a = pressure at A, P_b = pressure at B, g = gravitational constant and ϱ = fluid density.

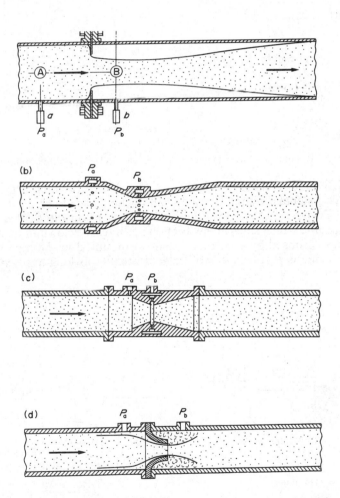

Figure 2 Restriction-type flow measuring devices: (a) orifice meter, (b) venturi tube, (c) Dall flow tube, (d) flow nozzle

Equations similar to equation (2) are used to describe the operation of the venturi meter, the flow nozzle and the Dall flow tube, which are also restriction devices similar in operation to the orifice meter. Thorough discussions of the use of these and other flowmeters appear in works by Doebelin (1975) and Miller (1983).

Flowmeters of the flow-restriction type are widely used in the bioprocessing industry. Like the pitot tube, however, these meters use pressure measurement devices which are usually connected to the flowing fluid by a leg of stagnant liquid. This can be particularly troublesome when solids are present in the fluid and when it is necessary to sterilize the flow system.

Rotameters. Rotameters are widely used in the bioprocessing industries. Their principle of operation is shown in Figure 3. Fluid flowing upward in a vertical tube of tapered cross-section lifts a float to a point in the tube where the annular flow area is such that the downward force of gravity on the float is balanced by the float's buoyancy plus the drag forces of the flowing fluid. Advantages of the rotameter are its almost 10:1 range of flow indication, the fact that there are no stagnant fluid columns or pressure sensors, and the ease with which it can be sterilized. Rotameters are of limited use in liquids where solids are present, or where the density of the liquid varies.

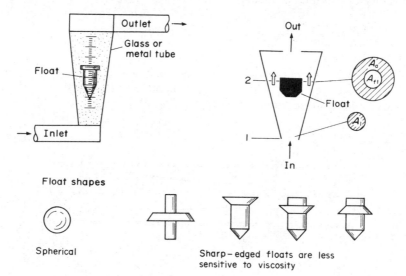

Figure 3 Rotameter design and operating principles and several different float shapes

Rotameters which have steel tubes in which the position of the float is sensed electrically can also be obtained. This signal can then be used to allow the device to be used as a flow controller.

Turbine meters. Turbine meters sense fluid velocity by using the fluid momentum to turn a radially centered turbine which rides in low friction bearings as shown in Figure 4a. The rotation of the turbine vanes is sensed by a magnetic element mounted in the pipe wall. The rotational frequency of the turbine is proportional to the fluid velocity as long as viscosity and density are relatively constant.

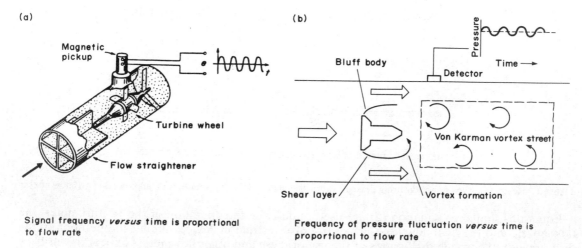

Figure 4 (a) Turbine and (b) vortex meters for liquid flow measurement

The rotational signal frequency signal from the sensor can be fed to a pulse rate meter to generate a digital signal proportional to the flow rate. The meter can be used as the rate sensing device in a flow rate control loop.

Turbine meters can be sterilized and find considerable use in clean streams, but normally cannot be used in particulate streams or liquids of high viscosity.

Vortex meters. Like the turbine meter, the vortex meter generates a digital signal proportional to fluid velocity. The principle of operation is shown in Figure 4b. A nonmoving body of unique shape is centered in the flow field; flow patterns around the edges of the body result in the formation of vortices in the liquid flow. The rate of formation of the vortices is proportional to flow rate. By mounting a detector on the pipe wall and counting vortex formation *versus* time, a digital signal proportional to fluid velocity is obtained.

Vortex meters have the advantage of having no moving parts, are relatively independent of fluid density, and have operating ranges of over 20:1. This type of meter is easily sterilizable, but is not normally used in streams containing particulates.

Electromagnetic flowmeters. Magnetic flowmeters use the principle of induced voltage to obtain a signal proportional to fluid velocity in an unobstructed flow path as shown in Figure 5. Liquids must have a conductivity of at least 2 μS cm^{-1} for proper sensing so many hydrocarbons are excluded.

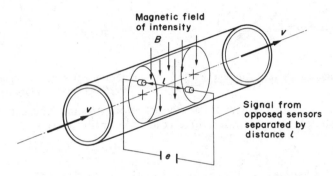

Figure 5   Magnetic flowmeter principle of operation

The flow of a conductive liquid at an average velocity v through a perpendicular magnetic field of intensity B generates a voltage e across a path of length l, thus:

$$e = B l v \qquad (3)$$

where e = induced voltage, B = magnetic field intensity, l = conductor length and v = conductor velocity.

Magnetic flowmeters are sterilizable and are ideally suited for fluids with solids such as slurries and fiber stocks. They are fabricated in all corrosion resistant materials and in a variety of line sizes.

The analog signal obtained from the meter can be used to position valves or control pumps.

Ultrasonic flowmeters. Ultrasonic flowmeters like magnetic flowmeters present an unobstructed flow path to the fluid and are ideal for sterilization and handling of sterile fluids. The principle of operation derives from the fact that an induced pressure pulse travels through a fluid at the speed of sound in the particular fluid. If the fluid itself is moving then the measured velocity of the impulse between two fixed points is the sum of two velocities, and the fluid velocity component of this velocity can be isolated by the design of the ultrasonic signal generator(s) and sensor(s) and associated circuitry. Figure 6 shows several possible configurations for an ultrasonic metering device.

Time-of-flight meters beam a pulse across the flow path and at an angle to the flow direction. The fluid velocity is then calculated from a measurement of the time it takes the pulse to traverse this known distance. Doppler effect meters depend on the reflection of the pulse back to a sensor by particles within the fluid. Thus, time-of-flight devices are usually used on clean streams while Doppler devices need some particulate matter in the fluid to be useful.

Target meters. Drag force or 'target' meters suspend an unmoving bluff body into the flow path of a fluid and measure the force that the fluid imparts to the body due to drag. This force is proportional to fluid velocity if density is constant.

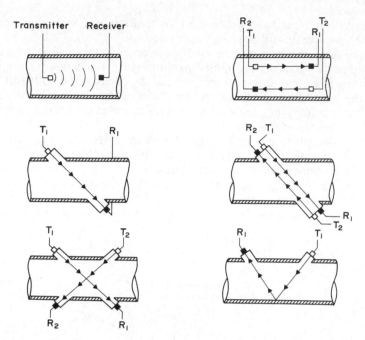

Figure 6 Several different configurations for ultrasonic flowmeter systems

Target meters are well suited for viscous and dirty flow streams, and can be used for streams containing small particles if they do not tend to collect on or foul the target.

15.2.2.3 Flowmeter selection

Characteristic operating parameters and applications of a number of flowmeters are shown in Table 2.

15.2.2.4 Direct mass flow measurement

In most instances in a chemical or biochemical process where flow information is desired, one wishes to know mass flow rates rather than volumetric rates. Two techniques have been developed for direct mass flow rate measurement. The first method incorporates a volumetric flow rate measuring device coupled with a density monitor. The two signals are processed in a small computer to yield mass flow rate information. Almost any of the volumetric rate measuring devices described above which give an analog electrical signal can be coupled with density monitors such as a buoyancy gauge, a balanced U-tube element or an ultrasonic density meter.

The second mass flow measuring technique is used by meters which monitor parameters directly related to mass flow. Such devices as angular-momentum elements are often used to generate a signal proportional to mass flow.

Both types of mass flowmeters are finding greater use as quality and cost of commercial devices become more acceptable. More information on mass flow measurement is given by Doebelin (1975) and Miller (1983).

15.2.2.5 Special considerations in bioprocesses

In most instances where it is desired to measure flow rates of streams in the bioprocessing industries, the same criteria apply as are used in the conventional chemical process industry applications. However, often sterile streams must be monitored, or the measuring device itself must be sterilizable along with the rest of the flow system. Many of the newer devices such as magnetic

Table 2 Selection Chart for Liquid Flowmeters (Miller, 1983)

Flowmeter	Pipe size (mm)	Liquids — Clean	Viscous	Dirty	Slurries — Corrosive	Fibrous	Abrasive	Sterilization	Temperature [°F(°C)]	Pressure [p.s.i. (kPa)]	Accuracy, uncalibrated (including transmitter)	Reynolds number
Square root scale; maximum range 4:1												
Orifice									Process temperature to 1000 °F (540 °C); transmitter limited to −20 to 250 °F (−30 to 120 °C)	To 6000 (41 000)		
square edged	>1.5 (40)	A	C	B	B	C	C	C			±1-2% URV	>2000
honed meter run	0.5–1.5 (12–40)	A	B	C	B	C	C	C			±1% URV	>1000
integral	<0.5 (12)	A	A	C	B	C	C	C			±2-5% URV	>100
quadrant, conic edge	>1.5 (40)	A	A	A	B	C	C	C			±2% URV	>200
eccentric	>2 (50)	B	B	C	A	B	C	C			±2% URV	>10 000
segmental	>4 (100)	B	C	A	B	C	C	C			±2% URV	>10 000
annular	>4 (100)	B	C	A	A	B	C	C			±2% URV	>10 000
Target	>0.5–4 (12–100)	A	A	A	A	A	C	C			±1.5–5% URV	>100
Venturi	>2 (50)	A	B	B	B	B	C	B			±1-2% URV	>75 000
Flow nozzle	>2 (50)	A	B	B	B	B	C	C			±1-2% URV	>10 000
Lo-loss	>3 (75)	A	C	C	A	C	C	C			±1.25% URV	>12 500
Pitot	>3 (75)	A	B	C	C	C	C	C			±5% URV	No limit
Annubar	>1 (25)	A	C	C	B	B	C	C			±1.25% URV	>10 000
Elbow	>2 (50)	A	C	C	B	B	B	C			±4.25% URV	>10 000
Linear scale; typical range 10:1												
Magnetic	1–72 (25–1800)	A	A	A	A	A	A	A	360 (180)	≤1500 (10 000)	± 5% of rate	No limit
Positive displacement	>12 (300)	A	C	C	B	C	C	B	Gases 250 (120) Liquids 600 (315)	≤1400 (10 000)	Gases ±1% URV Liquids ±0.5% of rate	≤8000 cS
Turbine	0.25–24 (6–600)	A	C	C	C	B	C	A	−450 to 500 (−268 to 260)	≤3000 (21 000)	Gases ±0.5% of rate Liquids ±1% of rate	≤2–15 cS
Ultrasonic time of flight	>0.5 (12)	A	B	C	A	C	C	A	−300 to 500 (180 to 260)	Pipe rating	±1% of rate	No limit
doppler	>0.5 (12)	C	B	A	A	A	A	A	−300 to 250 (−180 to 120)	Pipe rating	±5% URV	No limit
Variable area	≤3 (75)	A	A	A	C	B	C	B	Glass 400 (200) Metal 1000 (540)	Glass 350 (2400) Metal 720 (5000)	±0.5% of rate	To highly viscous fluids
Vortex	1.5–16 (40–400)	A	C	B	B	C	C	A	400 (200)	≤1500 (10 500)	±0.75–1.5% of rate	>10 000

A = Designed for this application. B = Normally applicable. C = Not designed for this application.

flowmeters and ultrasonic meters which need no direct, liquid connection to the flow stream and which present an unimpeded flow path to the fluid are ideal for this type of application.

15.2.2.6   Fluid dynamics and liquid transport

The majority of liquid transport systems in the chemical and biochemical process industries are composed of cylindrical piping with associated valves and fittings using some type of pumping device to move the fluid. In this section we will concentrate on the theory and practice of the proper design of piping systems for moving fluids in the biochemical industries, and on the selection of appropriate pumps.

15.2.2.7   Flow of liquids in cylindrical pipes

Liquids in steady flow inside cylindrical pipes develop a velocity profile in the axial direction which is characteristic of the pipe size, the fluid properties and the fluid velocity. Liquids in laminar flow develop parabolic velocity profiles where velocity is zero at the pipe wall and increases to a maximum at the pipe center line. In turbulent flow the fluid has well-developed eddies and intermixing and has a much flatter velocity profile.

In either laminar or turbulent flow the existence of a velocity profile indicates that there is uneven movement of the concentric fluid layers in the pipe. If the fluid is a viscous liquid some friction will be caused between the adjacent fluid layers and this will ultimately be transferred to the pipe wall. A mechanical energy balance for an incompressible liquid flowing in a conduit is

$$\frac{\Delta v^2}{\alpha} + g\Delta Z + \frac{\Delta P}{\varrho} + \Sigma F + W_S = 0 \qquad (4)$$

where v = average velocity of liquid at a point along the length of the pipe, α = a correction factor for using the average velocity; $\alpha = 1.0$ for turbulent flow and 0.5 for laminar flow, g = gravitational constant, Z = height of pipe relative to some plane of reference, P = pressure of the liquid inside the pipe at a point, ϱ = density of the liquid, F = friction forces caused by fluid flow in the system, and W_S = shaft work exchanged between the system and its surroundings.

For systems where a pump is used to move liquids the shaft work term can be written as

$$W_S = -\eta W_P \qquad (5)$$

where η = pump efficiency and W_P = pump power.

This balance expresses the basis used to size pumps and pipes in given flow systems. Usually we know the necessary mass flow rate and the start and finish points of the flow system. It is left for us to select the most economical system of pumps and piping. This is a trial and error procedure whereby one selects a pipe size (and designs a suitable flow system). This sets the velocity, the liquid head to be overcome and the overall system pressure drop. One can then calculate the friction losses and determine the pump size necessary for the system. Selection of a different pipe size for another trial will give another pump requirement. Assignment of values to pipe and pump capital costs and pump operating costs will allow computation of a minimum cost system. Such calculations are given in detail by many texts and handbooks such as Geankoplis (1978) and Peters and Timmerhaus (1980), among others.

15.2.2.8   Design of flow systems for biochemical process industries

Pumps and piping systems for most liquid transport applications in the biochemical process industries are the same as those in the conventional chemical industry applications. However, several unique problems occur in some biochemical processes which must be carefully dealt with by the system designer to assure successful operation. These unique situations are:
   liquids which are sterile;
   flow systems which must be repeatedly sterilized;
   liquids with microorganisms or biochemicals which give non-Newtonian flow characteristics;
   liquids with organisms or biochemicals (enzymes, other proteins, polysaccharides, *etc.*) which are shear sensitive;
   systems with sensitivity to metals contamination such as fermentation or enzyme substrates.

Sufficient materials of construction and pump designs exist to allow successful handling of each of these problems. The main points of importance are to properly identify the potential problems, and to specify a least-cost solution.

15.2.3 Pumps and Piping Systems

The design of pumps to move liquids can be divided into two major types, kinetic and positive displacement. These two classes can be further subdivided into some 30 different pump designs as shown in Figure 7.

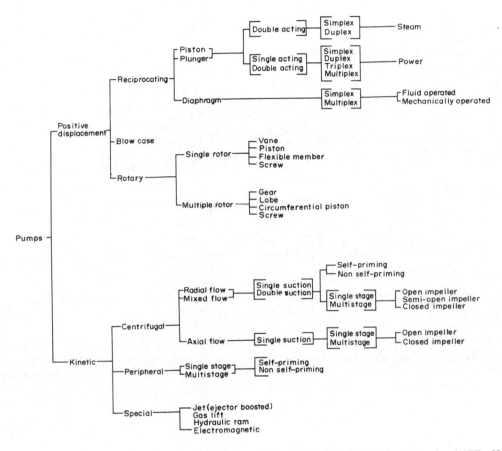

Figure 7 Pump classification (used by permission from *Hydraulic Institute Standards*, 13th edn. (1975), Hydraulic Institute, Cleveland, Ohio)

The general criteria for selection of a pump for a specific service are: (1) flow rate or capacity, (2) outlet pressure necessary for flow, (3) temperature and pressure at inlet, (4) material of construction needs, (5) metering ability, (6) slurry or viscous liquid service, (7) aseptic service, (8) shear sensitive service.

Attention to these points will allow the proper selection of pump design, pump size and other qualities such as seal design and materials of construction.

15.2.3.1 *Centrifugal pumps*

Centrifugal pumps are used in the majority of applications in the chemical and biochemical process industries. They have only one moving part and can operate at a steady, synchronous speed. Since centrifugal pumps can be made in almost any material of construction, are generally of low cost, and are made for a wide range of capacities, they are usually used for a service unless excluded by one of the following criteria:

(1) where very abrasive material is present in the stream;
(2) where flow rates are lower than 10 g.p.m.;
(3) where shear sensitive material is present in the stream;
(4) where liquid viscosity is above 2000–3000 cP;
(5) when absolute sterility must be maintained.

The main variable in the design of centrifugal pumps is the shape of the impeller. Impeller profiles shown in Figure 8 range from thin, fully shrouded radial vane types to axial flow turbines with a number of types in between. The efficiency of each type of impeller is a maximum at some different specific speed. Specific speed is defined as

$$N_S = \frac{N\sqrt{Q}}{H^{3/4}} \tag{6}$$

where N_S = pump specific speed, N = pump speed (r.p.m.), Q = flow in g.p.m. at optimum efficiency and H = total head in feet.

The specific speed scale in Figure 8 shows the speeds where each type of impeller operates at maximum efficiency.

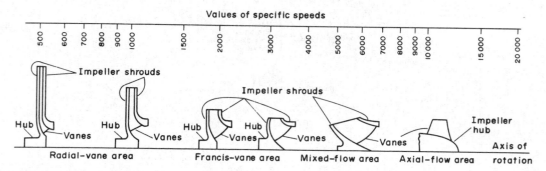

Figure 8 Comparison of centrifugal pump impeller profiles (used by permission from *Hydraulic Institute Standards*, 13th edn. (1975), Hydraulic Institute, Cleveland, Ohio)

The operation of centrifugal pumps is usually described by a characteristic pump curve as shown in Figure 9. These plots show the pressure generated at the pump outlet (the pump 'head') *versus* the flow rate. Often pump efficiency, net positive suction head (NPSH) and pump horsepower are also plotted.

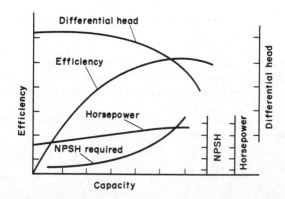

Figure 9 Typical operating characteristic curves for a centrifugal pump

Characteristic operating curves may be plotted for a series of centrifugal pumps of different impeller sizes and pump horsepower as shown in Figure 10. Charts of this nature are supplied by all centrifugal pump manufacturers and are useful for sizing and selection of an appropriate pump for a given service.

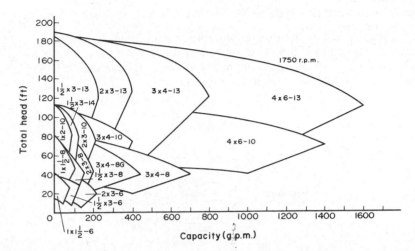

Figure 10 Operating ranges of a series of centrifugal pumps. Each labeled area represents the head/capacity region in which a specific size pump can operate. The first number is the diameter of the discharge line, the second is the diameter of the suction line, and the third is the maximum diameter of impeller that will fit in the pump casing. All diameters are in inches

Centrifugal pumps are manufactured in sizes to handle capacities from 5 to over 100 000 g.p.m. and viscosities up to about 2000 cP. Wetted parts are manufactured in all types of metals and a number of plastics. Pumps for clean or aseptic service are designed with easily removable heads for quick and thorough cleaning.

The operating characteristics of centrifugal pumps are affected significantly by the viscosity of the fluid being pumped. If the intended service involves a fluid with a viscosity greater than water, it is necessary to correct the operating characteristics so a workable pump is selected. Methods of making this correction are given by manufacturers or can be found in Hydraulic Institute Standards (1975) or Thurlow (1971).

15.2.3.2 *Rotating shaft seals*

One of the weakest points of centrifugal and all other rotating element pumps is the necessity to provide a seal around the shaft which drives the impeller. Some small centrifugals in low pressure, low capacity service evade the seal altogether by using a magnetic drive to turn the impeller. However, most pumps suitable for industrial service have shaft driven impellers and some type of rotary seal.

The stuffing box with packing is the most used type of rotary seal. The stuffing box is a cylindrical area surrounding the shaft with a movable plunger for applying restraining pressure to the packing.

The packing can be of any number of materials which are inert to the fluid being pumped and supply lubrication to the rotating shaft. If the fluid being pumped is dirty, gritty or corrosive, some sealing liquid must be supplied to the stuffing box. Clean liquids often act as their own sealing liquid.

In applications where leakage through the seal must be minimized, as in aseptic services, a mechanical seal is often used. The mechanical seal consists of one or two pairs of seal rings. Each pair consists of one soft and one hard ring with very smooth mating faces which rotate against each other. The seal is provided by a spring which forces the rotating surfaces together. Some seal liquid must also be supplied to mechanical seals, but leakage is much reduced from that expected in stuffing boxes.

All rotary seals permit some fluid leakage and are thus suspect when included in a piping system which must remain sterile for long periods of time. It is good design practice to place necessary pumps upstream of the sterilizing device and to have no pumps in the sterile piping section. In cases where this cannot be avoided, a mechanical seal is specified and either the process fluid itself or sterile steam condensate is used as the seal liquid.

15.2.3.3 Kinetic pump metering systems

Centrifugal pumps are not by themselves suitable for use in fluid metering applications. However, flow systems as shown in Figure 11 are often used to permit flow control when a centrifugal pump, or other constant speed pump, is the motive device. Such systems, however, are not suitable for streams which contain solids or must remain aseptic.

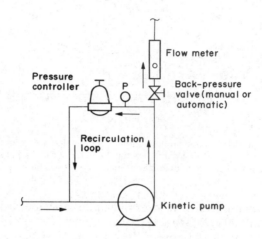

Figure 11   Flow metering system for a kinetic pump

15.2.3.4 Rotary pumps

Rotary pumps are positive displacement pumps which use gears, screws or lobes to provide a moving cavity which carries the process fluid from the pump inlet to the outlet. The size of the pump and the speed at which the pump is operated determine the flow rate, and rotary pumps driven through some variable shaft speed devices are often used as metering pumps.

The major types of rotary pumps are listed in Figure 12, and are shown in Figure 13.

15.2.3.5 Gear pumps

Gear pumps are the most used of rotary pumps. They are supplied in units which are designed for fluids with viscosities to 20 000 000 cP, pressures to 3000 p.s.i., and capacities to over 100 g.p.m. Pumping efficiency is usually from 80% to 95%.

Gear pumps are good for applications involving high viscosity fluids, for metering applications, for high pressure service, and for some dirty streams. They are not appropriate for applications where a low rate of shear is desirable.

15.2.3.6 Screw pumps

Screw pumps use two or three helical machined screws to create a positive displacement pumping action. Screw pumps are produced with capacities to 9000 g.p.m., discharge pressures to 700 p.s.i., and can handle fluids with viscosities to 150 000 SSU.

A special type of screw pump uses only one rotating screw which turns inside a stator to provide a moving cavity. These pumps are often specified when the service involves pumping fluids

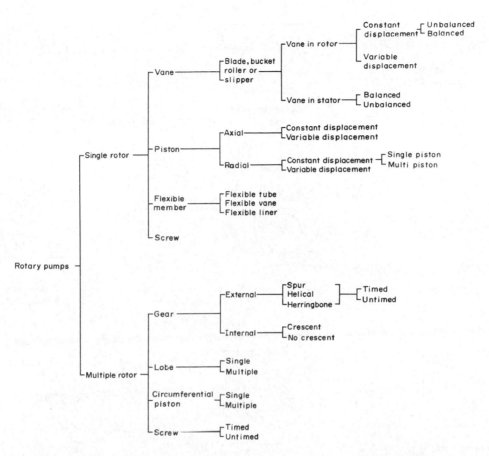

Figure 12 Classification of rotary pumps (used by permission of Hydraulic Institute, Cleveland, Ohio)

with suspended solids, and are used for starch suspensions, food purees, peanut butter and other such two-phase streams. Such pumps are often useful where no other pump can be applied, but care must be taken to select proper materials of construction or wear of the stator will be rapid.

15.2.3.7 Other rotary pumps

Rotating vane pumps are often used in food applications where no crushing or shearing can be tolerated. The rotating vanes simply serve as paddles which move the fluid through the pump body. This type of pump, however, has limited speed and head development.

Peristaltic pumps also supply pumping action with low shear. They are also the only pump type which can guarantee total asepsis and are often used on small fermentation applications. They are of limited size, however, and head development is low. Additionally, the abrasion and stressing of the flexible tubing eventually cause it to fail, so such pumps cannot be used where failure would be catastrophic.

The effect of shear on the integrity of protein precipitates has been studied by Hoare *et al.* (1982). The effects of several different pumps on the size of precipitates is shown in Figure 14. Gear and centrifugal pumps showed considerable destruction of the flocs while a rotor/stator screw pump was somewhat better. A peristaltic pump provided the least floc disruption.

15.2.3.8 Pipes, valves and fittings

Piping systems in the biochemical process industries must often be designed for services not usually found in normal chemical process applications. Systems in the fermentation industry

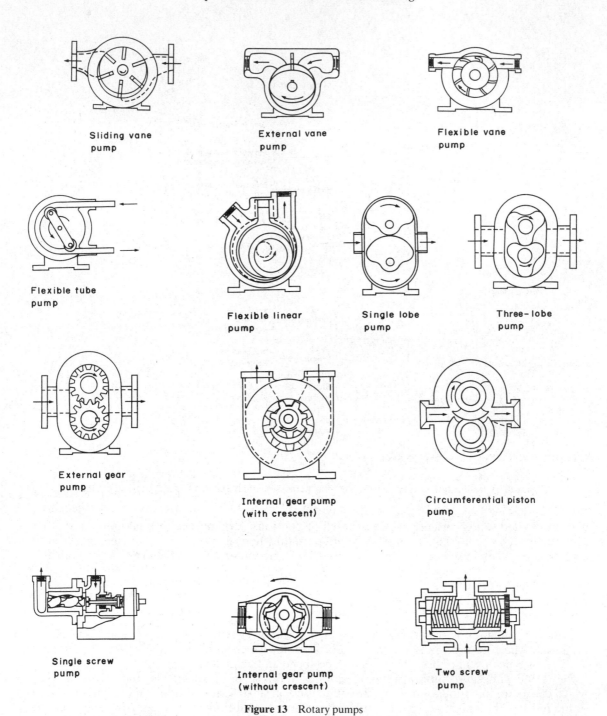

Sliding vane pump

External vane pump

Flexible vane pump

Flexible tube pump

Flexible linear pump

Single lobe pump

Three-lobe pump

External gear pump

Internal gear pump (with crescent)

Circumferential piston pump

Single screw pump

Internal gear pump (without crescent)

Two screw pump

Figure 13   Rotary pumps

often handle sterile fluids or liquids with high levels of mycelial or other cell solids. Streams in systems with enzyme-catalyzed reactions often must be free of certain trace metals. Food and brewing industry applications often deal with systems which must be thoroughly cleaned between uses.

Detailed treatment of the design of piping systems for process plants is given by Rase (1963) and Genereaux *et al.* (1973), and piping applications in biochemical industries are discussed by Solomons (1969) and Aiba *et al.* (1973).

Basic criteria for the design of a process piping system are: (1) material of construction, (2) most economic pipe size, (3) pressure levels of service, (4) temperature levels of service, (5) insulation or tracing, (6) support systems, (7) fabrication style, (8) aseptic or clean service, (9) maintenance, and (10) safety and system integrity.

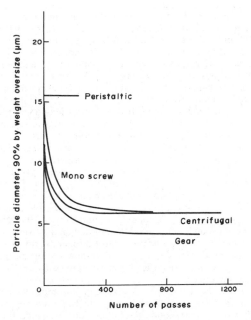

Figure 14 Flow of protein precipitates through various pumps. Total protein concentration = 2.5% by weight. Mono pump flow rate = 1.67×10^{-5} m^3 s^{-1}. Gear pump flow rate = 2.5×10^{-6} m^3 s^{-1}. Centrifugal pump flow rate = 5×10^{-4} m^3 s^{-1}. Peristaltic pump flow rate = 2.5×10^{-6} m^3 s^{-1} (reprinted with permission from *Ind. Eng. Chem. Fundam.*, **21**, 402–406 (1982))

15.2.3.9 *Materials of construction*

All common utility services and most process streams of noncorrosive, nonaseptic nature can easily be handled in conventional mild steel pipe of the proper wall thickness for the temperature and pressure of the stream. In cases where hot acid or alkali, sterile streams or metal-sensitive streams occur, one must select an appropriate material other than mild steel.

Pipe suitable for use in biochemical process flow systems is produced in metals, plastics or glass, or steel lined with one of these previous materials. Cast iron and clay pipe are not usually specified for process flow service. Service requirements in enzyme and fermentation process streams are usually of low pressure (less than 25 p.s.i.) and moderate temperatures (less than 250 °F) compared to normal service in the chemical process industries. On the other hand, pH can often fall to the range of 1.0–2.0 which can severely attack metals like copper and mild steel. Sensitivity to metal contamination often rules out copper and its alloys and sometimes many of the ferrous alloys. The need for repeated steam sterilization disqualifies many plastics and rubbers.

Metal pipe materials commonly used in biochemical applications are given in Table 3. Detailed information about corrosion resistance and application suitability of any metal is available from suppliers and manufacturers. It is always a good idea to test samples of candidate metals in the actual fluid at process conditions if at all possible. Often mixed streams with low levels of gases, salts and solids behave differently to pure chemicals for which corrosion data exist.

Glass is often used in pipes up to 6 inches in diameter where cleanliness, asepsis and chemical inertness are important. Glass piping systems are rather expensive and must be carefully anchored and supported. Working pressure limits are usually moderate as shown in Table 4. But head losses due to friction are lower than for equivalent steel pipe due to the interior surface smoothness of glass.

A great number of plastic piping materials have been introduced recently and polyethylene, polypropylene, PVC and fluorinated plastics are widely used where chemical corrosion is severe and temperatures and pressures are moderate.

Plastic pipe is designed for considerably lower service pressures than most metal pipe, and the effects of increased temperature are much more pronounced. Some plastic pipe gains increased strength and rigidity by using fibrous windings in the pipe wall or around the outside.

Table 3   Characteristics of Metals for Use in Process Piping

Aluminum 2S	Low in strength, most readily welded aluminum type. Used in food plants. Resistant to formaldehyde, ammonia, dyes, phenol, hydrogen sulfide
Aluminum alloy 3S	Better weldability and mechanical properties because of manganese content
Bronze, silicon	Used for processing tanks and equipment because of high strength and toughness. Resists brines, sulfite solutions, sugar solutions and organic acids
Copper	Resistant to corrosive waters. Used for condenser and heat-exchanger tubing
Durimet 20	Extremely resistant to sulfuric acid at all concentrations except between 60 to 90% at boiling temperatures
Hastelloy B	Used for handling boiling hydrochloric acid and wet hydrochloric acid gas.
Hastelloy C	Withstands strong oxidizing agents such as nitric acid and free chlorine. Also resistant to phosphoric, acetic, formic and sulfurous acids
Inconel	NiCr alloy with small percentage Fe. Chromium content makes it resistant to oxidizing as well as reducing solutions at high temperatures. Prevents contamination and tarnishing by substances encountered in the soap and food industries.
Monel 400	High-strength Ni–Cu alloy used primarily for handling alkaline solutions whenever copper contamination is not a problem. It has excellent resistance to corrosion by many airfree acids, caustic solutions, alkalies, salt solutions, food products and other organic substances. Generally recommended — like commercial nickel — under reducing conditions, rather than those that are oxidizing
Nickel 200	Highly resistant to corrosion. Most frequently used where copper contents of Monel is undesirable. Handles high concentrations of hot caustic and neutral salts
Nickel, silver, 18%	Called German silver. Highly resistant to corrosion. Used for foodstuffs with atmosphere contact
NI-Resist, type 1	For sulfuric acid service. Available in pipe and valve bodies
Steel, stainless; AISI type 304 or ASTM A312–64, Gr. TP–304	The most common of the stainless steels. Resists corrosion by many materials; provides sanitary conditions for food and drug industries
Steel, stainless; AISI type 316 or ASTM A312–64, Gr. TP–316	Most corrosion-resistant of the stainless steels; also higher in price
Tantalum	Resistant to nitric and other acids
Titanium, Ti-50A	High corrosion resistance in oxidizing media. Resists hypochlorites, 30% sulfuric acid, perchlorates. Resists also abrasion and erosion by cavitation
Worthite	Used for handling sulfuric acid and concentrated solutions of acetic and phosphoric acids

Table 4   Characteristics of Glass Piping Systems

Type of piping system			Capped pipe with fiberglass cladding	Capped pipe	Flange with soft gasket	Beaded end with clamp and soft gasket
Ratings	Maximum working pressure (p.s.i.)	1–3 in	150	75	50	15 (no 1″)
		4 in	100	50	35	15
		6 in	60	30	20	15
Measurements	Temperature, all sizes (°F)		0–350	0–400	cryogenic to 450	200 (continuous)
	Pipe sizes (in)		1–6 in	1–6	1–6	1½–6
	Deflection/joint		3°	3°	0.25° to 0.03°	3°

Plastic piping systems can be considerably less expensive than metal with similar corrosion resistance. However, the designer must make every effort to gain all necessary information about the temperature, pressure and corrosivity of the stream of interest and design for the most severe service contemplated.

More common than solid plastic pipe is lined pipe, which uses a carbon steel or other rigid shell to strengthen and support a thin, corrosion-resistant liner. Early attempts to use lined pipe saw many failures due to the separation of the lining from the shell, holes in the lining allowing corrosive media to attack the shell, and the limited availability of instrument taps and fittings. Current lined pipe is considerably more rugged and better designed. It offers advantages of lower cost than an equally corrosion-resistant metal. However, like plastic pipe, lined pipe must be carefully matched to the service intended. Severe temperature excursions cause failure in many lined piping systems, and a careful definition of all temperatures the system may experience during start-up, service, shutdown, or emergency shutdown must be made.

A number of pipe linings have failed because condensing vapors caused a vacuum in a system designed for pressure service only. Linings can separate from the shell under such conditions and destroy the integrity of the lining. Some lined pipe can resist full vacuum within certain temperature limits, and this feature must be checked if vacuum is a possibility.

Some common pipe lining materials are:

Rubber—Hard rubber linings resist strong acids and alkali, and soft rubber linings can be used for abrasive materials. Rubber lining has a long history of service and applications data are available for a number of chemicals.

Glass—Glass-lined pipe is available in a wide range of sizes and compositions. It has excellent chemical stability and good resistance to high temperature. However, thermal shock and impact resistance are low.

Titanium, tantalum—Expensive metals can be used as liners in carbon steel pipes.

Plastics—Probably more common in recently introduced systems, such lining materials as PVC, Teflon, Kynar, polypropylene, Saran and other plastics offer excellent corrosion resistance at moderate cost if moderate temperatures and pressures are contemplated.

15.2.3.10 Methods of fabrication

Piping systems are fabricated by threaded fittings, flanged fittings or welding. Tubing can be connected with compression or flared fittings, or with special quick-connecting clamp fittings. Each technique has advantages and disadvantages and all are commonly used successfully in biochemical process industry applications. Texts such as Rase (1963) and Littleton (1951) discuss these fabrication techniques in detail.

15.2.3.11 Design of biochemical piping systems

Optimum sizing and structural design of all piping systems are similar. Selection of the most economic line diameter is presented by Peters and Timmerhaus (1980) and Perry *et al.* (1963) among others. Design shortcuts and programs for small computers which permit optimum line sizing are presented by Garrett (1983). Standard practice for pipe supporting and anchoring structures must also be followed.

Several rules exist for the design of piping systems in some typical biochemical industry applications:

(1) Systems carrying sterile liquids should have no dead end runs and no undrained vertical drops.

(2) Systems carrying sterile liquids at moderate temperatures (where contamination could occur) should have no sample valves which are not steam sealed.

(3) Systems carrying biodegradable solids should have no internal crevices, dead areas around flanges, internal open threads, abrupt reducers or expansions, or other internal device which could allow the solids to accumulate.

(4) Systems which are aseptic or clean should be designed to drain dry with no trapped pools of liquid.

15.2.3.12 Piping costs

Piping systems are a major cost item in the construction of chemical and biochemical plants. The cost of pipes, valves and fittings, fabrication and erection labor, testing, insulation or tracing, supports and painting is usually between 3% and 20% of the total plant capital cost. Table 5 gives estimated piping costs for plants with primarily solid, mixed or fluid type processes. Two techniques which are used to estimate plant piping costs are the use of a percentage of plant capital equipment as indicated in Table 5, or the estimation of materials and labor to construct a piping system taken from detailed process layout drawings.

The use of percentage estimates is less accurate for small plants and for reconstruction and renovation work. If accuracy of ±10% is needed, a careful labor and materials estimate is necessary. Costs for all piping components useful for estimating purposes are given in Peters and Timmerhaus (1980) and Littleton (1951) as well as being available from manufacturers.

Total installed costs for a typical complex piping system containing a number of valves, short runs and fittings are shown in Table 6. Costs are given for a number of different materials of con-

Table 5 Estimated Cost of Piping (Peters and Timmerhaus, 1980)

Type of process plant	Percent of purchased equipment			Percent of fixed-capital investment
	Material	*Labor*	*Total*	*Total*
Solid[a]	9	7	16	4
Solid–fluid[b]	17	14	31	7
Fluid[c]	36	30	66	13

[a] A coal briquetting plant would be a typical solid-processing plant. [b] A shale oil plant with crushing, grinding, retorting and extraction would be a typical solid–fluid processing plant. [c] A distillation unit would be a typical fluid-processing plant.

struction and for pipe sizes of 2, 4 and 6 in. Costs for a much simpler system consisting largely of straight pipe run roughly one-half of the cost of the complex system.

15.2.3.13 *Valves*

Valves to be used in the bioprocessing industries are normally the same as those used in typical process industry applications, except that particular attention should be given to valves which must protect sterile systems against contamination. Valves which have complex internal flow paths, or dead pockets where liquids or solids can collect, present a danger of contamination in clean systems as do valves with moving stems which contact the process fluid.

In pipeline applications, diaphragm, pinch and plug or ball valves are often used where aseptic sealing is necessary. Figure 15 shows diagrams of these valves. Diaphragm valves offer the advantage of an unbroken interior surface which allows no contamination as long as the sealing diaphragm is intact. Plug and ball valves offer uninterrupted liquid flow paths with low pressure drop, but have a seal on the rotating shaft or plug which must be maintained to prevent leakage or contamination.

The materials of construction of valves should be similar to the material of the piping system both to present the same resistance to corrosion and to prevent electrochemical activity due to two dissimilar metals. Elastomers for diaphragms should be selected to meet the service intended. Butyl and Viton rubber diaphragms are usually used in fermentation service where repeated sterilization by wet steam is necessary. If hydrocarbons are present in the stream, then Viton is usually the material of choice.

Several special sanitary valves have been designed for use in sealing exit lines of sterile vessels. One type is fitted to the bottom of the vessel and when the seal element is seated it presents an almost flush interior surface on the inside of the vessel. Another uses an element with an O-ring seal and has an integral steam inlet line into the flow chamber to permit sterilization of the valve prior to flow of the process liquid.

15.3 TRANSPORT AND HANDLING OF SOLIDS

The transport and handling of solid materials is very important in the bioprocessing industries, and much of the equipment available today was developed for handling bioprocess raw materials such as cereal grains, sugar, plant fibers and other agricultural and animal products. Such materials vary widely in size, abrasiveness, flowability, shear-sensitivity and shape; commercially available solids handling equipment matches this diversity.

A thorough treatment of the storage and flow of solids is given by Jenike (1976) and Raymus and Steymann (1973), and a guide to solids handling has been published by the American Institute of Chemical Engineers (1969b). Peters and Timmerhaus (1980) give cost figures and an extensive list of references for solids handling techniques and equipment.

Although suitable equipment exists for most solids handling problems encountered in the bioprocessing industries, the use of some processes has been held back due to the lack of dependable solids conveyors and feeders. The recent development of equipment like the plug-flow, positive displacement feeder for fibrous cellulosics has greatly enhanced the prospects for using cellulose and other hard-to-handle solids as raw materials. Some new developments in fermentation technology such as the use of cell pellets or flocs, or cell or enzyme immobilization media such as gels, polymer sponges, ceramic beads or synthetic fibers have created solids handling applications for which equipment must yet be developed.

Table 6 Compared Costs of Installed Corrosion Resistant Piping Systems (ranked by total installed cost, first quarter, 1982) (taken from *Comparing Installed Costs of Corrosion Resistant Piping*, by permission of the Dow Chemical Co., Midland, MI)

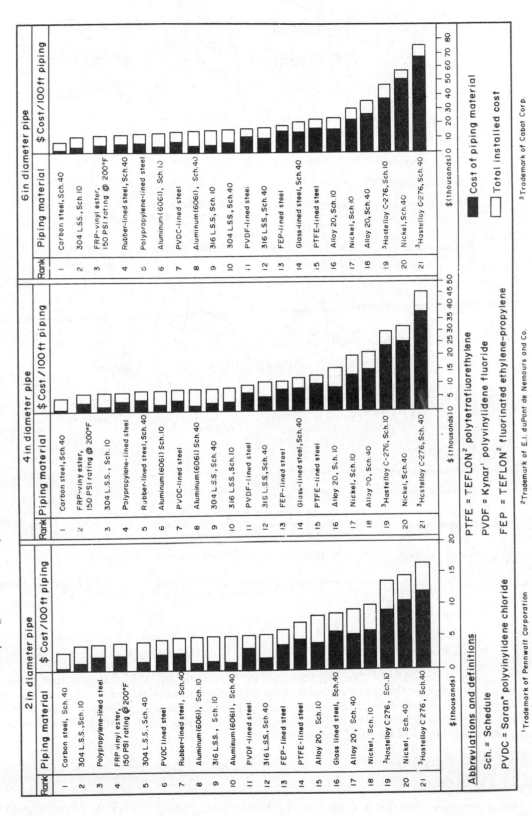

Rank	2 in diameter pipe — Piping material
1	Carbon steel, Sch. 40
2	304 L.S.S., Sch. 10
3	Polypropylene-lined steel
4	FRP vinyl ester, 150 PSI rating @ 200°F
5	304 L.S.S., Sch. 40
6	PVDC lined steel
7	Rubber-lined steel, Sch. 40
8	Aluminum (6061), Sch. 10
9	316 L.S.S., Sch. 10
10	Aluminum (6061), Sch. 40
11	PVDF-lined steel
12	316 L.S.S., Sch. 40
13	FEP-lined steel
14	PTFE-lined steel
15	Alloy 20, Sch. 10
16	Glass lined steel, Sch. 40
17	Alloy 20, Sch. 40
18	Nickel, Sch. 10
19	[3]Hastelloy C 276, Sch. 10
20	Nickel, Sch. 40
21	[3]Hastelloy C 276, Sch. 40

Rank	4 in diameter pipe — Piping material
1	Carbon steel, Sch. 40
2	FRP-vinyl ester, 150 PSI rating @ 200°F
3	304 L.S.S., Sch. 10
4	Polypropylene-lined steel
5	Rubber-lined steel, Sch. 40
6	Aluminum (6061) Sch. 10
7	PVDC-lined steel
8	Aluminum (6061) Sch. 40
9	316 L.S.S., Sch. 40
10	316 L.S.S., Sch. 10
11	PVDF-lined steel
12	316 L.S.S., Sch. 40
13	FEP-lined steel
14	Glass-lined steel, Sch. 40
15	PTFE-lined steel
16	Alloy 20, Sch. 10
17	Nickel, Sch. 10
18	Alloy 20, Sch. 40
19	[3]Hastelloy C-276, Sch. 10
20	Nickel, Sch. 40
21	[3]Hastelloy C-276, Sch. 40

Rank	6 in diameter pipe — Piping material
1	Carbon steel, Sch. 40
2	304 L.S.S., Sch. 10
3	FRP-vinyl ester, 150 PSI rating @ 200°F
4	Rubber-lined steel, Sch. 40
5	Polypropylene-lined steel
6	Aluminum (6061), Sch. 10
7	PVDC-lined steel
8	Aluminum (6061), Sch. 40
9	316 L.S.S., Sch. 10
10	304 L.S.S., Sch. 40
11	PVDF-lined steel
12	316 L.S.S., Sch. 40
13	FEP-lined steel
14	Glass-lined steel, Sch. 40
15	PTFE-lined steel
16	Alloy 20, Sch. 10
17	Nickel, Sch. 10
18	Alloy 20, Sch. 40
19	[3]Hastelloy C-276, Sch. 10
20	Nickel, Sch. 40
21	[3]Hastelloy C-276, Sch. 40

■ Cost of piping material □ Total installed cost

Abbreviations and definitions

Sch. = Schedule

PVDC = Saran* polyvinylidene chloride

PTFE = TEFLON[2] polytetrafluoroethylene

PVDF = Kynar[1] polyvinylidene fluoride

FEP = TEFLON[2] fluorinated ethylene–propylene

[1]Trademark of Pennwalt Corporation

[2]Trademark of E.I. duPont de Nemours and Co.

[3]Trademark of Cabot Corp.

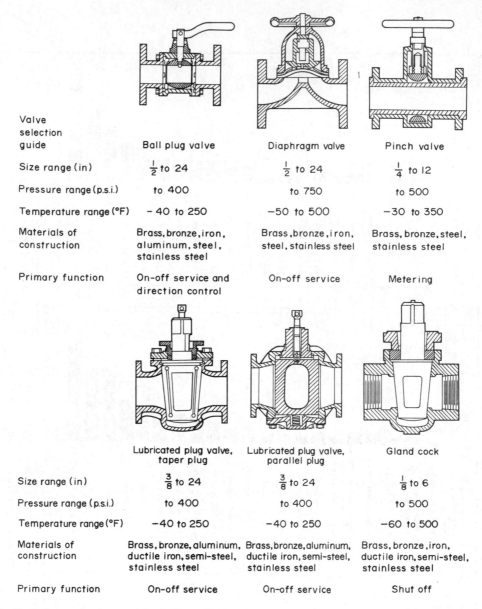

Valve selection guide	Ball plug valve	Diaphragm valve	Pinch valve
Size range (in)	$\frac{1}{2}$ to 24	$\frac{1}{2}$ to 24	$\frac{1}{4}$ to 12
Pressure range (p.s.i.)	to 400	to 750	to 500
Temperature range (°F)	– 40 to 250	–50 to 500	–30 to 350
Materials of construction	Brass, bronze, iron, aluminum, steel, stainless steel	Brass, bronze, iron, steel, stainless steel	Brass, bronze, steel, stainless steel
Primary function	On-off service and direction control	On-off service	Metering

	Lubricated plug valve, taper plug	Lubricated plug valve, parallel plug	Gland cock
Size range (in)	$\frac{3}{8}$ to 24	$\frac{3}{8}$ to 24	$\frac{1}{8}$ to 6
Pressure range (p.s.i.)	to 400	to 400	to 500
Temperature range (°F)	–40 to 250	–40 to 250	–60 to 500
Materials of construction	Brass, bronze, aluminum, ductile iron, semi-steel, stainless steel	Brass, bronze, aluminum, ductile iron, semi-steel, stainless steel	Brass, bronze, iron, ductile iron, semi-steel, stainless steel
Primary function	On-off service	On-off service	Shut off

Figure 15 Typical manual valves used in sterile applications (reprinted with permission from *Chem. Eng. Deskbook*, **76** (8), 134 (1969))

15.3.1 Properties of Solids

The ability to design a suitable handling system for a particular solid material depends upon a thorough knowledge of certain physical properties of the material and the effects that these properties have on its flow. Most solids handling equipment manufacturers have developed material information sheets similar to that shown in Figure 16 which permit the classification of a material according to its handling properties.

15.3.1.1 Hardness

Harder materials usually flow more easily than soft materials given similar particle geometry. This is because hard materials will deform less under pressure and will have fewer points of con-

ANALYSIS OF SOLID MATERIAL FOR MATERIALS HANDLING APPLICATION

Chemical information:
Formula :_____

pH of soln. _____ at _____ % concentration

Solubility in water_____ at _____ °F

Toxicity_____

Flammability_____

Other hazards_____

Handling properties:
Color_____Hardness_____

Melting point_____Sp. gr._____

Form_____Sphericity_____

Bulk density: aerated_____ packed_____

average_____ working_____ fluid_____

Compressibility_____%

Screen analysis: Number   % Retained
                 _____   _____
                 _____   _____
                 _____   _____
                 _____   _____

Moisture, level as received_____ %

            Rel. humidity   Equil. % moisture in solids
               20   _____
               40   _____
               60   _____
               80   _____
            Critical moisture level_____%

Angles of : repose _____° spatula _____°

            fall _____° slide _____°

Flow properties:
      Arches_____ Cohesive_____

      Packs_____ Bridges_____

      Channels_____ Other _____

Remarks:_____
        _____

Figure 16   Example of an information sheet as used by a solids handling equipment manufacturer

tact with neighboring particles. Sliding friction between particles will be lower and thus flow is easier. Hardness values for standard minerals and materials of commerce are given by Carr (1969) and in other handbooks.

15.3.1.2   Melting point

Although most solids of interest have melting points well above temperatures usually encountered in materials handling systems some, such as waxes, asphaltics and tars, can melt at relatively low temperatures. Since materials close to their softening or melting points exhibit drastic changes in surface properties, it is imperative to know this information.

15.3.1.3   Bulk density and specific gravity

Specific gravity refers to the density of one particle of the solid material and is the true density of the solid. The bulk density of a solid refers to the mass of a unit volume of the material in one of several states. The aerated bulk density is measured after the material has been allowed to assume its natural packing arrangement without external pressure or vibration. Packed bulk density is the density measured after tapping or vibrating the material until a minimum volume per unit mass is achieved. Average, working and fluid bulk densities are defined by Carr (1969) and are useful in sizing bins, hoppers, feeders and conveyors, but do not reflect the flow characteristics of the material. The compressibility of a particulate solid is defined as the packed bulk density minus the aerated bulk density and this divided by the packed bulk density. The figure is usually multiplied by 100 and expressed as percent compressibility. Table 7 shows the correlation between compressibility and the characteristic type of material and its typical flowability.

Table 7 Compressibility and Flow Characteristics of Different Classes of Solids
(Carr, 1969)

% compressibility		Class of dry solids	Flow
5	15	Free flowing granules	Excellent
12	18	Free flowing; powdered granules	Good
18	22	Flowable, powdered granules	Fair to passable
22	28	Very fluid powders	Poor–unstable
28	33	Fluid, cohesive powders	Poor
33	38	Cohesive powders	Very poor
38	40	Very cohesive powders	Very, very poor

15.3.1.4 Moisture

Both organic and inorganic solids have some degree of attraction for water in ambient air. Materials normally establish some equilibrium moisture content with respect to the relative humidity of the surrounding air. At some critical moisture level solids begin to adsorb moisture very rapidly and can agglomerate and clump causing great difficulties in their handling. This is especially important with materials like protein powders, sugars, starches and salts like calcium chloride and zinc chloride. The angle of repose is the angle to the horizontal assumed by the base of a conical pile of material being poured from above. This is an easily measured property and is strongly correlated with the flowability of the material.

The lower the value of the angle of repose the more flowable is the material. Hard, granular particles which flow easily have angle of repose values under 35°, powdered granules with medium flowability between 35 and 45°, while very cohesive powders and fibrous materials can have angle of repose values up to 70°. The angle of repose is a function of particle size, shape, hardness, bulk density, surface area and moisture level and is an excellent measure of the flowability of a solid. Angles of fall, internal friction, spatula and slide are defined by Carr (1969) and are used to further correlate a material's flow characteristics.

Mass flow phenomena such as arching, bridging, channeling and packing are descriptive of a material's flow in a given physical system. The terms are not quantitative, and the phenomena are often measured with rather arbitrary techniques.

It is important to establish a material's safety prior to designing a handling system. Flours and dusting organic solids present an explosion hazard and handling systems must be explosion proof and non-sparking. Some biological materials present health hazards and dust and particulates must be contained.

15.3.2 Solids Transport

The design of a complete solids handling and transport system requires specification of the following pieces of equipment:

bin or hopper for bulk storage of the material to be transported;
feeder system to transfer the material from the bin to the carrier system;
carrier system: proper conveyors and elevators to move the material from the starting point to the final delivery point;
discharge equipment necessary to disengage the solid from the carrier system and to deposit it into the desired delivery point.

System configuration depends on the starting and ending points of the system, the degree of rise or fall, the flow characteristics of the material and the amount of material to be transferred per unit time. Safety considerations affect motors and dust suppression equipment, and corrosiveness affects materials of construction.

Table 8 shows how to narrow the selection of solids handling equipment on the basis of material characteristics and paths of travel necessary for the application.

15.3.2.1 Bins and feeders

Proper design of bins and hoppers to hold solids in bulk is dependent on those properties of the solids that affect their flow such as cohesiveness, angle of repose, angle of flow and others as dis-

Table 8   Selection Guide for Bulk Handling Equipment (used by permission of FMC Corporation, Material Handling Systems Division)

Type of equipment	Material characteristics						Abrasiveness			Path of travel				
	Size			Flowability						Horiz-ontal (H)	Inclined (I)	Vertical (V)	Com-bined (HI)	Com-bined (HV)
	Fine	Gran-ular	Large lumps	Very free	Free	Slug-gish	Zero	Mild	High					
Feeding														
Apron feeders	•	•	•	•	•	•	•	•	•	•	• •		•	
Belt feeders	•	•	•	•	•	•	•	•	•	•	• •			
Reciprocating feeders	•	•	•	•	•		•	•	•	•				
Vibratory feeders	•	•	•	•	•	•	•	•	•	•				
Gravimetric weight feeders	•	•		•	•		•	•	•	•				
Mechanical vibrating feeders	•	•	•	•	•		•	•	•	•				
Bar flight feeders	•	•	•	•	•	•	•	•	•	•	• •		•	
Screw feeders	•	•		•	•		•	•	•	•				
Rotary table feeders	•	•	•	•	•		•	•		•				
Rotary plow feeders	•	•		•	•		•	•	•	•				
Rotary vane feeders	•	•		•	•		•	•	•	•				
Conveying and elevating														
Belt conveyors	•	•	•	•	•	•	•	•	•	•	•		•	
Oscillating conveyors	•	•	•	•	•	•	•	•	•	•	• • •		• •	
Apron conveyors	•	•	•	•	•		•	•	•	•	•			
Screw conveyors	•	•		•	•	•	•	•		•	•	•	• • •	•
Flight conveyors	•	•		•	•		•	•		•	•			
Wide chain drag conveyors	•	•		•	•		•	•	•	•	•			
Sidekar-Karrier				•	•		•	•	•	•				
Circular carrier				•	•		•	•	•	•				
Weigh larries				•	•		•	•		•				
Bucket elevators	•	•	•	•	•	•	•	•	•		• •	•	•	•
Skip hoists	•	•		•	•	•	•	•	•	•	• •	• • •	•	•
Bulk-Flo		•		•	•		•	•	•		•	• •		• •
Rotor lifts				•	•		•	•				•		
Gravity-discharge conv. elev.	•	•	•	•	•	•	•	•	•	•	•	•	•	•
Peck carrier	•	•	•	•	•	•	•	•	•	•	•	• •		• •

cussed in the previous section. Feeders are attached to bin outlets to meter or insure a steady flow of solids from the bin on to or into a carrier system. The bin and feeder are usually designed as a unit since the operation of one depends a great deal on the operation of the other. Design of bins and feeders is discussed by Johanson (1969) and Jenike (1976).

In the processing of a number of natural and biochemical products, problems occur with feeding and hoppering dry powders or fibrous solids. Powders can arch or rathole in bins, or can flush erratically from bin openings. Fibers can bridge over very large openings and compress to form blockages in bins and feeders. The only way to insure the success of a given bin and feeder combination is to test it with the material of interest. However, certain design criteria may be used to narrow down design choices. These are discussed by Johanson (1969) and Raymus and Steymann (1973).

15.3.2.2 Mechanical conveying

Almost all solids which cannot be fluidized are transported by mechanical conveying. A number of different types of mechanical conveyors are available. Buffington (1969) discussed the design of mechanical conveyors, Bates (1977) presents design information on screw conveyors, and Peters and Timmerhaus (1980) give extensive literature references and costs of conveying equipment.

Selection of a proper conveyor begins with the analysis of a material's flow characteristics and a comparison with the capabilities of available conveying equipment. Table 9 gives such a comparison for apron, belt, drag, en masse, screw and vibrating conveying systems.

15.3.2.3 Pneumatic conveying

Pneumatic conveying is the transfer of fluidized solids in a flow of gas in an enclosed pipeline. Dry granules or powders are transported in this manner, but wet or clumping solids cannot be handled. Advantages of pneumatic over mechanical conveying are the fact that the pneumatic system can flow up, down or horizontally with no special units. It is totally enclosed and dust tight, so contamination and dusting are at a minimum, and the materials can be conveyed great distances with incremental costs of only the carrier pipe.

The design of pneumatic systems is discussed by Kraus (1969), Caldwell (1976) and Gluck (1968) among others, and Peters and Timmerhaus (1980) give costs and references.

Pneumatic transport systems consist of a primary gas mover which may be a compressor or fan, devices to meter and fluidize solids into the gas, the pipeline conveyor and equipment to disengage the air and solids at the end. The system may be operated as either a positive pressure system, a vacuum system or a combination pressure–vacuum system.

An example of each of these types of pneumatic transport system is shown in Figure 17. Any of these systems can be used with an inert gas or dried air by connecting the inlet gas connection with the system gas outlet and operating the gas system as a closed loop. This technique can be used where the solids must be protected from contamination or an oxidizing or moist atmosphere. Closed systems normally require more power per unit of solids than open systems.

15.3.2.4 Expression and extrusion

Transporting solids by expressing or extruding them through a solids pumping device is used in instances where the solids must be fed over a significant pressure gradient. Equipment capable of this service with relatively good reliability and reasonable maintenance has become available only recently.

In general, free flowing solids such as flours, powders and granules are handled in single or twin screw extruders. Extruders can handle up to several tons per hour of solids and can heat or cool them while in the extruder barrel. They can be set up to increase or minimize the amount of shear introduced into the solid while traversing the machine, and they can provide a seal against a pressure gradient of several hundred atmospheres.

Fibrous solids are difficult to handle in extrusion devices due to their extreme resistance to flow when compacted. Recently a plug-type solids feeder has been developed which creates a seal by a

Table 9 Conveyor Selection Guide (used by permission of Rapistan, Inc.)

	Material transfer							Conveyor profile														Material properties																												
	Loading		Discharge					Angle of incline					Complexity		Horizontal carry			Vertical lift				Material size						Flowability — Angle of repose				Flowability — Specific material						Friability			Temperature				Corrosivity			Abrasiveness		
	Controlled feed	Uncontrolled feed	Head end	Head or foot end	Single intermed.	Multiple intermed.	Variable intermed.	Horizontal	10° max.	18° max.	30° max.	Up to 45° max.	Straight line	Compound	Up to 100 ft	Up to 250 ft	Over 250 ft	Up to 30 ft	Up to 60 ft	Up to 80 ft	Over 80 ft	Silt (under 250 M)	Very fine (250–100 M)	Granule (100 M to ¼ in)	Pebble (¼–2 in)	Small cobble (2–4 in)	Medium cobble (4–6 in)	Flushing (0–15°)	Free flowing (15–35°)	Sluggish (35–60°)	Sticky (60–90°)	Sludge and filtercake	Setting (trisuper phos.)	Packing (sodium bicarb.)	Metal turnings	Wood chips	Silverstick and pulpwood	Non-friable	Friable	Dusts excessively	Cold (below 32 °F)	Ambient (32–150 °F)	Hot (150–300 °F)	Very hot (300–900 °F)	Non-corrosive (pH>7)	Mildly corrosive (pH 5–7)	Very corrosive (pH 1–5)	Non-abrasive (1–3 moh)	Mildly abrasive (3–5 moh)	Very abrasive (>5 moh)
Apron — deep pan	•	•	•	•				•	•	•	•	1	•	2	•	•		•	•					•	•	•	•		•	•	•	•	•	•		•	•	•	•		•	•	•	•	•	•	1	•	•	•
Apron — hinged pan	•	•	•	•				•	•	•	•	1	•	2	•	•		•	•					•	•	•	•		•	•	•	•	•	•		•	•	•	•		•	•	•	•	•	•	1	•	•	•
Apron — shallow pan	•	•	•	•				•	•	•	•	1	•	2	•	•		•	•					•	•	•	•		•	•	•	•	•	•		•	•	•	•		•	•	•	•	•	•	1	•	•	•
Belt — rubber belt	•	•	•	•	•	•	•	•	3	3	3		•	2	•	•		•	•	•	•			•	•	•	•	1	•	•			•	•	•		•	•	•	•	1	•	•	•	•	•	•	1	•	•	•
Belt — shuttle	•	•	•	•	•	•	•	•					•		•	•		•	•	•	•			•	•	•	•	1	•	•			•	•	•		•	•	•	•	1	•	•	•	•	•	•	1	•	•	•
Belt — with plows	•	•	•	•	•	•	•	•					•		•	•		•	•	•	•			•	•	•	•		•	•			•	•	•		•	•	•	•		•	•	•	•	•	•	1	•	•	•
Belt/mobile trip.		•		•	•	•	•	•	3	3	3		•	2	•	•		•	•	•	•			•	•	•	•	1	•	•	1		•	•	•		•	•	•	•	1	•	•	•	•	•	•	1	•	•	•
Belt/static trip.		•		•	•	•	•	•	3	3	3		•	2	•	•		•	•	•	•			•	•	•	•	1	•	•	1		•	•	•		•	•	•	•	1	•	•	•	•	•	•	1	•	•	1
Drag chain	•	•	•	•	•	•		•	•	•		•	•	2	•	•		•	•	•	•	•	•	•	•				•	•	•	•	•	•	•	•	•	•	•		1	•	•	•	•	•	1	1	•	•	1
En masse	•	•	•	•	•	•		•	•	•	1	1	•	2	•	•	•	•	•	•	•	•	•	•				•	•	•		•	•	•				•	•	•	•	•	•	•	•	1	1	•	•	1
Flight	•	•	•	•	•	•		•	•	•	1	1	•		•	•	•	•	•	•		•	•	•	•				•	•	1	•	•	•				•	•	•	•	•	•	•	•	1	1	•	•	1
Screw	•	•	•	•	•	•		•	3	3	3	1	•		•	•	•	•	•			•	•	•				1	•	•	1	•	•	•				•	•	1	•	•	•	•	•	1	1	•	•	1
Vibrating	•	•	•	•	•	•		•	1	1	1	1	•		•	•		•				•	•	•	•	•	•		•	•			•	•	•	•	•	•	•	•	1	•	•	•	•	•	1	1	•	•	1

Key: 1. Special considerations required; consult examples such as special steels, special belts, enclosures, etc.
2. In combination of inclined and horizontal units only
3. Investigate max. operating angle for material handled

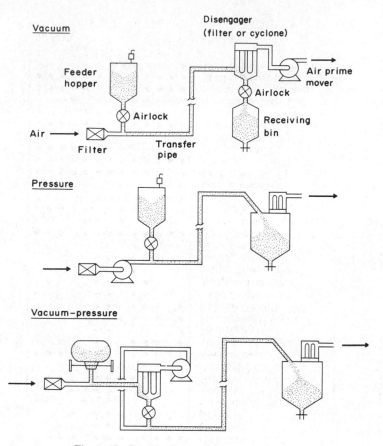

Figure 17   Pneumatic conveying system designs

compacted plug of solids forced into a transporting screw by a reciprocating ram. Figure 18 shows such a device connected to a continuous treatment section designed to digest cellulose fibers.

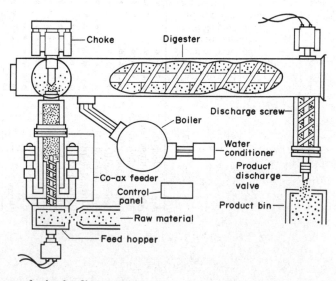

Figure 18   Reciprocating ram feeder for fibrous solids connected to a continuous fiber digestion section (used by permission from Stake Technology Inc.)

15.4 MIXING OF SOLIDS AND LIQUIDS

Mixing liquids, solids and liquid–solid mixtures is most important in applications where mass or heat transfer must be enhanced, streams must be homogenized prior to reaction or further blending or splitting, or products must be made homogeneous prior to packing and shipment. Excellent treatment of all mixing applications exists in the literature and will be referenced as appropriate.

15.4.1 Mixing in Tanks

Most liquid mixing applications are carried out in tanks which are agitated by mechanically driven impellers. Design considerations depend on what one wishes to accomplish in the application. Three general types of tank mixing applications can be characterized as follows:

(1) Blending of liquids: where two or more miscible liquids must be intimately mixed; or very slow settling solids must be suspended; or liquids must be moved past a heat transfer surface.

(2) Suspension of solids: where solid particles with settling velocities over 0.5 ft min^{-1} must be suspended in a liquid for reaction, dissolution or homogenization.

(3) Gas dispersion: where a gas must be dispersed in a liquid phase to promote mass transfer between phases.

Each of the three types of mixing can be characterized by a different system parameter. Liquid viscosities characterize blending applications, particle settling velocities characterize solids suspension problems, and gas dispersion is characterized by gas superficial velocity and mass transfer rates. These applications are quite different and require individual design solutions to assure proper operation.

Good design procedures exist for applications of turbine agitators in tanks. The procedure is presented by Gates *et al.* (1975) and consists of:

(1) *Classification.* Is the application blending, solids suspension or gas dispersion?

(2) *Magnitude of agitation necessary.* The size and difficulty of the problem are assessed and the required process result is quantified.

(3) *Equipment selection.* Agitator power and shaft speed, impeller design, and shaft and seal design are specified.

(4) *Economic evaluation.* All suitable mechanical designs are costed to permit selection of the most cost effective system.

The definition of the magnitude of the agitation necessary for a given application is a key step in which agitation parameters are quantified so that equipment specification can be done. Dimensional analysis is often used to permit modeling of agitated tank systems and its theory and application are described by Dickey and Fenic (1976) and Langhaar (1951).

One of the most important applications of gas dispersion is for mass transfer of oxygen in aerated fermenters. Since this is discussed in other parts of this work, it will not be covered in this section.

15.4.1.1 Blending of liquids

Theoretical aspects of the agitation of liquids in tanks for the purposes of mixing liquids, enhancing heat transfer and suspending slow settling particles are presented by Dickey and Hicks (1976). Actual design applications are discussed by Hicks *et al.* (1976) as well as by Uhl and Gray (1966) and Nagata (1975).

It should be remembered that smaller impellers must be rotated at higher speeds to obtain the same mixing values as larger impellers and usually create higher shear which may be undesirable in some biochemical processing applications. A system for evaluating the effects of shear in fermenters and other agitated vessels is described by Midler and Finn (1966). The technique uses shear-sensitive protozoa and effects of shear stresses in model systems and protozoan disruption are described.

Normally used design techniques are best for Newtonian systems of low to medium viscosity. Effects of non-Newtonian fluids and high viscosity liquids on agitator design are discussed by Uhl and Gray (1966) and Nagata (1975).

15.4.1.2 Solids suspension

Design of turbine agitation systems to suspend solids in tanks is described by Gates *et al.*
(1976). The design techniques presented are recommended where the solids have settling velocities greater than 0.5 ft min^{-1}.

The size and difficulty of agitation for solids suspension are described by the system equivalent
volume.

The characteristic parameter for a solids suspension system is the terminal settling velocity of
the particles in the slurry. Gates *et al.* (1976) then used a design procedure similar to that used in
liquid blending applications to specify an agitation system suitable for the problem.

15.4.1.3 Gas dispersion

Procedures for the design of turbine agitators to disperse gases in liquids in tanks are presented
by Hicks and Gates (1976). More detailed treatment of the design aspects of dispersing gas for
enhanced mass transfer in fermenters is presented elsewhere in this work.

15.4.1.4 Equipment

Typical applications of liquid agitation usually occur in baffled tanks as shown in Figure 19.
There are a number of manufacturers of agitation systems of this sort, and usually the tank and
baffles are supplied by a tank fabricator with a standard opening for the agitator. This opening is
usually a standard pipe flange size. Agitator manufacturers usually supply the motor drive, the
gear reduction system, the shaft, seal and impellers.

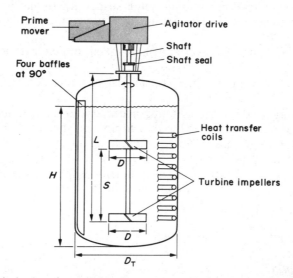

Figure 19   Typical agitated tank with baffles (may or may not have internal heat transfer coils)

Agitators are usually driven by electric motors although all kinds of prime movers may be used.
The agitator is usually of a top-entering type but bottom-entering designs are often used in
smaller fermenters to permit a less cluttered head. Seal design is more important with this
arrangement since the seal is always submerged in the broth.

Side-entering agitators are used on some very large tanks to reduce the length of the shaft and
to permit support of the agitator system by a framework not anchored to the tank. Paper pulp
stock tanks often use these systems.

The mechanical design of drive trains for turbine agitators is discussed by Hill and Kime
(1976); Ramsey and Zoller (1976) describe the design of shafts, seals and impellers; and scale-up
aspects are described by Hicks and Dickey (1976).

Costs of agitator systems can be estimated by methods presented by Meyer and Kime (1976)
and Peters and Timmerhaus (1980).

A number of systems other than a mechanically driven impeller in a baffled tank are used to agitate liquids and suspend solids. Draft tubes are often used with or without turbine impellers to promote top-to-bottom mixing in a vessel. Air-lift fermenters have draft-tube internals. Simple pump loops can also serve to agitate and mix liquids in vessels. Such systems usually leave considerable dead volume in the tank, however, and should not be used when rapidly settling solids are present or when the time of mixing is critical.

15.4.2 Mixing in Lines

When two or more flowing streams of relatively constant composition must be mixed, an in-line mixing device can often be used with considerable savings in cost over a tank and agitator system. Some types of these mixers are also used to introduce a brief exposure to an area of intense shear to a material flowing through a pipe as needed in some protein processing operations.

There are three basic types of in-line mixing devices:

(1) *Nozzles or jets*. There are often used to mix a gas or vapor with a liquid stream and may or may not use one stream to generate a suction to entrain the other fluid. The mixing action depends upon the generation of a turbulent region to create a mixing zone in a short section of the line.

Normal uses of nozzle or jet mixers are for direct steam injection heating applications, and in cases where two or more easily miscible streams of low viscosity and roughly equal volumetric flows are to be mixed. They do not work well where there is a highly viscosity difference between the liquids or where the flow of one stream is a small fraction of the other. Since there are no external connections, the mixer will remain sterile once sterilized.

(2) *Static mixers*. Static mixers are used to mix two or more streams which have been injected into the same pipe. Some of these mixers use elements composed of slanted fingers or spirals to split and re-mix the fluid flowing in the pipe. The length of the section and the number of internal units in the mixer control the degree of mixing obtained.

These units can mix streams of any composition and can effectively distribute even trace components into the stream, but are not effective on streams with too great a difference in viscosity. These units also will remain sterile once sterilized since there are no external connections.

(3) *Mechanical mixers*. Several different types of mechanical in-line mixers are available for applications where a more intense shear is desired and where two liquids of very different viscosities are to be mixed. Specially designed blenders of this type are available to mix liquids where the viscosity difference may be one million to one or more.

The mechanically agitated mixers use a motor to drive some type of proprietary mixing device within a body which can be mounted in a pipeline, usually between flanges. The design of the blender internals and the speed at which they are driven determine the degree of mixing and the degree of shear. Since the rotating shaft penetrates the process fluid these mixers need a shaft seal and present the same contamination hazard as a shaft-driven pump.

15.4.3 Solids Blending

Blending or mixing of particulate solids is a common unit operation in biochemical processing. Feed and fertilizer ingredients are mixed, solids to be pelleted or extruded are conditioned, reactants are blended, and finished products are coated and sized in equipment that is designed to blend solids. Liquids are often added to solid ingredients and when the resulting product remains a nonflowing solid the blending is done in some type of solids mixer.

A thorough treatment of solids mixing and sampling theory and practice is presented by Weidenbaum (1958). Peters and Timmerhaus (1980) give cost figures for mixing equipment and present a good list of references. Goldberger *et al.* (1973) discuss mixing theory and applications, and a testing procedure for solids mixing equipment is published by the American Institute of Chemical Engineers (1966b).

15.4.3.1 *Characteristics of particulate solids*

Different types of solids can exhibit very different mixing characteristics; some flow and mix easily while many biochemical solids cling and agglomerate excessively. These properties can be characterized by the description of several particle characteristics such as:

(1) *Shape*. The shape of regular and irregular particles is often expressed as sphericity ϕ_s, which is:

$$\phi_s = \frac{6V_p}{D_p S_p} \tag{7}$$

where ϕ_s = sphericity, V_p = volume of one particle, D_p = equivalent diameter of one particle, and S_p = surface area of one particle. Sphericity is independent of particle size and is equal to unity for spheres and regular cubes and cylinders where height equals diameter. Irregular particles have sphericity values of less than one, commonly between 0.6 and 0.7. Sphericity is often used in correlation between shape and particle size and surface area as shown by McCabe and Smith (1976).

(2) *Size*. Sizes and size distribution of particles are usually described by screen analyses. McCabe and Smith (1976) also show the use of screen analyses in correlations for the specific surface area, the mean and median particle diameter and the number of particles in a mixture.

(3) *Cohesiveness*. Particles which readily flow are called noncohesive; examples are sand, grain, pellets or hard chips. Materials such as fibers, sludges, clays or some fine powders are characterized by their reluctance to flow, and are termed cohesive solids.

(4) *Bulk density*. The density of a mass of the particulate solids when resting in a natural packing arrangement.

(5) *Friability*. The tendency of the particles to break or crumble upon abrasion or impact.

(6) *Moisture content*. The moisture content of solids can have dramatic effects on the cohesiveness of the solid mixture.

(7) *Angle of repose*. The maximum angle which the surface of a pile of the solids can attain without flow. A function of the particles' shape, bulk density and cohesiveness.

15.4.3.2 *Mixing equipment*

Two general types of mixing equipment are available; equipment that can handle very cohesive materials like pastes, plastics or heavy slurries, and equipment designed for free-flowing particulate solids. Since moisture often adds to the cohesiveness of a solid, most mixers for service on moist materials must be capable of handling cohesive materials. The major difference between the two classes of mixers is that in mixers designed for use with cohesive solids the material to be mixed is carried to the mixing impeller or the mixing impeller moves through the entirety of the mixture. In a blender designed for noncohesive materials, the mixture is allowed to tumble, fall or flow into the path of the mixing impeller. Some mixers can handle any type of material, although not always in the most economic manner, and the definitions of the capabilities of various blenders are not clear cut, so there are often several types of blender which will handle a given application.

Mixers designed to handle cohesive solids are can mixers, mullers, kneaders, extruders, mixing rolls, pugmills and pan mixers. A can mixer as shown in Figure 20(a) is usually limited in size to 100 gallons. The charge is added in a batch to the can and the blender is run until the desired blending is achieved. These blenders often have several different beaters or paddles which can be selected to match the viscosity and cohesiveness of the charge. They often have several speed selections and are useful for a variety of applications.

Kneaders and masticators are heavy duty mixers which use two parallel blades which rotate in a close-fitting trough to create a folding action in the solid mixture charge. They are batch machines and are designed with several choices of blades as shown in Figure 20(b). These machines can handle charges of up to several hundred gallons, and are used for the most cohesive and tough mixing tasks such as rubbers and plastics. They are seldom needed for mixing conventional bioprocessing materials, but may be necessary to properly condition polymer products.

Muller mixers as shown in Figure 20(c) use one or more rotating wheels to mash and smear materials inside a pan. Solids in the pan are guided into position by plows which turn the material and force it out of a bottom opening when mixing is finished. These mixers are often used to blend a small amount of liquid into a batch of solids. Single and twin screw extruders are often used to blend and process cohesive solids.

Blenders to handle noncohesive materials usually consume much less power and are constructed with much lighter bodies and materials. Such mixers as ribbon blenders, cone mixers, vertical screw mixers and tumbling blenders are designed for free-flowing materials.

Ribbon blenders, as shown in Figure 20(d), use a 'ribbon'-shaped impeller driven by a longitu-

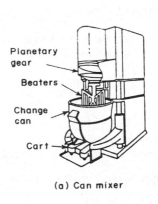

(a) Can mixer

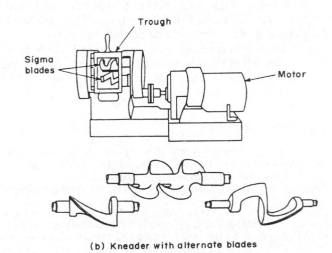

(b) Kneader with alternate blades

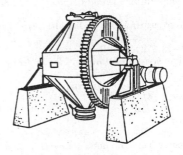

(c) Muller mixer

(d) Ribbon mixer

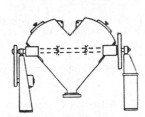

(e) Double cone mixer

(f) Twin shell mixer

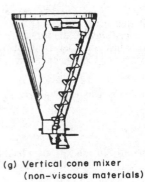

(g) Vertical cone mixer
(non-viscous materials)

(h) Vertical cone mixer
(viscous materials)

Figure 20 Solids mixers

dinal central shaft located in a U-shaped trough to lift and turn the charge. Sizes range up to 9000 gallons and the shell can be supplied with a jacket if heat transfer is needed. Ribbon blenders are used extensively for dry mixing and blending liquid into solids in which the viscosity does not get too high.

Tumbling mixers, such as the double cone and twin shell mixers shown in Figures 20(e) and 20(f) are used for free-flowing mixtures. They are batch machines. Some can be used as dryers if fitted with jackets and vacuum operation.

Vertical screw mixers, such as shown in Figures 20(g) and 20(h), lift and turn solids in a batch by a vertical flighted screw which traverses the perimeter of the cone-shaped vessel as it rotates. The mixed batch is then discharged from the bottom of the vessel.

15.5 REFERENCES

Aiba, S., A. E. Humphrey and N. F. Millis (1973). *Biochemical Engineering*, pp. 303–316. Academic, New York.

American Institute of Chemical Engineers (1966). *Solids Mixing Equipment—Equipment Testing Procedure*. American Institute of Chemical Engineers, New York.

American Institute of Chemical Engineers (1969a). Process Piping. In *Liquids Handling, Chem. Eng. Deskbook*, **76** (8), p. 95.

American Institute of Chemical Engineers (1969b). *Solids Handling, Chem. Eng. Deskbook*, **76** (22).

Bates, L. (1977). Using helical screws for solids handling, *Chem. Eng.*, **84** (5), 183.

Bird, R. B., W. E. Stewart and E. N. Lightfoot (1960). *Transport Phenomena*, Wiley, New York.

Blanch, H. W. and S. M. Bhavaraju (1976). Non-Newtonian fermentation broths: rheology and mass transfer. *Biotechnol. Bioeng.*, **18**, 745–790.

Buffington, M. A. (1969). Mechanical conveyors and elevators. *Chem. Eng.*, **76** (22), 33–49.

Caldwell, L. G. (1976). Pneumatic conveying materials — a pneumatic conveying primer, *Chem. Eng. Prog.*, **72** (3), 63.

Carr, R. L. (1969). Properties of solids. *Chem. Eng.*, **76** (22), 7–16.

Chang, H. T. and D. F. Ollis (1982). Extracellular microbial polysaccharides: generalized power law for biopolysaccharide solutions. *Biotechnol. Bioeng.*, **24**, 2309–2318.

Charles, M. (1978). Technical aspects of the rheological properties of microbial cultures. *Adv. Biochem. Bioeng.*, **8**, 1.

Deindoerfer, F. H. and J. M. West (1960). Rheological examination of some fermentation broths. *J. Biochem. Microbiol. Technol. Eng.*, **2** (2), 165–175.

Dickey, D. S. and J. C. Fenic (1976). Dimensional analysis for fluid agitation systems. *Chem. Eng.*, **83** (1), 139–145.

Dickey, D. S. and R. W. Hicks (1976). Fundamentals of agitation. *Chem. Eng.*, **83** (3), 93–100.

Doebelin, E. O. (1975). *Measurement Systems*. McGraw-Hill, New York.

Garrett, J. R. (1983). Published calculator programs for chemical engineers. *Chem. Eng.*, **90** (5), 149–160.

Gates, L. E., T. L. Henley and J. G. Fenic (1975). How to select the optimum turbine agitator, *Chem. Eng.*, **82** (26), 110–114.

Gates, L. E., J. R. Morton and P. L. Fondy (1976). Selecting agitator systems to suspend solids in liquids. *Chem. Eng.*, **83** (11), 144.

Geankoplis, C. J. (1978). In *Transport Processes and Unit Operations*. pp. 59–98. Allyn and Bacon, Boston.

Genereaux, R. P., P. P. O'Neill, W. D. Webb and R. W. Nolan (1973). Transport and storage of fluids. In *Chemical Engineers Handbook*, ed. R. H. Perry and C. H. Chilton, 5th edn., pp. 6.1–6.105. McGraw-Hill, New York.

Glickman, M. and A. H. Hehn (1969). Valves. In *Liquids Handling, Chem. Eng. Deskbook*, **76** (8), p. 134.

Gluck, S. E. (1968). Design tips for pneumatic conveyors. *Hydrocarbon Process.*, **47** (10), 88.

Goldberger, W. M. *et al.* (1973). Liquid–liquid and solid–solid systems. In *Chemical Engineers Handbook*, ed. R. H. Perry and C. H. Chilton, 5th edn., pp. 21.30–21.69. McGraw-Hill, New York.

Hicks, R. W. and D. S. Dickey (1976). Applications analysis for turbine agitators. *Chem. Eng.*, **83** (24), 127.

Hicks, R. W. and L. E. Gates (1976). How to select turbine agitators for dispersing gas into liquids. *Chem. Eng.*, **83** (15), 141.

Hicks, R. W., J. R. Morton and J. G. Fenic (1976). How to design agitators for desired process response. *Chem. Eng.*, **83** (9), 102–110.

Hill, R. S. and D. L. Kime (1976). How to specify drive trains for turbine agitators. *Chem. Eng.*, **83** (16), 89.

Hoare, M., T. J. Narendrathan, J. R. Flint, D. Heywood-Waddington, D. J. Bell and P. Dunhill (1982). Disruption of protein precipitates during shear in couette flow and in pumps. *Ind. Eng. Chem. Fundam.*, **21**, 402–406.

Hydraulic Institute (1975). *Hydraulic Institute Standards*. pp. 100–103. Hydraulic Institute, Cleveland, Ohio.

Jenike, A. W. (1976). *Storage and Flow of Solids*. Bulletin No. 123, Utah Engineering Experiment Station, University of Utah, Salt Lake City.

Johanson, J. R. (1969). Feeding. *Chem. Eng.*, **76** (22), 75–83.

Kabre, S. P., H. G. DeKay and G. S. Banker (1964). Rheology and suspension activity of pseudoplastic polymers. *J. Pharm. Sci.*, **53**, 492.

Kraus, M. N. (1969). Pneumatic conveyors. *Chem. Eng.*, **76** (22), 59–65.

Langhaar, H. L. (1951). *Dimensional Analysis and Theory of Models*. Wiley, New York.

Littleton, C. T. (1951). *Industrial Piping*. McGraw-Hill. New York.

McCabe, W. L. and J. C. Smith (1976). *Unit Operations of Chemical Engineering*, 3rd edn., p. 804. McGraw-Hill, New York.

Meyer, W. S. and D. L. Kime (1976). Cost estimation for turbine agitators. *Chem. Eng.*, **83** (20), 109.

Midler, M. and R. K. Finn (1966). A model system for evaluation shear in the design of stirred fermenters. *Biotechnol. Bioeng.*, **8**, 71–84.

Miller, R. W. (1983). *Flow Measurement Engineering Handbook*. McGraw-Hill, New York.

Myers, R. R. and C. J. Knauss (1965). Mechanical properties of starch pastes. In *Starch: Chemistry and Technology*, ed. R. L. Whistler and E. F. Paschall, pp. 393–406. Academic, New York.

Nagata, S. (1975). *Mixing, Principles and Applications*. Halstead, New York.

Perry, R. H., C. H. Chilton and S. D. Kirkpatrick (eds.) (1963). *Chemical Engineers Handbook*, 4th edn. McGraw-Hill, New York.

Peters, M. S. and K. D. Timmerhaus (1980). *Plant Design and Economics for Chemical Engineers*, 3rd edn. McGraw-Hill, New York.

Ramsey, W. D. and G. C. Zoller (1976). How the design of shafts, seals, and impellers affects agitator performance. *Chem. Eng.*, **83** (18), 101.

Rase, H. F. (1963). *Piping Design for Process Plants*. Wiley, New York.

Raymus, G. J. and E. H. Steymann (1973). Handling of bulk and packaged solids. In *Chemical Engineers Handbook*, ed. R. H. Perry and C. H. Chilton, 5th edn., pp. 7.3–7.50. McGraw-Hill, New York.

Roels, J. A., J. Van Den Berg and R. M. Voncken (1974). The rheology of mycelial broths. *Biotechnol. Bioeng.*, **16**, 181–208.

Sherman, P. (1970). *Industrial Rheology*. Academic, New York.

Solomons, G. L. (1969). *Materials and Methods in Fermentation*. pp. 150–229. Academic, New York.

Thurlow, C. (1971). Centrifugal pumps. In *Pump and Valve Selector, Chem. Eng.*, **78** (23), p. 29.

Uhl, V. W. and J. B. Gray (eds.) (1966). *Mixing Theory and Practice*, vol. 1. Academic, New York.

Van Wazer, J. R., J. W. Lyons, K. Y. Kim and R. E. Calwell (1963). *Viscosity and Flow Measurement*. Interscience, New York.

Washburn, E. W. (1926). *International Critical Tables*, vol. 5, pp. 10–30. McGraw-Hill, New York.

Weidenbaum, S. S. (1958). Mixing of solids. In *Advances in Chemical Engineering*, ed. T. B. Drew and J. W. Hooper, vol. 11, pp. 209–324. Academic, New York.

Whistler, R. L. and J. N. BeMiller (1959). *Industrial Gums*. Academic, New York.

16
Gas Compression

R. T. HATCH
University of Maryland, College Park, MD, USA

16.1 BACKGROUND

The gas flow rates through fermenters are determined by the requirement for adequate mass transfer and mixing. This is accomplished in stirred tank fermenters by the use of specific gas flow rates in the range of 0.2 to 2.0 volumes of gas per volume of fermenter per minute depending upon the height of the vessel. Other designs, such as the airlift (Hatch, 1973), can operate at higher specific gas flow rates. The actual upper limit is the gas flow rate at which flooding occurs at the top of the fermentation broth. This occurs at a superficial gas velocity of less than 30 cm s^{-1}. The more surface active the fermentation broth becomes, the lower the superficial gas velocity must be to minimize excessive foaming.

16.2 COMPRESSION THEORY

The compression of perfect gases is described by the perfect gas law:

$$PV = NRT \qquad (1)$$

where P is the absolute pressure, V is the volume, N is the number of moles of gas, R is the gas law constant, and T is the absolute temperature. A commonly used modification of the perfect gas law is the use of a correction factor, known as the compressibility factor (z). This corrects for deviations from the perfect gas law (Nelson and Obert, 1954):

$$PV = zNRT \qquad (2)$$

The compressibility factor is generally a function of temperature and pressure, however, most gases have the same compressibility factor at the same reduced temperature and pressure. This allows the use of generalized compressibility charts for a wide range of gases.

Gases may be compressed either adiabatically (no thermal energy input or output) or isothermally. Low to moderate compression ratios generally do not justify the use of external cooling for isothermal operation. The compression is therefore allowed to approximate to an adiabatic process. The change in enthalpy of the gas is then the work done by the compressor and the process is isentropic, *i.e.* the entropy change is zero. For isentropic compression the following relationships are true for perfect gases (Genereaux *et al.*, 1973):

$$P_2/P_1 = (V_1/V_2)^k \tag{3}$$

$$T_2/T_1 = (P_2/P_1)^{(k-1)/k} \tag{4}$$

where k = the ratio of specific heats (C_p/C_v).
The work of compression is simply:

$$\text{Work} = \Delta H = V dP \tag{5}$$

which becomes for a perfect gas compressed isentropically:

$$\Delta H = (k/k-1)wRT_1[(P_2/P_1)^{(k-1/k)} - 1] \tag{6}$$

The power of compression is defined as:

$$\text{Power} = w\Delta H \tag{7}$$

$$\text{or} \quad \text{Power} = (k/k-1)wRT_1[(P_2/P_1)^{(k-1/k)} - 1] \tag{8}$$

where w = the mass flow rate of gas.
The power of compression can also be expressed in terms of the volumetric flow rate of gas into the compressor by:

$$\text{Power} = C_1 Q_1 P_1 (k/k-1)[(P_2/P_1)^{(k-1/k)} - 1] \tag{9}$$

For power in units of horse power (HP), $C_1 = 0.00436z$, $Q_1 = \text{ft}^3 \, \text{min}^{-1}$ and $P_{1,2}$ = input, output p.s.i. absolute. The compressibility constant, z, is incorporated into the constant C_1 to account for deviations from the perfect gas law.

Operation at moderate to high compression ratios usually requires the use of multiple stage compression which justifies interstage cooling. The compression can then approximate to an isothermal process. The power of compression for an isothermal process is:

$$\text{Power} = C_1 Q_1 P_1 \ln(P_2/P_1) \tag{10}$$

This equation also approximates to the power dissipated during isothermal expansion of the gas as it passes through the fermenter.

16.3 COMPRESSOR SELECTION

16.3.1 Power Requirement

Once the gas flow rate is specified, the total pressure head must be determined. This is a combination of the pressure losses through the piping, filters, sparger, hydrostatic head of the fermentation broth, exhaust filter, and the exhaust gas valve. The pressure drop through the filters is generally specified to be less than 5 p.s.i. at the operating gas flow rate. As the filter becomes loaded with solids, the pressure drop rises but is not allowed to rise more than an additional 1 to 2 p.s.i. before the filter is serviced. This is necessary to ensure that particulates do not break through the filter. The major pressure drops are the hydrostatic head of the fermentation broth and the sparger pressure drop. The exhaust gas valve may be set to provide an overpressure in the vessel for increased gas partial pressures and corresponding higher gas–liquid mass transfer rates. This overpressure is generally no greater than 20 p.s.i. unless the fermentation is specifically designed for operation at elevated pressures. The hydrostatic head is simply determined by the ungassed liquid height in the fermenter (0.4335 p.s.i. per foot of water).

The remaining pressure drop is due to the sparger design. The general approach to sparger design is the specification of a small bubble size distribution created by the sparger with minimum frictional pressure losses. This results in the requirement to balance the sparger design with the overall design of the fermenter. For column heights above 1 m it has been found that the resulting bubble size distribution is independent of sparger design (LeFrancois, 1963). The spargers in industrial scale fermenters are therefore designed to minimize sparger pressure drops. In pilot scale or smaller fermenters the spargers are designed to provide a small bubble size distribution without contributing to an excessive pressure drop. For these cases careful consideration of the sparger design becomes important.

The equations defining the bubble size created by the sparger orifice depend upon the orifice Reynolds number. For an orifice Reynolds number less than 300, one bubble is created at a time and the bubble size is described in CGS units by (Leibson *et al.*, 1956):

$$d_b = (6d_o\gamma/\Delta\varrho g)^{1/3} \tag{11}$$

where d_b is the bubble size, d_o is the orifice diameter, γ is the interfacial surface tension, and $\Delta\varrho$ is the gas–liquid density difference. In this low flow range increasing gas velocities result in greater bubble frequencies without changing the bubble size. At a higher Reynolds number the bubble size increases while the frequency of bubble formation remains constant. The bubble size is then described in CGS units by:

$$d_b = 0.286d_o^{1/2} Re^{1/3} \tag{12}$$

This equation applies up to the onset of turbulence or an orifice Reynolds number of approximately 2000. As the orifice velocity is further increased, a jet of gas is formed at the orifice. Random bubble sizes are formed from this jet and the larger bubbles are unstable, leading to further breakup above the sparger. The bubble size in this regime can be defined in CGS units for Reynolds numbers above 2000 by:

$$d_b = 0.71/Re^{0.05} \tag{13}$$

The specification of the optimum orifice size is dependent upon the optimization of power input per unit of oxygen transferred. The power input is a function of gas flow rate and pressure drop. The pressure drop across the sparger can be described in CGS units by:

$$\Delta P = (2\gamma/r) + 0.28(U/C)^2(\varrho g/\varrho_l) \tag{14}$$

where r is the orifice radius, U is the gas velocity through the orifice, and C is a constant (approximately 0.7). The first term is the pressure drop due to surface tension and the second term is the frictional pressure drop across a sieve plate. Since the pressure drop across the orifice is virtually independent of the hydrostatic head, the contribution of the sparger to the overall power requirement per unit liquid volume decreases with increasing hydrostatic head. It is also apparent that as the orifice radius is decreased below 1 mm, the pressure drop due to surface tension becomes significant. Conversely, as the orifice diameter is increased above 10 mm, the pressure drop becomes virtually independent of orifice Reynolds number. For commercial scale fermenters, the sparger is therefore designed with orifice diameters of the order of 10 mm. Small scale fermenters may utilize much smaller orifice diameters in order to improve mass transfer. For liquid heads less than 1 m, bubble coalescence is not as significant and the bubble size distribution may be more determined by the sparger than the agitator. As long as the sparger pressure drop is not an important consideration, the sparger may be designed primarily for high oxygen transfer rates.

16.3.2 Compressor Design Considerations

The primary factors influencing the compressor selection process are the output pressure required and the gas flow rate required. Two general classes of compressors are the dynamic type and the positive displacement type (Evans, 1971; Neerken, 1975). The compressor types have been grouped by Neerken (1975) as shown in Table 1.

Table 1 Types of Compressor

Dynamic (centrifugal)	Positive displacement	Positive displacement
Radial flow	Reciprocating	Rotary
Single stage	Air cooled	Lobe
Multistage	Water cooled	Screw
Axial flow		Vane
Multistage		Liquid ring

The most frequently used high power compressors for process industries consist of the multistage radial flow types. Exhaust pressures as high as 10 000 p.s.i.g. and inlet flow rates in the range of 1000 to 100 000 ft^3 min^{-1} are available. When compression ratios less than five are required, single stage designs can be used (Peters and Timmerhaus, 1980). For higher outputs, multistage axial flow compressors are used. Since the boost in pressure per stage is lower than achieved in radial flow compressors, many more stages are required. Although the capital cost is higher, the compression efficiency is greater, which may justify the axial flow design in some cases. This is particularly the case in lower pressure, high mass flow output applications.

The reciprocating compressors cover a very broad range of pressures and flow rates from the laboratory scale to large process flows (Neerken, 1975; Peters and Timmerhaus, 1980; Bresler, 1970). The primary application tends to be high pressure, low flow rate gas flows. As is the case for the dynamic compressors, multiple stages are used for higher pressure services. The compression ratio per stage is generally no greater than 3.5 and may be limited by the temperature increase.

The second general category of positive displacement compressors includes the rotary designs. This class of compressors utilizes a rotating mechanical element to displace a fixed volume of gas per rotation. The earliest design to become established in the process industries was the lobe design. This compressor utilizes figure-eight shaped rotors which intermesh to displace the gas. This compressor is used primarily in low pressure applications (less than 30 p.s.i.g.) and secondarily in vacuum pump applications.

A second rotary design is the sliding vane, which utilizes an offset rotor in a cylindrical chamber. The vanes slide in and out of the rotor and push the gas through a constantly decreasing cavity volume, thereby increasing the pressure. At the end of the revolution the gas exits through an exhaust port. The sliding vane may produce slightly higher pressures per stage (approximately 50 p.s.i.g.). A second stage can be added to produce discharge pressures greater than 100 p.s.i.g.

A third design of rotary compressors is the screw design, which is capable of higher pressures and is available in large sizes. This design can be either oil cooled or constructed to run dry. The discharge pressures can range up to 400 p.s.i.g. with intake gas flow rates up to 20 000 ft^3 min^{-1}.

The fourth rotary compressor design is the liquid ring, which utilizes water or another fluid to act as a seal at the circumference of the casing. An off-center, vaned rotor maintains the fluid at the circumference. The spinning fluid forces the gas through a progressively decreasing volume which compresses the gas to the discharge port at the rotor center. The liquid ring design is used most frequently in vacuum pumps. It is also used for corrosive gas compression in low pressure applications.

Gas compression for aeration of fermenters usually requires discharge pressures less than 100 p.s.i.g. Careful design for minimum pressure losses can permit the use of compressors with discharge pressures less than 50 p.s.i.g. This greatly reduces both the capital cost and power consumption. Centrifugal compressors, such as the one or two stage radial design, and rotary compressors, such as the lobe design, are frequently selected for this application.

16.3.3 Operating Efficiency

Although the theoretical work of compression can be calculated depending upon the mechanical design, the actual work of compression depends upon the efficiency of the specific compressor (Lady, 1970). The isothermal efficiency is defined as:

$$\eta_{iso} = W_{iso}/W_a \tag{15}$$

and the isentropic efficiency is:

$$\eta_{isen} = W_{isen}/W_a \tag{16}$$

where W_a, W_{isen} and W_{iso} are the actual, isentropic, and isothermal works of compression, respectively. Extensive relationships between compressor efficiencies and output pressures have been developed (Evans, 1971). Ultimately, however, the efficiency depends upon the specific compressor design. The relationships of operating parameters to compressor efficiency are generally available from the compressor manufacturers. Usually the isentropic efficiency is quoted, since most compressors are not operated isothermally. The isentropic efficiency can be as high as 90% but is usually greater than 70%.

16.4 REFERENCES

Bresler, S. A. (1970). Guide to trouble-free compressors. *Chem. Eng.*, **77**, 161–170.
Evans, Jr., F. L. (1971). *Equipment Design Handbook for Refineries and Chemical Plants*, vol. 1. Gulf Publishing Co., Houston.

Genereaux, R. P., P. P. O'Neill, W. D. Webb and R. O. Nolan (1973). In *Chemical Engineers' Handbook*, ed. R. H. Perry and C. H. Chilton, section 6, pp. 15–21. McGraw-Hill, New York.

Hatch, R. T. (1973). *Experimental and Theoretical Studies of Oxygen Transfer in the Airlift Fermentor*, Ph.D. Thesis, Massachusetts Institute of Technology.

Lady, E. R. (1970). Compressor efficiency: definition makes a difference. *Chem. Eng.*, **77**, 113–116.

LeFrancois, L. (1963). *34th International Conference of Industrial Chemistry, Belgrade*, Section 14: Fermentation Industries.

Leibson, I., E. G. Holcomb, A. G. Cacoso and J. J. Jacmic (1956). Rate of flow and mechanics of bubble formation from single submerged orifices. *AIChEJ*, **2**, 296–306.

Neerken, R. F. (1975). Compressor selection for the chemical process industries. *Chem. Eng.*, **82**, 78–94.

Nelson, L. C. and E. F. Obert (1954). *Chem. Eng.*, **61**, 203–208.

Peters, M. S. and K. D. Timmerhaus (1980). *Plant Design and Economics for Chemical Engineers*, 3rd edn., pp. 552–554. McGraw-Hill, New York.

17

Selection Criteria for Fermentation Air Filters

R. S. CONWAY

Pall Corporation, Glen Cove, NY, USA

17.1 INTRODUCTION

The criteria for selecting a fermentation air filtration system include the efficacy of organism retention provided by the air filters, the economy of the system, the ease with which the filter assembly can be maintained and the service provided by the filter manufacturer. Of these criteria the most important is the filter's ability to remove viable contaminants from the fermentation inlet air and to control escape of organisms from the fermenter in the exhaust air.

Filtration of compressed air used in aerobic fermentation has been employed for many years as the most practical and economical solution to contamination control. Examples of filtration devices used in the past, and still in use today, include depth type filters such as granulated carbon beds and vessels containing packed fiberglass or packed cotton. These filtration devices will provide a fair statistical probability of organism retention as long as they are properly maintained and care is taken to control the quality and velocity of the inlet air. Depth type air filters, however, fail to perform with high efficiency if the influent air is not dried or if the inlet air velocity is not optimum. Because organism retention is dependent on the depth of the filter medium used, these filters have high pressure drops which restrict air flow to the fermenter and increase compressor costs. In addition, depth filters trap moisture and organic contaminants and can provide an environment conducive to microbial growth.

Sterilizing grade microporous membrane filters are replacing depth type filter devices as fermentation air filters. Membrane air or gas filter cartridges typically contain a pleated, hydrophobic filter medium having a fixed submicron pore structure of uniform size distribution. Use of a hydrophobic filter medium prevents transmembrane pressure drop increase due to filter wetting and ensures reliable organism retention by excluding moisture which can decrease filtration efficiency. The submicron pore structure allows quantitative microbial retention and the pleated filter medium results in a high filter area in a small unit volume which can be easily handled for servicing and installation.

The factors affecting filter selection criteria are discussed with emphasis placed on selection of industrial scale fermentation air filter systems.

17.2 MECHANISMS OF AEROSOL FILTRATION

The primary function of a fermentation air filter is to remove viable microbial contaminants from the inlet (or exhaust) air. For this reason the most important criterion of filter selection should be reliable organism retention.

When selecting an air filter it is important to understand the principal mechanisms of aerosol filtration responsible for microorganism and particle retention. This subject has been extensively investigated and many excellent reviews have been published (Chen, 1955; Davies, 1964; Davies, 1973; Aiba *et al.*, 1973; Gaden and Humphrey, 1956; Bruckshaw, 1973).

The following discussion of the mechanisms of air filtration apply to depth type hydrophilic air filters and hydrophobic membrane air filters. Hydrophilic filters can provide reliable particle retention efficiency if the filtered air remains dry, however, if moisture is present in the air stream the filter medium will become wet and decrease significantly in filtration efficiency. Because air moisture content is often difficult to control, hydrophobic air filters which are not affected by moisture provide more reliable filtration efficiency.

Particle retention by air filters is generally accepted to involve the following four mechanisms: direct interception, inertial impaction, diffusional interception and electrostatic effects. The net effects of these four factors can be quite complex. For example, where retention by inertial impaction predominates, capture efficiency increases as flow rate through a filter is increased. If retention by diffusional interception predominates, the opposite is the case. The degree to which particles are retained by a filter due to electrostatic interactions has been neglected in most investigations of filtration efficiency because the contribution to particle retention by this mechanism is difficult to measure and is dependent upon the filter medium employed and the amount of moisture and entrained oil droplets present in the influent air. Where present, however, electrostatic interactions will serve to increase filtration efficiency.

Removal of particles by direct interception is a sieve phenomenon and occurs when the particle diameter exceeds the limiting pore diameter of the filter medium. Retention efficiency by direct interception is independent of incident air velocity. Bacteria and bacterial spores have a size range of approximately 0.3 to 1 μm or greater. The limiting pore diameter of depth type air filters such as packed fiberglass or cotton is typically 2–10 μm. Microporous membrane filters used for air filtration typically have pore size ratings of 0.45 μm or less. For this reason, membrane air filters can retain particles such as bacteria and bacterial spores by direct interception.

In addition to retention by direct interception, particles with an average diameter of several micrometers or larger are removed by air filters due to inertial impaction. The particles entrained in air have a density considerably greater than that of the carrying gas. Thus their trajectory can deviate from the main air flow path as the gas flows through the filter medium. As a result of such deviation, particles inevitably strike a fiber of the filter structure. Particles impacting upon a structural fiber will be held in place by van der Waals forces. Once captured by the filter, particles will not be re-entrained into the air stream unless the (incident) air velocity is sufficient to overcome the attractive forces. Air flow rates required to generate re-entrainment are considerably higher than air flow rates typically encountered in fermentation air systems.

It has been determined by laboratory experiments and actual field experience that small submicron particles such as viruses and bacteriophages are retained by air filters having pore diameters that are much larger than these particles (Harstad, 1967; Decker, 1962; Fitzgerald, 1957; Decker, 1957; Conway, 1984). The mechanism of filtration responsible for submicron particle retention is diffusional interception. Diffusional interception occurs because gas molecules are constantly moving in random directions. A small particle suspended within the gas will be impacted by the gas molecules, causing it to move in a random fashion characterized by Brownian motion. The random movement of submicron particles increases the probability that they will strike the matrix of the filter and be retained by van der Waals forces. The rate of particle capture is inversely proportional to the square root of the particle diameter. Particle retention by diffusional interception is also dependent on the air velocity and viscosity of the fluid (gas) stream. Retention is optimal at low air velocities and in dry air.

A summary of the mechanisms of aerosol filtration can be seen in Figure 1 (Davies, 1964). This figure compares percentage penetration of dioctyl phthalate particles as a function of particle radius and incident air velocity. Experiments were conducted using a glass fiber filter, impregnated with resin, composed of fibers having a radius of 0.75 μm. The filter thickness was 1.2 cm. Homogeneous mists of dioctyl phthalate particles having a defined radius were produced and supplied to the test filter at air velocities ranging from 0.094 to 0.94 cm s^{-1}.

An examination of the figure shows that the most penetrating particle size ranges from 0.25 to

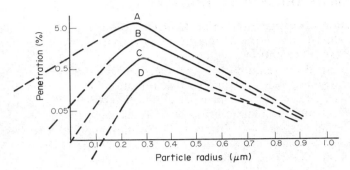

Figure 1 Mechanisms of aerosol filtration. Percent penetration of dioctyl phthalate particles through a resin impregnated glass fiber filter is plotted as a function of particle radius and incident air velocity (cm s^{-1}): A, 0.94; B, 0.42; C, 0.21; D, 0.094

0.35 μm. Particles having smaller or larger radii are removed by the filter with greater efficiency. Extrapolation of the curves for particles smaller than 0.25 μm shows that percentage penetration approaches zero, however, the lines do not converge, indicating that particle retention remains a strong function of incident air velocity. This result would be expected based on a diffusional mechanism of retention. For particles with a radius greater than 0.35 μm, extrapolation of the curves also demonstrates that percentage penetration approaches zero, however, the lines converge at a point where particle retention becomes independent of incident air velocity. This indicates that the combined effects of direct interception and inertial impaction mechanisms predominate over diffusional interception.

17.3 MICROORGANISM RETENTION TESTS OF A MEMBRANE AIR FILTER

It is well established that the particle retention efficiency of a filter is higher when it is used to filter air as opposed to a liquid. This observation was tested by examining the retention efficiency of a membrane air filter with respect to bacteria removal from liquid and bacteriophage removal from (wet and dry) air.

Figure 2 shows a scanning electron micrograph of polytetrafluoroethylene (PTFE) filter medium used in a membrane air filter. It is clear from this photograph that a membrane filter does not contain regularly arranged straight through pores but rather is characterized by a random pore structure resulting in a tortuous pore pathway. The filter medium has approximately 80% void volume. Particle retention efficiency by this type of filter has been reported to be >1 × 10^{12} for bacteria and >1 × 10^{10} for bacteriophages (Conway, 1984). The retention efficiency ratings were obtained by supplying a known quantity of either *Pseudomonas diminuta* (0.3 × 0.8 μm) or T$_1$ bacteriophage (0.05 × 0.1 μm) to the filter, collecting the filter effluent, and examining the effluent for the presence of the challenge organism by culture technique. Retention efficiency (or titer reduction) is calculated by dividing the number of organisms incident on the filter by the number recovered in the effluent. Where no organisms are recovered, the titer reduction can be assumed to be greater than the number of incident organisms.

The results of filter challenge tests are shown in Tables 1 and 2. These data were obtained using the liquid challenge test stand or aerosol challenge test stand shown in Figures 3 and 4, respectively. A liquid challenge test is a rigorous test of filtration efficiency as retention by diffusional and inertial impaction mechanisms are limited or nonexistent in the viscous (aqueous) medium. A liquid challenge test also simulates service conditions where significant quantities of fluid (which may contain organisms) are incident on the air filter. This can occur as a result of steam condensation and condensed water vapor, or in the case of exhaust gas filters exposed to a fermenter foam-out. The titer reduction obtained by a liquid challenge test for a given filter can only be enhanced when the filter is used in air service.

In order to measure titer reduction for the same filter in air, aerosol challenge tests were conducted. The results in Table 2 show that quantitative retention of T$_1$ bacteriophages was obtained based on challenge tests conducted at low and high relative humidity. Challenge tests at high relative humidity were conducted to simulate use of the filter in fermentation systems where the incident air is below the dewpoint or contains entrained moisture.

This type of filter testing demonstrates that membrane air filters can provide quantitative

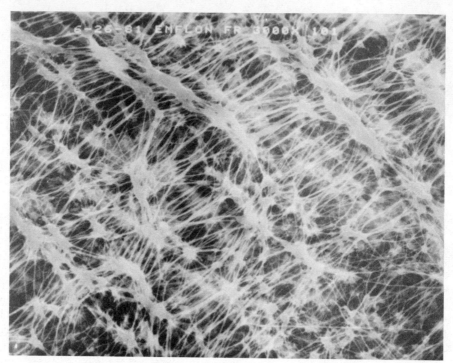

Figure 2 Photomicrograph of the surface of Emflon™ (PTFE) air filter membrane; 3000×

Table 1 Liquid Challenge Test Results[a]

Filter cartridge serial number	Total Pseudomonas diminuta challenge ($\times 10^{12}$)	Filter cartridge serial number	Total Pseudomonas diminuta challenge ($\times 10^{12}$)
82780Z21	1.2	N7780462	1.3
102180AG	1.2	N2090024	1.1
110480AE	1.2	92580B	1.7
110480AK	1.2	112180U	1.0
102180AI	1.2	N2080837	1.3
110480AB	1.0	N2070793	1.1
925801	1.0	112180Z16	1.3
N2090177	2.8	92580AF	2.3
92580AG	2.3	102480C	4.1
102180V	1.0	92580AZ	2.3
N2090094	1.3	N7780470	1.1
110480O	1.0	112180L	1.1
110480D	1.2	102480L	4.8
N2090003	1.3	92580K	1.7
N7780464	1.3	110480F	1.2

[a] Emflon air filters were challenged with liquid suspensions of *Pseudomonas diminuta*. The filter effluent was examined for the presence or absence of the challenge organism. The effluent was sterile in all cases. The Filters were challenged at a liquid flow rate of 1 l min^{-1} for 60 min. Test challenge parameters: fluid, sterile water; flow rate 1 l min^{-1}; total time, 60 min. Organisms were recovered in all cases, producing sterile effluents.

organism retention in conditions of actual service. When absolute organism exclusion or containment is required in fermentation air systems, membrane filters should be used.

17.4 ECONOMIC CONSIDERATIONS

In calculating air filter system economics one must consider costs associated with loss of fermentation batches and production downtime if the air filters fail to provide reliable organism retention. This factor is often difficult to quantify, but can be estimated on the basis of previous

Table 2 T₁ Bacteriophage Aerosol Challenge Test Results[a]

	Test no.				
	1	*2*	*3*	*4*	*5*
Actual bacteriophage challenge level	2.33×10^{10}	9.9×10^6	2.5×10^{10}	8.8×10^7	1.04×10^9
Effluent bacteriophage recovery	0	0	0	0	0
Titre reduction	$>2.33 \times 10^{10}$	$>9.9 \times 10^6$	$>2.5 \times 10^{10}$	$>8.8 \times 10^7$	$>1.04 \times 10^9$

[a] Emflon membrane air filters were challenged with aerosolized suspensions of T₁ bacteriophage. Tests 1–3 were performed at relatively low face velocity and dry air conditions (flux 0.13 SCFM/ft²). Tests 4–5 were performed at high face velocity and moist air conditions (flux 7.8 SCFM/ft²)

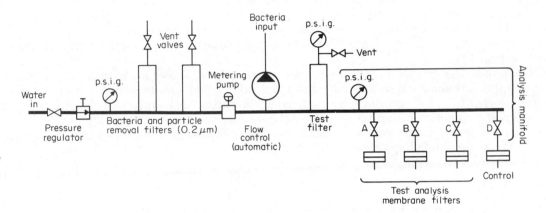

Figure 3 Bacteria (liquid) challenge test stand

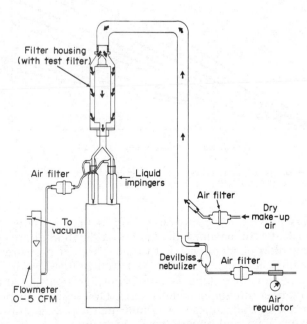

Figure 4 Bacteriophage (aerosol) challenge test stand

operating experience. This applies particularly where depth type or hydrophilic air filters are currently used. Absolute rated, fixed pore membrane air filters provide the most reliable organism retention, minimizing or eliminating risk of fermenter batch loss due to airborne contamination.

As stated earlier, a hydrophilic air filter medium can be wet by moisture present in the inlet (or exhaust) air supply resulting in decreased filtration efficiency and increased pressure drop. Selecting hydrophobic membrane air filters eliminates these concerns and reduces the controls which must be taken to remove moisture from the air supply. It is not unusual for the air incident on a hydrophilic air filter to be heated by steam tracing (or otherwise heating) the pipes leading to the filter vessel and often, the filter vessel itself. The expense of maintaining air temperature above dewpoint is significant and complicates system piping requirements. Moisture should also be removed from air lines when hydrophobic air filters are used to prevent accumulation of liquid in the filter vessel, however, this can be easily accomplished by using a coalescing type filter upstream of the final air filter. A coalescing filter reduces air moisture content and acts as a prefilter to lengthen final filter life by removing debris such as rust particles present in piping.

A second consideration affecting air filter system economics is cost associated with pressure drop across the filter assembly. This factor can be readily calculated by determining the cost of producing compressed air which is lost due to pressure drop. These calculations should be made for the filter assembly currently used or those under consideration. By selecting air filters which have a high effective filtration area per unit size, pressure loss is minimized. Reducing the pressure drop by only 1 to 2 p.s.i. can result in savings which may be applied in determining filter system payback time.

Results obtained by calculating the cost of pressure drop are shown in Figure 5. Kilowatt-hours expended on an annual basis are plotted as a function of pressure drop for air systems ranging from 5 to 75 p.s.i.g. inlet pressure. By multiplying kilowatt-hours by local electric rates, cost estimates can be obtained. Figure 5 is based on a reciprocating compressor supplying 1000 SCFM air. Compressor costs will vary depending on the type and efficiency of compressor used, local power rates and the operating requirements of the air system.

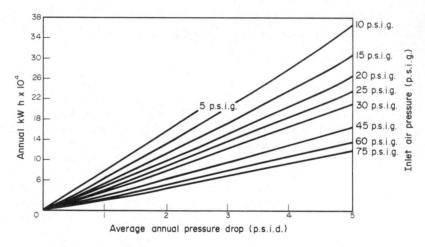

Figure 5   Annual power consumption due to pressure drop. The annual power consumption due to pressure drop for air systems operating at 1000 SCFM flow rate and inlet pressures ranging from 5 to 75 p.s.i.g.

Membrane air filters provide a low pressure drop at high air flow rates because the filter medium is thin and is pleated to maximize the filter area. High filter area per unit filter cartridge reduces the number of filter elements needed to meet system pressure loss requirements and reduces the size of the filter housing required to contain the filters.

When sizing a fermentation air system assembly, consideration should be given to minimizing pressure loss and the number of filter elements required to obtain a low pressure drop. As shown in Figure 5 there are economic advantages to minimizing the pressure drop, however, this saving must be balanced against filter and housing cost. An example of such an analysis is shown in Figure 6 by plotting system pressure drop *versus* total system cost. Pressure drop, plotted on the abscissa, for an air system operating at a fixed inlet air flow rate varies as a function of the number of filter elements used. As more filters are installed, pressure drop decreases and filter life increases. Total system cost, plotted on the ordinate, is the sum of the pressure drop cost and filter assembly cost. The point of maximum air filter system economy will be realized where the cost savings due to low pressure drop and the filter assembly size are optimum.

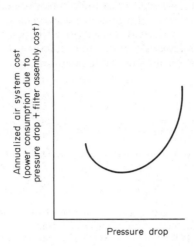

Figure 6 Annualized air system cost *vs.* pressure drop. Air system cost (cost due to pressure drop + filter assembly cost) is plotted as a function of system pressure drop

17.5 FILTER SYSTEM MAINTENANCE

In the previous section the cost of maintaining inlet air above the dewpoint was stated to be a factor affecting filter system economy. This consideration applies to filter assembly maintenance. Different filters have different operating requirements, the effects of which should be evaluated prior to selecting a particular type of air filter.

Selecting hydrophobic membrane filters eliminates the need to constantly maintain incident air temperature above the dewpoint. These filters also do not require extensive drying time following steam sterilization as is required for hydrophilic depth filters.

Membrane air filter cartridges which contain pleated filter medium provide a high filter area and low pressure drop. The use of a cartridge configuration enables ease of handling resulting in rapid installation time and minimal downtime. In contrast, replacing the filter medium of a fiberglass tower filter may require several days. The bed depth and packing density of depth filters is not reproducible. Membrane air filters contain a fixed submicron pore structure which is consistent from lot to lot provided strict quality control is enforced by the filter manufacturer.

A final consideration of system maintenance is an ability to test the integrity of the filter elements (or device) at installation or at any time following installation. A method to detect filter damage or proper installation in the filter housing is called an integrity test. An integrity test must be nondestructive, of sufficient sensitivity to detect minor filter system damage or leaks, and closely correlated with destructure live organism challenge tests of the type described earlier.

17.6 SUMMARY

The criteria involved in selecting a fermentation air filter system for inlet or exhaust gas filtration have been presented. Selection criteria include: filter retention efficiency, economy of operation, ease of filter use and service provided by the manufacturer. The most important selection criteria are filter efficiency and reliability of organism retention. In this regard fixed submicron pore size membrane filters provide the highest level of filtration efficiency. Use of a hydrophobic membrane minimizes or eliminates concerns of filter wetting due to air moisture content.

Filtration efficiency can be measured by (destructive) organism challenge tests such as those described here. An equally valid filter evaluation procedure is to install test filter elements in fermentation pilot plant applications. This type of testing allows the user to test filter efficiency and reliability at conditions of actual use, measure average filter system pressure drop, and gain experience with operating requirements which apply to specific filters.

Economy of operation and ease of filter use should be evaluated prior to filter selection. Costs associated with the need to heat the air above the dewpoint, pressure drop loss, filter replacement cost and labor cost of filter maintenance should be projected.

Finally, service and reliability of the filter manufacturer is a criterion for filter selection. Engineering consultation for filter system design and data relating to filter performance should be requested from manufacturers. Technical service once the filter system has been installed should be available. Quality standards applied to filter manufacture should be rigidly maintained to provide filter elements which consistently meet specification.

17.7 REFERENCES

Aiba, S., A. E. Humphrey and N. F. Millis (eds.) (1973). *Biochemical Engineering*, p. 270, chap. 10. Academic, New York.

Bruckshaw, N. B. (1973). Removal of contamination from compressed air. *Filtr. Sep.*, May/June, 296.

Chen, C. Y. (1955). Filtration of aerosols by fibrous media. *Chem. Rev.*, **55**, 595.

Conway, R. S. (1984). State of the art in fermentation air filtration. *Biotechnol. Bioeng.*, in press.

Davies, C. N. (1964). In *The Mechanics of Aerosols*, ed. N. A. Fuchs. Macmillan, New York.

Davies, C. N. (1973). *Air Filtration*. Academic, New York.

Decker, H. M., L. M. Buchanan, L. B. Hall and K. R. Goddard (1962). Air filtration of microbial particles. *Public Health Service Publication 953*. Government Printing Office, Washington DC.

Decker, H. M., J. B. Harstad and F. T. Lense (1957). Removal of bacteria from air streams by glass fiber filters. *J. Air Pollut. Control*, **7**, 15–16.

Fitzgerald, J. J. and C. G. Detweiler (1957). Collection efficiency of filter media in the particle size range of 0.005 to 0.1 micron. *Am. Ind. Hyg. Assoc. Q.*, **18**, 47–54.

Gaden, E. L. and A. E. Humphrey (1956). Fibrous filters for air sterilization design procedure. *Ind. Eng. Chem.*, **48**, 2172.

Harstad, J. B., H. M. Decker, L. M. Buchanan and M. E. Filler (1967). Air filtration of submicron virus aerosols. *Am. J. Public Health*, **57**, 2186–2193.

18
Media Sterilization

C. L. COONEY
Massachusetts Institute of Technology, Cambridge, MA, USA

18.1 INTRODUCTION

One of the first and most critical unit operations required for a successful fermentation is medium sterilization. The objective is to prevent the growth of undesired microorganisms during the course of a fermentation, bioconversion, enzyme catalyzed reaction, or medium storage. Sterilization may be achieved through the removal or destruction of any organism that will adversely affect the process or product. It is difficult, however, to remove selectively those organisms that may adversely affect a specific process. For this reason, medium sterilization is approached from the point of view of removing or destroying all microorganisms present in the material. While two strategies for sterilization are considered here, *e.g.* filtration for removal and thermal treatment for destruction, it is the latter that is of primary importance in processing most raw materials and media used in biochemical processes.

The choice of the sterilization method depends upon several factors including effectiveness in achieving an acceptable level of sterility, reliability, effect (positive or negative) on medium quality, and cost, including operating and capital expense, to achieve sterilization. The intention of this chapter is to provide the reader with a basic understanding of the fundamental principles in the use of filtration for removal and thermal pretreatment for destruction of microorganisms in the preparation of media for use in a biochemical reaction process. The reader is also referred to Richards (1968), Solomon (1969), Bailey and Ollis (1977) and Aiba *et al.* (1973) for additional discussion of this topic.

18.2 FILTRATION

18.2.1 Physical Characteristics of Microorganisms

In order to design an adequate procedure for removal of microorganisms from liquid media by filtration, it is important to understand the physical properties of these microorganisms. Filamentous, mycelial organisms, such as molds, typically have a minor dimension of greater than 7 μm and often a major dimension of several millimeters. As a consequence, these organisms are relatively easy to remove by filtration. Of more concern are the single-cell organisms. The largest are yeasts with typical dimensions of 3.5 μm and bacteria whose size ranges from 0.5 to 2 μm. Lastly, there are the microbial viruses or bacteriophages which are typically 0.04 to 0.1 μm. Fortunately, not all fermentation processes are susceptible to phage contamination and, as a consequence, one usually designs a filtration process to remove the smallest organisms, namely the bacteria. Although it is possible to assign a size to most organisms, they should not be considered as rigid spheres or rods. Cell size will change with the specific growth rate; slowly growing cells are much smaller. In addition, under pressure, many cells exhibit flexibility and are extruded through small spaces which are typically less than the minor dimension of the organism.

18.2.2 Filter Type

18.2.2.1 Depth filters

Depth filters are usually constructed from fairly porous or fibrous materials such that the typical pore or dimension between fibers in these filters is greater than the minimum size of the materials to be removed. Particle removal is based on the probability that a particle will be retained in the filter. The mechanisms of retention are interception, inertial impaction, diffusion, sedimentation and electrostatic interaction; the first three of these are most important and are shown graphically in Figure 1. Inertial impaction shown by trajectory B is based on the fact that a particle in motion will tend to stay in motion in a straight line. As a consequence, as the fluid passes around the fiber, a particle will leave the flow stream and tend to impact upon the fiber. Not surprisingly, the larger the organism and the greater its velocity V_0, the higher the probability that it will impact on a fiber.

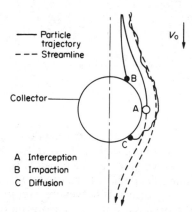

Figure 1 Schematic representation of particle removal in a fiberous depth filter (modified from Yao *et al.*, 1971)

The second method, interception, is illustrated by trajectory A. This is based on the fact that, if a particle attempts to move around a fiber or too close to a surface, it enters the stagnant liquid boundary layer where the liquid velocity is zero. If it comes in contact with the particle or the fiber itself, it will usually 'stick'. This removal method is important in all flow regimes; however, if the velocity of the liquid becomes too high, it is possible to reentrain intercepted or impacted particles into the fluid stream.

The third removal method is diffusion. As a particle enters the stagnant boundary layer, it exhibits diffusion or Brownian motion. If the fluid velocity is not too high, the retention time of the organism is sufficiently long for the particle to diffuse and there is some finite probability that it will 'bump' into and adhere to the fiber. Other methods for removal, sedimentation and electrostatic interaction have a relatively minor effect. For very small particles, diffusion is rate controlling. The efficiency of each mechanism is dependent on fluid velocity; there exists some optimal velocity for maximum particle retention. Thus, in the evaluation of alternative media for depth filters, it is important to identify the critical velocity at which the filtration efficiency is minimal. By using this value for design, any increase or decrease from the critical velocity will only improve filtration efficiency (Yao *et al.*, 1971).

The types of filter media used in depth filters include porous ceramics, sintered metals, diatomaceous earth, glass wool and cellulose fiber.

18.2.2.2 Absolute filters

An alternative filtration mechanism is absolute organisms removal based on size exclusion. It is possible to use membranes such as ultrafiltration, microporous or even macroporous membranes whose maximum pore size is less than the minimum size of the particles to be removed. Thus, the mechanism of filtration is primarily absolute size exclusion. In addition there is also bridging, achieved by particle accumulation around an open pore. These particles provide a filtration medium that is more efficient than the original filter itself. Lastly, there can be non-specific adsorption of particles to the solid portion of the filter.

18.2.3 Filtration Strategy

Media filtration is, in general, more expensive than thermal methods. However, it has the primary advantage that it is applicable to heat labile materials. Thus, in situations where it is essential to process nutrients or products in liquids, sterilization by filtration is quite common. Examples include sterilization of complex organic media and final product sterilization. Another increasingly important use of microporous and ultrafiltration as means for media sterilization includes the preparation of pyrogen free process water for use in bioreactors. Depth filters can be used quite effectively with media containing suspended solids, whereas absolute filters cannot. If the suspended particle level is above 10^6 ml^{-1}, it is not feasible to use absolute filters. As a consequence, for media sterilization one often employs a prefiltration with a depth filter followed by an absolute sterilizing filter.

18.3 THERMAL DESTRUCTION

18.3.1 Thermal Death Kinetics

Microbial death is a probabilistic phenomenon that follows first order kinetics as shown in Figure 2; as shown, it is possible to achieve a level of sterility that is lower than one organism/ml. This means that if a sterilization level is 10^{-3} then there is a probability that less than one batch in a thousand will become contaminated as a consequence of inadequate sterilization.

Microbial death is defined operationally; it is the inability of an organism to replicate itself in a given environment. For this reason, it is important to measure cell viability in an environment that is representative of the medium to be sterilized. Furthermore, death does not mean that all enzymatic activity in the cell is destroyed; even though a cell may not replicate, it may still have the potential for catalyzing one or more reactions. Usually, this is not of importance in media sterilization; however in food processing, the presence of active enzymes in a moist food for long periods of time can promote degradation.

Death kinetics can be described by a simple first order rate equation:

$$dN/dt = -kN \tag{1}$$

where N is the number of viable cells, t is time and k is the death rate constant. This equation may be integrated with respect to time, during which the number of viable organisms is reduced from an initial value N_0 to N. The resulting equation is:

$$\ln (N/N_0) = kt = \nabla_{total} \tag{2}$$

From this relationship, we define a sterilization criterion, ∇, which represents the extent of death that occurs during some time interval. As one might expect, the specific rate constant k is a function of temperature. This is seen by the kinetic relationships in Figures 3a and 3b for a typical

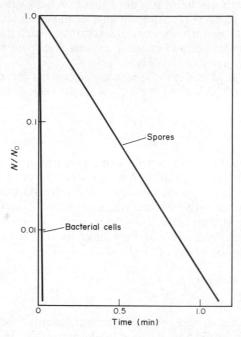

Figure 2 Kinetics of thermal death for typical bacterial vegetative cells and bacterial spores at 121 °C; *N* is the number of viable cells or spores at time *t*, and N_0 is the initial number

bacterial cell and a bacterial spore, respectively (Wang *et al.*, 1979). From these curves, it is clear that bacterial spores are much more resistant to thermal degradation than vegetative cells. In fact, the relative resistance differs by approximately six orders of magnitude. As a consequence, sterilization processes need to be designed around the most resistant organism. For the case of media preparation for fermentation or bioconversion or for the sterilization of foods, the criterion is based on thermal destruction of microbial spores such as from the genus *Bacillus* or *Clostridium*.

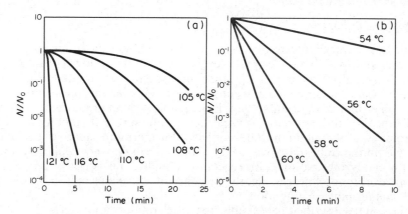

Figure 3 (a) Typical death rate at various temperatures for *E. coli* in buffer; (b) typical death rate for spores of *Bacillus stearothermophilus* in distilled water

18.3.2 Effect of Temperature on Death Kinetics

The effective temperature on thermal destruction may be described by the Arrhenius relationship shown by Figure 4 and equation (3):

$$K = Ae^{-E_a/RT}$$

(3)

where K is the thermal death constant, A is the Arrhenius constant, E_a is the activation energy, R is the universal gas constant and T is the absolute temperature (K). Typical E_a values for microorganisms are 250 to 290 kJ mol^{-1}.

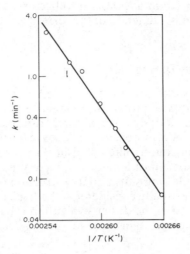

Figure 4 Correlation of isothermal death rate data for *Bacillus stearothermophilus*, where k = reaction rate constant and T = absolute temperature; value of E (activation energy) = 287.2 kJ/g mol.

During thermal treatment of the medium, there is not only the possibility of killing microbial spores and cells, but also of destroying ingredients in the medium. The sensitivity of nutrients to thermal destruction is illustrated in Table 1. The characteristic activation energy for vitamins and amino acids is typically 84 to 92 kJ mol^{-1}. By comparison, protein inactivation is about 165 kJ mol^{-1}. An observation from Table 1 is that the high activation energy for thermal destruction of microorganisms, *e.g.* 250–290 kJ mol^{-1}, means that a small increase in temperature has a relatively greater effect on cell death than on nutrient destruction. This fact becomes the basis for the use of high temperature short time sterilization discussed later in the chapter.

Table 1 Some Values of Activation Energy for Vitamins and Other Nutrients

Compound	Activation energy (kJ mol^{-1})
Folic acid	70.2
d-Panthothenyl alcohol	87.8
Cyanocobalamin (B$_{12}$)	96.6
Thiamine·HCl (B$_6$)	92
Riboflavin (B$_2$)	98.7
Hemoglobin	321.4
Trypsin	170.5
Peroxidase	98.7
Pancreatic lipase	192.3

18.4 DESIGN OF A BATCH STERILIZATION CYCLE

With an understanding of thermal death kinetics and their dependence on temperature, it is possible to proceed with the design of a batch sterilization cycle for prevention of contamination while minimizing overheating and destruction of the medium components. The required sterilization criterion may be calculated by a modification of equation (2) which expresses the initial number of spores in terms of a contamination load N_i (spores/ml) and a vessel volume, V_F:

$$\nabla_{\text{total}} = \ln \frac{N_i V_F}{N} \qquad (4)$$

The initial number of viable organisms, N_i, is assumed to be the initial spore contamination. Because of the difficulty in measuring spore levels rapidly, one usually assumes a conservative value, typically 10^6 bacterial spores/ml. This is a high level of spore contamination but adds an important safety margin in the design. The final level of contamination, N, represents some acceptable risk in sterilization. Typically, a value of 10^{-3} spores per fermenter is used. This suggests that one will tolerate one batch in a thousand to become contaminated. It is important to note that the sterilization criterion is a function of the fermenter volume.

18.4.1 Temperature–Time Profile During Sterilization

The design objective of batch sterilization is to calculate a temperature–time profile that will ensure adequate sterilization. A typical profile is shown in Figure 5. The heat up and cool down periods are dependent on equipment design, steam temperature, cooling water temperature and flow rate (Aiba *et al.*, 1973). In a new bioreactor, it is possible to manipulate the various design parameters. More often, however, one is interested in designing a more effective sterilization cycle for an existing piece of equipment or is simply trying to understand why a bioreactor is not being adequately sterilized or is allowing substantial destruction of essential nutrients. Thus, the heat up and cool down periods are usually considered to be fixed by the equipment and operating conditions in a given plant and most of the design evaluation is based on calculating the required holding time at the maximum temperature in order to assure adequate sterilization.

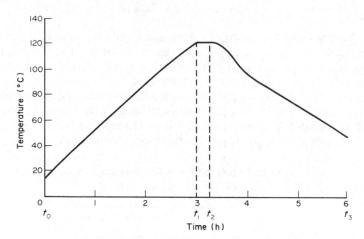

Figure 5 Effect of scale-up on medium sterilization: temperature–time profile for a batch sterilization of a 60 000 l industrial fermenter

The equations that may be used to calculate the temperature/time profiles during heat up and cool down are described by Aiba *et al.* (1973). For the purpose of the discussion here, it will be assumed that the temperature profile is experimentally measured for the equipment of interest.

The overall sterilization criterion, ∇_{total}, may be broken down into three component parts: heating, holding and cooling periods in the cycle shown in equation (5):

$$\nabla_{total} = \nabla_{heat} + \nabla_{hold} + \nabla_{cool} \tag{5}$$

Because the heating and cooling temperature/time profile is fixed by the equipment, it is possible to calculate from this profile the extent of cell death using the Arrhenius relationship. From data such as in Figure 5, it is possible to construct a plot of thermal death constant k as a function of time; the area under the curve is:

$$\ln (N/N_0) = \nabla_{total} = \int_0^{t_f} k \, dt \tag{6}$$

Thus, by graphical integration, one can readily calculate the extent of thermal death during the heat up and cool down period. Rearrangement of equation (5) gives the following equation for sterilization during the holding period:

$$\nabla_{hold} = \nabla_{total} - \nabla_{heat} - \nabla_{cool} \tag{7}$$

Since the temperature during this holding period is constant, by choosing a holding temperature k is fixed, and ∇_{hold} is calculated by equation (8):

$$\nabla_{hold} = k(t_2 - t_1) \tag{8}$$

This holding time $(t_2 - t_1)$ is usually much smaller than the time required for heat up or cool down (see Figure 5). However, because it is at the maximum temperature, most of the sterilization is achieved at this point. The impact of holding temperature on sterilization can be seen by the simple calculation illustrated in Table 2, which shows the time required to achieve a 15 log reduction in viable spores.

Table 2 Typical Values of K for *B. stearothermophilus* Spores

Temperature (°C)	K (min^{-1})	Holding time (min)[a]
100	0.02	1730
110	0.21	164
120	2.0	17
130	17.5	2
140	136	0.25
150	956	0.04

[a] For $N_0 /N = 10^{15}$.

18.4.2 Scale Up of Batch Sterilization

During the scale up of a batch sterilization cycle, an increase in the liquid volume affects both the overall sterilization criterion as well as the equipment design leading to alterations in the temperature–time profile for heat up and cool down. This is illustrated in Figure 6. Of particular concern is the effect of scale up on nutrient destruction (Mulley *et al.*, 1975). When the fermenter volume is increased from V_1 to V_2, it is necessary to hold the broth for a longer period of time; as a consequence, there is greater destruction of nutrients in the broth. If this nutrient is critical to product formation, oversterilization may lead to reduced product titers upon scale up. This phenomenon is often seen in fermentations utilizing complex nutrients. As a consequence, medium optimization for large scale fermentation processes must be done using sterilizing conditions that reflect those in larger vessels. The problem of oversterilization of media for citric acid fermentation has been described by Chopra *et al.* (1981).

18.5 CONTINUOUS STERILIZATION

Continuous medium sterilization is based on the concept of high temperature short time (HTST) treatment. This takes advantage of the fact that an increase in temperature has a relatively greater effect on thermal destruction of cells than on nutrients. The rationale for using continuous *versus* batch sterilization is based not only on the beneficial effect of heat sensitive media, but also on improved steam economy, easier and more reliable scale up, ease of interfacing with continuous processes and more efficient fermenter use. The steam usage in continuous sterilization is perhaps 20 to 25% of the requirements for a batch cycle. In addition, the total time required to sterilize media may be only two to three hours compared with a typical sterilization cycle of five to six hours for a large, *i.e.* 100 000 m^3, fermenter.

18.5.1 Equipment for Continuous Sterilization

18.5.1.1 *Continuous steam injection*

Continuous steam injection, as illustrated in Figure 7, is used to achieve rapid heat up of the medium without the use of a heat exchanger. The temperature–time profile of the medium in the sterilizer is shown in Figure 7b. This approach is particularly effective with media that tend to foul heat exchange surfaces. Pick (1982) described some approaches to steam injection. In the use of steam injection, it is important to avoid two phase flow. Therefore, one operates a continuous

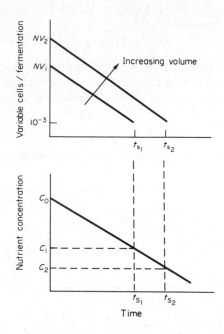

Figure 6 Effect of scale up on medium sterilization

sterilizer at 5 to 10 p.s.i. above the bubble point. A disadvantage of this approach is the dilution of the medium with condensed steam and the difficulty in controlling pressure and temperature due to variation in medium viscosity. It is also important to use clean steam to avoid deposition of materials from the steam into the medium. Heat exchangers are used for preheating and cooling of the medium in order to achieve efficient energy usage.

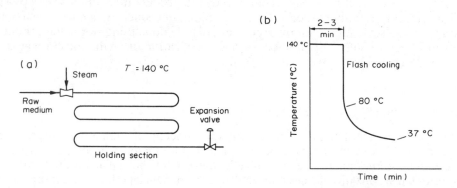

Figure 7 (a) Schematic view of continuous steam injection for media sterilization; (b) the temperature–time profile

18.5.1.2 *Continuous plate or spiral heat exchangers*

Continuous plate or spiral heat exchangers are used for media sterilization to avoid direct steam injection. The principles of operation and the temperature–time profile are shown in Figures 8a and 8b respectively. The disadvantages of this approach are increased capital cost over direct steam injection, potential for fouling of hot heat exchange surfaces and potential for leaks around the exchanger gaskets and seals. The spiral heat exchanger is particularly effective with media containing suspended solids. The wide gap, curved walls and high liquid velocity minimize surface fouling. In addition, such heat exchangers make efficient use of available space in the plant and minimize gasket area that could potentially leak. Plate and frame heat exchangers are often used for media with low suspended solids. They are very flexible in their design since the heat transfer area can be easily altered by adjusting the number of plates. It is also possible to achieve high heat transfer coefficients in these heat exchangers.

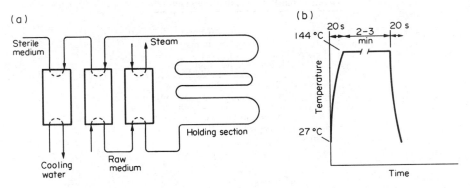

Figure 8 (a) Continuous plate or spiral heat exchangers used for media sterilization, avoiding direct steam injection; (b) the temperature–time profile

18.5.1.3 Sterilizer systems

The essential components of the continuous sterilizer system include a heat exchanger for recovery of heat from the sterilized fluid and preheating fresh medium, a heat exchanger or steam injector for heating the medium to sterilizing temperatures, a holding section for sterilization, a heat exchanger for cooling the medium to the fermentation temperature and a number of peripheral pieces of equipment including a medium mixing tank, sterile medium receiver, which is often the fermenter, and a set of pumps, valves and controllers. These components are typically arranged as shown in Figure 9. This type of system is started up by passing water through the system at sterilizing temperatures (dotted lines) in order to avoid any regions in the equipment below the critical temperature for sterilization. The time required to achieve medium sterilization for batch fermentation is typically two to three hours. The value as well as knowledge of the total amount of liquid to be sterilized will determine the flow rate through the continuous sterilizer.

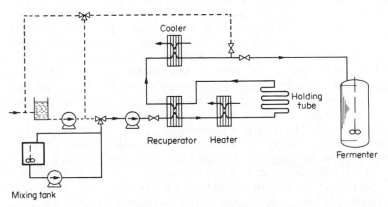

Figure 9 Continuous sterilization system illustrating recovery of heat from the sterilized medium. The bypass line (dotted) is used to recirculate water during start-up of the sterilizer

Because the time in a continuous sterilizer is much less than in batch sterilization, there is a potential problem associated with the presence of particles. Shown in Table 3 are the typical times to reach 99% of the final temperature for particles ranging in diameter from 1 μm to 1 cm. As a consequence, one can see that it is important to exclude particles with a dimension greater than 1 or 2 mm from continuous sterilizers (see de Ruyter and Brunet, 1973).

Table 3   Effect of Particle Size on
Heat-up Time of Solids

Diameter (cm)	Time to reach 99% of final temperature[a]
1×10^{-4}	1 μs
1×10^{-3}	0.1 ms
1×10^{-2}	10 ms
1×10^{-1}	1 s
1	100 s

[a] The particle is assumed to have the following property: $(\varrho C_p)/k = 250$, where ϱ = density, C_p = heat capacity and k = thermal conductivity. By comparison, values for water, rubber and wood are 182, 358 and 256, respectively.

18.5.2 Design of a Continuous Sterilizer

18.5.2.1 Fluid flow

The objective in design of a continuous sterilizer is to calculate the residence time of the medium in the sterilizer to assure adequate sterilization. For this reason, it is important to understand the type of fluid flow through a continuous sterilizer. Ideally, one would like perfect plug or piston flow. In reality, the flow will be somewhere between viscous and fully turbulent flow such that the average velocity is 0.5 to 0.82 times the maximum velocity. In order to minimize overheating of the medium, it is desirable to approach as closely as possible fully turbulent flow. This occurs when the Reynolds number, Re, is at least 2.5×10^3 and preferably above 2×10^4. Furthermore, this will minimize the extent of axial dispersion in the sterilizer. The Reynolds number in a tube is:

$$Re = (D\varrho U)/\mu \tag{9}$$

where D is the tube diameter, U is the average velocity, ϱ is the specific gravity, and μ is the medium viscosity.

The degree of axial dispersion is described by the Peclet number which is equal to:

$$Pe = (UL)/E_z \tag{10}$$

where U is the average fluid velocity, L is the length of the sterilizer, and E_z is the axial dispersion coefficient. The effect of the degree of dispersion as indicated by the Peclet number on residence time in the continuous sterilizer is shown in Figure 10. This figure shows the effect of dispersion on the residence time in a continuous sterilizer. If a step change is made on fluid entering the reactor and the reactor operates with perfect plug flow, *e.g.* the Peclet number is infinite, there is no change in outlet concentration until precisely one nominal residence time when the concentration falls precipitously to zero. Realizing that this cannot be achieved but, rather, Peclet numbers of 3 to 600 are more typical, one sees that with a nominal residence time of 0.9, the cells or tracer concentration begin to decline for the initial value C_0. This deviation from plug flow must be taken into account in the design of the sterilizer.

Keeping in mind that the objective in sterilizer design is to achieve adequate sterilization to minimize the extent of overheating, the first step in continuous sterilizer design is to determine the type of flow. The Reynolds number should be greater than 2×10^4. By choosing a type diameter, D, one can calculate the required liquid velocity, U, to achieve this Reynolds number. By calculation (Levenspiel, 1958) or measurement, one can derive the Peclet number and hence the degree of axial dispersion at this Reynolds number or in a given piece of equipment (Aiba *et al.*, 1973). The next step is to choose the sterilization criterion N/N_0. The value of N should reflect the desired tolerance for contamination, *e.g.* one batch in a thousand, thus $N_0/N = (V_F) 10^{12}$ for a batch fermentation. In a continuous fermentation, the value F is the medium flow rate and T is the operation time for fermentation. In the case of continuous fermentation, N reflects the number of fermentations that one would tolerate to become contaminated. Thus, having chosen the sterilization criterion, one can use Figure 11 along with knowledge of the Peclet number to calculate the reaction number N_r, which is equal to the product of the thermal death constant

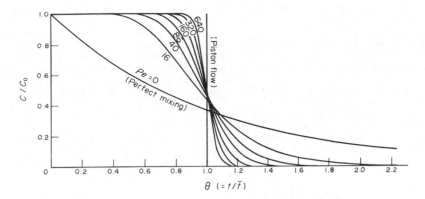

Figure 10 Effect of different types of flow as shown by different *Pe*, values of the Peclet numbers, and holding time, ▽, in continuous medium sterilization (modified from Aiba *et al.*, 1973)

times the length of the sterilizer divided by the average velocity in the sterilizer. Since the fluid velocity was chosen earlier to meet a Reynolds number requirement for turbulent flow, it is now possible to choose a holding temperature to calculate the value of K. Thus, from the reaction number, N_r, one can calculate L, the length of the sterilizer.

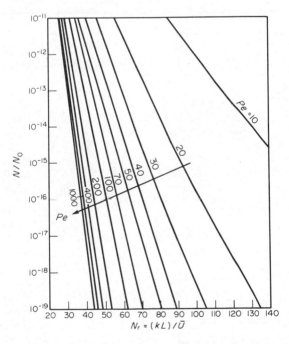

Figure 11 Effect of different types of flow (as shown by different *Pe* values) on the destruction of organisms (N/N_0) at different rates of destruction indicated by the reaction number $N_r = KL/U$

It is possible to see that a tradeoff can be made between the length of the sterilizer and the operating temperature. This represents a tradeoff between capital investment and operating costs.

The theory of continuous medium sterilization has been developed in more detail by Lin (1975) and additional data for HTST destruction of spores is given by Jacobs *et al.* (1973).

18.6 REFERENCES

Aiba, S., A. E. Humphrey and N. F. Millis (1973). *Biochemical Engineering*. University of Tokyo Press, Tokyo.
Bailey, J. E. and D. F. Ollis (1977). *Biochemical Engineering Fundamentals*. McGraw-Hill, New York.

Chopra, C. L., G. N. Qazi, S. K. Chaturvedi, C. N. Gaind and C. K. Atal (1981). Production of citric acid by submerged fermentation. Effect of medium sterilisation at pilot-plant level. *J. Chem. Technol. Biotechnol.*, **31**, 122–126.

de Ruyter, P. W. and R. Brunet (1973). Estimation of process conditions for continuous sterilization of foods containing particulates. *Food Technol.*, **7**, 46–51.

Jacobs, R. A., L. L. Kempe and N. A. Milone (1973). High temperature short time (HTST) processing of suspensions containing bacterial spores. *J. Food Sci.*, **38**, 168–172.

Levenspiel, O. (1958). Longitudinal mixing of fluids flowing in circular pipes. *Ind. Eng. Chem.*, **50**, 343.

Lin, S. H. (1975). A theoretical analysis of thermal sterilization in a continuous sterilizer. *J. Ferment. Technol.*, **53** (2), 92–98.

Mulley, E. A., C. R. Stumbo and W. M. Hunting (1975). Kinetics of thiamine degradation by heat. *J. Food Sci.*, **40**, 985–996.

Pick, A. E. (1982). Consider direct steam injection for heating liquids. *Chem. Eng.*, **6**, 87–89.

Richards, J. W. (1968). *Introduction to Industrial Sterilization*. Academic, New York.

Solomons, G. L. (1969). *Materials and Methods in Fermentation*. Academic, New York.

Wang, D. I. C. *et al.* (1979). *Fermentation and Enzyme Technology*. Wiley, New York.

Yao, K. M., M. T. Habibian and C. R. O'Melia (1971). Water and waste filtration: concepts and applications. *Curr. Res.*, **5**, 1105–1112.

19

Heat Management in Fermentation Processes

J. R. SWARTZ
Genentech Inc., San Francisco, CA, USA

19.1 INTRODUCTION

Whether complicated or simple, nearly all fermentation processes benefit from temperature control. In order to achieve this control, the addition and removal of heat must be carefully managed. That is the topic of this chapter.

Heat management is usually quite simple in small reactors. However, as processes are scaled up to larger and larger volumes, heat removal and temperature control often become a serious design and operational limitation. As a result, the need for heat management can add substantially to capital and operating costs. In this chapter, we will discuss overall heat balances as well as methods to estimate the magnitudes of heat generated and the capacity of the bioreactor to remove that heat. This understanding is crucial to the design of processes and their equipment. Although the primary focus will be on mechanically agitated, liquid fermentations, considerations for air agitated designs and for solid substrate fermentations will also be suggested.

19.2 OVERALL HEAT BALANCE

All biological processes generate heat and exchange it with their environment. The maintenance of an isothermal state then requires that these heat sources and sinks be dynamically balanced. The following equation presents a general description of this balance:

$$mC_p \, dT/dt = Q_{ox} + Q_{ag} - Q_{evap} - Q_{sen} - Q_{exch} \tag{1}$$

where mC_p = the heat capacity of the fermentation (J °C^{-1}); T = the temperature of the fermentation (°C); t = time (h); Q_{ox} = rate of heat generation by substrate oxidation (J h^{-1}); Q_{ag} = rate of heat generation by agitation (J h^{-1}); Q_{evap} = rate of heat loss by evaporation (J h^{-1}); Q_{sen} = rate of sensible heat gain by equipment or feeds (J h^{-1}); and Q_{exch} = rate of heat exchange with the surroundings (J h^{-1}).

It can be seen that the right hand side of the equation must equal zero in order to maintain a constant fermentation temperature. Q_{ox} describes the heat associated with the growth and maintenance of the organism as well as with product formation. In large, intense fermentations this term can become the major heat source. Q_{ag} accounts for the heat which is added to the system when the fermentation is agitated to promote mixing and mass transfer. This agitation is usually imposed mechanically, but may also be effected by sparging air into the vessel as in airlift or bubble column fermentations.

The other three terms indicate effects which are usually heat sinks. Q_{evap} and Q_{sen} are normally the more minor influences; the first describes heat lost as water vaporizes into the sparged gas and the second includes sensible heat effects as either the fermentation equipment changes in temperature or as gas or liquid feeds enter at temperatures different than the fermentation temperature. Finally, Q_{exch} includes the influence of the surrounding environment. While the evaluation of this term is somewhat dependent upon where the boundaries are drawn in the analysis, it usually describes the rate of heat gained or lost by the temperature control fluids or the air surrounding the fermenter. It is the term which is usually controlled to counteract the other influences so that a constant temperature can be maintained.

19.3 EVALUATION OF HEAT SOURCES

19.3.1 Metabolic Heat

The first and often most important heat source in equation (1) is the heat produced by the organism itself. This rate can be measured directly using calorimeters or with specially equipped fermenters (Cooney *et al.*, 1968; Ho, 1979; Ishikawa and Shoda, 1983). However, it can also be estimated quite accurately by correlation with expected oxygen uptake rates in aerobic fermentations or by energy balances around the organism itself. The second is usually most easily accomplished by a stoichiometric balance using heats of combustion of the substrates, cells and products.

Cooney *et al.* (1968) showed that, for aerobic fermentations, approximately 460 J of heat were evolved for each mmole of oxygen consumed. Later, Minkevich and Eroshin (1973) and Erickson *et al.* (1978) showed theoretically that 113 J are evolved as each mmole of electrons is transferred from the substrates to the electron acceptor(s). Since 1 mmol of molecular oxygen can receive 4 mmol of electrons, the theoretical number agrees well with the observations of Cooney *et al.* Thus, if one knows the expected oxygen uptake rate of a fermentation, the expected rate of heat evolution associated with aerobic metabolism can be calculated. Likewise, if one knows the maximum oxygen transfer rate of a bioreactor, the maximum rate of metabolic heat evolution can be calculated for aerobic fermentations.

If the fermentation uses electron acceptors other than oxygen, or if oxygen uptake rates cannot be anticipated, then a stoichiometric analysis can be used to estimate heat evolution. Such an analysis begins with the following equation:

$$A\,C_aH_bO_cN_d + B\,O_2 + C\,NH_3 \rightarrow D\,C_eH_fO_gN_h + E\,CO_2 + F\,H_2O + G\,C_iH_jO_kN_l + \text{heat} \qquad (2)$$

where: A to G = stoichiometric rates (mmol h^{-1}); $C_aH_bO_cN_d$ = molecular formula of the carbon-energy source(s); $C_eH_fO_gN_h$ = 'molecular formula' of cells; $C_iH_jO_kN_l$ = molecular formula of product(s); and heat = rate of metabolic heat evolution (J h^{-1}). The molecular formulae for the substrate and product can represent stoichiometric averages if there is more than one of either. The molecular formula for cells can be determined by C, H, O and N determinations (note that this does not include the other elements, usually about 5 to 10% of dry cell mass). If elemental analysis of the cells is not available, the formula $C_6H_{10}NO_3$ can often be used with good results (Wang *et al.*, 1976).

Now an energy balance can be written for cell growth and product formation by using the heats of combustion of the components:

$$AH_{c,S} + CH_{c,A} = DH_{c,C} + GH_{c,P} + \text{heat} \qquad (3)$$

where $H_{c,S}$ = heat of combustion of carbon-energy source(s) (J mmol^{-1}); $H_{c,A}$ = heat of combustion of ammonia (aq) (= 347 J mmol^{-1}); $H_{c,C}$ = heat of combustion of cells (J mmol^{-1}); and $H_{c,P}$ = heat of combustion of product(s) (J mmol^{-1}). The heats of combustion of substrates and products can be found in chemical handbooks. Again, a stoichiometric average can be used if there is more than one of either. The heat of combustion of cells can be measured directly or can be estimated from the molecular formula and the reduction level (RL) of the cells (Wang *et al.*, 1976). The reduction level is calculated as:

$$RL = \frac{2nc + 1/2nh - no}{2nc} \qquad (4)$$

where: nc = the number of carbon atoms per molecule; nh = the number of hydrogen atoms per molecule; and no = the number of oxygen atoms per molecule. Now the heat of combustion is calculated by the formula:

$$H_{c,C} = RL \times 460 \times nc \qquad (5)$$

The rate of heat evolution can now be calculated using equation (3). First, values for the cell and product yields from substrate must be measured or estimated. If A is assumed to be equal to 1, the yields can be used to calculate the values of D and G. Finally, a nitrogen balance is applied to equation (2) to determine a value for C, and the only remaining unknown in equation (3) is heat which will be expressed as joules evolved per mmole of substrate consumed. This second method is somewhat longer and, perhaps, not quite so precise as the correlation with oxygen uptake, but both will provide estimates of suitable accuracy for process and equipment design.

19.3.2 Heat of Agitation

The other major heat source is the heat of agitation. Although good correlations exist for relating ungassed power to gassed power (Pollard and Topiwala, 1976), methods for calculating this term for mechanically agitated vessels are beyond the scope of this chapter. Often the power input into an agitated vessel is known or is specified by the agitator vender. If it is not known for an existing vessel, the value can be determined by measuring the power draw of the agitator motor with an empty vessel and subtracting that value from the power draw when the vessel is in operation. If the vessel is being designed, the power input is usually calculated to deliver a desired oxygen transfer coefficient. These methods are also beyond the scope of this chapter.

Roels and Heijnen (1980) address the topic of how to estimate the heat of agitation in air agitated fermenters. Using nonequilibrium thermodynamics they show that energy transferred to the liquid in the form of work is nearly balanced by the energy absorbed by the expansion of the sparged gas. Although this does not account for changes in the potential and kinetic energy of the gas, these latter terms are generally negligible. Thus the net heat input from air agitation is usually quite small in relation to other effects.

19.4 EVALUATION OF HEAT SINKS

19.4.1 Sensible and Evaporative Heat Losses

As equation (1) suggests, the usual heat sinks are heat loss by evaporation, heat lost as sensible heat gained by the equipment or feeds, and heat transferred to the surroundings. The heat lost by evaporation can be significant, especially when the sparged air is dried during the compression process. The effect can be evaluated by calculating the enthalpy change of the air as it passes through the fermenter. Usually the air will become saturated with water, but this should probably be checked experimentally. The enthalpy change is then multiplied by the sparging rate to determine the net heat lost by evaporation as well as the sensible heat gained by the sparged gas.

Enthalpy balances are also used to determine the heat effects of adding feeds at temperatures different than the fermentation temperature. This term is usually small in comparison to other effects. The sensible heat gained by the equipment is also usually negligible when the fermentation, and therefore the equipment in contact with the fermentation, is maintained at a constant temperature.

19.4.2 Heat Exchange with the Surrounding Environment

The final term in equation (1) is the one over which we generally have the most control, that is the rate of heat exchange with the fermentation's surroundings. It is predominantly this term which balances the heat sources and allows us to control temperature. The magnitude of the exchange must be carefully regulated in existing equipment and accurately estimated for effective equipment design. The capacity to remove heat from the fermentation must exceed the value of the potential heat sources.

Occasionally fermentations are air cooled or cooled by external liquid film evaporation. For very active fermentations such as those for single cell protein, some designs also circulate the fermentation broth through external heat exchangers (Kanazawa, 1975). Usually however, temperature control is effected by flowing a temperature control fluid through a jacket installed around the fermentation vessel, through coils installed within the vessel, or both.

The rate of heat exchange is calculated by the general heat transfer equation according to the treatment of Bailey and Ollis (1977):

$$Q_{\text{exch}} = h_{\text{oa}} \times A \times \Delta T \tag{6}$$

where: h_{oa} = overall heat transfer coefficient (W m^{-2} °C^{-1}); A = surface area for heat transfer (m^2); and ΔT = temperature driving force between bulk fermentation fluid and temperature control fluid (°C). ΔT is calculated as the difference between the bulk fluid temperature and the log mean temperature of the temperature control fluid. For transfer to the vessel jacket, we assume the vessel wall approximates a planar surface, and the overall heat transfer coefficient can be estimated by:

$$1/h_{\text{oa}} = (l/h_{\text{W1}}) + (l/k_{\text{s}}) + (l/h_{\text{W2}}) \tag{7}$$

where: h_{W1} = heat transfer coefficient between inside bulk fluid and vessel wall (W m^{-2} °C^{-1}); l = thickness of vessel wall (m); k_{s} = thermal conductivity of the vessel wall (W m^{-1} °C^{-1}); and h_{W2} = heat transfer coefficient between outside vessel wall and temperature control fluid (W m^{-2} °C^{-1}). For transfer through the walls of cooling coils, the equation is modified:

$$1/h_{\text{oa}}d_{\text{o}} = 1/h_{\text{o}}d_{\text{o}} + [\ln (d_{\text{o}}/d_{\text{i}})/2k_{\text{s}}] + (l/h_{\text{i}}d_{\text{i}}) \tag{8}$$

where d_{o} = the outside diameter of the tube wall (m); h_{o} = heat transfer coefficient between vessel side bulk fluid and outside tube wall (W m^{-2} °C^{-1}); d_{i} = the inside diameter of the tube (m); and h_{i} = heat transfer coefficient between inside tube wall and the cooling fluid within the tube (W m^{-2} °C^{-1}). When h_{oa} from equation (8) is used in equation (6), A represents the outside surface area of the tube.

Methods for calculating heat transfer coefficients for turbulent flow within the tubes and jacket are given by Perry and Chilton (1973). The vessel side coefficients are more difficult to estimate. Good methods for transfer to single phase fluids are also described by Perry and Chilton (1973), but Pollard and Topiwala (1976) point out that these do not apply well to gas sparged liquids. They found tank-side heat transfer coefficients for a sparged agitated fermenter to be in the range of 4–9 kW m^{-2} °C^{-1}, and conclude from their study that more work is needed in this area. Finally it should also be pointed out that overall heat transfer coefficients can be negatively affected by the formation of films on the heat transfer surfaces. To summarize, the *a priori* estimation of heat transfer coefficients for sparged fermentation is somewhat difficult, and it may be wise to over-design heat removal capacities to allow for errors. For existing equipment, however, heat transfer coefficients may be estimated by measuring the rate of cooling which can be achieved when the other heat effects are known. Care should be taken to cool a fluid with similar rheological properties to the fermentation broth expected.

19.5 SUMMARY

In conclusion, this chapter has described methods for evaluating the various terms which influence the heat balance in a fermentation process. The same balance is applicable to both liquid and solid fermentations and to either air or mechanically agitated fermenters. With solid substrate fermentations, however, Rathbun and Shuler (1983) point out that temperature gradients within the fermentation merit additional consideration.

Still, in all fermentation processes, effective heat management and process design depend upon an adequate understanding of the overall balance. This becomes especially true with larger fermentation processes.

19.6 REFERENCES

Bailey, J. E. and D. F. Ollis (1977). *Biochemical Engineering Fundamentals*. McGraw-Hill, New York.

Cooney, C. L., D. I. C. Wang and R. I. Mateles (1968). Measurement of heat evolution and correlation with oxygen consumption during microbial growth. *Biotechnol. Bioeng.*, **11**, 269–281.

Erickson, L. E., I. G. Minkevich and V. K. Eroshin (1978). Application of mass and energy balance regularities in fermentation. *Biotechnol. Bioeng.*, **20**, 1595–1621.

Ho, L. (1979). Process analysis and optimal design of a fermentation process based upon elemental balance equations: generalized semitheoretical equations for estimating rates of oxygen demand and heat evolution. *Biotechnol. Bioeng.*, **21**, 1289–1300.

Ishikawa, Y. and M. Shoda (1983). Calorimetric analysis of *Escherichia coli* in continuous culture. *Biotechnol. Bioeng.*, **25**, 1817–1827.

Kanazawa, M. (1975). The production of yeast from *n*-paraffins. In *Single-cell Protein II*, ed. S. R. Tannenbaum and D. I. C. Wang. MIT Press, Cambridge, MA.

Minkevich, I. G. and V. K. Eroshin (1973). Productivity and heat generation of fermentation under oxygen limitation. *Folia Microbiol.* (*Prague*), **18**, 376–385.

Perry, R. H. and C. H. Chilton (1973). *Chemical Engineer's Handbook*, 5th edn. McGraw-Hill, New York.

Pollard, R. and H. H. Topiwala (1976). Heat transfer coefficients and two-phase dispersion properties in a stirred-tank fermentor. *Biotechnol. Bioeng.*, **18**, 1517–1535.

Rathbun, B. L. and M. L. Shuler (1983). Heat and mass transfer effects in static solid-substrate fermentations: design of fermentation chambers. *Biotechnol. Bioeng.*, **25**, 929–938.

Roels, J. A. and J. J. Heijnen (1980). Power dissipation and heat production in bubble columns: approach based on nonequilibrium thermodynamics. *Biotechnol. Bioeng.*, **22**, 2399–2404.

Wang, H. Y., D. G. Mou and J. R. Swartz (1976). Thermodynamic evaluation of microbial growth. *Biotechnol. Bioeng.*, **18**, 1811–1814.

20

Disruption of Microbial Cells

C. R. ENGLER
Texas A&M University, College Station, TX, USA

20.1 INTRODUCTION

Techniques for the disruption of microbial cells have had an instrumental part in unlocking mysteries of cellular composition and structure, enabling a better understanding of how life processes function to be gained. The ability to manipulate genetic information contained in microbial cells promises to open whole new industries based on microbial products. Whether these products are to be used for medical diagnostic or therapeutic applications, as catalysts for production of industrial chemicals or food ingredients, or as precursors or end products in the chemical and food industries, isolation and purification of these products generally will be required. While some products will be extracellular, many will be located in interior parts of the cells and require disruption as an early step in isolating the product.

The main purpose of this chapter is to review methods for disruption of microbial cells, both in the laboratory and on an industrial scale. Although disruption can be approached in an empirical manner, an understanding of cell wall structures, which are the target of any disruption process, can aid in developing more efficient methods. Therefore, a brief description of cell wall structures has been included. In addition, a discussion of analytical methods for measuring the extent of disruption has been included.

20.2 COMPOSITION AND STRUCTURE OF CELL WALLS

For a microorganism to be disrupted, the envelope surrounding the cell must be damaged to the extent that it no longer separates the intracellular components from the fluid in which the organism is suspended. Cell envelopes of microorganisms for which disruption processes are of interest generally consist of a cytoplasmic membrane and a cell wall which provides a rigid outer support. The cytoplasmic membrane maintains concentration gradients between the interior and exterior of the cell and is composed primarily of proteins and lipids. Without the cell wall, the membrane is readily susceptible to osmotic shock, therefore the cytoplasmic membrane generally is of little concern in disruption processes.

Since the primary resistance to disruption is provided by the cell wall, an understanding of its composition and structure is important for analysis of cell disruption. Structures and compositions of cell walls depend on both genetic and environmental factors, giving rise to great diversity among structures when examined in close detail. However, similarities in gross structural aspects can be observed within groups of microorganisms. Brief descriptions of general structure and composition are given below for bacteria (with differences between Gram-positive and Gram-negative types noted), yeasts and other fungi.

20.2.1 Bacterial Cell Walls

Composition, structure and function of bacterial cell walls have been the subjects of intensive research efforts for over three decades since cell wall material from mechanically disrupted cells was first isolated (Salton, 1964). Several recent review articles (Costerton *et al.*, 1974; Ghuysen and Shockman, 1973; Glauert *et al.*, 1976; Sutherland, 1975) provide a comprehensive picture of the general structures and compositions of both Gram-positive and Gram-negative bacteria.

The rigid matrix of the walls of nearly all bacteria, except *Mycoplasma* and some L-forms and halophiles, is composed of peptidoglycan, a network of glycan chains cross-linked by short peptides (Ghuysen and Shockman, 1973). The peptidoglycan forms a nearly continuous network around the cell and provides both shape and strength.

The glycan chains are composed almost exclusively of alternating residues of *N*-acetylglucosamine and *N*-acetylmuramic acid in β-(1→4) linkage. The *N*-acetylmuramic acid may have variations in the side group substitutions in different species, none of which significantly affect the three-dimensional organization of the glycan chains. The lactyl groups of the *N*-acetylmuramic acid residues provide the points at which peptides are linked to the glycan strands.

At least a portion of the lactyl groups of the *N*-acetylmuramic acid residues are substituted by tetrapeptide units (Ghuysen and Shockman, 1973). The units on adjacent glycan chains are, in turn, cross-linked by peptide bridges as shown in Figure 1. The degree of cross-linking varies considerably among different organisms. In *Escherichia coli*, about 50% of the tetrapeptide units are uncross-linked and the others are linked only as dimers (Ghuysen, 1968; Weidel and Pelzer, 1964). On the other hand, nearly 90% of the tetrapeptide units are cross-linked in *Lactobacillus acidophilus* with about 30% of the units cross-linked as trimers (Coyette and Ghuysen, 1970).

Although nearly all bacteria contain a basic peptidoglycan network, there are significant differences in the wall structures formed by Gram-positive and Gram-negative bacteria. The walls of Gram-positive bacteria are relatively thick (15–50 nm) and contain 40–90% peptidoglycan with the remainder being primarily polysaccharides and teichoic acids (Ghuysen and Shockman, 1973). Some species have regularly arranged protein subunits on the outer surface as shown in Figure 2a (Thornley *et al.*, 1974). In Gram-negative bacteria the wall consists of a much thinner peptidoglycan layer (1.5–2.0 nm) and an outer membrane similar in appearance in electron micrographs to the cytoplasmic membrane. The peptidoglycan layer apparently has lipoproteins covalently attached to it. In some species, an additional layer of regularly arranged protein subunits occurs outside of the outer membrane as shown in Figure 2b (Glauert and Thornley, 1971).

From the available information, the major resistance to disruption of bacterial cells appears to be the peptidoglycan network. The tightness and perhaps strength of this peptidoglycan network depends on both the frequency with which peptide units occur on the glycan chains and also the

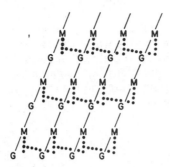

Figure 1 Schematic representation of bacterial peptidoglycan structure. Glycan strands contain alternating residues of *N*-acetylmuramic acid (M) and *N*-acetylglucosamine (G). Vertical dots from M represent amino acid residues of the tetra-peptide units and horizontal dots are peptide cross-linking bridges (reprinted from Ghuysen *et al.*, 1968, by courtesy of Elsevier/North-Holland, New York)

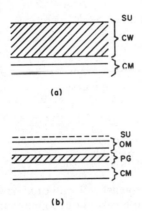

Figure 2 Schematic cross-sectional views of (a) Gram-positive and (b) Gram-negative bacterial cell walls: CM represents the cytoplasmic membrane; CW, the cell wall structure; SU, regularly arranged subunits; PG is the peptidoglycan layer and OM the outer membrane for Gram-negative bacteria; the subunits may or may not be present in either type

frequency with which peptide units are cross-linked. In addition, the network will be tighter if the peptide units are cross-linked to more than one other unit.

20.2.2 Yeast Cell Walls

Cell wall structures of yeasts apparently are much more difficult to elucidate than those of bacteria. The basic structural components have been identified as glucans, mannans and proteins, but the way in which these are combined to form the structure is not known. The overall structure is somewhat thicker than Gram-positive bacteria. Moor and Muhlethaler (1963) report cell walls of bakers' yeast to be approximately 70 nm thick and that the thickness increases with age. However, because the structural organization of yeast cell walls is not completely understood, it is possible that only a portion of the thickness contributes significantly to the rigidity and strength of the wall.

In contrast to bacterial cell walls, which are composed primarily of peptidoglycans, yeast cell walls contain large fractions of glucans, mannans and proteins. An extensive review of the chemical structures of these cell wall components is given by Phaff (1971). Although the structure of glucan varies from species to species and definitive structural analyses are difficult to make, glucans appear to be at least moderately branched molecules composed of glucose residues primarily in β-(1→3) and β-(1→6) linkages. Mannans from bakers' yeast cell walls have been studied extensively and are characterized by a backbone of mannose residues in α-(1→6) linkage having short oligosaccharide side chains composed of mannose residues in mostly α-(1→2) linkage with a small amount of α-(1→3) linkage. Phosphodiester links also occur in the mannan. The majority of proteins found in yeast cell walls are complexed with mannans and many are enzymes rather than structural components.

A schematic picture of the yeast cell wall consistent with the bulk of experimental data is presented by Lampen (1968) and shown in Figure 3. The innermost part of the cell wall is composed of glucan fibrils which constitute the rigid matrix that gives the cell its shape. Covering the fibrils is a layer of glycoprotein, and beyond that is a mannan mesh covalently linked by 1,6-phosphodiester bonds. Within the mannan mesh are the mannan–enzyme complexes which may or may not be covalently attached to the mesh.

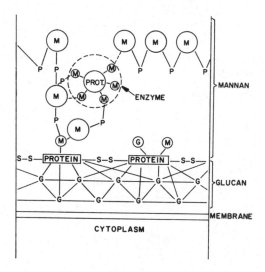

Figure 3 Schematic structure of yeast cell wall: the outer portion contains mannans (M) which may be linked by phosphodiester bridges (P); the inner portion contains cross-linked glucans (G) (reprinted from Lampen, 1968, by courtesy of the British Mycological Society)

A somewhat different picture of the yeast cell wall organization is presented by Ballou (1976). The major point of difference is in the location of the mannan–enzyme complexes which Ballou hypothesized to be located between the cytoplasmic membrane and the glucan–mannan layer. However, the primary structural component is similar in both models. As with bacterial cell walls, the resistance of yeast cell walls to disruption appears to be a function of how tightly cross-linked and how thick the structural portion is.

20.2.3 Cell Walls of Other Fungi

Because of the diversity in fungal cell wall composition and construction and the few examples for which detailed information is available, generalization from one species to another cannot be done at present. However, viewing wall structure from the perspective of how it may affect disruption, some general comments can be made. These comments will relate only to walls of hyphal growth since those have been most widely studied and also are most likely to be of interest for disruption.

In most fungi, the cell wall is composed primarily of polysaccharides with lesser amounts of proteins and lipids. Several recent reviews (Bartnicki-Garcia, 1968; Burnett, 1979; Griffin, 1981; Wessels and Sietsma, 1979) indicate there is wide variation in cell wall composition among the fungi, and wall structures have been studied in detail for only a few. As with bacteria and yeasts, it appears that shape and strength of the wall are provided by polysaccharides. The polysaccharides apparently occur in distinctive pairs, *e.g.* chitin and glucan, which can be related to taxonomic classification (Bartnicki-Garcia, 1968) as shown in Table 1. The most numerous category is that in which the wall polysaccharides are primarily chitin and glucan.

As with yeasts, walls of fungi in the chitin–glucan group appear to be constructed in layers. A detailed structure for the walls of *Neurospora crassa* has been proposed by Burnett (1979). Three major polymers occur: glucan with predominantly β-(1→3) linkages and some β-(1→6), chitin in a microfibrillar state, and a glycoprotein. In the mature wall, these polymers predominate in con-

Table 1 Cell Wall Composition and Fungal Taxonomy[a]

Key polysaccharides	Taxonomic group
Cellulose, glycogen	Acrasiomycetes
Cellulose, β-glucan	Oomycetes: Saprolegnicales, Peronosporales, Leptomitales (Rhipidiaceae)
Cellulose, chitin, β-glucan	Oomycetes: Leptomitales (Leptomitaceae)
Cellulose, chitin[b]	Hyphochytridiomycetes
Chitosan, chitin	Zygomycetes
Chitin, β-glucan	Chytridiomycetes
	Ascomycetes (except Hemiascomycetidae)
	Basidiomycetes (except Sporobolomycetaceae)
Mannan, β-glucan	Hemiascomycetidae
Mannan, chitin	Sporobolomycetaceae
	Rhodotorulaceae

[a] From Griffin (1981). [b] Presence of other polysaccharides, *e.g.* β-glucan, not tested.

centric layers as shown in Figure 4. The wall is thinner at the hyphal apex and the glycoprotein does not form a recognizable network in this region.

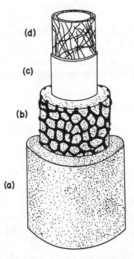

Figure 4 Schematic structure of the cell wall of *Neurospora crassa*. Compositions of the layers: (a) outer mixed α- and β-glucans; (b) the glycoprotein reticulum, glucans merging into proteinaceous material; (c) primarily proteinaceous material; (d) inner chitinous region, chitin microfibrils imbedded primarily in proteinaceous material (reprinted from Burnett, 1979, by courtesy of the British Mycological Society)

Although its composition is different, the layered pattern also is evident in walls of *Schizophyllum commune* (Wessels and Sietsma, 1979). Three distinct polymers comprise about 70% of the dry weight of the wall: (S)-glucan, (R)-glucan and chitin. The (S)-glucan contains α-(1→3) linkages and comprises the outer layer of the water-insoluble portion of the wall. The (R)-glucan is a highly branched polymer with β-(1→3) and β-(1→6) linkages and occurs complexed with chitin microfibrils in the inner layer of the wall.

The strength of fungal cell walls appears to be related to polymer networks, as in both yeast and bacterial cell walls. In addition, at least some fungal walls contain fibrous structures of either chitin or cellulose which may give added strength.

20.2.4 Cell Wall Structure and Disruption

The shape and strength of microbial cell walls appear to depend on the structural polymers within the wall and the degree to which they are cross-linked to one another and to other wall components. To disrupt a cell, the major resistance which must be overcome is the covalent bonding of this structural network. Wall structure and composition have been studied in detail for only a few organisms, but these show great diversity. Not only is the structure dependent on genetic information, but it is also dependent on the growth environment (McMurrough and Rose, 1967; Sutherland, 1975) and, in fungi, on the developmental stage (Bartnicki-Garcia, 1968). The wall structure of fungi also has been shown to be altered by mechanical effects of mixing in a fermenter (Musilkova *et al.*, 1981).

For mechanical disruption, size and shape of the cell and degree of cross-linking of the structural polymers are important factors in determining the ease of disruption. Although it should be possible to alter cell wall structures by changes in either genetic coding or environmental factors, there is not enough information available at the present time to use this as a means of increasing susceptibility to mechanical disruption. In addition, there is not enough information available to predict *a priori* the relative resistances of various organisms to mechanical disruption.

Knowledge of cell wall structure and composition is particularly important for enzymatic and chemical methods of cell lysis. Since structural networks may be covered by protective coatings, detailed knowledge of both structural and non-structural components would be useful for selecting lytic enzymes and chemical methods. However, with present limitations of detailed knowledge, the reverse procedure of using lytic enzymes and chemicals to elucidate fine structure is more likely to occur.

20.3 ANALYSIS OF DISRUPTION

Accurate analytical techniques are required for determining the extent of cell disruption if quantitative results are to be obtained for isolation of cellular components from microbial cells. The fraction of cells disrupted can be found either directly, by counting numbers or mass of whole cells, or indirectly, by measuring the release of a particular cell component. Although direct methods may be useful for standardizing other methods, indirect estimation of disruption generally is more precise, can be applied to a greater variety of situations, and is less tedious and time-consuming.

20.3.1 Direct Measurement

Direct counting of numbers of whole cells can be done by microscopy or by use of an electronic particle counter. Both methods generally are suitable for counting cells prior to disruption. However, materials released during disruption, *e.g.* DNA and other polymeric constituents, may interfere with either counting method (Hughes *et al.*, 1971). Staining can be used to differentiate disrupted cells from undamaged cells for microscopic counting, for example damaged Gram-positive bacteria frequently stain as Gram-negative (Hughes *et al.*, 1971), and using Gram's method, yeast cells appear purple if undamaged and light red if damaged (Engler, 1979).

For large numbers of samples, microscopic counting is too tedious and time-consuming. The extent of disruption of yeast cells has been determined using an electronic particle counter (Magnusson and Edebo, 1974). However, large particles of cell debris may interfere with this method. In addition, the method is not sensitive enough to apply to bacteria.

20.3.2 Indirect Measurement

Indirect measurements of the fraction of cells disrupted are based on the increase of cytoplasmic materials released into the medium used to suspend the cells for disruption, *e.g.* measurement of soluble protein or enzyme activities. For dilute cell suspensions, concentration or activity in the aqueous phase can be compared directly to a standard obtained for 100% disruption. However, care must be taken when using enzyme activities since kinetics may be different in crude cell extracts and in intact cells or for purified enzymes (Hughes *et al.*, 1971).

For accurate results when using more concentrated cell suspensions, a correction must be made for the increase in aqueous fraction which occurs as cytoplasmic contents from disrupted cells mix with the suspending medium (Hetherington *et al.*, 1971). However, the dilution procedure is not suitable for samples containing partially denatured materials which may result from more severe disruption conditions. For such cases, Engler and Robinson (1979) have developed a method based on mass balances.

20.3.2.1 Dilution technique

When denaturation is not a problem, the extent of disruption can be obtained by measuring soluble protein contents of clear supernatants from the disrupted sample (C_u) and from a diluted sample (C_d). The aqueous-phase volume fraction (F_d) of the disrupted sample can then be obtained from the following equation developed by Hetherington *et al.* (1971):

$$F_d = (V_d/V_s)[C_d/(C_u - C_d)] \qquad (1)$$

where V_d is the volume of diluent added to a volume V_s of disrupted sample. Soluble protein can be determined by either the Folin–Lowry method (Lowry *et al.*, 1951) or the biuret method (Gornall *et al.*, 1949; Layne, 1957). The aqueous-phase fraction can then be used to obtain protein released per unit cell mass in the disrupted suspension (R_p) from

$$R_p = F(C_u/x) \qquad (2)$$

where x is the mass concentration of cells on a dry weight basis. To calculate the fraction of cells disrupted, the value of R_p corresponding to 100% disruption must be determined.

Hetherington *et al.* (1971) found that aqueous-phase fractions were independent of the amount of sample dilution. This indicates that protein solubility was not affected by sample dilution in their study. However, Engler and Robinson (1979) found that protein solubility was affected by dilution for samples disrupted under more severe conditions, therefore effects of dilution on protein solubility should be checked before proceeding with the dilution method.

20.3.2.2 Mass balance technique

An alternative method for determining the extent of disruption is based on mass balances as described by Engler and Robinson (1979). The aqueous-phase volume fraction for a disrupted sample is related to the aqueous-phase fraction for the sample prior to disruption (F_o) and the internal moisture content (mass fraction, M) of whole cells through the equation

$$F = F_o + (1-F_o)MR(\varrho_c/\varrho_{aq}) \qquad (3)$$

where R is the fraction of cells disrupted, ϱ_c is the density of the cells and ϱ_{aq} is the density of the aqueous suspending medium. In deriving equation (3), it is assumed that only the internal moisture content of cells contributes to increasing the volume fraction of the aqueous phase.

An independent equation can be written to relate F and R using Kjeldahl nitrogen contents of the disrupted sample aqueous phase (C_N) and undisrupted cells (C_{NO}):

$$R = FC_N/C_{NO}\, x \qquad (4)$$

Equations (3) and (4) may be combined to eliminate F giving

$$R = F_o\, C_N/[C_{NO}\, x - (1-F_o)\, C_N\, M\, (\varrho_c/\varrho_{aq})] \qquad (5)$$

Accurate measurement of cell concentration (x) by direct methods is difficult when x is large, however, x can be calculated from physical properties and the mass fraction of solids (W) in the sample:

$$x = \varrho_c\, \varrho_{aq}\, W(1-M)/[W(\varrho_{aq}-\varrho_c) + \varrho_c(1-M)] \qquad (6)$$

Because of several assumptions made in using mass balances to calculate the fraction of cells disrupted, the method probably is somewhat less accurate than the dilution technique. However, it does allow close estimation of the extent of disruption when the dilution technique cannot be used because of dilution effects on protein solubility.

20.3.2.3 Conductivity measurements

For rapid estimation of the extent of disruption in a large number of similar samples, Luther (1980) has developed a procedure based on changes in conductivity caused by the release of cytoplasmic material into the aqueous phase. He reported a linear increase in conductivity with fraction of cells disrupted. However, the procedure requires standardization by another method because conductivity readings were found to depend on the type of organism and the conditions under which it was handled, cell concentration, temperature, and electrolyte content of the suspending medium. Thus, the method appears to be useful only if all conditions, except extent of disruption, are held constant.

20.4 LABORATORY-SCALE DISRUPTION TECHNIQUES

Disruption of microorganisms on a laboratory scale has been practiced for many years, and a variety of methods have been developed for efficient disruption while minimizing damage to cellular materials. Laboratory disruption techniques have found widespread use in the study of cell structure and function and for preparing purified enzymes and other cell components. Reviews of laboratory disruption methods have been given by Edebo (1969) and Hughes *et al.* (1971).

20.4.1 Mechanical Methods

Mechanical methods for disruption of microbial cells frequently are chosen because they allow processing of larger quantities of cells very quickly. With these methods, cells are subjected to high stresses produced by high pressure, rapid agitation with glass beads, or ultrasonics. In most cases, cooling is provided to remove excessive heat generated by dissipation of the mechanical energy.

One of the most widely used laboratory disruption methods is sonication. Ultrasonic disintegrators generally operate at frequencies of 15 to 25 kHz, with cell disruption resulting from cavitation effects. Doulah (1977) has suggested that disruption is caused by shear stresses developed by viscous dissipative eddies arising from shock waves produced by imploding cavitation bubbles. It has been further suggested that ultrasonic disruption follows the statistical theory of reliability, indicating that wear-out damage to cell walls is caused by applied stresses (Doulah, 1978). During disruption, the cell suspension is cooled by packing the sonication cup in ice or circulating coolant through a jacketed cup. While ultrasonic disruption can be adapted for continuous operation by placing the probe tip in a flowing stream of cell suspension (Neppiras and Hughes, 1964), it does not appear suitable for large-scale processing because of difficulties in providing adequate cooling at high power input (Lilly and Dunnill, 1969). A disadvantage of sonication is that it appears to cause significant degradation of enzymes (Augenstein *et al.*, 1974; Coakley *et al.*, 1974).

Use of high pressure to force suspended cells through a needle valve, originally described by Milner *et al.* (1950), is another frequently-used method for disruption. Pressures up to 210 MPa are applied to a sample contained in a steel cylinder by means of a tight-fitting piston. The sample is bled through a needle valve while keeping the pressure constant. To prevent heat denaturation from occurring in the needle valve, Ribi *et al.* (1959) devised a system to inject compressed CO_2 into the valve for cooling. Presses with either manual or automatic pressure control are available. A highly refined model, using precooled nitrogen to chill the needle valve and incorporating an automatic refilling cycle, was marketed as the Sorvall-Ribi fractionator (Hughes *et al.*, 1971) but is no longer available. Although scale-up of this method has not been studied directly, the principle of operation is similar to that of high pressure homogenizers and probably the mechanisms causing disruption are the same.

Another high pressure method has been developed in which a frozen suspension of cells is forced through a small opening (Hughes *et al.*, 1971). Pressures up to 550 MPa are required for operation. Abrasion by ice crystals (Hughes *et al.*, 1971), changes in ice crystal states (Edebo, 1960) and plastic flow through the discharge gap (Scully and Wimpenny, 1974) have been suggested as mechanisms of disruption. A combination of these phenomena is likely, as results from a study by Magnusson and Edebo (1976) indicate that internal friction (shear), sustained by a high concentration of cells, low temperatures and a high mean pressure, is an effective promoter of disruption. In addition, ice crystals may be present at passage through the orifice thereby increasing disruption. Advantages of freeze pressing include its effectiveness for disruption of a wide variety of cell materials, retention of biological activity and less comminution of cell fragments (Edebo, 1969; Hughes *et al.*, 1971). This method is not suitable for preparation of materials sensitive to freezing and thawing (Hughes *et al.*, 1971).

Rapid agitation of a cell suspension with glass beads is used in several disruption devices. Some disruption can be obtained by processing the cell suspension and glass beads in a blender (Edebo, 1969; Hughes *et al.*, 1971). Better results are obtained in devices made especially for cell disruption such as the Mickle tissue disintegrator (Mickle, 1948) and Braun homogenizer. A major disadvantage of these devices is the rapid increase in sample temperature during disruption, but this can be partly offset by cooling the container with CO_2. Somewhat larger samples can be processed in laboratory colloid mills (Garver and Epstein, 1959), and a laboratory-scale high speed ball mill that can be operated either for batch or continuous flow processing is now available (Limon-Lason *et al.*, 1979). The mill is provided with a cooling jacket to provide temperature control in the grinding chamber. Although the relative influence of mechanisms causing disruption may vary among the different devices, the mechanisms discussed below for high speed ball mills (Section 20.5.1) generally are expected to apply in the laboratory devices as well.

20.4.2 Non-mechanical Methods

A wide variety of non-mechanical techniques are available for disruption of microbial cells including enzyme lysis, chemical lysis, osmotic shock, heat treatment, freezing and thawing, cell wall deficient mutation and phage lysis. Many of these methods have been reviewed by Edebo (1969) and Hughes *et al.* (1971). Some are of limited usefulness, particularly when biologically active materials are being prepared.

One of the most promising and widely studied methods of cell disruption is enzyme lysis. Numerous enzymes have been identified which attack specific bonds in the cell wall structure of microorganisms. Application of enzyme lysis requires selection of an appropriate enzyme or enzyme system and determination of specific reaction conditions for efficient lysis. Some organisms may be susceptible to enzyme lysis only in a particular stage of growth (Hughes *et al.*, 1971) or when grown under specific conditions (Engler, 1979). Organisms seemingly resistant to enzyme lysis may be made more susceptible by treatments such as irradiation (Watanabe *et al.*, 1981), addition of salts in high concentration (Kitamura and Yamamoto, 1981) or use of biological factors to promote activity (Takahara *et al.*, 1976). Combinations of lytic enzymes have been shown to be more effective in some cases than the individual enzymes alone (Knorr *et al.*, 1979). Major advantages of enzyme lysis are the very specific action of the enzyme and the mild conditions under which lysis occurs. At present, costs of lytic enzymes are high and many are not available commercially.

Autolysis is another method of enzyme lysis in which the lytic enzymes are produced by the organism itself. Most microorganisms have the capability of producing enzymes which hydrolyze polymeric structures in the cell wall to allow normal growth processes to occur. However, changes in an organism's environment may trigger overproduction of these enzymes or activate production of other autolytic enzymes. Parameters which affect the autolytic process include temperature, time, pH, molarity of buffer, and the metabolic state of the cells (Hughes *et al.*, 1971). Generally, long incubation times of 2 to 20 h are required for autolysis and yields of disrupted cells may be low (Tannenbaum, 1968). In addition, considerable protein denaturation may occur (Dunnill and Lilly, 1975; Edebo, 1969).

Although the mode of action is different, inhibition of cell wall synthesis can give results similar to enzyme lysis. Antibiotics such as penicillin, which act by blocking synthesis of new cell wall material (Ghuysen and Shockman, 1973), may be added late in the growth phase to initiate lysis. Conditions for lysis will be favorable only if biosynthesis and reproduction continue to occur after addition of the inhibitor (Hughes *et al.*, 1971) so that deficient cell walls are present at the time of division.

A less-explored area of enzymatic lysis is phage infection. Upon infection, an enzyme carried by the phage hydrolyzes the host cell wall to allow phage nucleic acid to enter the cell (Hughes *et al.*, 1971). If infection is heavy, lysis may occur before the nucleic acid can penetrate (lysis from without). Otherwise, enzymes produced in the phage reproduction cycle will lyse the host cell to free new phage material (lysis from within). Cellular constituents may be significantly altered with this method, particularly if lysis occurs from within.

Chemical treatment may be used either to lyse cells or to extract cellular components. Treatment with alkali may solubilize most components except for the cell wall (Edebo, 1969). However, Lee *et al.* (1979) reported that alkali did not solubilize protein from dried *Candida lipolytica* but that 6 N HCl was effective in hydrolyzing the cells. The acid treatment apparently hydrolyzed proteins to free amino acids as well. Detergents often cause either cell lysis or leakage of some components out of the cell although biological activity may be destroyed (Edebo, 1969). Other chemical treatments such as with butanol or urea have been used, but these create problems of separation and/or recovery of the chemical and toxicity (Tannenbaum, 1968).

One of the gentlest methods of disruption is osmotic shock. Cells are allowed to equilibrate briefly in a medium of high osmotic pressure such as 1 M glycerol or 1 M sucrose. When the medium is suddenly diluted, water enters the cells rapidly increasing the hydrostatic pressure which causes disruption. This method is good only for very fragile organisms or when the cell wall is first weakened by enzyme treatment or by inhibition of cell wall synthesis (Hughes *et al.*, 1971). Osmotic shock also has been reported as a means of enhancing autolysis (Leduc and van Heijenoort, 1980).

Physical treatments, such as freezing and thawing in repeated cycles, and heating and drying, have been used successfully for disruption or release of several cell constituents (Edebo, 1969; Hughes *et al.*, 1971). Freezing and thawing frequently gives only low yields even after a large number of cycles. In addition, some components are sensitive to freezing and thawing and may denature. Heating and drying usually require conditions severe enough to cause denaturation of proteins and other components. Hedenskog and Mogren (1973) found that drum drying gave higher protein accessibility than did spray drying or lyophilization when measured by PER, but extraction of protein from dried whole cells was low regardless of drying method.

Selection of an appropriate laboratory disruption method requires consideration of several factors such as quantity of cells, sensitivity of desired products to disruption conditions (temperature, chemicals, proteolytic enzymes, *etc.*), extent of disruption needed, and speed with which disruption must be carried out. In most cases, it is desirable to use the gentlest method possible. However, when products being studied have potential large-scale applications, disruption techniques that are suitable for scale-up should be chosen.

20.5 LARGE-SCALE DISRUPTION TECHNIQUES

Availability of large quantities of relatively pure intracellular components from microorganisms would create a wide range of new products and processes for industrial, therapeutic, diagnostic and food applications. A key factor in producing these microbial components economically is an efficient, large-scale disruption process. With currently available methods, disruption is a high-cost, energy-intensive unit operation (Engler, 1979; Mogren *et al.*, 1974), therefore disruption costs may be a significant part of total production costs for these new products.

Although there have been some attempts at scaling up laboratory disruption methods for use on a large scale, greater success has been achieved in adapting commercially available high speed ball mills and high pressure homogenizers for disruption. An advantage to this approach is the relatively low cost of commercial equipment having already established markets.

20.5.1 High Speed Ball Mills

In high speed ball mills, suspensions of cells and glass beads are agitated by discs rotating at high speed. Cell disruption apparently is caused by collisions between shear force layers and also by rolling of the grinding elements (Rehacek and Schaefer, 1977). Cell disruption in these devices can be affected by a number of operating parameters including bead diameter and loading, cell concentration in the feed to the disintegrator, flow rate of feed, agitator speed and configuration, and temperature. In addition, the grinding chamber can be oriented either horizontally or vertically. These parameters affect not only the degree of disruption obtained but also power requirements.

Effects of bead size reported in the literature have not been consistent. Currie *et al.* (1972) reported that for a range of bead diameters studied (0.5 to 2.8 mm diameter) the smaller beads were more effective for disruption of bakers' yeast. Marffy and Kula (1974) found beads having diameters of 0.25 to 0.50 mm were more effective for disruption of brewers' yeast than either very small beads (0.1 to 0.25 mm) or larger beads (0.75 to 1.00 mm). Schutte *et al.* (1983) indicate that location of an enzyme within a cell influences the optimum bead size for release of enzyme activity. In addition, they indicate that for disruption of bacteria, better results are obtained with smaller beads than those found optimum for yeast. Rehacek and Schaefer (1977) suggest that bead size must be selected in relation to cell size and cell concentration. Another consideration in selecting bead size is retention of beads in the grinding chamber during continuous flow operations. Separator designs have been developed to prevent clogging of the outlet while retaining small particles (Marffy and Kula, 1974; Rehacek and Schaefer, 1977; Schutte *et al.*, 1983).

The volume fraction of the grinding chamber filled with glass beads affects both disruption and power requirements. Currie *et al.* (1972) reported increased disruption with increased bead content in the grinding chamber. Rehacek and Schaefer (1977) observed similar behavior but indicated that filling to greater than 88% was not suitable for disruption. On the other hand, Schutte *et al.* (1983) found 80% loading of beads to be optimum for disruption of bakers' yeast. At a given agitator speed, an increase in bead loading causes a greater increase in temperature during disruption (Rehacek and Schaefer, 1977; Schutte *et al.*, 1983).

Although increasing agitator speed generally increases disruption (Currie *et al.*, 1972; Marffy and Kula, 1974; Mogren *et al.*, 1974), several studies have indicated a leveling off of disruption at higher speeds (Rehacek *et al.*, 1969; Schutte *et al.*, 1983) or even a decrease in disruption at very high speeds (Limon-Lason *et al.*, 1979). Limon-Lason *et al.* (1979) have suggested that the actual rate constant for disruption continues to increase at higher speeds but a simultaneous increase in dispersion in the grinding chamber negates the effect.

Agitator design appears to have a significant effect on efficiency of operation. Dunnill and Lilly (1975) used two agitator designs with the principal difference being the fraction of chamber cross-section occupied by the agitator discs. They found a more open design was less efficient at low flow rates, apparently due to back-mixing. At higher speeds, the open design gave greater agitation and had increased power consumption. Rehacek *et al.* (1969) compared smooth and grooved agitator discs and found that the grooved design reduced slippage in more concentrated cell suspensions. Commercial agitator mills having circular discs mounted eccentrically to form a helical array were studied by Currie *et al.* (1972) and Schutte *et al.* (1983). The helical arrangement causes movement of beads opposite to the direction of flow of suspension thereby preventing compaction of the beads (Schutte *et al.*, 1983). While most ball mills have agitator discs mounted perpendicular to the drive shaft, Rehacek and Schaefer (1977) investigated discs mounted obliquely. They found that inclining the discs further from perpendicular gave better results, particularly at lower agitator speeds, but at a cost of higher power consumption and a greater increase in product temperature.

In addition to agitator design, orientation of the grinding chamber may affect disruption. Rehacek *et al.* (1969) compared both vertical and horizontal orientations of single-disc prototype models and found the horizontal arrangement gave better results. Most commercial mills tested for disruption have had horizontal arrangements, however, a vertical mill tested by Currie *et al.* (1972) was reported to be very effective. The vertical mill had several disadvantages which are related to other design considerations: it was unsealed, the outlet sieve plugged easily and it was not constructed of stainless steel (Dunnill and Lilly, 1975).

Operating temperature was reported by Currie *et al.* (1972) to have little effect on disruption within the range 5 to 40 °C. Of greater importance is the increase in temperature which occurs in the grinding chamber. Although jacketing the grinding chamber allows adequate temperature control in smaller machines, heat removal must be an important consideration in scaling up to large grinding chambers.

Effects of cell concentration in the feed are not consistent. Mogren *et al.* (1974) found no effect of cell concentration within the range 40 to 200 kg m^{-3} (dry wt). Using a single-disc prototype device, Rehacek *et al.* (1969) reported slippage occurred with concentrations of 175 kg m^{-3}. Marffy and Kula (1974) reported an increase in disruption with increasing cell concentration, whereas Schutte *et al.* (1983) reported an optimum concentration for low speeds but little effect of concentration at higher speeds. Limon-Lason *et al.* (1979) found disruption rate to be independent of concentration at higher concentrations for one type of impeller. However, for a different impeller, disruption rate was reported to decrease continuously with increasing concentration. They attributed their observations to variations in rheological behavior of the suspension.

Disruption in ball mills generally is a first order process with respect to time (t):

$$\ln [1/(1-R)] = kt \tag{7}$$

where k is the first order rate constant. For batch processing with cell suspension recirculated through the grinding chamber, the rate constant decreases with an increase in flow rate (Currie *et al.*, 1972; Marffy and Kula, 1974). Apparently higher flow rates disturb patterns of bead movement thus reducing disruption efficiency. On the other hand, Mogren *et al.* (1974) obtained a first order relation between disruption and the inverse of flow rate in continuous flow experiments, which indicated the rate constant was independent of flow rate. An analysis of the reactor characteristics of high speed ball mills has been reported by Limon-Lason *et al.* (1979). They found disruption in a 5 l mill was adequately described by a first order reaction and a CSTR-in-series model with the number of tanks corresponding to the number of impellers. However, the model had to be extended to incorporate backflow to obtain good agreement with experimental data from a smaller (0.6 l) mill. They concluded that an increase in dispersion or backflow decreases disruption efficiency, particularly at low throughput. Increasing agitator speed leads to increased disruption rates, but it also increases the backflow which may partially or fully negate effects of the increased rate constant.

Total power consumption of ball mills increases with increasing agitator speed, bead loading and flow rate through the grinding chamber. Agitator design also has a significant effect on power requirements (Dunnill and Lilly, 1975; Rehacek and Schaefer, 1977). However, the more important consideration is the power requirement per unit mass disrupted. This is influenced by a number of factors and does not necessarily follow the same trends as total power or rate of disruption (Mogren *et al.*, 1974). Efficiency also may not follow trends expected on the basis of dispersion effects. Limon-Lason *et al.* (1979) reported higher efficiency for a mill having greater dispersion than for one with little or no dispersion, with the most likely reason for the higher efficiency being a greater number of agitator blades per unit volume.

In addition to the operating parameters discussed above, disruption is dependent on the type of organism and its conditions of growth. Rehacek *et al.* (1969) reported that *Escherichia coli* was more difficult to disrupt than *Aspergillus niger* or a *Basidiomyces* species. *Candida utilis* was reported to be more difficult to disrupt than bakers' yeast (Mogren *et al.*, 1974; Rehacek and Schaefer, 1977) or brewers' yeast (Rehacek and Schaefer, 1977). Schutte *et al.* (1983) reported several different bacteria to be more difficult to disrupt than yeast, most likely because of the smaller size of bacteria.

Because many microbial products are easily denatured, operating conditions for ball mill disintegrators must be carefully studied to optimize disruption in a given application. Marffy and Kula (1974) found that enzymes were detectably inactivated by the disruption process, even when operating temperatures were kept below 5 °C. Apparently this denaturation is caused by shear or other mechanical effects. Such denaturation could be expected to increase as conditions are adjusted to give higher amounts of disruption.

20.5.2 High Pressure Homogenizers

High pressure homogenization is an alternative method for large-scale cell disruption. In high pressure homogenizers, disruption is accomplished by passing cell suspension under high pressure through an adjustable, restricted orifice discharge valve. There are fewer operating parameters to consider than with high speed ball mills. The major parameters are operating pressure, temperature and number of passes through the valve. In addition, valve design may play an important role.

20.5.2.1 Effects of operating parameters

Disruption in high pressure homogenizers generally has been found to be a first order process with respect to the number of passes through the valve, with the rate constant being a function of temperature and pressure (Brookman, 1974; Dunnill and Lilly, 1975; Gray *et al.*, 1972; Hetherington *et al.*, 1971). However, Whitworth (1974b) found that the maximum amount of protein released during disruption increased with increasing pressure which is at variance with the simple first order relationship found by others. The general first order equation is given below:

$$\ln\left[1/(1-R)\right] = kN_p P^a \tag{8}$$

where N_p is the number of passes through the homogenizer valve and P is the operating pressure. In equation (8), k has been found to depend on temperature (Hetherington *et al.*, 1971). The value of the power (a) to which pressure is raised varies with different organisms (Gray *et al.*, 1972) and may depend on the range of pressures covered (Dunnill and Lilly, 1975; Engler and Robinson, 1981a). Typical data for disruption of bakers' yeast, obtained by Hetherington *et al.* (1971), are plotted according to equation (8) in Figure 5.

Disruption has been found to be independent of cell concentration in the feed over a wide range. Brookman (1974) reported no significant effect of concentration from 28 to 224 kg m^{-3} (dry weight). However, above approximately 168 kg m^{-3}, Hetherington *et al.* (1971) found that disruption was reduced.

Release of enzyme activities during disruption of bakers' yeast has been studied by Follows *et al.* (1971). They reported that release of enzyme activity relative to soluble protein was not affec-

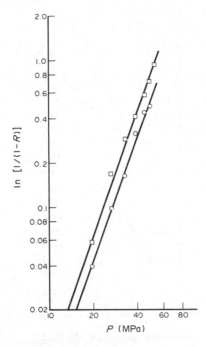

Figure 5 Relationship between amount of disruption and operating pressure for disruption of bakers' yeast in a high pressure homogenizer; data shown for temperatures of 30 °C (□) and 5 °C (○) with cell suspension recycled through the homogenizer (redrawn from Hetherington *et al.*, 1971, by courtesy of the Institution of Chemical Engineers)

ted by disruption pressure. However, rates of release of different enzymes were not the same, with enzymes located in the cell wall released at a faster rate than soluble protein, those in the cytoplasm at about the same rate, and those in subcellular particles at a slower rate. For most enzymes, no loss in activity was found after prolonged recycling through the homogenizer at temperatures below 30 °C.

Typical pressures used for disruption in commercial homogenizers cover a range up to 55 MPa. As shown in Table 2 for pressures of approximately 55 MPa, amounts of disruption in one pass through the valve vary from 12 to 67% depending on the type of organism and other factors. Gray *et al.* (1972) reported that *Escherichia coli* grown on a simple synthetic medium were more easily disrupted than when grown on a complex medium. They also found the disruption rate constant to be dependent on pressure to the power 2.2 in contrast to the power of 2.9 found by Hetherington *et al.* (1971) for bakers' yeast. Engler and Robinson (1981b) also reported differences in disruption behavior for different organisms and for a given type of organism grown under different conditions as shown in Figure 6.

Table 2 Single-pass Disruption of Various Organisms Using High Pressure Homogenizers

Organism	Pressure (MPa)	Disruption (%)	Reference
Bakers' yeast	53	62	Hetherington *et al.*, 1971
Bakers' yeast	55	12	Brookman, 1974
Brewers' yeast	55	61	Whitworth, 1974b
Candida lipolytica	55	43	Whitworth, 1974a
Escherichia coli	53	60[a]	Gray *et al.*, 1972
Escherichia coli	53	67	Higgins *et al.*, 1978

[a] This includes 17% disruption which occurred during centrifugation prior to processing through the homogenizer.

To achieve 90% disruption, at least two passes through the homogenizing valve would be required based on the best results reported in Table 2. Because of the strong dependence of disruption on operating pressure, it may be more efficient to increase the operating pressure and use fewer passes. Adaptations of homogenizing valves for operation at higher pressures have been investigated. Brookman (1974) fitted a homogenizer valve to an extreme pressure pump and

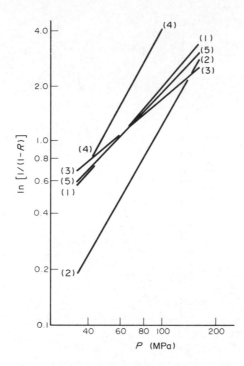

Figure 6 Effects of cell type and growth conditions on disruption by impingement: (1) *C. utilis*, batch culture, $\mu = 0.5$ h^{-1}; (2) *C. utilis*, continuous culture, $D = 0.1$ h^{-1}; (3) *S. cerevisiae*, aerobic continuous culture, $D = 0.1$ h^{-1}; (4) spent brewery yeast; (5) *B. subtilis*, continuous culture, $D = 0.2$ h^{-1}

reported 100% disruption at a pressure of approximately 175 MPa. In testing an experimental homogenizer capable of operating at up to 118 MPa, Dunnill and Lilly (1975) reported that disruption increased somewhat more slowly with pressure above 70 MPa, the normal maximum pressure for commercial homogenizers. Although Dunnill and Lilly (1975) attributed the reduced effect to non-optimal design of the valve for the higher pressures, Engler and Robinson (1981a) have suggested the pressure dependency may not be constant over an extremely wide range of operating pressures.

Valve design also has been shown to affect disruption in high pressure homogenizers. Hetherington *et al.* (1971) found that a knife-edge valve gave a higher rate of disruption than a flat valve, and Gray *et al.* (1972) indicate that the disruption rate is less if the knife edge is worn. To provide a firm basis for further improvement in valve design, greater knowledge of disruption mechanisms and how to enhance them is required.

20.5.2.2 Disruption mechanism

Although only a few operating parameters affect disruption in high pressure homogenizers, with pressure being the most significant, the mechanisms causing disruption are difficult to elucidate. Several mechanisms have been suggested on the basis of homogenizer data, but the hydrodynamic phenomena occurring in the valve are too complex to allow definitive conclusions from that data alone. A more fruitful approach has been to isolate the various phenomena for individual study. The flow pattern in a homogenizing valve may create high shear stress, normal stress, turbulence and stress caused by impact of a jet on a stationary surface.

From his study of a homogenizer valve adapted to fit an extreme pressure pump, Brookman (1974) concluded that major factors causing cell disruption are the magnitude of the pressure drop and the rate at which it occurs. For further study of the disruption mechanism, a needle valve or hypodermic needle tubing were fitted to the pump discharge (Brookman, 1975). Results for the hypodermic needle tubing using suspensions of bakers' yeast showed that for a constant pressure peak of 124 MPa, the amount of disruption increased as tube length decreased. Those results were interpreted as support for the mechanism relating disruption to the magnitude and rate of pressure drop. However, it is equally valid to interpret the results on the basis of a shear stress mechanism (Engler, 1979).

A hydrodynamic mechanism for cell disruption in high pressure homogenizers was proposed by Doulah *et al.* (1975). The authors suggested that turbulent eddies, of dimensions smaller than a cell, cause the cell liquid to oscillate with enough kinetic energy to disrupt the cell wall. Because of the relationship between eddy and cell sizes, cells having a diameter smaller than some minimum value are not disrupted. Although the theory is in reasonably good agreement with many experimental results, it does not yield a first order response to multiple passes without the additional consideration of a time factor. In addition, the theory does not account for experimental results showing nearly complete disruption after several passes unless a weakening of the walls of smaller cells occurs on each pass.

The pulsating nature of extreme pressure pumps makes it difficult to analyze data even for simple hydrodynamic phenomena, therefore Engler and Robinson (1981a) designed a disruption system based on a pressure intensifier capable of processing up to 150 ml of cell suspension at constant pressure within a range from 50–300 MPa. By selecting suitable discharge nozzles, effects of normal stress, shear stress and impingement could be studied independently. Results for disruption of *C. utilis* by normal stress and impingement, shown in Figure 7, indicate normal stress is only about 20% as effective for disruption as impingement. Results for shear stress were less conclusive but again indicated impingement was more effective. Impingement data for *C. utilis* and *S. cerevisiae* show similar trends to data for disruption of bakers' yeast obtained by Dunnill and Lilly (1975) using an experimental higher pressure homogenizer as seen in Figure 8. Those results clearly indicate that impingement of a high velocity jet of suspended cells is a major cause of disruption in high pressure homogenizers. In addition, Engler and Robinson (1981a) report there is some evidence that turbulence near the point of impact may enhance disruption.

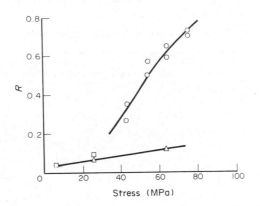

Figure 7   Disruption of *C. utilis* by (O) impingement and (Δ,□) normal stress; cells grown in (Δ) cyclic batch culture and (O,□) continuous culture (reprinted from Engler and Robinson 1981a by courtesy of Wiley, New York)

20.5.3   Energy Requirements

Mechanical disruption of microbial cells is an energy intensive process regardless of the method used. Therefore, a major consideration in studying mechanical disruption processes should be the efficiency of energy utilization. Although power consumption data have been reported in several studies, results are not always given on the same basis. For comparison of different machines or results from different laboratories, a common basis for efficiency such as dry mass of cells disrupted per unit energy input is needed.

Energy efficiency data for ball mills are summarized in Table 3 on a basis of disrupted dry cell mass per unit energy input. Unfortunately, cell concentrations and efficiencies have been expressed on either a wet weight basis (Limon-Lason *et al.*, 1979) or a dry weight basis (Mogren *et al.*, 1974; Rehacek and Schaefer, 1977) but not both. Data of Limon-Lason *et al.* (1979) were converted to a dry mass basis for comparison in Table 3 by assuming packed wet yeast contained 70% water. These data show efficiency varies with operating parameters such as cell concentration, agitator design, agitation per unit volume and type of organism as discussed earlier.

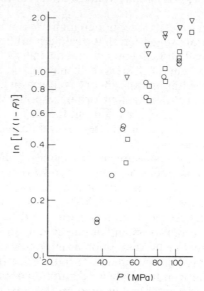

Figure 8 Comparison of impingement and high pressure homogenizer disruption data: (○) homogenizer, bakers' yeast (Dunnill and Lilly, 1975); (□) impingement, *C. utilis*, continuous culture, $D = 0.1 \text{ h}^{-1}$; (▽) impingement, *S. cerevisiae*, aerobic continuous culture, $D = 0.1 \text{ h}^{-1}$

Although data were not given, Mogren *et al.* (1974) indicated that both *C. utilis* and *Scenedesmus obliquus* had lower disruption efficiencies than *S. cerevisiae*.

Table 3 Disruption Efficiencies for Ball Mills

Organism	Disruption (%)	Concentration (% mass/volume)	Efficiency (mg J⁻¹)	Notes
S. cerevisiae	80	6	0.51	a
	80	11	1.11	a
	80	16	1.11	a
	85	10–20	0.93	a
	95	17	1.26–1.39	b
	65	13.5	2.32	c,f
	42	13.5	1.25	d,f
	40	13.5	5.40	e,f
S. carlsbergensis	95	17	1.26–1.39	b
C. utilis	85–90	17	0.84–1.07	b

ᵃ Mogren *et al.* (1974). ᵇ Rehacek and Schaefer (1977). ᶜ Limon-Lason *et al.* (1979). Polyurethane impellers in 5 l mill. ᵈ Limon-Lason *et al.* (1979). Stainless steel impellers in 5 l mill. ᵉ Limon-Lason *et al.* (1979). 0.6 l mill. ᶠ Concentration reported on wet weight/volume basis, converted to dry mass basis assuming fresh packed yeast contains 70% water. Efficiencies reported in reference are too low by factor of 10^3.

Power consumption data for disruption of bakers' yeast and *E. coli* in a high pressure homogenizer have been reported by Dunnill and Lilly (1975) on a basis of percent disruption per unit energy input. Since disruption of bakers' yeast was reported to be independent of concentration over the range 8.5 to 17% dry mass per volume, the highest concentration should give the highest efficiency. Based on a rated throughput for the homogenizer of 280 l h⁻¹ (Hetherington *et al.*, 1971), energy efficiency at 90% disruption is 1.65 mg J⁻¹. Engler (1979) reported efficiencies for disruption of *S. cerevisiae* and *C. utilis* using a high pressure impingement device. Those data are shown in Table 4 along with the data of Dunnill and Lilly (1975) for comparison. The data in Table 4 again indicate *C. utilis* is more difficult to disrupt than *S. cerevisiae*. They also show that employing several passes at lower pressure is less efficient than a single pass at a high pressure.

Although operation under more severe conditions to obtain high amounts of disruption in a single pass through either ball mills or high pressure devices appears to be more energy efficient, those conditions may cause denaturation or degradation of desired products. Engler (1979) reported that essentially all of the energy input to high pressure devices is converted to heat almost instantaneously. Thus at higher pressures, cell material is subjected to higher tempera-

Table 4 Disruption Efficiencies for High Pressure Devices

Organism	Disruption (%)	Concentration (% mass/volume)	Efficiency (mg J^{-1})	Notes
S. cerevisiae	90	17	1.65	a
	94	20	1.25	b
C. utilis	90	20	1.26	c
	87	20	0.83	d

[a] Dunnill and Lilly (1975). High pressure homogenizer. [b] Engler (1979). Impingement, 3 passes at 50 MPa. [c] Engler (1979). Impingement, 1 pass at 143 MPa. [d] Engler (1979), Impingement, 3 passes at 70 Mpa.

tures, at least for a short period of time. Similarly, most of the energy input to ball mills will be dissipated as heat; however, the rate of energy dissipation may not be as high.

20.5.4 Non-mechanical Methods

Because of the high capital and operating costs for mechanical disruption processes, interest has been shown in various non-mechanical methods. These include development of mutant organisms with weakened cell walls, mutants programmed to lyse under non-growth conditions, enzymatic lysis, chemical treatment and autolysis. Although most of these methods are at present suitable only for laboratory use, chemical treatment and autolysis are used to some degree on an industrial scale. However, those methods generally are accompanied by significant denaturation or degradation of protein which may make them impractical for enzyme production or food uses where functionality of the proteins is important (Dunnill and Lilly, 1975).

Yeast autolyzates, plasmolyzates and hydrolyzates are the most common products obtained from non-mechanical cell rupture. These processes have been reviewed by Peppler (1970) and Reed and Peppler (1973). Yeast autolysis requires 12 to 24 h reaction time at a temperature of 45 to 50 °C. Small amounts of sodium chloride, ethyl acetate or chloroform may be added to accelerate the reaction. Undigested cell walls may be removed to form a clear extract.

Plasmolysis involves contact of viable cells with high concentrations of salt, sugar or certain acetate esters (Peppler, 1970). This extracts cellular materials without involvement of catabolic enzymes. Resulting products are limited in application by the presence of the plasmolyzing substance.

Acid hydrolysis is the fastest of these non-mechanical methods requiring 6 to 12 h for the reaction. Concentrated yeast suspensions are acidified and refluxed in a wiped-film evaporator. After hydrolysis, the slurry is neutralized, filtered, decolorized and concentrated. While hydrolysis gives higher yields, there is some loss of vitamins, protein and flavor (Peppler, 1970).

Numerous lytic enzymes have been identified that are able to hydrolyze microbial cell walls, but most are used only for small-scale laboratory studies. Recent studies indicate that large-scale production of yeast-lytic enzymes may be possible. Asenjo *et al.* (1981) produced yeast-lytic enzyme from *Cytophaga* species in a 900 l fermentation with results that compared favorably to those of smaller fermentations. Isolation and immobilization of the enzymes were further reported by Asenjo and Dunnill (1981). Enzyme bound to a linear soluble carbohydrate polymer (Dextran 300) retained lytic activity, but when bound to a soluble globular carbohydrate polymer (Ficoll 400), activity was lost. They indicated that effective use of free enzyme represented the best technology currently available. A process for production of yeast-lytic enzyme and disruption of whole yeast with free enzyme was described by Asenjo (1981).

Another approach to large-scale lysis of organisms is to develop mutants programmed to lyse under specific conditions. Boudrant *et al.* (1979) studied several mutants of *S. cerevisiae* which would start to lyse when the temperature was increased from 27 to 37 °C. Cell density and physiological state were found to have significant effects on ability to lyse at the higher temperature.

20.6 SUMMARY

A wide variety of mechanical, physical and chemical techniques have been developed for disruption of microorganisms. Although most are suitable only for laboratory use, several have been developed to the pilot-plant stage for large-scale disruption. In selecting an appropriate disruption technique consideration must be given to the amount of cells to be processed, sensitivity of desired products to disruption conditions and the required yield of disrupted cells. Because of the many factors that affect the ease with which microbial cells can be disrupted and also the wide range of operating parameters for most disruption processes, the most effective disruption method in a given situation must be determined empirically. An understanding of cell wall structure and effects of various environmental factors on that structure can be useful not only in selecting an appropriate disruption technique, but also in determining conditions for production of organisms more susceptible to disruption.

Disruption processes that have been tested on a pilot-plant scale use mechanical methods such as high speed ball milling or high pressure homogenization. Energy requirements, as well as capital investment, for these processes are high. In addition, since most of the energy input is converted to heat, effective cooling is required to prevent heat denaturation or inactivation of cellular components.

Although enzymatic lysis methods have not been scaled-up to the pilot plant stage, these offer considerable promise for large-scale disruption if low-cost sources of enzymes or effective immobilization procedures can be developed. Alternatively, genetic mutation to produce cells with weaker walls or cells programmed to lyse under certain conditions may be a useful technique for simplifying disruption requirements.

20.7 REFERENCES

Asenjo, J. A. (1981). Process for the production of yeast-lytic enzymes and the disruption of whole yeast cells. In *Advances in Biotechnology*, ed. C. Vezina and K. Singh, vol. 3, pp. 295–300. Pergamon, Toronto.

Asenjo, J. A. and P. Dunnill (1981). The isolation of lytic enzymes from *Cytophaga* and their application to the rupture of yeast cells. *Biotechnol. Bioeng.*, **23**, 1045–1056.

Asenjo, J. A., P. Dunnill and M. D. Lilly (1981). Production of yeast-lytic enzymes by *Cytophaga* species in batch culture. *Biotechnol. Bioeng.*, **23**, 97–109.

Augenstein, D. C., K. Thrasher, A. J. Sinskey and D. I. C. Wang (1974). Optimization in the recovery of a labile intracellular enzyme. *Biotechnol. Bioeng.*, **16**, 1433–1447.

Ballou, C. (1976). Structure and biosynthesis of the mannan component of the yeast cell envelope. *Adv. Microbiol. Physiol.*, **14**, 93–158.

Bartnicki-Garcia, S. (1968). Cell wall chemistry, morphogenesis, and taxonomy of fungi. *Annu. Rev. Microbiol.*, **22**, 87–108.

Boudrant, J., J. DeAngelo, A. J. Sinskey and S. R. Tannenbaum (1979). Process characteristics of cell lysis mutants of *Saccharomyces cerevisiae. Biotechnol. Bioeng.*, **21**, 659–670.

Brookman, J. S. G. (1974). Mechanism of cell disintegration in a high pressure homogenizer. *Biotechnol. Bioeng.*, **16**, 371–383.

Brookman, J. S. G. (1975). Further studies on the mechanism of cell disruption by extreme pressure extrusion. *Biotechnol. Bioeng.*, **17**, 465–479.

Burnett, J. H. (1979). Aspects of the structure and growth of hyphal walls. In *Fungal Walls and Hyphal Growth*, ed. J. H. Burnett and A. P. J. Trinci, chap. 1, pp.1–25. Cambridge University Press, Cambridge.

Coakley, W. T., R. C. Brown, C. J. James and R. K. Gould (1974). Optimization of the release of undegraded material extracted from cells by ultrasound. *Biotechnol. Bioeng.*, **16**, 659–673.

Costerton, J. W., J. M. Ingram and K.-J. Cheng (1974). Structure and function of the cell envelope of gram-negative bacteria. *Bacteriol. Rev.*, **38**, 87–110.

Coyette, J. and J. M. Ghuysen (1970). Structure of the walls of *Lactobacillus acidophilus* Strain 63 AM Gasser. *Biochemistry*, **9**, 2935–2943.

Currie, J. A., P. Dunnill and M. D. Lilly (1972). Release of protein from bakers' yeast (*Saccharomyces cerevisiae*) by disruption in an industrial agitator mill. *Biotechnol. Bioeng.*, **14**, 725–736.

Doulah, M. S. (1977). Mechanism of disintegration of biological cells in ultrasonic cavitation. *Biotechnol. Bioeng.*, **19**, 649–660.

Doulah, M. S. (1978). Application of the statistical theory of reliability to yeast cell disintegration in ultrasonic cavitation. *Biotechnol. Bioeng.*, **20**, 1287–1289.

Doulah, M. S., T. H. Hammond and J. S. G. Brookman (1975). A hydrodynamic mechanism for the disintegration of *Saccharomyces cerevisiae* in an industrial homogenizer. *Biotechnol. Bioeng.*, **17**, 845–858.

Dunnill, P. and M. D. Lilly (1975). Protein extraction and recovery from microbial cells. In *Single-Cell Protein II*, ed. S. R. Tannenbaum and D. I. C. Wang, chap.8, pp. 179–207. MIT Press, Cambridge, MA.

Edebo, L. (1960). A new press for the disruption of micro-organisms and other cells. *J. Biochem. Microbiol. Technol. Eng.*, **2**, 453–479.

Edebo, L. (1969). Disintegration of cells. In *Fermentation Advances*, ed. D. Perlman, pp. 249–271. Academic, New York.

Engler, C. R. (1979). *Disruption of Microorganisms in High Pressure Flow Devices*. Ph.D. Thesis, University of Waterloo, Waterloo, Ontario, Canada.

Engler, C. R. and C. W. Robinson (1979). New method of measuring cell-wall rupture. *Biotechnol. Bioeng.*, **21**, 1861–1869.

Engler, C. R. and C. W. Robinson (1981a). Disruption of *Candida utilis* cells in high pressure flow devices. *Biotechnol. Bioeng.*, **23**, 765–780.

Engler, C. R. and C. W. Robinson (1981b). Effects of organism type and growth conditions on cell disruption by impingement. *Biotechnol. Lett.*, **3**, 83–88.

Follows, M., P. J. Hetherington, P. Dunnill and M. D. Lilly (1971). Release of enzymes from bakers' yeast by disruption in an industrial homogenizer. *Biotechnol. Bioeng.*, **13**, 549–560.

Garver, J. C. and R. L. Epstein (1959). Method for rupturing large quantities of microorganisms. *Appl. Microbiol.*, **7**, 318–319.

Ghuysen, J. M. (1968). Use of bacteriolytic enzymes in determination of wall structure and their role in cell metabolism. *Bacteriol. Rev.*, **32**, 425–464.

Ghuysen, J. M. and G. D. Shockman (1973). Biosynthesis of peptidoglycan. In *Bacterial Membranes and Walls*, ed. L. Leive, chap. 2, pp. 37–130. Dekker, New York.

Ghuysen, J. M., J. L. Strominger and D. J. Tipper (1968). Bacterial cell walls. In *Comprehensive Biochemistry*, ed. M. Florkin and E. H. Stotz, vol. 26A, chap. 2, pp. 53–104. Elsevier, Amsterdam.

Glauert, A. M. and M. J. Thornley (1971). Fine structure and radiation resistance in *Acinetobacter*: a comparison of a range of strains. *J. Cell Sci.*, **8**, 19–41.

Glauert, A. M., M. J. Thornley, K. J. I. Thorne and U. B. Sleytr (1976). The surface structure of bacteria. In *Microbial Ultrastructure*, ed. R. Fuller and D. W. Lovestock, pp. 31–47. Academic, New York.

Gornall, A. G., C. J. Bardawill and M. M. David (1949). Determination of serum proteins by means of the biuret reaction. *J. Biol. Chem.*, **177**, 751–766.

Gray, P. P., P. Dunnill and M. D. Lilly (1972). The continuous-flow isolation of enzymes. In *Fermentation Technology Today*, ed. G. Terui, pp. 347–351. Society for Fermentation Technology, Japan.

Griffin, D. H. (1981). *Fungal Physiology*. chap. 3, pp. 40–72. Wiley, New York.

Hedenskog, G. and H. Mogren (1973). Some methods for processing of single-cell protein. *Biotechnol. Bioeng.*, **15**, 129–142.

Hetherington, P. J., M. Follows, P. Dunnill and M. D. Lilly (1971). Release of protein from bakers' yeast (*Saccharomyces cerevisiae*) by disruption in an industrial homogenizer. *Trans. Inst. Chem. Eng.*, **49**, 142–148.

Higgins, J. J., D. J. Lewis, W. H. Daly, F. G. Mosqueira, P. Dunnill and M. D. Lilly (1978). Investigation of the unit operations involved in the continuous flow isolation of β-galactosidase from *Escherichia coli*. *Biotechnol. Bioeng.*, **20**, 159–182.

Hughes, D. E., J. W. T. Wimpenny and D. Lloyd (1971). The disintegration of micro-organisms. In *Methods in Microbiology*, ed. J. R. Norris and D. W. Ribbons, vol. 5B, chap. 1, pp. 1–54. Academic, New York.

Kitamura, K. and Y. Yamamoto (1981). Lysis of yeast cells showing low susceptibility to zymolyase. *Agric. Biol. Chem.*, **45**, 1761–1766.

Knorr, D., K. J. Shetty and J. E. Kinsella (1979). Enzymatic lysis of yeast cell walls. *Biotechnol. Bioeng.*, **21**, 2011–2021.

Lampen, J. O. (1968). External enzymes of yeast: their nature and formation. *Antonie van Leeuwenhoek*, **34**, 1–18.

Layne, E. (1957). Spectrophotometric and turbidimetric methods for measuring proteins. In *Methods in Enzymology*, vol. 3, pp. 447–454. Academic, New York.

Leduc, M. and J. van Heijenoort (1980). Autolysis of *Escherichia coli*. *J. Bacteriol.* **142**, 52–59.

Lee, C.-H., S. K. Tsang, R. Urakabe and C. K. Rha (1979). Disintegration of dried yeast cells and its effect on protein extractability, sedimentation property, and viscosity of cell suspension. *Biotechnol. Bioeng.*, **21**, 1–17.

Lilly, M. D. and P. Dunnill (1969). Isolation of intracellular enzymes from microorganisms—the development of a continuous process. In *Fermentation Advances*, ed. D. Perlman, pp. 225–247. Academic, New York.

Limon-Lason, J., M. Hoare, C. B. Orsborn, D. J. Doyle and P. Dunnill (1979). Reactor properties of a high-speed bead mill for microbial cell rupture. *Biotechnol. Bioeng.*, **21**, 745–774.

Lowry, O. H., N. J. Rosebrough, A. L. Farr and R. J. Randall (1951). Protein measurement with the Folin phenol reagent. *J. Biol. Chem.*, **193**, 265–275.

Luther, H. (1980). Characterization of the mechanical disintegration of micro-organism cells with the aid of conductivity measurement. *Die Nahrung*, **24**, 265–272.

Magnusson, K.-E. and L. Edebo (1974). Estimation of the disruption in freeze-pressed *Saccharomyces cerevisiae* by an electronic particle counter. *Biotechnol. Bioeng.*, **16**, 1273–1282.

Magnusson, K.-E. and L. Edebo (1976). Influence of cell concentration, temperature and press performance on flow characteristics and disintegration in the freeze pressing of *Saccharomyces cerevisiae* with the X-press. *Biotechnol. Bioeng.*, **18**, 865–883.

Marffy, F. and M.-R. Kula (1974). Enzyme yields from cells of brewers' yeast disrupted by treatment in a horizontal disintegrator. *Biotechnol. Bioeng.*, **16**, 623–634.

McMurrough, I. and A. H. Rose (1967). Effect of growth rate and substrate limitation on the composition and structure of the cell wall of *Saccharomyces cerevisiae*. *Biochem. J.*, **105**, 189–203.

Mickle, H. (1948). Tissue disintegrator. *J. R. Microsc. Soc.*, **68**, 10–12.

Milner, H. W., N. S. Lawrence and C. S. French (1950). Colloidal dispersion of chloroplast material. *Science*, **111**, 633–634.

Mogren, H., M. Lindblom and G. Hedenskog (1974). Mechanical disintegration of microorganisms in an industrial homogenizer. *Biotechnol. Bioeng.*, **16**, 261–274.

Moor, H. and K. Muhlethaler (1963). Fine structure in frozen-etched yeast cells. *J. Cell Biol.*, **17**, 609–628.

Musilkova, M., E. Ujcova, J. Placek, Z. Fencl and L. Seichert (1981). Release of protoplasts from a fungus as a criterion of mechanical interactions in the fermenter. *Biotechnol. Bioeng.*, **23**, 441–446.

Neppiras, E. A. and D. E. Hughes (1964). Some experiments on the disintegration of yeast by high intensity ultrasound. *Biotechnol. Bioeng.*, **6**, 247–270.

Peppler, H. J. (1970). Food yeasts. In *The Yeasts*, ed. A. H. Rose and J. S. Harrison, vol. 3, chap. 8, pp. 421–462. Academic, New York.

Phaff, H. J. (1971). Structure and biosynthesis of the yeast cell envelope. In *The Yeasts*, ed. A. H. Rose and J. S. Harrison, vol. 2, chap. 5, pp. 135–210. Academic, New York.

Reed, G. and H. J. Peppler (1973). *Yeast Technology*, chap. 12, pp. 355–366. AVI, Westport.

Rehacek, J., K. Beran and V. Bicik (1969). Disintegration of microorganisms and preparation of yeast cell walls in a new type of disintegrator. *Appl. Microbiol.*, **17**, 462–466.

Rehacek, J. and J. Schaefer (1977). Disintegration of microorganisms in an industrial horizontal mill of novel design. *Biotechnol. Bioeng.*, **19**, 1523–1534.

Ribi, E., T. Perrine, R. List, W. Brown and G. Goode (1959). Use of pressure cell to prepare cell walls from mycobacteria. *Proc. Soc. Exp. Biol. Med.*, **100**, 647–649.

Salton, M. J. R. (1964). *The Bacterial Cell Wall*. Elsevier, Amsterdam.

Schutte, H., K. H. Kroner, H. Hustedt and M.-R. Kula (1983). Experiences with a 20 litre industrial bead mill for the disruption of microorganisms. *Enzyme Microb. Technol.*, **5**, 143–148.

Scully, D. B. and J. W. T. Wimpenny (1974). Thermodynamics and rheology of the Hughes press. *Biotechnol. Bioeng.*, **16**, 675–687.

Sutherland, I. W. (1975). The bacterial wall and surface. *Process Biochem.*, **10** (3), 4–8.

Takahara, Y., Y. Hirose, N. Yasuda, K. Mitsugi and S. Murao (1976). Effect of peptide lipids produced by *Bacillus* on the enzymatic lysis of gram negative bacterial cells. *Agric. Biol. Chem.*, **40**, 1901–1903.

Tannenbaum, S. R. (1968). Factors in the processing of single-cell protein. In *Single-Cell Protein*, ed. R. I. Mateles and S. R. Tannenbaum, pp. 343–352. MIT Press, Cambridge, MA.

Thornley, M. J., A. M. Glauert and U. B. Sleytr (1974). Structure and assembly of bacterial surface layers composed of regular arrays of subunits. *Philos. Trans. R. Soc. London, Ser. B.*, **268**, 147–153.

Watanabe, H., H. Iizuki and M. Takehisa (1981). Radiation induced enhancement of enzymatic cell lysis of *Micrococcus radiodurans*. *Agric. Biol. Chem.*, **45**, 2323–2328.

Weidel, W. and H. Pelzer (1964). Bagshaped macromolecules—a new outlook on bacterial cell walls. *Adv. Enzymol.*, **26**, 193–232.

Wessels, J. G. H. and J. H. Sietsma (1979). Wall structure and growth in *Schizophyllum commune*. In *Fungal Walls and Hyphal Growth*, ed. J. H. Burnett and A. P. J. Trinci, chap. 2, pp. 27–48. Cambridge University Press, Cambridge.

Whitworth, D. A. (1974a). Hydrocarbon fermentation: protein and enzyme solubilization from *Candida lipolytica* using an industrial homogenizer. *Biotechnol. Bioeng.*, **16**, 1399–1406.

Whitworth, D. A. (1974b). Assessment of an industrial homogenizer for protein and enzyme solubilization from spent brewery yeast. *C. R. Trav. Lab. Carlsberg*, **40**, 19–32.

21

Centrifugation

H. A. C. AXELSSON
Alfa-Laval Separation AB, Tumba, Sweden

21.1 INTRODUCTION

The continuously operating centrifugal separator was invented in 1878 by the Swedish engineer Gustaf de Laval. It was used for separating cream from milk. In 1896 the centrifugal separator made its entry into the fermentation industry. That year a machine was introduced which recovered bakers' yeast (Figure 1). This application was the first one outside the milk field and is still an important part of the use of centrifuges.

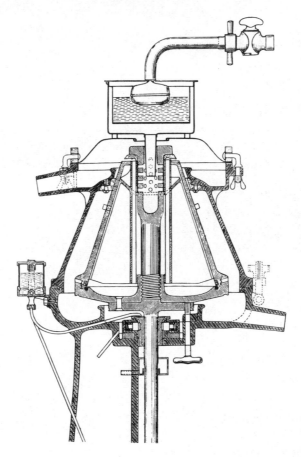

Figure 1   Cross-section of the first bakers' yeast separator from 1896 (Courtesy of Alfa Laval Separation AB, Sweden)

From this starting point the use of the centrifuge in the fermentation industry has become very diversified. In terms of product value, the biggest part of the fermentation industry today is the brewery industry. It is therefore not very surprising that for many years, centrifugal separators were primarily developed for that industry and for the bakers' yeast industry, and applied afterwards to other fermentation industries.

With the increasing interest to use the rDNA techniques, large-scale centrifugation equipment must meet new requirements, *e.g.* asepsis and containment. Especially in those cases where the product is intracellular, many process steps are necessary (Fish and Lilly, 1984). Rosén and Datar (1983) have estimated that equipment costs for down-stream processing are three times higher than for fermentation.

The step yield must be as near to 100% as possible. These two factors call for high separation efficiencies.

21.2 FLUID AND PARTICLE DYNAMICS

21.2.1 Stokes' Law

The sedimentation velocity for a solid particle settling under the influence of gravity is given by Stokes' law.

$$v_g = \frac{\varrho_p - \varrho_f}{18\mu}d_p^2 g \tag{1}$$

where ϱ_p = density of the particle, ϱ_f = density of the surrounding medium, μ = viscosity of the suspension, d_p = diameter of the particle, and g = gravitational acceleration. This relation is valid for spherical particles with a Reynolds number Re $(= d_p v_g \varrho_p/\mu) < 1$ and for low particle concentration, where hindered settling is absent (Sokolov, 1971).

21.2.2 Settling in the Centrifugal Field

The equations of continuity and motion for centrifugation have been derived and presented most recently by Hsu (1981). Assuming that the particle travels at its terminal velocity and neglecting the Coriolis effect, the settling velocity v_c in the centrifugal field may be written as

$$v_c = \frac{\varrho_p - \varrho_f}{18\mu} d_p^2 \omega^2 r \tag{2}$$

where ω = angular velocity and r = radial position of particle.

Equation (2) is valid for a particle $Re < 0.4$. For higher Re, equations have been derived by Sokolov (1971) and Hsu (1981). By dividing equation (2) by equation (1) one obtains

$$Z = \frac{\omega^2 r}{g} \tag{3}$$

where Z is usually called 'relative centrifugal force' or 'centrifuge effect' or 'g-number'. This parameter is often used to characterize centrifuges, but cannot be used for capacity estimations. The range of Z for industrial centrifuges is of the order of 300 to 16 000. For laboratory swinging-bucket and fixed angle-head rotors Z can be up to 500 000, and for special zonal rotors up to 1 000 000 (Hsu, 1981)

21.2.3 Centrifuge Configurations

The centrifuge designer must solve a number of conflicting requirements. From equation (2) it can be seen that $\omega^2 r$ must be made as large as possible which means high bowl speed or large radial distances from the center of rotation. The designer must also design the bowl so that particles are captured at a short distance from where they enter the centrifugal force field and provide for a long residence time. For continuous centrifuges the handling of the sedimented captured solids phase often conflicts with these requirements and has led to the development of a number of geometries: (1) tubular bowl (Figure 2) with large length/diameter (L/D) ratio; (2) one-chamber bowl with small L/D ratio; (3) multichamber bowl (Figure 3); (4) decanter with interior screw conveyor (Figure 4); (5) disc type (Figure 5a–e).

21.2.4 Tubular Bowl Centrifuges

The tube centrifuge (Figure 2) has the simplest configuration of all centrifuges. In spite of this, theoretical flow models have so far proved inadequate. Sokolov (1971) reviewed a number of velocity distributions, *e.g.* plug flow, a parabolic distribution and a quasi-parabolic velocity profile, also used by Svarovsky (1977). He compared the actual separation performance with those calculated from the first and the third profile. The plug flow model gave the best fit, although it overestimated separation performance. The quasi-parabolic profile, however, underestimated it.

The phenomena that make the flow models invalid are end effects, vortices between wing inserts, velocity lag and non-ideal feed accelerators (Bell *et al.*, 1983). Zeitsch (1978) has related the feed effect on the surface lag to three dimensionless numbers and found that the lag is of the order of 3–10%.

21.2.5 Chamber Centrifuges

The one-chamber centrifuge can be considered a tube centrifuge with a small L/D ratio and the multichamber centrifuge (Figure 3) as a battery of tube centrifuges coupled in series. Sokolov's (1971) analysis for the multichamber machine is founded on a very simplified model, and does not take into account the changing conditions along the liquid path during a run, *e.g.* regarding solids accumulation and concentration.

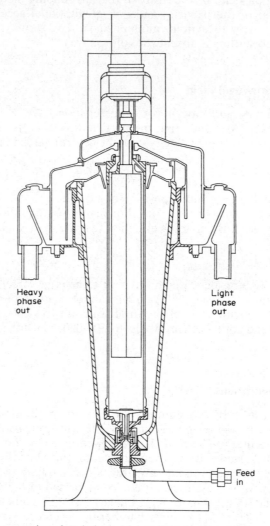

Heavy
phase
out

Light
phase
out

Feed
in

Figure 2   Cross-section of a tubular bowl centrifuge (courtesy of Pennwalt Ltd, UK)

Very little theoretical analysis of this type of machine has been made.

21.2.6   Decanter Centrifuges

The objective of decanters is usually to concentrate the solids in process liquids with high solids content. In most cases, the solids particles settle easily. Therefore, Stokes' analysis is not valid. Schnittger (1970) has found experimentally that the bowl speed exponential in the flow rate/bowl speed relationship is 1.1–1.5, whereas it is 2 in the case of Stokes free settling. Bell *et al.* (1983) describe some recent theoretical work.

The hydrodynamics of the decanter are complicated. The conveyor adds turbulence and forces the clarified liquid into a helical path. Below the top layer, vortices are formed (Sokolov, 1971).

The rheology of the concentrated solids is often a problem. Many thickened solids can be conveyed under the liquid surface, but when they are transported up the beach the radial force quite suddenly increases and the solids collapse (Hsu, 1981).

21.2.7   Disc-stack Centrifuges

The discs in disc-stack centrifuges split the liquid into thin layers. The flow between the discs becomes laminar, which makes particle settling easier. The settling distance is drastically reduced. The disc stack also keeps the liquid rotating at or near to the bowl speed, especially if

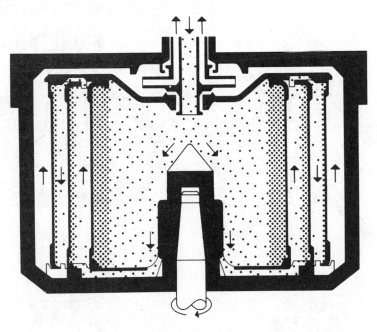

Figure 3 Multichamber bowl centrifuge

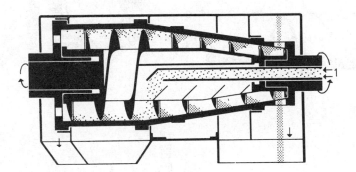

Figure 4 Decanter centrifuge with interior screw conveyor

the discs have spacer strips (Brunner and Molerus, 1979). The critical Reynolds number for transition to turbulent flow increases with decreasing disc spacing and increasing bowl speed (Sokolov, 1971).

The question if the incoming liquid distributes itself evenly between the discs has been investigated by Rachitskii and Skvortsov (1980). They found that the distribution is even, as was assumed by Carlsson (1979) in his dissertation.

The velocity distribution across the disc spacing has been demonstrated to depend on a dimensionless number λ (Sokolov, 1971; Bohman, 1974; Brunner and Molerus, 1979; Carlsson, 1979):

$$\lambda = h\sqrt{\frac{\omega \sin \alpha}{\nu}} \tag{4}$$

where h = disc spacing, ω = radial velocity of bowl, ν = kinematic viscosity, and α = half cone angle of disc.

If λ is smaller than π, the velocity profile is parabolic. The value of λ in industrial centrifuges is usually between 5 and 28. Sokolov (1971), Brunner and Molerus (1979) and others have calculated profiles for a number of different λ values. The relative circumferential velocity increases from the disc surfaces to the middle of the disc spacing, and becomes—for large λ values—λ times larger than the average radial velocity. This velocity component splits at large λ values into two thin layers, one near to each disc surface. The velocity in these layers increases and their thick-

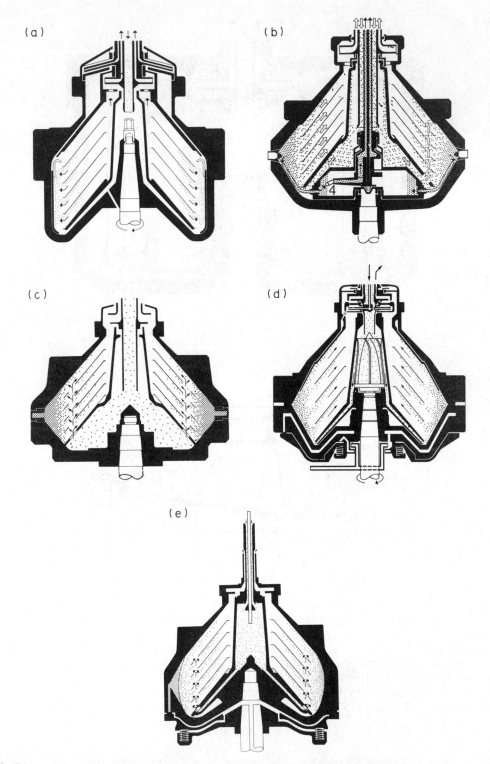

Figure 5 Disc-type bowl configurations: (a) solids-retaining centrifuge; (b) nozzle machine with pressurized discharge of concentrate; (c) nozzle machine with peripheral nozzles; (d) periodically solids-ejecting centrifuge with radial slits; (e) periodically solids-ejecting centrifuge with axial channels

ness decreases with increasing λ. The Reynolds number increases and transition to turbulence can take place. The reason for this behaviour is an interplay between Coriolis, friction and centrifugal forces.

In some centrifuges, the discs have distribution holes, from which a large part of the feed liquid enters the disc stack. Around each distribution hole a vortex is formed due to Coriolis forces. The size of the vortex increases with bowl speed and feed flow rate (Sokolov, 1971). When leaving the distribution hole, the liquid will flow in a direction counter to the bowl rotation.

Willus and Fitch (1973) have studied the flow pattern in a device simulating a disc without distribution holes but with six long radial spacing ribs. A rotating camera observed vortices between each spacing rib, occupying a large part of the disc area.

The flow field between discs has been computed by several workers using the linearized Navier–Stokes equations (Brunner, 1956; Goldin, 1971; Fitch, 1965; Bohman, 1974; Carlsson, 1979). They have all used some simplifying assumptions, *e.g.* axi-symmetry. No solution to the Navier–Stokes equations for non-axi-symmetrical conditions has been published. One of the unanalyzed conditions is the technically important case of long radial spacing ribs, which have a positive effect on separation efficiency (Brunner and Molerus, 1979).

The flow fields obtained from the Navier–Stokes equations are one of the necessary bases for the calculation of particle trajectories. Simplifying assumptions are necessary if the numerical integration shall not become too time consuming for computers, *e.g.* low solids concentration and constant drag coefficient (Brunner and Molerus, 1977, 1979; Carlsson, 1979; Gupta, 1981).

A particle is usually considered to be separated when it reaches the disc surface. Brunner (1956) and Carlsson (1980) have postulated that this is the case only within a region where the centrifugal force exceeds the counteracting shear force. Stokes' law requires settling in a fluid with no boundaries. Carlsson (1980) has included an empirical factor in his model to account for the wall effect and has found good fit with experimental data from Brunner and Molerus (1979) for small particles, which are considered to be able to keep to the disc surface more easily than large particles during the countercurrent movement of the solids layer and, because of this, avoid the shear forces better. The optimal spacing between discs has been found experimentally to decrease with increasing bowl speed and decreasing solids concentration (Carlsson, 1984).

21.2.8 Hindered Settling

Aiba *et al.* (1965) made tests with *Serratia marcescens* and found that the rate of settling decreased by a factor of 2 at a volume fraction of cells of 0.2 and by a factor of 19 at 0.5. It has been found that the volume fraction of yeast cells is 0.8 at the edge of the disc stack of a disc bowl centrifuge when it is running at its maximum capacity with an acceptable loss of yeast in the beer. This means that hindered settling must be an important factor governing the centrifugation efficiency, especially since the volume fraction of yeast cells in the feed and in distribution holes often is greater than 0.15.

21.3 SCALE-UP

21.3.1 General

A prerequisite for a successful scale-up is a true mathematical model of the process taking place in the equipment. Preferably it should define one or several quantities, typical for the equipment, independent of the process parameters. The most used quantity to characterize centrifuges, the Σ concept, fulfills the second condition. Ambler (1952, 1959, 1961) has shown that the volumetric flow rate, at which half the solids particles of diameter d_p will be removed from the liquid, is given by

$$Q = \frac{(\varrho_p - \varrho_f)d_p^2 V \omega^2 r_e}{9\mu s_e} \qquad (5)$$

where Q = volumetric flow rate, V = volume of liquid in bowl, s_e = average value of sedimentation distance and r_e = average value of the radius. The way of calculating s_e and r_e depends on the geometry of the bowl.

Equation (5) may be written

$$Q = 2\frac{(\varrho_p - \varrho_f)d_p^2 g}{18\mu} \frac{V\omega^2 r_e}{gs_e} = 2v_g\Sigma \qquad (6)$$

where Σ is the area equivalent of the centrifuge, with the dimension (length)^2.

The derivation is based on the assumptions that viscous drag is determining the particle movement, the flow between the discs is laminar and symmetrical, the liquid rotates at the same speed as the bowl, the particle concentration is low (no hindered settling), the particle must all the time be in equilibrium, *i.e.* move at its final settling velocity, and that

$$v_c/v_g = \omega^2 r/g \tag{7}$$

These assumptions are not fulfilled, which the discussion above in Section 21.2 shows.

Ambler uses the particle which is separated to 50% in his formulae. Therefore, his analysis is a special case of the grade efficiency function which Svarovsky (1977) has developed and where the ratio of sedimented mass to the mass in the feed material for each particle size is determined and integrated over all particle sizes to get a more realistic measure of the separation efficiency. Murkes (1969) worked with a simple Gaudin particle size distribution, verified his equations with a kaolin clay suspension in a disc-bowl centrifuge and reached a number of practically useful conclusions. The formulae obtained were found to be valid also for monodispersed particles (Murkes and Carlsson, 1978). Other particle size distributions are also used. Higgins *et al.* (1978) plotted separation data in a log/probability diagram and found straight lines for a suspension of whole and disrupted cells of *Escherichia coli*.

21.3.2 Tubular Bowl Centrifuges

The expression for Σ can be found in Table 1. On comparing tube centrifuges of different sizes and bowl speeds it was found that they performed according to the Σ model (Lavanchy *et al.*, 1964). Compared to test tube centrifugation the efficiency of the tube centrifuge is given as 90% (Lavanchy *et al.*, 1964) or 97–98% (Ambler, 1959). Frampton's figure of 37.5% is based on different definitions of Σ and the critical particle.

Table 1   Equations for Σ for Various Bowl Geometries

Machine type	Equation for Σ	Symbols used	Ref.
Tubular bowl	$\dfrac{\pi\omega^2}{g} L \dfrac{r_2^2 - r_1^2}{\ln\left(\dfrac{2r_2^2}{r_2^2 + r_1^2}\right)}$ or $\dfrac{\pi\omega^2}{g} 2L\left(\dfrac{3}{4}r_2^2 + \dfrac{1}{4}r_1^2\right)$	r_1 = inner radius of liquid r_2 = inner radius of bowl L = inner length of bowl	Ambler (1959) Ambler (1952)
Multichamber bowl	$\dfrac{\pi\omega^2}{g} \dfrac{L}{3} \sum\limits_{i=0}^{i=n} \dfrac{r_{2i+1}^3 - r_{2i+2}^3}{r_{2i+1} - r_{2i+2}}$	r = radius of liquid Indices with even numbers = inner radius of chamber Indices with odd numbers = outer radius of chamber $n + 1$ = number of chambers	Sokolov (1971)
Decanter	$\dfrac{\pi\omega^2}{g}\left[L_1\left(\dfrac{3}{2}r_2^2 + \dfrac{1}{2}r_1\right) + \right.$ $\left. L_2\left(\dfrac{r_2^2 + 3r_2 r_1 + 4r_1^2}{4}\right)\right]$	L_1 = length of cylindrical part L_2 = length of conical part	Ambler (1959) Svarovsky (1977)
Disc bowl	$\dfrac{\pi\omega^2}{g} \dfrac{2}{3} N\left(r_2^3 - r_3^1\right) \cot\alpha$	N = number of discs r_2 = max. radius of disc r_1 = min. radius of disc α = half cone angle of disc	Ambler (1952)

21.3.3 Decanter Centrifuges

The Σ value of the decanter centrifuge is found in Table 1 (Ambler, 1959). In decanters, it is important to perform the scale-up work with more than separation performance according to the Σ theory in mind. It is recommended to make test runs in a pilot machine where bowl speed and differential speed can be varied in addition to the feed flow rate. Gösele (1980) recommends a scale-up procedure where turbulent power per unit volume, driving forces for dewatering (compression or drainage) and residence time are important parameters. Records (1977) describes a pilot machine test procedure from a more practical viewpoint.

Various authors agree that the efficiency of a decanter centrifuge compared to test tube centrifugations is 54–67% (Ambler, 1959; Morris, 1966; Sokolov, 1971). Most of the reduction is due to the solids that occupy a large part of the liquid volume, reducing the residence time (Ambler, 1959).

21.3.4 Disc-stack Centrifuges

21.3.4.1 Σ value

The Σ value for this type of machine can be found in Table 1 (Ambler, 1952). It is the most used relationship and has been modified several times, *e.g.* by Jury and Locke (1957). Their analysis bears some similarity to the grade efficiency analysis by Svarovsky (1977) but is based on mostly the same assumptions as Ambler's. The analyses of Jury and Locke, and of Ambler were compared by Gupta (1980) who found very small differences.

Compared to test tube centrifugation, the efficiency of disc-stack centrifuges has been measured by Murkes and Carlsson (1978) who found 73% for a dilute monodisperse suspension. Various quoted values are 55% (Ambler, 1959, 1961; Frampton, 1963), 40% (Purchas, 1981) and 45% (Morris, 1966).

21.3.4.2 KQ value

Based on tests on milk skimming, E. A. Forsberg of AB Separator (now Alfa-Laval AB), Sweden, found a possible relationship at the beginning of this century, taking into account the irregularities of the flow:

$$v_c/v_g = \left(\frac{\omega^2 r}{g}\right)k \tag{8}$$

He looked at the worst possible case: a particle on the upper side of one disc at r_2 (= outer radius of the disc) being separated out on the underside of the next disc at r_1 (= inner radius of the disc). He arrived at the following formula:

$$KQ = \frac{3.6}{100^k}\frac{2\pi}{k+2}\omega^{2k}N\cot\alpha(r_2^{2+k} - r_1^{2+k}) \tag{9}$$

Forsberg found the best correlation with test data with $k = 0.75$. r_1 and r_2 are expressed in cm.

A practical and traditional form of this equation is that of Sullivan and Erikson (1961).

$$KQ = 280\left(\frac{n}{1000}\right)^{1.5}N\cot\alpha(r_2^{2.75} - r_1^{2.75}) \tag{10}$$

where n = bowl speed (r.p.m.), and r_2, r_1 = radii of disc (cm).

Much later, Zastrow (1976) has found $k = 0.71$ for 1% PVAC/water suspension in a solids-ejecting disc separator. This form of the KQ formula is dimensionally incorrect, but it is possible to make it correct, expressing the KQ value as an entity with the dimension $(\text{length})^2$.

One can regard KQ as the product of a feed flow rate Q and a particle-related constant. Comparing with Ambler, K is inversely proportional to v_g, and found by experiments on a disc bowl with known geometry (Bergner, 1957). The reduced exponents in the KQ formula may be explained by the shear force or the pressure gradient (Carlsson, 1979).

21.3.4.3 Dimensionless parameters

Gupta (1981) has worked out a scale-up method based on grade-efficiency plots and three dimensionless parameters. Similar uniformity and equivalence relationships have been proposed by Bohman and Murkes (1969).

21.4 TYPES OF SEPARATOR

21.4.1 General

The factor determining the design of a centrifuge is the method by which the solid phase in the process material is handled. The first two types developed were applied to duties where the products were in liquid form (cream from milk and yeast cream), which could flow out of the bowl. A comprehensive description of centrifuges can be found in the book by Sokolov (1971). Shorter treatises are made by Records (1977), Svarovsky (1977), Ambler (1979) and Purchas (1981).

Brochures and technical publications from centrifuge manufacturers, *e.g.* Alfa-Laval Separation AB (1984b), also provide information on machine types and sizes.

21.4.2 Solid Bowl Machines

21.4.2.1 *Solid bowl disc centrifuges*

In Figure 5a, a solid bowl machine is shown. This version is equipped with two liquid outlets, each having stationary centripetal pumps (paring discs) reducing foaming. The machine is also equipped with a basket for easy removal of accumulated solids. The feed to the bowl may also be introduced from below through a hollow spindle and the liquid is allowed to fill the bowl completely. On the outlet(s), centrifugal pump wheel(s) are fitted. The machine is thus completely enclosed, with rotating mechanical face seals at the bottom of the bowl spindle and at the liquid outlets. In this way the operation is fully hermetic. The solids space volume can be up to about 50 l.

21.4.2.2 *Tube centrifuges*

In Figure 2, a tube centrifuge is shown, equipped for separation of two liquid phases. The liquid(s) are discharged from the top *via* overflow of ring dam(s). In order to remove the solids, the machine must be stopped and dismantled.

The bowl can develop up to 18 000*g* in industrial models and up to 65 000*g* for laboratory models. The accumulated solids volume can be up to 2–4 l.

21.4.2.3 *Zonal centrifuges*

This type of machine is intended for purification of viruses or isolation of RNA, DNA, membranes or other cell constituents. The bowl is filled with liquid which can establish a density gradient. The feed is introduced into the machine and the particles are allowed to sediment into a band where their density is equal to that of the liquid (isopyknic banding). Hsu (1981) and Rickwood (1978) have described equipment and methods. The machines have different *L*/*D* ratios. The bowl volume can be up to 7 l. The construction material is usually an aluminum alloy but titanium alloys are used more and more. The bowls rotate in vacuum, and usually a maximum *g* number of 100 000 is developed. There is, however, at least one model which develops about 1 000 000*g*.

21.4.2.4 *One-chamber and multichamber bowl centrifuges*

These machines are suitable for recovery of valuable solids and are used where the scale of production is too large for the tube centrifuges. Figure 3 shows a multichamber bowl. The chamber volume can be up to 64 l. Smaller one-chamber bowl centrifuges are used in protein fractionation from blood plasma (Foster and Watt, 1980). In this process, a very close temperature control is necessary. Therefore, machines have been developed with cooling channels in the bowl and paring discs and a cooling jacket in the frame.

21.4.2.5 *Decanter centrifuges*

In the 1950s a separator with scroll discharge was developed for very large quantities of solids. The machine is called the decanter (Figure 4). It is equipped with a screw conveyor, which rotates at a speed slightly higher or lower than the bowl. The differential speed is obtained *via* a gearbox rotating at the bowl speed, and can be varied by rotating the so-called 'sun-wheel' of the gearbox which usually is of planetary type.

Since the decanter has found its largest use in the dewatering of flocculated sludges, a number of different feed arrangements have been developed because of the susceptibility of the flocs to shear (Bell *et al.*, 1983). Special baffle plates are also provided to make it possible to use a 'negative beach' for those sludges which are difficult to convey because of low inner friction. In order to increase retention time, decanters with very steep cones have been developed. Three-phase decanters have recently been introduced into the biotechnology industry (Katinger *et al.*, 1981).

Vesilind (1974) has applied thickener theory to estimate decanter capacity. He also gives relationships for the conveyor capacity.

21.4.3 Solids-discharging Disc Separators

21.4.3.1 *Nozzle machine with pressurized discharge of concentrate*

The first yeast separator had no intermediate discs (Figure 1), but from 1901 discs were used. In yeast separators, the concentrate is flowing through channels in the bowl wall from the periphery towards the center. At the end of each channel is a nozzle, the size of which determines the flow rate of the cream. The reason why the nozzles can be situated at reduced diameter (retracted nozzles) is the favorable rheological characteristics of yeast concentrates: they are pseudoplastic, and so are many other microbial cell suspensions. Since the flow rate through a given nozzle is proportional to its distance from the center of rotation, the nozzle size can be bigger for a given flow in this type of separator than in the conventional nozzle machine, which means lower risk of clogging the nozzles by oversized particles.

A further development of this type of machine was made in 1959 when the nozzles were made to discharge into a central chamber in the bowl, instead of discharging into the collecting cover (Figure 5b). In the central chamber is a stationary tube which picks up the concentrate and discharges it under pressure in a closed pipe, which is of advantage for reasons of hygiene and plant layout. The peripheral nozzles are open only during cleaning. In a very large model of this type of nozzle machine the nozzles are rectangular slits, which can be blocked off partially by radial plungers in an internal slide in the paring chamber. The slide can be moved up and down, controlling the nozzle opening. Therefore the solids concentration can be closely controlled.

Since the pressure drop over the nozzles in this type of machine is smaller than in all other nozzle machines, the risk for cell break-up is smaller. The latest development for separators with retracted nozzles is the application of a discharge mechanism like the one use for solids-ejecting separators with axial channels (Figure 5e). The bowl must have a star-shaped inner contour, *i.e.* integral filler pieces. The discharge mechanism is used for a very efficient cleaning-in-place procedure.

21.4.3.2 *Nozzle machine with peripheral nozzles*

Sediments which easily pack together or have adverse flow properties cannot be discharged through a nozzle machine with retracted nozzles. Therefore, early in the 1930s, when the requirement for a centrifuge which could continuously discharge large amounts of solids arose, the machine in Figure 5c was developed. It is equipped with peripheral nozzles. It can be equipped with a recirculation device for the sediment—a separate central tube which is divided into tubes along the bowl bottom ending just in front of the nozzle. This makes it possible to increase the concentration of the solids phase, without having to decrease the nozzle size leading to increased risk of nozzle clogging. Hsu (1981) presents some Japanese investigations on nozzle flow rates and gives equations for flow rate as a function of nozzle and bowl geometry.

21.4.3.3 *Number of nozzles*

In all nozzle machines, solids accumulate between the nozzles or nozzle channels, forming cones. The height of the cone depends on the angle of repose of the solids. This cone must never build up into the disc stack. When this happens, the separation is seriously impaired. Therefore, the number of nozzles must be increased when the bowl size is increased. The number varies from four on laboratory size machines up to 20 on the very biggest production machines. The solids cone can be replaced by filler pieces made of plastic or machined out of the bowl.

21.4.3.4   *Solids-ejecting separator with radial slits*

With increasing cost of labor and the introduction of continuous processes, the cleaning of the solid-bowl separators became more and more troublesome. Together with the foremost limitation of the nozzle machine—the low concentration of solids—this led to the development of the solids-ejecting separator (Figure 5d). It was introduced at the beginning of the 1950s, and is now the most common type of disc centrifuge. With an automated, periodical partial discharge of the sediment it is possible to obtain a considerably higher concentration of the solids than with the nozzle machine. The machine is discharged in about 0.1 second by pressing a so-called sliding bowl bottom in the bottom part of the bowl downwards, thereby opening big discharge ports in the outer bowl wall. In order to make the discharge at the right moment, the discharge can be initiated by a mechanism which hydraulically senses the solids level in the solids space (so-called self-triggering) (Barker 1975, Broadwell 1974).

This type of centrifuge has been developed much along the same lines as chamber centrifuges: cooling of bowl, centripetal pump and sediment catcher. Hermetic seals have been introduced between the bowl casing and the bearing and gear housing (drive system) enabling sterilization with steam at 120 °C and above. Figure 8 shows a modern solids-ejecting separator developed for the biotechnology industry. The bowl casing is a pressure vessel which can withstand a pressure of 300 kPa. Solids-ejecting machines can also be equipped with hermetic seals and a hermetic inlet from below in a hollow bowl spindle.

Figure 6   Yeast separation and washing plant with three machines of type FEUX 512T-31C. From Swedish Yeast Corporation, Rotebro, Sweden (Courtesy of Alfa Laval Separation AB, Sweden)

21.4.3.5 *Solids-ejecting separator with axial channels*

The separation efficiency of a centrifugal separator may be expressed as a function of the dimensions of the disc stack and the rotational speed. The disc stack of the solids-ejecting machine (Figure 5d) will be limited due to a big sludge room and an interior sliding bowl bottom. The lock ring joint, which connects the upper and bottom parts of the rotor, must also have a diameter close to the outer diameter of the rotor in order to allow fitting of the sliding bowl bottom. As the solids outlets are shaped like axial channels, tightened from the outside, they allow a big disc stack and the lock ring joint being placed nearer the center, where the centrifugal force is lower. In this way, the rotational speed can be increased (Lagergren, 1980). The new solids-ejecting nozzle separator (Figure 5e) is an example of what is achievable in this way with an industrial, centrifugal separator of medium size. At a rotor diameter of 600 mm 14 000g is obtained at the periphery, which is more than twice the g-force of solids-ejecting machines with radial slits of the corresponding size.

This machine has the appropriate seals (*cf.* solid-bowl machines) to make it suitable for incorporation in a sterile system. It is jacketed for cooling of the frame. The discharge mechanism is operated with pressurized air instead of water, which is an advantage in sterile systems. The pressurized air forces a ring-slide, which is equipped with valve seats, downwards, opening the axial channels. The bowl is machined in such a way that the solids are forced towards the channels, as described above in Section 21.4.3.1. In Figure 7, the machine is shown installed in an insulin production plant.

Figure 7   Solids-ejecting centrifuge type AX 213S-31 installed in an insulin production plant (Courtesy of Alfa Laval Separation AB, Sweden)

21.4.4   Summary

A summary of some data on centrifugal separators is given in Table 2. The figures for solids content in the feed material are only a guide. The largest solids-ejecting separators of today can handle a solids flow maximum of 2700 l h^{-1} (12 g.p.m.). The corresponding figure for nozzle machines is above 50 m^3 h^{-1} (220 g.p.m.).

Bell *et al.* (1983) have listed some of the advantages and disadvantages of various types of separator; see Table 3.

Table 2   Basic Types of Centrifugal Separators

	Transport of sediment	*Solids content in feed material (% by vol.)*	Maximum throughput in largest machine [m^3 h^{-1} (g.p.m.)]
Solids bowl separators	Stays in bowl	0–1	150 (650)
Solids-ejecting nozzle separator	Intermittent discharge through axial channels	0.01–10	200 (880)
Solids-ejecting separators	Intermittent discharge through radial slot	0.2–20	100 (440)
Nozzle separator	Continuous discharge through nozzles at or near bowl periphery	1–30	300 (1320)
Decanter	Internal screw conveyor	5–80	200 (880)

Table 3   Advantages and Disadvantages of Various Separator Types

System	*Advantages*	*Disadvantages*
Tubular bowl	a) High centrifugal force b) Good dewatering c) Easy to clean d) Simple dismantling of bowl	a) Limited solids capacity b) Foaming unless special skimming or centripetal pump used c) Recovery of solids difficult
Chamber bowl	a) Clarification efficiency remains constant until sludge space full b) Large solids holding capacity c) Good dewatering d) Bowl cooling possible	a) No solids discharge b) Cleaning more difficult than tubular bowl c) Solids recovery difficult
Disc centrifuges	a) Solids discharge possible b) Liquid discharge under pressure eliminates foaming c) Bowl cooling possible	a) Poor dewatering b) Difficult to clean
Scroll discharge	a) Continuous solids discharge b) High feed solids concentration	a) Low centrifugal force b) Turbulence created by scroll

21.5   METHODS FOR SELECTION OF CENTRIFUGES

Many papers have discussed the design of solid/liquid separation systems (Davies, 1965; Tiller, 1974; Fitch, 1974). Several books have been written on this subject (Svarovsky, 1977; Records, 1977; Purchas, 1981). To complete the picture, cross-flow filtration should also be considered among the possible methods (Rosén and Datar, 1983).

Before choosing the equipment, it is important to define the process requirements. The next step is usually a test-tube centrifugation, which can give an approximate Σ or KQ value for the full-scale centrifuges, employing the efficiencies mentioned in Section 21.3.

Apart from the settling characteristics, a number of other parameters must be considered, including possible pretreatment (heating or flocculation), corrosion properties, particle susceptibility to shear forces, sterility and containment possibilities and explosion hazards (Day, 1974; Moyers, 1966).

The final step in the selection of machine is usually pilot-machine testing. The purpose of this is at least twofold: to obtain a more reliable Q/Σ or Q/KQ than test-tube centrifugation can give and to assess the solids flow behavior in the bowl and in the receptacle.

21.6 DESCRIPTION OF APPLICATIONS

21.6.1 Introduction

Centrifuges in the biotechnical industry can be found in many stages of the process. For the sake of this presentation the process can be divided into the following stages: substrate pretreatment; biomass separation; further processing. For the biomass separation two cases are possible: biomass or supernatant valuable.

21.6.2 Separators in Substrate Preparation

21.6.2.1 Wort production for beer

When boiling the wort in the copper, hot break is formed. The amount of trub is between 0.5 and 1% by volume. If powder hops or hop-pellets are used, the total solids content increases up to about 2% by volume. The solids must be removed, and four methods are in use: settling, whirlpool treatment, filtration and centrifugation. Here solids-ejecting centrifuges with self-triggering are used. To recover wort from whirlpool, decanters can be used. Of the four clarification methods, centrifugation gives the right clarification, low extract losses and low effluent load.

21.6.2.2 Molasses pretreatment

For many fermentation products, *i.e.* bakers' yeast, ethanol and organic acids, it has been found advantageous to clarify the molasses in conjunction with sterilization or pasteurization. The reason is better quality of the yeast or easier further processing (*e.g.* lower scaling by Ca salts).

The clarification is performed in solids-ejecting machines or nozzle machines, the latter when the solids loading is high, which often is the case with cane molasses. When the solids concentration is high, the loss of molasses with the solids from the centrifuge can become so high that a molasses recovery system with a solids-ejecting machine or preferably a decanter can be profitable, especially when the cost of the molasses is high (Alfa Laval, 1984a).

21.6.3 Biomass Separation

21.6.3.1 Introduction

The range of settling characteristics of various microorganisms can be seen in Table 4, together with the type of centrifuge used. The low relative throughput of *Actinomyces* in spite of their large size is due to the very low density difference.

Table 4 Biomass Separation

Product	Microorganism Type	Size (μm)	Relative throughput in centrifuge	Type of separator
Bakers' yeast	*Saccharomyces*	7–10	100	Nozzle
Brewers' yeast	*Saccharomyces*	5–8	70	Nozzle, solids-ejecting
Alcohol yeast	*Saccharomyces*	5–8	60	Nozzle, solids-ejecting
SCP	*Candida*	4–7	50	Nozzle, decanter
Antibiotics	Mould	—	10–20	Decanter
Antibiotics	*Actinomyces*	10–20	7	Solids-ejecting
Citric acid	Mould	—	20–30	Solids-ejecting, decanter
Enzymes	*Bacillus*	1–3	7	Nozzle, solids-ejecting
Vaccines	*Clostridia*	1–3	5	Solid bowl, solids-ejecting

For the supernatant some requirements are:
  practically sterile (enzymes);
  absolutely clear (beer, antibiotics);
  somewhat cloudy (SCP, antibiotics);
  turbid (alcohol).

For the biomass:
    highest possible dry matter content (SCP);
    right consistency (antibiotics);
    cleaned, washed (bakers' yeast, SCP);
    minimal amount of entrained valuable product (antibiotics, beer, organic acids).

The separation efficiency needed in biomass separation varies from 70% up to 99.99% and higher. In order to illustrate what this difference means to the throughput of a given separator, the results found by Higgins *et al.* (1978) can be used. They separated *E. coli* and plotted unsedimented solids fraction against Q/Σ in a normal distribution/log diagram. In order to increase efficiency from 70% to 99.99% the feed flow rate had to be decreased by factor of seven. For microorganisms with a wider particle size distribution the factor is larger.

21.6.3.2 *Supernatant valuable*

After the separation of the biomass, the supernatant is subjected to further processing, *e.g.*

adsorption	glutamic acid
liquid/liquid extraction	penicillin
crystallization/precipitation	tetracycline
distillation	ethanol
ultrafiltration	enzymes
bottling	beer, wine

It is not very surprising that the separation systems are fairly diverse and employ a whole range of separation equipment.

The clarification of beer after fermentation is one of the oldest applications of solids-ejecting separators. About $20\text{--}80 \times 10^6$ yeast cells ml^{-1} (0.5–2% by volume) are reduced to $0.2\text{--}0.4 \times 10^6$ cells ml^{-1}. Today self-triggering is employed. The only alternative to centrifugation is gravity settling, which is a very slow process, and is less and less employed. After this centrifugation the beer is conditioned, and protein precipitates. In order to remove this, kieselguhr and sheet filters are used before bottling. To reduce the trouble with the filtration, the beer is often preclarified in solids-ejecting machines. All machines in beer clarification are equipped with a hydro-hermetic seal to avoid air pick-up or CO_2 losses.

In the production of alcohol from molasses, the Melle–Boinot method is used (Lagomasino, 1949). Here the yeast from one batch is recovered, conditioned and used for inoculating the next fermenting tank. This is the method used today in Brazil for its alcohol-for-gasoline program. Machines with retracted nozzles, peripheral nozzles (Figure 5c) or with pressurized yeast discharge (Figure 5b) are used. The peripheral nozzle type is preferred for the cases where the solids content of molasses or process water is high. For economic and environmental reasons it is important to keep the liquid volumes as low as possible in grain alcohol plants. One possibility is to use distillery slop as mashing-in liquid. Ronkainen *et al.* (1978) have shown that this is feasible provided that suspended solids are removed. This is done in a decanter with a solids-ejecting separator as a polisher.

Alcohol from spent sulfite liquor is produced by continuous fermentation with yeast recycling, accomplished with nozzle machines of either of the three above-mentioned types (Ericsson, 1947).

In the continuous Biostil fermentation and distillation process (Cook, 1983), nozzle machines are used to recover yeast before distillation and recycle it to the fermenter. From the beer still, an intermediate fraction is recycled back to the fermenter, which operates under limitation of osmotic pressure.

In glutamic acid and lysine production, machines with pressurized concentrate discharge are used. In order to recover glutamic acid from the cells, a three-stage countercurrent washing system is used.

In the production of *Actinomyces* antibiotics, the most common harvesting method is vacuum drum filtration with precoat. The suspension is often very difficult to filtrate, and the kieselguhr cost is high. Some producers use centrifuges instead. *Erythromycin mycelium* is harvested in solids-ejecting mchines. The mycelium is washed with water once (Larsson, 1974).

In vaccine production (Hepple, 1971; Lawrence and Barry, 1982) solid bowl and solids-ejecting machines of both types are used for harvesting of *Clostridia* bacteria before filtration. The machines are hermetically sealed to avoid the creation of aerosols. Examples of processes using other bacteria have been published (Walker and Foster, 1982; van Hemert, 1982a).

In extracellular enzyme manufacture (Aunstrup, 1979), disc-bowl centrifuges are often used for harvesting, especially if a bacterial enzyme is manufactured. When a fungus is used for production, filtration may be advantageous. When the substrate contains large and/or fibrous particles, decanter centrifuges can precede the disc-stack machine.

The clarified liquid from biomass separation in bacterial enzyme manufacture must in some cases have a very low cell count. One way of reducing the bacterial count is to use a machine common in dairies: a bactofuge (Thurell, 1977) which has a hermetic inlet and a cooled frame. In the bactofuge a bacterial concentrate of 3% by volume of the feed is continuously drawn off. A two-stage process typically reduces the cell count from 10^8 to 10^3 per ml (Larsson, 1974).

21.6.3.3 Biomass valuable

Some examples of process steps after biomass harvesting are:

evaporation/drying	SCP, lysine for fodder
packaging	bakers' yeast
cell break-up	endoenzymes
autolysis	yeast extract

The oldest application of centrifugal separation in the fermentation industry is in the production of bakers' yeast. The separation is always a series separation with countercurrent water washing in three steps (Figure 6). Separators with retracted nozzles or with pressurized yeast discharge are used. The latter has the advantage that the yeast is always under pressure in closed pipes which gives better hygiene than external pumps. It is also an advantage from the cleaning-in-place point of view.

In the production of single-cell protein it is of highest importance to concentrate the biomass as much as possible. The final process stage is spray, flash or drum drying, sometimes preceded by evaporation. The thermal dewatering is much more expensive per kg of removed water than mechanical dewatering (Labuza, 1975). Most of the single-cell protein produced today is made from spent sulfite liquor. Nozzle-machines are used. Petroleum yeast plants are in operation in Comecon countries. One is using gas–oil as a substrate. Such a process, which needs at least three separation stages, has been described by Lainé and du Chaffaut (1975). A demonstration plant using the bacterium *Methylophilus methylotrophus*, feeding on methanol, has been built in Great Britain (King, 1982). After fermentation the broth is subjected to a pH and heat shock, causing flocculation (Mayes, 1982). The protein mass ('Pruteen') is then concentrated by flotation (Atkinson and Mavituna, 1983), and before drying is further dewatered by nozzle centrifuges with pressurized concentrate discharge (Figure 5b).

Some kinds of antibiotics, *e.g.* bacitracin and tetracycline, are used as growth stimulants. Therefore, the cell mass can also be included in the fodder. The active ingredient can be precipitated on the cell mass and recovered in solids-ejecting or nozzle machines before drying.

In intracellular enzyme production, solids-ejecting centrifuges of both kinds are used (Figures 5d and e). Examples of this are glucose isomerase from *Bacillus coagulans* (Hemmingsen, 1979), β-galactosidase (Higgins *et al.*, 1978) and penicillin acylase from *E. coli*.

So far, the latter organism is the most widely used in recombinant DNA processes; solids-ejecting machines with axial channels are often employed. These machines are also used in steroid processes. Zamola *et al.* (1981) describe the use of various machines for recovery of microbial insecticides. Atkinson and Mavituna (1983) have reviewed various processes for vitamin B_{12}. Very little is known of what types of centrifuge are used, however.

After making initial batch centrifugations, Patrick and Freeman (1960) verified the estimated flow rates for continuous processing of virus in a tube centrifuge.

21.6.4 Separators in Further Processing

21.6.4.1 Supernatant valuable

The production of penicillin has been described in detail elsewhere (Rehm, 1967; Sylvester and Coghill, 1954). The extraction is usually performed in centrifugal extractors. Various such machines are described by Gebauer *et al.* (1982). The two most used machines can be compared to a rotating sieve tray column and a countercurrent integrated mixer–settler, respectively.

The three-phase decanter centrifuge has recently been introduced for penicillin extraction from mycelium-containing broth. The process is a countercurrent extraction in two steps (Katinger *et al.*, 1981). In extracellular enzyme production, solids-ejecting centrifuges are often used to recover the precipitated protein fractions (Aunstrup, 1979). A review article on protein precipitation methods and parameters has recently been published (Bell *et al.*, 1983). Precipitation is often used to isolate and concentrate antibiotics, *e.g.* rifamycin, erythromycin and tetracycline, with the use of the same machine type for recovery.

Precipitated calcium citrate and crude monosodium glutamate (MSG) crystals are often recovered in decanters.

21.6.4.2 Biomass valuable

The isolation and purification of intracellular molecules is a much more challenging task than for extracellular products. It is however more and more important with the advent of recombinant DNA technology.

After autolysis or homogenization of the cells, the cell debris must be removed in most cases. This is usually a very difficult separation, since the particles are small and of irregular shape. At the same time, RNA can give high viscosity to the suspension (Higgins *et al.*, 1978; Mosqueira *et al.*, 1981). Machines with high *g*-numbers are to be preferred on an industrial scale, *e.g.* the solids-ejecting centrifuge with axial channels. For *E. coli* debris, the flow rate for acceptable separation in a given machine is often lower by a factor of two to three compared with whole cells, and by a much bigger factor for large cells, *e.g.* bakers' yeast. It is also difficult to eliminate all the turbidity because of the large fraction of fines and colloidal material.

An important similar application is the production of yeast extract, where cell debris is separated and washed in a counter-current system (Peppler, 1982).

A new extraction method could make it possible in some cases to isolate the wanted substance in the presence of cell debris. Kula *et al.* (1982) and Kroner *et al.* (1984) have employed two immiscible aqueous phases in enzyme recovery. They have used laboratory scale nozzle and solids-ejecting centrifuges for the separation. An economic analysis (Kroner *et al.*, 1984) shows that the method is promising, but that the chemicals cost may be too high in some cases.

21.7 CONTAINMENT AND STERILITY

Incidents of exposure to biological materials leading to infections, allergic reactions and sensitization have been reported (Dunnill, 1982; Mayes, 1982; Wedum, 1973). In many cases, these incidents have been caused by aerosols created by the centrifuge involved.

For small centrifuges, zonal or tube, cabinets with varying degrees of isolation have been designed. The simplest of these are laminar air flow cabinets, whereas the more complicated ones have physical barriers between the operator and the work area and other arrangements, *e.g.* negative air pressure in the cabinet and HEPA (High Efficiency Particulate Air) filters on exit air (Henke, 1973; Evans *et al.*, 1974; van der Groen *et al.*, 1980; West, 1981; Flickinger and Sansone, 1984). For these small machines, the cabinet is regarded as the primary containment barrier. For larger units, totally enclosed steam-sterilizable systems have been developed (van Hemert, 1982b; Lawrence and Barry, 1982; Walker and Foster, 1982) for which the frame of the machine is considered to be the primary barrier (Figure 8).

In contrast to many small centrifuges, the industrial types do not run in vacuum. Therefore, the bowl heats up the air round it through friction. This makes it necessary to have good ventilation of air in the piping system and cooling of the upper and lower part of the frame and the solids catcher. In Figure 9 the principle of an installation of a solids-ejecting centrifuge with axial channels may be seen. The version shown is designed without pumps. On discharging, both types of solids-ejecting machines produce a pressure shock wave that propagates into the concentrate catcher (here a cyclone) and into the concentrate tank. Therefore, the volume of this must be above a certain minimum. It must be ventilated through a sterile air filter. When the discharged volume of suspension leaves the bowl, a vacuum is created, which is eliminated by a circulation pipe from the top of the cyclone *via* a valve to the frame. The valve opens automatically for a short time at discharge. In steam-sterilizable machines, mechanical seals are necessary (*cf.* Section 21.4.2.1). They must be cooled and lubricated with sterile water; the spent water is collected in the slop tank. The discharge mechanism is operated by sterile air, sterile water or saline. The

Figure 8 Solids-ejecting centrifuge type BTPX 205SGD-34CDLP in a sealed and sterile system. Control panel in the foreground, solids and slop tanks in the background (Courtesy of Alfa Laval Separation AB, Sweden)

sterile water or saline is collected in the slop tank. The cleaning-in-place liquids are also collected in this tank.

A method for risk studies, HAZOP, has been developed for use in the chemical industry (Lawley, 1976). It is a systematic analysis of the consequences of process and operational deviation from the normal with a given plant design. It can also be used to advantage for fermentation processes.

The large centrifuge manufacturers develop, together with centrifuge users, cleaning and sterilization programs. These are usually based on a number of basic rules, but the details are largely dependent on process requirements and the specific installation design.

21.8 SUMMARY

The fluid and particle dynamics, starting from Stokes' law, are described for different centrifuge types. Especially in the case of disc-stack centrifuges the internal flow patterns are very complicated and some of the latest research is presented. As a basis for scale-up, the Σ model is described and its limitations and simplifying assumptions are presented. The KQ formula is derived. The salient design features of different types of centrifuge are described. The methods for selection of centrifuges are discussed briefly. A description of applications of centrifuges in the biotechnological industry is made, presenting the use of the machines for substrate pretreatment,

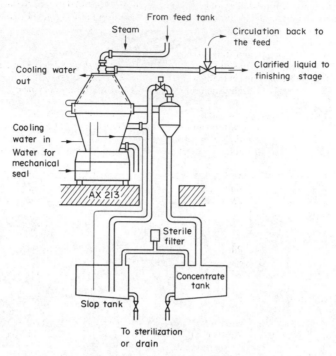

Figure 9 Separation of pathogenic bacteria cells. Installation of AX 213S-31 (Courtesy of Alfa Laval Separation AB, Sweden)

biomass separation and further processing. Finally, sterility and containment problems are discussed.

21.9 REFERENCES

Aiba, S., A. E. Humphrey and N. Millis (1965). *Biochemical Engineering*. Academic, New York.
Alfa Laval (1984a). Equipment for the fermentation industries. Technical brochure IB 40601 E2. Tumba, Sweden.
Alfa Laval (1984b). A complete range of centrifuges for the pharmaceutical industry. Technical brochure IB 41045E. Tumba, Sweden.
Ambler, C. M. (1952). The evaluation of centrifuge performance. *Chem. Eng. Progr.*, **48**, 150–158.
Ambler, C. M. (1959). The theory of scaling up laboratory data for the sedimentation type centrifuge. *J. Biochem. Microbiol. Technol. Eng.*, **1**, 185–205.
Ambler, C. M. (1961). Centrifugation equipment—theory. *Ind. Eng. Chem.*, **53**, 430–433.
Ambler, C. M. (1979). Centrifugation. In *Handbook of Separation Techniques for Chemical Engineers*, ed. P. A. Schweitzer, pp. 4-55–4-104. McGraw-Hill, New York.
Atkinson, B. and I. S. Daoud (1976). Microbial flocs and flocculation in fermentation process engineering. In *Advances in Biochemical Engineering*, ed. A. Fiechter, vol. 4, pp. 41–124. Springer-Verlag, Berlin.
Atkinson, B. and F. Mavituna (1983). *Biochemical Engineering and Biotechnology Handbook*. Nature Press, New York.
Aunstrup, K. (1979). Production, isolation and economics of extracellular enzymes. In *Applied Biochemistry and Bioengineering*, ed. L. B. Wingard, Jr. *et al.*, vol. 2, pp. 27–69. Academic, New York.
Barker, T. A. (1975). Control systems for centrifuges and centrifugal processes. *Filtr. Sep.*, **12**, 33–36.
Bell, D. J., M. Hoare and P. Dunnill (1983). The formation of protein precipitates and their centrifugal recovery. In *Advances in Biochemical Engineering/Biotechnology*, ed. A. Fiechter, vol. 26, pp. 1–72. Springer-Verlag, Berlin.
Bergner, N. (1957). The importance of particle size distribution for technical centrifugal separation. *Trans. Inst. Chem. Eng.*, **35**, 181–194.
Bohman, H. (1974). Hydrodynamics of the liquid flow in the disk stack of a centrifugal separator. In *Proc. Eur. Conf. Mixing Centrifugal Sep.*, *1st, 1974*, paper F3. BHRA Fluid Engineering, Cranfield.
Bohman, H. and J. Murkes (1969). Centrifuges—uniformity and equivalence conditions. *Chem. Process Eng. (London)*, **50** (7), 97–99.
Broadwell, E. (1974). New separation developments in protein processes. *Filtr. Sep.*, **11**, 509–515.
Brunner, A. (1956). *Über das Reinigen von Schweröl mittels der Zentrifuge*. Dissertation, Technische Hochschule, Zürich.
Brunner, K.-H. and O. Molerus (1977). Partikelbewegung im Tellerseparator. *Verfahrenstechnik*, **11**, 538–541.
Brunner, K.-H. and O. Molerus (1979). Theoretical and experimental investigation of separation efficiency of disc centrifuges. *Ger. Chem. Eng.* (Engl. Transl.), **2**, 228–233.
Carlsson, C.-G. (1979). *Separation Efficiency and the Flow of Two Liquid Layers in a Disk Stack Centrifuge*. Dissertation, Chalmers University of Technology, Gothenburg, Sweden.

Carlsson, C.-G. (1980). Mathematical modelling of centrifugal separation including shear force effects. Paper presented at the International Symposium on Solids Separation Processes. Institution of Chemical Engineers, Irish branch, Dublin, Ireland.

Carlsson, C.-G. (1984). Personal communication.

Cook, R. (1983). Biostil—a breakthrough in distillery design. *Sugar Azucar Yearbook*.

Davies, E. (1965). Selection of equipment for solid–liquid separations. *Trans. Inst. Chem. Eng.*, **43**, T256–T259.

Day, R. W. (1974). Techniques for selecting centrifuges. *Chem. Eng.*, **81**, May 13, 98–104.

Dunnill, P. (1982). Biosafety in the large-scale isolation of intra-cellular microbial enzymes. *Chem. Ind. London* (22), November 20, 877–879.

Ericsson, E. O. (1947). Alcohol from sulphite waste liquor. *Chem. Eng. Progr.*, **43**, 165–167.

Evans, C. G. T., R. Harris-Smith, J. E. D. Stratton and J. Melling (1974). Design and construction of a ventilated cabinet for a continuous flow centrifuge. *Biotechnol. Bioeng.*, **16**, 1681–1687.

Fish, N. M. and M. D. Lilly (1984). The interactions between fermentation and protein recovery. *Bio/Technology*, **2**, 623–627.

Fitch, B. (1965). Separating power of disc centrifuges. Paper presented at AIChE 55th National Meeting, Preprint 11a.

Fitch, B. (1974). Choosing a separation technique. *Chem. Eng. Progr.*, **70**, December, 33–37.

Foster, P. R. and J. G. Watt (1980). The CSVM fractionation process. In *Methods of Plasma Protein Fractionation*, ed. J. M. Curling, pp. 17–31. Academic, London.

Frampton, G. A. (1963). Evaluating the performance of industrial centrifuges. *Chem. Process Eng. (London)*, **44**, 402–412.

Flickinger, M. C. and E. B. Sansone (1984). Pilot- and production-scale containement of cytotoxic and oncogenic fermentation processes. *Biotechnol. Bioeng.*, **24**, 860–870.

Gebauer, K., L. Steiner and S. Hartland (1982). Zentrifugalextraktion—Eine Literaturübersicht. *Chem.-Ing.-Tech.*, **54**, 476–496.

Goldin, E. M. (1971). Hydrodynamics of clarifying separators with guiding fins and peripheral feed. *Theor. Found. Chem. Eng. (Engl. Transl.)*, **5**, 246–252.

Gösele, W. (1980). Scale-up of helical conveyor type decanter centrifuges. *Ger. Chem. Eng. (Engl. Transl.)*, **3**, 353–359.

Gupta, S. K. (1980). Comparative study of Ambler's and Jury's and Locke's analyses of a disc-stack centrifuge. *Indian Chem. Eng.*, **22**, 39–42.

Gupta, S. K. (1981). Scale-up procedures for disc-stack centrifuges. *Chem. Eng. J.*, **22**, 43–49.

Hemmingsen, S. H. (1979). Development of an immobilized glucose isomerase for industrial application. In *Applied Biochemistry and Bioengineering*, ed. L. B. Wingard, Jr. *et al.*, vol 2, pp. 157–183. Academic, New York.

Henke, C. B. (1973). Containment recommendations and prototype equipment for ultracentrifuges in cancer research. In *Proceedings of the National Cancer Institute Symposium on Centrifuge Biohazards*. Cancer Research Safety Monograph Series, vol. 1, pp. 111–136. DHEW Publ. no. (NIH) 78-373.

Hepple, J. R. (1971). Filtration in the manufacture of vaccines and anti-sera. Paper presented at the Conference on Filtration in Process Plant Design and Development. Filtech/71, London.

Higgins, J. J., D. J. Lewis, W. H. Daly, F. G. Mosqueira, P. Dunnill and M. D. Lilly (1978). Investigation of the unit operations involved in the continuous flow isolation of β-galactosidase from *E. coli*. *Biotechnol. Bioeng.*, **20**, 159–182.

Hsu, H.-W. (1981). *Separations by Centrifugal Phenomena*. Wiley, New York.

Jury, S. H. and W. L. Locke (1957). Continuous centrifugation in a disc centrifuge. *AICE J.*, **3**, 480–483.

Katinger, H., F. Wibbelt and H. Scherfler (1981). Kontinuerliche direkte Extraktion von Antibiotika mit Separatoren und Dekantern. *Verfahrenstechnik*, **15**, 179–182.

King, P. P. (1982). Biotechnology. An industrial view. *J. Chem. Tech. Biotechnol.*, **32**, 2–8.

Kroner, K. H., H. Hustedt and M.-R. Kula (1984). Extractive enzyme recovery: economic considerations. *Process Biochem.*, **19**, 170–179.

Kula, M.-R., K. H. Kroner and H. Hustedt (1982). Purification of enzymes by liquid–liquid extraction. In *Advances in Biochemical Engineering*, ed. A. Fiechter, vol. 24, pp. 73–118. Springer-Verlag, Berlin.

Labuza, T. P. (1975). Recovery and drying for SCP manufacture. In *Single-cell Protein II*, ed. S. R. Tannenbaum and D. I. C. Wang, pp. 69–104. MIT Press, Cambridge, MA.

Lagergren, B. (1980). Recent advances in separation technology. In *Food Process Engineering*. ed. P. Linko *et al.*, vol. 1. pp. 653–666. Applied Science Publishers, Barking.

Lagomasino, J. M. (1949). The Melle–Boinot alcoholic fermentation method using the yeast repeatedly. *Int. Sugar J.*, **51**, 338–339.

Lainé, B. M. and J. du Chaffaut (1975). Gas–oil as a substrate for single-cell protein production. In *Single-cell protein II*, ed. S. R. Tannenbaum and D. I. C. Wang, pp. 424–437. MIT Press, Cambridge, MA, USA.

Larsson, Å. (1974). Various approaches to the separation process for harvesting of the products of fermentation in the field of antibiotics. *Biotechnol. Bioeng. Symp.*, **4**, 917–931.

Lavanchy, A. C., F. W. Keith and J. W. Beams (1964). Centrifugal separation. In *Kirk–Othmer Encyclopedia of Chemical Technology.*, 2nd ed., vol. 4, pp. 710–758. Wiley, New York.

Lawley, H. C. (1976). Size up plant hazards this way. *Hydrocarbon Process*, **55**, 247–252.

Lawrence, A. and A. Barry (1982). Potential hazards associated with the large scale manufacture of bacterial vaccines. *Chem. Ind. (London)* (22), November 20, 880–884.

Lilly, M. D. (1979). Production of intracellular microbial enzymes. In *Applied Biochemistry and Bioengineering.*, ed. L. B. Wingard, Jr. *et al.*, vol. 2, pp. 1–26. Academic, New York.

Mayes, R. W. (1982). Lack of allergic reactions in workers exposed to Pruteen (bacterial single-cell protein). *Br. J. Ind. Med.*, **39**, 183–186.

Morris, B. G. (1966). Application and selection of centrifuges. Part 1. *Br. Chem. Eng.*, **11**, 347–351.

Mosqueira, F. G., J. J. Higgins, P. Dunnill and M. D. Lilly (1981). Characteristics of mechanically disrupted bakers' yeast in relation to its separation in industrial centrifuges. *Biotechnol. Bioeng.*, **23**, 335–343.

Moyers, C. G., Jr. (1966). How to approach a centrifugation problem. *Chem. Eng.*, **73**, June 20, 182–189.

Murkes, J. (1969). The effect of suspension characteristics in centrifugal separation. *Br. Chem. Eng.*, **14**, 1692–1697.

Murkes, J. and C.-G. Carlsson (1978). Mathematical modelling and optimisation of centrifugal separation. *Filtr. Sep.*, **15**, 18–22.

Patrick, W. C., III and R. R. Freeman (1960). Continuous centrifugation in virus processing. *J. Biochem. Microbiol. Technol. Eng.*, **2**, 71–80.

Peppler, H. J. (1982). Yeast extracts. In *Economic Microbiology*, ed. A. H. Rose, vol. 7, pp. 293–312. Academic, London.

Purchas, D. B. (1981). *Solid/Liquid Separation Technology*. Uplands Press, Croydon.

Rachitskii, V. A. and L. S. Skvortsov (1980). Method for determining the load of a packet of trays of a centrifugal separator. *J. Appl. Chem. USSR (Engl. Transl.)*, **53**, 367–370.

Records, F. A. (1977) Sedimenting centrifuges. In *Solid/Liquid Separation Equipment Scale-up*, ed. D. B. Purchas, pp. 199–240. Uplands Press, Croydon.

Rehm, H.-J. (1967). *Industrielle Mikrobiologie*. Springer-Verlag, Berlin.

Rickwood, D. (1978). *Centrifugation: A Practical Approach*. Information Retrieval Ltd, London.

Ronkainen, P., O. Leppänen and K. Harju (1978). The re-use of stillage water in the mashing of grain as a means of energy conservation. *J. Inst. Brewing (London)*, **84**, 115–117.

Rosén, C.-G. and R. Datar (1983). Primary separation steps in fermentation processes. In *Proc. Biotech '83, London*, pp. 201–224. Online Publications, Northwood.

Rosen, K. (1978). Continuous production of alcohol. *Process Biochem.*, **13** (5), 25–26.

Schnittger, J. R. (1970). Integrated theory of separation for bulk centrifuges. *Ind. Eng. Chem., Process Des. Dev.*, **9**, 407–413.

Sokolov, V. I. (1971). *Moderne Industriezentrifugen*. VEB Verlag Technik, Berlin.

Sullivan, F. E. and R. A. Erikson (1961). Centrifugation equipment-design. *Ind. Eng. Chem.*, **53**, 434–438.

Svarovsky, L. (1977). Separation by centrifugal sedimentation. In *Solid–Liquid Separation*, ed. L. Svarovsky, pp. 124–147. Butterworths, London.

Sylvester, J. C. and R. D. Coghill (1954). The penicillin fermentation. In *Industrial Fermentations*, ed. L. A. Underkofler and R. J. Hickey, vol. 2, pp. 219–263. Chemical Publishing, New York.

Thurell, K.-E. (1977). Bactofugation of cheese milk. *Nordisk Mejeriindustri*, **4**, 239–241.

Tiller, F. M. (1974). Bench-scale design of SLS systems. *Chem. Eng.*, **81**, April 29, 117–119.

van der Groen, G., P. C. Trexler and S. R. Pattyn (1980). Negative-pressure flexible film isolator for work with class IV viruses in a maximum security laboratory. *J. Infection*, **2**, 165–170.

van Hemert, P. (1982a). Strictly aseptic techniques for continuous centrifugation. *Antonie van Leeuwenhoek*, **46**, 510.

van Hemert, P. (1982b). Biosafety aspects of a closed-system Westfalia-type continuous centrifuge. *Chem. Ind. (London)* (22), November 20, 889–891.

Vesilind, P. A. (1974). Estimating centrifuge capacities. *Chem. Eng.*, **81**, April 1, 54–57.

Walker, P. D. and W. H. Foster (1982). Containment in vaccine production. *Chem. Ind. (London)* (22), November 20, 884–887.

Wedum, A. G. (1973). Microbiological centrifuging hazards. In *Proceedings of the National Cancer Institute Symposium on Centrifuge Biohazards*. Cancer Research Safety Monograph Series, vol. 1, pp. 5–16. DHEW Publ. no. (NIH) 78-373.

West, D. L. (1981). Special design considerations: biohazard facilities for infectious microorganisms, recombinant DNA materials and oncogenic viruses. In *Proceedings of the National Cancer Institute Symposium on Design of Biomedical Research Facilities*. Cancer Research Safety Monograph Series, vol. 4, pp. 147–168. NIH Publ. no. 81–2305.

Wiesmann, U. and H. Binder (1982). Biomass separation from liquids by sedimentation and centrifugation. In *Advances in Biochemical Engineering*, ed. A. Fiechter, vol. 24, pp. 119–171. Springer-Verlag, Berlin.

Willus, C. A. and B. Fitch (1973). Flow patterns in a disc centrifuge. *Chem. Eng. Progr.*, **69**, 73–74.

Zamola, B., P. Valles, G. Meli, P. Miccoli and F. Kajfez (1981). Use of the centrifugal separation technique in manufacturing a bioinsecticide based on *Bacillus thuringiensis*. *Biotechnol. Bioeng.*, **23**, 1079–1086.

Zastrow, J. (1976). Theoretical calculation of the performance of disc centrifuges. *Int. Chem. Eng.*, **16**, 515–518.

Zeitsch, K. (1978). Effect of the feed rate on the active acceleration of overflow centrifuges. *Trans. Inst. Chem. Eng.*, **56**, 281–284.

22
Filtration of Fermentation Broths

P. A. BELTER
The Upjohn Company, Kalamazoo, MI, USA

22.1 INTRODUCTION

Filtration is a basic operation for separating solids from a liquid phase by forcing the liquid through a filter medium consisting of a solid support, a woven textile and a deposited layer of the solids. It is a straightforward procedure for well-defined crystalline materials, however, the physical characteristics of fermentation broths complicate the operations for a variety of reasons. The nature, sliminess, morphology and size of the microorganisms, pH and viscosity of the broth, temperature history of the broth, possible contamination by undesired microorganisms, and the inherent nature and physical properties of the insoluble portion of residual substrates are factors that contribute to filtration difficulties with fermentation broths. Non-Newtonian behavior of the broth and compressibility of the mycelia or cells, which in turn decrease the permeability of the accumulated cake on the filter surface, add additional complications for the filtration sequence.

The desired product may be found in either the soluble portion of the fermentation broth or in the biomass, *i.e.* the resultant filter cake. In either case separation of the two phases permits an appreciable upgrading of the product. Consequently, this unit operation is usually conducted early in the recovery sequence. If the desired product remains in the filter cake either as an intracellular constituent or because of solubility properties, subsequent treatment to solubilize the product will be necessary. This treatment will be followed by another separation operation which can, of course, be a second filtration step.

The major problem associated with filtration of fermentation broths is the compressibility of the accumulated solids on the filter surface and the resultant decrease of permeability during the filtration cycle. This phenomenon decreases filtration rates and washing efficiency, which in turn results in losses of product and undesirable overall economics. The problem has been minimized by the judicious use of precoat materials and the addition of filter aids to the fermentation broths prior to filtration. This approach coupled with the selection of appropriate equipment has resulted in the widespread use of filtration as a primary separation tool for the fermentation industry.

22.2 FILTER AIDS

Only two of the numerous filter aids proposed have achieved major commercial importance in the fermentation industry. These are the diatomaceous earths, the skeletal remains of tiny aquatic plants deposited centuries ago, and the perlite types, processed from volcanic rock. These materials are processed to give a wide range of particle sizes and flow characteristics. Typical properties of various grades of diatomaceous filter aids are presented in Table 1. This information is a summary extracted from brochures of Johns–Manville. Although the perlite materials do not resemble the diatomaceous earths microscopically, they have been processed to give high porosity. The selection of a combination of precoat material and suitable filter aid material to give the desired degree of clarity, filtration and washing rates are factors that must be considered in optimizing the filtration process for a given broth.

Table 1 Typical Properties of Diatomaceous Earths[a]

Grade	Density (lb ft^{-3}) Dry	Wet	pH	Water adsorption (%)	Relative flow rate
Filter Cel	7.0	15.9	7.0	235	100
Standard Super Cel	8.0	17.9	7.0	255	200
512 Hyflo	8.0	17.9	7.0	250	300
Super Cel	9.0	17.9	10.0	245	500
501	9.5	16.9	10.0	250	750
535	12.0	17.6	10.0	245	1350
545	12.0	18.0	10.01	240	2160

[a] Summary of data presented in Johns–Manville brochures.

Ground wood pulp and starch have been advocated as filter aids in applications where silica-containing materials are undesirable. These materials have been reserved for specialty uses only. In addition, the reported ability of aminoglycoside antibiotics and proteins to bind to diatomaceous earths and cellulosic filter aids (Wagman *et al.*, 1975) is a source of loss and must be considered in the final selection of a filter aid.

22.3 EQUIPMENT

The type of equipment selected for filtration of the broth will depend on the production rate, slurry concentration, particle size, physical properties of the two phases, and other unique properties of the broth such as toxicity and/or physiological activity.

Continuous rotary vacuum filters are usually the choice for large scale filtration of fermentation broths. Materials that tend to settle slowly and may be described as having slow to medium filtration rates are handled adequately by this type of equipment with the help of precoat and filter aid materials. Because of its general flexibility the rotary vacuum drum filter has the greatest variation in design and has become the work-horse of the fermentation industry.

This filter consists of a horizontally rotating cylinder whose outer surface is divided into individually segmented compartments. These compartments are connected internally by drainage tubes to a terminal plate located in one head of the cylinder. The terminal plate revolves against a stationary plate which is compartmented to permit the separation of filtrate and wash water and to permit application of different vacuum levels to independent drum compartments if needed. The cylinder is supplied with a central shaft and trunnion bearings so that the drum can be rotated and its speed varied. The cylinder is suppported within a trough tank equipped with a rake agitator to maintain the solids suspended in the slurry or broth for filtration. In addition wash headers are supported over the cylindrical drum to supply wash liquor to remove adhering broth from the filter cake.

Various methods of discharging the processed cake have been devised throughout the years. Selection of the method of discharge depends on the nature of the cake. The methods are: (1) simple doctor blade assisted by an air blowback through the filter medium; (2) a string discharge also assisted by the release of vacuum and a small blowback; (3) a continuous belt discharge where the filter medium is not caulked to the filter drum but travels as a continuous loop and lies against the rotating drum during the filtration and washing sequences in the cycle but leaves the drum and moves over a series of small diameter rolls for discharge; and (4) the continuous rotary precoat filter. In method (4), a layer of precoat material, such as diatomaceous earth, perlite materials, shredded cellulose or starch, is applied to the drum prior to filtration of the actual slurry. As the drum rotates during the filtration cycle, a thin layer of biomass and filter aid accumulates on the surface of the precoat and builds up during the filtration cycle. Upon further dewatering followed by washing of the formed cake, a doctor blade removes a thin layer of the cake and filter aid and exposes a fresh surface of precoat for the next cycle. The depth of the cut is also controlled to remove any penetrated particles. In this operation vacuum is maintained throughout the entire cycle. When the bed of precoat is depleted, a new layer of precoat material is applied and further filtration of the broth proceeds.

22.4 THEORY

The operation of continuous rotary filters, whether under vacuum or pressure, consists of a cyclic sequence containing the following stages: (1) cake formation; (2) cake dewatering; (3) cake washing to remove retained solubles; (4) cake dewatering; and (5) cake discharge.

The last stage, cake discharge, does not affect the size of the filter. However, the other segments of the cycle are interrelated by the geometry of the system and must be considered as a whole to understand and interpret actual operating results and/or to design a suitable filter for a given application.

Modification of the Poiseuille relationship results in the following correlation of the pertinent variables:

$$\frac{dV}{A\,dt} = \frac{\Delta p}{\mu_c[\alpha(\frac{W}{A}) + r]}$$

A = area of filtration surface; V = volume of filtrate collected; t = time; Δp = differential pressure across the filter; μ_c = broth viscosity; α = average specific cake resistance; W = mass of accumulated biomass; and r = resistance of filter medium.

Integration of this relationship, assuming a non-compressible cake, results in a linear relationship between t/V and V with a slope and intercept dependent on the conditions employed. However, cakes from fermentation broths are seldom non-compressible and actual filtration rates must be measured for ultimate design. Plots of this type showing filtration results for a *Streptomyces griseus* broth were published originally by Shirato and Esumi (1963). These rates may be obtained in the laboratory using the 'leaf-test' approach (Smith, 1976) or from pilot plant studies. In either case it is important that the slurry tested be typical of the broth to be filtered ultimately. Typical filtration rates for a variety of applications are given in Table 2. Dahlstrom and Purchas (1957) have published scale-up methods whereby the various elements of the filtration operation can be studied and incorporated into a suitable design for the overall filtration operation.

Table 2 Typical Filtration Rates

Product	Organism	Filter type	Filtration rate ($l\,h^{-1}\,ft^{-2}$)
Kanamycin	*S. kanamyceticus*	Vacuum precoat	8
Lincomycin	*S. lincolnensis*	Vacuum precoat	35
Neomycin	*S. fradiae*	Vacuum precoat	12
Penicillin	Unknown	Vacuum precoat	130–170
Protease	*B. subtilis*	Vacuum precoat	10–40
Disrupted yeast cake	Unknown	Rotary vacuum	28

Okabe and Aiba (1974) have presented an algorithm based on mass balances, theoretical filtration rate and washing rate relationships that simulates the operation of a continuous filter for the purpose of minimizing cost of the filtration operation. Rotation speed and amount of filter aid added were used as independent variables in their optimization study. It was shown that an optimal combination of these variables exists which gives a minimum cost for the filtration evaluated.

The same authors (1976) have demonstrated an overall optimization approach for a hypothetical two stage recovery process involving an antibiotic filtration step followed by a solvent extraction operation. Their approach consists of combining the algorithms for each segment and calculating the interactions involved. Simulations of this nature will help to develop optimal design and operation of antibiotic recovery processes.

Although filtration has been the major primary separation tool for antibiotics in the past, this unit operation is slowly being replaced by other approaches such as centrifugation and ion exchange. These alternatives are covered in other chapters in this volume.

22.5 SUMMARY

Filtration is one of the processing alternatives used to separate the insoluble mycelia or cells from the solubles in the liquid fermentation broth. The desired product may be found in either the soluble portion of the broth or in the biomass (the resultant filter cake).

The major problem associated with filtration of fermentation broths is the compressibility of the accumulated solids on the filter surface and the resultant decrease of permeability during the filtration and washing cycles. The problem has been minimized by the judicious use of precoat materials and the addition of filter aids to the broth prior to filtration. This approach coupled with the selection of appropriate equipment has resulted in the widespread use of filtration as a primary separation tool for the fermentation industry.

The general flexibility of the rotary vacuum drum filter has permitted the greatest variation in design and, consequently, it has become the work-horse of the industry.

The general theory for this type of equipment and methods for scale-up were covered briefly.

Although filtration has played a major role as a primary separation tool in the past, this unit operation is slowly being replaced by other approaches.

22.6 REFERENCES

Aiba, S. and M. Okabe (1976). Simulation of filtration and extraction, followed by coordinating optimization in an antibiotic recovery process. A demonstration. *Process Biochem.*, **11** (3), 25–6, 28–30.

Dahlstrom, D. A. and D. B. Pruchas (1957). *Joint Symposium: Scaling-up Chemical Plant and Processes (London)*, pp. 91–102.

Okabe, M. and S. Aiba (1974). Simulation and optimization of cultured-broth filtration. *J. Ferment. Technol.*, **52**, 759–777.

Shirato, S. and S. Esumi (1963). Filtration of the culture broth of *Streptomyces griseus*. *Hakko Kogaku Zasshi*, **41**, 86–92.

Smith, G. R. S. (1976). Filtration: advances and guidelines. 2. How to use rotary, vacuum, precoat filters. *Chem. Eng.*, **83** (4), 84–90.

Wagman, H., V. Bailey and M. Weistein (1975). Binding of antibiotics to filtration materials. *Antimicrob. Agents Chemother.*, **7**, 316–319.

23

Cell Processing Using Tangential Flow Filtration

F. R. GABLER
Millipore Corporation, Bedford, MA, USA

23.1 INTRODUCTION

This chapter will deal with membrane tangential flow filter applications for manipulating and processing cells. The types of cells that can be processed with membranes include viruses, bacteria, streptomyces, yeasts and mammalian cells. What will be presented is a brief summary of a relatively new way of performing some traditional processes on both the laboratory and industrial scales. Membrane filter processing of cells offers a number of distinct advantages over classical techniques in the way of increased speed and recovery. The purpose of this chapter is to present a basic understanding of how these membrane filter applications work and to illustrate their versatility and effectiveness.

23.2 OVERVIEW OF TANGENTIAL FLOW FILTRATION APPLICATIONS

Filtration as a method for performing coarse separations of components from complex mixtures has had a long history (Dudka, 1981), however, during this century, and particularly the last 20 years, extensive use of filtration has become routine in performing critical separations of biological components. The introduction of synthetic polymeric membrane filters has opened a vast array of applications that were previously awkward, expensive, time consuming, *etc.* (Brock, 1983; Kesting, 1971). There are basically three types or classes of membrane filters: microporous (MF), ultrafiltration (UF), and reverse osmosis (RO). Each class of membrane has its own characteristics, and some of these are noted in Table 1. Cell processing with membranes involves both UF and MF types.

351

Table 1   General Classes of Membrane Filters and their Properties

	Pore size (μm)	Entity retained	Membrane material	Major uses
Microporous	0.1–10	Bacteria	Cellulose esters, polyvinylidene fluoride, polycarbonate	Sterile filtration, cell processing
Ultrafiltration	0.001–0.05	Cells, macrosolutes	Polysulfone, cellulosics	Cell and macrosolute processing, pyrogen removal
Reverse osmosis	0.0005–0.001	Ions	Cellulose acetate, polyamides	Water purification

The specific applications of cell processing that will be covered in this chapter are as follows: cell harvesting, cell washing and cell recycle in fermentation.

Cell harvesting is that process in which cells from a fermenter are concentrated and separated from the liquid phase of the fermentation fluid. If the product of interest is intracellular, then it will be found in the solid or cell mass phase in a volume that is much smaller than that of the original fermenter. If the product of interest is extracellular, it will be located in the filtrate, and the volume here is essentially the same as the fermenter. In any case, the object of the first few post fermentation steps is to perform a solid/liquid separation and reduce the bulk volume of the phase where the product is found, thus facilitating further processing. Working with small volumes is easier than with large volumes.

In cell washing, the solvent conditions of the cell suspensions are changed. For instance, changes in pH or ionic strength can be effected with membrane techniques faster than the traditional dialysis methods. Cell washing with membranes also has application in separating cellular mass and debris from component proteins, for instance *E. coli* from a recombinantly derived protein, or hemoglobin from a red blood cell. In these latter applications, the alternative procedure has always involved cumbersome steps with differential centrifugation.

Cell recycle is a form of fermentation which can actually change the dynamics of the normal batch fermentation process by increasing cell mass and product yield. In cell recycle, the fermentation broth is continuously passed tangentially across a membrane filter where a liquid/solid separation takes place. The cells are returned to the fermenter and commercially valuable products and/or metabolic inhibitors are removed from the fermenter.

23.3 TANGENTIAL FLOW FILTRATION

There are two modes of flowing fluids through membranes, dead-ended and tangential. In dead-ended flow, the fluid passes perpendicular to the plane of the filter, whereas in the tangential regime, fluid passes essentially parallel to the filter surface. Common sterile filtration procedures are operated in the dead-ended flow pattern, however, for processing cells, dead-ended filtration is somewhat limiting in terms of total throughput. In dead-ended filtration, the material not passing through the filter remains close to the upstream surface of the filter and forms a layer of retained material that can severely limit subsequent flow rates and throughput. In tangential flow, this material is swept along with the fluid which acts as a cleaning force to keep the upstream surface of the membrane relatively clear of retained material. For the same cell suspension, tangential flow procedures can produce flow rates and throughputs 100 to 1000 times higher than corresponding dead-ended procedures (Henry, 1972, 1972a). All applications in this chapter are by tangential flow filtration (TFF).

23.4 PRACTICAL ASPECTS AND THEORY OF TFF

Figure 1 is a scanning electron microscope (SEM) side view of a UF membrane. Several characteristics of UF membranes are illustrated here. The active portion of the membrane which actually rejects molecules or cells is extremely thin. In the SEM, this rejection layer is seen as the thin, wavy, horizontal line at the top of the membrane. The remainder of the membrane below is very open in structure and acts as a support for the active layer. Resistance to fluid flow occurs mainly at the rejecting layer at the upper surface. As the SEM illustrates, the membrane is very asymmetric, and fluid flow proceeds from the top down only (Porter and Michaels, 1971). Microporous membranes, on the other hand, are more symmetrical than UF membranes as seen in Figure 2. Flow can proceed in either direction, and resistance to flow is encountered throughout the entire volume of the filter. Regenerating or cleaning of UF membranes is relatively easy compared to MF membranes because only the top surface needs to be cleaned, not the entire filter volume. Both MF and UF membranes are used in processing cells, and each generic type of membrane has distinct advantages in particular applications.

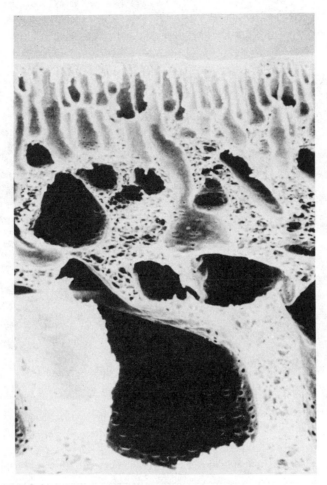

Figure 1   A side view scanning electron microscope photograph of an ultrafiltration membrane; note the asymmetric structure from top to bottom

One criterion in choosing a particular type of membrane for an application is its chemical compatibility with the fluids to be processed. Another reason relates to the compatibility with the cleaning materials. Table 2 shows some representative polymeric materials that are used to construct membrane filters and the chemical compatibilities of these materials with various cleaning solvents. Normal fermentations are aqueous based so there are seldom solvent incompatibilities seen here.

Microporous membranes are rated in terms of a mean pore size, usually measured in microns, whereas UF filters are rated by their molecular weight retention characteristics, that is the molecular weight of globular protein which will be rejected at the upstream surface of the filter. Representative pore size MF membranes used in cell processing are 0.22 μm and 0.45 μm. A 100 000 nominal molecular weight limit (NMWL) UF filter is commonly used in cell processing, although a broad range of NMWL membranes are commercially available.

Membranes are commercially fabricated in several geometrical configurations. Flat sheets or cassette and spiral cartridges are two examples. Both types of configurations can be used in cell processing applications. Figure 3 illustrates a flat sheet and also a spiral configuration. In the flat

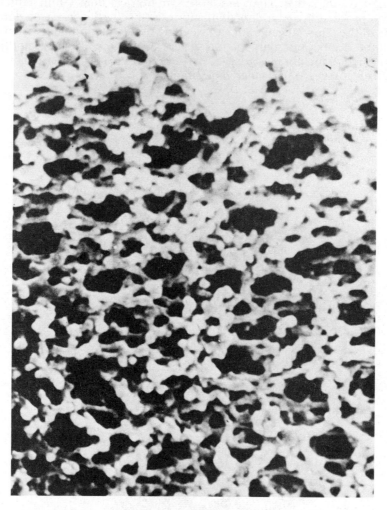

Figure 2   A side view scanning electron microscope photograph of a microporous membrane of pore size 0.22 μm; note the uniform structure throughout

Table 2   Chemical Compatibilities of Cleaning Solutions with Different Membranes

Filter material	pH range	NaOH	NaOCl	Ethanol	Formalin
Cellulosics	4–8	None	10 p.p.m. (0.5 h)	20%	2%
Polysulfone	1–14	1 M (50 °C)	0.1%	70%	4%
Polyvinylidene fluoride derivative	2–13	0.1 M	0.1% (0.5 h)	100%	4%

sheet configuration, layers of membrane are stacked on top of one another to increase filter surface area and the edges are sealed with an adhesive. Holes along opposite sides allow fluid to enter and exit. Fluid flow is parallel to the plane of the stacked sheets. In a spiral configuration, the flat sheets are layered as in the cassette configuration, but the layers are then wound around a central perforated cylinder. Fluid exits from this hollow, central core. Both flat sheet and spiral devices can have separator screens separating and maintaining a defined space between membrane layers preventing the various filter sheets from collapsing on each other.

The theory of tangential flow filtration is covered in detail elsewhere (Blatt *et al.*, 1970) and only the highlights will be presented here. When a cell suspension is passed tangentially across the upstream surface of a membrane filter, macrosolutes in the suspension rapidly form a thin layer immediately adjacent to the filter surface. This thin layer is termed a concentration polarization or gel layer, and it always serves to limit the flow of fluid through the membrane, in effect,

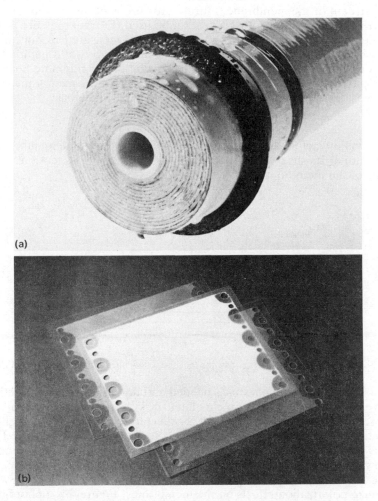

Figure 3 (a) A spiral membrane device: flat sheets are wound around a central, perforated, hollow core through which the filtrate exits; large surface areas are possible in relatively small volumes. (b) A flat sheet membrane configuration which fits into a plate and frame holding device

the gel layer becomes a secondary membrane itself. Under conditions of equilibrium, the rate of solvent flow through a membrane under laminar flow conditions is given as:

$$J = B \left(\ln C_g / C_B \right) \left(\dot{\gamma} \frac{D^2}{L} \right)^{1/3} \tag{1}$$

where J = solvent flux in volume/area of membrane; B = a constant depending on wall boundary conditions; C_g = concentration of macrosolute in the polarization layer; C_B = concentration of bulk species retained upstream of the membrane; $\dot{\gamma}$ = fluid shear rate at the membrane surface; D = diffusion coefficient of macrosolute species in the polarization layer; and L = length of the flow channel. From equation (1), it can be seen why tangential flow has such an advantage over dead-ended flow. In dead-ended flow, $\dot{\gamma}$ is very low or zero, and the polarization layer will build up to become eventually the limiting factor in the flow. With tangential flow, $\dot{\gamma}$ can become an appreciable factor in keeping J at a high value.

23.5 CELL HARVESTING

23.5.1 Operating Parameters

A schematic of a tangential flow filtration device used for cell harvesting is shown in Figure 4. The main parameters the operator has to control the performance of this device are the inlet pressure and outlet pressure. These parameters control the recirculation rate and transmembrane pressure which in turn controls the filtration rate. It should be noted that both inlet and outlet pressures are upstream of the membrane. Filtrate pressure is normally kept at zero, although to minimize polarization effects, non-zero values can be used. Filtrate pressure is measured downstream of the membrane. Flow through the membrane is measured as bulk flow (volume/unit time) or flux (volume/unit membrane area/unit time). Flux is a useful parameter in comparing performances from one experiment to another. Recirculation, or retentate flow rate, is that flow rate of fluid that passes tangentially across the upstream side of the membrane and returns to the cell reservoir. Average transmembrane pressure (TMP) is calculated as:

$$\text{Average TMP} = [(P_{\text{in}} + P_{\text{out}})/2] - P_{\text{filtrate}} \tag{2}$$

and it is TMP that is the actual driving force pushing fluid through the membrane. The value $P_{\text{in}} - P_{\text{out}}$ is proportional to shear rate or the velocity of the cell suspension flowing across the upstream surface of the membrane.

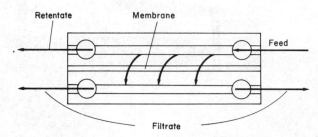

Figure 4 Schematic of a typical plate and frame tangential flow filtration device used for cell harvesting

Figure 5 shows what type of apparatus is needed and how it is plumbed together to perform a typical cell harvest operation. The pump circulates the cell suspension to the filter device where the liquid/solid separation takes place. Filtrate is collected separately and the cells are returned to the original container where they then start the whole process all over again. Thus, the path of any single cell is a circular one upstream of the membrane separator. The pump used should be one that has low internal shear, no local hot spots, is easily sanitized and cleaned, has no air/liquid interfaces, and has the capacity necessary to provide the flow rates which will give the shear rates that minimize polarization effects on the membranes. Figure 6 is a photograph of a typical laboratory size system used for concentrating cells. In Figure 6, the cell reservoir contains yeasts in a concentration of about 10^6 ml^{-1}, and they can be reduced in volume from 6 l to 2 l in about 9 min under normal operating conditions. Normal operating conditions means the filtration device has 5 ft^2 of 0.45 μm membrane and P_{in} of typically 20 p.s.i.

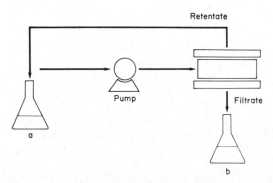

Figure 5 Schematic of how various pieces of equipment are plumbed together for the purpose of performing a cell harvest procedure

For cell harvesting, either UF or MF membranes can be used. Comparative data for concentrating *Bordetella pertussis* (Valeri *et al.*, 1979) from around 20 l to 1 l is shown in Figure 7. Flow decay data *versus* time is shown for a 10^6 NMWL UF membrane and also a 0.22 μm MF membrane. The flow decay curves shown in Figure 7 are typical in that flow decreases rapidly at first

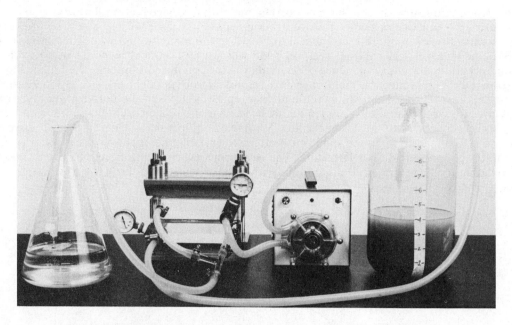

Figure 6 Photograph of a typical laboratory-scale system used for harvesting cells

and eventually reaches a steady value as the polarization layer comes to an equilibrium. Figure 7 also indicates that one-half the area of a 0.22 μm membrane gives the same flow compared to a UF membrane. However, one should not generalize that MF membranes are always preferable and always give greater fluxes. For repeated use applications, UF membranes are easier to clean and the original fluxes can be regenerated more easily than with MF membranes. In the long run, UF membranes can give better performances and this is especially true for industrial applications. There are some applications which will be discussed later where MF membranes are appropriate. It should also be mentioned that in comparison to traditional centrifugation techniques, tangential flow recovered significantly more biomass (85% *vs.* 65%) for the *Bordetella* harvesting example. Greater recovery of biomass by TFF compared to centrifugation methods is one of the general advantages of the filtration technology. Cells that remain suspended in the supernatant fluid are lost in centrifugation. Centrifugation separations of cells are based on density differences, whereas filtration methods are based on size.

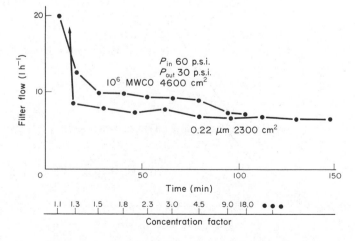

Figure 7 Flow decay curves for concentrating *Bordetella pertussis* suspensions with a UF and an MF membrane; membranes used and areas are as shown

Figure 8 is a graph of filtration rate *vs.* average TMP for the concentration of an industrial antibiotic producing streptomycete suspension. The membrane device is a spiral cartridge. The figure illustrates two practical aspects of concentrating cells. First, as TMP increases, so too does filtration rate, however, there is a practical limit where the increase in flow begins to peak out as TMP is increased, so a further increase in TMP will not give a linear response in flow rate. The second point illustrates the influence of recirculation rate on flux. In general, the higher the recirculation rate, the higher the membrane flux because J is dependent on $\dot{\gamma}^{1/3}$. By increasing the recirculation rate, the shear will increase and sweep away more of the polarization layer and its respective flow restrictions. As seen in Figure 8, varying the recirculation flow rate from 4 to 7 to 10 gal min^{-1} at any TMP, increases the flux. With fixed equipment and membranes, the crossflow rate across the membrane is the single most effective variable the operator has in order to reduce the effects of concentration polarization. The streptomycete concentration in Figure 8 shows that it is possible to concentrate cell suspensions with initial viscosities in the 300–500 cP range.

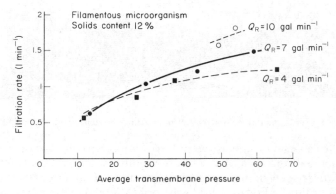

Figure 8 Flow rate *vs.* average transmembrane pressure for a cell concentration process using an industrial streptomycete culture which produces an antibiotic; recirculation flow rates and TMP influence flux

Another method of reducing the influence of concentration polarization is to change membrane configurations. Figure 9 shows a plot of calculated shear rate *vs.* flux for two different membrane devices, a plate and frame device (Pellicon®) and a spiral configuration. In both cases, the same *E. coli* solution was being concentrated. For the higher shear rates, higher fluxes are realized. The difference in shear rate between these two devices is governed by the spacial separation of the membranes and the geometry of the flow channel. The major operating principle to remember is the higher the shear rate, the lower is the effect of the concentration polarization layer.

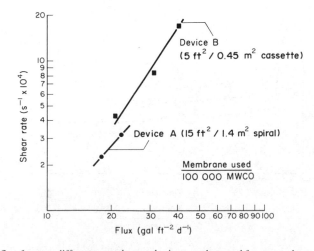

Figure 9 Shear rate *vs.* flux for two different membrane devices: a plate and frame, and a spiral processing identical suspensions of *E. coli* (10 gal ft^{-2} d^{-1} = 17 l m^{-2} h^{-1})

Table 3 is a brief condensation of some typical operating and performance parameters for concentrating solutions of *E. coli*. The crossflow mentioned is a measure of the recirculation rate per unit of membrane area, and it is a major factor in determining the flux. Crossflows for spirals in general are lower than for cassette (plate and frame) devices, however, spirals have their advantages in processing larger volumes of dilute solutions. Spirals usually cannot support such large values of ΔP across them as can plate and frame devices. By increasing the crossflow parameter above that which is normally used, it is possible to obtain increased fluxes. This effect is shown in the column marked HPUF^{TM} (High Performance Ultrafiltration). Raising the crossflows by factors of 3 and over, allows increases in flux as shown in Table 3. A specific example of using higher crossflow operating conditions is shown in Figure 10 where the same recombinant *E. coli* suspension has been concentrated under two sets of operating conditions. For the HPUF^{TM} conditions, the fluxes are double those found under more ordinary or normal operating conditions throughout the entire concentration procedure. The main difference between the two concentration procedures is the shear. For the HPUF^{TM} conditions, shear is some 5 times greater, and theoretically, one would expect a 70% increase in flux because of it. In actual fact for the data shown in Figure 10, flux for HPUF^{TM} conditions was double normal operating conditions, which indicates that the dependence of flux on shear was closer to the one-half power. Another very important aspect of using HPUF^{TM} conditions is the design of the filter device itself. It must be capable of withstanding the high inlet pressures generated. For the case shown in Figure 10, P_{in} was 100 p.s.i. This restriction rules out the use of hollow fibers, because the individual fibers start breaking in the 20–25 p.s.i. range. To illustrate HPUF^{TM} operating conditions with one more example, we have concentrated 16 l of *Streptococcus faciens* to 1 l in 1 min 10 s using 5 ft^2 of a 0.45 μm membrane.

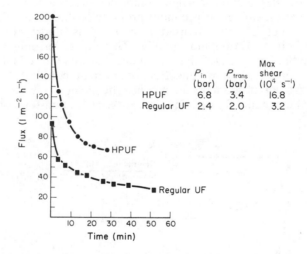

	P_{in} (bar)	P_{trans} (bar)	Max shear (10^4 s^{-1})
HPUF	6.8	3.4	16.8
Regular UF	2.4	2.0	3.2

Figure 10 Flux decay *vs.* time for harvesting an *E. coli* suspension in which normal and HPUF operating conditions were separately used

23.5.2 Membrane Fouling

Fouling of membranes is a distinctly different phenomenon from that of concentration polarization, although both effects manifest themselves in the same manner, mainly a reduction in flux with time (Zahka and Leahy, 1982). There are basically two forms of fouling: adsorption and instability. Adsorption fouling is caused by a chemical constituent in fermentation media actually adsorbing to the membrane itself. Instability, on the other hand, is caused by a gradual condensation and solidifying of the concentration polarization layer with time. In the latter case, components in the polarization layer begin to adhere to each other, and also to the membrane. Fouling is differentiated from true concentration polarization by the fact that changes in flux due to fouling are not sensitive to crossflow rate, shear or velocity of fluid on the upstream side of the membrane, whereas polarization effects are. Membranes whose performance is limited by true fouling are responsive to TMP only. If TMP increases, then flux will increase. In many cases where a

Table 3 Summary of Typical Operating Conditions for Harvesting
Suspensions of *E. coli* Using Different Membrane Devices and
Operating Conditions[a]

	Cassette (Pellicon®)	Spiral	HPUF cassette (Pellicon®)
Average flux (ml/5 ft²)			
100K UF	200–500	100–600	1000+
0.45 μm MF	300–1000	—	2000+
Inlet pressure (p.s.i.)	10–30	50	20–200
Recirculation rate (l min⁻¹)	1–2	40	16
Crossflow (ml min⁻¹ ft²)	300	<100	1000+

[a] High performance ultrafiltration (HPUF), also called high performance crossflow filtration, is an extension of normal operating conditions.

membrane flux decreases with time, a combination of polarization and fouling effects are usually responsible. Deciding how much of which pure effect is responsible for a flow decay is usually an academic exercise. What is of practical importance is how to increase the flux. It is difficult to decide *a priori* which component in a fermentation fluid medium will cause fouling of a membrane except by experimentation. One component that is commonly found in all fermentation broths that can cause fouling is an antifoam. Antifoams added to fermentations can be one of a wide variety, however, they have the common element of being hydrophobic. Both antifoam type and concentration are variables to manage when membranes are used to concentrate cells directly from a fermenter. Some antifoam agents are more prone to fouling than others and, in general, the lower the concentration of the antifoam used the better. It is important to optimize the whole process, not just each unit operation individually. This means picking an antifoam plus its respective concentration that optimizes both the fermentation and the cell harvesting, not just the fermentation step alone. Figure 11 shows the effect of antifoam concentration on the flux for an *E. coli* cell suspension growth in defined salts media. The clean water flux line is included as a reference to show typical tangential flow performance of a clean, non-fouling fluid.

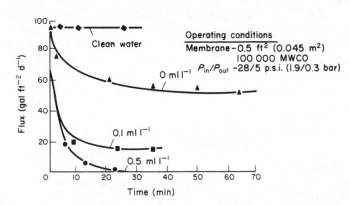

Figure 11 The effect of concentration of one type of antifoam on flux for an *E. coli* suspension grown in defined salts solution (10 gal ft⁻² d⁻¹ = 171 l m⁻² h⁻¹)

If antifoaming agents are a problem with a particular cell harvest application, there are basically four approaches that can be tried to alleviate the flux reductions. First, use less antifoam, or as little antifoam as possible, which will still reduce the foaming problem. Second, change to an antifoam that does not foul the membrane. Experiments performed at Millipore® have evaluated a number of different antifoams for their fouling characteristics. Results show there are certain antifoams that have minimal effects on membrane performance. Third, change the membrane that is being used. If adsorptive fouling is the problem, then by using a membrane made with a different polymer, adsorptive effects may be reduced. Many UF type membranes are made from polysulfone which is hydrophobic. If an antifoam problem occurs with polysulfone, try a hydrophilic membrane. Durapore®, a hydrophilic derivative of polyvinylidene fluoride, has been shown to have less non-specific adsorption characteristics than most membranes. Fourth and last, membranes can have the antifoams washed or cleaned off by a suitable cleaning agent at the end of a run. Membrane fluxes, in many cases, can be restored to their original fluxes after such a cleaning. For severe fouling conditions, cleaning of membrancs will be the most practical approach. By using one, or a combination, of the above strategies, membrane fouling can be managed.

23.5.3 Particulate Loading

Heavy concentrations of particulates in fermentation broths can occasionally influence membrane device performance. The problems usually occur with industrial type media that have been chosen for their low cost and improved product yield, as opposed to their solubility characteristics. Heavy particulate loading will cause plugging of membrane devices, rather than the membranes themselves. Blockage of inlet ports causes high operating inlet pressures without the corresponding high crossflows, and also extra strain on pumps. Typical laboratory media are usually quite soluble and present minimal particulate problems.

The best approach to solving particulate loading problems is prefiltration. In prefiltration, a depth or other type of high dirt holding filter removes particulates and thus protects both the pump and the membrane device. The amount of prefiltration depends on the particular device used. For instance, in spiral configuration, we recommend prefiltration down to at least 100 μm. In general, the smaller the spacing between membrane surfaces, the smaller the particle that must be prefiltered out. Narrowly spaced flow channels give better shear rates, but they also present a problem in blockage. The type of prefilter used also depends on the concentration of particles. For lower concentrations, an in-line screen filter is adequate. For moderate concentrations, a larger dirt handling prefilter, like a bag filter, can be used, and for large quantities of particles, a vibrating table or a self-cleaning filter is appropriate.

23.5.4 Industrial Applications

Classical methods that have been used to concentrate cells on either a laboratory or industrial scale are: filter presses, precoat vacuum filters, and centrifuges. Table 4 compares these techniques with TFF based on mode of action and the limit of separation. For a filter press, removing organisms less than 10 μm is usually not practical and the corresponding figure for a precoat vacuum filter is 1 μm. Centrifuges separate cells based on density differences between the cells and the suspending fluid. TFF techniques can concentrate basically any size cell by just changing the membrane to one with a lower NMWL cutoff. Also, TFF can produce cell free extracts, whereas this is not usually found in other techniques, especially centrifugation. The speed of a TFF procedure, especially one operating in the HPUF mode, is unmatched by other techniques.

Table 4 Comparison of Various Technologies used for Separating Cells from Broth

Technique	Basis of separation	Limit of separations (μm)
Filter press	Size	10
Precoat vacuum filter	Size	1
Centrifugation	Density	Density related
TFF	Size	<0.1

Tangential flow filtration is not limited to small or laboratory-scale use (Reid and Adlam, 1976; Henry, 1972). Although the technique has many advantages for concentrating cells in the laboratory, it is also finding increasing adoption for routine large-scale processing. Figure 12 shows some typical operating data for a 400 l batch concentration of *E. coli* that is routinely carried out by one industrial firm. The *E. coli* is harvested *via* TFF as part of the normal process in preparing a veterinary vaccine. Figure 13 shows several performance parameters for the concentration. Of particular note is the exact match between the volume reduction and the corresponding increase in titer. No viability is lost as a result of concentration. The company uses a plate and frame device (Pellicon®) that is mounted on a cart which can be wheeled around from one fermenter to another. A screw action pump ensures the cells are handled in a reasonably gentle fashion. One added benefit for the company over the previously used centrifugation procedure was the elimination of aerosols that were causing allergic reactions among employees. With the closed membrane system, the allergy problem was no longer an issue. Cleaning the membranes was carried out on a routine basis. Routine cleaning was done after each use, intermittent regeneration to restore flux was done after every 10 uses, and after every three months, a backflush wash was performed. With this type of cleaning regime, membrane lifetime is reported by the company to be at least one year.

Batch size	400 l	P_{in}	40 p.s.i.
Membrane area used	25 ft² (2.3 m²)	P_{out}	25 p.s.i.
Membrane type	100 000 NMWL	Recirculation rate	800 l h⁻¹

Figure 12 Typical operating data for an industrial-scale *E. coli* harvesting procedure

Initial volume	400 l	Recovery	
Final volume	20 l	Initial titer	1×10^9
Concentration factor	20 ×	Final titer	2×10^{10}
Average flux	26 l h⁻¹ m⁻²	Processing time	6 h

Figure 13 Typical performance data for an industrial *E. coli* harvesting procedure

23.6 CELL WASHING

23.6.1 Practical Aspects and Theory

Cell washing applications using TFF are similar to cell harvesting in many respects. However, while filtrate is being collected, new solvent is simultaneously being added to the cells. The net result is a changing of solvent conditions such as pH or ionic strength, or the removing of a cellular component such as a protein. Cell washing procedures can be used to remove an unwanted protein or to separate a valuable intracellular protein from lysed cells and cell debris.

A schematic representation for a cell washing setup is shown in Figure 14. As filtrate is collected in the appropriate container, a vacuum is created in the head space of the reservoir holding the cells. Because this container is closed to the atmosphere, fluid from the wash container is drawn into the cell reservoir *via* the tubing connection. At equilibrium, the filtrate rate and the flow of fluid from the wash reservoir to the cells are identical. As a consequence, the cell volume remains constant. Individual cells are still recirculated from the cell reservoir through the pump and filtration device and back to the reservoir. As a function of time, the old cell suspension medium is gradually changed to that of the new solvent. If the membrane is chosen properly, large molecular weight components can pass through while cells and debris remain upstream.

Cell washing techniques are very efficient in terms of the volume of wash solution needed and also the amount of time for processing. Theoretically, the exchange process in Figure 14 can be described mathematically as:

$$F = 1 - \exp(-V_F/V_0) \tag{3}$$

where V_0 is the initial volume of the cells, V_F is the volume of filtrate and F is the fraction of the wash completed assuming that the component being washed through the membrane has no hindrance. Table 5 gives values of F for various filtrate volumes.

The figures in Table 5 indicate that after only five volumes of wash solution, the old solvent conditions are 99.3% changed. After only three wash volumes, 95% of the old solvent conditions are changed. Figure 15 shows a graph of alcohol removed from an albumin solution as a function of filtrate volume where the removal rate follows equation (3) (Schmitthauesler, 1977). The albumin solution shown was isolated by the Cohn method used commercially to fractionate protein from plasma.

Cell washing *via* the TFF system has been used to remove hemoglobin from lysed red blood cells where the purpose was to isolate protein in cell membrane fragments (Rosenberry *et al.*, 1981). The basic procedure used is as shown in Figure 14. An isotonic buffer was used to wash the

Table 5 The Percentage Completion of
a Wash Procedure *vs.* Filtrate Volume

Completion (%)	Filtrate volume[a]
63.2	1
86.5	2
95	3
98.2	4
99.3	5

[a] In units of original cell suspension volume V_0.

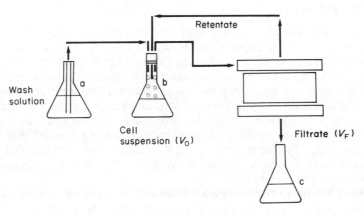

Figure 14 A schematic diagram illustrating how equipment is plumbed together to perform a constant volume cell wash

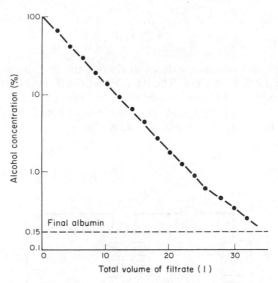

Figure 15 Decrease of alcohol concentration in an albumin solution as a function of wash volume; initial albumin volume was 5 l; PTGC 10 000 MW cassettes (Schmitthauesler, 1977)

cells. The isotonic solution lysed the cells, and since a 100 000 NMWL membrane was used, hemoglobin passed through. Rosenberry reported 99.9% of the hemoglobin could be removed that way. The normal method of using differential centrifugation for washing the cells and isolating the cell membrane fragments took over 20 h, whereas similar results were obtained *via* TFF in 2.5 h. As part of his work, Rosenberry also circulated whole red blood cells through the TFF system for 16 h in an effort to determine if the hydrodynamic forces of recirculation would lyse the cells. After this period, only 3.0% of the RBC had been lysed indicating that even fragile mammalian cells without cell walls can be processed by TFF (Radlett, 1972; Gardon and Mason, 1955).

Using a similar arrangement to isolate Ca^{2+} transport ATPase from red blood cells, Gietzen and Kolandt (1982) found considerably higher enzymatic activities of material isolated using TFF methods compared to methods used in other laboratories. Centrifugation methods also were reported to take five times longer than the TFF procedure, and enzyme yield was 40% higher with the membrane method compared to centrifugation steps. Other cellular components have also been isolated with the aid of TFF techniques (Howlett *et al.*, 1979; Silvestri *et al.*, 1978; Iida and Nussenzweig, 1981).

23.6.2 Industrial Applications

Cell washing techniques can also be used to isolate a valuable protein. When *E. coli* is used as a recombinant organism in genetic engineering work, the protein of interest is intracellular. After fermentation, the cells can be concentrated using TFF to a small, workable volume. The next step is to lyse the cells. If a 0.45 μm membrane was used in the concentration step, then it can be used again for the cell washing step where the protein is collected in the filtrate. Using essentially the same apparatus, the cells can be both concentrated and the protein separated from the debris. Several industrial companies are now using this procedure. The membrane used in the cell washing step should have low, non-specific protein adsorption and a pore size that will easily accommodate protein passage. In these types of cell washing applications, MF membranes are appropriate because their relatively large pore size allows free flow of macrosolutes through the membrane compared to UF membranes.

One further example where cell washing has been used industrially to save time is in the area of influenza virus processing. After the virus is collected from the allantonic fluid and the major debris is removed, the virus is usually isolated using density gradient centrifugation. This step is then followed by a dialysis step to remove the material that made up the density gradient. By using TFF to replace this process, the dialysis procedure which normally took overnight, could be replaced by a washing which now took only 4 h. Also, loss of product due to dialysis bag breakage was completely eliminated. Overall, virus isolation time was reported as being reduced by a factor of 2.5. Again, this use of TFF is in routine industrial use at several companies.

23.7 CELL RECYCLE

Cell recycle using TFF is a technique that can change the dynamics and productivity of a fermentation compared to a batch process. The basic idea of cell recycle is to circulate fermentation broth continuously through a TFF device, and to return the cells back to the fermenter. A schematic of this process is shown in Figure 16. As filtrate is collected, fresh nutrient broth has to be added to the fermenter to maintain a constant volume, so it is necessary to couple the filtrate rate to that of the addition rate.

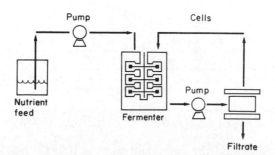

Figure 16 Schematic illustrating how cell recycle is performed

Cell recycle techniques have a number of advantages. First, product can be collected on a continuous basis if it is small enough to permeate the filter. Second, productive cells are returned to the fermenter so they can generate more product. Returned cells use nutrients only to maintain their metabolism and to produce product, not to grow to maturity. Third, if the product is part of a feedback inhibition cycle, then by removing the product and keeping its concentration below a critical level, it is possible to 'fool' the organisms into continuing to produce, which they would not ordinarily do (Abbott and Gerhardt, 1970). Increased productivity can result compared to regular batch fermentation. Fourth, if the cells produce growth inhibitory products, these can be removed continuously and hence, growth conditions are improved. Last, larger biomasses are possible because cells are receiving fresh nutrients and having waste products removed (Goto, 1979). Literature references have shown in excess of one order of magnitude increases in cell density compared to regular batch fermentations (Gallup and Gerhardt, 1963; Landwall and Holme, 1977).

Rogers *et al.* (1980) have shown that both biomass and productivity of ethanol by *Zymomonas mobilis* can be significantly increased with cell recycle. Using PVC membranes of 0.6 μm pore size and filtration rates between one and three fermenter volumes per hour, biomass was increased by a factor of 12 and alcohol productivities by a factor of 10 compared to batch fermentations.

In a novel use of cell recycle, Van Reis *et al.* (1982) recycled human leukocytes in order to increase their production of α-interferon. Using two tangential flow devices, they separated the interferon from the recycled cells in one step and concentrated the interferon in the second step. Experiments were performed to determine just how effective cell recycle was as a function of how long the recycle process continued and when recycle was started after the induction of the leukocytes with Sendai virus. By using recycle for half an hour, and starting 3 h after induction, a 200% increase in interferon was observed compared to controls.

23.8 SUMMARY

Tangential flow filtration is a technique that can complement, supplement, or completely replace some classical techniques for processing cells. This technology offers a number of distinct advantages. First, it is a closed system so no aerosols are generated in the workplace, which is especially important when processing pathogens or recombinant organisms. Second, the techniques are flexible enough to allow a person to perform a variety of procedures with the same equipment. Examples of the different procedures that can be performed by TFF include concentration, clarification and constant volume washing. Third, compared to classical methods, TFF can offer definite advantages in time savings and product recovery. Fourth, and last, TFF can be operated continuously; it is not necessary to stop in the middle of a cell concentration, for instance, as is the case for a batch centrifuge. It should also be emphasized that TFF is not limited to small or laboratory scale, but can be scaled up to handle pilot or production size quantities.

ACKNOWLEDGEMENTS

The author would like to thank Joe Zahka, Charlie Stoddard, Hank Lane and Tim Leahy for the new experimental work reported here and for help in preparing this manuscript.

23.9 REFERENCES

Abbott, B. J. and P. Gerhardt (1970). Dialysis fermentation: enhanced production of salicylic acid from naphthalene by *Pseudomonas fluorescens*. *Biotechnol. Bioeng.*, **12**, 577–589.
Blatt, W. F., A. David, A. Michaels and L. Nelsen (1970). Soluble polarization and cake formation in membrane ultrafiltration: causes, consequences and control techniques. In *Membrane Science and Technology*, ed. E. Flinn. Plenum, New York.
Brock, T. (1983). *Membrane Filtration: A User's Guide and Reference Manual*. Science Tech Inc., Madison, WI.
Dudka, B. J. (1981). *Membrane Filtration Application, Techniques and Problems*. Dekker, New York.
Gallup, D. M. and P. Gerhardt (1963). Dialysis fermentor system for concentrated culture of microorganisms. *Appl. Microbiol.*, **11**, 506–512.
Gardon, J. L. and S. G. Mason (1955). *Can. J. Chem.*, **33**, 1625.
Gietzen, K. and J. Kolandt (1982). Large-scale isolation of human erythrocyte Ca^{+2}-transport ATPase. *Biochem J.*, **207**, 155–159.
Goto, S., T. Kuwajima, R. Okamoto and T. Inui (1979). Separation of biomass by membrane filtration and continuous culture with filtrate recycling. *J. Ferment. Technol.*, **57**, 47–52.
Henry, J. D. (1972). Concentration of bacterial cells by crossflow filtration. *Dev. Ind. Microbiol.*, **13**, 177.

Henry, J. D. (1972a). Crossflow filtration. In *Recent Developments In Separation Science*, ed. N. Li, vol. 2, pp. 205–225. CRC, Cleveland, OH.

Howlett, A. C., P. C. Sternweis, B. A. Macik, P. M. Van Arsdale and A. G. Gilman (1979). Reconstitution of catecholamine-sensitive adenylate cyclase. *J. Biol. Chem.*, **254**, 2287–2295.

Iida, K. and V. Nussenzweig (1981). Complement receptor is an inhibitor of complement cascade. *J. Exp. Med.*, **153**, 1138–1150.

Kesting, R. E. (1971). *Synthetic Polymeric Membranes*. McGraw-Hill, New York.

Landwall, P. and T. Holme (1977). Removal of inhibitors of bacterial growth by dialysis culture. *J. Gen. Microbiol.*, **103**, 345.

Porter, M. C. and A. S. Michaels (1971). *Chem Technol.*, Jan.

Radlett, P. J. (1972). The concentration of mammalian cells in a tangential flow filtration unit. *J. Appl. Chem. Biotechnol.*, **22**, 495–499.

Reid, D. E. and C. Adlam (1976). Large-scale harvesting concentration of bacteria by tangential flow filtration. *J. Appl. Bacteriol.*, **41**, 321–324.

Rogers, P. L., K. J. Lee and D. E. Tribe (1980). High productivity ethanol fermentations with *Zymomonas mobilis*. *Process Biochem.*, Aug./Sept., 7–11.

Rosenberry, T. L., J. F. Chen, M. M. L. Lee, T. A. Moulton and P. Onigman (1981). Large-scale isolation of human erythrocyte membranes by high volume molecular filtration. *J. Biochem. Biophys. Methods*, **4**, 39–48.

Schmitthauesler, R. (1977). Molecular filtration in human plasma fractionation. *Process Biochem.*, Oct., 13–16.

Silvestri, L. J., R. A. Craig, L. O. Ingram, E. M. Hoffman and A. S. Bleiweis (1978). Purification of lipoteichoic acids by using phosphatidyl chlorine vesicles. *Infect. Immun.*, **22**, 107–118.

Valeri, A., G. Gazzeri and G. Genna (1979). Tangential flow filtration of *Bordetella pertussis* submerse cultures, *Experientia*, **35**, 1535–1536.

Van Reis, R., R. R. Stromberg, L. I. Friedman, J. Kern and J. Franke (1982). Production and recovery of leukocyte-derived alpha interferon using a cascade filtration system. *J. Interferon Res.*, **2**, 533–541.

Zahka, J. and T. Leahy (1982). *Practical Aspects of Tangential Flow Filtration in Cell Separations, American Chemical Society Meeting.*

24

Cell Separations with Hollow Fiber Membranes

R. S. TUTUNJIAN,
Amicon Corporation, Danvers, MA, USA

24.1 INTRODUCTION

The recovery of bacteria and cells after fermentation and culture can be tedious, inefficient and costly when using traditional processes such as centrifugation. Other problems include aerosol formation, poor product recovery and slow processing. These difficulties are more pronounced when dealing with large volumes of genetically altered bacteria, which must be contained and completely recovered.

Cross-flow membranes have been demonstrated to be an attractive alternative to centrifugation for cell separations. As discussed in Chapter 26 of Volume 2, cross-flow techniques in ultrafiltration move solutes away from the membrane and improve filtration rates. This same principle can be applied to bacteria and cells, even though they are discrete particles instead of solutes. Both ultrafiltration and microporous membranes have been used in cellular applications; these are available in flat sheet and hollow fiber configurations. This chapter will focus on the latter type of membranes.

24.2 THEORY

24.2.1 Membranes

Hollow fiber membranes are self-supporting porous polymer tubes which have a cylindrical channel. Channels can vary from 0.2 to 1.1 mm diameter in commercially available cartridges.

Unlike hollow fibers, flat-sheet systems, such as spiral cartridges, use spacer screens to separate membranes (see Figure 1). Problems can arise during cell processing operations in that these spacer screens can become clogged at high cell densities (Lukaszewicz *et al.*, 1981). As shown in Figure 2, flow through hollow fibers is unobstructed. Although fibers have lower pressure limits than flat-sheet systems, this does not substantially affect mass transfer rates due to higher velocities and larger channels (see Table 2, Chapter 26, Volume 2).

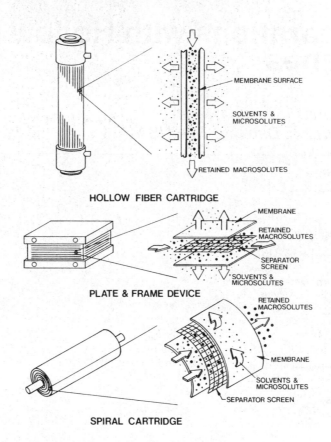

HOLLOW FIBER CARTRIDGE

PLATE & FRAME DEVICE

SPIRAL CARTRIDGE

Figure 1 Examples of different membrane configurations

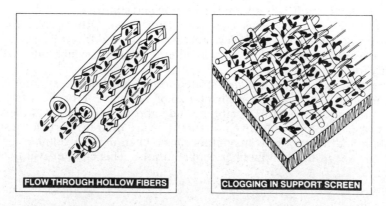

FLOW THROUGH HOLLOW FIBERS

CLOGGING IN SUPPORT SCREEN

Figure 2 Comparison of flow through hollow fibers and membrane support screen

In comparing membranes, a distinction must also be made based on filter matrix structure. Traditional microporous filters possess an open, tortuous path structure which causes irreversible entrapment within the matrix (Reti *et al.*, 1979; Howard and Duberstein, 1980). This can create difficulties in cleaning and reuse of the filters. Ultrafilters and some newer types of microfilter have a 'skinned' or anisotropic structure which allows retention on the filter surface (see Figure 3).

Figure 3 Comparison of anisotropic and conventional microporous filters

Both types of microporous filter are available in hollow fiber configuration. These are fabricated either in polypropylene with an open structure (Membrana, Inc.) or in polysulfone with an anisotropic structure (Amicon Corp.). Since bacteria are typically no smaller than 0.3 μm in size, use of microporous filters is generally suitable for cell recovery.

24.2.2 Fluid Mechanics

The use of cross-flow filtration for cell recovery necessitates recirculation of suspensions through membrane channels. Recirculation velocity across the filter surface is of great importance to the overall filtration rate (see Sections 26.2.3.2 and 26.3.2, Volume 2). Maintaining a high cross-flow rate reduces filter blockage allowing high flux rates.

The concentration of cells during harvesting causes substantial changes in the viscosity of the whole broth. As shown in Figure 4, the apparent viscosity of yeast broth increases dramatically after the cell concentration exceeds 18% by weight. It is also important to note that the flow properties are pseudoplastic; the viscosity decreases as shear rate is increased. In other words, as the broth is recirculated at a faster rate, the shear rate increases and viscosity drops.

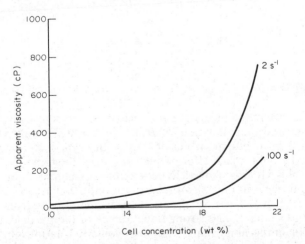

Figure 4 Viscosity of yeast broth as a function of cell density, varying shear rates

Viscosity of cell broth is also strongly dependent on temperature. In the case of yeast broth shown before, increasing the temperature to 55 °C reduces the apparent viscosity for a 22% suspension to about 10 cP. At this temperature, the flow properties are close to Newtonian in nature so that shear rate is not a major factor (Labuza *et al.*, 1970).

24.2.3 Flux Rate Determination

With cross-flow ultrafiltration of dissolved macrosolutes, the filtration flux rate will depend on: (1) the transmembrane pressure (ΔPTM), (2) the gel layer concentration (C_G), (3) the mass transfer coefficient (K) and (4) the bulk solute concentration (see Section 26.2.3.2 Volume 2). Flux will tend to increase linearly with ΔPTM until gel polarization occurs. At this point, the latter three variables will determine flux rate (J) according to

$$J = K\ln(C_G/C_B) \tag{1}$$

and K is governed by the Sherwood equation:

$$\frac{Kd}{D_v} = ARe^B Sc^{1/3} \tag{2}$$

where d is the fluid channel height above the membrane, D_v is the diffusivity, Re is the Reynolds number (dV/v), Sc is the Schmidt number (v/D_v), v is the viscosity, V is the fluid velocity, and A and B are constants.

The value of B will vary from 0.5 in laminar flow to about 1.0 in turbulent flow. These values are slightly higher than the theoretical figures of 0.33 and 0.8, respectively. This key parameter determines the relationship of fluid velocity and flux rate.

The cross-flow filtration rates of cell suspensions behave differently to those of macromolecular solutions. Most cell harvesting processes will exhibit relatively little decline in flux at lower cell densities (see Figure 5). When a near-packed condition is approached, the filtration rates will drop sharply in logarithmic fashion. This behavior is similar to that of a protein solution before and after gel polarization occurs (see Section 26.2.2, Volume 2). However, the 'unpolarized' condition during cell concentration may still exist at moderately high densities; ultrafiltration of macromolecules will show polarized behavior at even very low concentrations.

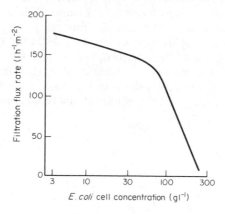

Figure 5   Filtration flux rate as a function of concentration, *E. coli* suspension

A second major difference in the flux characteristics of cell suspensions is in the velocity dependence. Porter (1972) showed that the value of B in equation (2) would vary from about 0.8 in laminar flow to over 1.3 in turbulent, when dealing with such suspensions. He attributed this behavior to the so-called tubular pinch effect, which enhances movement of particles away from a fluid boundary layer. This creates a particle-deficient layer at the fluid wall, which has been reported and observed by several investigators (Brandt and Bugliarello, 1961; Segre and Silberberg, 1961; Karnis *et al.*, 1966). The rate of particle movement away from the wall is dependent on the square of the fluid velocity. The strong dependence of the particle migration on the recirculation rate would explain the abnormally high flux–velocity relationship found in cell processing.

Flux rate will also be impacted by one other factor: viscosity. As viscosity increases, the fluid recirculation rate will have to drop, assuming that the channel pressure drop is at a maximum. Decreases in recirculation rate will have major effects on filtration rate due to the strong flux–velocity relationship mentioned above. As shown in Section 24.2.2, the viscosity of cell suspensions will increase sharply at high densities. Since most cell suspensions also exhibit pseudoplastic behavior at these levels, a decrease in flow will cause further increases in viscosity. As a result, fluid velocity will decrease substantially causing flux rate to drop rapidly, as shown in Figure 5.

In summary, during cross-flow filtration of cells, flux rate will be a direct function of recirculation velocity. This dependence is much stronger than that observed in ultrafiltration of dissolved macromolecules. Flux will also tend to decline only slightly at low to moderate cell densities. At higher levels, filtration rate will drop rapidly as viscosity increases markedly.

24.3 PROCESS CONSIDERATIONS

24.3.1 Operation

As discussed in Section 26.3.2, Volume 2, ultrafiltration fluxes are maximized by variation of recirculation rate and transmembrane pressure; inlet and outlet pressures are used for monitoring. However, the filtration rate during cross-flow filtration of cells will depend strongly on the fluid velocity. Generally speaking, flux rates are maximized by maintaining the highest possible recirculation rate. This is achieved by using an outlet pressure of zero and a maximum inlet pressure. With hollow fiber systems, this pressure is 30–40 p.s.i. A slight backpressure may be used in some cases where solids levels are low and the tubular pinch effect is negligible.

Cell processing by hollow fiber filtration can be categorized in a manner similar to ultrafiltration (see Section 26.3.3, Volume 2): concentration, diafiltration and purification. The three processes were defined as follows:

(1) Concentration and recovery where the cells are or contain the desired product (intracellular) also known as 'cell harvesting'.
(2) Diafiltration involves removal of low molecular weight solutes in the medium by the addition and removal of water/buffer; also known as 'cell washing.'
(3) Purification processes use the membrane to retain the cells and allow an extracellular product to be recovered in the filtrate.

Greater details on the general material balances and recovery calculations are discussed in this earlier chapter.

24.3.2 Combined Operations: Optimization

Most cell processes combine concentration and diafiltration for cell harvesting/washing or for purification of an extracellular product. In these situations, the point at which diafiltration is carried out can be critical in minimizing total process time. With macromolecular ultrafiltration, this point can be defined by the logarithmic flux–concentration profile (see Section 26.3.3.5, Volume 2). However, during cell concentration, flux will decline only slightly until relatively high cell densities (see Figure 5). For cell harvesting and washing, it is generally best to concentrate to a point slightly less than the cell density where rapid flux decay takes place. An example of a typical process is shown in Table 1. In this case, flux remains relatively constant until a density of 60 g 1^{-1} is reached. Two process arrangements are shown: in case A, the washing step is done at the final desired cell density, while in case B, washing is carried out at 60 g 1^{-1}, where flux is still high. As shown, the latter case gives a 26% lower overall process time. This is accomplished in spite of the fact that the diafiltrate volume (3000 1) is 2.5 times higher than case A.

It should be noted that, in cases where initial cell density is high, flux rates can decline logarithmically from the start. In these situations, the optimum concentration for washing (C^*) can be determined by the relation defined by Ng *et al.* (1976):

$$C^* = C_G/e \qquad (3)$$

where C_G is the maximum cell density (zero flux extrapolated) and e is the natural logarithm base.

In the purification of extracellular products, such as antibiotics, minimizing process time may not be the only consideration. In the cases shown in Table 1, the more rapid process resulted in a total permeate volume 38% larger than the slow process. This would further dilute an extracellular product and may not be desirable. The choice would depend on the cost to recover the product.

Table 1 *E. coli* Harvesting/Washing: Comparison of Different Diafiltration Concentrations

	Cell volume (l)	Cell density (g l^{-1})	Filtration rate (l h^{-1})	Average rate (l h^{-1})	Filtrate volume (l)	Filtration time (h)
Case A						
(1) Initial	4000	15	2200			
				2000	3000	1.5
(2) Concentration 4 ×	1000	60	1840			
				980	600	0.6
(3) Concentration 2.5 ×	400	150	420			
				460	1200	2.6
(4) Diafiltration 3 ×	400	150	480			
				Totals	4800	4.7
Case B						
(1) Initial	4000	15	2200			
				2000	3000	1.5
(2) Concentration 4 ×	1000	60	1840			
				2000	3000	1.5
(3) Diafiltration 3 ×	1000	60	2100			
				1120	600	0.5
(4) Concentration 2.5 ×	400	150	480			
				Totals	6600	3.5

24.3.3 System Design and Cleaning

The general considerations for system design have been discussed in detail in Section 26.3.4, Volume 2. A basic system requires a recirculation pump, backpressure valve, pressure gauges and process reservoir (see Figure 6). Hollow fiber systems also usually incorporate an automatic shutoff to prevent overpressuring cartridges.

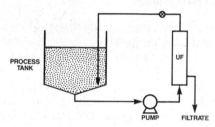

Figure 6   Flow schematic of batch cross-flow filtration system

The design of the fluid reservoir may be quite important in cell processing operations. Tanks should have conical bottoms with outlets at the lowest point for easy draining and complete recovery. These outlets must also be large enough to allow recirculation flow rates without pump cavitation. The return, or retentate, line from the UF system should run below the fluid level near the tank bottom (see Figure 6); this minimizes foaming during operation.

For genetically altered bacteria, complete containment is a critical factor. Hollow fiber cartridges are supplied in self-contained housings, complete with sanitary, clamp-type fittings. As a result, they are ideal for use with biohazardous products (Hanisch *et al.*, 1982; Trudel and Payment, 1980). For such use, reservoirs must be closed and vented with sterilizing air filters. The filtrate should also be collected in a similar tank for subsequent purification or disposal after treatment.

Hollow fiber systems are available in a wide variety of sizes to suit any process need. Membrane area may be added to increase filtration rates as required. A laboratory-scale system is shown in Figure 7; this unit is supplied complete with variable-speed pump and 20 l reservoir. With a single 5 ft^2 hollow fiber cartridge, cell harvesting rates of 90 l h^{-1} can be achieved. The larger process-scale system shown in Figure 8 uses a centrifugal pump with sanitary stainless steel piping. Cell processing can be carried at over $2000 \, l \, h^{-1}$ using $120 \, ft^2$ of hollow fiber area.

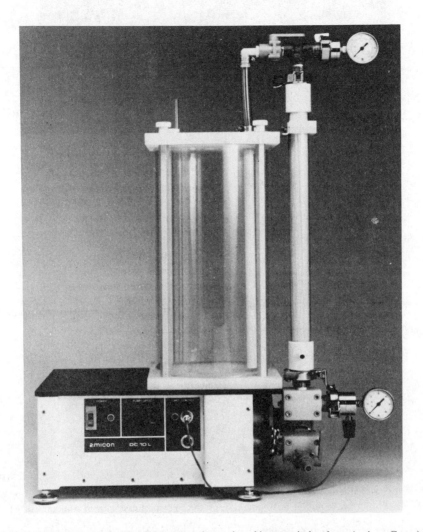

Figure 7 Laboratory hollow fiber system (reproduced by permission from Amicon Corp.)

Cleaning of hollow fiber membranes is described in Section 26.3.5, Volume 2. In addition to normal chemical cleaning (with sodium hydroxide, sodium hypochlorite, *etc.*), hollow fibers may be backflushed to force foulants off the filter surface. Recycle cleaning by closing the filtrate ports and reversing flow allows backflushing during operation. This in-process cleaning has been particularly useful with high solids cell suspensions. As a result, most hollow fiber systems for cell processing are designed for easy recycle/flow reversal; in larger systems, this can be automated for timed, intermittent cycling.

24.4 APPLICATIONS

24.4.1 General

Cell processing applications have been categorized into three basic areas (see Section 24.3.1): concentration (cell harvesting), diafiltration (cell washing) and purification (of extracellular products). It should be noted that these distinctions are generally applied to batch fermentations. The use of hollow fibers for continuous fermentation and cell culture is a separate, general category. For the purposes of this discussion, hollow fiber separations will be divided into batch and continuous processes. The former group will be further broken down into cell harvesting/

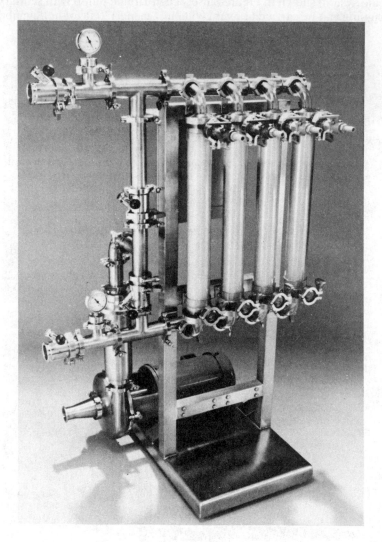

Figure 8   Pilot scale hollow fiber system (reproduced by permission from Amicon Corp.)

washing and product purification; the latter will describe continuous bacteria fermentation as well as mammalian cell culture.

24.4.2   Batch Processes

24.4.2.1   *Cell harvesting/washing*

The recovery of intracellular products generally begins with concentration of the bacterial broth in order to facilitate cell disruption. In doing so, it is also desirable to remove medium constituents which would be impurities in the final product. As mentioned previously, cross-flow, hollow fiber filtration has been used successfully for cell harvesting/washing. Hollow fiber filtration offers many advantages over centrifugation, the most widely used traditional method.

The use of centrifugation on an industrial scale requires continuous removal of centrate. Cells are removed continuously or intermittently. The throughput rate for complete particle removal (J_c) is a function of several variables:

$$J_c = \frac{d_P^2(\varrho_P - \varrho_F)}{18v} \frac{\omega^2 rV}{S} \tag{4}$$

where d_P is the particle diameter, ϱ_P is the particle density, ϱ_F is the fluid density, v is the kinematic viscosity, ω is the angular velocity, r is the rotation radius, V is the centrifuge liquid volume and S is the liquid layer thickness. This relationship leads to the following key factors:

(1) throughput is strongly dependent on cell diameter and density; *E. coli* is about one-half the size of the most yeast cells and would have only about 25% of the throughput; (2) cell recovery also affects throughput (due to the liquid layer thickness *S*); increasing recovery from 90% to 98% can reduce throughput by over 55% (Hanisch *et al.*, 1982).

Continuous centrifuges also have several other operating limitations (Wang *et al.*, 1979). Most types produce aerosols which contain wet solids. This can present problems when dealing with biohazards and genetically altered bacteria, where containment is required. Heat generation can be substantial in some cases and requires complicated cooling systems. Tubular bowl units (*e.g.* Sharples Model 1P and 6P, Pennwalt Corp.) have limited capacity (about 4 kg wet cell weight) and must be stopped to recover cells. Cell washing must be carried out by repeated centrifugation and redilution steps.

Hollow fiber filters have been utilized for cell processing without the drawbacks of centrifuges. Since membranes are housed in self-contained cartridges, aerosols are not generated. Cell washing is also facilitated since a diafiltration step may be done as part of the overall process.

Throughput of hollow fiber filters is not a function of particle size as shown in centrifugation (see equation 4). Filtration rates will not drop significantly as smaller bacteria are harvested. Since the membrane physically retains the cells, recovery is essentially 100% at all times and is not affected by throughput. Finally, capacity is not limited since scale-up may be carried out by merely adding surface area.

Direct comparison of the two processes shows an economic advantage to hollow fiber filters in addition to the operating benefits. Table 2 presents data for smaller-scale, laboratory processes. In case 1, the centrifuge gave poor recovery even at a low process rate; the hollow fiber system output was five times higher with about the same capital expense. In case 2, a higher process rate was achieved with a capital cost only 30% that of the centrifuge.

Table 2 Comparisons of Hollow Fiber Filtration and Continuous Centrifugation

	Case 1[a]		Case 2[b]	
	Hollow fiber filtration	*Continuous centrifugation*	*Hollow fiber filtration*	*Continuous centrifugation*
Initial concentration	4×10^9 cells cm^{-3} (0.7 vol %)		1×10^7 cells cm^{-3}	
Concentration factor	10 ×	10 ×	100 ×	500 ×
Cell recovery	98%	<70%	98%	90%
Process rate (l h^{-1})	30[c]	6	40[d]	30
System cost ($)	9000[c]	8000	6000[d]	20 000

[a] Personal communication; Codon, Inc., Brisbane, California.
[b] Personal communication; University Micro Reference Lab, Ann Arbor, Michigan.
[c] Process rate can be increased fivefold with extra cost of $5000.
[d] Process rate can be increased twofold with extra cost of $1000.

More detailed cost information is presented in Tables 3 and 4. In the former case, a comparison is made for harvesting given bacteria, such as *E. coli*. The data are based on a continuous disc-type centrifuge (Alfa-Laval Model AX 213 with 37 HP motor). As shown, the hollow fiber system (based on Amicon 0.1 μm microporous-type) has about 70% lower capital cost and about 25% lower operating cost. In Table 4, where yeast is used for comparison, a smaller sized centrifuge now has twice the output. To provide equivalent process rates, a larger filtration system is required. Operating and capital costs now are higher with the hollow fiber system.

In brief summary, hollow fiber filters provide several operating advantages over centrifuges; these include absence of aerosols, higher recovery, simple scale-up and easy cell washing. The economic advantages of hollow fibers are also substantial when dealing with smaller bacterial cells. For yeasts and large cells, centrifuge outputs will increase significantly and harvesting with hollow fibers will be less attractive economically. A listing of cells which have been harvested/ washed with hollow fibers is shown in Table 5 with filtration performance data.

24.4.2.2 Product purification

The purification of extracellular fermentation products generally involves removal of bacteria from the broth. This has traditionally been carried out by such equipment as centrifuges, filter presses and rotary drum vacuum filters. Since the use of centrifugation has been discussed in detail (see Section 24.4.2.1), this section will focus on comparisons of hollow fibers with other types of filter.

Table 3 Comparison of Hollow Fiber Filtration and Centriguation for Bacteria Harvesting

| | Hollow fiber filtration | | Centrifugation[a] | |
	Annual ($)	$ per 1000 l	Annual ($)	$ per 1000 l
Depreciation	4500	0.22	15000	0.75
Maintenance	1800	0.09	6000	0.30
Labor	7500	0.38	7500	0.38
Power	1500	0.08	6200	0.31
Water, chemicals	1400	0.07	1200	0.06
Membranes	10500	0.52	—	—
Total operating costs	27200	1.36	35900	1.80
Capital cost	45000		150000	

Assumptions

(1) System output: 5000 l h^{-1} (water removal rate)
(2) Operation 250 d y^{-1}, 16 h d^{-1} plus 2 h d^{-1} cleanup
(3) Depreciation: 10 y S/L
(4) Maintenance: 4% of capital cost annually
(5) Labor at $15.00 h^{-1} (2 h d^{-1})
(6) Power at $0.05 kWh^{-1}
(7) Membrane life of 1 y, $157 m^{-2}
(8) Membrane average flux rate: 75 l m^{-2} h^{-1}
(9) Broth: *E. coli*, simple medium, 20 g l^{-1} feed, 180 g l^{-1} product
(10) Centrifuge type: Alfa-Laval Type AX213

[a] Personal communication, Alfa-Laval AB, Tumba, Sweden (1984).

Table 4 Comparison of Hollow Fiber Filtration and Centrifugation for Yeast Harvesting

| | Hollow fiber filtration | | Centrifugation[a] | |
	Annual ($)	$ per 1000 l	Annual ($)	$ per 1000 l
Depreciation	7500	0.19	5000	0.12
Maintenance	3000	0.07	2000	0.05
Labor	7500	0.19	7500	0.19
Power	2400	0.06	2900	0.07
Water, chemicals	2300	0.06	1500	0.04
Membranes	17400	0.43	—	—
Total operating costs	40100	1.00	18900	0.47
Capital cost	75000		50000	

Assumptions

(1) System output: 10 000 l h^{-1} (water removal rate)
(2) Operation 250 d y^{-1}, 16 h d^{-1} plus 2 h d^{-1} cleanup
(3) Depreciation: 10 y S/L
(4) Maintenance: 4% of capital cost annually
(5) Labor at $15.00 h^{-1} (2 h d^{-1})
(6) Power at $0.05 kWh^{-1}
(7) Membrane life of 1 y, $157 m^{-2}
(8) Membrane average flux rate: 90 l m^{-2} h^{-1}
(9) Broth: *S. cerevisiae*, simple medium, 40 g l^{-1} feed, 180 g l^{-1} product
(10) Centrifuge type: Alfa-Laval Type SFDX–209S

[a] Personal communication, Alfa-Laval AB, Tumba, Sweden (1984).

The use of filter presses and rotary drum filters depends on forming a filtration 'cake' on a cloth, plastic or metal mesh. The effective retention characteristics will actually depend on the porosity of this cake. As the filtration proceeds, the cake thickens and hence its resistance to flow will increase. Theoretically, the flow rate (dV/dt) will decrease in inverse proportion to the total volume (V) throughput:

$$\frac{dV}{dt} = \frac{a}{V} + b \tag{5}$$

where the constant a is dependent on the filter cake resistance and the constant b is dependent on the resistance of the supporting mesh. Practically speaking, however, this equation is limited by several factors (Wang *et al.*, 1979). In some cases, the b term will decline due to blockage of the filter mesh. With cells and biological products, compression of the filter cake solids can be a

Table 5 Filtration Flux Data for Cell Harvesting/Washing[a]

Application	Bacteria type (cell density) [g l^{-1}]	Fiber cut-off	Average filtration rate (l m^{-2} h^{-1})
Cell harvesting	*E. coli* (10–100)	0.1 μm	170
	E. coli (10–250)	0.1 μm	110
	B. subtilis (20–200)	100 000 D	150
	B. subtilis (20–200)	0.1 μm	190
	S. cerevisiae (20–100)	0.1 μm	260
	S. cerevisiae (20–240)	0.1 μm	190
	Streptomyces sp. (complex medium)	100 000 D	110
	Corynebacterium sp. (30–150)	100 000 D	54
	Corynebacterium sp. (30–260)	100 000 D	32
	Pertussis cells (10 ×; nonserum media)	100 000 D	130
	Algae (10–100)	100 000 D	93
	Pseudomonas sp.	100 000 D	71
	Leptospira sp.	100 000 D	86
	Streptococcus sp. (2–15)	100 000 D	110
	Pediococcus sp. (20–160)	100 000 D	54
	L. acidophilus (15–150)	100 000 D	71
Cell washing	*S. aureus* (protein A in filtrate)	0.1 μm	170
	B. thuringiensis (40)	100 000 D	79
	C. perfringens (16) (exotoxin in filtrate)	0.1 μm	30
	Red cell ghosts	0.1 μm	130
	Bacillius sp. (α-amylase in filtrate)	0.1 μm	54
	Algae (polysaccharides in filtrate)	100 000 D	64
	S. cerevisiae (150)	0.1 μm	130
	Corynebacterium sp. (amino acid in filtrate)	100 000 D	43
	V. cholerae (cholera toxin in filtrate)	100 000 D	79

[a] Amicon Corporation, Danvers, MA, USA; Publication No. 494, 'Hollow Fibers for Fermentation and Cell Culture' (1984).

major factor. Above a certain pressure, the cake may actually collapse to a very compressed form with a much lower flow rate.

In dealing with bacteria and cell debris, the use of filter aids will improve solids retention. Diatomaceous earth is the most common type of filter aid. Unfortunately, use of filter precoats does lead to certain problems. Waste cells are more difficult to dispose of; they cannot be used for animal feed. Filter aids will also inhibit complete passage of product in the liquor. If protein precipitation occurs, recovery is very difficult.

As shown in the previous section, hollow fiber filters offer advantages over traditional cell processing methods. Hollow fibers would completely retain even small cells, obviating the need for filter aids. Even with a filter aid, some cells will be found in the filtrate of a press or drum type filter (Wang *et al.*, 1979). In some cases, an ultrafilter may be selected to remove high molecular weight contaminants. This type of purification may be carried out if the extracellular product is of low molecular weight, such as with an antibiotic (Beaton, 1980). Even such larger proteins as interferon have been purified using a 100 000 D cutoff hollow fiber (Leuthard and Schuerch, 1980).

In addition to the above operating advantages, hollow fiber filters can provide large reductions in operating expense. As shown in Table 6, a rotary vacuum filter is less than one-half the capital cost. However, the savings in filter aid are large enough to allow a one year payback in this case. These economics do not include the extra cost for disposal of filter aid. It should be noted that the hollow fiber flux rate is quite low in this case; the rate is about half of that shown in Tables 3 and 4. This reduction is due to the high solids level and viscosity found in the mycelial broth containing the antibiotic. In spite of this low flow rate, the hollow fiber system gives a rapid payback, complete removal of cells and simpler disposal.

24.4.3 Continuous Processes

24.4.3.1 *Fermentation*

Batch fermentation has been the traditional means for growth of bacteria. Research on the use of membranes in small-scale dialysis fermentation systems with cell recycle shows substantial

Upstream and Downstream Processing

Table 6 Comparison of Hollow Fiber Filtration and Rotary Vacuum Filtration for Antibiotic Purification

| | Hollow fiber filtration | | Rotary vacuum filtration[a] | |
	Annual ($)	$ per 1000 l	Annual ($)	$ per 1000 l
Depreciation	14000	0.47	6000	0.20
Maintenance	8400	0.28	3600	0.12
Labor	7500	0.25	7500	0.25
Power	4000	0.13	3700	0.12
Water, chemicals	3600	0.12	—	—
Membranes	31500	1.05	—	—
Filter aid	—	—	144000	4.80
Total operating costs	69000	2.30	164800	5.49
Capital cost	140000		60000	

Assumptions
(1) System output: $7500 \, l \, h^{-1}$ (water removal rate)
(2) Operation $250 \, d \, y^{-1}$, $16 \, h \, d^{-1}$ plus $2 \, h \, d^{-1}$ cleanup
(3) Depreciation: $10 \, y$ S/L
(4) Maintenance: 4% of capital cost annually
(5) Labor at $15.00 \, h^{-1}$ ($2 \, h \, d^{-1}$)
(6) Power at $0.05 \, kWh^{-1}$
(7) Membrane life of $1 \, y$, $157 \, m^{-2}$
(8) Filter aid used: $24 \, kg \, m^{-3}$ at $0.20 \, kg^{-1}$
(9) Membrane average flux rate: $37.5 \, l \, m^{-2} \, h^{-1}$
(10) Vacuum filtration rate: $400 \, l \, m^{-2} \, h^{-1}$

increases in cell density and total product yield (Abbott and Gerhardt, 1970; Friedman and Graden, 1970). Continuous removal of inhibitory by-products allows growth to continue for weeks.

Hollow fiber membranes offer the same advantage as dialysis fermentation but allow easy use in an industrial environment. They have proved successful in increasing total cell densities to substantially higher levels than those found in normal batch fermentations: up to 10^{15} cells l^{-1} in certain cases (Michaels *et al.*, 1980). These bacteria can be withdrawn continuously to extract intracellular products or to test cell viability. Where extracellular products are of interest, filtrates are removed for further purification.

Typical process arrangements use the hollow fiber cartridge in a normal cross-flow mode. Cells are recirculated through the fibers and returned to the fermenter (see Figure 9). In some cases, bacteria are immobilized on the outside of the fibers and remain productive in a stationary phase (Kan and Shuler, 1978; Mattiason *et al.*, 1981).

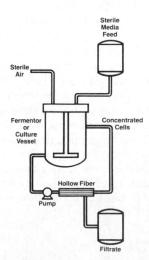

Figure 9 Flow schematic for continuous fermentation/recycle system

The use of hollow fiber cartridges in a high solids fermentation system has been reported (Chang and Furjanic, 1981). The membrane cartridge was connected to a fermenter to remove waste effluent, while fresh substrate was added to the broth. Production efficiency was increased by

100%. It was also determined that at high cell density, lactic acid biomass production was controlled by substrate limitations rather than by product inhibition.

In a more detailed report, a hollow fiber filter was utilized for continuous ethanol production by *Zymomonas mobilis* (Charley *et al.*, 1983). Earlier work utilized a microporous (0.45 μm) cassette-type system for recycle; this system was found to be difficult to operate on a long-term continuous basis due to membrane and support screen blockage. A two-stage continuous stirred tank reactor (CSTR) was used with recycle on the second stage. Steady state operation was attained about 36 hours after reaching the approximate operating biomass concentration. A bleed from the second stage gave a mean cell residence time of 27.8 hours; this was somewhat higher than the desired optimum of 24 hours, which was determined in a simulation. The specific rate of product formation (0.68 g g^{-1} h^{-1}) was about 82% of that shown in a two-stage system without recycle. Very high effluent alcohol concentrations (over 100 g l^{-1}) and product yield factors (0.46) were obtained throughout steady state operation. Overall volumetric productivity (18.1 g l^{-1} h^{-1}) was higher than a single-stage CSTR without recycle (11.3 g l^{-1} h^{-1}), but much lower than one with recycle (151 g l^{-1} h^{-1}). Thus, the hollow fiber recycle gave a 60% increase in productivity compared to a single-stage CSTR and allowed product concentrations of over 100 g l^{-1}.

24.4.3.2 *Cell culture*

Large-scale growth of animal or plant cells in culture can be difficult with such techniques as roller bottles, microcarrier beads and agitated vessels. These approaches have been examined for suspension cultures as well as anchorage-dependent cells.

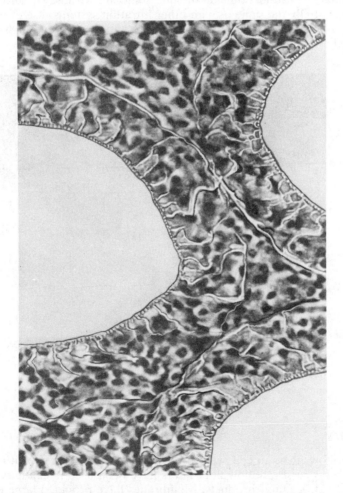

Figure 10 Cells growing within and around hollow fibers (reproduced by permission from Amicon Corp.)

Hollow fiber cartridges have been evaluated extensively for growth of animal cell lines. By allowing growth on the outside of the membrane, the fibers actually simulate capillaries. Examination of cultures shows that the cells grow into the porous fiber to the membrane wall itself and completely occupy all available intercapillary space (see Figure 10). Growth is truly tissue-like in nature as opposed to monolayers found in roller bottles and microcarrier beads (Ku *et al.*, 1981). Densities have been shown to be in the range of 10^5–10^6 cells per cm^2 fiber area and 10^7–10^8 cells cm^{-3} (Knazek *et al.*, 1974; Fike *et al.*, 1977; Ehrlich *et al.*, 1978). These levels are about 10 times higher than those found in roller bottles.

Although cell division eventually ceases, the other cell functions continue for weeks and even months thereafter. Significantly, the *per cell* production of viruses and macromolecules is often greater and lasts longer than in other systems. Superior performance has been observed with virus (Johnson *et al.*, 1978; Ratner *et al.*, 1978), hormone (Knazek *et al.*, 1974), antigen (David *et al.*, 1978; Rutzky *et al.*, 1979) and interferon (Strand *et al.*, 1982).

Recent research reports the use of hollow fibers to grow hybridomas for production of monoclonal antibodies (Calabresi *et al.*, 1981; Weimann *et al.*, 1983). In one case, mouse hybridomas were constructed by fusing myeloma cells with spleen cells from mice immunized with β-galactosidase. Hollow fibers with 50 000 D cutoff were inoculated with the hybridoma cells. Solid masses of cells were observed to fill the interstices of the fibers. Daily output of monoclonal antibody was equivalent to a suspension culture 50 times larger in volume and containing about 5×10^5 cells cm^{-3}. Weekly analysis showed that antibody production increased with time. After two days of culture, the antibody concentration was 33 times higher than the control medium. After 29 days of continuous culture, the concentration increased to 136 times the control level.

An actual hollow fiber culture system is shown in Figure 11. Medium is pumped from a spinner reservoir to the inside of the fibers where it perfuses the extracapillary space and the cells. Oxygen and fresh medium may be introduced periodically into the reservoir. The products, which are typically extracellular, will collect outside the fibers and are withdrawn periodically. Scale-up of these systems is essentially the same as with other filtration systems: use of more cartridges with larger surface area. In summary, hollow fiber culture systems achieve high cell density, output and lifetime as well as allow simple process scale-up.

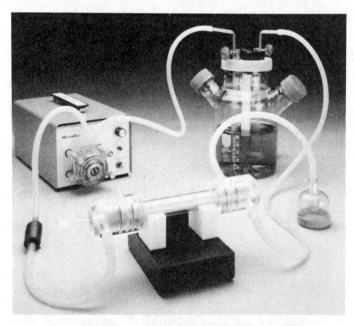

Figure 11 Laboratory hollow fiber cell culture system (reproduced by permission from Amicon Corp.)

24.4.4 SUMMARY

The utility of hollow fibers in cell separation processes has been shown to be extensive. For batch fermentation processes, hollow fiber filters have been used in cell harvesting, cell washing and purification of extracellular products. Membranes have provided large economic and operational advantages over such equipment as continuous centrifuges, filter presses and rotary drum vacuum filters. With continuous fermentation and cell culture processes, hollow fibers have allowed high cell productivity, long-term growth and easy scale-up. The use of hollow fibers for such separation processes will become significant as biotechnology creates more and more products from cells.

24.5 REFERENCES

Abbott, B. S. and P. Gerhardt (1970). Dialysis fermentation I. Enhanced production of salicylic acid from naphthalene by *Pseudomonas fluorescens*. *Biotechnol. Bioeng.*, **12**, 557–589.
Beaton, N. C. (1980). The application of ultrafiltration to fermentation products. In *Ultrafiltration Membranes and Applications*, ed. A. R. Cooper, pp. 373–404. Plenum, New York.
Brandt, A. and G. Bugliarello (1961). *Trans. Soc. Rheol.*, **10**, 229–251.
Calabresi, P., K. L. McCarthy, D. L. Dexter, F. J. Cummings and B. Rotman (1981). Monoclonal antibody production in artificial capillary cultures. *Proc. AACR ASCO*, p. 302.
Chang, W. T. H. and J. J. Furjanic (1981). Production of biomass with hollow-fibre ultrafiltration–fermentation system. *SIM News*, July, 24–25.
Charley, R. C., J. E. Fein, B. H. Lavers, H. G. Lawford and G. Lawford (1983). Optimization of process design for continuous ethanol production by *Zymomonas mobilis* ATCC 29191. *Amicon News*, No. 474, p. 1. Amicon Corp., Danvers, MA.
David, G. S., R. A. Riesfeld and T. H. Chino (1978). Continuous production of CEA in hollow fiber cell culture units. *J. Natl. Cancer Inst.*, **60**, 303–306.
Ehrlich, K. C., E. Stewart and E. Klein (1978). Artificial capillary perfusion cell culture: metabolic studies. *In Vitro*, **14**, 443–448.
Fike, R. M., J. L. Glick and A. A. Burns (1977). Propagation of human lymphoid cell lines in hollow fiber capillary units. *In Vitro*, **13**, 170.
Friedman, M. R. and E. F. Gaden (1970). Growth and acid production by *L. delbrueckii* in a dialysis culture system. *Biotechnol. Bioeng.*, **12**, 961–974.
Hanisch, W. H., S. Fuhrman, D. Harano and M. Pemberton (1982). Separation and purification techniques applicable to the biological processes. *Proc. Aust. Biotechnol. Conf.*, pp. 153–170.
Howard, G. and R. Duberstein (1980). New separation technique for the chemical process industries. *J. Parent. Drug Assoc.*, **34** (2), 95–102.
Johnson, A. D., G. A. Eddy, J. D. Gangemi, H. R. Hamsburg and J. F. Metzger (1978). Production of Venezuelan equine encephalitis virus in cells grown on artificial capillaries. *Appl. Environ. Microbiol.*, **35**, 431–434.
Kan, J. K. and M. L. Shuler (1978). Urocanic acid production using whole cells immobilized in a hollow fiber reactor. *Biotechnol. Bioeng.*, **20**, 217–230.
Karnis, A. (1966). *Can. J. Chem. Eng.*, **44**, 181.
Knazek, R. A., P. O. Kohler and P. M. Gullino (1974). Hormone production by cells grown in vitro on artificial capillaries. *Exp. Cell Res.*, **84**, 251–254.
Ku, K., M. J. Kuo, J. Delente, B. S. Wildi and J. Feder (1981). Development of a hollow-fiber system for large-scale culture of mammalian cells. *Biotechnol. Bioeng.*, **23**, 79–95.
Labuza, T. P. (1975). Cell collection: recovery and drying for SCP manufacture. In *Single-Cell Protein II*, ed. S. R. Tannenbaum and D. I. C. Wang, pp. 69–104. MIT Press, Cambridge, MA.
Leuthard, P. and A. R. Schuerch (1980). A simple and rapid method for concentration of interferon and removal of concentrated inducing virus. *Experientia*, **36**, 1447.
Lukaszewicz, R. C., A. Kurin, D. Hauk and S. Chrai (1981). Functionality and economics of tangential flow microfiltration. *J. Parent. Sci. Technol.*, **35**, 231–236.
Mattiason, B. and M. Ramstorp (1981). Comparison of the performance of a hollow-fiber microbe reactor with a reactor containing alginate entrapped cells. *Biotechnol. Lett.*, **3**, 561–566.
Michaels, A. S., C. R. Robertson and S. N. Cohen (1980). Hollow membrane fiber bioreactor. A novel approach to continuous, immobilized whole cell biochemical synthesis. Paper presented at Second Chem. Congr. of the North American Continent, San Francisco, CA.
Ng, P. K., J. L. Lundblad and G. Mitra (1976). *Separ. Sci. Technol.*, **11**, 499–502.
Porter, M. C. (1972). Concentration polarization with membrane ultrafiltration. *Ind. Eng. Chem. Prod. Res. Dev.*, **11**, 234–248.
Ratner, P. L., M. L. Cleary and E. James (1978). Production of rapid harvest moloney murine leukemia virus by continuous cell culture on synthetic capillaries. *J. Virol.*, **26**, 536–539.
Reti, A. R., T. Leahy and P. M. Meier (1979). The retention mechanism of sterilizing and other submicron high efficiency filter structures. *Proc. Second World Filt. Congr.*, pp. 427–435.
Rutzky, L. P., J. T. Tomita, M. A. Calenoff and B. D. Kahan (1979). Human colon adenocarcinoma cells III. In vitro organoid expression and carcino-embryonic antigen kinetics in hollow fiber culture. *J. Natl. Cancer Inst.*, **63**, 893–898.
Segre, G. and A. Silberberg (1961). *Nature (London)*, **189**, 209–210.
Strand, J. M., J. M. Quarles and S. H. Black (1982). Human fibroblast interferon production in a matrix perfusion microcarrier bead system. *Proc. ASM 82nd Annual Meet.*, Paper No. T20.
Trudel, M. and P. Payment (1980). Concentration and purification of rubella virus hemagglutinin by hollow fiber ultrafiltration and sucrose density centrifugation. *Can. J. Microbiol.*, **26**, 1334–1339.
Wang, D. I. C., C. L. Cooney, A. L. Demain, P. Dunnill, A. E. Humphrey and M. D. Lilly (1979). Enzyme isolation. In *Fermentation and Enzyme Technology*, ed. C. R. Heden, pp. 238–310. Wiley, New York.
Weimann, M. C., E. D. Ball, M. W. Fanger, D. L. Dexter, O. R. McIntyre, G. Bernier, Jr. and P. Calabresi (1983). Human and murine hybridomas: growth and monoclonal antibody production in the artificial capillary culture system. *Clin. Res.*, **31**, 511.

25

Ultrafiltration

M. S. LE and J. A. HOWELL
University College, Swansea, UK

25.1 INTRODUCTION

To those who are involved in biotechnology or who are concerned with the isolation and purification of macromolecular products, ultrafiltration needs no introduction. However, in the authors' experience, a surprising number of people are still strangers to the word ultrafiltration. With rapid development of new membranes which will stand harsh cleaning regimes and higher pressures there will be increasing use of ultrafiltration in the downstream processing of fermentation products. Yet the process is poorly understood fundamentally and suffers from a number of problems that have not yet been properly solved. A lengthy explanation and discussion of the technique to the newcomer is, therefore, justified. On the other hand it would be unnecessary and wasted effort to merely reproduce material already available in good reviews or from membrane manufacturers, although the latter often underplay the problems in ultrafiltration and present the process in very simple terms whilst extolling the virtues of their products. Hence the authors of the chapter will endeavour to present information of interest to beginners as well as being useful to experts.

Recent advances in polymer technology have made available a wide range of materials from which membranes of desired properties can be fabricated to suit particular applications. Consequently there is now a large choice of membranes from which the user can make his selection. We shall outline the methods of membrane manufacturing, the membrane characteristics and performance.

For a long time it has been felt that the existing theory in ultrafiltration is both inadequate and misleading. In the following pages we shall examine this theory closely and discuss some of the latest ideas on membrane theory. The successful application of ultrafiltration depends largely on a semi-permeable membrane; it would be an incomplete discussion if the membrane itself were omitted.

Undoubtedly, anyone who has any experience with ultrafiltration would agree that the stumbling block which hinders the commercial exploitation of the process is the phenomenon of membrane fouling. The subject of fouling has been under extensive investigation by numerous research workers in the past few years and their progress will be reviewed here.

We shall briefly discuss the concept of module design and the advantages and disadvantages of various systems. Finally, by way of an example we shall show how to scale up a UF plant from research data plus an analysis of the running costs of such a plant.

25.2 MEMBRANES

25.2.1 Development of the Semi-permeable Membrane

The history of the semi-permeable membrane can be traced back more than 200 years. It began in 1748 when Abbe Nollet observed the swelling of an animal bladder containing alcohol when it was immersed in water. Over one hundred years passed until this observation of osmosis was formulated into a law by Van't Hoff. Artificial membranes from collodion or cellulose nitrate were first prepared by Fick in the middle of the last century. In 1939 Dobry showed that porous films of cellulose acetate could be made by dissolving cellulose acetate in saturated aqueous magnesium perchlorate solution, casting it and then precipitating the cellulose acetate by adding water. Although these films had very high water flux they had no ability to reject salts. The most significant breakthrough in membrane technology occurred in the mid 1950s with the finding of Reid and Breton (1959; and Breton, 1957) that cellulose acetate membranes exhibit a remarkable selectivity between salts and water, and the discovery by Loeb and Sourirajan (1960) of a method for preparing high flux cellulose acetate membranes that have an extremely thin selective layer backed by a highly porous substrate. The application of these processes that has received the greatest attention has been the desalination of brackish water or sea water by reverse osmosis (RO). Much of the support for membrane development came from the Office of Saline Water of the US Department of the Interior in the early 1960s (Lonsdale *et al.*, 1965). At about the same time research activity at MIT led to the discovery of the polyelectrolyte complex hydrogels and their anomalous water transport properties (Michaels, 1965, 1967). An early commercial venture in ultrafiltration was undertaken by a collaboration between Amicon Corporation and Dorr-Oliver Inc. in 1963 for the purpose of developing an economical process for large scale removal of colloidal and macromolecular impurities from secondary sewage effluent. These first generation membranes were shown to be asymmetric in structure and were several orders of magnitude more permeable than the homogeneous gel membrane structures of comparable solute retentivity. They took an important place in laboratory and medical practice due to their high retentivity for small molecules and their inertness to biochemicals.

At this stage it may be appropriate to make a distinction between the various types of membranes used in separation processes. Membranes are classified in three general groups according to their operation: microfiltration (MF); ultrafiltration (UF) and reverse osmosis (RO). There are no fundamental differences between the three types of membranes. In fact, some are actually made by the same process involving the precipitation of the polymer from a polymer–solvent casting solution by a non-solvent (usually water). Reverse osmosis is usually operated at 500–1000 p.s.i. while the other two processes operate in the pressure range of 10–100 p.s.i. Figure 1 shows the pore size spectrum of the different types of membranes and, as can be seen, there is considerable overlap in the pore size range. It is believed that the chief separating mechanism in RO is charge repulsion and diffusion, whereas in UF and MF sieving predominates. However, the differences between the three processes are subject to considerable speculation and for the present they must remain arbitrary.

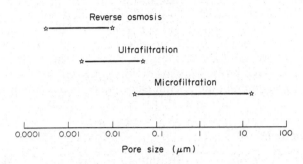

Figure 1 Membrane filtration size spectrum

25.2.2 Membrane Preparation

25.2.2.1 Introduction

Traditionally cellulose has been the most common material for membrane fabrication. The casting technique generally involves first rendering the cellulose soluble, then evaporating or precipitating it to form a membrane. Cellulose can be made soluble by several methods. (a) By nitration of the OH groups then dissolving in an alcohol or ether solvent. The membrane can be formed by evaporation or precipitation with water after which it is denitrated. (b) By treatment with a strong caustic solution followed by carbon disulfide. Sodium cellulose xanthate is produced; this is dissolved in a weak caustic solution to form a viscose which is left to ripen for several days. The polymer is coagulated in sulfuric acid which regenerates the cellulose. (c) Purified cellulose can be dissolved in copper sulfate/ammonia solution which forms a complex with the cellulose. The solution is coagulated in water and washed in acid. (d) By reaction with acetic acid to form cellulose acetate. In this process the OH groups are acetylated and the polymer product is soluble in acetone.

Membranes from all these preparations can be produced by several methods. For example, a cuprammonium dope can be spread onto a glass plate then dipped into cold water to coagulate. The membrane produced by this technique is suitable for use in artificial kidneys. Asymmetric membranes can be made of different blends of cellulose nitrate and cellulose acetate from different mixtures of acetone/propanol/glycerine/water. Surface evaporation of the more volatile components is allowed to take place prior to coagulation in water. The top surface of the membrane produced this way has a finer pore structure. A skinned RO membrane can be prepared from a solution of cellulose acetate in acetone. The thin film is evaporated for 3–4 min, coagulated in cold water and then annealed in water at 80 °C. The top layer forms a thin homogeneous membrane with very high permselectivity and the lower layer yields a supporting matrix of a highly porous structure.

Despite its availability and many useful characteristics which make it suitable for use in the manufacture of a good membrane, cellulose suffers chemical and biological attack and it has low resistance to temperature. For these reasons cellulose based membranes are now being phased out to be replaced by physically and chemically more rugged and durable membranes made from synthetic polymers. Some of the most commonly used are the aromatic amides. Others include polycarbonates, polyacrylonitrile–polyvinyl chloride copolymer and polysulfone. Many new materials have also been tried.

25.2.2.2   Composite membranes

Composite membrane production is an important area in membrane technology at the present. This type of membrane consists of an extremely thin layer (typically <1 mμ) of very high perm-selectivity; it is supported on a much thicker (20 μm–1 mm) layer of microporous open-celled sponge. If desired, these membranes can be further supported by a fibrous sheet (*e.g.* paper) to provide greater strength and durability. The support provided to the skin by the spongy sublayer is adequate to prevent film rupture even at high pressures and the hydraulic resistance to liquid flow through the layer is small compared to the skin resistance. In the earliest method of membrane production a film of cellulose acetate was deposited by dipcoating the supporting layer with a dilute polymer solution. More recently composite membranes have been made by the so-called phase inversion process or interfacial polymerization. In this process a homogeneous polymer solution changes into a two-phase system in which a solidified polymer phase forms the porous membrane structure while a liquid phase, poor in polymer, fills the pores. Various researchers have reported quite a variety of membrane morphologies produced by the phase inversion process. Depending on the precipitating technique employed the resulting membranes may be skinned or unskinned, or have porous sublayers with an even distribution of pore size or with increasing pore size over the membrane thickness, or have porous sublayers with cavities.

25.2.2.3   Precipitation methods

(i)   Precipitation from the vapour phase (Strathmann and Koch, 1977, Maier and Scheuermann, 1960)

In this technique, membrane formation is accomplished by penetration of the precipitant through a polymer film from the vapour phase which is saturated with the solvent used. A porous membrane is produced without a skin and with an even distribution of pores over the membrane thickness. Such membranes are suitable for bacteria filters.

(ii)   Precipitation by evaporation (Maier and Scheuermann, 1960; Kesting, 1973)

The polymer is dissolved in a mixture of a volatile and a less volatile solvent. As the more volatile component evaporates, the polymer precipitates and forms a skinned membrane.

(iii)   Immersion precipitation (Guillotin *et al.*, 1978; Koehnen *et al.*, 1978)

This technique was used successfully by Loeb and Sourirajan (1960) for their well-known RO membrane. It has been exhaustively studied and exploited for commercial production of skinned membranes. The main feature of the technique is of course the immersion of the cast film in a bath of non-solvent for coagulation of the membrane.

(iv)   Thermal precipitation (Tanny, 1974)

Precipitation of the polymer solution is brought about by a cooling step. If the evaporation of the film surface was allowed prior to cooling a skinned membrane would result.

25.2.2.4   Other methods

Membranes have also been prepared by sintering of powders, *e.g.* PTFE, or by extrusion and stretching at right angles to the direction of extrusion to produce controlled tears. The production of membranes with an extremely narrow pore size distribution is now possible with the use of a laser beam, or collimated charged particles. A homogeneous film is exposed to the radiation which breaks the chemical bonds, then it is placed in an etching bath where the damaged material is preferentially removed. Such membranes are suitable for use in microfiltration. Although membrane production steps are quite simple, the exact polymer additives, solvent formulations and the timing are critical; this tends to be a complex process and the exact preparation procedures are currently very empirical.

25.2.3   Factors Affecting Membrane Structure

The performance of a membrane to a large extent is dependent on the process by which it is manufactured. It is appropriate, therefore, to examine the factors which control the final structure of the membrane. Every step in the casting process has important effects on the average pore size and the distribution of pores on the surface of the membrane (Sourirajan and Kunst, 1978). The composition and temperature of the gelation media are important film casting variables which can be exploited for the production of useful membranes. Sourirajan *et al.* (Sourirajan and Kunst, 1978; Kunst and Sourirajan, 1974; Kutowy and Sourirajan, 1975) showed that a decrease in solvent/polymer ratio, or an increase in non-solvent/polymer ratio in the casting solution composition, or a decrease in the casting temperature all tend to increase the average pore size. Furthermore, increases in solvent/polymer ratio and temperature also increase the effective number of pores on the membrane surface. However, often during such variations an increase in the size of the pores on the membrane surface results in a simultaneous decrease in the number of pores. These criteria offer a wide choice for composition of the casting solution and casting conditions for making ultrafiltration membranes suitable for many different applications.

It has been shown (Kutowy *et al.*, 1978; Tweddle and Sourirajan, 1978) that the gelation process could provide an effective means to control the average pore size on the membrane surface. For example, by using alcohol/water mixtures as gelation media, at different temperatures and in a wide range of alcohol concentrations, cellulose acetate RO and UF membranes with different surface porosities have been obtained. The success of this technique led to the development of a differential gelation process for achieving asymmetric porosity in the making of tubular UF membranes (Thayer *et al.*, 1978). In this technique gelation media of different composition and/or temperature are cast consecutively onto a porous support tube. The casting solution penetrates into the voids of the porous support during film casting by capillary action and forms the anchoring points.

The mechanism of pore formation is of fundamental importance in porous membrane technology. The subject is very complex and science has not yet progressed far enough in this field to help in producing better membranes. Meares (1966) estimates that 99% of the water flow in cellulose acetate membranes probably takes place by viscous flow in tortuous channels 6–10 Å in diameter. These channels occupy about 4% of the membrane volume, but how they are formed is still uncertain. According to Meares there are several possible explanations for the existence of pores. They may be morphological artifacts introduced by the membrane preparation method or they may be an inherent property of the cellulose acetate. They may be a consequence of the natural packing of the molecules or they may result from non-uniform swelling on a molecular scale when the polymer absorbs water. Kesting (1973) critically studied the role of certain inorganic salts in the gelation process and concluded that hydrated salt ions are capable of forming complexes with certain sites on the polymer and cause additional water to be incorporated into the gel network. The inclusion of such salts in a casting solution containing polymer and water results in salt–polymer interactions and the formation of a more open network than otherwise in the absence of the salts. The same people also suggested that the formation of a cellulose acetate membrane is a gelation process involving the coagulation of the polymer solution into a comparatively rigid mass incorporating a large amount of water. In the dilute solutions the polymer molecules are random coils enclosed in an envelope of solvent which acts as a barrier between the macromolecules. As the polymer concentration is increased molecules of the polymer come into contact and aggregation occurs. Coacervation and desolvation give rise to an open cell structure in the resulting membrane. Desolvation may be effected by evaporation or by immersion of the polymer solution into a gelation medium such as water. Coacervation is the coalescence of the colloidal particles whose solvent envelopes diffuse into one another. Based on diffusion studies in cellulose acetate gels prepared from cellulose acetate–benzyl alcohol, Klemm and Friedman (1932) concluded that the gel structure was porous and that the average pore radius decreased with an increase in cellulose acetate concentration. They suggested that gel formation was the result of incomplete precipitation of the dispersed medium and the resulting pore openings consisted of minute tubes of varying lengths randomly arranged but each intersecting others to form continuous passages through the gel. The pore size in the membrane can also be changed permanently by heating (Wasilewski, 1965), however, for a more lengthy discussion interested readers should consult the references.

25.3 UF MEMBRANE CHARACTERISTICS AND PERFORMANCE

The formal definition of ultrafiltration is, briefly, that it is a process whereby a solution containing a solute of molecular size significantly greater than the solvent is depleted of the solvent by the application of a hydraulic pressure which forces only the solvent to flow through a suitable membrane. Figure 2 illustrates the principle of ultrafiltration.

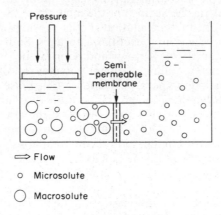

Figure 2 The principle of ultrafiltration

The heart of the process is, of course, the semi-permeable membrane and for successful and useful ultrafiltration the membrane must possess the following characteristics: (1) high hydraulic permeability to solvent (water), that is the membrane must be capable of passing water at high rates per unit membrane area under modest pressure; (2) sharp molecular weight cutoff, that is the membrane must be able to retain all molecules above a specific size and pass all smaller molecules; (3) good mechanical durability, chemical and thermal stability; (4) high fouling resistance, that is no tendency to adsorb solutes or to retain solutes in the pore networks thereby reducing the hydraulic permeability; (5) easily cleaned and sterilized; and (6) long membrane life.

25.3.1 Membrane Permeability

Table 1 is a list of commercially available UF membranes. Taken collectively, all the different types of membrane now comprise a family covering an extremely broad range of permeability and retention characteristics. These new generations of membranes are highly reproducible compared with the collodion membrane of the 1930s. In most cases the water fluxes in Table 1 have been normalized to 55 p.s.i., which is a typical operating pressure for many commercial UF systems, although laboratory filters tend to operate at lower pressure. The majority of commercial membranes currently available have an anisotropic structure which has been found to resist plugging and is more permeable to water. The pure water flux of a clean membrane bears no relationship to the membrane permeation behaviour in solutions of different solutes. Hancher and Ryon (1973) tested a number of early membranes with nominal molecular weight cutoff ranging from 500–300 000 in water and in salt solutions. They found that half of the membranes showed a significant decrease in flux during the first 24 h and two-thirds of them reached steady state flux values less than 50% of the manufacturers' ratings, especially those with a more open structure. The reduction in flux cannot be readily explained because of the complex interaction of different fouling processes inside the membrane structure and on its surface. For example, the contaminants in the salts or bacterial cells in the water may collect on the membrane surface and block the pores underneath, or such particles may be carried into the pore network and become trapped. In this way the membrane can act as a depth filter. It has been suggested (Swaminathan, 1980) that the salt ions may interact with the polymer matrix and cause the membrane to swell or contract reversibly with concomitant flux decrease or increase. Generally, flux in a macromolecular solution such as protein can be expected to be as low as 10% of the flux value in pure water. Interestingly, although the water permeability of various type of membranes in Table 1 differ by more than 100 fold, in protein solutions the difference is less marked. For instance, the pure water flux of a PM30 membrane is 360–600 cm h^{-1} in water and the corresponding figure for a T6/B membrane is 3.2–5 cm h^{-1}, but when used for concentrating cheese whey the flux ratio of the PM30 to T6/B is less than 4. One hypothesis is that, all other factors such as the membrane thickness, the pore size, the porosity and the degree of anisotropy being equal, the flux will be determined solely by the surface property of the membrane. Clearly, the differences in surface properties will be masked when the membranes are covered with a monolayer of protein. In view of the unpredictable flux performance of a membrane with any particular solute it is highly recommended that a small laboratory unit be used to evaluate all the possible membranes against the solutes of interest before committing oneself to a large membrane system.

Table 1 Characteristics of Some Commercial Membranes

Membrane series	Manufacturer	Composition	Flux at 55 p.s.i. (cm h^{-1})[a]	Molecular wt. cutoff ×1000	Working temp. maximum (°C)	pH limits[b]
PEM	Gelman	Cellulose	1.2	60	—	—
UM Diaflo	Amicon	Polyelectrolyte	0.6–1.8	0.5–10	50	2–12
XM Diaflo	Amicon	Substituted olefins	60–120	50–300	75	2–12
PM Diaflo	Amicon	Aromatic polymer	90–300	10–30	115–121	2–12
YM Diaflo	Amicon	—	1.2–120	1–30	75	2–12
PSED	Millipore	Cellulose ester	8	50	—	—
HFA	Abcor	Cellulose acetate	50–80	10–30	50–63	2–9
CA	DDS	Cellulose acetate	8–12	6–60	50	2–8
GR, FS	DDS	Copolymer	4–15	6–60	80	2–14
T2/A–T5/A	PCI	Cellulose acetate	2–3	4–10	30	2–8
BX1, BX2, T6/B	PCI	Polyacrylonitrile copolymer	3.2–5	30–100	70	2–13

[a] Actual water permeation rate varies widely with water quality.   [b] The operating pH depends on the working temperature.

25.3.2 Solute Rejection

The rejection characteristic is perhaps the second most important criterion in the membrane selection process. It is a unique property of the semi-permeable membrane and, by definition, it is the fraction of solute retained by the membrane, *i.e.* rejection = 1 − concentration in the permeate / concentration at the membrane surface.

In most cases the solute concentration at the membrane surface is greater than the concentration in the bulk solution due to concentration polarization, but as the surface concentration cannot be determined by simple measurement techniques, for the purpose of defining the rejection coefficient the surface concentration is taken to be the same as the bulk concentration. The rejection value obtained in this way is known as the apparent rejection and it is smaller than the actual coefficient. Usually the membrane manufacturer specifies the rejection characteristics of a membrane by indicating its molecular weight cutoff values. Basically the specification procedure involves passing solutes of different molecular weight through the membrane and the cutoff is the value at which a 99% rejection occurs. Such figures are intended only as a rough guide since manufacturers are aware that the molecular diameter of a molecule is not directly related to its mass, but on what is termed steric hindrance, that is its tertiary structure; a globular protein molecule has less steric hindrance than a linear chained polymer of the same mass, the protein is expected, therefore, to be able to pass through the pore more easily than its counterpart.

One popular misconception which still exists at present is that if the membrane retains a solute of certain size then its pore size must be smaller than the solute dimensions. Particle bridging, which is the result of many particles arriving at the pore entry at the same time, is a possibility that exists when the size of the solute is within one order of magnitude of the pore size. The rejection is effected by mutual exclusion between the solute molecules.

Another process which can greatly influence the rejection characteristic of a membrane is adsorption. Solute adsorption on the pore wall will reduce the effective pore size and enhance rejection considerably. At an extreme the adsorbed layer may form a dynamic membrane with its own rejection characteristics. Figure 3 illustrates the flux and rejection performance of a UF membrane backing paper in a crossflow filtration unit. The increase in protein rejection accompanying a decrease in the flux indicate the formation of a dynamic membrane with a high rejection coefficient. Membrane solute double-layer interactions will have a strong influence on adsorption and it is important to select a membrane for which such forces will tend to prevent adsorption. This may be difficult when ultrafiltering complex mixtures such as cheese whey, which contains several proteins, fat, lipoproteins, sugar and salts.

Other factors which influence the apparent rejection of UF membranes are the operating pressure, temperature and the fluid velocity tangential to the membrane, and the bulk concentration. Nakao and Kimura (1981) showed that the apparent rejection decreases slightly with increasing pressure. The mechanism for this effect is not clear, but it is probable that the high pressures

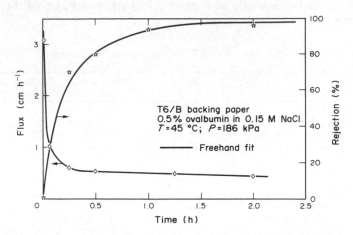

Figure 3 Time variation of flux and rejection for the membrane backing paper

unblock some of the pores by forcing the trapped particles through them. The effect of tangential velocity is illustrated by Figure 4 which shows the rejection of PEG20 000 by the T6/B membranes which were coated with films of ovalbumin. Membranes with a loose pore structure show a dramatic variation in rejection as the feed velocity changes while membranes with high rejection seem little affected. The variations shown in Figure 4 are due entirely to concentration polarization and they can be accounted for accordingly (Nakao and Kimura, 1981). The temperature and the solute concentration of certain materials have a significant effect on the rejection of a membrane. Figures 5 and 6 show a T6/B membrane membrane rejection for PEG as a function of the operating temperature and solute concentration respectively. Rejection decreases with increasing temperature and increases with increasing concentration. All these effects are attributed to a reversible type of adsorption (Michaels, 1980; Howell and Velicangil, 1980; Le, 1982) and the increase in rejection is due to a reduction in the pore radius as a result of solute adsorption on the pore wall. Figure 5 also shows the reduction in the effective pore radius as a consequence of adsorption. The slight decrease in rejection at high bulk concentration is due to the reduction in the mass transfer coefficient which tends to enhance the effect of concentration polarization.

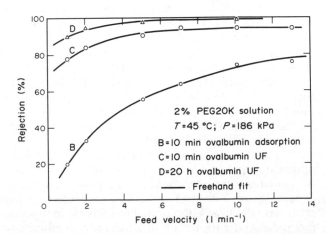

Figure 4 Effect of feed velocity on apparent rejection

25.3.3 Molecular Fractionation

We have discussed all the factors that are known to influence the membrane's rejection ability; perhaps it is appropriate now to rethink how UF membranes could be used for effective molecu-

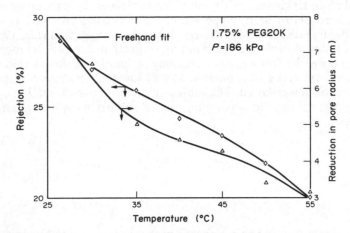

Figure 5   Effect of temperature on effective pore radius and membrane rejection

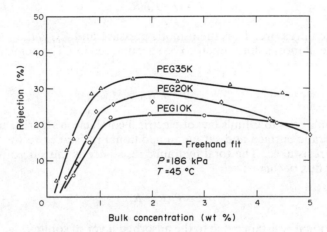

Figure 6   Effect of solute bulk concentration on the rejection coefficient

lar fractionation which was a great hope for membrane technology in the late 1960s. The potentially exciting utilization of membranes for large scale complex mixture separations, which are currently performed by costly and slow processes such as gel and ion exchange, chromatography, selective precipitation or electrophoresis, is a real possibility once all the problem areas have been defined and ways to control them have been found.

The basic problem is the interaction between the solutes and the membrane itself, therefore, the answer must be the prevention of that interaction or to build into the membrane an allowance for it. For example, membrane materials can be chosen that have no affinity for the solutes or the membrane surface can be precoated (Le, 1982) with substances that will prevent solute adsorption. The alternative would be to make use of the adsorbed solute layer to effect the separation, thus by choosing a membrane with a very open pore structure one relies on one of the solute components to form a monolayer on all exposed surfaces and reduce the pores to a workable size. The difficulty arises when one encounters the wide pore size distribution within the pore network underneath the membrane surface where depth filtration is present.

There is a need for a very thin membrane, not in order to maximize the flux, but to have a good molecular fractionating capability. It should be noted, however, that certain molecular mixtures are amenable to fractionation, namely hydrophilic macromolecules, such as dextran, hydroxyethyl cellulose and polyethylene glycol, which are well solvated and do not adsorb strongly to the membrane. A clear understanding of the interaction mechanism between the solutes and the membrane and also between the solutes themselves in this particular application is essential for successful results. Consider proteins which form the largest and most important group of materials requiring purification; protein can adsorb to almost any known surface (Dillman and Miller, 1973; Norde and Lyklema, 1978) and when subjected to denaturing conditions, such as shear stresses, extreme pH or high temperature, the molecules crosslink (Grant, 1980) *via* an intermolecular sulfhydryl-disulfide interchange. It has been suggested (Le, 1982) that this cross-linking mechanism is responsible for the build up of protein deposits on membranes in cheese whey ultrafiltration with the fouling layer reaching 30 μm in thickness (Glover and Brooker, 1974). For such labile materials fractionation at very low temperature with gentle recirculation and dilute concentration is warranted. The subject of molecular fractionation by membranes is an important and fascinating area in separation, and a subject to which greater research effort should be devoted.

25.4 THEORY

25.4.1 Water flux

Since 99% of the water flow in the membrane probably takes place by viscous flow in tortuous channels, the rate of permeation or the flux (J) may be described by a modified Kozeny–Carmen equation

$$J = (P - P_o)/R_m \tag{1}$$

where P is the applied pressure, P_o is the osmotic pressure and R_m is the membrane resistance. The osmotic pressure of most solutes likely to be encountered in UF is negligible and hence flux is often expressed as

$$J = P/R_m \tag{2}$$

Equation (2) is analogous to Ohm's law of electrical conduction. During ultrafiltration solutes adsorb to the membrane surface and set up an additional resistance to the water flow in series with the membrane resistance. The total hydraulic resistance is the sum of the resistances. The expression for water flux becomes:

$$J = P/(R_m + R_s) \tag{3}$$

where R_s is the additional resistance due to the adsorbed layer of solutes.

25.4.2 Concentration Polarization

The separation of solute and solvent takes place at the membrane surface where the solvent is allowed to pass through the membrane, but the solute is retained and causes the local concentration to increase; this is termed concentration polarization. A fraction of the solute diffuses back into the bulk solution where the concentration is lower according to Fick's law of diffusion and a concentration profile is established at the membrane as illustrated by Figure 7. At steady state the rate of solute arriving at the membrane is balanced by the sum of the solute leakage and the rate of back diffusion, *i.e.*

$$JC = D dC/dx + (1 - E)C_s J \tag{4}$$

where E is the true rejection coefficient, C is the solute concentration at x, C_s is the solute concentration at the membrane, and D is the solute diffusivity.

The rate of solute leakage can also be expressed as

$$C_p J = (1 - E)C_s J \text{ or } C_p = (1 - E)C_s \tag{5}$$

where C_p is the solute concentration in the permeate. Combining (4) and (5) we obtain

$$J(C - C_p) = D dC/dx \text{ or } J = K_m \ln[(C_s - C_p)/(C_b - C_p)] \tag{6}$$

where $K_m = D/X$, the mass transfer coefficient. For a totally retentive membrane, $E = 1$ and the expression for flux under concentration polarization is

$$J = K_m \ln(C_s/C_h) \tag{7}$$

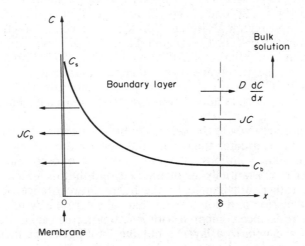

Figure 7 Concentration polarization in ultrafiltration

25.4.3 Gel Polarization Model

The plots of flux *versus* applied pressure in Figure 8 are typical of most macromolecular solution UF operations. The flux profile for distilled water follows equation (2), however, the flux for the polysaccharide solution becomes less dependent on the applied pressure as the latter is increased. This effect is caused by concentration polarization, because as flux increases by increasing the pressure, solute buildup at the membrane increases accordingly and so raises the additional hydraulic resistances in the adsorbed layer. The effects of changing the bulk solute concentration and the mass transfer coefficient are as predicted by equation (7). However, the existence of the flux plateau, *i.e.* the region in which increase in the applied pressure yields no flux increase, cannot be explained from basic principles. Michaels (1965) and Blatt *et al.* (1970) forwarded an hypothesis that as the concentration at the membrane increases the solute reaches its solubility limit and precipitates on the membrane surface. The layer of precipitated solute is commonly referred to as the gel layer and sometimes as a 'slime' or 'cake' layer. The solubility limit is known as the gel concentration. According to Michaels' model an increase in pressure produces a temporary increase in flux which brings more solute to the gel layer and increases its thickness thereby reducing the flux to the original level.

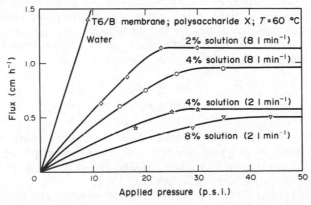

Figure 8 The effects of pressure, recirculation rate and solute concentration on the rate of ultrafiltration

25.4.3.1 Limitations of the gel model

The gel polarization model is successful in so far as it correctly predicts the effects of changing the bulk concentration and the hydrodynamic regime and the presence of the flux plateau. However, it bears two implications which put its foundation in doubt. Firstly, one would expect to find a constant flux in the gel-polarized region irrespective of the membrane permeability for the same hydrodynamic condition. Secondly, the gel concentration should be independent of the membrane type and the flow regime. Neither of these is observed in practice. In fact, the fluxes of different membranes in the so-called gel region differ by many fold and the gel concentration of any solute is never constant (Blatt *et al.*, 1970; Goldsmith, 1971; Porter, 1972; Nakao *et al.*, 1979), but is dependent on the membrane type and system parameters such as the bulk concentration or the feed velocity. Curiously enough, hydrophilic materials, such as polyethylene glycol (PEG15 500), which has infinite solubility in water, were found to have limiting concentration as low as 5.3%.

Colloidal UF is another area where the gel model breaks down. The fluxes of colloidal solutions are usually many times greater than that predicted by the gel model and often no flux plateau is observed with dilute solutions. Blatt *et al.* (1970) proposed that either back diffusion from the polarized layer is substantially augmented beyond that of a Fickian diffusion mechanism or the permeation rate is not limited by the hydraulic resistance of the gel layer. They favoured the latter explanation and pointed out that cake formed by micron size particles has a relatively low specific resistance compared with macromolecular cake. Porter (1972) disputed this and argued that if the polarized layer is not the limiting factor then the flux should be independent of the bulk concentration and proportional to the applied pressure. However, in practice he found the opposite to be true since he observed decreasing flux with increasing bulk concentration. Porter used the tubular pinch effect, which is sometimes known as the radial migration phenomenon, to explain his data, but Madsen (1978) pointed out that this effect can only account for a very small part of the high flux and that Porter's model would under-predict flux of colloidal solutions by a factor of 50 in some cases. Another difficulty which confronts the gel model is the fact that flux dependence of the feed velocity varies with solute types and membrane system configurations (Henry and Porter, 1972); this means that mass transfer rate at the membrane surface deviates from any accepted theory of diffusive convective transport, but Porter pointed out that this does not include the non-diffusive tubular pinch effect.

25.4.4 Developments in Theoretical Modelling

In the light of the gel model's severe shortcomings in the areas discussed above, several attempts have been made to relate flux to the operating parameters in other ways. Howell and Velicangil (1980), for example, showed that the time taken for the build up of the membrane wall concentration to the limiting or gel concentration is of the order of a few seconds after the application of pressure. It is expected that once concentration polarization is established, the flux should reach a steady state value. In practice, they found that for protein ultrafiltration the flux never attained a steady state value. Flux decay occurs in three distinct stages corresponding to concentration polarization during the first few seconds followed by an adsorption process which lasts no more than 10 minutes, from then on flux loss is due to a polymerization process. The polymerization mechanism increases the thickness of the fouling layer and thereby reduces the flux. The rate of solute deposition is expressed as

$$\mathrm{d}l/\mathrm{d}t = K_\mathrm{r} C_\mathrm{s}^n \tag{8}$$

$$R_\mathrm{s} = P_\mathrm{g} l \tag{9}$$

where l is the thickness of the gel or fouling layer, C_s is the wall concentration, n is the order of the polymerization reaction, K_r is a rate constant, and P_g is the gel permeability. Equation (8) produces a reasonable fit to many different sets of data. It has been found that the parameter n varies markedly with the mass transfer coefficient (Le, 1982).

The scour model (Fane *et al.*, 1980) is another independent approach to flux correlation. It was derived for an ultrafiltration system with suspended solids ($>0.5~\mu$m) in the solution. According to this model the suspended solids modify the surface of the membrane by protecting some of the pores which would otherwise become obstructed by the macromolecules. The suspended solids could also augment the rate of back diffusion of the macromolecules by scouring, turbulence promotion and radial migration. Data on activated sludge were found to be correlated by the relationship $J = AC_b^{-m/U_b}$, where U_b is the feed velocity, A is a function of the feed concentration and the nature of the solids, m is a parameter which is dependent on the feed velocity. The authors of the scour model assume that long term flux decline is due to an aging process in the gel layer. The gel resistance is expressed as

$$R_s = R_o + R_a \tag{10}$$

where R_o is the initial resistance and R_a is the increase in resistance due to aging. It is assumed that R_a rises from zero at time $= 0$ and approaches a maximum asymptotic value which is proportional to the initial resistance, this leads to the following expression

$$G = R_a/R_o = G_{max} \, t/(K+t) \tag{11}$$

where K is an aging constant, G_{max} is the value of G as t approaches infinity.

In general, all the models considered so far are very system specific and cannot be applied to a wide range of solutes and membrane equipment. The model presented below is more versatile and it avoids all the pitfalls that caused problems with earlier models.

25.4.5 Limiting Flux Model

While in dilute solutions of polymer chains in a good solvent the chains are swollen and the size of the molecule increases as $N^{3/5}$, where N is proportional to the molecular weight, in a concentrated solution or melt the chain becomes ideal and has a size R which varies as $N^{1/2}$ and many chains overlap to build up the total concentration (de Gennes, 1979). This total concentration in mass terms is then likely to become gelling at a constant value irrespective of the molecular weight. Table 2 shows the wall concentration at which the flux plateau occurs for different molecular weight fractions of dextran (Goldsmith, 1971). The limiting wall concentration increases with increasing molecular weight and thus it would seem unlikely that the limiting concentration (C^*) could be equated with the gelling concentration (C_g). Molecules close to the membrane surface will be in a concentrated solution and behave as ideal chains whose radius of gyration R scales as $N^{1/2}$. The area of the surface where a segment of the molecule could adsorb will scale as R^2 and the chance of a segment being in any particular region of space such as the surface will scale as N/R^3. The average number of attachments per molecule will then scale as $R^2N/R^3 = N^2/N^{3/2} = N^{1/2}$. As the concentration at the membrane surface C_s is in mass terms, the total number of molecules in the region will scale as C_s/N and thus the total number of adsorbed sites will scale as $N^{1/2} C_s/N = C_s N^{-1/2}$.

Table 2 also shows the results of calculation of the total number of adsorbed sites. It appears that the flux becomes limiting at some fixed level of adsorption of segments of the polymer chains on the surface. This would relate directly to the probability of blocking a pore. When a molecule with a fixed tertiary structure such as protein is involved, intermingling of the chains is no longer likely even at high concentrations. The local density of chain adsorption sites is thus not going to exceed N/R^3, but the number of sites per molecule will still be $N^{1/2}$ and the total density of sites $C_s N^{-1/2}$ as before.

Table 2 Limiting Concentration and the Number of Adsorption Sites for
Different Dextran Fractions

Molecular weight (N)	Limiting concentration (C*, %)	Number of adsorption site (CN$^{-1/2}$)
21 800	14.2	0.0962
39 500	20.3	0.1021
69 000	25.5	0.0981
100 500	33.0	0.1041

The limiting flux model then postulates that flux can be increased by increasing the applied pressure only up to some limiting value which will depend on the bulk solute concentration, its molecular weight and the hydrodynamic regime above the membrane. In all cases the probability of pore blockage for a particular membrane system is constant. Increasing the pressure results in a transitory increase in flux, an increase in the solute concentration at the membrane and a consequent increase in the probability of pore blockage. An additional number of pores become blocked reducing the flux and the blocking probability to its original value. Reducing the applied pressure will have the opposite effects. It is assumed that adsorption is completely reversible. Irreversible adsorption or fouling would modify the membrane characteristics and alter the blocking probability function accordingly.

In actual practice some pores may become blocked and others unblocked in a dynamic fashion. This dynamic process is equivalent to cyclic blocking and unblocking of all the pores simultaneously. While a pore is blocked it is non-permeating and while unblocked it always tends to pass a flux close to that for a pure solvent. The fraction of pores which may be blocked is C_s/C^*. Thus in the prelimiting flux region a fraction of the pores $(1 - C_s/C^*)$ is not blocked at all, and the flow through this fraction will be proportional to the applied pressure. The flux contribution from the fraction which is not blocked at all is

$$J_1 = \left(1 - \frac{C_s}{C^*}\right) J_p \tag{12}$$

where J_p is the pure solvent flux at the applied pressure P, $J_p = P/R_m$. If the net prevailing flux is J, then the flux contribution by the remaining partially blocked pores will be $J_2 = J - J_1$. The flow through a partially blocked pore, which for the present we call a pore in an occupied state, is smaller than the flow through an unoccupied pore, so $J_2 < (JC_s)/C^*$ or $J_2 = (XJC_s)/C^*$ where $0 \leqslant J \leqslant J^*$; if $J = J^*$ then $C_s = C^*$, $J_1 = 0$ and $J_2 = J^*$. Thus X is such that $X = 0$ when $J = 0$, $X = 1$ when $J = J^*$ and $0 \leqslant X \leqslant 1$. X will be a function of C_s (related to J) and membrane solute interaction characterized by C^* (related to J^*). Of a number of functional forms tried, $X = J/J^*$ was both simple and gave satisfactory results. Hence

$$J = \frac{C_s}{C^*} \frac{J^2}{J^*} + \left(1 - \frac{C_s}{C^*}\right) J_p \tag{13}$$

This equation assumes that a limiting flux exists corresponding to a limiting concentration at the membrane surface. Noting that as C_b tends to zero, J tends to J_p and as C_s tends to C^*, J tends to J^*; thus the pressure flux profile must be as in Figure 9.

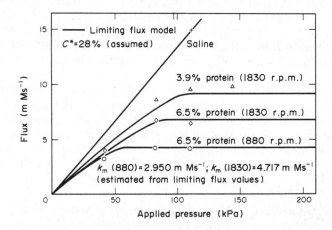

Figure 9 Dependence of flux on the applied pressure (Porter, 1972)

In Figure 9 the lines are those predicted by this model using best values of C^* and K_m for the data points by Porter (1972). The value of the mass transfer coefficient used in the prediction was calculated at the limiting condition, *i.e.* for $C_s = C^*$, and it was assumed constant as in the linear/log-plot of the gel polarization equation in the literature. Clearly at low bulk concentration the mass transfer coefficient is greatest and discrepancy arising from the assumption of a constant coefficient is also greatest as reflected very well by the plots. The assumption of a constant mass transfer coefficient for a constant feed velocity results in deviation from linearity in flux *versus* log concentration plots in many systems (Fane, 1983).

Colloidal solutions behave effectively as ordinary solutions except that a true threshold pressure or true limiting flux is never reached. In practice, for instance, with a 1% colloidal polymer latex (average size 0.1 μm) flux was still increasing strongly with pressure above 620 kPa, but for similar membranes and human plasma or BSA at 1–6.5% concentration limiting flux was found between a pressure of 35–100 kPA (Porter, 1972). Consider a unit area of membrane. For an ideal membrane with hexagonally close pore arrangement the pore density N will be $N \propto 1/d$, where d is the pore diameter. Assume that only particles in direct contact with the pores can cause the flux to be limited and any one particle can cause only one pore to be fully flux limited, but by virtue of its larger size may decrease flux in adjacent pores to some extent by its projection over them. The maximum number of particles of radius D_p that may adsorb to the surface at any time will be $M^* \propto 1/D_p$, where $D_p > d$. Hence the maximum pore fraction that could be put into an occupied state is M^*/N. Then for $J \geqslant J^*$, $J_2 = \gamma J^*$, where $\gamma = M^*/N$, flow through these pores is fully limited and J_2 is the flux contribution by this pore fraction.

Flow through the remaining pore fraction will be $J_1 = J - J_2$. Since the remaining pore fraction $(1 - \gamma)$ is never fully blocked at any time, flow through it is nevertheless reduced to some extent by the particles' projection from the occupied pores.

Clearly $J_1 < J_p (1 - \gamma)$, where J_p is the pure solvent flux, or $J_1 = X J_p (1 - \gamma)$. Total prevailing flux is then $J = \gamma J^* + X J_p (1 - \gamma)$ for $J > J^*$. Note that for $\gamma = 1$ or for normal solution UF $J = J^*$ for pressure > threshold p.

For convenience the area γ which is fully blocked will be referred to as the blocking zone. In this zone the flux is fully limited and is completely independent of pressure. The term exclusion zone will apply to the remaining area $(1 - \gamma)$. This zone is affected by the colloid but is never completely blocked and the flux is pressure dependent. Within the exclusion zone at any instance a fraction of the area may be covered and therefore does not permeate and the remaining area would permeate freely. Thus one can write

$$\begin{array}{cccc} \text{Solute} & + & \text{free area} & \rightleftharpoons & \text{covered area} \\ C_s & & 1-\theta & & \theta \end{array} \qquad (14)$$

If one assumes Langmuir type adsorption then

$$\theta = \frac{C_s}{K + C_s} \qquad (15)$$

where K = adsorption constant. The flux contribution by the exclusion zone will be

$$J_1 = (1 - \gamma)(1 - \theta)J_p \qquad (16)$$

and the total prevailing flux is given by

$$\begin{aligned} J &= J_1 + J_2 \\ &= (1 - \gamma)\frac{KJ_p}{K + C_s} + \gamma J^* \end{aligned} \qquad (17)$$

Note that when $J = J^*$, $C_s = C^*$ and $J_p = J_p^*$. C^* can be interpreted as the colloid concentration which can cause the adsorption of a complete monolayer on the membrane. J_p^* is the pure solvent flux at the pressure when $J = J^*$. Solve for K and on substitution the above equation becomes

$$J = \gamma J^* + (1-\gamma)J_p \frac{C^*}{(\sigma - 1)C_s + C^*} \qquad (18)$$

where $\sigma = J_p^*/J^*$, for $J \geqslant J^*$. Figures 10 and 11 show the plots of equation (12) and equation (18). The model predictions for bacterial cells and colloidal latex polymer were good in both cases. The values of γ and C^* were obtained by a parameter estimation method and J^* values were estimated from the break points in the curves. It is quite likely that the slight deviation of the data from the model in Figure 11 is contributed by the non-uniformity in the particle size of the polymer suspension and also by the fact that the same mass transfer coefficient was employed for three different colloid concentrations.

25.4.6 Membrane Fouling and Treatment

The most serious problem in ultrafiltration is membrane fouling which is manifested by a progressive reduction in the permeation rate. The effect of concentration polarization, which

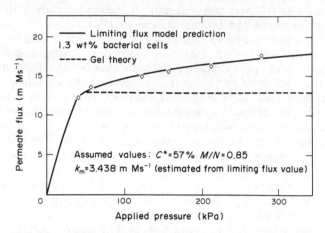

Figure 10 Dependence of flux on the applied pressure for bacterial cell suspension (Henry and Porter, 1972)

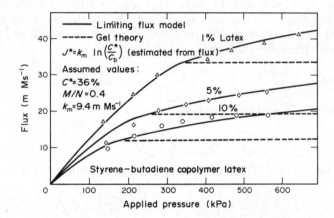

Figure 11 Dependence of flux on pressure for a colloidal latex (Porter, 1972)

accounts for a large proportion of the flux loss, is not regarded as fouling. It is now recognized that fouling occurs in two stages: the first is due to the adsorption of the solute to the membrane and is generally completed within 10 minutes of contact between the protein or solute with the membrane. This causes a flux loss varying from 30%, in a low flux membrane, to a ten-fold reduction in the membrane hydraulic permeability relative to the measured pure water permeation rate (Ingham, 1980; Michaels, 1980; Velicangil, 1979). The second stage of fouling is manifested by a relatively slow, continuous decline in permeation rate that is substantially independent of feed solute concentration and upstream hydrodynamic conditions. This fouling process, which has variously been ascribed to membrane pore plugging or to the formation of a slowly consolidating, gelatinous layer on the membrane surface (Michaels, 1968) or protein polymerization (Velicangil, 1979), is very unpredictable, and appears to vary markedly in severity depending on the membrane composition, the nature of the retained solute, and other variables such as solution pH, ionic strength, electrolyte composition, temperature and operating pressure. Under the worst circumstances, this type of fouling can cause a 90% irreversible flux loss (relative to the initial flux observed with a clean membrane) within a period of a few days. While in most cases the foulants can be removed from the membrane by the appropriate cleaning and scouring procedures which do not necessitate module disassembly and can yield a large restoration of the original performance, there are cases where such cleaning procedures are relatively ineffective (Michaels, 1980) or where the rigorousness of the scouring process required to remove deposits, such as one of the procedures adopted by Velicangil (1979), is so severe that it compromises the membrane flux and its lifetime. In all events, membrane cleaning and regeneration procedures reduce the operating time, consume costly reagents and in many cases degrade the membranes, thereby contributing to the operating cost. Another undesirable aspect which results from the fouling process is the loss of the fractionating capability of the ultrafiltration membrane for large-scale resolution of macromolecular mixtures such as plasma proteins. Such loss is most certainly due to the formation of a deposit layer on the membrane surface and membrane pore obstruction.

The problems of reduced throughput capacity, increased power consumption, compromised separation capability and reduced membrane service life associated with membrane fouling, stubbornly resisted attempts to provide an adequate solution despite 10 years of experience with the systems both on a laboratory scale and on an industrial scale. Methods for circumventing the ultrafiltration process limitations to date can be classified under two broad categories, namely physical and chemical. Kopecek and Sourirajan (1969) noted significantly high flow rates with anisotropic reverse osmosis membranes when they were first used upside down for distilled water filtration. This is the 'back pressure' or 'backwash' treatment which was also adopted for ultrafiltration membranes by van Oss (1972). Lefebvre *et al.* (1980) reported the most comprehensive application of this technique with their novel polyamide membranes. When they ultrafiltered distilled water with the membranes properly placed, a 60% flux decline was observed in 40 minutes. When the position of the membranes was reversed at this stage a sudden recovery of up to 85% of the initial flux value was noticed which remained constant for a further 80 minutes. When the experiment was repeated with the reversed order of configurations, first a nearly constant flux was noted with the upside down position, and then a sharp 60% drop when the configuration was changed. Note that this technique is not applicable to large scale UF modules, because of the risk of membrane rupture. Prefiltration of the process material to remove lumps or centrifugation to remove oil and fats are known to alleviate fouling (Lee *et al.*, 1976). Several workers (Goldsmith *et al.*, 1971; Hiddink *et al.*, 1970; Fane, 1982), including the present author, have tried to reduce concentration polarization and by inference expected to reduce the rate of fouling through the addition of small particles into the feed. Although the flux improvement was good, membrane scratching made this method an unattractive proposition. Similarly, foam balls have been used to remove deposits from tubular reverse osmosis and ultrafiltration membranes, but they tend to reduce the membrane life considerably.

One of the most appealing features of the hollow fibre membrane modules is their ability to withstand a negative pressure difference across the fibre wall, a property which permits them to be operated in a 'backwash' mode, which significantly aids unplugging of membrane pores and detachment of surface deposits. Chemical treatment methods suggested to reduce flux decline in cheese whey ultrafiltration cannot always be extended to other protein sources because of the differences in their fouling characteristics. It has been demonstrated that by intermittent and periodic flushing of the feed channels and membrane surfaces with cleaning solutions of suitably adjusted temperature and pH, frequently containing proteases to facilitate degradation of protein deposits, it is possible to restore a large portion of the flux loss. By treating whey with acid to a low process pH, by addition of calcium sequestering agents and compounds like EDTA or urea to modify specific protein side chains, or by increasing the ionic strength permeation rate increases between 16 and 72% were obtained (Lee and Merson, 1976). Hayes *et al.* (1974) reported 50 to 100% flux improvement in cheese whey permeation rate due to combined pH adjustment and heat treatment. Various attempts in the use of proteases immobilized on the membrane to reduce protein build up have been reported. The idea underlying this technique is the exploitation of the active proteases to hydrolyse the fouling proteins as they adhere to the membrane so that the rate of flux decline is retarded. Gillespie (1978) claimed a 93% improvement in terms of flux increase for the use of immobilized industrial grade proteases in the ultrafiltration of dried milk solution. The significance of this claim is questionable since dried milk which has been sprayed and heat dried would have been considerably denatured. Jenq (1980) obtained a 12% flux enhancement with the use of such techniques for the treatment of raw sewage. A most comprehensive study of the use of protease-coupled membranes in the UF of whey and other proteins was undertaken by Velicangil (1979). He reported that initial flux losses due to the enzyme immobilization procedure amounted to between 35 and 49% of the original water flux, but these membranes exhibited only 1 to 2% further losses during a standard 22 h run. Flux improvement in the range 25 to 78% was obtained in the ultrafiltration of 0.5% BSA or haemoglobin. For reconstituted cheese whey (spray dried) improvement between 60 and 270% was obtained, but there was no significant improvement with fresh cheese whey and hence the technique has only limited application.

Michaels (1980) discussed the problem of fouling and suggested that techniques for altering the hydrophobicity or electrostatic charge on the ultrafiltration membrane may provide the answer to irreversible membrane fouling. This seems to be a logical approach since Feder (1980) has shown that a negatively charged surface under certain conditions can greatly reduce the protein loading on that surface. Yet another approach to the fouling problem which may provide a convenient and versatile solution to the problem is the pretreatment of the membranes with dilute aqueous solutions of selected water soluble polymers which form highly hydrated, high hydraulic permeability films on the membrane by filtration or adsorption. These films should prevent direct adsorption of fouling macromolecules onto the membrane surface. Such pretreatment has the advantage of being able to be performed easily by perfusing the dilute solution of the selected polymer through the ultrafiltration module prior to production runs and can be reapplied *in situ* in the modules without serious interference with normal plant operations.

The present authors applied this technique to cheese whey ultrafiltration using polyethylene glycol as the precoating material. They found that such a technique greatly facilitated membrane cleaning and as result membrane flux could be maintained at almost a constant level for periods up to a week before rigorous treatment with hydrogen peroxide to remove the protein deposit was necessary.

Table 3 summarizes some of the techniques applied successfully against membrane fouling in ultrafiltration.

Table 3   Techniques for Fouling Reduction in Ultrafiltration

Technique	Processed material	Ref.
Pretreatment		
Heat treatment and pH adjustment	Cheese whey	Hayes *et al.* (1974)
pH and ionic adjustment	Cheese whey	Fane *et al.* (1982), Lee and Merson (1976)
Ca sequestering by EDTA	Cheese whey	Lee and Merson (1976)
Prefiltration	Cheese whey, oily wastes	Lee and Merson (1976), Bansal (1975)
Membrane characteristic		
Charged membrane	Electrodeposition paint	Breslau *et al.* (1980)
Enzyme coating	Protein	Howell and Velicangil (1980)
Hydrophilic precoat	Cheese whey	Le (1982)
Cleaning		
Backflush	Several	Breslau *et al.* (1980)
Foam balls	Sea water	Kuiper *et al.* (1973)
Biological detergent	Cheese whey	Howell and Velicangil (1980), Matthiasson and Sivik (1980)
Oxalic acid	Oily wastes	Bansal (1975)
Hydrogen peroxide	Proteins	Le and Howell (1982)
Hypochlorite	Cheese whey	Le and Howell (1983)

25.5 ULTRAFILTRATION EQUIPMENT

25.5.1 Module Design

A wide range of UF modules is available. In general, modules are designed to provide an appropriate tangential velocity to minimize concentration polarization and to provide adequate support for the membrane, to provide a large membrane area to volume ratio. A means of permeate collection must be available and the system should be easily cleaned or sanitized as well as being readily remembraned. The various industrial scale UF equipment design concepts differ mainly in the size and shape of the flow channels. Common designs include plate and frame, pressure sustaining porous tubes, membrane lined and self-supporting hollow fibres. In other designs flat sheet membranes are incorporated either into a supporting fabric and spirally wound like a Swiss roll or into flat leaf cartridges. Such elements are then built into separate pressure sustaining housings.

Ultrafiltration modules can be classified into two main groups. The first of these consists of the laminar flow systems which rely on the high shear rate at the membrane surface to minimize concentration polarization to provide adequate flux. This can be achieved by using very small diameter tubes or channel height. Turbulent flow systems are either tubular (6–25 mm in diameter) or thin channel. They rely on much larger volumetric pumping rates for concentration polarization control. Table 4 summarizes the advantages and disadvantages of various systems.

Table 4 Comparison of Ultrafiltration Systems

System	Advantages	Disadvantages
Turbulent flow systems		
Tubes 6–25 mm dia.	Easily cleaned, no dead space, well-developed equipment, individual tube may be replaced	High hold up volume, large pressure drops in joints
Flat plate	Well-developed, small hold up volume	Large area of dead space, entire module must be replaced on failure
Spiral wound	High area/volume ratio, easy to replace membrane	Large hold up volume, difficult to clean
Laminar flow systems		
Plate and frame 0.7 mm channel rectangular or spiral	Economic for viscous solution, low hold up, low pressure drops, well-developed high area/volume ratio	Difficult to clean, large area of dead space
Hollow fibre	Low hold up, large area/volume ratio, low cost, low pressure drops	Less well developed, whole unit must be replaced on failure

25.5.2 Design Considerations

In theory the permeate size of the membrane has no effect on the performance of a UF system, however, in practice, for ease of operation the design of this part of the module requires careful consideration. The permeate compartment should have as small a volume as possible and it should be filled with liquid throughout. Such measures are necessary to ensure that the cleaning or sanitizing solutions and the flushes reach all parts of the system. For example, in large cheese whey plants it is not uncommon to find colonies of moulds or bacteria where flushing fails to wash away the lactose deposits. In all cases where the volume is large it would be almost impossible to wash out all the permeate hold up without great time loss and large amounts of water or sanitizing solution.

The prevention of dead spots on both sides of the membrane is essential. There is a tendency for deposits to accumulate and for scale formation where flushing is ineffective. Extra care must be given where cellulosic membranes are in use, because very often membranes are biodegraded behind such scale. Although modern polymeric membranes do not suffer microbial attacks, the risk of product contamination is too great to ignore such precautions.

Generally, a great deal of time and effort are given to the fine detailed design of the flow channel to obtain the desired hydrodynamic for concentration polarization control, however, inadequate attention is devoted to the pressure distribution on the membrane. Very often the feed solution is over-pressurized near the inlet and under-pressurized near the outlet; this is particularly true of thin channel equipment with a long flow path. While the over-pressured regions are liable to serious irreversible pore blockage, the under-pressured regions do not give the maximum flux that the membrane is capable of yielding. For optimal operation the pressure over the whole membrane should be fairly uniform and approximate to the threshold pressure. The following expressions have been used with success by various authors for the calculations of channel dimensions.

25.5.2.1 Circular channel (Goldsmith, 1971)

For turbulent flow

$$K_m = K_1 D Re^{0.913} Sc^{0.346} \tag{19}$$

For laminar flow

$$K_m = K_2 D (Re/dL)^{0.5} Sc^{0.33} \tag{20}$$

25.5.2.2 Rectangular channel (Porter, 1972)

For laminar flow

$$K_m = 0.816(6QD^2/wLb^2)^{0.33} \tag{21}$$

For turbulent flow

$$K_m = 0.02Q^{0.8}D^{0.67}/bw^{0.8}v^{0.47} \tag{22}$$

where K_1, K_2 are constants, D is solute diffusivity, d is tube diameter, L is channel length, b is channel height, w is channel width, $v = X/Y =$ kinematic viscosity, Q is volumetric flow rate, $Sc = A/D =$ Schmidt number, and Re is Reynolds number. The above equations give good predictions of the mass transfer coefficients in general, however, they should be used with caution since it has been shown (Goldsmith, 1971; Porter, 1972) that on many occasions they produced results which deviated many times from the actual values.

25.5.3 Plant Configuration

UF systems can be operated either on a batch or continuous basis. For a given amount of membrane, the process mode selected will affect the capacity of the system. As the retentate concentration increases the flux decreases, therefore, it is advantageous to operate under conditions such that the processing time at high concentration should be as short as possible. The batch approach is suited for small to moderate quantities of feed solution, where intermittent operation is either desirable or not difficult. In this system (Figure 12) fluid is withdrawn continuously from the feed tank, introduced into a recirculation loop and recycled back to the tank. The concentration of solids in the system increases with time as the volume of the permeate increases. The feed rate is maintained at a high level to ensure that the concentration of solids in the recirculation loop is not significantly larger than that in the feed tank. Typically the feed rate to the permeation rate ratio is about 20. In this way ultrafiltration takes place continuously at the lowest possible retentate concentration and provides an optimal time-average flux.

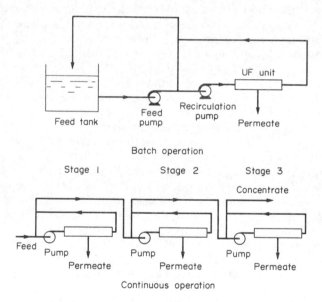

Figure 12 Ultrafiltration process modes

The continuous UF approach is most suitable for large volume flow and where intermittent operation is not feasible. Due to the modular nature of UF equipment a whole range of arrays is possible. The simplest form is series flow with equal membrane area in each stage, however, tapered cascade with more membrane area at the head end may be more efficient; calculation of the most efficient cascade is a complex optimization problem. Generally, it is more economical to operate a multistage system in cascade configuration (Figure 12). In such a system several recirculation loops are connected in series, with the concentrate from one stage becoming the feed for the next. The advantage of higher fluxes obtained in the front stages of the cascade compensate for the lower fluxes at the cascade end. The average process flux is usually greater than that of a batch operation, being typically 80% better in a four-stage system (Beaton, 1980).

25.6 DESIGN EXAMPLE

The following example is based on an actual industrial problem. A polysaccharide is available in solution at a concentration of approximately 2% w/w. Current technology isolates the polysaccharide by a solvent precipitation method which is expensive due to the diluteness of the polysaccharide solution. The alternative would be to deploy UF to concentrate the solution five-fold, then spray dry the product. It is decided, therefore, to build a pilot unit to process 10 m^3 of the fermented broth per day.

The quantity of the feed is relatively small, a batch system would, therefore, be appropriate. The problem now remains to determine the module type and the membrane area, the operating temperature and pressure, and the recirculation rate. In this case the choices of membrane and module types can be quickly narrowed down to the final selection. For total retention of the polysaccharide a membrane of molecular weight cutoff of 30 000 is suitable. Hollow fibre is rejected because the flow channel is too small (0.5 mm) to accommodate the high flowrates and it is susceptible to blocking by large lumps of solids. Plate and frame or spiral type filters are difficult to flush following pump failure. The tendency is not to go for thin channel equipment, it is decided to employ a tubular type membrane (12 mm). A laboratory unit consisting of 300 cm^2 of this type of membrane is set up to obtain the data essential for the plant sizing. Flux as a function of pressure for a series of flowrates is determined at 60 °C. The operating temperature is chosen to be 60 °C since at this temperature microbial activities are almost non-existent and the compound is most stable in the range 55–100 °C. Also high temperature means higher fluxes which will reduce membrane area requirements.

Figure 13 shows the plots of the threshold pressure against concentration for the polysaccharide. The threshold pressure is that pressure at which flux begin to plateau. According to the limiting flux model, the threshold pressure decreases with increasing bulk concentration, the plot for the 2 1 min^{-1} recirculation rate seems to obey the model. However, the pressure profile for the 8 1 min^{-1} recirculation rate seems to behave in the opposite way. In fact, the recorded pressure is the average pressure between the outlet and the inlet to the test unit and since a constant flux is observed only when the whole of the membrane area is exhibiting a limiting flux, this means in effect, the actual threshold pressure is the outlet pressure and not the average pressure which includes the pressure loss along the flow channel. The optimal operating pressure, as shown, varies with the retentate concentration and recirculation rate and should, therefore, be adjusted accordingly as ultrafiltration proceeds.

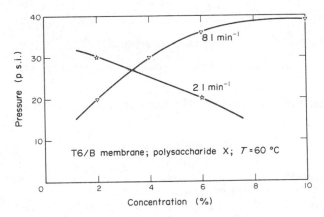

Figure 13 Effect of concentration on the threshold pressure

25.6.1 Membrane Area Requirement

The membrane area requirement will depend on the rate of recirculation to be operated. A high pumping rate means costly energy demands, but a smaller membrane requirement and *vice versa*. It is necessary to optimize the total running cost. The pressure drop per unit channel length is illustrated by Figure 14 and the flux *versus* concentration data for the polysaccharide is presented by Figure 15.

Suppose the recirculation rate to be employed is 8 1 min^{-1}. The values of C^* and K_m corresponding to this flowrate are then read off from Figure 15. The system is to be operated 20 h d^{-1},

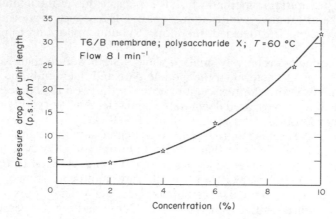

Figure 14   Effect of concentration on pressure drops

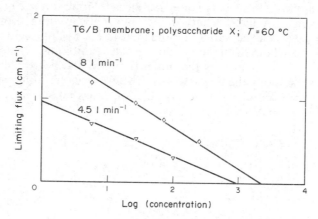

Figure 15   Effect of bulk concentration on flux

the remaining 4 h being use for cleaning. The membrane requirement is then determined as follows:

$$V = V_0 - A\int_0^t J^* \mathrm{d}t \tag{23}$$

where $t = 20$ h, A is membrane area, V_0 is original batch volume, V is batch volume at time t.

By assuming a value for the membrane area and a small time step, calculate the time taken to concentrate the fermented broth from 2 to 10%. If the time is too large or too small then assume a new area value and repeat the calculations. The results shown in Figure 16 are obtained with a desk top computer and from the plots it is clear that 45 m² of membrane is sufficient for the pilot plant.

25.6.2 Power Demand

Rate of work = mgh, where h = pressure drop (m of water), g = acceleration due to gravity (m s⁻²), m = mass flowrate (kg s⁻¹). Tube length = 1.2 m, tube diameter = 0.012 m, number of tubes required = 995.
Total pumping rate required = 995 × 8 / 60 = 133 kg s⁻¹.
Average pressure drop (across tube over 20 h) = 14.78 m of water (21 p.s.i.).
Work rate = 133×9.81×14.78/1000 = 19.3 kW.

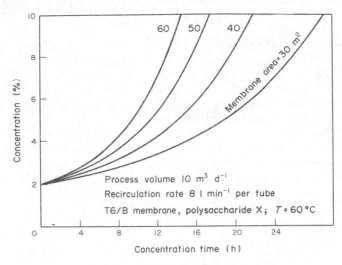

Figure 16 Membrane requirement chart

25.6.3 Cost Optimization

Energy cost per year at 5 p/kW h (260 d/year) = $19.3 \times 20 \times 260 \times 0.05$ = 5018
Cost of membrane replacement at £120 m^{-2} = £45 $\times$ 120 = £54 000
Now, flux $\propto K_m \propto Q^{0.913}$, where Q is the pumping rate; membrane area and 1/flux $\propto Q^{-0.913}$, therefore

$$\text{Membrane cost} \propto Q^{-0.913} \tag{24}$$

$$\text{Pressure drop} \propto Q^2 \tag{25}$$

$$\text{Work rate} \propto Q^3 \tag{26}$$

The total work load also depends on the number of tubes, which is proportional to the membrane area, therefore

$$\text{Energy cost} \propto Q^3 \times Q^{-0.913} \propto Q^{2.09} \tag{27}$$

The energy cost and membrane replacement cost as a function of the pumping rate is shown in Figure 17. The optimal recirculation rate is about 6 l min^{-1}. Note that this rate depends on the unit cost of electricity and the membrane which may change independently. Many production processes are designed and optimized by the same economic principle, it would be worthwhile to reappraise existing processes from time to time to ensure that they operate at the best possible conditions.

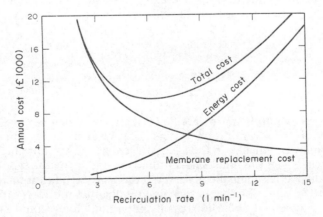

Figure 17 Effects of pumping rate on operating costs

25.6.4 Scale-up

The problems involved in scaling up from pilot plant operations to full scale systems are similar to those already discussed. However, the size of the plant and the huge capital expenditure entailed in such a venture means there is no room for errors. Matthews (1979) indicated the questions that need to be answered before a scale-up can be carried out as follows: (1) volume of product to be handled; (2) the end product compositions; (3) characteristics of the material to be processed; (4) will the characteristics of the final product meet client's requirement; and (5) floor space and services availability.

Although batch mode studies are useful, large scale UF plants are rarely operated batchwise. Data obtained from batchmode pilot plants can be used for the design of large continuous systems provided that certain rules are observed. The feed composition must be the same. The operating conditions of the pilot plant, *i.e.* the temperature, pH, recirculation rate and the average pressure drop must be employed in each stage of the multistage plant.

25.6.5 Pump System Design

Processed materials are often non-Newtonian, with viscosity being dependent on shear rate. Many are also thixotropic, *i.e.* their tertiary structure or stabilizing crosslinks are broken by a combination of temperature and pH extremes and the shear forces. In the case of proteins this is manifested in denaturation such as the loss of enzyme activity or whipping tendency; with polymeric materials this may be the loss of gelling power or the ability to form an emulsion. In batchwise operations the exposure time of the materials to the processing conditions can be minimized by splitting a large volume into smaller batches. The total process time on the same UF unit is the same but additional time may be required for emptying and refilling the batches or changing the feed tank. However, the best strategy would be in the selection of the right pump for the job. Generally, positive displacement pumps such as the rotary types have a more gentle action on the fluids than centrifugal pumps. They are also more suitable for the viscous suspensions often encountered with fermentation products.

The increase in viscosity during concentration in ultrafiltration can be quite dramatic, with values as high as 500–800 cP being attained. This increase in viscosity imposes certain requirements on the design of the hydraulic systems. As the viscosity increases the pressure difference across the membrane increases as well as the pressure gradient along the flow channel in the recirculation loop.

The recirculation flow will also decrease in accordance with the flow–head characteristics of the pump. Eventually, if the viscosity becomes sufficiently high, the maximum head that can be developed by the pump will be reached. Centrifugal pumps generally have a relatively low maximum head. Pump cavitation problems are likely to occur if over-concentration takes place, therefore, limits on the degree of concentration must be imposed. In a system where the frictional pressure drop due to concentration is large, the recirculation loop must be as short as possible. Thus within a recirculation system, greater usage of parallel-connected modules is warranted in order to reduce the cost of the pumps and ancillary equipment.

25.6.6 Process Economics

For a particular application the permeate flux is a critical parameter in establishing the capital and operating costs of an ultrafiltration system. The level of fluxes obtained with fermented broths of various origins is typically in the range 1–5 cm h^{-1} at the initial feed concentration and decrease to 0.5–1.0 cm h^{-1} at the final concentration depending on the type of membrane and the products. A small flux decrease may be observed if there are significant flux–time decay effects as illustrated in Figure 18. While fouling is specific to individual products, a 10–40% flux decay in 20 h is typical. For commercial operation, plants are usually interrupted every 20 h for cleaning and sanitization. Plants which operate in a batchmode are often cleaned twice daily.

Cleaning has two objectives, the first is to eliminate microbial contamination, and the second is to restore the membrane flux to its original level.

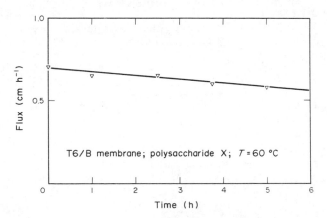

Figure 18 Rate of flux decay

Other cost factors which have not been discussed so far are labour, general maintenance, overheads and plant depreciation. Table 5 shows a complete cost analysis for the UF plant in our example.

Table 5 Assumptions and Operating costs for a UF Plant

Parameters	Assumptions	Annual cost in 1982 (£)
Throughput	$10 \, \text{m}^3 \, \text{d}^{-1}$	—
Capital cost	£100 000	—
Membrane area	$45 \, \text{m}^2$	—
Membrane life	1 year	—
Membrane replacement	—	5400
Recirculation rate per tube	$8 \, \text{l min}^{-1}$	—
Process duty	$260 \, \text{d year}^{-1}$	—
Pump duty	19.3 kW	—
Energy cost	$5 \, \text{p kW}^{-1} \, \text{h}^{-1}$	5018
Labour cost	$£10 \, \text{d}^{-1}$	2600
Annual maintenance	—	1000
Cleaning chemicals	—	150
Depreciation (SL)	10 year	10 000
Overheads	10% of total	2422
Total		26 590

While depreciation cost is the most costly item in this example, in a full scale plant it is the cost of membrane replacement (Le, 1982). The capital cost of a plant is related to its size by an index of 0.6–0.8. For example, a plant with $300 \, \text{m}^2$ of membrane would probably cost about £312 000. Labour cost is also related to plant size by an index $n < 1.0$, but it is expected that power consumption will be directly proportional to the plant size.

25.7 NOMENCLATURE

C	solute concentration	(kg m^{-3})
D	molecular diffusivity	$(\text{m}^2 \, \text{s}^{-1})$
E	rejection efficiency	$(—)$
J	flux	$(\text{m}^2 \, \text{s}^{-1})$
K_{m}	mass transfer coefficient	(m s^{-1})
M	number of molecules in a monolayer	$(—)$
N	molecular weight	(Daltons)

P	transmembrane pressure	(N m^{-2})
Q	pumping rate	$(\text{m}^3\,\text{s}^{-1};\,\text{l min}^{-1})$
R_{m}	membrane resistance	(N s m^{-3})
V	batch volume	(m^3)
X	empirical function of the blocking probability	
A	fraction of the surface covered	
B	ratio of the water flux at threshold pressure or limiting flux	
C	fraction of the pores being blocked	

Subscripts and superscripts

*	limiting condition
s	at the membrane surface
g	gelling condition
b	bulk concentration
p	in the permeate, or pressure dependent
0	initial condition

25.8 REFERENCES

Bansal, I. K. (1975). UF of oily wastes from process industries, *AIChE Symp. Ser.*, **151** (71), 93.

Beaton, N. C. (1980). *Polym. Sci. Technol.*, **13**, 373–404.

Blatt, W. F., A. Dravid, A. S. Michaels and A. Nelson (1970). Solute polarisation and cake formation in membrane ultrafiltration: causes, consequences and controlling techniques. In *Membrane Science and Technology*, ed. J. E. Flinn, p. 47. Plenum, New York.

Breslau, B. R., A. J. Testa, B. A. Milnes and G. Medjanis (1980). *Polym. Sci. Technol.*, **13**, 109–128.

Breton, E. J., Jr. (1957). Water and ion flow through imperfect osmotic membrane. *US Office Saline Water Res. Dev. Prog. Rep.*, **16**. University of California.

Dillman, W. J. and I. F. Miller (1973). Adsorption of serum proteins on polymer membrane surfaces. *J. Colloid Interface Sci.*, **44**, 221–241.

Fane, A. G., C. J. D. Fell and M. T. Nor (1980). Ultrafiltration/activated sludge system-development of a predictive model. *Polym. Sci. Technol.*, **13**, 631–658.

Fane, A. G., C. J. D. Fell and A. Suki (1983). The effects of pH and ionic environment of the UF of protein solutions. *J. Membr. Sci.*, **16**, 195.

Fane, A. G. (1982). *Membrane Technology for Industry*. Notes for a short course at Imperial College, London.

Feder, J. and I. Giaever (1980). Adsorption of ferritin on lexan polycarbonate and carbon. *J. Colloid Interface Sci.*, **78**, 144–154.

de Gennes, A. B. (1979). *Scaling Concept in Polymer Physics*. Cornell University Press, Ithaca, NY.

Gillespie, C. (1978). M. Sc. Thesis, Rutgers University, New Brunswick, NJ.

Glover, F. A. and B. E. Brooker (1974). The structure of the deposit formed on the membrane during the concentration of milk by reverse osmosis. *J. Dairy Res.*, **41**, 89.

Goldsmith, R. L. (1971). Macromolecular UF with microporous membranes. *Ind. Eng. Chem. Fundam.*, **10**, 113.

Goldsmith, R. L., P. P. de Fillippi and S. Hassain (1971). New membrane process applications *AIChE Symp.Ser.*, **120**, 7–14.

Grant, R. A. (ed.) (1980). *Applied Protein Chemistry*. Applied Science, London.

Guillotin, M., C. Lemoyne, C. Noel and L. Monnerie (1978). *Desalination*, **22**, 205.

Hancher, C. W. and A. D. Ryon (1973). *Biotechnol. Bioeng.*, **15**, 677–692

Hayes, J. F., J. A. Dunkerley, L. L. Muller and A. T. Griffin (1974). Studies on whey processing by ultrafiltration II. Improving permeation rates by preventing fouling. *Aust. J. Dairy Technol.*, **29**, 132–140.

Henry, J. D. and M. C. Porter (1972). In *Recent Developments in Separation Science*, ed. N. N. Li. CRC Press, Ohio.

Hiddinck, J., R. de Boer and P. F. C. Nooy (1970). *J. Dairy Sci.*, **63**, 204–214.

Howell, J. A. and O. Velicangil (1980). Ultrafiltration of protein solutions. *Polym, Sci. Technol.*, **13**, 217–230.

Jenq, C. Y., S. S. Wang and B. Davidson (1980). Ultrafiltration of raw sewage using an immobilised enzyme membrane. *Enzyme Microb. Technol.*, **2**, 145–149.

Kesting, R. E. (1971). *Synthetic Polymeric Membranes*. McGraw-Hill, New York.

Kesting, R. E. (1973). *J. Appl. Polym. Sci.*, **17**, 1771.

Kesting, R. E., M. K. Barsh and A. L. Vincent (1980). *J. Appl. Polym. Sci.*, **9**, 1873.

Klemm, K. and L. Friedman (1932). *J. Am. Chem. Soc.*, **54**, 2637.

Koehnen, D. M., M. H. V. Mulder and C. A. Smolder (1978). *J. Membr. Sci.*, **21**, 199.

Kopocek, J. and S. Sourirajan (1969). *J. Appl. Polym. Sci.*, **13**, 637.

Kuiper, D., C. A. Bom, J. L. van Hezel and J. Verdouw (1973). *Proceedings of the 4th International Symposium on Fresh Water From the Sea, Heidelberg.* pp. 207–215.

Kunst, B. and S. Sourirajan (1974). *J. Membr. Sci.*, **18**, 3423.

Kutowy, O. and S. Sourirajan (1975). *J. Membr. Sci.*, **19**, 1449.

Kutowy, O., W. L. Thayer and S. Sourirajan (1978). *Desalination*, **26**, 195.

Le, M. S. and J. A. Howell (1982). The fouling of ultrafiltration membranes and its treatment. In *Symposium of Progress in Food Engineering.* Springer Verlag AG, Zurich.

Le, M. S. (1982). *Membrane ultrafiltration: fouling and treatment.* Ph.D. thesis, University of Wales.

Le, M. S. and J. A. Howell (1983). A model for the effects of adsorbents and cleaners on UF membrane structure. *Chem. Eng. Res. Des.*

Lee, D. N. and R. L. Merson (1976). Chemical treatment of cottage cheese whey to reduce fouling of ultrafiltration membranes. *J. Food Sci.*, **41**, 778–786.

Lefebvre, M. S., C. J. D. Fell, A. G. Fane and A. G. Waters (1980). *Polym. Sci. Technol.*, **13**, 79–98.

Loeb, S. and S. Sourirajan (1960). Sea water desalination by means of a semipermeable membrane. *UCLA Water Resources Center Report WRCC 34.* UCLA, CA.

Lonsdale, H. K., U. Merten, R. L. Riley and K. D. Vos (1965). R. O. for Water Desalination, US Office for Saline Water, Res. Dev. Prog. Rep. 150, General Dynamics, General Atomics Division.

Madsen, R. E. (1978). *Hyperfiltration and Ultrafiltration in Plate and Frame Systems.* Elsevier, New York.

Maier, K. and S. Scheuermann (1960). *Kolloid Z.*, **171**, 122.

Matthews, M. E. (1979). *Proceedings of the Whey Research Workshop II, Palmerston, North, NZ.*

Matthiasson, E. and B. Sivik (1980). *Desalination*, **35**, 59–103.

Meares, P. (1966). *Eur. Polym. J.*, **2**, 241.

Michaels, A. S. (1965). *Ind. Eng. Chem.*, **57**, 32.

Michaels, A. S. (1967). *US Pat.* 3 276 598.

Michaels, A. S. (1968). New separation techniques for the CPI. *Chem. Eng. Progr.*, **64** (12), 31–43.

Michaels, A. S. (1980). *Polym. Sci. Technol.*, **13**, 1–20.

Nakao, S. I., T. Normura and S. Kimura (1979). Characteristics of macromolecular gel layer formed on ultrafiltration tubular membrane. *AIChEJ*, **25**, 615–622.

Nakao, S. I. and S. Kimura (1981). Analysis of solute rejection in ultrafiltration. *J. Chem. Eng. Jpn.*, **14**, 32–37.

Norde, W. and J. Lyklema (1978). *J. Colloid Interface Sci.*, **66**, 257–302.

van Oss, C. J. (1972). *Technol. Surf. Colloid Chem. Phys.*, **1**, 89.

van Oss, C. J. (1981). Detergent treatment of Uf membrane. *J. Colloid Interface Sci.*, **79**, 590.

Pepper, D. (1982). The basis of reverse osmosis. In *A Short Course in Membrane Technology for Industry.* Imperial College, London.

Porter, M. C. (1972). Concentration polarisation with membrane ultrafiltration. *Ind. Eng. Chem. Prod. Res. Dev.*, **11**, 234–248.

Reid, C. E. and E. J. Breton (1959). *J. Appl. Polym. Sci.*, **1**, 133.

Sourirajan, S. and B. Kunst (1978). Cellulose acetate and other cellulose ester membranes. In *Reverse Osmosis and Synthetic Membranes.* NRCC, Ottawa.

Strathmann, H. and K. Koch (1978). *Desalination*, **21**, 24.

Swaminathan, T. (1980). Flux enhancement in ultrafiltration by detergent treatment of membrane, *J. Colloid Interface Sci.*, **76**, 573–579.

Tanny, G. B. (1974). *J. Membr. Sci.*, **18**, 2199.

Thayer, W. L., L. Pagean and S. Sourirajan (1978). *Desalination*, **21**, 209.

Tweddle, T. A. and S. Sourirajan (1978). *J. Appl. Polym. Sci.*, **22**, 2265.

Velicangil, O. (1979). Ph.D. thesis, University of Wales.

Wasilewski, S. (1965). Microwave investigation of temperature induced transient changes in CA osmosis membranes. Report no. 65, p.10. Department of Engineering, University of California.

26

Ultrafiltration Processes in Biotechnology

R. S. TUTUNJIAN
Amicon Corporation, Danvers, MA, USA

26.1 INTRODUCTION

Ultrafiltration (UF) is a pressure-driven membrane separation technique for dissolved and suspended materials based on size and molecular scale. Substances smaller than the pore size of the filter are driven through with the solvent while larger solutes are retained. This is a very simple procedure and requires no phase change, no chemical addition and little energy.

The separation involved in ultrafiltration can deal with species ranging in molecular weight from 500 to 1 000 000, or approximately 10 to 100 Å. This distinguishes the process from reverse osmosis (RO), where even salts are retained, and from microporous filtration, where pore size ranges from 0.01 to 10 μm (see Figure 1).

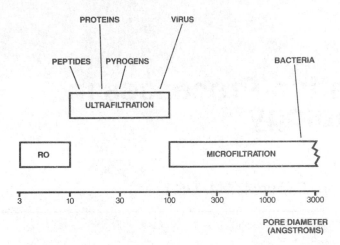

Figure 1   Comparison of ultrafiltration to other membrane processes

UF membranes are also differentiated from most microporous filters by their unique 'skinned' or anisotropic structure (see Figure 2). The tight pore structure opens to a more porous network below. This allows retention to take place on the membrane surface rather than within the filter structure. Filtration rates are also greatly increased since the major flow resistance, the membrane surface structure, is extremely thin. Conversely, typical microporous filters have an open, tortuous path structure which causes particle entrapment within the filter matrix (Reti *et al.*, 1979; Howard and Duberstein, 1980). This is not true of radiation-etched membranes and some newer anisotropic microfilters. Flow rates gradually decrease as filter clogging takes place.

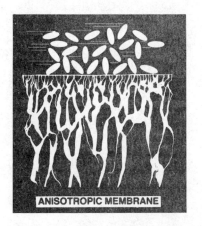 

Figure 2   Comparison of anisotropic and conventional microporous filters

It should also be noted that ultrafiltration membranes are produced in a wide variety of polymers which exhibit high resistance to acids, bases, alcohols and temperature, allowing very effective membrane cleaning. In this manner, ultrafilters can be reused without deterioration of flow rates and without cross contamination. In most cases, filter lifetimes last one to four years.

During the past decade, ultrafiltration has become established as a useful separation process for the pharmaceutical, chemical and food processing industries. Present applications are found in the plasma products, vaccines, fermentation products, enzymes, whey proteins and related products. Ultrafiltration has been used to replace and/or supplement many conventional separation techniques, such as centrifugation, rotary drum vacuum filtration, distillation, salt precipitation, lyophilization, evaporation and chromatography. The results have shown dramatic savings in cost and time as well as improvements in product purity.

The recent activity in the field of biotechnology has also utilized ultrafiltration as an integral part of purification processes. Present production of recombinant DNA products and of monoclonal antibodies generally requires extensive isolation techniques. Bacteria and cells require harvesting and washing prior to product recovery. Products are also typically dilute and highly impure directly after cellular extraction. Ultrafiltration has allowed economical, efficient concentration and purification of cells and biologicals as well as simple scale-up of production processes. The advantages of ultrafiltration should have increased impact on this emerging industry as process research continues in the future.

26.2 THEORY

26.2.1 Mass Transfer and Concentration Polarization

During concentration, pressure is exerted on a solution in contact with the UF membrane. This causes a flow of solutes and water toward the ultrafilter. However, the concentration of retained macrosolutes will build up at the membrane surface due to the removal of water. This results in a concentration gradient with the maximum solute level at the membrane (see Figure 3). This phenomenon is known as concentration polarization (Michaels, 1968).

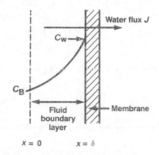

Figure 3 Concentration gradient during ultrafiltration

As a result of the increased concentration at the membrane surface, there is a tendency for solute to diffuse away from this point. Under steady state conditions, the convective mass transfer due to filtration is balanced by the diffusive movement in the opposite direction. This condition can be expressed as follows:

$$JC - D_v \frac{dC}{dx} = 0 \tag{1}$$

where JC and D_v (dC/dx) are the convective and diffusive terms respectively (J is filtration flux rate, C is the solute concentration at point x and D_v is the solute diffusivity). Equation (1) can be integrated across the solute boundary layer (see Figure 3) to give:

$$J = (D_v/\delta) \ln (C_W/C_B) \tag{2}$$

where C_W is the solute concentration at the membrane surface, C_B is the bulk solute concentration and δ is the boundary layer thickness. Since this latter term is typically unknown, a mass transfer coefficient, K, usually replaces D_v/δ:

$$J = K \ln (C_W/C_B) \tag{3}$$

It is important to note that the mass transfer coefficient is generally not a function of the solute concentration. However K is dependent on the driving pressure and any fluid flow across the membrane. These forces affect K by changing the thickness of the boundary layer, δ. (Fluid flow across the membrane is used to improve filtration flux; this will be discussed further in Section 26.2.3.)

26.2.2 Gel Polarization

The filtration flux rate may be increased by raising the solute concentration at the membrane, C_W (see equation 3). However, the value of C_W can be increased only to a certain extent. Under actual operating conditions, a point is reached where the retained solute forms a so-called gel layer (Michaels, 1968; Blatt *et al.*, 1970). This gel concentration, C_G, is the maximum value of C_W and may be substituted into equation (3):

$$J = K \ln (C_G/C_B) \qquad (4)$$

The concentration at which gel formation takes place will depend on several variables including pressure, temperature, solubility and pH. Ingham *et al.* (1980) suggest that C_G is actually the concentration at which osmotic back-pressure is high enough to prevent flux.

Once gel polarization takes place, C_G becomes constant and flux (J) will decline linearly with ln C_B (see Figure 4); this assumes K to be constant by maintaining fixed pressure/flow conditions. The gel layer also offers a hydraulic resistance to flow and solute permeation, somewhat like a secondary membrane (see Figure 5). Blatt *et al.* (1970) showed the gel layer actually could interfere with solute passage through the membrane.

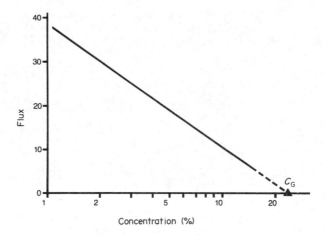

Figure 4   Filtration flux as a function of solute concentration

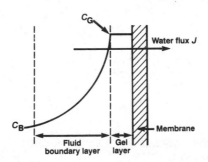

Figure 5   Concentration gradient during gel polarization

Gel polarization will actually take place under all but the most dilute solute concentrations. Figure 6 shows how the flux rate at very low concentrations is relatively constant. At these levels, flux is determined by equation (3); the wall concentration is increasing with the bulk concentration. When the gel layer forms, flux will decline logarithmically as shown. Ingham and Busby (1980) showed that albumin displayed this behavior after reaching a concentration of only about 1 mg ml^{-1} (0.1%). It should also be noted that an increase in pressure would cause gel formation at even lower concentrations.

To summarize briefly, gel polarization places a limit on the solute concentration at the membrane. Gel formation takes place at very low solute levels and is affected by operating pressure. In most actual operating conditions, flux rates are determined by gel polarization conditions.

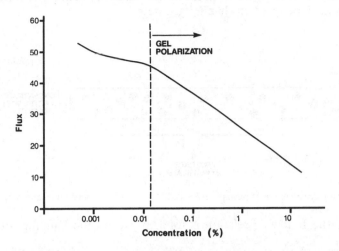

Figure 6 Filtration flux as a function of concentration before and after gel polarization

26.2.3 Cross-flow Filtration

26.2.3.1 Pressure characteristics

The effects of gel polarization can be reduced by various methods. The principle behind these centers on maximizing movement of solute away from the membrane and reducing the thickness of the gel layer. This is usually accomplished by cross-flow filtration where retained fluid is recirculated over the membrane surface (See Figure 7). Other approaches involving mechanical agitation, with stir bars and vibrators, are used only in small-scale laboratory equipment.

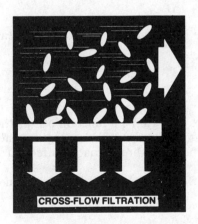

Figure 7 Comparison of cross-flow and conventional filtration

Characterization of cross-flow ultrafiltration is usually accomplished by monitoring pressures. A schematic diagram of a general process is shown in Figure 8. The feed stream flows across the ultrafilter and creates a pressure differential from the inlet (P_i) to the outlet (P_o). The so-called cross-flow ΔP is simply:

$$\Delta P = P_i - P_o \qquad (5)$$

This pressure drop can be related to the flow rate (Q) or velocity (V) across the membrane according to the Poiseuille equation for laminar flow, modified as follows:

$$\Delta P = (C_1 \mu L V)/d^2 = (C_2 \mu L Q)/d^4 \qquad (6)$$

where μ is the viscosity, L is the filter length, d is the fluid channel height above the membrane, C_1 and C_2 are constants dependent on channel geometry.

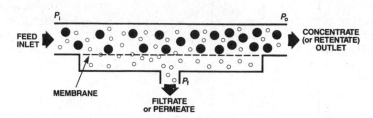

Figure 8   Cross-flow filtration pressure relationships: cross-flow $\Delta P = P_i - P_o$; transmembrane $\Delta PTM = [(P_i + P_o)/2] - P_F$

In turbulent flow, the following relation is used, based on the Fanning, or Darcy, equation modified as follows:

$$\Delta P = (C_3 f L V^2)/d = (C_4 f L Q^2)/d^5 \tag{7}$$

where f is a factor based on the Reynolds number, and C_3 and C_4 are constants dependent on channel geometry.

The driving force through the membrane is also determined by pressure. This transmembrane pressure is the difference between the pressure on the feed side and on the filtrate side of the ultrafilter. The differential will be highest at the inlet and reduce to a minimum at the outlet. An average driving force (ΔPTM) is characterized as follows:

$$\Delta PTM = \frac{P_i + P_o}{2} - P_f \tag{8}$$

Generally, the filtrate pressure is negligible and P_f is taken as zero.

It should also be noted that the transmembrane pressure and the cross-flow pressure drop can be related by (assuming zero permeate pressure):

$$\Delta PTM = P_i - (\Delta P/2) \tag{9}$$

This shows that for fixed inlet pressures, a change in the cross-flow velocity as measured by ΔP will also affect ΔPTM.

26.2.3.2   Flux rate determination

With cross-flow ultrafiltration, the filtration flux rate will be a function of: (1) the transmembrane pressure (ΔPTM), (2) the gel layer concentration (C_G), (3) the mass transfer coefficient (K), and (4) the bulk solute concentration (C_B). The latter three variables affect flux rate according to the gel polarization equation (4).

Blatt *et al.* (1970) showed that K is a function of the fluid velocity (V) across the membrane; the following Sherwood number (Sh) relationship is generally used:

$$Sh = (Kd)/D_v = A\,Re^B Sc^{1/3} \tag{10}$$

where d is the fluid channel height on top of the membrane, Re is the Reynolds number (dV/v), Sc is the Schmidt number (v/D_v), A and B are constants. Theoretically, the value of the constant B should be 0.33 in laminar flow conditions and 0.8 in turbulent flow (Leveque, 1928). However several researchers (Blatt *et al.*, 1970; de Filippi and Goldsmith, 1970; Cheryan, 1977) have shown that the value of the constant B is actually 0.5 in laminar flow and about 1.0 in turbulent flow. Thus, K will increase with the square root of V in laminar flow conditions. In turbulent flow, K will increase almost linearly with velocity in many cases.

The gel concentration (C_G), as mentioned previously, is a function of several variables including temperature, solubility and pH. The value of C_G can also be dependent on the recirculation flow velocity and flow channel geometry. This fact is important in that velocity changes may affect C_G as well as the mass transfer coefficient.

The third key parameter governing flux rate is the transmembrane pressure (ΔPTM). The flux will be a function of ΔPTM as defined by:

$$J = \Delta PTM/(R_{\mathrm{M}} + R_{\mathrm{G}}) \tag{11}$$

where R_{M} and R_{G} are the hydraulic resistances created by the membrane and the gel layer, respectively. The membrane resistance is considered to be constant during normal operation (membrane compaction may take place at very high pressures).Therefore, in a situation where no gel polarization occurs, flux will increase linearly with ΔPTM. Using water as an example, this linear flux behavior can be verified (see Figure 9).

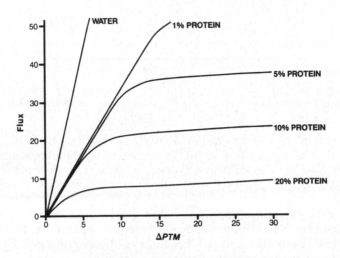

Figure 9 Filtration flux as a function of transmembrane pressure, varying solute concentrations

In the presence of a retained solute, gel formation will create an additional resistance to flow. If R_{G} were constant, the flux rate would still vary linearly with transmembrane pressure. However, many researchers (Michaels, 1968; Blatt *et al.*, 1970; Cheryan, 1977) have demonstrated that R_{G} is highly variable. As shown in Figure 9, flux rate will increase linearly up to a certain transmembrane pressure. Beyond this point, the gel layer will form and increase overall resistance. Further increases in pressure will increase the thickness and the resistance of the gel layer. Consequently, at high ΔPTM the flux rate will reach a maximum and become relatively constant with pressure.

The onset of gel polarization and its pressure independent flux behavior will vary according to the solute concentration and fluid recirculation velocity. At higher solute concentrations, flux will level off at lower transmembrane pressures (see Figure 9). Also, as velocity increases, gel polarization occurs at higher ΔPTM values (see Figure 10).

Under gel polarization conditions, flux is a function of velocity and concentration as determined by equations (4) and (10). Flux will fall logarithmically with solute concentration under constant velocity/pressure conditions. As shown in Figure 9, the flux rates achieved decline as the bulk solute concentration increases. In this case, the flux curves for concentrations 5% and above exhibit gel polarized behavior. The values at a ΔPTM of 15 p.s.i. for C_{B} (%) and J ($\mathrm{l\,m^{-2}\,h^{-1}}$) are: $C_{\mathrm{B}} = 5, J = 36$; $C_{\mathrm{B}} = 10, J = 22$; and $C_{\mathrm{B}} = 20, J = 8$. These rates may be correlated according to equation (4). In this case, K and C_{G} are determined to be $20\,\mathrm{l\,m^{-2}\,h^{-1}}$ and 30% respectively; equation (4) becomes:

$$J = 20 \ln (30/C_{\mathrm{B}}) \tag{12}$$

At solute levels less than 5%, gel polarization is not observed and flux will not increase logarithmically according to equation (12).

If the transmembrane pressure is reduced, equation (12) would also not apply. At a low ΔPTM of 5 p.s.i. gel polarization does not even occur at a solute concentration of 10% (see Figure 9). In fact, above this level, flux will be constant as concentration is increased.

Flux will also depend on the velocity according to the mass transfer coefficient. The value of K will increase with the square root of velocity in laminar flow and approximately directly in turbu-

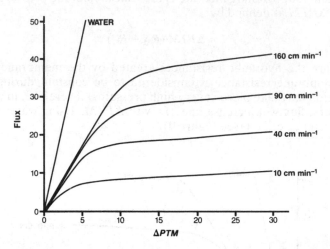

Figure 10 Filtration flux as a function of transmembrane pressure, varying recirculation velocities

lent flow. This velocity dependence is shown in Figure 10. As always, flux increases linearly at low ΔPTM, but becomes relatively constant at higher levels after gel polarization occurs. The final flux rate increases with velocity as discussed previously. The case shown is under laminar flow conditions; at a ΔPTM of 15 p.s.i., the flux rate doubles when velocity is increased by a factor of 4.

In brief summary, cross-flow ultrafiltration is controlled by adjustment of operating pressures and recirculation flow rate. The filtration rate will increase linearly with the transmembrane pressure until gel formation occurs and flux becomes relatively constant. The flux rate can be improved, however, by increasing velocity across the membrane.

26.2.4 Membrane Rejection

The ability of an ultrafiltration membrane to retain a given species is defined by the rejection coefficient, σ.

$$\sigma = 1 - (C_P/C_B) \tag{13}$$

where C_P is the concentration of the species in the permeate side of the membrane at a given instant of time. If the membrane completely retains the species, its concentration in the permeate would be zero ($C_P = 0$) and its rejection coefficient would be one ($\sigma = 1$). If, on the other hand, the membrane was completely permeable to the solute, as would be the case for a salt, its concentration in the permeate would be unaffected by the membrane ($C_P = C_B$) and its rejection coefficient would be zero ($\sigma = 0$). This point is of particular importance because it shows that the membrane can selectively concentrate some solutes while allowing others to pass. This selectivity allows an ultrafiltration membrane to purify as well as concentrate. Thus if a macromolecule such as an enzyme was prepared in the presence of inorganic salts or other low molecular weight impurities, these could be removed by ultrafiltration. This type of operation could not be accomplished with a reverse osmosis membrane which would retain the microsolutes as well as macrosolutes.

The molecular weight rating of an ultrafiltration membrane is determined experimentally by challenging with known test molecules. The membrane should have a sharply defined molecular weight cutoff, as illustrated in Figure 11. Ultrafilters with more diffuse cutoff characteristics are limited in purification use since low molecular weight solutes are retained to a greater extent.

Fractionation of solutes relatively similar in molecular weight can be difficult with any ultrafiltration membrane. Rejection can be a function of such parameters as solute concentration and transmembrane pressure. In fact, retained solutes can cause rejection of solutes which normally would pass through a membrane. This effect has been demonstrated in albumin–γ-globulin and lysozyme–albumin mixtures (Blatt *et al.*, 1970; Ingham and Busby, 1980).

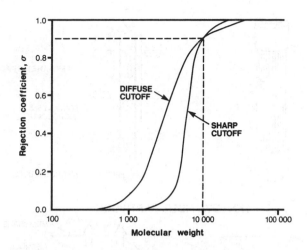

Figure 11 Rejection coefficient as a function of molecular weight

26.3 PROCESS CONSIDERATIONS

26.3.1 Membrane Configurations

There are two basic configurations of ultrafiltration membranes, namely hollow fibers and flat sheets. Filtration equipment with these membranes utilize a variety of operational modules (see Figure 12).

Hollow fibers are generally supplied in self-contained cartridge housings. Fibers are easy to clean and allow good product recovery with their unique capability for backflushing. They also allow a large amount of membrane area in a compact space. Fibers are somewhat limited in their overall pressure capability (up to 30–40 p.s.i.).

Large diameter (0.25 – 1 in) membrane tubes are basically a variant of hollow fibers. High recirculation rates allow turbulent flow to take place. Due to their large hold-up volume and high space requirements, tubular membranes are generally not well-suited for biological purification.

Flat sheet membranes have been fabricated in several types of modules. Spiral wound cartridges use a rolled membrane design with spacer screens separating the filters. Certain plate-and-frame type devices also use separator screens except that the membranes are stacked vertically. These devices allow high pressure operation (to 100 p.s.i.) but can have cleaning and product recovery problems due to clogging of the screens (Lukaszewicz *et al.*, 1981).

True plate-and-frame devices are also available which are easier to clean. These tend to have higher hold-up volumes and require more space. These systems do offer certain advantages in cases where prefiltration is not desired and/or high pressures are required.

26.3.2 Fluid Mechanics

As discussed previously, ultrafiltration systems use cross-flow filtration to reduce gel polarization and associated reductions in flux rate. This recirculating flow creates a pressure differential between the inlet and outlet of the system on the retentate side of the filter. The filtration flux is a function of the transmembrane pressure (ΔPTM) as well as the recirculation flow velocity (see Section 26.2.3). Theoretically, flux could be increased almost indefinitely merely by raising velocity and pressure.

In actual system use, however, pressures are constrained by the operating limits of the equipment. Most UF systems operate at under 100 p.s.i. maximum pressure; hollow fiber membranes use less than 40 p.s.i. as a limit.

During operation, flux rates are optimized by varying the recirculation rate (measured by ΔP), while inlet pressure is usually maintained at the maximum allowed by the system. Equation (9) shows that ΔP and ΔPTM are directly related assuming a fixed inlet pressure; as ΔPTM is

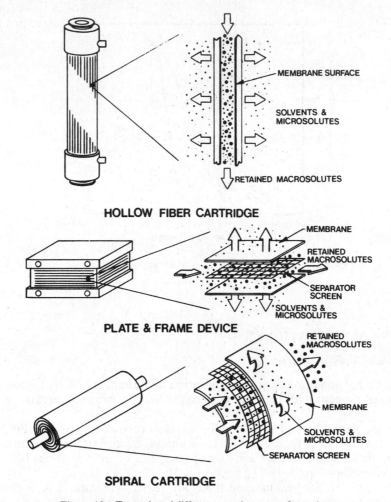

HOLLOW FIBER CARTRIDGE

PLATE & FRAME DEVICE

SPIRAL CARTRIDGE

Figure 12   Examples of different membrane configurations

increased, ΔP (and thus velocity) is decreased. Flux rates normally increased with higher ΔPTM may actually decrease since velocity is reduced.

In a case shown in Table 1, inlet pressure was limited to 25 p.s.i. and flow was laminar. The optimum flux rate was reached with a ΔP of 15 p.s.i. Further reduction in velocity caused a drop in filtration rate. Higher velocity reduced the driving force, ΔPTM, to the point where flux decreased.

Table 1   Filtration Flux Rate as a Function of Recirculation Rate
and Pressure Drop

P_i	P_o	ΔP	ΔPTM	Recirculation rate $(1\ min^{-1})$	Flux rate $(1\ min^{-1})$
25	0	25	12.5	50	2.3
25	5	20	15	40	2.8
25	10	15	17.5	30	2.9
25	15	10	20	20	2.4
25	20	5	22.5	10	1.7

This example may be shown graphically as a function of ΔPTM and ΔP (see Figure 13). As shown previously in Figure 10, flux will tend to level off at high ΔPTM values, depending on the velocity (or ΔP). When the constraint of a maximum pressure is imposed, operating conditions are limited to a certain range as shown in Figure 13. At low ΔPTM, conditions are constrained to where ΔP equals the maximum allowable system pressure (25 p.s.i. here); this means $P_o = 0$. When ΔPTM exceeds one-half the maximum pressure, further increases will cause a drop in velocity. The ΔP associated with a value of ΔPTM may be calculated directly from equation (9), for example, if $\Delta PTM = 20$, then ΔP will be 10 ($P_i = 25$). Flux will be optimized at a particular pair of operating pressures. This will generally occur at the point where gel polarization starts.

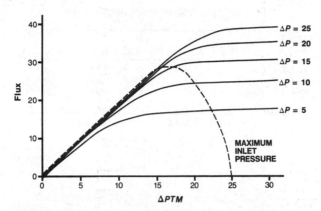

Figure 13 Filtration flux as a function of transmembrane pressure, varying recirculation rates and fixed maximum inlet pressure

It should also be noted that the optimum operating conditions are a strong function of product concentration; a higher ΔP is required at elevated solute concentrations. The optimum flux curve shown in Figure 13 can be expanded to show results at various product concentrations (see Figure 14). As concentration increases, higher velocity is required and outlet pressure is reduced; above a certain point no back pressure is used, allowing maximum velocity.

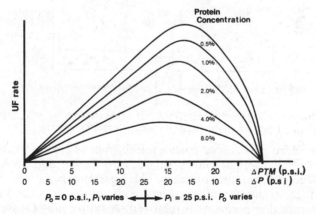

Figure 14 Filtration flux optimized as a function of transmembrane pressure, varying solute concentrations and fixed maximum inlet pressure

System geometry can also have a large impact on filtration flux rate. As mentioned in Section 26.3.1, certain systems can operate at higher pressures and thus higher velocities. However, most of these use spacer screens which reduce flow channel size; higher pressure drops (ΔP) are necessary to achieve a given flow velocity. The effect of system geometry can be examined using equations (6) and (10) rewritten as follows (assuming laminar flow where $B=0.5$ in equation 10):

$$\Delta P = C_5 \frac{VL}{d^2} \tag{14}$$

$$K = \frac{AD_v}{d} \frac{(dV)^{1/2}}{v^{1/2}} Sc^{1/3} = C_6(V/d)^{1/2} \tag{15}$$

where C_5 and C_6 are constants. Equations (14) and (15) can be combined to give:

$$K = C_7(d\Delta P/L)^{1/2} \tag{16}$$

where C_7 is a constant. This equation can be used to compare three basic types of membrane configurations. As shown in Table 2, hollow fiber devices have about the same mass transfer coefficient as higher pressure systems. This means that flux rates will be comparable despite lower operating pressures.

Table 2 Comparison of Mass Transfer Coefficients For
Various Types of Membrane Configurations

Membrane type	d (mm)	ΔP (p.s.i.)	L (mm)	K[a]
Hollow fiber	1.1	30	500	26
Cassette type	0.2	100	300	26
Spiral cartridge (open channel)	0.5	100	900	24

[a] Assumes equation (16) constant $C_7 = 100 \, l \, m^{-2} \, h^{-1} \, p.s.i.^{-0.5}$

26.3.3 Operation

26.3.3.1 General

The applications of ultrafiltration can be broadly categorized as follows; concentration, diafiltration or purification.

Concentration of suspended particles or macromolecules is where the retained stream is the product (see Figure 15).

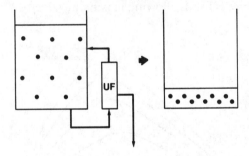

Figure 15 Flow schematic for concentration process

Diafiltration of suspended particles or macromolecules is where the retained stream contains the product and low molecular weight solutes, such as salts, sugars and alcohols, pass through the membrane by the addition and removal of water. This operation is shown schematically in Figure 16. The permeate that leaves the system is replaced with deionized water, ideally through a level controller at the same rate that permeate is removed. A buffer may be used instead of water if salt exchange is desired.

The third type of UF process involves the purification of solvents and solutions of low molecu-

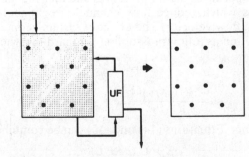

Figure 16 Flow schematic for diafiltration process

lar weight solutes where the permeate stream contains the product; the retained species may also contain a product of interest (see Figure 17).

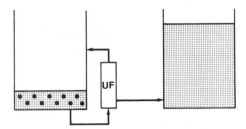

Figure 17 Flow schematic for purification process

26.3.3.2 *Concentration*

In a batch system concentration, an initial volume V_0 is ultrafiltered to produce a volume of permeate V_P. The final volume of concentrate V_F is determined by a simple material balance:

$$V_F = V_0 - V_P \tag{17}$$

It is important to note that V_F may not exactly equal the final volume in the process tank since a certain volume may be held up in the UF system.

The concentration factor (CF) is defined as:

$$CF = V_0/V_F \tag{18}$$

Thus if the feed stream was a 1000 l batch and the system was allowed to run until the final volume was 100 l, the concentration factor would be 10. A macromolecule that was completely retained by the membrane would have been increased by a factor of 10 in concentration. The concentration of a completely permeable species such as a salt will not be affected by ultrafiltration.

In cases where the rejection coefficient (σ) is not simply 1 or 0, a more general relation must be used to determine product concentrations:

$$C_F = C_0 (CF)^\sigma \tag{19}$$

where C_F and C_0 are the final and initial product concentrations. Using this equation, the final concentration may be calculated for various values of CF and σ (see Table 3).

Table 3 Final Product Concentration as a Function of Concentration Factor and Rejection Coefficient[a]

| | *Rejection coefficient* | | | | |
CF	*1.0*	*0.9*	*0.5*	*0.1*	*0.0*
1.0	1.00	1.00	1.00	1.00	1.00
2.0	2.00	1.87	1.41	1.07	1.00
3.0	3.00	2.69	1.73	1.12	1.00
5.0	5.00	4.26	2.24	1.17	1.00
10.0	10.00	7.94	3.16	1.26	1.00
20.0	20.00	14.82	4.47	1.35	1.00

[a] Initial concentration 1%.

The product recovery (R) can be determined by comparing the total product in the final concentrate to the initial feed, as follows:

$$R = \frac{C_F V_F}{C_0 V_0} = \frac{C_F}{C_0} \frac{1}{CF} \tag{20}$$

By substituting equation (19) into equation (20), this may be rewritten as:

$$R = CF^{\sigma-1} \tag{21}$$

A material balance for the product can be used to determine the concentration in the permeate (C_P):

$$V_P C_P = V_0 C_0 - V_F C_F \tag{22}$$

For example, if 2000 l of a 1% product is concentrated twenty-fold, then V_P is 1900 and V_F is 100. If rejection is 90%, C_F is 14.82% (see Table 3). The amount of product recovered in the final concentrate is about 74% of the total feed. The concentration of product in the permeate is 0.27% (see equations 20–22).

26.3.3.3 Diafiltration

As stated previously, diafiltration processes use the ultrafilter's ability to selectively pass low molecular weight solutes, usually salts, while retaining larger product molecules. By removing these solutes and replacing them with an equal volume of deionized water, the retained product is purified. Since the volume is constant, product concentration is constant, assuming complete retention.

The concentration of microsolute (C_S) will decline as the total volume of water added (V_D) increases. The following relation holds if the microsolute is completely permeable:

$$C_S = C_{S0} e^{-V_D/V_0} \tag{23}$$

where C_{S0} is the initial microsolute concentration and V_0 is volume of retained product. This is shown schematically in Figure 18.

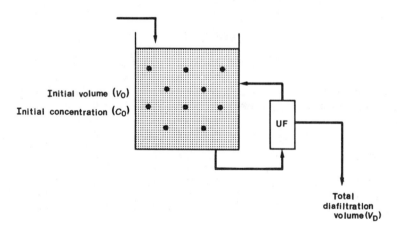

Figure 18 Diafiltration flow schematic: salt concentration $(C) = C_0 e^{-V_D/V_0}$

The amount of solute removal may also be shown graphically (see Figure 19) as a function of the multiple of diafiltration volumes (V_D/V_0). As shown, a four-fold diafiltration would remove 98% of a microsolute. In other words, adding 4000 l of deionized water continuously to 1000 l of product would reduce an initial 1% salt level to 0.02%.

In cases where the rejection coefficient (σ) is greater than 0, a correction must be made in equation (23) as follows:

$$C_S = C_{S0} e^{-(1-\sigma)V_D/V_0} \tag{24}$$

The effect of partial retention of the microsolute is shown in Table 4. The greater the rejection coefficient, the more diafiltration volume is required to effect a desired removal. For example, a three-fold diafiltration removes 95% of a completely permeable solute; almost four volumes are needed if the solute is rejected by only 20%.

It should be noted that diafiltration can also be carried out in step-wise, or discontinuous, fashion. This procedure involves repeated concentration and dilution with deionized water or buffer. If, for example, a product is concentrated twenty-fold and then diluted back to its original volume, the salt concentration would be reduced by 95% (assuming zero retention). Discontinuous diafiltration requires less water or buffer to effect a given salt removal. However, since

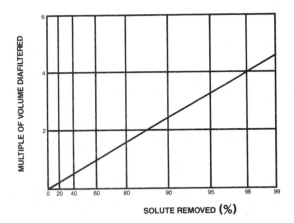

Figure 19 Microsolute removal as a function of diafiltration volume

Table 4 Microsolute Removal (%) as a Function of
Diafiltration Volumes and Rejection Coefficient

V_D/V_0	Rejection coefficient				
	0.0	*0.1*	*0.2*	*0.3*	*0.5*
1.0	63.2	59.3	55.1	50.3	39.3
2.0	86.5	83.5	79.8	75.3	63.2
3.0	95.0	93.3	90.9	87.8	77.7
4.0	98.2	97.3	95.9	93.9	86.5
5.0	99.3	98.9	98.2	97.0	91.8
10.0	100.0	100.0	100.0	99.9	99.3

product concentration is taking place, flux rate is dropping steadily and overall process time may be longer. Process optimization will be reviewed in Section 26.3.3.5.

26.3.3.4 Purification

Purification processes are similar to those using diafiltration in that large and small molecules are separated. In purification uses, the filtrate contains the product solute. Generally, these processes combine an initial concentration step (concentrating the high MW impurities) with a diafiltration step.

The concentration of product in the permeate (C_P) can be determined by a material balance on the retentate. The initial and final concentration of product in the retentate (C_0 and C_F) are calculated by using equations (19) and (23) or (24). Then a material balance yields:

$$V_P C_P = V_0 C_0 - V_F C_F \tag{25}$$

where V_P, V_0 and V_F are the volumes of permeate, initial and final retentate, respectively. The overall recovery (R) of product in this case is:

$$R = \frac{V_P C_P}{V_0 C_0} = 1 - \frac{V_F C_F}{V_0 C_0} \tag{26}$$

An example of a purification process is shown in Table 5. In this case, the product is completely permeable ($\sigma = 0$). The data show that the initial concentration step allows 90% of the product to be recovered in the permeate with no dilution. Further diafiltration steps improve product recovery, but cause dilution of the product in the permeate. By concentrating to near maximum levels initially, product recovery can be high with little dilution. In this way, the retentate is also highly purified; this is important in some fractionation processes where both solutes are of interest.

In situations where the product rejection is not zero, equations (25) and (26) are still appropriate. However, the values of C_0 and C_F must be calculated using the correct rejection coefficient in equations (19) and (24).

Table 5 Purification of Product Showing Recovery and Permeate Concentration: No Product Rejection

Condition	Retentate			Permeate		
	Volume (l)	Retentate concentration (%)	Product concentration (%)	Volume (l)	Product concentration (%)	Product recovery (%)
Initial	1000	1.0	5.0	0	—	0
10 × concentration	100	10.0	5.0	900	5.0	90.0
1 × diafiltration	100	10.0	1.8	1000	4.8	94.4
2 × diafiltration	100	10.0	0.68	1100	4.5	98.6
3 × diafiltration	100	10.0	0.25	1200	4.0	99.5

Table 6 Purification of Product Showing Recovery and Permeate Concentration: 20% Product Rejection

Condition	Retentate			Permeate		
	Volume (l)	Retentate concentration (%)	Product concentration (%)	Volume (l)	Product concentration (%)	Product recovery (%)
Initial	1000	1.0	5.0	0	—	0
10 × concentration	100	10.0	7.9	900	4.7	84.2
1 × diafiltration	100	10.0	3.6	1000	4.6	92.9
2 × diafiltration	100	10.0	1.6	1100	4.4	96.8
3 × diafiltration	100	10.0	0.72	1200	4.1	98.6
4 × diafiltration	100	10.0	0.32	1300	3.8	99.4

An example shown in Table 6 assumes a rejection of 20%. As expected, in comparison to the completely permeable product (Table 5), product recoveries and permeate concentrations are lower. A 98.6% recovery is now achieved after 3× diafiltration where only 2× was required previously; product concentration drops to 4.1% (from 4.5%) for this case.

26.3.3.5 *Combined operations—optimization*

Many processes combine concentration and diafiltration to purify the retained solutes. In purification processes (see Section 26.3.3.4) these steps are carried out to give high recovery and minimal product dilution. When the retained solute is the product of interest, however, conditions are generally optimized relative to process time.

In a combined concentration/diafiltration process, a decision must be made as to when the diafiltration should be carried out. If implemented at the starting protein (or macrosolute) concentration, the flux rate will be maximized but the total amount of diafiltration fluid required will also be maximum. Equation (23) showed that salt, or microsolute, removal depends on the ratio of diafiltration volume to process volume. So, as the product is concentrated and process volume reduced, the diafiltration volume required will decline proportionately.

Unfortunately, during concentration the flux rate will decline according to equation (4), so if the diafiltration step was carried out at the final protein concentration, the diafiltration volume and the flux rate would be at their lowest points. The optimum process may, in fact, require carrying out the diafiltration at a point in between the initial and final protein concentrations.

An example of a typical process is shown in Table 7. A 1000 l volume of 2% protein is to be concentrated ten-fold to 20%. In addition, it is desired to reduce an initial salt concentration of 5% to 0.05%; this is a 99% removal which requires a 4.6× diafiltration. The flux–concentration relation was experimentally determined to be:

$$J = 500 \ln (27/C_B) \tag{27}$$

which means $K = 500$ l h^{-1} and $C_G = 27\%$ at these operating conditions.

The filtration rates for four protein levels are calculated using this relation (see Table 7). The diafiltration volume (always 4.6× the batch volume) is divided by the filtration rate to yield the time required for diafiltration. As shown, the minimum time is achieved at 10% protein. So the optimal process should be carried out in three steps: (1) concentrate five-fold to 10% protein, (2) diafilter 4.6×, and (3) concentrate two-fold to final 20% protein level. Usually, it is reasonable to assume that the total time required for concentration will not be affected by diafiltration.

Table 7 Diafiltration Time as a Function of Protein Concentration and Batch Volume

Protein concentration (%)	Batch volume (l)	Diafiltration volume (l)	Diafiltration rate (l h^{-1})	Diafiltration time (h)
2	1000	4600	1300	3.5
5	400	1840	840	2.2
10	200	920	500	1.8
20	100	460	150	3.1

The optimum point for continuous diafiltration, as described above, may be determined mathematically. Ng *et al.* (1976) showed that this optimum protein concentration (C^*) is related to the gel concentration (C_G) by the equation:

$$C^* = C_G/e \tag{28}$$

where e is the natural logarithm base. This relation does assume complete protein rejection although other conditions have been examined (Cooper and Booth, 1978). The example given in Table 7 had a C_G value of 27%; when substituted into equation (25), C^* is found to be about 10% which agrees with the previous results.

26.3.4 System Design

The various types of filter configurations have been discussed (see Section 26.3.1). All designs incorporate the membranes into separate, modular elements or cartridges. These can be manifolded together increasing membrane area to provide the necessary filtration rate. The elements are contained inside external pressure-sustaining housings.

The need for flow across the UF membrane surface has also been defined. This recirculation flow is achieved by connecting a pump to the inlet of the ultrafiltration unit. The pump must be sized to provide the flow velocity and pressure as dictated by the membrane configuration. The transmembrane pressure (ΔPTM) is controlled by a restricting process valve on the UF outlet. The pump recirculation rate determines the ΔP. Pressure gauges are incorporated to monitor these parameters. These basic components of a UF system are shown in Figure 20.

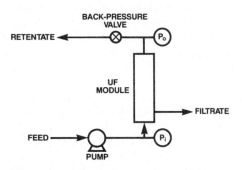

Figure 20   Flow schematic of basic ultrafiltration/cross-flow system

Ultrafiltration systems may be designed to operate either on a batch or a continuous basis (see Figure 21). In a batch system, fluid is removed from a process tank, pumped through the UF unit and returned to the tank. The concentration of product in the system gradually increases as filtrate is withdrawn. In this manner, the filtration flux rate is high initially and decreases steadily during the process.

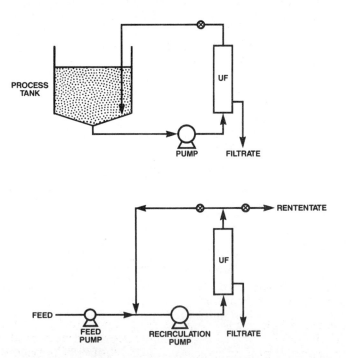

Figure 21   Comparison of batch and single-stage continuous system

In a continuous process, the fluid is introduced to a recirculation loop by means of a feed pump. A recirculation pump maintains the proper velocity and pressure at the membrane surface. The concentrate is bled continuously from the system so that the ratio of feed to concentrate flow equals the specified concentration factor. When the steady state is achieved, the system will contain fluid at the final and maximum concentration. Consequently, the flux rate will be at a minimum throughout the run.

If a continuous process is used, it is usually more economical to use a multistage system (see Figure 22). In a staged continuous process, several recirculation loops are connected with the concentrate of one feeding the next. The concentration of product in the loop increases from stage to stage. In this manner, the filtration flux will be higher in the initial stages and be at the minimum level only in the final stage. As more stages are added, the average system flux rate increases and approaches the flux rate of the batch system. In this manner, less filter area is required to provide a desired process rate. However, additional stages add some capital costs (pumps, flow controls, *etc.*). Beyond a certain point, additional stages do not decrease membrane area considerably and increase overall capital costs. Table 8 compares a batch and several continuous processes, showing the impact of adding stages. As shown here, it is generally most economical to use three to five stages.

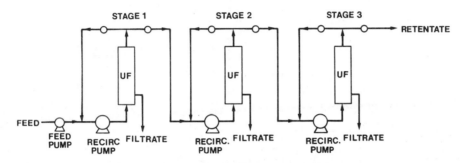

Figure 22 Multistage continuous system

Table 8 Comparison of Batch and
Continuous/Multistage Systems[a]

System	Flux ($l\,m^{-2}\,h^{-1}$)	Total area (m^2)
Batch	33.1	136
Continuous		
One-stage	8.1	555
Two-stage	31.1	243
	8.1	
Three-stage	38.7	194
	23.4	
	8.1	
Five-stage	44.7	165
	35.6	
	26.4	
	17.3	
	8.1	

[a] System design for $10 \times$ concentration factor and feed rate of 5000 l h^{-1}. Flux rate from: $J = 20 \ln (30/C)$

The basic choice between batch and continuous operation is usually dictated by the overall purification process. Continuous systems allow minimum residence time of product with the UF unit. This may be important if the product is heat or shear sensitive. Continuous processing is used extensively in processing dairy and food proteins; in these cases, the entire recovery process is continuous. Batch systems, on the other hand, provide maximum average flux rates and minimum membrane area. They are also much simpler and less expensive. Recovery of most pharmaceutical/biological products are carried out on a batch basis.

26.3.5 Membrane Cleaning

Another consideration in system operation is that of maintaining flux rates and cleaning after use. Most membranes are highly resistant to pH and are cleaned with sodium hydroxide, sodium hypochlorite, phosphoric acid or other suitable chemical agents. This is normally done in the standard ultrafiltration mode of operation. In cases where severe membrane fouling has occurred, hollow fiber membranes offer an extra advantage since they can be backflushed with cleaning solution (see Figure 23). This will force residual proteins off the filter surface and flush them away.

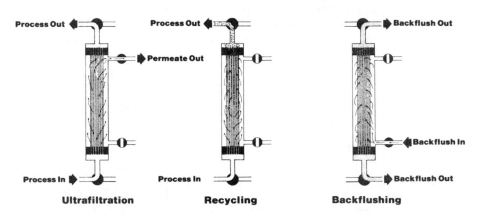

Figure 23   Operational modes for hollow fiber cartridge

A method useful for cleaning during product processing is that of recycling (see Figure 23). In this case, the permeate outlets are closed while the cartridge is filled with ultrafiltrate. This causes backflushing on the outlet half of the cartridge, where the permeate pressure now exceeds the retentate pressure. By then reversing the direction of the process flow, the other end of the filter can be similarly cleaned. This procedure requires only a few minutes to perform and has proven very useful in regenerating flux rates in process. In fact, the process can be automated for easy operation. Recycling is particularly effective in dealing with process streams with high suspended solids, such as bacterial cells and protein precipitates. As with backflushing, recycle operation is unique to hollow fiber membranes.

Another important consideration in a membrane cleaning cycle is that of protein adsorption to the membrane. In some cases, product may remain bound to the ultrafilter after the system has been drained. If the product is very valuable, it should be recovered prior to cleaning. This may be carried out by flushing with a minimal amount of water or suitable buffer and pooling this solution with the final concentrate. Backflushing is particularly effective for recovery since the volume used can be reduced. The problems created by adsorption can be avoided in many cases by proper choice of membrane material; some membrane manufacturers produce low adsorptive membranes for use with proteins.

26.4 BIOTECHNOLOGY APPLICATIONS

26.4.1 General

As described previously, ultrafiltration process applications fall into three categories: concentration, diafiltration and purification (see Section 26.3.3). In actual use and for the purposes of discussion, the former two are generally combined.

The types of products which are processed can also be separated into two basic groups: dissolved macromolecules and cellular products. The former includes such classes of biochemicals as enzymes, plasma proteins, viruses/viral antigens, hormones, peptides, lymphokines and interferons.

Cellular products are defined as bacteria and cells themselves as well as extracellular products which may be extracted during the process. Antibiotics and alcohols are examples of extracellular products which have been purified in this fashion. Membrane processing of cells has been carried out for continuous fermentation as well as batch; the latter recovery procedure is known as cell harvesting. Microporous filters operating in a cross-flow mode have also been used successfully for cell separations. The use of membranes for cell processing is discussed in full detail in Chapter 24.

26.4.2 Biochemical Concentration/Diafiltration

26.4.2.1 Processes

Ultrafiltration has been incorporated into many biochemical recovery processes as a means of concentration and purification. UF has allowed improved product recovery, faster processing and reduced costs when compared to traditional operations (*e.g.* lyophilization, centrifugation, precipitation). Although most data concerning genetic engineering processes are highly proprietary, information about UF uses in other biological applications is available. Several of these are discussed in more detail below.

26.4.2.2 Albumin

Human serum albumin is a protein isolated from plasma. This purification is normally carried out by a process known as Cohn fractionation (Cohn *et al.*, 1946), which is based on solvent precipitation. The final steps of this procedure involve concentration of protein and removal of alcohols and salt. Originally, this had been carried out by lyophilization and redilution to the desired protein level.

Ultrafiltration is currently used for final preparation of albumin by essentially all manufacturers. Using concentration and diafiltration, the product can be purified by a single piece of equipment. Typically, protein is concentrated from 6–8% initially to about 28–30% and rediluted slightly to yield a 25% final product. Alcohol concentration will drop from 7–10% to as low as 0.05% with a five-fold diafiltration. Diafiltration is usually done at 10–12% protein. Using UF, the freeze-drying and subsequent reconstitution steps are eliminated.

The benefits of ultrafiltration have been shown to be several-fold in this process. First of all, the final product is virtually free from albumin dimers. Lyophilization and other methods of drying all produce varying levels of dimerization which contaminate the product (Mercer, 1977). It has also been shown that diafiltration can provide more complete alcohol removal than freeze drying (Ng *et al.*, 1978). Finally, ultrafiltration is substantially less expensive in capital and operating (labor and energy) costs, as shown in Table 9. As a result of these clear advantages, UF has become a standard unit operation in plasma fractionation.

Table 9 Comparison of UF and Other Processes for Concentration/Purification of Albumin

Method	Dimer (%)[a]	Energy costs	Labor (man-h)[b]	Capital cost ($)[b]
UF	Trace	Low	20	60 000
Lyophilization	3.3–4.7	High	180	300 000
Vacuum distillation	1.2–1.5	High	100	180 000
Wiped film evaporation	1.8–2.0	High	N.A.	200 000
Solvent drying	4.6–5.2	Medium	N.A.	150 000

[a] Mercer (1977). [b] Figures for preparation of 1000 l of 25% albumin; N.A. = not available.

26.4.2.3 Viruses

Viruses or viral antigenic proteins must often be concentrated to obtain workable volumes for final purification. Centrifugation or precipitation can create such problems as aerosols and loss of infectivity. Large scale production, required in virus vaccine production and in cancer virus studies, can be difficult using such techniques.

Ultrafiltration has gained wide acceptance as a method for concentration of large volumes of virus. The use of UF has been reported for such viruses as polio, hepatitis A, influenza, herpes simplex and others (van Wezel, 1975; Thornton, 1976; Webster, 1976).

The use of ultrafiltration has allowed substantial improvements in recovery of viral activity. Weiss (1980) reports that hollow fiber UF membranes allowed over 90% recovery of a fragile C-type virus; these viruses have been very difficult to isolate using continuous-flow zonal centrifuges. Various methods of concentrating foot and mouth virus have also been compared (Morrow *et al.*, 1974). The results (see Table 10) showed that centrifugation gave highly variable recoveries, as low as 24%. Precipitative methods gave somewhat better yields, but are very difficult to scale-up. Ultrafiltration, however, allowed recoveries consistently above 92% with easy processing of large volumes.

Table 10 Comparison of UF and Other Processes For Concentration of Foot and Mouth Virus[a]

Method	Initial volume (l)	Concentration factor	Recovery (%)
UF	100	24×	92–100
Centrifugation	6	25×	24–77
PEG precipitation	4	20×	63–85
Sulfate precipitation	4	20×	80

[a] Morrow *et al.* (1974).

High concentration factors are also important, particularly when processing large quantities of dilute virus. Most viruses are grown in tissue culture in the presence of serum proteins. These proteins can limit the degree of concentration allowed by UF since they may be retained along with the virus. However more porous membranes (100 000 MW cutoff and higher) will allow passage of albumin, the major protein in tissue culture medium. Using this type of ultrafilter, rubella virus was concentrated 600-fold, even though the starting medium contained 2% serum (Trudel and Payment, 1980).

26.4.2.4 Enzymes

The production of enzymes and biochemicals involves many purification steps following initial fermentation or extraction from animal tissues/glands. These techniques include such processes as chromatography, salt precipitation and solvent extraction. Ultrafiltration has also been used extensively in enzyme purifications, complementing and/or replacing some of these other methods.

Since most of these isolation processes have been developed in industry, little has been published on details. Examples of how UF has been utilized are: (1) concentration and diafiltration of final product, generally after gel filtration chromatography; this reduces volume substantially prior to lyophilization; (2) concentration prior to salt precipitation and reducing the amount of salt required; this cuts costs directly and indirectly; a smaller volume is easier to separate after precipitation and salt disposal expenses are reduced; (3) concentration of products in solution following precipitation; this concentrate may then be easily processed further, or, if normally discarded, it can be recycled; (4) diafiltration to remove alcohols following solvent precipitation and redilution of the precipitate; and (5) initial diafiltration to remove such low molecular weight contaminants as peptides, amino acids, sugars and salts.

The extensive use of UF in enzyme purification has allowed good definition of operating costs for these processes. A typical economic analysis is shown in Table 11 with a comparison to freeze concentration. Freeze concentration is somewhat less expensive than lyophilization since the product is not completely dried in the former. These processes are used in enzyme and other biological production rather than evaporators due to the heat sensitivity of the product. In comparing the two processes, it is clear that ultrafiltration is substantially less expensive, particularly in capital costs. It should also be noted that the figures shown for freeze concentration are well below those reported in the literature (Renshaw *et al.*, 1982). In addition to capital-related operating costs, UF offers a large saving in energy consumption; this is more than enough to offset membrane replacement costs. Even though ultrafiltration cannot provide a completely dried product, its use to concentrate prior to lyophilization provides significant economic benefits.

Table 11 Cost Comparison of Ultrafiltration and Freeze Concentration[a]

	UF		Freeze concentration	
	Annual cost ($)	Cost per 1000 l ($)	Annual cost ($)	Cost per 1000 l ($)
Depreciation	14 000	0.78	170 000	9.44
Maintenance	8 400	0.47	102 000	5.67
Labor	28 100	1.56	12 500	0.69
Power	4 000	0.22	62 700	3.48
Water, chemicals	3 600	0.20	14 000	0.78
Membranes	31 500	1.75	—	—
Total operating costs	89 600	4.98	361 200	20.06
Capital cost	140 000		1 700 000	

[a] Assumptions: (1) system output—4500 l/h (water removal rate); (2) operation 250 d/year, 16 h/d; for UF system add 2 h/d cleaning; (3) depreciation—10 years s/l; (4) maintenance—6% of capital cost annually; (5) labor at $12.50/h (UF—9 h/d, F.C.—4 h/d); (6) power at $.05/kW h; (7) membrane life of 1 year, $188/m²; (8) membrane average flux rate—27 l/m²/h.

26.4.2.5. Interferon

The use of ultrafiltration has been extensive in the recovery of interferon. Interferon is a naturally occurring protein which was prohibitively expensive to produce prior to recombinant-DNA techniques. Typical isolation processes utilize several chromatography steps. Ultrafiltration has proven invaluable in concentration and diafiltration of interferon following these steps (Georgiades, 1981).

In some cases, protein adsorption to the membranes can cause incomplete yield (Zoon *et al.*, 1980). With hollow fiber membranes, a backflush with a small quantity of filtrate has shown some success in improving recovery. In one case, yield was improved from 80% to 98% with this technique. In many cases, special, non-adsorptive membranes must be used to allow complete recovery of activity (Georgiades, 1981).

26.4.3 Biochemical Purification

26.4.3.1 Processes

Ultrafiltration has also been used extensively in a purification mode of operation where the filtrate contains the product of interest. In these situations, solutes which are of relatively high molecular weight are removed by the ultrafilter. These contaminants have included cellular debris, proteins, endotoxins, viruses and viral antigens. Examples of these are described below.

26.4.3.2 Pyrogens

The need to ensure the absence of pyrogens from water or solutions can be a significant problem for pharmaceutical manufacturers and clinical and research laboratories. Though defined as any substances which cause a temperature rise when injected, most pyrogens are lipopolysaccharides (LPS) from bacterial cell walls. The sizes of these endotoxins can range from subunits of molecular weight about 20 000 to aggregates with diameters of about 0.1 μm. Their threshold of detectability ranges from 0.01 to 0.1 ng ml^{-1}.

Ultrafiltration membranes have been used successfully to remove pyrogens from solutions at even very high feed concentrations (Tutunjian, 1982). As shown in Table 12, an ultrafilter can have a removal efficiency (C_F/C_P) of over 10^5. The filtrate concentrations in this case were below the limit of detectability of standard LAL tests.

It should be noted that pyrogens cannot be eliminated by autoclaving or by typical microporous filtration (Nelsen, 1978). Processes which have been used traditionally are distillation, reverse osmosis and adsorption by asbestos or other media. The disadvantages of these methods are varied. Distillation equipment requires large capital expenditures; operating costs are reduced only with more expensive thermocompression or multi-effect stills. Reverse osmosis systems have

Table 12 Pyrogen Removal from Water by Ultrafiltration

Day	*Pyrogens* (ng ml^{-1}) Before ultrafiltration	After ultrafiltration
1	300	<0.050
2	100	<0.050
3	300	<0.050
4	10 000	<0.050
5	1 000	<0.050
10	1 000	<0.050
15	1 000	<0.050

been difficult to maintain in sanitary condition. Asbestos and charged filter media have relatively low removal efficiency (Gerba *et al.*, 1980).

A final important point is that neither distillation nor reverse osmosis can be used to depyroge-nate solutions with low molecular weight solutes (antibiotics, saline or sugar solutions, amino acids, *etc.*). These processes will remove the solute along with the pyrogens; UF membranes will allow free permeation of such products. This type of purification has been reported for a variety of solutions including small proteins (Henderson and Beans, 1978; McGregor *et al.*, 1978).

A comparison of the various methods of pyrogen removal is shown in Table 13. This is based on the process considerations discussed above.

Table 13 Comparison of Ultrafiltration and Other Methods of Pyrogen Removal

	Hollow fiber UF	*Distillation*	*Thermocompression distillation*	*Reverse osmosis*	*Charged filters*
Usability					
For water	Yes	Yes	Yes	Yes	Yes
For product	Yes	No	No	No	Yes
Sanitary design	Good	Good	Good	Fair	Good
Removal efficiency	>10^5	>10^5	>10^5	>10^5	10^2–10^3
Approximate operating cost, U.S. $ per 1000 l	0.50–0.80	10–15	1–2	1–2	2.50–10
Approximate capital cost (500 l/h output)	$20 000	$150 000	$350 000	$50 000	$15 000

26.4.3.3 *Cell debris and proteins*

Intracellular products of cell culture and fermentation must be free of cellular debris for purifi-cation after lysis. While centrifugation has traditionally been used for such operations, ultrafil-tration is gaining acceptance as a suitable alternative, in many cases. Due to the large size of most protein molecules, achieving satisfactory permeation through an ultrafilter is difficult at times. However, new high porosity membranes are now available which allow filtration of such solutes. Cell debris has been removed successfully from such products as interferon with yields of 80–100% (Leuthard and Schuerch, 1980).

Removal of proteins by UF has been applied to the purification of fermentation media. These broths frequently incorporate hydrolysates of soy, casein and other proteins. However, in most cases, residual proteins are present in the medium which do not improve growth. These proteins actually are contaminants which must be removed in later purification steps. Purification by UF removes these proteins thereby simplifying product recovery (Deeslie and Cheryan, 1981). This filtration step also removes bacteria from the medium prior to use (Cox, 1975). The same prin-ciple has been applied to purification of cell culture fluid. In one case, albumin was removed from a smaller MW product using hollow fiber UF (Lachman *et al.*, 1978).

26.5 SUMMARY

Ultrafiltration has been shown to be a useful process step which may be applied in a variety of ways. This versatility can perhaps be best summarized by examining a product recovery scheme in a typical biotechnology process (see Figure 24). In this case, it is assumed that the product is intracellular. As described previously, ultrafiltration may be used in the following ways: (1) purification of culture medium prior to fermentation, removing unhydrolyzed proteins; (2) cell harvesting after fermentation; (3) removal of cell debris after lysis; (4) concentration/diafiltration of product after intermediate chromatography or other purification steps; (5) concentration/diafiltration of final product; and (6) providing non-pyrogenic water for final product makeup and vial washing.

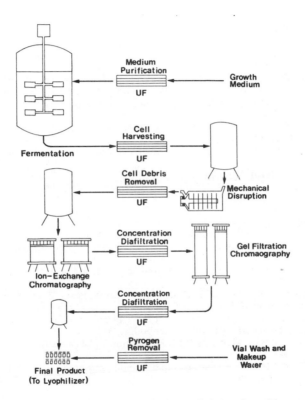

Figure 24 Typical recovery process for intracellular product of fermentation

It is clear that the advantages of ultrafiltration will have a growing impact as biotechnology moves from laboratory to full-scale production.

26.6 REFERENCES

Abbott, B. S. and P. Gerhardt (1970). Dialysis fermentation I. Enhanced production of salicylic acid from naphthalene by *Pseudomonas fluorescens*. *Biotechnol. Bioeng.*, **12**, 557–589.
Beaton, N. C. (1980). The application of ultrafiltration to fermentation products. In *Ultrafiltration Membranes and Applications*, ed. A. R. Cooper, pp. 373–404. Plenum, New York.
Blatt, W. F., A. Dravid, A. S. Michaels and L. Nelson (1970). In *Membrane Science and Technology*, ed. J. E. Flinn, pp. 47–97. Plenum, New York.
Brandt, A. and G. Bugliarello (1961). *Trans. Soc. Rheol.*, **10** (1), 229–251.
Calabresi, P., K. L. McCarthy, D. L. Dexter, F. J. Cummings and B. Rotman (1981). Monoclonal antibody production in artificial capillary cultures. *Proc. AACR ASCO*, 302.
Chang, W. T. H. and J. J. Furjanic (1981). Production of biomass with hollow-fibre ultrafiltration–fermentation system. *SIM News*, July, 24–25.
Charley, R. C., J. E. Fein, B. H. Lavers, H. G. Lawford and G. Lawford (1983). Optimization of process design for continuous ethanol production by *Zymomonas mobilis* ATCC 29191. *Amicon News* (Amicon Corp., Danvers, MA) **474**, 1–2.
Cheryan, M. (1977). Concentration of soy protein isolates. *J. Food Proc. Eng.*, **1**, 269–281.

Cohn, E. J., E. L. Strong, W. L. Hughes, D. J. Mulford, J. N. Ashworth, M. Mein and H. L. Taylor (1946). *J. Am. Chem. Soc.*, **68**, 459–475.

Cooper, A. R. and R. G. Booth (1978). Ultrafiltration of synthetic polymers. Part 1. Optimization of solvent flux during diafiltration. *Sep. Sci. Technol.*, **13**, 735–744.

Cox, J. C. (1975). New method for the large-scale preparation of diphtheria toxoid. *Appl. Microbiol.*, **29**, 464–468.

David, G. S., R. A. Riesfeld and T. H. Chino (1978). Continuous production of CEA in hollow fiber cell culture units. *J. Natl. Cancer Inst.*, **60**, 303–306.

Deeslie, W. D. and M. Cheryan (1981). A CSTR hollow fiber system for continuous hydrolysis of proteins. *Biotechnol. Bioeng.*, **23**, 2257–2271.

de Filippi, R. P. and R. L. Goldsmith (1970). In *Membrane Science and Technology*, ed. J. E. Flinn, pp. 33–46. Plenum, New York.

Ehrlich, K. C., E. Stewart and E. Klein (1978). Artificial capillary perfusion cell culture: metabolic studies. *In Vitro*, **14**, 443–448.

Fike, R. M., J. L. Glick and A. A. Burns (1977). Propagation of human lymphoid cell lines in hollow fiber capillary units. *In Vitro*, **13**, 170.

Friedman, M. R. and E. F. Gaden (1970). Growth and acid production by *L. delbrueckii* in a dialysis culture system. *Biotechnol. Bioeng.*, **12**, 961–974.

Georgiades, J. A. (1981). Production and purification of the human interferon-gamma. *Tex. Reports Biol. Med.*, **41**, 179–183.

Gerba, C. P., K. Hon, R. Babineau and J. Fiore (1980). Pyrogen control by depth filtration. *Pharm. Technol.*, **4** (6), 82–88.

Hanisch, W. H., S. Fuhrman, D. Harano and M. Pemberton (1982). Separation and purification techniques applicable to the biological processes. *Proceedings of the Australian Biotechnology Conference*, pp. 153–170.

Henderson, L. W. and E. Beans (1978). Successful production of sterile pyrogen-free electrolyte solution by ultrafiltration. *Kidney Int.*, **14**, 522–525.

Henry, J. D. (1972). Cross flow filtration. In *Recent Developments in Separation Science*, ed. N. N. Li, vol. II, pp. 205–225. CRC Press, Cleveland, OH.

Howard, G. and R. Duberstein (1980). New separation technique for the chemical process industries. *J. Parenter. Drug Assoc.*, **34** (2), 95–102.

Ingham, K. S. and T. F. Busby (1980). Methods of removing PEG from plasma fractions. In *Ultrafiltration Membranes and Applications*, ed. A. R. Cooper, pp. 141–152. Plenum, New York.

Inloes, D. S., W. J. Smith, D. P. Taylor, S. N. Cohen, A. S. Michaels and C. R. Robertson (1983). Hollow-fiber membrane bioreactors using immobilized *E. coli* for protein synthesis. *Biotechnol. Bioeng.*, **25**, 2653–2681.

Johnson, A. D., G. A. Eddy, J. D. Gangemi, H. R. Hamsburg and J. F. Metzger (1978). Production of Venezuelan equine encephalitis virus in cells grown on artificial capillaries. *Appl. Environ. Microbiol.*, **35** (2), 431–434.

Kan, J. K. and M. L. Shuler (1978). Urocanic acid production using whole cells immobilized in a hollow fiber reactor. *Biotechnol. Bioeng.*, **20**, 217–230.

Knazek, R. A., P. O. Kohler and P. M. Gullino (1974). Hormone production by cells grown *in vitro* on artificial capillaries. *Exp. Cell Res.*, **84**, 251–254.

Ku, K., M. J. Kuo, J. Delente, B. S. Wildi and J. Feder (1981). Development of a hollow-fiber system for large-scale culture of mammalian cells. *Biotechnol. Bioeng.*, **23**, 79–95.

Labuza, T. P. (1975). Cell collection: recovery and drying for SCP manufacture. In *Single-Cell Protein II*, ed. S. R. Tannenbaum and D. I. C. Wang, pp. 69–104. MIT Press, Cambridge, MA.

Lachman, L. B., J. O. Moore and R. S. Metzgar (1978). Preparation and characterization of lymphocyte-activating factor. *Cell Immunol.*, **7386**, 226–233.

Leuthard, P. and A. R. Schuerch (1980). A simple and rapid method for concentration of interferon and removal of concentrated inducing virus. *Experientia*, **36**, 1447.

Leveque, M. A. (1928). *Ann. Mines*, **13**, 1–12.

Lukaszewicz, R. C., A. Kurin, D. Hauk and S. Chrai (1981). Functionality and economics of tangential flow microfiltration. *J. Parenter. Sci. Technol.*, **35**, 231–236.

McGregor, W. C., H. L. Newmark and A. H. Ramel (1978). Stabilized thymosin composition and method. *US Pat.* 4 082 737.

Mattiason, B. and M. Ramstorp (1981). Comparison of the performance of a hollow-fiber microbe reactor with a reactor containing alginate entrapped cells. *Biotechnol. Lett.*, **3**, 561–566.

Mercer, J. E. (1977). Alternative methods of solvent removal. *Proceedings of the International Workshop on Technology for Protein Separation of Blood Plasma Fractionation*, pp. 160–174. US Department of H. E. W, Washington, DC.

Michaels, A. S. (1968). *Chem. Eng. Prog.*, **64**, 131–142.

Morrow, A. W., C. J. Whittle and W. A. Eales (1974). A comparison of methods for the concentration of foot-and-mouth disease virus. *Bull. Off. Int. Epiz.*, **81**, 1155–1167.

Nelsen, L. (1978). Removal of pyrogens from parenteral solution by ultrafiltration. *Pharm. Technol.*, **2** (5), 46–49.

Ng, P. K., J. L. Lundblad and G. Mitra (1976). *Sep. Sci. Technol.*, **11**, 499–502.

Ng, P. K., G. Mitra and J. L. Lundblad (1978). Simultaneous salt and ethanol removal from human serum albumin. *J. Pharm. Sci.*, **67** (3), 431–433.

Porter, M. C. (1972). Concentration polarization with membrane ultrafiltration. *Ind. Eng. Chem . Prod. Res. Dev.*, **11**, 234–248.

Ratner, P. L., M. L. Cleary and E. James (1978). Production of rapid harvest moloney murine leukemia virus by continuous cell culture on synthetic capillaries. *J. Virol.*, **26**, 536–539.

Renshaw, T. A., S. F. Sapakie and M. C. Hanson (1982). Concentration economics in the food industry. *Chem. Eng. Prog.*, **5**, 33–40.

Reti, A. R., T. Leahy and P. M. Meier (1979). The retention mechanism of sterilizing and other submicron high efficiency filter structures. *Proceedings of the 2nd World Filtration Congress*, pp. 427–435.

Rutzky, L. P., J. T. Tomita, M. A. Calenoff and B. D. Kahan (1979). Human colon adenocarcinoma cells III. *In vitro* organoid expression and carcino-embryonic antigen kinetics in hollow fiber culture. *J. Natl. Cancer Inst.*, **63**, 893–898.

Segre, G. and A. Silberberg (1961). *Nature (London)*, **189**, 209–210.

Strand, J. M., J. M. Quarles and S. H. Black (1982). Human fibroblast interferon production in a matrix perfusion micro-carrier bead system. *Proceedings of the ASM 82nd Annual Meeting*, Paper no. T20.

Thornton, H. (1976). Concentration of herpes simplex viruses. *Virus Concentration*, Publication no. 463:3. Amicon Corp., Danvers, MA.

Trudel, M. and P. Payment (1980). Concentration and purification of rubella virus hemaglutinin by hollow fiber ultrafil-tration and sucrose density centrifugation. *Can. J. Microbiol.*, **26**, 1334–1339.

Tutunjian, R. S. (1982). Pyrogen removal by ultrafiltration. In *Endotoxins, and their Detection with the Limulus Amebo-cyte Lysate Test*, ed. S. W. Watson, J. Levin and T. J. Novitsky, pp. 319–328. Alan R. Liss, New York.

van Wezel, A. L. (1975). Concentration of biological products in vaccine production by hollow fiber ultrafiltration. *Amicon Dialog* (Amicon Corp., Danvers, MA) **7** (1), 1–2.

Wang, D. I. C., C. L. Cooney, A. L. Demain, P. Dunnill, A. E. Humphrey and M. D. Lilly (1979). Enzyme isolation. In *Fermentation and Enzyme Technology*, ed. C. R. Heden, pp. 238–310. Wiley, New York.

Webster, R. (1976). Concentration of swine influenza virus. *Virus Concentration*, Publication no. 463:3. Amicon Corp., Danvers, MA.

Weiss, S. A. (1980). Concentration of baboon endogeneous virus in large-scale production by use of hollow fiber ultrafil-tration technology. *Biotechnol. Bioeng.*, **22**, 19–31.

Wiemann, M. C., E. D. Ball, M. W. Fanger, D. L. Dexter, O. R. McIntyre, G. Bernier, Jr. and P. Calabresi (1983). Human and murine hybridomas: growth and monoclonal antibody production in the artificial capillary culture system. *Clin. Res.*, **31**, 511.

Zoon, K. C., M. E. Smith, P. J. Bridgen, D. Z. Nedden, D. M. Miller and C. B. Anfinsen (1980). Human lymphoblas-toid interferon: purification, amino acid composition, and amino-terminal sequence. *Ann. N.Y. Acad. Sci.*, 390–398.

27

Liquid–Liquid Extraction of Antibiotics

T. A. HATTON
Massachusetts Institute of Technology, Cambridge, MA, USA

27.1 INTRODUCTION

An important step in the production of antibiotics is the recovery and concentration of the product from the fermentation broth in which it is produced. The physicochemical properties which distinguish the desired solute from other fermentation products must be exploited in the selection of an appropriate separation technique for this step, and physical sorption, ion-exchange and liquid–liquid extraction have all been employed successfully in the isolation of different antibiotics.

Impetus for the large-scale use of solvent extraction for antibiotics recovery was provided by the critical need for increased supplies of penicillin during the Second World War. Since that time, refinements in extractor design and improvements in fermentation technology have yielded considerable increases both in equipment throughput capacities and in overall product yields, although the basic technology itself has changed but little over the years. It is the purpose of this chapter to outline the general features of antibiotic purification *via* solvent extraction techniques as they are practised today. The principles of solvent extraction are discussed briefly in Section 27.2, and this is followed by a discourse on solvent and equipment selection in Sections 27.3 and 27.4, respectively. In Section 27.5, common extraction processes, as typified by the extraction of penicillin from its fermentation broth, are discussed, and Section 27.6 concludes with a summary of the major points brought out in the chapter.

27.2 PRINCIPLES OF LIQUID EXTRACTION

The successful implementation of liquid extraction in the separation and concentration of antibiotics depends both on thermodynamic equilibrium constraints and on process considerations such as the relative flow rates and the contacting patterns of the two phases. A discussion of the factors controlling the effectiveness of an extraction process is given in this section.

27.2.1 The Partition Coefficient

When antibiotics in solution are brought into contact with a second, immiscible phase, they tend to distribute themselves between the two phases. It is this phenomenon that is exploited in liquid–liquid extraction operations, since it provides a method whereby the desired solute can be removed selectively from a solution in which there are other, undesirable impurities present, thereby allowing some purification of the solute to be achieved. In addition the solute accumulated in the extractant is usually in a more concentrated form than in the original feed solution, and this facilitates further downstream processing.

The degree to which solute redistribution occurs is determined by thermodynamic considerations, and depends on the molecular interactions between the solute and the individual solvents. It it usually quantified in terms of the partition or distribution coefficient:

$$m = C_{solvent}/C_{feed} \tag{1}$$

where $C_{solvent}$ and C_{feed} are the antibiotic concentrations at equilibrium in the solvent and feed phases, respectively. In general, the partition coefficient is a function of the actual concentration levels in the system, although frequently the variations with concentration can be neglected to facilitate extraction calculations.

The partitioning behaviour of antibiotics can be influenced strongly by solution properties (*e.g.* pH, ionic strength, *etc.*) and temperature, and can also be modified through the introduction of additives which interact with either the solute or the solvent to cause effective changes in the solute distribution between the phases. An example is given in Figure 1, which depicts the effect of pH on the distribution of penicillins and impurities between an organic solvent and an aqueous solution (Queener and Swartz, 1979). Clearly, low aqueous phase pH values favour selective partitioning of the penicillins to the organic solvent phase, while for pH values greater than about 6.0, the distribution of the penicillins will be biased toward the aqueous phase. This is because penicillins are strong organic acids. Erythromycin, on the other hand, is an organic base, and is extractable with organic solvents only under basic, rather than acidic, conditions. Such control of the partitioning behaviour by pH adjustment is readily exploited in the extraction and subsequent recovery and concentration of these antibiotics, as is emphasized in Section 27.5

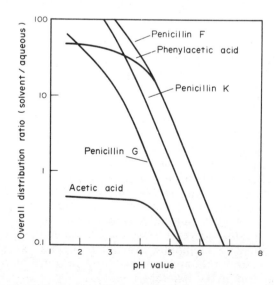

Figure 1 The effect of pH on the distribution of solutes between the solvent and aqueous phases (Queener and Swartz, 1979)

Other antibiotics, such as the streptomycins, are highly polar solutes and under normal circumstances do not show appreciable partitioning toward organic solvents. However, the addition of long-chain fatty acids to the solution can greatly enhance their organic phase solubility because they associate strongly with the solute to reduce its polarity and increase its hydrophobicity. Sulfonic acids, other detergents and trichloroacetic acid are also known to exhibit similar effects on the solubility of polar solutes, probably by ion pairing (Craig and Sogn, 1975).

For non-polar solutes, it is frequently necessary to influence the partition ratio in the reverse manner, *i.e.* to increase the aqueous phase solubility of the solute to enable effective extraction from the organic solvent, and this can be achieved through additives which change the water structure. For this purpose urea, formamide, or salts which have a so-called 'salting in' effect, can be used (Craig and Sogn, 1975).

27.2.2 The Extraction Factor and Degree of Separation

While the partition coefficient can be used to determine the relative concentrations of an antibiotic in two immiscible phases at equilibrium, it does not in itself determine the actual degree of separation of the solute; this is governed also by the relative volumes of the two phases. Clearly, as the solvent volume is increased, a greater proportion of the original solute will appear in the solvent, and the aqueous phase will become more depleted of the solute. The degree of separation can then be defined as:

$$G = (m\ V_{\text{solvent}})/V_{\text{feed}} \tag{2}$$

which, in effect, gives the relative capacities of the two phases for the solute under equilibrium conditions. Here, V_{solvent} and V_{feed} are the solvent and aqueous feed volumes, respectively. From this relation it is evident that the higher m is, the smaller will be the solvent volume required to obtain a desired degree of separation, and, more importantly, the greater will be the antibiotic concentration in the extractant. This concentrating effect is, of course, desirable if a pure material is to be the final product. Frequently, if insufficient contact time between the phases is permitted, equilibrium will not be attained and the actual degree of separation will therefore be less than that given by equation (2).

In continuous operations, the two phases are usually introduced countercurrently to one another in a multistaged extraction device and it is the rates at which these phases are fed to the system rather than the relative volumetric hold-ups within the device that are important in characterizing the extraction conditions. It is therefore customary to define the extraction factor by:

$$E = (m\ v_{\text{solvent}})/v_{\text{feed}} \tag{3}$$

which is again a measure of the relative capacities of the two phases for the solute under the given operating conditions. The solvent and aqueous flow rates are v_{solvent} and v_{feed}, respectively, and for dilute systems it is often acceptable to assume that neither m nor the flow ratio $v_{\text{solvent}}/v_{\text{feed}}$ changes significantly over the extraction device. As is shown in the next section, the parameter E is important in determining the overall extraction rates in a multistaged extraction operation.

27.2.3 Extraction Calculations

In the design of a solvent extraction unit it is necessary to determine either the number of stages required to effect a given solute recovery under specified feed conditions or, given the number of stages available, one must determine the solvent feed rates necessary to attain the desired separation. If it can be assumed that neither the distribution ratio m nor the feed ratio $v_{\text{solvent}}/v_{\text{feed}}$ changes in going from one stage to the next, then the change in concentration over an extraction unit can be predicted from the simple relation (Treybal, 1980).

$$\frac{X_{\text{F}} - X_{\text{N}}}{X_{\text{F}} - X_{\text{N}}^*} = \frac{E(E^N - 1)}{E^{N+1} - 1} \tag{4}$$

Here, N is the number of theoretical (equilibrium) stages required to perform the desired separation, and X_{F} and X_{N} are the feed and effluent solute mole fractions, respectively. The quantity X_{N}^* is that effluent concentration that would be in equilibrium with the fresh solvent feed. Equation (4) is shown plotted in Figure 2, where it is strongly evident that the extraction factor plays an important role in determining the overall performance of the extraction unit.

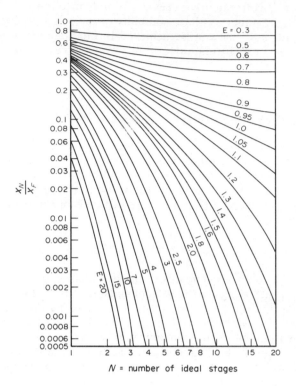

Figure 2   Number of ideal stages for extraction operations with constant extraction factor

Since equilibrium between the phases is seldom attained within any given stage, because of mass transfer limitations, incomplete contacting of the phases and backmixing, the actual number of stages required is generally larger than N. It is difficult to estimate precisely the effects of the factors contributing to non-ideal mass transfer behaviour under extraction conditions, and it is necessary to resort to the use of overall and/or stage efficiencies to correct for these non-idealities.

The overall efficiency is defined by

$$\eta_o = N_{theoretical}/N_{actual} \tag{5}$$

and must generally be determined empirically. With correlations for η_o on hand, the design of an extraction unit will involve determination of the theoretical number of stages required for a given separation, assuming ideal equilibrium conditions in each stage, and then correction of this value using equation (5) to determine the actual number of stages needed. Unfortunately, data on overall efficiencies are scarce in the literature, and particularly for new applications, it is probably necessary to obtain the required information from pilot plant runs, preferably in collaboration with equipment manufacturers.

Stage efficiencies can also be used in the design of extraction units. These are defined in terms of the fractional approach to equilibrium at any given stage, and can be related in a reasonably fundamental way to the conditions existing within the stage. However, this degree of refinement in the calculational procedure is usually not warranted. Further details are given in the standard texts of Treybal (1980) and King (1981). Laddha and Degaleesan (1976) also mention the utility of overall and stage efficiencies in comparing the performance of different contactor types for any specific extraction duty.

Under conditions where the distribution coefficient or the flow ratio varies in going from one stage to another, equation (4) will no longer be valid and it is necessary to perform either numerical or graphical calculations to determine the number of stages required for a given extraction problem. These procedures are discussed adequately by Treybal (1980), King (1981) and Laddha and Degaleesan (1976).

27.3 SOLVENT SELECTION

The successful implementation of an antibiotic liquid extraction process depends, of course, on the solvent selected for recovery of the solute from its fermentation broth. Treybal (1980) and King (1981) have documented several criteria that should be considered in the selection of solvents for extraction processes in general, and these are summarized in Table 1. Often, this selection represents a compromise between the desired physicochemical properties for the system and economic considerations. Fortunately, in the antibiotics industry these trade-offs are minimal, and the solvents in general use fare well in most, if not all, of the categories listed in Table 1.

Table 1 Criteria for Solvent Section

1. Cost and availability
2. Selectivity for the product
3. Immiscibility with the aqueous solution
4. Density differential
5. Physical properties such as viscosity and boiling point
6. Toxicity and corrosiveness
7. Ease of recovery
8. Sensitivity to the extracted product

Popular solvents employed in antibiotic extractions are alcohols (*e.g. n*-butanol), ketones (especially methyl isobutyl ketone) and acetates, such as pentyl or butyl acetate. These solvents are all inexpensive and readily available, are immiscible with aqueous solutions, and exhibit the desirable physical properties of low viscosity and densities significantly different from that of water. Selectivity of the solvent for the product can usually be influenced by adjustment of the broth pH, or by the introduction of other additives to the aqueous phase. The formation of emulsions is sometimes a problem, particularly in the presence of solids, but this is related more to the condition of the feed broth than to any undesirable features of the solvent itself.

For laboratory scale fractionation of antibiotics other solvents or solvent mixtures can be used. Craig and Sogn (1975) have prepared an extensive compilation of such solvents, and the antibiotics which can be extracted using them in countercurrent distribution runs.

27.4 EXTRACTION EQUIPMENT SELECTION

Over the years, a number of different extractor configurations have been developed to meet a variety of process needs. These include spray and packed towers, pulsed columns, mechanically agitated columns of the rotating disc and Oldshue–Rushton types, horizontal and centrifugal contactors, and mixer–settler units. Hanson (1968) has discussed the various factors that should be considered in selecting extraction equipment for a given task, and has summarized his analysis in the form of an easily used selection guide. An excellent discussion on equipment selection has been given recently by Pratt and Hanson (1983).

In the antibiotics industry, the contactor of choice is usually of the centrifugal type, of which there are a number of different configurations available (Todd and Podbielniak, 1965; Perry and Chilton, 1973; Hafez, 1983). The reasons for this selection are severalfold. Antibiotics are generally unstable to heat, to wide pH ranges and to enzyme reaction, and are often decomposed in solution. Penicillin extraction, for example, must be undertaken under acidic conditions, and prolonged exposure to low pH environments causes degradation and loss of potency of the product. Consequently, upon acidulation of the fermentation broth, extraction of the penicillin by the solvent should be as rapid as possible and short phase contact times are required. These can be realized in centrifugal extractors, which are characterized by extremely low liquid hold-ups per stage. Other problems can also arise due to the presence of solids in the aqueous feed and the nature of the broth itself, which can lead to easy emulsion formation and thus to problems in separating the two phases after contacting. The development of high centrifugal forces in centrifugal contactors is important in overcoming these otherwise intractable difficulties and again points to the logical selection of these devices for antibiotic extraction separations.

A description of the more common centrifugal extractors is given below. Because of the predominance of the Podbielniak extractor in the antibiotic industries, it is pertinent to describe its operation in more detail than is done for other contactor types.

27.4.1 The Podbielniak Extractor

This extraction unit is illustrated in Figure 3. It consists of a horizontally rotating drum mounted on a rigid shaft through which the solvent and feed streams enter, and the effluent extract and raffinate streams leave. The drum itself is comprised of a number of perforated concentric shells which behave much in the same way as do the trays in a conventional sieve tray column. The heavier liquid is introduced near the shaft, and is flung outward by centrifugal force to displace the lighter fluid fed to the unit near the periphery, causing it to move inward. The net effect is for the two liquids to pass countercurrently to one another through a series of mixing zones, or stages. The driving force for this motion is due to centrifugal action, rather than to gravity, and in this respect the Podbielniak contactor has an added flexibility over conventional gravity-fed columns; the mixing and settling forces can be varied over wide ranges by simple adjustment of the rotor speed. For fermentation broth extractions, capacities of up to 260 gal (US) min^{-1} are attainable.

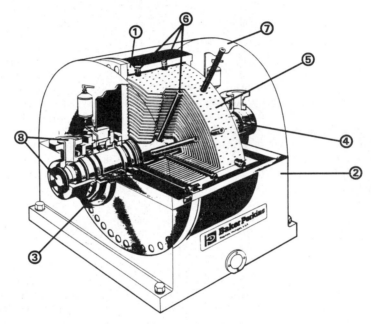

Figure 3 The Podbielniak contactor: (1) rotor and shaft assembly, (2) base, (3) bearings, (4) multi-V-belt drive, (5) concentrically assembled perforated elements, (6) liquid distribution tubes, (7) 'Asco' tubes to permit access for cleaning and flushing, (8) inlet and outlet ports (Courtesy of Baker–Perkins)

Operation of the Podbielniak contactor is reasonably straightforward and relies primarily on the regulation of the feed pressures, and of the light phase effluent backpressure (Todd and Podbielniak, 1965; Todd, 1972; Hafez, 1983). The main consideration here is which is to be the dispersed phase and which phase is to be continuous. The line of demarcation separating the regions where the two phases are respectively continuous is called the principal interface and is illustrated schematically in Figure 4. Increases in the light liquid backpressure cause this interface to move outwards and to fill the rotor with a continuous phase of light liquid, while the reverse happens when the back pressure is reduced. Todd (1972) describes a straightforward procedure to ascertain under which conditions the contactor is operating and the interested reader is referred to this work for the appropriate details on how to control the interface position.

Increases in extraction efficiency of up to 40% can result with proper positioning of the interface. If it is not immediately evident from the process which phase is to be continuous, it is generally recommended that the major phase flow be dispersed in the minor one and that the principal interface be maintained at the radial entry point of the minor phase flow (Todd, 1972).

Other operational characteristics of importance to the operation of the Podbielniak extractor include the rotor speed, the flow ratio and the total throughput, all of which should be adjusted to optimize a specific extraction process. Increases in rotor speed tend to provide for easier emulsion breakup because of the higher centrifugal forces developed in the unit, but they also reduce the effective contact times between the two phases, with a consequent reduction in the effective

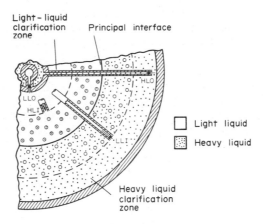

Light–liquid
clarification
zone

Principal interface

HLO

LLO

HLI

LLI

☐ Light liquid

▨ Heavy liquid

Heavy liquid
clarification
zone

Figure 4 Distribution of the phases in a Podbielniak contactor

number of theoretical stages available for the extraction. Clearly, these two effects should be balanced to obtain optimum performance. The selection of the flow ratio is determined in part by the magnitude of the distribution coefficient and the required extraction factor, as discussed in Section 27.2. However, too high or too low a ratio can lead to emulsion formation and once again a compromise must be struck between two opposing effects. In a like manner, optimization of the total throughput must be based on consideration of a number of different factors. For instance, a high throughput ensures that turbulent conditions will be set up in the extractor, thus preventing solids build-up, creating a better dispersion and increasing mass transfer rates. At the same time, however, phase contact times will be decreased and there will be an enhanced risk for the formation of unwanted emulsions. Again, compromises must be sought if the unit is to perform optimally.

There are no fundamental relationships on which to base the choice of operating conditions. Optimization is essentially a trial and error process, relying on the trying out of various combinations of flow rates, rotor speed and backpressure to determine the optimal operating conditions. Empirical correlations, based on pilot plant data, can be used to assist in this optimization process. As mentioned in Section 27.2, a convenient quantitative measure of the extraction efficiency is given by the number of theoretical stages, which can be estimated using experimental results together with Figure 2. Under whole-broth extraction conditions, values of N ranging betwen 0.4 and 2.4 have been reported (Anderson and Lau, 1955).

27.4.2 The Alfa-Laval Extractor

This extractor is used mainly in Western Europe and differs from the Podbielniak in that it is mounted vertically. The light phase is introduced near the periphery of the rotating bowl, while the heavy phase is fed near the centre. The bowl itself consists of concentric channels containing spirally wound baffles to direct the flow paths of the two phases. Within any channel, a distinct interface is established, as the two phases move as layers in a spiral countercurrently to one another. Movement of the phases from one channel to the next occurs at orifices located alternately at the top and bottom of the channels, where intense mixing of the phases occurs as the lighter fluid moves inwards and the heavier liquid moves outwards. The heavy liquid leaves the extractor at the top by way of the outermost channel, while the light fluid exits through the central channel in the shaft.

27.4.3 Other Centrifugal Extractors

Many other centrifugal extractors have been developed over the years, although not all have found widespread commercial application. Hafez (1983) gives a good overview of the different types available, which fall in three distinct categories. The Quadronic extractor, a modified version of the Podbielniak, is classified as a differential contact unit, and is joined in this classification by the Podbielniak and Alfa-Laval extractors discussed above. Multistage contactors in common use are the Luwesta extractor and the Robatel SGN unit, while the Robatel BXP and the Westfalia TA extractors are representative of the single stage devices that are commercially available. These extractors have not been used to any great extent in the antibiotics industry and need not be considered in more detail here.

27.5 PROCESS CONSIDERATIONS

In this section, details are presented of typical processes for penicillin recovery from fermentation broths using liquid extraction technology. The principles discussed here will also carry over to the extraction of other antibiotics, although the details will differ to some extent when going from one fermentation product to another. Brief descriptions of two other representative antibiotic extractions are given at the end of this section.

27.5.1 Broth Pretreatment

A typical broth contains about 20–35 g penicillin l^{-1}, while the solids content can approach one-third of the total feed. In conventional extraction operations, the solids are removed from the broth by filtration before being introduced to the extraction sequence. This controls the rheological properties of the aqueous phase and reduces the tendency for emulsion formation within the extractor. However, the filtration step can be bypassed, and whole-broth extractions are frequently carried out successfully in the pharmaceutical industries.

Even with filtered broth, additional proteinaceous solids may be formed during the broth pretreatment and these may build up in the extraction unit over time leading to a gradual impairment in extraction efficiency. Moreover, these solids may stabilize broth–solvent emulsions formed in the extractor, and cause unacceptable loss of solvent with the spent broth phase. For this reason, small traces of wetting agents and de-emulsifiers are added to the filtered broth. These are effective in reducing solids accumulation rates in the rotor, in ensuring that the rich extract is more readily clarified, and in decreasing the problems associated with emulsion formation. It should be emphasized, though, that addition of too much treating agent can aggravate the emulsion problem rather than ameliorate it (Todd and Davies, 1973), and care should be exercised in formulating the recipe for the treating agent.

It is customary also to cool the fermentation broth to 0–4 °C to minimize chemical and enzyme degradation during the solvent extraction process.

27.5.2 Penicillin Extraction

In conventional penicillin extraction processes, the neutral broth is acidulated either with phosphoric acid or, more commonly, with sulfuric acid to lower the pH to about 2. This converts the penicillin to the acid form, which is soluble in several organic solvents; commonly used solvents are pentyl acetate, butyl acetate and methyl isobutyl ketone. Todd and Davies (1973) point out that this acidity is a compromise between better extraction distribution coefficients at lower pH values (see Figure 1) and the degradation and loss of potency of the penicillin which is inevitably caused by the lowering of the pH.

The acidulated broth is immediately contacted with the solvent in a multistage centrifugal extractor, as illustrated schematically in Figure 5. Typically, solvent-to-broth ratios of 0.1 to 0.2 are employed in this step, and better than 90% recovery is achieved. The loaded solvent phase leaving the extractor is usually crystal clear, although there may be some solvent entrainment, ostensibly due to the presence of proteinaceous solids, in the spent broth.

A second extraction step follows wherein the rich solvent phase is now contacted countercurrently with an aqueous buffer or alkali solution (potassium or sodium hydroxide) of pH approximately 7.5. Under these conditions the distribution of the penicillin favours the aqueous phase (Figure 1) and the solute is very effectively stripped from the solvent. Since solvent to aqueous ratios of about 5 to 10 are typical in this extraction step, a further concentration effect is accomplished. Emulsion formation in this stage is not a problem.

The concentrated penicillin can be precipitated from the resulting aqueous phase. However, in many plants facilities exist for final purification by crystallization from the solvent phase, and in

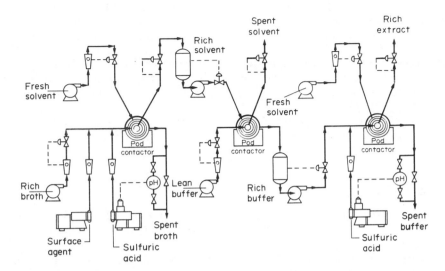

Figure 5 A multiple solvent extraction process for antibiotics recovery (Courtesy of Baker–Perkins)

these cases a third stage extraction is performed. The concentrated aqueous solution is once again acidified and the penicillin re-extracted with fresh solvent to obtain a further enhancement in the penicillin concentration. Various procedures are used to crystallize the penicillin and the resulting salt is washed and dried before being subjected to rigorous, government mandated tests. The spent solvents are recovered for re-use.

Process modifications have been implemented to improve the penicillin extraction efficiency or to eliminate some of the pretreatment steps in the product recovery sequence (Todd and Davies, 1973). One source of antibiotic loss can be attributed to the stable emulsion which is often formed within the extractor and which exits with the spent broth. As much as 5–10% of the penicillin tied in with this emulsion is not recovered during the extraction process, and is therefore lost. A method for overcoming this problem is to connect two centrifugal contactors in series, operated in the countercurrent mode as shown in Figure 6. Overall product recovery can be increased to about 98–99%, and it is frequently possible to reduce the solvent requirements significantly. This configuration has the advantage that the difficulties associated with emulsion formation, and loss of efficiency due to solids build-up in the contactor, are confined primarily to the feed stage. The second stage, which is not plagued by these problems, is responsible for maintaining the high extraction efficiency.

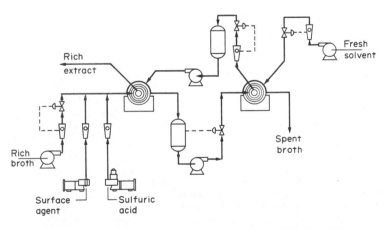

Figure 6 Centrifugal contactors in series (Todd and Davies, 1973)

Whole-broth processing, which circumvents the filtration pretreatment step, has also been used with some success in industrial antibiotic recovery operations. As pointed out by Todd and Davies (1973), the broth particulate material can both adsorb and occlude the desired product.

Under conventional extraction processes, where prefiltration is standard procedure, the product bound to the filtered solids would not be available for recovery by extraction. Provision for direct contact between the suspended solids and the solvent will make possible the recovery of this otherwise lost product. With whole-broth processing, series extraction is usually justified, and the feed extractor is generally larger and of more open design than normal. A disadvantage of whole-broth processing is that more frequent cleaning of the extractor is required than in conventional processing owing to a more rapid silting up of the device because of the higher concentration of suspended solids.

In most operations, be they conventional or whole-broth, the spent-broth and spent-solvent streams are processed for solvent recovery, usually by steam-stripping.

27.5.3 Examples of Other Antibiotic Extractions

Penicillin is only one of the many antibiotics that can be recovered from a fermentation broth using liquid extraction. Other representative systems amenable to separation using suitable solvents are discussed here.

Erythromycin is an organic base, and can be isolated by solvent extraction under basic conditions, using organic solvents such as pentyl acetate, rather than the acidic conditions necessary for penicillin extraction. The primary extraction is conducted at a pH of about 9.4 and the back-extraction into water occurs under slightly acidic conditions, corresponding to a pH of about 5.

Several procedures are available for the recovery of tetracyclines (Porter, 1976), some of which rely on solvent extraction techniques. In one process, precipitates formed with high molecular weight quaternary ammonium compounds are subsequently dissolved in solvents such as methyl isobutyl ketone or butanol, and are then extracted from the organic phase with aqueous acid. Alternatively, the filtrate can be extracted directly into one of these solvents at pH 8.5, following which the solution is concentrated and the antibiotic precipitated as the free base upon addition of Skellysolve C.

Craig and Sogn (1975) have compiled a list of available systems that have been of practical use in the isolation of individual solutes from small samples of mixtures.

27.6 CONCLUSIONS

Liquid–liquid extraction is an effecive and economical method for the recovery and concentration of a number of important antibiotics either from whole fermentation broths or from clean, filtered supernatant. These separations are usually carried out in centrifugal contactors, which offer special advantages over other contactor types in that they are able to handle both highly emulsified systems and liquids with small density differences. In addition, the short phase residence times characteristic of these contactors minimize the risk of solute degradation by heat, hydrolysis or enzyme reaction during the extraction operation.

27.7 REFERENCES

Anderson, D. W. and E. F. Lau (1955). Commercial extraction of unfiltered fermentation broths in the Podbielniak contactor. *Chem. Eng. Prog.*, **51**, 507–512.
Craig, L. C. and J. Sogn (1975). Isolation of antibiotics by countercurrent distribution. In *Methods in Enzymology*, ed. J. H. Hash, vol. XLIII, pp. 320–346. Academic, New York.
Hafez, M. (1983). Centrifugal extractors. In *Handbook of Solvent Extraction*, ed. T. C. Lo, M. H. I. Baird and C. Hanson, pp. 459–474. Wiley, New York.
Hanson, C. (1968). *Chem. Eng.*, August 26, 76.
King, C. J. (1980). *Separation Processes*, 2nd edn. McGraw-Hill, New York.
Laddha, G. S. and T. E. Degaleesan (1976). *Transport Phenomena in Liquid Extraction*. Tata McGraw-Hill, New Delhi.
Perry, R. H. and C. H. Chilton (eds.) (1973). *Chemical Engineers' Handbook*, 5th edn. McGraw-Hill, New York.
Porter, J. N. (1976). Antibiotics. In *Industrial Microbiology*, ed. B. M. Miller and W. Litsky, pp. 60–78. McGraw-Hill, New York.
Pratt, H. R. C. and C. Hanson (1983). Selection, pilot testing, and scale-up of commercial extractors. In *Handbook of Solvent Extraction*, ed. T. C. Lo, M. H. I. Baird and C. Hanson, pp. 475–495. Wiley, New York.
Queener, S. and R. Swartz (1979). Penicillins: biosynthetic and semisynthetic. In *Economic Microbiology*, ed. A. H. Rose, vol. 3, pp. 35–122. Academic, New York.
Todd, D. B. (1972). Improving performance of centrifugal extractors. *Chem. Eng.*, July 24, 152–158.

Todd, D. B. and G. R. Davies (1973). Centrifugal pharmaceutical extractions. *Filt. Sep.* **1**, 663–666.
Todd, D. B. and W. J. Podbielniak (1965). Advances in centrifugal extraction. *Chem. Eng. Prog.*, **61** (5), 69–73.
Treybal, R. E. (1980). *Mass-transfer Operations*, 3rd edn. McGraw-Hill, New York.

28

Liquid–Liquid Extraction of Biopolymers

M.-R. KULA
Gesellschaft für Biotechnologische Forschung mbH, Braunschweig-Stöckheim, Federal Republic of Germany

28.1 INTRODUCTION

Liquid–liquid extraction is a well known and extensively used unit operation in the chemical industry (Treybal, 1963; Hanson, 1971). It is considered to have special advantages compared to other separation techniques in cases where labile substances have to be handled, when distillation is impossible for material properties or economic reasons, when large volumes have to be treated and short reaction times in continuous processes are required. In recovery processes for natural products liquid–liquid extraction is employed, *e.g.* for separation of antibiotics like penicillin or erythromycin, from filtered or whole broth (Brunner *et al.*, 1981).

However, liquid–liquid extraction has until very recently not been considered for industrial recovery processes of biopolymers like proteins or nucleic acids. This neglect is somewhat surprising looking at the reasons cited above for the advantages of the technique, but on the other hand there are only very few proteins soluble in organic solvents commonly employed in the chemical industry, *e.g.* butanol. Other solvents like phenol lead to extensive denaturation of proteins, a fact commonly exploited for the purification of nucleic acids which has been performed up to pilot plant scale (Hancher *et al.*, 1969). To extract biologically active proteins obviously other ways and means to establish two immiscible liquid phases have to be considered. The properties of such phases must meet all the requirements necessary to maintain proteins in their natural active state in solution. It has been known for a long time that phase separation can be brought about by addition of hydrophilic polymers to aqueous solutions (Beijerinck, 1896). Each of the phases formed this way contains a high proportion of water, forming a favorable environment for biologically active proteins and even cell organelles as has been shown by Albertsson (1971, 1977, 1978, 1979, 1982). Aqueous two-phase systems can be exploited for the extraction of enzymes and other proteins from cell homogenates in large scale improving considerably down stream processing of intracellular products (Kula *et al.*, 1976, 1979). By additional single or multiple stage extractions further purification of the desired protein can be achieved. Several processes for the isolation of enzymes have been developed and are operated on pilot plant (Kula *et al.*, 1982; Kroner *et al.*, 1982) and industrial scale.

Besides application in recovery operations aqueous phase systems are currently investigated for analytical purposes, separation of nucleic acids and phase transfer catalysis employing enzymes and/or microbes. The following description will be centered on biochemical and technological aspects of protein extraction in aqueous phase systems. Other applications of such systems will be briefly reviewed.

28.2 BIOCHEMICAL ASPECTS OF PROTEIN EXTRACTION

28.2.1 General Considerations

Partition of a protein between two immiscible liquids is described by:

$$K = C_u / C_l \tag{1}$$

where K is the partition coefficient and C_u and C_l the concentration of the protein at equilibrium in the upper and lower phase respectively. The partition coefficient is a constant and independent of the protein concentration as long as molecular properties are identical in both phases. Surface properties of the protein or particle have a steering effect on partition. Surface energy and charge are the main forces contributing to the partition coefficient. Gerson (1980) has recently shown that for partition of cells in aqueous phase systems the following relations apply:

$$- \log K = \alpha \Delta \gamma + \delta \Delta \psi + \beta \tag{2}$$

where $\Delta \gamma$ is the difference in surface free energy in the two phases, $\Delta \psi$ the difference in the electrical potential and α, δ and β are constants which contain parameters difficult to quantify such as actual surface area, surface charge and standard chemical potential. According to equation (2) an exponential relation exists between the partition coefficient and $\Delta \gamma$ or $\Delta \psi$ respectively, or a linear combination of both. Since surface area is a function of the molecular weight it can be expected that large macromolecules, cell organelles or cells will show one-sided partition and minute changes in $\Delta \gamma$ produce large effects on K. Most enzymes with molecular weights in the range between 20 000 and 400 000 exhibit partition coefficients between 0.01 and 100, quite suitable for extractions. The degree of separation G is determined by:

$$G = K \frac{V_u}{V_l} \tag{3}$$

where V_u and V_l denote the volume of the upper and lower phase respectively.

Albertsson discusses the following increments to contribute to the experimentally observed partition coefficient of a protein

$$\ln K = \ln K_{el} + \ln K_{hphob} + \ln K_{hphil} + \ln K_{conf} + \ln K_{ligand} \tag{4}$$

where the subscript el, hphob, hphil, conf, ligand refer to electric (charge), hydrophobic or hydrophilic surface properties, conformation and ligand interactions of the protein respectively. Changes in the environment, *e.g.* ionic composition and pH, act in a complex manner on the carrier phase system as well as on the protein mainly influencing the first three increments in equation (4). The interdependence of many parameters makes quantitative correlation of partition coefficients with independently measurable, molecular properties of a given protein presently impossible. Experimentally a number of variables can be manipulated to achieve the desired extraction or separation, which are discussed below.

28.2.2 Contribution of the Polymers on the Partition of Proteins in an Aqueous Phase System

Most hydrophilic polymers show incompatibility in mixtures and form two liquid phases under suitable conditions. At low concentrations of polymers homogeneous solutions are obtained, but at discrete concentration ratios, measurable by, for example, the cloud point method, phase separation occurs. The line connecting these points in a phase diagram is called the binodal. The concentrations sufficient to split the solution into two immiscible phases will vary with the molecular weight of the polymers involved, decreasing with increasing molecular weight. Table 1 lists a number of two-phase systems according to Albertsson (1971). Other available natural and/or synthetic polymers, *e.g.* polysaccharides other than dextran and polyacrylamides, have not yet been systematically studied for partition of biopolymers. Biotechnological applications have been restricted to systems composed of polyethylene glycol and dextran or polyethylene glycol and salts, mainly for legal but also for technical reasons. Both polymers are extensively studied, the toxicology is known and they are included into the pharmacopoeias of most countries. Only a limited number of phase diagrams of aqueous phase systems have been published, Tables 2 and 3 list the polyethylene glycol/dextran and polyethylene glycol/salt systems, analyzed so far. Published data cover only the ternary systems, polymer A/polymer B/water or polymer/salt/water. In Figure 1 the phase diagram of a polyethylene glycol 1550/phosphate system is given. At concentration ratios presented by points above the binodal, *e.g.* point M, a polyethylene glycol rich upper phase and a salt rich lower phase are spontaneously formed. The composition of these phases in equilibrium is given by point T and B on the binodal. The connecting line from T to B passes through M and is called a tie-line. All mixtures represented by points on the same tie-line give rise to identical phase systems but with different volume ratios. The latter can be approximated by the distance between M and B (V_u) and M and T (V_l) respectively, on the tie-line. Addition of salt to a two polymer system or a second salt to a one polymer aqueous phase system will alter the carrier system in a not quite predictable way.

Table 1 Selected List of Two-phase Systems Formed by Incompatibility of Neutral Polymers in Water

Polypropylene glycol	– methoxypolyethylene glycol
	– polyethylene glycol
	– polyvinyl alcohol
	– polyvinylpyrrolidone
	– hydroxypropyldextran
	– dextran
Polyethylene glycol	– polyvinyl alcohol
	– polyvinylpyrrolidone
	– dextran
	– Ficoll
Polyvinyl alcohol	– methylcellulose
	– hydroxypropyldextran
	– dextran
Polyvinylpyrrolidone	– methylcellulose
	– hydroxypropyldextran
	– dextran
Methylcellulose	– hydroxypropyldextran
	– dextran
Ethylhydroxyethylcellulose	– dextran
Hydroxypropyldextran	– dextran
Ficoll	– dextran

Depending on the polymers involved aqueous phase systems show a very fine tuning of the hydrophobicity, which is important for van der Waals interactions between chain segments of a given protein as well as between solvent and protein, and also influences protein solubility. Hydrophobicity increases in the aqueous solutions of polymers in the order dextran sulfate < carboxymethyldextran < dextran < hydroxypropyldextran < methylcellulose < polyvinyl alcohol < polyethylene glycol < polypropylene glycol (Albertsson, 1971). For the same polymer, *e.g.* PEG, the hydrophobicity increases with increasing molecular weight, since the number of endgroups diminishes. The partition coefficient of a protein in PEG–dextran and PEG–salt systems will be strongly influenced by changes in the molecular weight of PEG. Depending on the desired direc-

Table 2   List of Two-phase Systems Composed of a Neutral
Polymer (P), a Salt (S) and Water

P	*S*
Polypropylene glycol	Potassium phosphate
Methoxypolyethylene glycol	Potassium phosphate
Polyethylene glycol	Potassium phosphate
Polyvinylpyrrolidone	Potassium phosphate
Polyethylene glycol	Magnesium sulfate
Polyethylene glycol	Ammonium sulfate
Polyethylene glycol	Sodium sulfate
Polyethylene glycol	Sodium formate
Polyethylene glycol	Sodium, potassium tartrate

Table 3   Selected List of Published Phase Diagrams and Binodales of Aqueous Two-phase Systems

P(1)	*P(2)*	*Salt*	*T(°C)*	*References*
PEG 20000	Crude dextran	—	20	Wennersten *et al.*, 1983
PEG 20000	Dextran-T 500	—	20	Schürch *et al.*, 1981
PEG 6000	Dextran D 68 (MW 2.2×10^6)	—	20	Albertsson, 1971
PEG 6000	Dextran D 68 (MW 2.2×10^6)	—	0	Albertsson, 1971
PEG 6000	Dextran-T 500	—	20	Albertsson, 1971
PEG 6000	Dextran-T 500	—	4	Albertsson, 1971
PEG 6000	Dextran-T 500	—	0	Albertsson, 1971
PEG 4000	Crude dextran	—	20	Kroner *et al.*, 1982a
PEG 4000	Dextran-T 500	—	20	Albertsson, 1971
PEG 4000	Dextran-T 500	—	0	Albertsson, 1971
PEG 20000	—	Potassium phosphate	20	Albertsson, 1971
PEG 6000	—	Potassium phosphate	20	Albertsson, 1971
PEG 6000	—	Potassium phosphate	0	Albertsson, 1971
PEG 4000	—	Potassium phosphate	20	Albertsson, 1971
PEG 4000	—	Potassium phosphate	0	Albertsson, 1971
PEG 1540	—	Ammonium sulfate	20	Albertsson, 1971
PEG 4000	—	Ammonium sulfate	20	Albertsson, 1971
PEG 4000	—	Magnesium sulfate	20	Albertsson, 1971

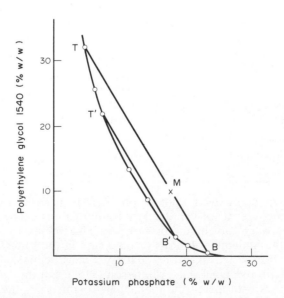

Figure 1   Phase diagram of the system polyethylene glycol 1540–potassium phosphate at 20 °C (data replotted from Albertsson, 1971)

tion of an extraction process, the average molecular weight of PEG should be lowered to obtain a higher yield in the top phase and should be increased to improve the yield in the dextran or salt rich phase. Besides the average molecular weight $\overline{MW}$, the molecular weight distribution of polyethylene glycol also contributes to the relative hydrophobicity of the phases. In the case of fumarase the partition coefficient was altered six orders of magnitude by changing the average molecular weight and the molecular weight distribution of PEG in a PEG–phosphate system (Kula *et al.*, 1982). The effect observed is very selective for fumarase since the apparent partition of total protein changed only 20-fold under these conditions. In two-phase systems incorporating a polymer with a large molecular weight distribution some fractionation will occur leading to a higher average molecular weight $\overline{MW}$ in the polymer rich phase. In such cases the partition coefficient will be changed with the concentration of the polymer even if two systems are on the same tie-line.

The length of the tie-line in any given system depends on the total concentration of the polymers or polymer/salt mixture and is a measure of the relative difference of the phases. At the critical point in the phase diagram, where the tie-line approaches zero, phase composition should be identical, therefore the partition coefficient should be 1. With increasing length of the tie-line, the partition coefficient deviates from 1 and proteins may be shifted into the polyethylene glycol rich upper phase in PEG/salt systems. With increasing tie-line length the interfacial tension increases also and may lead to adsorption to the interface (Albertsson, 1971). However, adsorption of soluble proteins at the liquid–liquid interface is rarely detected and normally does not represent a problem.

28.2.3 Contribution of the Ionic Environment on the Partition of Proteins in an Aqueous Two-phase System

28.2.3.1 *The ionic composition and ionic strength of salts constituting or added to the system*

In two polymer systems ions have individual partition coefficients (Albertsson, 1971; Johansson, 1974). These have been measured for a number of cations and anions and are listed in Table 4. For reasons of electroneutrality constraints exist on partition of single ions leading to the build-up of an electric potential across the interface, a major factor influencing especially the partition of charged macromolecules like proteins and nucleic acids.

Table 4 Logarithm of the Partition Coefficient (K_+ and K_-) of Ions in a System Containing 8% (w/w) Polyethylene Glycol 4000 and 8% (w/w) Dextran-T 500 at 25 °C at Zero Interfacial Potential[a]

Ion	K_-	K_+	Ion	K_-	K_+
K^+	—	−0.084	I^-	+0.151	—
Na^+	—	−0.076	Br^-	+0.083	—
NH_4^+	—	−0.036	Cl^-	+0.051	—
Li^+	—	−0.015	F^-	−0.040	—

[a] Data compiled from Johansson (1974).

The following equation has been derived by thermodynamic reasoning by Albertsson and co-workers (1971):

$$\ln K_p = \ln K_p^0 + (Z_p F/RT)\psi$$
$$\psi = RT/(Z^+ - Z^-)F \ln K_-/K_+ \tag{5}$$

where ψ is the interfacial potential, Z_p the net charge of the protein of interest, Z^+ and Z^- are the net charges and K_- and K_+ the partition coefficients of the cations and anions respectively. K_p is the partition coefficient of the protein, R the gas constant and F the Faraday constant, K_p^0 is the partition coefficient of the protein in the system at zero interfacial potential or if Z_p becomes zero, *e.g.* at the isoelectric point. ψ becomes very important if liquid ion exchangers are included in a two-phase system since these are expected to collect predominantly in one phase.

Due to the unusual low partition coefficient of the HPO_4^{2-} ion in PEG–dextran systems phosphate buffers above pH 7 can be used conveniently to alter the interfacial potential and thereby displace negatively charged proteins into the polyethylene glycol rich phase (Albertsson, 1971;

Kula, 1979). It should be noted that the ratio of the individual partition coefficients of the ions present will determine the interfacial potential and not the concentration. However, all changes in the ionic environment invariably perturb the carrier system and therefore K_p^0.

Also in PEG–salt systems the ionic composition is important presumably due to effects on the solvation of proteins. As an illustration of this up to now not very well documented field, Figure 2 shows the changes in phase composition when adding sodium chloride to a PEG–phosphate system. The differential response of individual proteins under these conditions has been used for the selective purification of a number of enzymes, *e.g.* fumarase (Hustedt *et al.*, 1981), formate dehydrogenase (Kroner *et al.*, 1982b), 1,4-α-glucanphosphorylase (Hustedt *et al.*, 1978). At very high salt concentration, solubility limits of proteins may be reached resulting in high apparent partition coefficients, which may no longer be independent from the concentration of the protein (Menge *et al.*, 1983).

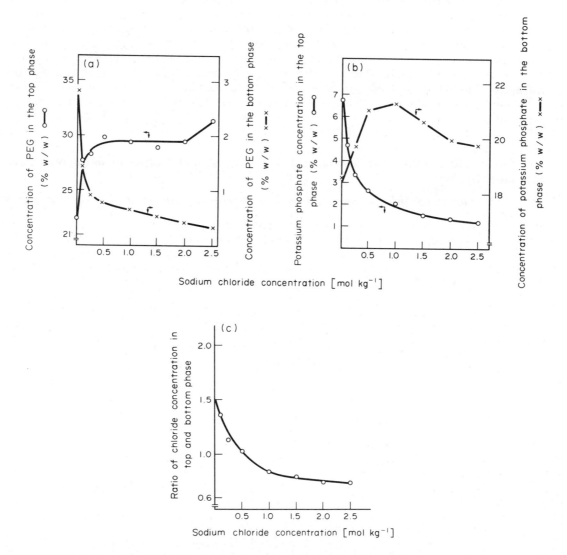

Figure 2 Perturbation of a polyethylene glycol–potassium phosphate system by the addition of sodium chloride into an aqueous two-phase system composed of 14% PEG 1540 and 12% potassium phosphate pH 7.0. The phases were brought to equilibrium at 20 °C and analyzed for the PEG content (a) and phosphate concentration (b) of the top and bottom phases and the apparent chloride partition (c)

28.2.3.2 The pH value

The pH influences the dissociation of ionizable groups of the protein which in turn will alter surface charges and therefore partition. Methods have been described to determine the isoelectric point of proteins, cell organelles and particles by cross partitioning (Albertsson, 1971). For preparative purposes changes in the dissociation of multivalent ions, *e.g.* phosphate, with pH are also important, since this will alter the electric potential between the phases and may give rise to significant changes in partition behaviour of proteins as discussed above. In PEG–salt systems the partition coefficients of proteins are also strongly influenced by pH; depending on the protein, changes over 2–3 orders of magnitude are observed at certain rather small pH intervals (Hustedt and Schütte, private communication). The underlying mechanism is presently under investigation.

28.2.3.3 *Temperature*

Fortunately the sensitivity of partition coefficients to changes in temperature is not very high. Large scale single stage extraction can be performed without extensive temperature regulation. Rises in temperature of 1–2 °C in the liquid during processing have only negligible consequences for recovery of the desired protein and performance of separation with the exception that the system must be far enough from the binodal to ensure phase formation over this temperature interval (Kroner *et al.*, 1978; Kula *et al.*, 1981). Large scale operations are usually carried out at room temperature to avoid expenditure for cooling devices and energy. Two facts contribute to such desirable operating conditions. The polymers introduced stabilize proteins and in general high activity yields are obtained operating at ambient temperatures. In addition the viscosity of the dispersion and phases will be lower at 20 °C compared to 4 °C improving the performance of the separation unit.

28.2.4 How to Find a Suitable System for the Extraction of Intracellular Proteins from Cell Homogenates

Because of the large number of interdependent parameters governing partitioning, a suitable system has to be found by trial and error. The time needed will depend on the peculiarities of the single case but a fast analytical determination for the desired product and a carefully designed strategy will contribute significantly to the success of a search program. Experiments are set up best in 10 ml graduated centrifuge tubes with a conical tip. Concentrated stock solutions of polymers and salts are used to prepare a series of phase systems, each varying in one parameter. To improve the reproducibility of experiments the rather viscous stock solutions are not measured by volume but weight into the centrifuge tube on a top balance. pH is usually set by addition of an appropriate buffer. Cell homogenate or protein solution is added and the system made up to 5.0 or 10.0 g by addition of water. After sealing the tubes the phase system is thoroughly mixed either using a Vortex for 20–60 s or inverting it with 6–10 r.p.m. for 5–10 min. Phases are separated by a short centrifugation in a swinging bucket rotor in a small laboratory table top centrifuge. 1800–2000 *g* are usually sufficient to separate the phases in 3–5 minutes. The volumes of upper and lower phase are noted and a sample of each phase is carefully removed using a syringe or pipet. The activity or concentration of the compound of interest is determined in both phases in an appropriate manner and the partition coefficient calculated. The total amount recovered in the upper and lower phases is also calculated and compared to the input. This way adsorption to the interface or precipitation under experimental conditions are easily detected, as well as serious errors in the analysis. It is advisable to check the influence of the phase components on the assay employed, especially for complex biological assays. In general, influences on biochemical procedures are small, but protein determination according to Lowry is impaired by precipitates formed from PEG and heteropolyacids in the Folin reagent (Lowry, 1951). This can be circumvented either by precipitation of protein with TCA removing most of the PEG before the assay or using the procedure of Bradford (1976) for protein determination after suitable dilution, since high concentrations of PEG or salt lead to erroneous readings with this assay too. Albertsson (1971) recommends variation of the length of the tie-line as a rationale to find a suitable extraction system that will work with cell organelles. The length of tie-line is also very important for partitioning of the cell debris into the bottom phase in PEG–salt systems (Hustedt *et al.*, 1978b). To obtain an additional purification already at this step, a PEG with a molecular weight as high as possible should be used. With increasing molecular weight of PEG less total protein will be extracted from dextran rich or salt rich phases. Such systems are therefore quite selective. Rather high initial purification of the desired protein can be obtained, if the protein can be extracted from cell homogenates in systems containing PEG 6000, *e.g.* the specific activity of β-galactosidase increased 12-fold by a single stage extraction in a 6.3% PEG 6000/10% potassium phosphate system (Veide *et al.*, 1983). Otherwise if PEG with lower average molecular weight is used as a phase forming polymer, purification factors tend to be much lower since the bulk of the proteins present will have a higher partition coefficient under such conditions. Selectivity can be improved by investigating different salts and ionic ratios, as shown in detail for formate dehydrogenase (Kroner *et al.*, 1982b) and 1,4-α-glucanphosphorylase (Hustedt *et al.*, 1978a). In PEG –dextran systems variation of pH in the presence of phosphate will increase partition coefficients above pH 7 of negatively charged proteins by the rising electric potential. Not well understood are changes in partition coefficients brought about by increasing the ionic strength to rather high values, which will in general lead to higher protein content of the PEG-rich upper phase which has a higher solvation power under these conditions.

28.2.5 Influence of the Cell Homogenate on the Initial Extraction

The concentration of biopolymers is an important parameter for the initial extraction of proteins from cell homogenates. For economic reasons the amount of cell homogenate should be as high as possible in the initial extraction, since this will lower the process volume and reduce expenses for the polymers employed. But under such circumstances the biopolymers introduced with the cell homogenate, cell wall fragments, nucleic acids and proteins, will reach concentrations comparable to the hydrophilic polymer employed to induce phase formation. The carrier system is more or less perturbed, which is evidenced by drastic changes in the volume ratio (Kula, 1979) and by apparent shifts of the binodal leading to phase formation at lower polymer concentrations than expected. The extraction of formate dehydrogenase from *C. boidinii* has been described at 9% PEG 4000, 0.5% dextran (Kula *et al.*, 1981), aminoacyl tRNA-synthetases from *E. coli* at 10% PEG 6000, 8% potassium phosphate (Kroner *et al.*, 1978), and glucose isomerase from *Streptomyces* species at 14% PEG 1540, 7.5% potassium phosphate (Hustedt *et al.*, 1978b). In these and many other cases polymer and salt concentrations are below the binodal of the carrier system, yet in the presence of large amounts of cell homogenate in the system, two liquid phases form. However, there are limits to the amount of cell homogenate which can be included in extraction systems. Depending on the microorganism involved or protein to be extracted, between 200 and 400 g of wet cells can be included after homogenization in a 1 kg extraction system. The upper limit is normally set by a decrease in the partition coefficient of the desired protein at high cell concentrations, which will lead, together with the decrease in the volume ratio, to a drop in yield by single step extractions (Kula, 1979). Reproducibility of partition coefficients, even in such highly complex systems, is very good. It has been shown that the mode of cell disintegration does not influence the results (Kroner *et al.*, 1978). Table 5 lists proteins extracted from a variety of microorganisms. At high concentrations of cell homogenate the amount of solid material present in the phase system reaches rather high levels too, which contributes to the viscosity and flow behaviour of the lower phase. We anticipate that 50% broken wet cells will be about the upper limit which can be handled in an extraction system.

28.2.6 Design of a Multistep Extraction Process

As can be seen from Table 5 the specific activity of proteins can often be considerably improved during the initial extraction by removing not only cell debris but more or less of the accompanying proteins. For an enzyme to be used as industrial catalyst a high purity is not generally required, but absence of interfering activities converting either substrate or product in an unwanted fashion is demanded. In addition a certain specific activity is desired to keep the reactor volume small. We have shown that such goals can be met by a series of 2–4 consecutive single stage extraction steps. Partition is also suited to remove nucleic acids and unwanted byproducts of fermentation such as undefined polysaccharide material and pigments. These byproducts are either more hydrophilic than proteins and can be preferentially extracted into a salt rich phase, or more hydrophobic and stay behind in a polyethylene glycol rich phase. If during the initial step a larger portion of nucleic acids and polysaccharide was extracted, together with the desired protein, into the polyethylene glycol rich phase, an extraction of this phase with a concentrated salt solution should follow as a second step, adjusting the conditions in a way that the protein remains in high yield in the polyethylene glycol rich phase. This step may be omitted if the initial partition

Table 5 Selected List of Enzymes Extracted from Cell Homogenates of Various Microorganisms[a]

Enzyme	Organism	Two-phase system	Purification factor	Yield (%)
Catalase	*Candida boidinii*	PEG dextran	—	81
Formaldehyde dehydrogenase		PEG dextran	—	94
Formate dehydrogenase		PEG salt	1.5	94
Isopropanol dehydrogenase		PEG salt	2.6	98
α-Glucosidase	*Saccharomyces cerevisiae*	PEG salt	3.2	95
Glucose-6-phosphate dehydrogenase		PEG salt	1.8	91
Hexokinase		PEG salt	1.6	92
Glucose isomerase	*Streptomyces* species	PEG salt	2.5	86
Leucine dehydrogenase	*Bacillus* species	PEG salt	1.3	98
Alanine dehydrogenase		PEG salt	2.6	98
Glucose dehydrogenase		PEG salt	2.3	95
β-Glucosidase	*Lactobacillus* species	PEG salt	2.4	98
D-Lactate dehydrogenase		PEG salt	1.5	95
L-Hydroxy isocaproate dehydrogenase		PEG salt	1.6	98
Glucose-6-phosphate dehydrogenase	*Leuconostoc mesenteroides*	PEG salt	7.3	94
Fumarase	*Brevibacterium* species	PEG salt	7.5	83
Phenylalanine dehydrogenase		PEG salt	1.5	99
Aspartate-β-decarboxylase	*Pseudomonas dacunhae*	PEG salt	1.1	100
Fumarase	*Escherichia coli*	PEG salt	3.4	93
Aspartase		PEG salt	6.6	96
Penicillinacylase		PEG salt	8.2	90
β-Galactosidase		PEG salt	12.0	75
Pullulanase	*Klebsiella pneumoniae*	PEG dextran	2.0	91
1,4-α-Glucanphosphorylase		PEG dextran	1.2	85

[a] With the exception of β-galactosidase (Veide *et al.*, 1983) all data from the Department of Enzyme Technology, GBF (H. Hustedt, K. H. Kroner, U. Menge, H. Schütte, published and unpublished results).

was selective enough for proteins. In the last extraction step the protein should be transferred into a salt rich phase. The strategy outlined allows the recovery of the major portion of the polyethylene glycol introduced into the process in the last extraction step. The protein can be easily recovered from the salt rich phase by ultrafiltration and/or diafiltration. During this procedure residual polyethylene glycol is effectively removed. A general outline for the process design is presented in Figure 3.

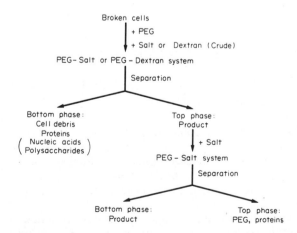

Figure 3 General scheme for the extraction of intracellular proteins

The parameters during the extraction steps can be adjusted to improve the specific activity of the protein of interest to the desired level and also to remove interfering activities. The selectivity of such subsequent partition step may be quite high, factors up to 4000 have been reported (Kula *et al.*, 1982). Current research activities focus on affinity partition as a rational approach to the selective separation and purification of a desired product (Kroner *et al.*, 1982c; Kopperschläger and Johansson, 1982; Johansson *et al.*, 1983). Even if the demands on the purity of the final product are very high as for instance for pharmaceutical or analytical use, it is worthwhile to carry out some extraction steps initially, since the speed of operation on a large scale is very high and yields are excellent. This will considerably improve performance of subsequent chromatographic steps and lower the total amount of protein to be handled. Unfortunately the perturbation of the carrier system by the cell homogenate makes it difficult to predict the composition of the enzyme containing phase obtained in the first extraction step. Even so an analysis of the phase forming components is relatively easy to conduct, the mixture is still very complex and secondary extractions are best developed on an experimental basis. The development is guided by general considerations discussed above and knowledge about the partition behaviour of the product accumulated during the design of the initial step.

A good illustration of the consequent application of such principles in a practical case is given by Hummel *et al.* (1983) in the process design for the isolation of D-lactate dehydrogenase. Table 6 lists several processes developed and carried out on pilot plant scale. A detailed analysis of the recovery process was published by Kroner *et al.* for formate dehydrogenase (1982) and Schütte *et al.* (1983) for leucine dehydrogenase. Overall yields of product are remarkably high. In two cases (pullulanase and formate dehydrogenase) a 70% purity of the enzyme in the final product was achieved. However, it should be kept in mind that in most processes listed in Table 6 a high enrichment of the catalyst was not of prime interest.

Table 6 Examples for the Isolation and Purification of Enzymes by One or Several Subsequent Partition Steps in Large Scale Production[a]

Enzyme	Organism	Number of steps	Overall purification	Overall yield (%)
Pullulanase	*Klebsiella pneumoniae*	4	6.3	70
Glucose isomerase	*Streptomyces*	1	2.5	86
Fumarase	*Brevibacterium ammoniagenes*	2	22	70
Aspartase	*E. coli*	3	18	82
Penicillin acylase		3	12	75
Glucose dehydrogenase	*Bacillus megaterium*	3	33	83
Leucine dehydrogenase	*Bacillus sphaericus*	2	3.1	87
Leucine dehydrogenase	*Bacillus cereus*	2	2.4	89
Formate dehydrogenase	*Candida boidinii*	4	3.8	70
D-Lactate dehydrogenase	*Lactobacillus confusus*	2	1.9	91
L-Hydroxy isocaproate dehydrogenase		3	20	66

[a] Experiments were carried out starting with >20 kg of centrifuged, wet cells, which were disintegrated in a suitable way. The first partition step was always designed to remove cell debris.

28.3 TECHNICAL ASPECTS OF PROTEIN EXTRACTION

28.3.1 Approach to Equilibrium

For the application of the method in industrial processes the following steps have to be considered: speed of formation of an aqueous phase system, mixing, phase transfer rates to reach equilibrium of partition and mechanical separation of the two liquid phases. The formation of a two-phase system by incompatibility between polymers occurs spontaneously. The phase diagram (Figure 1) shows that the components involved have a residual solubility in the excluded phase. Mixing is necessary to achieve equilibrium between the phases. Because of the low interfacial tension little energy is needed for thorough mixing and the time to reach equilibrium is apparently quite small and has not yet been studied in detail. In practice, formation of the phase system is incorporated in the first extraction step. Polymers and salts are added to the cell homogenate either as solids or concentrated solutions. If the mixture is stirred, *e.g.* by a turbine stirrer, overall equilibrium is usually reached within 15 minutes after complete solubilization of polymers and salts. The dissolution of polymer and/or salt seems to be the rate limiting step under these conditions. The density difference and the viscosity of the phases (which may be very high in the first step) influences the mixing and therefore also the approach to equilibrium. From experiments contacting concentrated solutions of PEG and/or salt with the cell homogenate in static mixers it can be concluded that it is possible to reach the coupled equilibrium of phase formation and partition in very short times depending mainly on the optimization of the mixing process (Hustedt, personal communication). In general the time needed to establish equilibrium will depend on the mass transfer area, the difference in concentration and the diffusion coefficient D. According to Fick's law the following equation applies:

$$\frac{dc}{dt} = D\frac{d^2c}{dx^2} \tag{6}$$

where c is the concentration of the compound of interest, t the time, D the diffusion coefficient and x the distance traveled. The rate of diffusion is influenced by viscosity and interfacial tension as well as temperature. Since for very large entities, like cell debris, with apparent molecular weights of several millions the diffusion coefficient is very small, equilibrium can only be reached quickly if the distance x to be traveled by a single particle or molecule also becomes very small.

As already pointed out the interfacial tension in aqueous phase systems is extremely small, of the order of 2.0–0.1 mN cm^{-1} for PEG salt systems and 0.1–0.0001 mN m^{-1} for PEG–dextran systems (Albertsson, 1971). Mechanical mixing will therefore produce very small droplets even at low energy input, thereby decreasing x and increasing the available surface area for exchange at the same time. Both facts contribute to a high transfer rate. The coalescence of the droplets formed apparently is also not inhibited, therefore it is no longer surprising that after removal of cell debris equilibrium in partition of proteins in well mixed aqueous phase systems takes place in seconds. It becomes difficult to measure phase transfer rates under such conditions and to resolve experimentally the time course in the approach to equilibrium. Experiments reported by Shanbhag (1973) using a cell with a defined interface show that the resistance of the liquid–liquid interface to phase transfer of proteins cannot be very high. The low interfacial tension in aqueous two-phase systems contributes not only significantly to the rapid establishment of equilibrium in phase formation and partition, it also minimizes denaturation of labile proteins at the interface and allows a high recovery of activity, even if extremely large surface areas between the phases are generated. The low interfacial tension will however make the mechanical separation of the phases more difficult.

28.3.2 Mechanical Separation of Phases

The mechanical separation of phases can be carried out under normal gravity or accelerated by centrifugation. In both cases one can work with or without superimposed flow. The sedimentation velocity is described by Stokes' law:

$$V_s = \frac{D^2\Delta\varrho}{18\eta}g \text{ (under gravity)} \tag{7}$$

$$V_z = \frac{D^2\Delta\varrho}{18\eta}r\omega^2 \text{ (centrifugal acceleration)} \tag{8}$$

where D is the diameter of the droplets, $\Delta\varrho$ the difference in density between the phases, η the dynamic viscosity, g the acceleration due to gravity, r the radius of the bowl and ω the angular velocity of the centrifuge. Dispersions can only be separated by sedimentation if the two phases exhibit a sufficient density difference. The density of aqueous solutions of dextran ($\overline{MW}$ 500 000), PEG 6000, methylcellulose ($\overline{MW}$ 140 000), polyvinyl alcohol (PVA 48/20) and sodium dextran sulfate at 20 °C have been tabulated by Albertsson (1971). As a first approximation the density of the phases can be calculated from the polymer composition and the densities of the individual polymer solutions. The density difference of pure phases in PEG–dextran systems reaches from 0.02–0.07 and in PEG–phosphate systems from 0.04–0.10. Albertsson (1971) describes one example where the density of the phases becomes equal at certain compositions of a ficoll dextran system. Such isopyknic phases can no longer be separated by sedimentation; including cell homogenates in a phase system leads to an increase in the density difference, which should improve separation. This effect is however counteracted by the rising viscosity. At higher concentrations the changes in viscosity dominate and a plot of $\Delta\varrho/\eta$ as a function of cell homogenate indicates the concentration range where sedimentation velocity deteriorates. This has been analyzed in detail by Kroner *et al.* (1978) for the extraction of pullulanase from *Klebsiella pneumoniae* and by Kroner and Kula (1978) for extraction of α-glucosidase from *Saccharomyces cerevisiae*. Results show that interferences are to be expected around 30% broken cells (wet) in the system. The time needed for separation will critically depend on the droplet size and therefore on the intensity of prior mixing and the rate of coalescence. Unfortunately both parameters are very difficult to measure experimentally especially in systems containing cell debris or precipitates. In general it can be expected that PEG–salt systems will separate faster than PEG–dextran systems. This behaviour can be attributed to the larger droplet size (higher interfacial tension), higher density difference and lower viscosity of PEG–salt systems. It will take 10–20 h to separate PEG–salt systems containing cell debris in a settling tank (Kula *et al.*, 1982b; Hustedt *et al.*, 1978b). The purity of the lower phase containing cell debris is only 75–85% under these conditions and does not increase any further, indicating that small droplets of upper phase are trapped and will not rise in a quite viscous lower phase without increasing gravity forces. While the viscosity of the dispersion in most practical cases will be dominated by the PEG-rich continuous phase and reaches values between 3 and 15 mPa, the lower phase during the initial extraction may exhibit a viscosity between hundred and several thousand mPa, depending on the volume ratio, the concentration of cell homogenate and the amount of precipitate present. A large viscosity difference between dispersion and the lower phase is also found if crude dextran is employed as a phase forming polymer (Kroner *et al.*, 1982). In further extraction steps, once all insoluble material is separated, the viscosity of the salt-rich phase becomes even smaller than the PEG-rich phase and conditions for separation under normal gravity are more favourable. Separation times between 45 and 90 min have been reported (Kula *et al.*, 1981; Kroner *et al.*, 1982b) in 120–150 1 scale experiments using a settling tank with a height/diameter ratio of the liquid column around 2.6. It can be expected that separation times could be decreased in settlers with improved geometry. Operation under normal gravity minimizes the energy consumption for the process and since the necessary equipment is also very simple, investment costs are small too.

If we consider sedimentation with superimposed flow the relation between flow and clearance can be described by the following equations:

$$Q = \frac{D_{\lim}^2 \Delta \varrho g}{18\eta} A \tag{9}$$

$$Q = \frac{D_{\lim}^2 \Delta \varrho g}{18\eta} \frac{r\omega^2}{g} A \tag{10}$$

$$Q = V_s \zeta A \tag{11}$$

where Q is the volumetric feed rate, $D_{\lim}$ the limiting diameter of the smallest particle separated and A the clarification area. The first term, V_s the sedimentation velocity, includes all product and dispersion properties. The last term A represents the clarification area of a settler or the effective clarification area, if additional surfaces have been incorporated in the flow path. The term ζ represents the centrifugal factor, the acceleration of the sedimentation by additional g-forces. In the case of a disc stack separator, commonly employed for liquid–liquid separation, the clarification area can be expressed as

$$A = \tfrac{2}{3} r t g \phi (r_{d_1}^3 - r_{d_2}^3) N \tag{12}$$

where ϕ is the angle of discs in the stack and N the number of discs, r_{d_1} is the outer radius of discs and r_{d_2} the inside radius of the discs. The product of ζ and A is often called the Sigma factor Σ, which is used to correlate and extrapolate performance of centrifuges with similar design.

Figure 4 illustrates the functioning of a disc stack separator. The dispersion is fed centrally into the separator and gets accelerated to the angular velocity as it flows towards the radius in the lower part of the bowl. The dispersion enters the disc stack and is distributed in the available space through the rising channels in the stack. Separation takes place in the disc stack. The heavier phase is displaced towards the rotor wall and leaves the separator ultimately at the radial position L. The lighter liquid collects at the core of the rotor and is discharged at the radial position U. In general the radial position of U is fixed in commercial separators, while the radial position of L is variable within certain limits to allow processing of liquids with different densities. Hydrostatic balance is obtained if:

$$\varrho_u (r_s^2 - r_u^2) = \varrho_1 (r_s^2 - r_1^2) \tag{13}$$

$$r_1 = \left[r_s^2 \left(\frac{\varrho_1}{\varrho_u} - 1 \right) + r_u^2 \right]^{\frac{1}{2}} / (\varrho_1/\varrho_u)^{\frac{1}{2}} \tag{14}$$

where ϱ_1 and ϱ_u are the density of the lower and upper phase respectively, r_1, r_u the radial position of the discharge of the lower and upper phase respectively, and r_s the radial position of the interface in the disc stack. In order to obtain both phases in high purity the interface line should be positioned within the rising channels in the disc stack. This can be accomplished by the correct choice of r_1. In open disc stack separators where the underflow is discharged freely into a cover, the inner diameter of the gravity disc has to be varied accordingly.

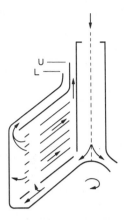

Figure 4 Flow path through a disc stack separator

In a closed separator the outflowing phases are collected under pressure by centripetal pumps. The diameter of the centripetal pump of the underflow and the back pressure in the discharge lines can be used to influence the positioning of the interface. With both types of separators balance conditions have to be met quite precisely to get optimal results with regard to the purity of the phases and the extraction yield. Calculation of r_1 according to equation (14) gives a reasonable approximation, which should be verified and if necessary modified by trial runs to establish the best position. It may become necessary to alter r_1 in smaller intervals than provided for with different gravity discs or centripetal pumps by the manufacturer of the separator. The missing parts can be machined easily in local shops from larger parts. With the exception of such a minor modification in accessory parts commercially available machines have been used very successfully to separate aqueous two-phase systems. Depending on the physical properties of the actual dispersion and the separator employed, nearly complete separation of the two liquid phases was obtained with residence times ranging between 2 and 120 s (Kula *et al.*, 1981, 1982). In the presence of cell homogenate, values below 40 s are commonly reached. A typical example is analyzed in Figure 5. Taken together with a rather high loading of cell homogenate in aqueous phase systems the short residence times indicate that the capacity of small machines is very high and scaleup of the technology for the extraction of ton quantities of microorganisms quite simple. The short residence times help to avoid extensive measures for temperature control. Differences in the temperature between feed and enzyme containing phase did not exceed 1 °C at ambient temperature during prolonged operation, as analyzed by Kroner *et al.* (1978) and Kula *et al.* (1981) in PEG–dextran systems.

During the fast radial acceleration within the separator redispersion of the phases occurs due to the low interfacial tension. Evidence comes from plots of the logarithmic ratio between the concentration of disperse phase in the effluent and in the feed *versus* the logarithm of the feed rate. From this plot maximal throughputs for optimal clearance can be determined as well as the limiting droplet size c which occurs in the separator under conditions of high flow velocities. Reliable independent determinations of the droplet sizes in the feed are difficult to obtain and presently not available. But estimates show that the droplet size in the dispersion entering the separator should be about one order of magnitude larger (Menge, personal communication). Such plots also reveal that open disc stack separators exhibit not only a maximal but an optimal throughput, since clearance will become less at lower feed rates. This behaviour is in contrast to the performance of closed separators where constant high clearance is observed up to the maximum feed rate (Kula *et al.*, 1981). The different performance has been attributed to the different response towards flow resistance by the two types of machines and to a better control of the interface line by regulating the back pressure in closed separators. The influence of the flow resistance in the separator bowl is also quite apparent handling aqueous phase systems with high viscosities of the lower phase during the initial extraction step. If the rheological properties of such phases become very unfavourable the discharge of the lower phase may be clogged. In such cases nozzle separa-

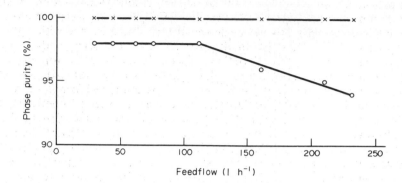

Figure 5 Continuous phase separation in a closed disc bowl centrifuge, model SAOH 205 from Westfalia. The purity of top ($\times$) and bottom phase ($\bigcirc$) is plotted as a function of the feed rate during the separation of fumarase from broken cells of *Brevibacterium ammoniagenes* using a PEG–salt system and the Westfalia separator SAOH 205 (bowl volume 0.25 1). Physical data of the phase system: volume ratio $V_u/V_1 = 1.5$, $\varrho_u = 1.17$ g cm^{-3}, $\varrho_1 = 1.26$ g cm^{-3}, $\eta_{\text{dispersion}} = 28$ cP, $\eta_u = 16$ cP, $\eta_1 = 255$ cP, 20% (w/w) wet cells included in the system (H. Hustedt, personal communication)

tors have been successfully utilized to separate the phases very quickly and with quite good separation efficiency. Separation efficiency (S_e) can be defined as:

$$S_e = P_u - \frac{1 - P_1}{1 + V_f'} \tag{15}$$

where P_u and the P_1 are the purity factors of overflow and underflow (% pure phase/100) and V_f' the ratio of overflow to underflow, $\dot{Q}_u/\dot{Q}_1$. The separation factor can reach a maximal value of 1. This factor can serve to characterize the performance of a liquid–liquid separation in separators of different design and also for different operating conditions. Kroner *et al.* (1982a,b) have shown that nozzle separators can indeed be employed for an efficient separation of aqueous phase systems, provided the interface line can be kept within the bowl. This will be possible if the density and viscosity of the heavier phase are high enough and the feed rate is sufficient so that a liquid seal is established. Otherwise loss of lighter phase through the nozzles cannot be avoided. Besides the viscosity and rheological properties of the heavier phase the number and geometry of the nozzles will also be important, influencing the back pressure of the underflow. In most commercial nozzle separators the number of nozzles can be varied, *e.g.* from six over four or three to two nozzles, keeping a symmetrical flow pattern. For the separation of aqueous phase systems nozzles with the smallest opening (0.4–0.5 mm) provided by the manufacturer are the most useful. In principle the length of the orifice in the nozzle could also serve as a modulator of the back pressure, but this has not as yet been tried. The separation efficiency in a nozzle separator is dominated by the purity of the heavier phase discharged through the nozzles. Stable operating conditions require a balance between the amount of discharged and delivered heavier phase per unit time.

$$\dot{Q}_n = \dot{Q}_F \frac{1}{V_f} \tag{16}$$

$$\dot{Q}_n = \dot{Q}_F \frac{V_1}{V_u} \tag{17}$$

where $\dot{Q}_n$ is the flow through the nozzles, $\dot{Q}_F$ the feed rate to the separator and V_f the volume ratio of the phases in the feed. The maximal feed rate is limited because a certain residence time is necessary to separate the dispersion in the disc stack and avoid flooding of the bowl. This leads to the conditions:

$$\dot{Q}_n = \dot{Q}_{\lim} \frac{V_1}{V_u} \tag{18}$$

In practice the choice of the nozzles should give $\dot{Q}_n < \dot{Q}_{\lim} (V_1/V_u)$. Also from mass balances the throughput at the nozzles can be described as:

$$\dot{Q}_n = \frac{\dot{Q}_T P_T}{(C_e/C_o) - 1} \tag{19}$$

where $\dot{Q}_T$ is the throughput of clarified top phase, C_e the emergent concentration of heavier phase in the nozzle product (%) and C_o the concentration of heavier phase in the feed (%).

Equation (19) provides a guide to optimize the operating conditions. Kroner *et al.* (1982a) and Kula *et al.* (1981) reported the separation of various PEG–dextran phase systems especially with crude dextran as phase components in nozzle separators. Performance with a PEG–salt system was analyzed by Kroner *et al.* (1982b). In all cases the cell debris was removed with the bottom phase. The opposite situation, when the cell debris is partitioned into the top phase, poses a more difficult separation problem, since now the lighter phase will be more viscous. The cell debris should remain suspended in the top phase during the separation and formation of a solid layer at the interface avoided. Fortunately this situation is quite rare since in general conditions can be found where the enzyme is extracted in the top phase and the cell debris is collected with the lower phase. However, for the purification of fumarase from *Brevibacterium ammoniagenes* fumarase exhibited a partition coefficient of 0.24 while cell debris had an apparent partition coefficient $\gg 1$. Hustedt *et al.* (1981) reported that a closed disc stack separator gave the best results in continuous operation with residence times of approximately 20 s. In order to maintain high yields a purity of the enzyme containing phase of 95% had to be accepted under these conditions.

28.3.3 Scale-up

Extraction of enzymes from up to 200 kg of cell homogenate has been carried out at the GBF without problems. Equilibrium of partition has always been reached in short times with adequate mixing. In large scale production, separation of the phases was performed either in nozzle separators or in closed disc stack separators, in continuous operation at ambient temperatures. Loss of enzyme due to the separation efficiency could be kept very low, between 1 and 2%. Two processes have been published in detail by Kroner *et al.* (1982a) and Schütte *et al.* (1983) for the isolation and purification of formate dehydrogenase from the yeast *Candida boidinii* and leucine dehydrogenase from bacteria, *Bacillus sphaericus* and *Bacillus cereus* respectively. Results of large scale processes could be accurately calculated from the partition coefficient of the enzyme and the volume ratio of the phase systems determined by laboratory experiments in 10 ml graduated tubes as described above. In scale-up all concentrations were kept and the proportions increased linearly according to the amount of cell homogenate included in the extraction system. Variations in the quality of the starting material, *e.g.* differences of as much as 50% in formate dehydrogenase activity in cells of *Candida boidinii*, could be accommodated without changes of the process parameters. Extraction of 200 kg cell homogenate can be performed during a single shift using one small separator with a volume of the bowl of only 820 ml (nozzle separator) or even 250 ml (disc stack separator) respectively. Larger production would require use of a larger disc stack or nozzle separator, which are readily available commercially. Selection of larger machines and adjustment of operating conditions can follow established practices in chemical engineering as discussed above.

Since the enzyme or any other interesting biologically active protein stays in solution at all times extraction lends itself to a continuous processing. This approach represents an interesting alternative to batchwise operation in larger scale and would allow in addition a drastic decrease in process time and improve the productivity considerably as pointed out by Dunnill and Lilly (1972). Liquid–liquid extraction technology in enzyme isolation follows a rather general sequence of mixing and phase separation as illustrated in Figure 3. This sequence of unit operations, together with the necessary measuring devices involved, should be quite flexible in application and allow different products to be handled in the same equipment. The biggest difficulty encountered in developing continuous processing so far is the high capacity of the method and the rather short residence times of separators which require large amounts of cells and material for experiments. Hustedt *et al.* (1983) reported encouraging results in the continuous extraction of a bacterial enzyme in a two stage process illustrated in Figure 6. The experiment had to be terminated after 5 h because the supply of cells was exhausted. During the progress of the experiment no changes in the relevant process data like enzyme concentration in the phases, volume ratio of the phases and phase purity were detected. The final yield and quality of the product was identical to a batch process.

28.3.4 Multistage Procedures

In previous chapters single step partitions have been dealt with. Under such conditions isolation of single components in high purity cannot be expected considering the complex nature of

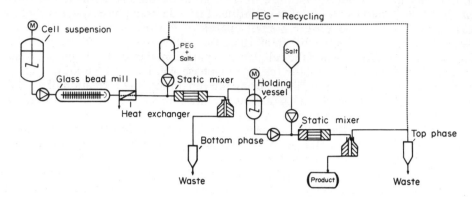

Figure 6 Flow diagram of a continuous enzyme extraction process

the starting material. However, in chemical engineering extraction technology has been developed for multistage continuous operation, which is also available in principle for protein purification by multistage partition in aqueous phase systems (Treybal, 1963; Hanson, 1971).

All the major chromatographic procedures in protein purification technology are batch processes. With multiple, automatically repeated partitions performed in a suitable extraction unit, continuous processing could be developed also with high resolving power as required. Progress in this special application has been slow so far, again due to the high capacity of the method and the high demand for materials in the experiments. Several common extractors of different design have been tested and found suitable for work with aqueous phase systems, *e.g.* mixer–settler batteries and a centrifugal separator, the Podbielniak extractor (Hustedt *et al.*, 1980), but performance has not yet been analyzed in detail. Of highest interest in this regard are extraction columns which can be operated either with countercurrent flow of both phases or by keeping the heavier phase stationary and the lighter phase mobile in the unit. Extraction columns differ in the way mixing and dispersion are carried out. It can be expected that mixing is of particular importance working with aqueous phase systems. As a consequence of the low interfacial tension mixing must be gentle enough to avoid flooding of the columns and at the same time support sufficient mass transfer. Blomquist and Albertsson (1972) studied performance of a small rotary extraction column for the separation of nucleic acids, proteins and particles in a PEG_{6000}–dextran-T 500 system. The efficiency of separation was very good but the feed rate had to be reduced to 0.12 ml min^{-1} for counter current flow operating at maximal stirrer speeds of 80 r.p.m. As discussed above PEG–salt systems exhibit faster separation rates. This allowed feed rates of 8–10 ml min^{-1} in a five stage rotary extraction column of 2 cm diameter and 200 ml total volume investigated by Kula *et al.* (1981, 1982a). The optimal stirrer speed also depends on the physical parameters of the system (average molecular weight of the PEG employed and the length of the tie-line) and varies between 100 and 300 r.p.m. The capacity of the 200 ml column for the purification of formate dehydrogenase or the separation of cytochrome *c* and catalase was judged to be 250–1000 mg h^{-1}. Also the Graesser contactor, a horizontal rotary extraction column with 36 chambers and a total volume of 7 l could be successfully operated with PEG–salt systems after minor modifications. Feed rates of 6–10 l h^{-1} were realized for several hours of continuous operation without flooding at approximately 2 r.p.m. In this case industrial dimensions are already reached, since a capacity for the purification of proteins of 40–50 g h^{-1} was anticipated from results obtained in the extraction of dyes. At least in principle scale-up of extraction technology utilizing aqueous phase systems appears possible. However, much more detailed investigations are necessary to evaluate the potential of this approach and compare it with competing techniques. Available knowledge in chemical engineering can be applied for future developments.

28.4 PARTITION OF NUCLEIC ACIDS IN AQUEOUS PHASE SYSTEMS

The partition of DNA in PEG–dextran systems depends on the ionic environment (Albertsson, 1971) and the size of the macromolecule (Favre and Pettijohn, 1967). Li^+ increases the partition coefficient while all other alkali metal cations show an opposite effect. The phase potential plays a dominant role which is also evident from the increase of the partition coefficient in the presence of di- and tri-valent anions like sulfate, phosphate and citrate. Since monovalent anions either decrease K or show no influence, the partition can be steered by proper selection of the salt employed. Partition of RNA follows a similar pattern (Albertsson, 1971). A separation of RNA from DNA and also of native from denatured DNA is possible, by one or a few partition steps. Including base specific ligands covalently attached to polyethylene glycol in an aqueous phase system increases the affinity of DNA for the top phase in a differential fashion depending on the AT or GC content respectively (Müller and Eigel, 1981). Binding of one PEG–dye molecule per 50 base pairs in the DNA will overcome a salt effect and raise the partition coefficient by about 3 logs in the presence of 30 mM KCl. A difference of 20% in the GC content of two DNA's would be sufficient to get 90% pure components by a single partition step assuming that the initial concentration of both DNA's was equal. Substantial purification of GC rich calf thymus satellite DNAs was achieved after 10 countercurrent extraction steps in preparative scale (Müller and Eigel, 1981). Higher resolution is possible by a partition chromatography, where the dextran-T 500 rich phase is physically bound to cellulose (Müller *et al.*, 1979; Müller and Kütemeier, 1982) and the PEG 6000 rich phase is employed for development of the column. By this liquid–liquid partition chromatography DNA fragments from 20 to 30 000 base pairs could be separated by size under optimized conditions; examples include a variety of restriction fragments of calf thymus DNA, phage DNA (λ, T 7) and plasmid DNA (pBR 322, Col E 1). Performance of the chromatographic process depended on the ratio of mobile phase and bound phase in the column, the temperature and the ionic strength. Gradient elution was carried out by changing the concentration of potassium acetate and/or lithium acetate, or lithium sulfate, respectively, in the mobile phase. Due to the viscosity of this system and the influence of the diffusion rate on the equilibrium partition of the DNA a rather low linear flow velocity, not exceeding 5 cm h^{-1}, is necessary in such columns. The same principles should be applicable to the separation of RNA species (Müller and Kütemeier, 1982).

In PEG–salt phases partition coefficients for high molecular weight DNA will be very small (Albertsson, 1971). Veide *et al.* (1983) report that during the extraction of β-galactosidase from an *E. coli* cell homogenate only < 0.1% of DNA was found in the PEG rich top phase. Since in such systems precipitates collect in the bottom phase the low recovery in the top phase may be attributed besides the partition also to precipitation of DNA under the extraction conditions. The recovery of RNA in the PEG rich phase is higher than of DNA and also depends more on the extraction conditions, reflecting the higher abundance of RNA species like tRNA (MW 25–30 000) or messenger RNA in the cell homogenate. Strategies to remove residual nucleic acids from proteins have been discussed above.

28.5 PARTITION OF CELLS AND CELL ORGANELLES IN AQUEOUS PHASE SYSTEMS

Partition is a very gentle fractionation method which has been applied to the separation of cells and cell organelles by many research groups after the pioneering work of Albertsson. The resolving power of a single partition can be enhanced by countercurrent distribution and further by adding specific ligands to the polymers forming the phase systems, *e.g.* hydrophobic ligands or liquid ion exchangers. Most work has been performed using a thin layer CCD unit with 60 transfer steps developed by Albertsson (1971). Early work is summarized in Albertsson's monograph on *Partition of Cell Particles and Macromolecules* (1971). Recent applications center on separation of chloroplasts, mitochondria from plant and animal cells, membranes and membrane vesicles, but also on separation and analysis of cell populations like erythrocytes, platelets, leucocytes and lymphoid cells. Recent work has been reviewed by Walter (1977), Albertsson (1979), Larsson and Andersson (1979), Eriksson and Johansson (1979), Fisher (1981) and Albertsson *et al.* (1982). While most of the work carried out with cell organelles and cell populations had quite other objectives than biotechnological goals, the method itself could be very useful in this context too, *e.g.* in combination with modern techniques in genetic engineering requiring only small amounts of DNA from specified cells, or for studies of cell adhesion (Schürch *et al.*, 1981; Gerson and Scheer, 1980).

28.6 EXTRACTIVE CONVERSION WITH BIOCATALYSTS IN AQUEOUS TWO-PHASE SYSTEMS

Cells or enzymes can be confined to one of the phases of an aqueous phase system by proper selection of parameters resulting in a one-sided partition. This provides a method of immobilizing

Table 7 Examples of Extractive Conversion

Carrier system	Reaction	Enzymes/cells	Operation and scale	References
PEG 20000 crude dextran	Starch → glucose	α-Amylase/glucoamylase	22 l, 0.24 l h^{-1}	Wennersten *et al.* (1983)
PEG 8000 dextran-T 40	Cellulose → alcohol	Cellulase/cellobiase/*S. cerevisiae*	60 ml batch	Hahn-Hägerdal *et al.* (1981)
PEG 8000 dextran-T 40	Cellulose → alcohol	Cellulase/cellobiase/*S. cerevisiae*	120 ml batch	Hahn-Hägerdal *et al.* (1982)
PEG 8000 dextran-T 40	Cellulose → alcohol	Cellulase/cellobiase/*S. cerevisiae*	120 ml batch	Hahn-Hägerdal *et al.* (1981)
PEG 20 000 potassium phosphate	Penicillin G → 6 APA	Penicillin/acylase	Repeated batch	Hahn-Hägerdal *et al.* (1983)
PEG 300 potassium phosphate	D, L-*N*-Acetylmethionin →L-methionin	Acylase	180 ml batch	Kuhlmann *et al.* (1980)

the catalyst in a simple and reversible way without any potentially harmful chemical treatment and keeping it in solution at all times. Since small molecules will have a partition coefficient close to one, substrate could be supplied and product removed with the opposite phase working with small molecules, while polymeric substrates would be introduced into the catalyst containing phase. The low interfacial tension allows the generation of large surface areas for fast and efficient mass transport. Demands on the partition coefficient of the catalyst are quite high to avoid washout in continuously operating systems. Such demands can be met working with intact cells, but soluble enzymes present a more difficult situation. An interesting approach is described by Wennersten *et al.* (1983), who studied the continuous conversion of starch into glucose using the substrate starch as an affinity ligand to keep the enzyme effectively in the bottom phase of a system composed of PEG 20 000–crude dextran and 14–17% starch at the beginning of the experiment. Approximately 90% conversion could be obtained. The process was carried out in a mixer–settler arrangement. Presumably due to the high viscosity of the bottom phase the throughput reported was rather low resulting in residence times of 92 h. Table 7 summarizes the systems studied so far by this approach. It is too early to judge the value of the extractive conversions, in comparison with established technology, for lack of more quantitative data. But those should be forthcoming in the near future.

28.7 ANALYTICAL APPLICATIONS OF LIQUID–LIQUID SEPARATION TECHNIQUES

Partition across the interface of an aqueous phase system is a process where equilibrium can be reached rapidly. If the components of an interacting system:

$$A + B \rightleftharpoons AB$$

such as enzyme–substrate, enzyme–inhibitor, antibody–antigen, or lectin–carbohydrate complexes differ sufficiently in their partition coefficient binding, studies can be performed by following the partition of one ligand as a function of the concentration. Gray and Chamberlin (1971) measured the binding of small ligands to an enzyme in this way, Hustedt and Kula (1977) extended the method to determine equilibrium binding between macromolecules, using aminoacyl-tRNA synthetases and tRNA as examples. If one component can be determined conveniently by radioactivity or spectroscopic measurements a competing assay can also be developed to quantify the interacting compound. A prerequisite for a sensitive assay of this kind is that the reactants should separate into opposite phases. In recent years Mattiasson *et al.* (1982) developed the partition affinity ligand assay (PALA). In order to achieve the desired partition into opposite phases of at least one reactant and the complex formed the concept of the 'separator molecule' was successfully introduced. It implies that the ligand can be modified selectively by covalent linkage with hydrophilic or hydrophobic groups to a nonbinding region, in order to steer the partition in the desired direction. Assay procedures for digoxin, triiodothyronine, concanavalin A, albumin and cells like *Staphylococcus aureus* and *Saccharomyces cerevisiae* have been reported (Mattiasson *et al.*, 1982). The principle should be generally applicable and simplify and speed up current analytical techniques such as RIA and ELISA.

ACKNOWLEDGEMENTS

I would like to thank Bo Mattiasson (University of Lund) and Anna-Lisa Smeds (Royal Institute of Technology, Stockholm) for providing me with preprints of recent work. I am indebted to my collaborators at the Department of Enzyme Technology at the Gesellschaft für Biotechnologische Forschung mbH, Braunschweig, who contributed so much to the progress described. The help of the staff in the pilot plant of the GBF is also greatfully acknowledged. Special thanks are due to Dr. H. Hustedt and Dipl.-Ing. K. H. Kroner for critical reading of the manuscript.

28.8 REFERENCES

Albertsson, P. Å. (1971). *Partition of Cell Particles and Macromolecules*. 2nd edn. Wiley, New York.
Albertsson, P. Å. (1977). Separation of particles and macromolecules by phase partition. *Endeavour* (new series), **1**, 69–74.
Albertsson, P. Å. (1978). Partition between polymer phases. *J. Chromatogr.*, **159**, 111–122.

Albertsson, P. Å. (1979). Phase partition of cells and subcellular particles. In *Separation of Cells and Subcellular Elements*, ed. H. Peeters, pp. 17–22. Pergamon, Oxford.

Albertsson, P. Å., B. Andersson, Ch. Larsson and H. E. Åkerlund (1982). Phase partition—a method for purification and analysis of cell organelles and membrane vesicles. In *Methods of Biochemical Analysis*, ed. D. Glick, vol. 28, pp. 115–150. Wiley, New York.

Beijerinck, M. W. (1896). Über eine Eigentümlichkeit der löslichen Stärke. *Centralblatt Bakt. 2. Abt.*, **2**, 697–699.

Blomquist, G. and P. Å. Albertsson (1972). A study of extraction columns for aqueous two-phase systems. *J. Chromatogr.*, **73**, 125–133.

Bradford, M. M. (1976). A rapid and sensitive method for the quantitation of microgram quantities of protein utilizing the principle of protein–dye binding. *Anal. Biochem.*, **72**, 248–254.

Brunner, K. H., H. Scherfler and M. Stölting (1981). Erfahrungen mit dem Gegenstrom-Extraktionsdekanter bei der Gewinnung von Antibiotika aus Fermentationsbrühen. *Vt-Verfahrenstechnik*, **15**, 619–622.

Dunnill, P. and M. D. Lilly (1972). Continuous enzyme isolation. *Biotechnol. Bioeng. Symp.*, **3**, 97–113.

Eriksson, E. and G. Johansson (1979). Partition of cells between aqueous phases. In *Cell Populations, Methodological Surveys (B) Biochemistry*, ed. E. Reid, vol. 8, pp. 81–90, Ellis Horwood, Chichester.

Favre, J. and D. E. Pettijohn (1967). A method for extracting purified DNA or protein–DNA complex from *E.coli. Eur. J. Biochem.*, **3**, 33–41.

Fisher, D. (1981). The separation of cells and organelles by partitioning in two-polymer aqueous phases. *Biochem. J.*, **196**, 1–10.

Gerson, D. F. (1980). Cell surface energy, contact angles and phase partition. I. Lymphocytic cell lines in biphasic aqueous mixtures. *Biochim. Biophys. Acta*, **602**, 269–280.

Gerson, D. F. and D. Scheer (1980). Cell surface energy, contact angles and phase partition. III. Adhesion of bacterial cells to hydrophobic surfaces. *Biochim. Biophys. Acta*, **602**, 506–510.

Gray, C. W. and M. J. Chamberlin (1971). Measurement of ligand–protein binding interactions in a biphasic aqueous polymer system. *Anal. Biochem.*, **41**, 83–104.

Hahn-Hägerdal, B., B. Mattiasson and P. Å. Albertsson (1981a). Extractive bioconversion in aqueous two-phase systems. A model study on the conversion of cellulose to ethanol. *Biotechnol. Lett.*, **3** (2), 53–58.

Hahn-Hägerdal, B., E. Andersson, M. Lopez-Leiva and B. Mattiasson (1981b). Membrane biotechnology, co-immobilization, and aqueous two-phase systems: alternatives in bioconversion of cellulose. *Biotechnol. Bioeng. Symp.*, **11**, 651–661.

Hahn-Hägerdal, B., B. Mattiasson, E. Andersson and P. Å. Albertsson (1982). Soluble temporarily immobilized biocatalysts. *J. Chem. Technol. Biotechnol.*, **32**, 157–161.

Hahn-Hägerdal, B., E. Andersson, M. Larsson and B. Mattiasson (1983). In *Biochemical Engineering III,* ed. K. Venkatasubramanian, *Ann. N.Y. Acad. Sci.*, **413**, 542–544

Hancher, C. W., E. F. Phares, G. D. Novelli and A. D. Kelmers (1969). Large scale production of transfer ribonucleic acids from *E.coli* K-12 MO7 *Biotechnol. Bioeng.*, **11**, 1055–1070.

Hanson, C. (1971). *Recent Advances in Liquid–Liquid Extraction.* Pergamon, Oxford.

Hummel, W., H. Schütte and M.-R. Kula (1983). Large scale production of D-lactate dehydrogenase for the stereospecific reduction of pyruvate and phenylpyruvate. *Eur. J. Appl. Microbiol. Biotechnol.*, **18**, 75–85.

Hustedt, H. and M.-R. Kula (1977). Studies of the interaction between aminoacyl-tRNA synthetases and transfer ribonucleic acid by equilibrium partition. *Eur. J. Biochem.*, **74**, 191–198.

Hustedt, H., K. H. Kroner, W. Stach and M.-R. Kula (1978a). Procedure for the simultaneous large scale isolation of pullulanase and 1,4-α-glucan phosphorylase from *Klebsiella pneumoniae* involving liquid–liquid separations. *Biotechnol. Bioeng.*, **20**, 1989–2005.

Hustedt, H., K. H. Kroner, U. Menge and M.-R. Kula (1978b). Aqueous two-phase systems for large-scale enzyme isolation processes. In *1st European Congress on Biotechnology*, part I, pp. 48–51. DECHEMA, Frankfurt.

Hustedt, H., K. H. Kroner, U. Menge and M.-R. Kula (1980). Enzyme purification by liquid–liquid extraction. In *Enzyme Engineering*, ed. H. H. Weetall and G. P. Royer, vol. 5, pp. 45–47. Plenum, New York.

Hustedt, H., K. H. Kroner and M.-R. Kula (1981). Large scale isolation of fumarase by partition. Poster presented at the *1st Engineering Foundation Conference on Advances in Fermentation Products Recovery Process Technology.* Banff, Canada.

Hustedt, H., K. H. Kroner, H. Schütte and M.-R. Kula (1983). Extractive purification of intracellular enzymes. In *3rd Rothenburger Fermentations Symposium 'Enzyme Technology'*, ed. R. M. Lafferty, pp. 135–145. Springer-Verlag, Berlin.

Johansson, G. (1974). Effects of salts on the partition of proteins in aqueous polymeric biphasic systems. *Acta Chem. Scand., Ser. B*, **28**, 873–882.

Johansson, G., G. Kopperschläger and P. A. Albertsson (1983). Affinity partitioning of phosphofructokinase from baker's yeast using polymer-bound Cibacron-Blue F3G-A. *Eur. J. Biochem.*, **131**, 589–594.

Kopperschläger, F. and G. Johansson (1982). Affinity partitioning with polymer-bound Cibacron-Blue F3G-A for rapid, large-scale purification of phosphofructokinase from baker's yeast. *Anal. Biochem.*, **124**, 117–124.

Kroner, K. H. and M.-R. Kula (1978). Extraction of enzymes in aqueous two-phase systems. *Process Biochem.*, **13** (4), 7–9.

Kroner, K. H., H. Hustedt, S. Granda and M.-R. Kula (1978). Technical aspects of separation using aqueous two-phase systems in enzyme isolation processes. *Biotechnol. Bioeng.*, **20**, 1967–1988.

Kroner, K. H., H. Hustedt and M.-R. Kula (1982a). Evaluation of crude dextran as phase forming polymer for the extraction of enzymes in aqueous two-phase systems in large scale. *Biotechnol. Bioeng.*, **24**, 1015–1045.

Kroner, K. H., H. Schütte, W. Stach and M.-R. Kula (1982b). Scale-up of formate dehydrogenase by partition. *J. Chem. Technol. Biotechnol.*, **32**, 130–137.

Kroner, K. H., A. Cordes, A. Schelper, M. Morr, A. F. Bückmann and M.-R. Kula (1982c). Affinity partition studied with glucose-6-phosphate dehydrogenase in aqueous two-phase systems in response to triazine dyes. In *Affinity Chromatography and Related Techniques*, ed. T. C. Gribnau, J. Visser and R. J. Nivard, pp. 491–501. Elsevier, Amsterdam.

Kuhlmann, W., W. Halwachs and K. Schügerl (1980). Racemat-Spaltung von Aminosäuren durch Immobilisierung von Acylase in wässrigen Zweiphasensystemen. *Chem. Ing. T.*, **52**, 607, Synopse 821/80.

Kula, M.-R., K. H. Kroner, H. Hustedt, S. Grandja and W. Stach (1976). Separation of enzymes. *Ger. Pat. 2 639 129* (1979), *US Pat. 4 144 130* (1979).

Kula, M.-R. (1979). Extraction and purification of enzymes using aqueous two-phase systems. In *Applied Biochemistry and Bioengineering*, ed. L. B. Wingard Jr., E. Katchalski-Katzir and L. Goldstein, vol. 2, pp. 71–95. Academic, New York.

Kula, M.-R., K. H. Kroner, H. Hustedt and H. Schütte (1981). Technical aspects of extractive enzyme purification. *Ann. N.Y. Acad. Sci.*, **369**, 341–354.

Kula, M.-R., K. H. Kroner and H. Hustedt (1982a). Purification of enzymes by liquid–liquid extraction. *Adv. Biochem. Eng.*, **24**, 73–118.

Kula, M.-R., K. H. Kroner, H. Hustedt and H. Schütte (1982b). Scale-up of protein purification by liquid–liquid extraction. In *Enzyme Engineering*, ed. I. Chibata, S. Fukui and L. B. Wingard Jr., vol. 6, pp. 69–74. Plenum, New York.

Larsson, Ch. and B. Anderson (1979). Two-phase methods for chloroplasts, chloroplast elements and mitochondria. In *Plant Organelles, Methodological Surveys (B) Biochemistry*, ed. E. Reid, vol. 9, pp. 35–46. Ellis Horwood, Chichester.

Lowry, O. H., N. J. Rosebrough, A. L. Farr and R. J. Randall (1951). Protein measurement with the folin phenol reagent. *J. Biol. Chem.*, **193**, 265–275.

Mattiasson, B., M. Ramstorp and T. G. I. Ling (1982). Partition affinity ligand assay (PALA): application in the analysis of hapteus, macromolecules and cells. *Adv. Appl. Microbiol.*, **28**, 117–147.

Menge, U., M. Morr, U. Mayr and M.-R. Kula (1983). Purification of human fibroblast interferon by extraction in aqueous two-phase systems. *J. Appl. Biochem.*, **5**, 75–90.

Müller, W., H. J. Schuetz, C. Guerrier-Takada, P. E. Cole and R. Potts (1979). Size fractionation of DNA-fragments by liquid–liquid chromatography. *Nucleic Acid Res.*, **7**, 2483–2499.

Müller, W. and A. Eigel (1981). DNA-fractionation by two-phase partition with aid of a base-specific macroligand. *Anal. Biochem.*, **118**, 269–277.

Müller, W. and G. Kütemeier (1982). Size fractionation of DNA-fragments ranging from 20 to 30 000 base pairs by liquid–liquid chromatography. *Eur. J. Biochem.*, **128**, 231–238.

Schürch, S., D. F. Gerson and D. J. L. McIver (1981). Determination of cell/medium interfacial tensions from contact angles in aqueous polymer systems. *Biochim. Biophys. Acta*, **640**, 557–571.

Schütte, H., K. H. Kroner, W. Hummel and M.-R. Kula (1983). Recent developments in separation and purification of biomolecules. *Ann. N.Y. Acad. Sci.*, **413**, 270–282.

Shanbhag, V. P. (1973). Diffusion of proteins across a liquid–liquid interface. *Biochim. Biophys. Acta*, **320**, 517–527.

Treybal, R. E. (1963). *Liquid–Liquid Extraction*. McGraw-Hill, New York.

Veide, A., A. L. Smeds and S. O. Enfors (1983). A process for large-scale isolation of β-galactosidase from *E. coli* in an aqueous two-phase system. *Biotechnol. Bioeng.*, **25**, 1789–1800.

Walter, H. (1977). Partition of cells in two-polymer aqueous phases: a surface affinity method for cell separation. In *Methods Cell Separation*, ed. N. Catsimpoolas, pp. 307–354. Plenum, New York.

Wennersten, R., F. Tjerneld, M. Larsson and B. Mattiasson (1983). Extractive bioconversions: hydrolysis of starch in an aqueous two-phase system. In *Proceedings of the International Solvent Extraction Conference, 1983, Denver*. In press.

29

Ion Exchange Recovery of Antibiotics

P. A. BELTER
The Upjohn Company, Kalamazoo, MI, USA

29.1 INTRODUCTION

The principle of ion exchange has been an extremely useful tool for many purposes in the medical and pharmaceutical industries over the years. Antibiotics are probably the most important pharmaceutical products processed extensively by this approach. The need for improved technology to separate, concentrate and purify antibiotics from fermentation sources has led to extensive evaluation of this adsorption approach.

In general, the sequence of operations for the recovery of antibiotics is similar to that used throughout the chemical industry. The total cycle consists of a series of segments: a contacting step to load the resin, a washing step to remove residual broth, an elution sequence with the appropriate solvent to displace the antibiotic, a wash to remove residual eluant, a regeneration step followed by a wash to remove the residual regeneration medium and return of the resin back to the contacting segment for a new cycle. The eluate solution is processed subsequently by a suitable recovery procedure for a given application. Antibiotics, because of their inherent instability and physical properties, present severe constraints for ion exchange that must be met for overall success.

Vast improvements in the nature of ion exchange materials and the ingenuity of engineers in using these improved exchange materials in economical processing sequences have contributed equally to the overall success of this processing approach.

29.2 MATERIALS

29.2.1 Resins

Modern ion exchange technology is based on the attachment of selected polar functional groups to an insoluble organic polymer matrix. The matrices which have been most useful in the processing of antibiotics are the styrene–divinylbenzene, methacrylic acid–divinylbenzene and acrylic acid–divinylbenzene copolymers. Fixed ionic groups are attached to the matrix by one or more chemical reactions, along with their mobile counterions. Although many functional groups have been tried, the useful groups for antibiotic adsorption are shown in Figures 1 and 2 for anionic and cationic resins respectively. The original resins were prepared as gels which had the advantages of high capacity, chemical stability under widely varying conditions and physical stability as discrete beads under adverse processing conditions. However, these resins contained no true porosity and it was necessary for the ionized antibiotics to diffuse through the bead structure to reach the exchange sites. The diffusion rates for materials as large as antibiotics are relatively small, which affected both the adsorption and elution kinetics of the prccess.

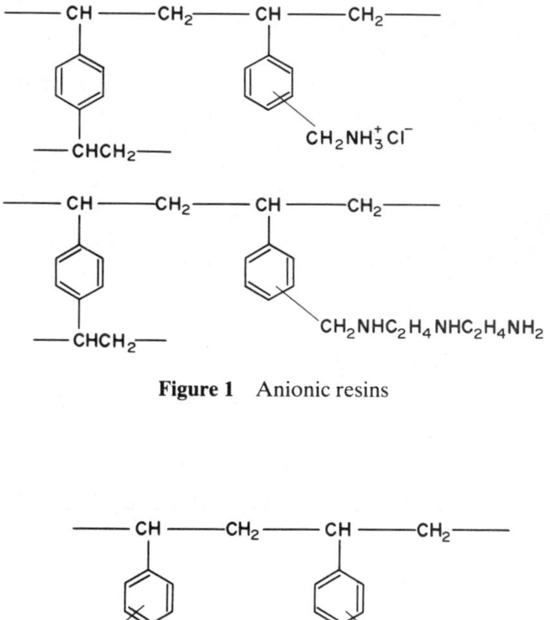

Figure 1   Anionic resins

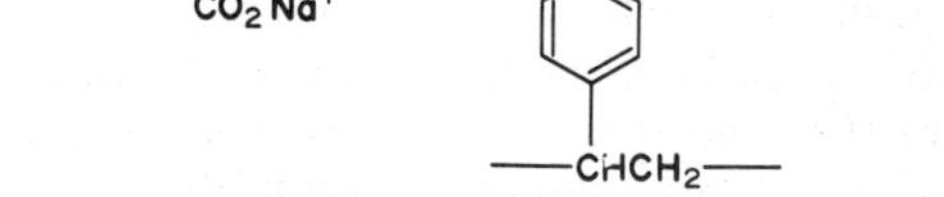

Figure 2   Cationic resins

The search for exchange materials with greater porosity led to the development of macroporous or macroreticular resins. These materials prepared by special polymerization methods have a sponge-like structure and, consequently, large discrete pores that facilitate maximum diffusion. In general, these types of resin have improved kinetics, better resistance to osmotic shock and can accommodate higher molecular weight adsorbates such as the antibiotics. The success of ion exchange technology for the recovery of antibiotics parallels the development of new and more useful exchange materials.

It is apparent that the selection of a suitable exchange resin for a given application is a compromise of numerous factors such as molecular dimensions of the adsorbate, resin capacity, equilibrium relationships, elution properties, hydrodynamics of the column during flow, attrition, resistance to osmotic shock, susceptibility to fouling and ease of regeneration.

29.2.2 Adsorbates—Antibiotics

The antibiotics are relatively complex organic compounds produced in fermentations as secondary metabolites of specific organisms. Their molecular weights range from about 400 to 750. Their importance as pharmaceutical products lies in their ability to inhibit the growth of pathogenic microorganisms and, consequently, they are used to fight infections. In general, these materials can be classified into four clinically active groups: (1) the β-lactams including the penicillins and cephalosporins; (2) the aminoglycosides; (3) the macrolides; and (4) the tetracyclines. Lincomycin and novobiocin are antibiotics that do not fit into these classifications. Recovery of the first two classes of antibiotics has benefited appreciably with the advance of ion exchange technology.

Although the use of ion exchange has been reported for laboratory separations of the macrolides, such as erythromycins, oleandomycins and rosamicin, most of the effort has been concentrated on solvent extraction or adsorption schemes using non-ionic resins for these materials. Also, commercial methods of recovery for the tetracyclines have been limited to solvent extraction and processes based on the precipitation of suitable salts. Since these processes are outside the scope of this chapter, there will be no further discussion of the recovery of these two classes of antibiotics.

29.2.2.1 β-Lactams

The penicillins are characterized by a four-membered lactam ring and a thiazolidine ring fused together as illustrated in Figure 3 where the basic nucleus of 6-aminopenicillanic acid (6-APA) and several different penicillins are shown. The penicillins, which are differentiated by the various side chain constituents, are prepared by the addition of monosubstituted acetic acid precursors to penicillin fermentations. Although these biosynthetic penicillins were recovered by solvent extraction techniques, the discovery of the presence of 6-APA, and its recovery by ion exchange techniques from fermentation broths to which precursors had not been added, provided the practical means of obtaining large quantities of 6-APA. The subsequent acylation of 6-APA permitted the introduction of various side chains and resulted in the semisynthetic penicillins.

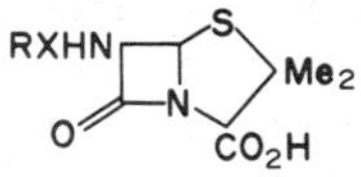

Figure 3 Structures of typical penicillins: 6–aminopenicillanic acid, R = —, X = H; benzylpenicillin, R = $PhCH_2$, X = CO; phenoxymethylpenicillin, R = $PhOCH_2$, X = CO; ampicillin, R = $PhCHNH_2$, X = CO

The ion exchange process for this important intermediate has been described by Batchelor *et al.* (1961). Because of the strongly acidic properties of 6-APA, it is absorbed on Dowex 1 resin in the H^+ cycle and eluted with dilute 0.1 N hydrochloric acid. The active fractions of the eluate are neutralized and concentrated. Crystallization is accomplished by the slow addition of 20% hydrochloric acid to a final pH of 4.3. The resultant crystals obtained by cooling are filtered, washed with water and dried. If a product of higher quality is desired, recrystallization following a similar procedure can be performed.

The cephalosporins constitute a second major class of antibiotics that are characterized by the β-lactam ring fused to a dihydrothiazine ring. The basic structure of this class of compounds is illustrated in Figure 4 where the structure of the compound cephalosporin C is given. This material occupies a unique place in the development of this class of antibiotics since it is the source of the 7-aminocephalosporanic acid (7-ACA) from which the numerous semisynthetic cephalosporins have been prepared.

Many approaches have been published for the recovery of this important intermediate. However, ion exchange has proved to be the most economical with practically complete recovery of

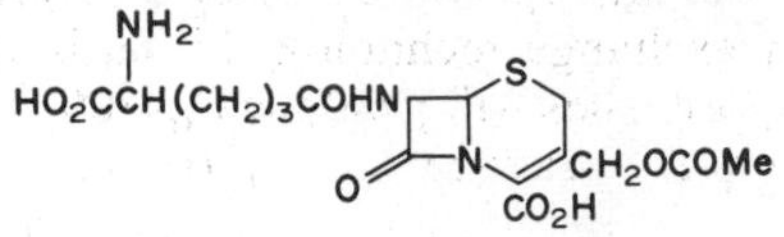

Figure 4 Cephalosporin C (MW 415.44)

this activity. Recovery processes for cephalosporin C uing ion exchange technology have been published by Trown *et al.* (1962) and Abraham *et al.* (1965).

29.2.2.2 Aminoglycosides

Aminoglycosides are characterized by carbohydrate moieties bound to complex cyclitol structures. Examples of this class of antibiotic are streptomycin, neomycins, gentamicins and kanamycins. The structure of streptomycin is shown in Figure 5 to illustrate this type of antibiotic. It is the strongly basic guanidine functionality of streptomycin that enters into the exchange reactions.

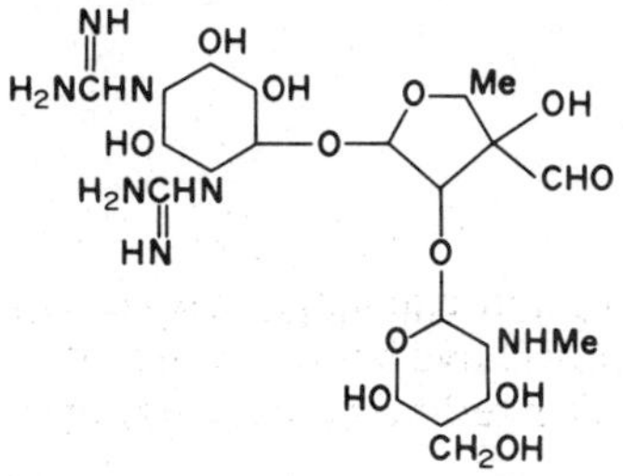

Figure 5 Streptomycin (MW 581.58)

Howe and Putter (1951) have patented an ion exchange process which uses a weakly acidic carboxylic acid resin in the salt form to obtain the desired loading. Once the resin is loaded, the antibiotic can be eluted with a dilute acid solution. After elution, the resin is washed with water to remove any excess acid, returned to the salt form with a stoichiometric amount of sodium hydroxide, washed with water and returned to the adsorption segment of the cycle.

Subsequent operations to remove inorganic salts and to decolorize the antibiotic solutions are necessary if a high quality product is desired. The desalting operation can be conducted by a two-stage exchange system consisting of contact with a cationic resin such as Amberlite IR-124 in the H^+ form, followed by contact with an anionic resin in the base form, such as Amberlite IR-45. The decolorization step has been accomplished by contact with Amberlite IRA 4015 in the chloride form. Sodium chloride is used for regeneration of this resin.

A similar ion exchange process for neomycin has been reported in a patent by Nager (1954). However, the basic amine groups in the neomycin molecule are considerably weaker than the guanidine grouping in the streptomycin molecule, therefore different conditions are employed to process this antibiotic. The carboxylic acid resin, such as Amberlite IRC-50, is adjusted to a pH of 7 to 8 prior to the contacting operation. The exchange takes place between the neutralized filtered fermentation broth and the partially neutralized carboxylic acid resin. The loaded resin, after washing with water to remove residual spent broth, is eluted with an alkaline solution. The eluant is usually a dilute ammonia solution of sufficient pH to repress the ionization of the amine groups on the neomycin molecule and overcome its affinity for the resin. Similar processes have been reported by Rosselet *et al.* (1964) and Lee *et al.* (1974) for the gentamicins and by Marquez and Kreshner (1978) for the kanamycins.

29.2.2.3 Lincomycin

The structure of lincomycin is shown in Figure 6. Miller and Gonzales have reported on a number of methods for the recovery of lincomycin in Volume 3 of this work. Most of these processes take advantage of the fact that the structure is basic, having a pK_a of 7.6. These methods include solvent extraction, carbon adsorption, azeotropic drying, and non-ionic macroporous resin adsorption as well as ion exchange technology. This is typical of fermentation product recovery, in that a number of approaches will produce a product but the final choice is based on economics and the intended use of the product.

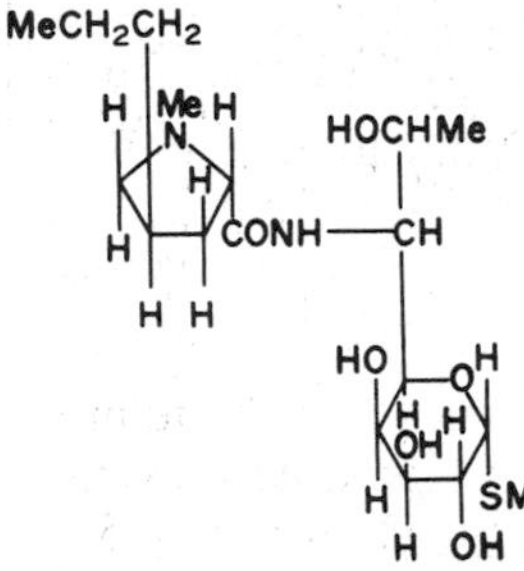

Figure 6   Lincomycin (MW 406.56)

29.3 PROCESSING METHODS

29.3.1 Fixed Bed Modes

Numerous methods of contacting ion exchange resins with antibiotic carrying solutions have been used over the years. Adsorptions have been conducted in both batch and continuous modes. Processing has been performed with fixed bed (column) contacting, fluidized bed contacting and countercurrent contacting.

The first step in the conventional ion exchange recovery sequence has been filtration to remove mycelia, spores and other insoluble impurities. Removal of these particulates was necessary to facilitate subsesquent concentration and purification steps such as solvent extraction, adsorption, ion exchange or direct precipitation of a crude material. Obviously, suspended cells and other particulates would interfere with any of these sequences.

The majority of ion exchange operations in the antibiotic industry have been conducted in the fixed bed mode. In this mode the clarified antibiotic solution is fed to the top of a static bed of ion cxchange resin which is sufficiently large to accomplish the desired adsorption. It has been reported by Wheaton and Seamster (1966) that industrial columns range in size from a few inches in diameter to over 20 feet with multiple column total heights of over 100 feet.

The entire sequence of operations consists of the adsorption step, washing with water to remove residual broth from the column, elution of adsorbed antibiotic with a suitable eluant, washing to remove the residual eluant, regeneration of the resin back to its original condition, washing of the resin to displace the residual regenerating solution and return of the resin to the initial operation of ion exchange. Each of these operations is performed as an unsteady state operation. Consequently, as discussed by Belter (1979), the performance of these columns is more difficult to evaluate because of the inherent transient conditions used.

29.3.1.1 Simulation models

(i) Continuum models

Simulation models for the adsorption (or ion exchange) of antibiotics in the fixed bed mode have been developed and reported. These models have used both of the conventional approaches: consideration of the fixed bed as a continuum of resin of uniform porosity (Chen *et al.*, 1968) and the 'tanks-in-series' approach (Chen *et al.*, 1972). The set of differential equations to be solved for the first approach is given in Table 1. The continuity equation for a differential element of a bed is a partial differential equation with independent variables of position and time. In this equation C represents the concentration of antibiotic in the solution surrounding the resin, Z is the position within the column, t is time, ε is the void volume of the column, U is the superficial velocity of the solution, ϱ is the bulk density of the resin and q is the concentration of antibiotic within or on the resin. The second relationship expresses the rate of mass transfer in terms of an overall transfer coefficient and a potential driving force. C^* is the concentration of antibiotic assumed to be in equilibrium with q, the resin concentration, and K_a is the overall mass transfer coefficient. Even though several resistances to mass transfer exist, this approach represents a simplification of the actual transient profiles established around and within the resin matrix. The third equation expresses the relationship between the value of the overall mass transfer coefficient and the loading on the resin. This relationship must be determined empirically for given operating conditions since the coefficient is dependent on both liquid phase diffusion and pore diffusion mechanisms. The fourth and final equation represents the surface or interfacial equilibrium between the solute in the liquid and that in the resin phase. Although the equilibrium relationship shown is of the Freundlich-type, other mathematical representations of the isotherm may be used. Solution of this set of simultaneous equations requires the help of a fairly powerful digital computer. Such calculations have proved extremely successful for the design and scaleup of transient state antibiotic–ion exchange processes.

Table 1 Adsorption Model for Fixed Beds

Continuity equation	$\dfrac{U}{\varepsilon}\dfrac{\partial C}{\partial Z} + \dfrac{\partial C}{\partial t} = \dfrac{-\varrho}{\varepsilon}\dfrac{\partial q}{\partial t}$
Exchange rate	$\dfrac{\partial q}{\partial t} = K_a(C - C^*)$
Transfer coefficient	$K_a = f(q)$
Equilibrium	$C^* = b(q)^a$

(ii) Tanks in series

The set of differential equations presented in Table 2 constitutes the simulation model for the second approach, the 'tanks-in-series' approach. In this approach the fixed bed is conceptually divided into a number of discrete stages and each treated as a completely mixed tank with feed and effluent streams. This concept simplifies the continuity equation to an ordinary differential equation by eliminating position within the column as an independent variable. In the continuity equation shown in Table 2 V_c represents the volume of the tank, V_R is the volume of resin in the tank and F is the volumetric flow rate of antibiotic solution. Other symbols have the same definitions as given previously. The set of simultaneous equations shown in Table 2 is applicable also to the fluidized bed method of contacting resin and broth to be discussed later.

Table 2 Adsorption Model for Fluidized Beds

Continuity equation	$V_c\dfrac{dC_n}{dt} = FC_{n-1} - FC_n - V_R\dfrac{dq}{dt}$
Exchange rate	$\dfrac{dq}{dt} = K_a(C - C^*)$
Transfer coefficient	$K_a = f(q)$
Equilibrium	$C^* = b(q)^a$

29.3.2 Fluidized Beds

The inherent problems associated with the filtration or centrifugation necessary to pretreat the fermentation broth have been covered by Belter (1979). Equipment, filter aid and labor costs are high for these operations and are usually accompanied by an appreciable loss of antibiotic activity in the discarded cake. In addition, several pH adjustments and/or a heating step may be required to render the whole broth suitable for filtration on commercial equipment.

Bartels *et al.* (1956) patented new technology for the adsorption of streptomycin on a cationic exchange resin from a whole unfiltered broth using specialized equipment. These authors reported in 1958 a description of the equipment and process. This approach eliminated the excessive losses, operations and costs of the filtration step. Belter *et al.* (1973) reported a mechanically similar process for the recovery of the acidic antibiotic, novobiocin, from an unfiltered fermentation broth.

29.3.2.1 Process descriptions

The sequence of operations in this process is as follows. The 'whole beer or broth', direct from the fermenter, is passed over a high capacity vibrating screen to remove hulls, chaff and other agglomerated insoluble materials from the media which might accumulate in the resin bed if not removed. The resulting screened broth is then contacted with the desired resin in a series of specially designed, well mixed columns. Each of the columns is equipped with a special head screen sized to permit passage of mycelia and other insolubles but to retain the resin particles.

Periodically, the lead column is removed from the train, washed free of insolubles, eluted and regenerated to its initial form. Subsequently, the resulting eluates are processed to a crystalline product of high quality by conventional purification methods. Periodic countercurrent operation is obtained in the sorption sequence by advancing each of the uneluted columns one position and placing a freshly eluted and regenerated column in the tail position of the train. The broth feed cycle is repeated.

29.3.2.2 Simulation models

The equations which govern the performance of each column in the adsorption train have been given in Table 2. Modelling of the exchange sequence consists of solving this set of simultaneous equations for each of the columns in the system. Such calculations permit the prediction of effluent concentrations as a function of time (loss curves) and resin concentrations as a function of time (resin loading curves). The loss and loading curves can, in turn, be converted to total or percentage loss for process evaluation studies. It should be remembered that these equations apply to any constituent of the beer as well as the antibiotic and that with additional assay information, resin loading and loss curves could be obtained for other components as well. The described model has been used for the study of a number of antibiotic recovery processes using ion exchange and the results of these studies have resulted in several production plants based on this technology.

29.3.3 Other Methods of Contacting

Although the previous processing technique gives a periodic countercurrent contacting, a second approach to 'whole broth' operation, which results in a true countercurrent process, is that developed by the Greerco Corporation. This equipment consists of a series of horizontal paddle-agitated conveyors assembled so that each of the operations, ion exchange, washing, elution, washing, regeneration and washing is conducted in a countercurrent fashion. Personal communications (1983) reveal an efficiency increase with this technique. Several other methods of contacting resins and liquids in a countercurrent mode, such as the Higgins contactor (1954), the Asahi (1963) and the Aconex exchange systems (1964), have been developed for the chemical industry. However, no references were found to suggest that any of these systems have been employed for antibiotic recovery.

29.4 SUMMARY

The principle of ion exchange has been an extremely useful tool for the recovery of antibiotics over a period of several decades.

In general, the sequence of operations for the recovery of antibiotics is similar to that used in ion exchange throughout the chemical industry. However, antibiotics, because of their inherent instability and physical properties, impose severe constraints for ion exchange that must be met for overall success.

The development of increasingly porous resins by the manufacturers have contributed appreciably to process improvements for the recovery of antibiotics. The selection of a suitable exchange resin for a given application is a compromise of many factors.

Ion exchange processes for the recovery of β-lactams, aminoglycosides, lincomycin and novobiocin are discussed briefly.

In addition to the contributions of the resin manufacturers the ingenuity displayed by process engineers to use these resins efficiently for the recovery of antibiotics has also contributed substantially to the success of modern ion exchange technology.

New methods of contacting fermentation broths with ion exchange resins have improved the recovery of antibiotics and the modelling of these processes has resulted in a better understanding of the recovery operations. Modelling of both fixed bed and fluidized bed contacting was discussed.

29.5 REFERENCES

Abraham, E. P., G. G. F. Newton and C. W. Hale (1965). US Pat. 3 184 454.
Bartels, C. R. (1958). *Chem. Eng. Prog.*, **54** (8), 49.
Bartels, C. R., B. Berk and W. L. Bryan (1956). US Pat. 2 765 302.
Batchlor, F. R., E. B. Chain, T. L. Hardy, K. R. L. Mansford and G. N. Robinson (1961). *Proc. R. Soc. London, Ser. B,* **154**, 498.
Belter, P. A. (1979). General procedures for isolation of fermentation products. *Microbial and Fermentation Technology,* chap. 16. Academic, New York.
Belter, P. A., F. L. Cunningham and J. W. Chen (1973). Development of a recovery process for novobiocin. *Biotechnol. Bioeng.*, **15**, 533.
Chen, J. W., J. A. Buege, F. L. Cunningham and J. I. Northam (1968). Scaleup of a column adsorption process by computer simulation, *Ind. Eng. Chem. Process Des. Dev.*, **7**, 26.
Chen, J. W., F. L. Cunningham and J. A. Buege (1972). Computer simulation of plant scale multicolumn adsorption processes under periodic countercurrent operation. *Ind. Eng. Chem. Process Des. Dev.*, **11**, 430.
Higgins, I. R. and J. T. Roberts (1954). *Chem. Eng. Prog., Symp. Ser. 14,* **50**, 87.
Howe, E. E. and I. J. Putter (1951). US Pat. 2 541 420.
Marques, J. A. and A. Kershner (1978). In *Antibiotics—Isolation, Separation and Purification*, ed. J. Weinstein and G. H. Wagman, p. 182. Elsevier, Amsterdam.
Nager, U. F. (1954). US Pat. 2 667 441.
Lee, B. K., R. G. Condon, G. H. Wagman, K. Byrne and C. Shaffner (1974). *J. Antibiot.*, **27**, 822.
Rosselet, J. P., J. Marquez, E. Meseck, S. Murawski, A. Hamdan, C. Joyner, R. Schmidt, D. Migliore and H. L. Herzog (1964). *Antimicrob. Agents Chemother.*, 14.
Trown, P. W., E. P. Abraham, G. G. F. Newton, C. W. Hale and G. A. Miller (1962). *Biochem. J.*, **84**, 157.
Wheaton, R. M. and A. H. Seamster (1966). Ion exchange. *Kirk-Othmer Encyclopedia of Chemical Technology*, 2nd edn., vol. 11, p. 871. Wiley, New York.

30

Ion Exchange Recovery of Proteins

D. E. PALMER and R. V. DOVE
Bio-Isolates Ltd., Swansea, UK
and
R. A. GRANT
Ecotech Systems Ltd., Poole, Dorset, UK

30.1 INTRODUCTION

Ion exchange as a technology for the recovery of materials such as metals is well known and has been employed industrially for many years. These exchangers are based on synthetic materials, usually of coal or oil origin and their use is limited to the recovery of small ions although attempts have been made to use them for the recovery of macromolecules. One of the main disadvantages of these exchangers when used for the recovery of organic macromolecules is that the delicate structure of the molecules is disrupted and proteins, for instance, cannot be recovered without denaturation. Additionally, adsorption capacity is generally small (Table 1).

The first attempts at producing ion exchangers suitable for the recovery of proteins and similar macromolecules from solution were restricted to hydrophilic exchangers based on natural cellulose. For a long time the technology was limited to laboratory use where it proved very valuable for protein analysis and laboratory scale isolation. A major obstacle to the scaling up of processes based on natural celluloses was that, because they were in the physical form of fibres, the exchangers had unsatisfactory hydraulic properties for use in large scale plants.

The second generation of hydrophilic ion exchangers was also based on cellulose, but on regenerated cellulose. Work carried out in New Zealand and in the UK led to the development of a range of hydrophilic ion exchangers which had a granular structure and, consequently, improved hydraulic properties. The improved flow rates obtained with the new exchangers enabled protein fractionations to be performed at rates of up to ten times those previously attainable and without loss of resolution efficiency. Some characteristics of ion exchangers are summarized in Table 1.

Table 1   Characteristics of Ion Exchangers

(i)	*'Small Ion' Exchangers employing Synthetic Resins*	
	Uses	De-ionization, water treatment, metal recovery
	Limitation	Resins unable to successfully adsorb macromolecules, *e.g.* protein, with reasonable uptake and without damage to protein molecules
(ii)	*Macromolecular (Hydrophilic) Ion Exchangers*	
	First generation	Based on natural cellulose Good capacity; protein undenatured but fibrous nature prohibited large-scale use
	Second generation	Based on regenerated cellulose Excellent capacity for protein; granular structure gives improved hydraulic properties and allows large-scale operation
	Third generation	Based on modified, regenerated cellulose High performance materials

30.2   TYPES OF HYDROPHILIC ION EXCHANGERS

High performance regenerated cellulose ion exchangers are now available in strongly and weakly acidic or basic forms corresponding to the main types of conventional ion exchange resins. The principal types so far developed are the carboxymethyl, diethylaminoethyl, sulfopropyl and quaternary base derivatives. The ionization characteristics for these resins, determined by titration using a glass electrode in the presence of salt, are shown in Figure 1. The effective working pH range covered by these types of resin is approximately 2–12.

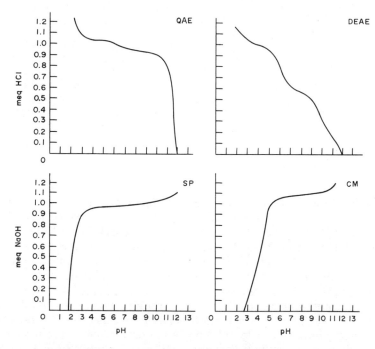

Figure 1   Resin ionization characteristics

30.3   ADSORPTION AND DESORPTION OF PURE PROTEINS

Figure 2 shows a typical adsorption–desorption cycle for the DEAE anion exchange regenerated cellulose resin and 0.1% aqueous bovine serum albumin solution. Protein concentration was monitored in the column effluent by UV absorption measurement at a wavelength of 280 nm. The completely flat baseline indicates total uptake of protein up to the breakthrough point; this was confirmed by the absence of turbidity when phosphotungstic acid was added to the column effluent. The uptake curve may be compared with a similar curve obtained with an identical column of a macroporous condensation type resin (Figure 3). It may be seen that in this case breakthrough occurred very rapidly after the solution was applied to the column, indicating negligible capacity of this resin for protein.

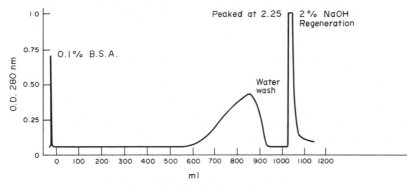

Figure 2 Protein adsorption curve

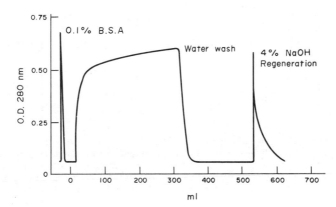

Figure 3 Protein adsorption curve

The UV absorbance measured at 280 nm is directly proportional to the protein concentration for solutions of pure proteins. In Figures 2 and 3 the baselines were initially adjusted using distilled water, and the height of the first peak in each case indicates the absorption due to the bovine serum albumin test solution. The volume during which the UV absorption falls to the blank (H_2O) level indicates the volume for which complete protein uptake was obtained in each case.

Following breakthrough the cellulose resin column was washed with water, applied by downflow, until the UV absorbance fell to zero. Regenerant solution consisting of 0.5 M sodium hydroxide was then applied, which stripped the protein from the bed in a concentrated solution, as shown by the second peak; it may be noted that the elution curve is tall and narrow and shows little evidence of tailing. Usually complete regeneration is achieved with one bed volume of regenerant followed by rinsing with two bed volumes of water. In contrast, the regeneration peak for the macroporous condensation type resin was small and asymmetric indicating minimal uptake of protein.

Repeated cycling of regenerated cellulose resins with pure protein over many cycles has shown no significant changes in adsorption efficiency, indicating that there is little irreversible binding of protein under these conditions.

In general the capacities for proteins have been found to lie within the range 300–500 mg protein (bovine serum albumin or lysozyme) per gram dry resin; exceptionally, capacities up to 1 g protein per gram resin have been recorded.

30.4 ION EXCHANGE SYSTEMS

Given a suitable ion exchange material, the other requirement for a successful ion exchange system is an effective means of separating solids from liquids. This may be achieved by using columns, fixed beds, fluidized beds, agitated columns or stirred tanks; the choice depends on the properties of the exchanger and/or the nature of the material to be treated. For instance, a stirred tank system is more successful than a fixed bed column for material with high suspended solids. When fat is present, an agitated column or stirred tank to assist in fat removal is more successful. More highly concentrated eluates may be obtained when columns are used.

30.5 APPLICATIONS OF HYDROPHILIC ION EXCHANGE

30.5.1 Abattoir Effluent Treatment

The new hydrophilic ion exchangers were first used in a column system in New Zealand for the treatment of abattoir effluent and in the UK in a continuous stirred tank reactor system consisting of three sections with settling cones (Figure 4).

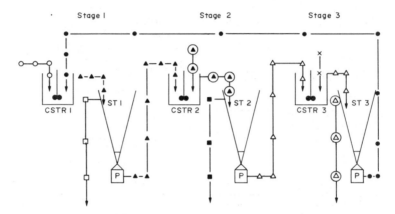

Figure 4 Flow diagram of continuous ion exchange system. Stage 1: protein adsorption. Stage 2: protein recovery and media regeneration. Stage 3: media washing. Key: ○, effluent containing protein; ●, cellulose adsorbent material; ▲, adsorbent plus protein stream; ◍, regenerant NaCl; ◭, desorbed protein plus adsorbent plus regenerant; ■, concentrated protein stream; □, deproteinized effluent; △, regenerant plus cellulose adsorbent; ×, wash-water; ⊗, regenerant plus wash-water; CSTR, continuous stirred tank reactors; P, pump; ST, settling tanks.

30.5.1.1 Column techniques

The columns or beds used were of the conventional type, provided with means for fluidization and backwashing.

Abattoir effluents are very polluting with high biological oxygen demand (BOD) values, the principal contaminants being fats and protein. In general such effluents contain large amounts of suspended insoluble matter as well as fat, and if applied directly to the resin bed will cause rapid clogging of the bed with a large fall-off of flow rate. Therefore, such effluents should be subjected to a pretreatment, preferably consisting of flocculation and air flotation to remove particulate matter and fat. Typical results obtained with this type of effluent are shown in Tables 2–4 where ion exchange has been combined with a pretreatment of the above type. Combined treatment of this type employing the DEAE cellulose anion exchanger in the second stage has given reductions in the BOD level in excess of 95%. The general forms of the adsorption and breakthrough curves are shown in Figures 5 and 6.

The resin columns were regenerated with a solution of 1% NaOH + 3.5% NaCl followed by a water backwash and downward rinse. Since the DEAE resin has a certain degree of strong base function and since the effluent contains salts, a high initial pH is produced which falls steadily thereafter. Usually, the pH remains at about 10 for the first 10–15 bed volumes (BV) after which it decreases to about 9 during the next 30–50 bed volumes and eventually falls to about 8. Under

Table 2 Residual COD (%) in Slaughterboard Effluent after Treatment

Treatment	Sample							Means
	1	2	3	4	5	6	7	
Pretreatment	19	40	31	37	35	31	31	32
Ion exchange	18	31	31	28	25	27	15	25
Pretreatment + ion exchange	0[a]	14	9	22	0[a]	11	0[a]	11

[a] Value too low for measurement by standard method.

Table 3 Residual Protein (%) after Treatment in Slaughterboard Effluent

Treatment	Sample					Means
	1	2	3	4	5	
Pretreatment	29	40	32	29	47	35
Ion exchange	36	21	18	23	45	28
Pretreatment + ion exchange	6	5	6	8	5	6
Initial level (mg ml^{-1})	0.63	0.70	0.39	0.18	0.43	0.47

Table 4 Residual BOD (%) in Treated Slaughterboard Effluent

Treatment	Sample					Means
	1	2	3	4	5	
Pretreatment	38	41	31	23	27	32
Ion exchange	10	13	23	19	18	17
Pretreatment + ion exchange	0[a]	5	12	7	0[a]	5

[a] Value too low for measurement by standard method.

normal conditions both the colour (optical density of 409 nm) and the protein concentration of the effluent are minimal until breakthrough occurs, usually at 40–100 BV depending on the strength of the effluent. Following breakthrough the values for protein and colour in the effluent often exceed the input values for a time before decreasing to a level equal to the input. In the case of slaughterhouse effluent, this overshoot appears to be due to haemoglobin accumulating in a band which travels down the column to emerge at the breakthrough point.

Chemical oxygen demand (COD) measurements result in a different type of curve in which a fairly constant residual value equal to 5–20% of the initial COD value is maintained up to the breakthrough point. This behaviour is explained by the presence in the effluent of uncharged organic compounds, such as carbohydrates, urea, *etc.*, which are not adsorbed by the resin, and also to low MW protein breakdown components produced by the action of bacteria in the effluent. It tends to become more pronounced the longer the effluent is kept before treatment. It is generally found that following breakthrough the protein curve tends to lag behind the optical density curve and the COD lags behind the protein curve with this type of effluent. Generally the adsorption cycle is terminated at the breakthrough point and the resin bed regenerated with alkaline brine solution (1 bed volume) which is allowed to stand in the bed for 30 min to 1 h before rinsing off with water. In cases where a small amount of colour and protein can be tolerated in the effluent, the adsorption cycle may be continued beyond the breakthrough point.

Since the spent regenerant from the ion exchange bed consists of concentrated protein solution, this should be treated to remove as much as possible of the organic matter and reduce the BOD level before discharge. It is possible to recover virtually all the desorbed protein by adjustment of the pH to about 4.5 followed by heat coagulation. The insoluble heat-coagulated protein is readily dewatered by screening or filtration and may be dried to yield a by-product suitable for inclusion in animal feeds. The nutritional value of effluent protein recovered by ion exchange has been evaluated by amino acid analysis and animal feeding trials; there was no indication of toxic effects arising from the treatment process.

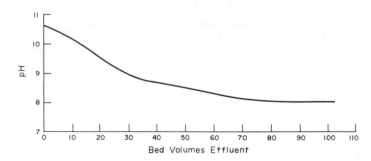

Figure 5   Column breakthrough curve

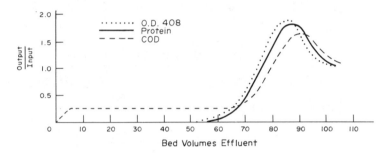

Figure 6   Column breakthrough curves

30.5.1.2   *Continuous method*

The effluent, together with the cellulose medium stream, is fed into a stirred tank where the protein is adsorbed and the suspension is allowed to flow into the first settling cone where the medium, with adsorbed protein, settles out and the treated liquor overflows. The medium with adsorbed protein is transferred to the second cone where the protein is desorbed by the addition of salt. The medium is allowed to settle and the protein solution overflows. The medium is transferred to the third cone where it is washed and regenerated, ready for recycling.

Most of the protein is removed and recovered and since the BOD in the case of abattoir effluent is due mainly to protein, the recovery system also serves as an efficient effluent treatment system, capable of removing about 98% of BOD.

However, the economics of the process for the treatment of abattoir effluent are doubtful unless effluent treatment charges increase. It seems likely that the application of the technique in abattoirs will, for some time, be restricted to the isolation of protein from hygienically collected blood, either for preparation of bulk quantities of albumin, or globulins, or for the isolation of high value proteins as reference proteins in biochemistry.

30.5.2   Whey Treatment

Cheese whey is an obvious source of high quality protein suitable for recovery by ion exchange. Cheese or casein wheys contain the albumins and globulins of milk which together amount to about 20% of the milk protein. Typically, whey contains: 0.6% protein, 0.6% minerals, 0.06% fat and 4.6% lactose. Total whey solids amount to 60 g l^{-1} of which 6 g is useful protein.

A separation method used for whey treatment which is based on a filter bottom stirred tank reactor has been developed.

A simplified flow sheet is shown in Figure 7. The reactor is a vessel fitted with a sieve bottom, the reaction taking place above the sieve. Whey is drawn into the reactor, the pH is adjusted and the protein is adsorbed onto the medium. Deproteinized whey is removed through the sieve and, depending on the purity required, the protein-laden medium is washed free of impurities. Water is added and the pH adjusted to release the protein into solution.

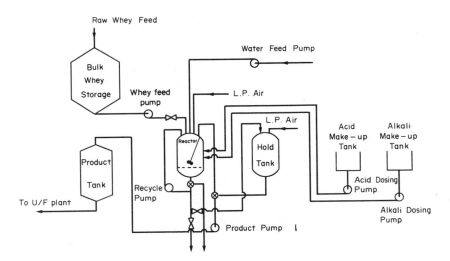

Figure 7 Simplified flow sheet of a whey protein recovery process

After complete desorption, the protein solution is concentrated and spray dried to produce a white protein powder. As extremes of pH are not used, little protein denaturation occurs. Invariably the protein recovered is of high quality.

30.5.3 Other Protein Wastes

Other waste streams that have been treated successfully by ion exchange include potato effluents, distillery effluents and animal by-product treatment effluents. The efficiency of protein removal is very good (Table 5) and at some future date it is likely that high quality, high value proteins may be isolated from these streams. The composition of protein products recovered from some of these wastes is summarized in Table 6. Other sources to be further examined include pea, sunflower, palm oil and leaf.

Table 5 Recovery of Proteins from some Food Processing Wastes by Ion Exchange

| | Protein N (g l^{-1}) | | Removal |
Waste stream	Before	After	of protein (%)
Potato effluent	0.78	0.035	95.5
Distillery effluent	0.48	0	100.0
Animal by-product	0.52	0.036	93.0

Table 6 Typical Analysis of Proteins Isolated by Ion Exchange from some Process Effluents

Process effluent	Protein (%)	Ash (%)
Soy	92	4
Potato	94	1.5
Rapeseed	95	3

30.6 CONCLUSION

Hydrophilic ion exchange, based on high performance cellulosic exchangers for the large scale recovery of proteins, is now a practical proposition. The economic viability of the processes, however, usually depends on achieving a recovered product of substantial resale value.

31

Molecular Sieve Chromatography

M. L. YARMUSH, K. ANTONSEN and D. M. YARMUSH
Massachusetts Institute of Technology, Cambridge, MA, USA

31.1 INTRODUCTION

Molecular sieve chromatography (MSC), also often called gel permeation chromatography, gel exclusion chromatography, gel filtration, or simply gel chromatography, is a form of partition chromatography where molecules are separated according to their effective size in solution using matrices with defined porosities. This concept is illustrated schematically in Figure 1. The smaller the hydrodynamic volume of a particular molecule the more effectively it may penetrate into the stationary phase thus removing itself from the solvent bulk flow. Consequently, components which are small compared to the matrix interstices traverse the column very slowly, and appear last in the chromatogram (Figure 1c). Molecules which are large compared to matrix porosity are completely excluded from the stationary phase and thus move swiftly through the column appearing first in the chromatogram (Figure 1a). Molecules which are intermediate in size enter the interstices but spend less time within the stationary phase than do the smallest particles (Figure 1b). These components travel at speeds consistent with their particular sizes.

The widespread use of MSC for separation of polymeric material, especially protein mixtures, can be attributed to several factors including relatively low cost, simplicity of operation and ability to process relatively large amounts of material. In addition to its preparative aspects, MSC

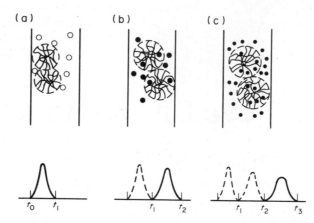

Figure 1 Schematic representation of MSC: (a) large molecules which do not penetrate the support appear first in the chromatogram (t_0-t_1); (b) molecules of intermediate size are retarded somewhat and thus appear second in the chromatogram (t_1-t_2); (c) the smallest molecules spend the most time within the stationary phase and thus appear last in the chromatogram (t_2-t_3)

can also be used for product characterization (*i.e.* molecular weight determination), estimation of product purity and characterization of a variety of molecular interactions.

The objectives of the chapter are to provide a short discussion of some of the theoretical and practical aspects of MSC. No attempt has been made to provide an exhaustive review of the voluminous literature which now exists on this subject. More extensive treatments can be found in numerous books and monographs (Determann, 1969; Giddings, 1965; Fischer, 1980; Provder, 1980; Kremmer and Boross, 1979; Morris and Morris, 1976). Laboratory handbooks dealing mainly with practical aspects of MSC are also available from several commercial concerns (Pharmacia, Bio-Rad).

31.2 MATERIALS

The most important component in the design of a molecular sieve separation is the column packing material since its characteristics have a major effect on column size, throughput and separation efficiency. Important variables in the choice of an appropriate packing material include chemical inertness, stability, rigidity and pore-size distribution.

31.2.1 Inertness

The ideal packing material is inert with respect to the proteins or other compounds being separated. In practice this means that the packing material should be hydrophilic because proteins tend to adsorb and denature irreversibly on hydrophobic surfaces (Porath, 1974). Packings carrying surface charge are generally considered undesirable because the charges introduce electrostatic effects. A given charged protein is thus either attracted to or repelled from the packing, which either decreases or increases, respectively, its rate of passage through the column. Highly charged packings cannot be used because proteins are irreversibly adsorbed.

The inertness requirement immediately eliminates many packings used for non-biological applications. For example, controlled-pore glass, although hydrophilic, is highly charged and can be used only if the surface is modified.

31.2.2 Rigidity

Packing materials can generally be classified into two distinct groups: compressible supports and non-compressible supports. As the pressure drop across a compressible support is increased from zero, the flow rate first increases, but then passes through a maximum and declines. This limit on the maximum flow rate leads to long separation times or, for large-scale separations, low throughput. Rigid particles, on the other hand, impose no such limit. Moreover, particle size limitations are not as severe when dealing with rigid materials. For example, smaller particles increase separation efficiency by reducing intraparticle mass transfer resistance and dispersion. Hence, a smaller column volume can be used for a given separation, leading to lower cost and shorter separation times. Because a larger pressure drop per unit length is necessary to push fluid through a column made up of smaller particles, use of small-particle compressible supports is precluded.

31.2.3 Stability

The packing material should be insoluble and stable over the pH and temperature range to which it is likely to be exposed. Moreover, it should not be degradable by the appropriate solvents, chaotropic salts or bacteria.

31.2.4 Pore-size Distribution

The required pore-size distribution of the packing material is determined by the particular application. For example, for desalting a particular solution of macromolecules, very small pores are needed so that all macromolecules are excluded from the pore matrix. Protein separations, on the other hand, require differences in the accessibilities of the individual proteins to the pores for adequate separation to occur.

No packing material available today rigorously meets all of these requirements, which to some extent compete with one another. Rather, those available represent a compromise, and as a consequence, relatively few have been exploited. Nevertheless, recent development has allowed introduction of some new materials. Presently, the most commonly used supports are gels of cross-linked dextran, polyacrylamide, agarose and combinations thereof. The characteristics of these materials are discussed below. A partial list of commercially available products and their separation characteristics is given in Tables 1 and 2. In Table 3, the chemical and physical properties of these supports are summarized.

31.2.5 Cross-linked Dextran

Dextran is a water-soluble polysaccharide made by the bacterium *Leuconostoc mesteroides*. It is transformed into an insoluble gel by cross-linking it with epichlorohydrin. The cross-linking reaction is normally performed in a water-in-oil emulsion so that gel beads of various sizes are produced (Determann, 1969). The pore-size distribution is controlled by varying the relative amounts of dextran and cross-linking agent.

Cross-linked dextrans (Sephadex) are stable and inert, which accounts for their popularity. They are stable in water, organic solvents and at high pH. However, the glycosidic linkages in the dextran chains hydrolyze below a pH of 1.7 (Pharmacia booklet, 1979). They are susceptible to bacterial attack, and thus must be stored in an antimicrobial agent. They are, however, very stable with respect to temperature and can therefore be autoclaved. Dextran beads with small pores are rigid, but as the pore size increases, the beads become more compressible. As shown in Table 1, the maximum flow rates for the beads with large pores are dramatically reduced.

Pharmacia Inc. makes a dextran gel cross-linked with N,N'-methylenebisacrylamide (Sephacryl). These gels are much more rigid than those cross-linked with epichlorohydrin, and flow rates ten times greater are possible. However, the acceptable range of pH is narrower (3–11), and both proteins and Blue Dextran (used as a tracer) have a greater tendency to adsorb to the gel.

31.2.6 Polyacrylamide

Polyacrylamide gels are entirely synthetic, formed by polymerizing acrylamide, $CH_2CHCONH_2$, in the presence of the cross-linking agent N,N'-methylenebisacrylamide. As with the dextran gels, the pore-size distribution is controlled by adjusting the relative amounts of monomer and cross-linking agent. Linear polyacrylamide is soluble in water, indicating that it is hydrophilic. By cross-linking the strands together, an insoluble network is formed. The network

Table 1 Compressible Packing Materials

Composition	Manufacturer	Product name	Fractionation range (MW)[a]	Maximum flow rate (cm h^{-1})	Maximum pressure drop (cm H$_2$O)	Notes
Dextran (cross-linked)	Pharmacia	Sephadex G-10	50–700	—	—	b, c
		G-15	50–1500	—	—	b, c
		G-25	1000–5000	—	—	b, c
		G-50	1500–30 000	—	—	b, c
		G-75 (except Superfine)	3000–70 000	77	160	c
		G-75 (Superfine)	3000–70 000	18	160	c
		G-100 (except Superfine)	4000–150 000	50	96	c
		G-100 (Superfine)	4000–150 000	12	96	c
		G-150 (except Superfine)	5000–300 000	23	36	c
		G-150 (Superfine)	5000–300 000	6	36	c
		G-200 (except Superfine)	5000–600 000	12	16	c
		G-200 (Superfine)	5000–600 000	3	16	c
Dextran/ bisacrylamide	Pharmacia	Sephacryl S-200 (Superfine)	5000–300 000	30	300	c
		S-300 (Superfine)	10 000–800 000	25	300	c
		S-400 (Superfine)	20 000–8 × 10^6	40	125	c
		S-500 (Superfine)	40 000–2 × 10^7	40	100	c
		S-1000 (Superfine)	500 000–>10^8	40	70	c
Polyacrylamide	Bio-Rad	Biogel P-2	100–1800	—	—	b
		P-4	800–4000	—	—	b
		P-6	1000–6000	—	—	b
		P-10	1500–20 000	—	—	b
		P-30	2500–40 000	—	—	b
		P-60	3000–60 000	d	100	e
		P-100	5000–100 000	d	100	e
		P-150	15 000–150 000	d	100	e
		P-200	30 000–200 000	d	75	e
		P-300	60 000–400 000	2.8	20	f
Agarose	Pharmacia	Sepharose 6B	10 000–4 × 10^6	14	90	c
		4B	60 000–20 × 10^6	12	60	c
		2B	70 000–40 × 10^6	10	30	c
	Bio-Rad	Bio-Gel A-0.5m	10 000–500 000	—	(>100)^g	c
		A-1.5m	10 000–1.5 × 10^6	—	(>100)^g	c
		A-5m	10 000–5 × 10^6	—	(>100)^g	c
		A-15m	40 000–15 × 10^6	d	90	
		A-50m	100 000–50 × 10^6	d	50	
		A-150m	1 × 10^6–>150 × 10^6	d	30	

Type	Product	Manufacturer	Fractionation range			
	Ultrogel A6	LKB	25 000–4 × 10^6	d	100	[h]
	A4		55 000–20 × 10^6	15	40	[h]
	A2		120 000–50 × 10^6	10	40	[h]
Agarose/polyacrylamide	Ultrogel AcA202	LKB	1000–20 000	(>50)[g]	—	[i]
	AcA54		5000–90 000	(>50)[g]	—	[i]
	AcA44		10 000–200 000	(>50)[g]	—	[i]
	AcA34		20 000–750 000	40	30	[i]
	AcA22		100 000–3 × 10^6	18	15	[i]
Cross-linked agarose	Sepharose CL 6B	Pharmacia	10 000–4 × 10^6	30	>120	[c]
	4B		60 000–20 × 10^6	26	120	[c]
	2B		70 000–40 × 10^6	15	50	[c]
Cross-linked polyether	Fractogel TSK HW40	EM Science (Toyo Soda)	100–10 000	b	b	
	HW50	(Toyopearl)	500–80 000	b	b	
	HW55		1000–70 000	(>200)[a]	—	
	HW65		50 000–5 × 10^6	b	b	
	HW75		500 000–50 × 10^6	b	b	

[a] Maximum number indicates the exclusion limit. [b] Obeys Darcy's law (*i.e.* is rigid) over expected operating conditions. [c] Flow data determined on a 2.5 (dia.) × 30 cm column. [d] Not available. [e] Flow data determined on a 2.5 (dia) × 41 cm column. [f] Flow data determined on a 2.5 (dia) × 15 cm column. [g] Obeys Darcy's law to specified flow rate or pressure drop. [h] Flow data determined on a 1.1 (dia) × 15 cm column. [i] Flow data determined on a 2.6 (dia) × 40 cm column. Flow data determined on a 1.8 (dia) × 15 cm column.

Table 2 Rigid Packing Materials

Composition	Manufacturer	Product name	Fractionation range (MW)[a]
Surface-modified glass (1,2-dihydroxypropyl glass)	Pierce	Glycophase CPG40	1000–8000
		CPG100	1000–30 000
		CPG240	2500–125 000
		CPG500	11 000–350 000
	Electro-Nucleonics	Glyceryl CPG- 75Å	3000–30 000
		120Å	7000–130 000
		170Å	12 000–400 000
		240Å	$22\ 000-1.2 \times 10^6$
		350Å	$40\ 000-4.0 \times 10^6$
		500Å	$70\ 000-1.0 \times 10^7$
		700Å	$130\ 000-3.0 \times 10^7$
		1000Å	$250\ 000-1.0 \times 10^8$
		1400Å	$400\ 000-3.0 \times 10^8$
		2000Å	$700\ 000-9.0 \times 10^8$
		3000Å	$1.2 \times 10^6-2.7 \times 10^9$
Surface-modified silica (1,2-dihydroxypropyl silica)	EM Science	LiChrosorb DIOL[a]	10 000–80 000
		LiChrosper[a] Si 100 DIOL	10 000–80 000
		Si 300 DIOL	70 000–300 000
		Si 500 DIOL	250 000–500 000
		Si 1000 DIOL	$400\ 000-1.0 \times 10^6$
Cross-linked poly(2-hydroxy-ethylmethacrylate)	LaChema Koch-Light KnauerAG	Spheron P-1	<1000[b]
		P-40	$<40\ 000$
		P-100	$<100\ 000$
		P-200	$<200\ 000$
		P-300	$<300\ 000$
		P-500	$<500\ 000$
		P-700	$<700\ 000$
		P-1000	$<1.0 \times 10^6$
		P-10 000	$<1.0 \times 10^7$

[a] Suitable for HPLC.   [b] From Janak *et al.* (1975). Lower exclusion limits not available.

Table 3 Chemical and Physical Stability of Packing Materials

Material	Safe pH range	Max. temp. (°C)	Special limitations
Cross-linked Dextran	>2	>120	Subject to microbial degradation
Dextran/bisacrylamide	3–11	>120	
Polyacrylamide	2–10	>120	Avoid strong oxidizing agents; organic solvents cause shrinkage
Agarose	4–9	40	Avoid strong oxidizing agents; guanidine HCl, urea and chaotropic salts (*e.g.* KSCN); not good in organic solvents
Agarose/polyacrylamide	3–10	40	Subject to microbial degradation; variable resistance to chaotropic salts and detergents, depending upon polyacrylamide content
Cross-linked agarose (w/ 2,3-dibromopropanol)	3–14	>120	Avoid strong oxidizing agents
Cross-linked polyether	1–14	>120	—
Silica and glass, coated	<9	>120	Avoid strong oxidizing agents
Cross-linked poly(2-hydroxy-ethylmethacrylate)	N/A	>120[a]	—

[a] From Mikes *et al.* (1978).

swells in the presence of water, producing an open, porous structure. Because polyacrylamide is synthetic, it is not subject to microbial degradation. Moreover, its thermal stability allows the gel to be autoclaved. It is also stable for long periods of time in the pH range of 2–10. Both polyacrylamide and the cross-linked dextrans are rigid in the low pore size range and compressible in the larger pore sizes (Bio-Rad, 1975). Polyacrylamide is advantageous when contamination with carbohydrates must be eliminated. Furthermore, at low ionic strength, adsorption mediated by surface charge is somewhat less than that observed with cross-linked dextran gels.

31.2.7 Agarose

Both polyacrylamide and dextran-based gels are of limited use when large proteins are to be fractionated. For fractionating proteins and other macromolecules with molecular weights above 500 000, agarose gels are generally used primarily for lack of a better alternative. Agarose gels are highly compressible, fragile, and stable only in a narrow temperature and pH range.

Agarose is the gelling agent in agar, a polysaccharide mixture extracted from red seaweed. It is an alternating copolymer of D-galactose and 3,6-anhydro-L-galactose. At 100 °C, it is soluble in water, but even a 1% solution will form a stiff gel when cooled to room temperature. Commercial agarose gels are manufactured in bead form by cooling in emulsion form. Pore sizes are controlled by adjusting the agarose concentration, usually between 2% and 6%. Bio-Gel A and Sepharose are commercially available forms. The pores are large enough to separate not only proteins but also cell organelles, viruses or intact cells.

Unmodified agarose redissolves at temperatures above 40 °C, and can only be used between pH values of 4 and 9. The manufacturers also do not recommend using it with chaotropic salts. It is, however, generally resistant to microbial degradation even though it is a natural product.

The high compressibility of agarose is a significant problem, and thus only very low flow rates are used (see Table 1). In an effort to improve the performance, Pharmacia Inc. also offers a cross-linked version (Sepharose-CL) with considerably improved physical and chemical properties.

31.2.8 Other Materials

The primary objection to the preceding column packings is that they are not sufficiently rigid. The development of acceptable rigid gels with suitable adsorption characteristics has indeed been a problem. Yet, recent progress in this area has produced a small set of rigid materials which are currently commercially available. Their separation characteristics are summarized in Table 2. One approach has been to prepare cross-linked organic gels. One such product is Toyopearl, also marketed as Fractogel TSK. Although its precise composition is not known (Barth, 1980; Dubin, 1981), it is probably a cross-linked hydroxylated polyether (Dubin, 1983). The gel is slightly compressible (hence listed in Table 1), but superficial velocities of up to 200 cm h^{-1} are attainable (Gurkin and Patel, 1982). It is stable with respect to temperature (120 °C or more), pH, chaotropic salts, detergents, organic solvents and bacterial action. However, some proteins, such as hemoglobin, are adsorbed slightly. At present, separations are possible only up to a molecular weight of 80 000.

Another cross-linked, synthetic organic gel (Spheron) is made by copolymerizing 2-hydroxyethylmethacrylate and ethylenedimethacrylate. The product is rigid and hydrophilic (Coupek, 1982) and appears to be stable with respect to pH and commonly used salts. A key advantage is that particles with very large pores can be manufactured, allowing separation of macromolecules with molecular weights of up to 5×10^6. Manufacturers are LaChema, Koch-Light and Knauer KG (Dubin, 1983). The methacrylate beads show great potential because of their inertness, rigidity and wide range of pore sizes.

Another approach in preparing rigid packings has been to treat the surface of a rigid, porous material in order to generate supports with more suitable chemical properties. This has been done with controlled-pore silica, the disadvantages of which have been noted earlier. The modifying agent is glycidoxypropyltrimethylsilane, which produces a 'glycol-like' surface (Dubin, 1983). One remaining drawback is that silica is soluble at high pH values and there is a tendency to lose the protective coating with time (Coupek, 1982). Commercially available products are Lichrosorb DIOL and Glycophase G. Glycophase G and Sephacryl have similar separation characteristics. For example, a 100 cm column of Glycophase G with 17 nm pores (Regnier and Noel, 1976) and a 94 cm Sephacryl S-300 column (Pharmacia booklet, 1979) can both amply resolve serum proteins. However, the Glycophase column can do so at much higher velocities (580 cm h^{-1}, compared to 2.4 cm h^{-1}) using particles of similar size.

31.3 EQUIPMENT

Equipment necessary for MSC includes three basic items: (1) the eluting solvent, (2) a packed column and (3) tubes for collecting fractions. Modern setups not only allow these components to be interconnected in a continuous fashion (*via* tubing) but also include other devices which automate and monitor various aspects of the system. Even more sophisticated arrangements can be applied whereby the entire separation process is automated, including data processing of the resulting chromatogram.

A typical laboratory setup is shown in Figure 2. The eluting solvent (1) is pumped through the column with a peristaltic pump (3). Alternatively, one may use gravity feed instead of a pump. The sample is introduced through one port of the three-way valve (2) and pumped through the column either in an upward or downward direction. The separated sample is identified *via* a detector (5) connected to a recorder (7) and collected at predetermined volumes in tubes (6) in a fraction collector.

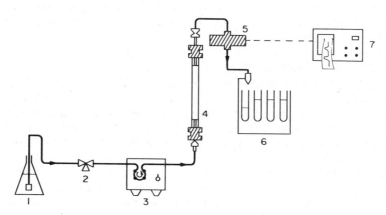

Figure 2 Laboratory MSC setup. Elution solvent (1) is pumped through a three-way valve (2) by a peristaltic pump (3). The desired volume of sample is introduced through one port of the valve while the solvent flow is temporarily blocked. The valve is repositioned to allow elution of the sample through the packed column (4). The eluting solution is monitored by a detector (5) and collected in a fraction collector (6). A permanent record of the chromatogram is obtained from a recorder (7)

Columns should be precision bore tubes (glass) of appropriate length. Commercially available columns are generally equipped with precisely molded or machined fittings which contain the gel at the bed ends. The fittings enable the solvent to move from the gel bed to tubing with a minimum of dead volume. It is important to remember that the dead volume or 'mixing volume' is not only within the netting or support fittings of the column but is actually the entire volume within the system beyond the pump output which is not taken up by the gel bed. Dead volume should be minimized because Taylor dispersion and diffusional processes may alter the composition of the liquid found in the pre- and post-column volumes, thereby giving rise to slightly broader elution profiles.

The column is secured and adjusted in a vertical position using a level. The gel suspension is thoroughly deaerated to prevent air pocket discontinuities in the column. The softer gels (*e.g.* Sephadex) are carefully poured, preferably in one application. If necessary, an additional reservoir connected to the top of the column can be used to ensure a single pouring of the gel. The more rigid gels can be poured more rapidly and are often packed with a pump at relatively high flow rates (*e.g.* for Sephacryl S-200, 3.5 ml min^{-1} for a column 2.5 cm in diameter). Regardless of which gel is selected, packing flow rates should exceed the rate to be used for sample fractionation. To attain stabilization of the gel bed and proper equilibrium, 2–3 bed volumes of buffer must be run through the column prior to use. Various gel bed lengths can be attained conveniently using adapters sold commercially. Besides defining a precise length, the adapters (primarily that which is fitted on the top) help reduce dead volume. A well-packed column is visually homogeneous throughout lacking any boundaries or air bubbles.

Generally, the pump employed in MSC (excluding high pressure) is the peristaltic pump. The most popular type is the roller type in which regulated revolving rollers compress elastomer tubing, such as silicone rubber tubing, which in turn advances the solvent in the specified direction. Silicone rubber tubing is superior to most other tubing in its ability to withstand continuous deformation by compression. A minor but noticeable drawback of peristaltic pumps is the pulsating delivery of the solvent (quite noticeable at low flow rates (0.5–1 ml min^{-1}) even with properly positioned tubing). One possible remedy is to use a pulse damper. Nonetheless, the contributions to substandard separations by other factors (*e.g.* improperly packed column or large dead volume) far outweigh the consequences of pulsating solvent flow. Tubing in the interconnected system outside the pump is generally made of teflon or polyethylene (18 gauge AWG or smaller).

The detection system is placed between the column and the fraction collecting device. In order to minimize dead volume and hence avoid peak broadening due to additional diffusional processes, one should connect the detector as close as possible to the column outlet. Means of detection include UV–visible absorbance, fluorescence, refractive index and conductance. Radioactivity measurement is usually performed after fractions have been collected.

The fraction collector usually operates in three different collection modes. This is either collection by volume, drops or time. Additional features on some of the more recent fraction collection systems enable individual peaks to be collected in one tube and allow solvent not containing a threshold value of separated sample to be directed to waste.

31.4 THEORY

31.4.1 Column Parameters

MSC is considered a form of partition chromatography where components to be separated are distributed between two solvent phases: the stationary phase and the mobile phase. The separation process requires a gel matrix which is packed in a column and surrounded by solvent. The total bed volume (gel particles and solvent) can be divided into two distinct parts: (1) the solvent molecules within the gel structure or inner volume of the bed, V_i (stationary phase), and (2) the interstitial or outer volume of the gel bed, V_o, through which the solute flows (mobile phase).

The behavior of a particular solute is usually represented by its elution volume, V_e, which is the volume of carrier solvent required to carry the solute through the column from the time sample penetration begins until it emerges at a maximum concentration. Molecules totally excluded from the inner volume of the pores (V_i) elute with the void or outer volume, $V_e = V_o$. If the inner volume is fully accessible to the solute then an elution volume of $V_e = V_o + V_i$ is required for solute emergence. Solutes which are partially excluded from the pores elute with a volume described by equation (1):

$$V_e = V_o + K_d V_i \tag{1}$$

where the partition ratio K_d is defined as the fraction of the internal volume available to the solute. Rearrangement of equation (1) provides the following expression for K_d:

$$K_d = (V_e - V_o)/V_i \tag{2}$$

When $V_e = V_o$, $K_d = 0$ and when $V_e = V_i + V_o$, $K_d = 1$. A K_d value greater than one indicates adsorption or other interfering effects. These relationships between the elution volume and other volumetric parameters are illustrated in Figure 3.

K_d is a convenient parameter when comparing separation results obtained with different columns, since it is independent of bed size and packing density. However, the value obtained for K_d may be imprecise since it is dependent on V_i which, in some cases, is an estimated value. For example, V_i can be computed from the following expression (Gelotte, 1960; Granath and Flodin, 1961):

$$V_i = aW_r \tag{3}$$

where a = quantity of dry gel (grams) and W_r = water regain of gel (grams of water per gram of dry gel divided by the density of water).

Such a determination of V_i includes, with the internal volume, the water of hydration associated with the gel which is generally inaccessible to large solute molecules. In addition, a, the exact amount of dry gel actually packed into the column, is usually unknown. One can circumvent part of the problem by using the density of the wet gel, d (Kremmer and Boross, 1979), in which case:

$$d = \frac{a + aW_r \varrho_{H_2O}}{V_m + V_i} = \frac{a(1 + W_r \varrho_{H_2O})}{V_t - V_o} \tag{4}$$

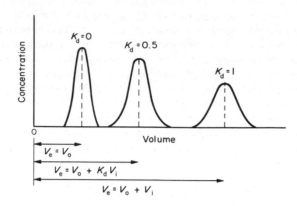

Figure 3 Parameters (discussed in text) measured upon elution of a narrow band of sample; V_e for the respective peaks
appears at the center of the elution profile

where $V_t = V_o + V_i + V_m$ and V_m = volume of the solid gel matrix and ϱ_{H_2O} is the density of water.
Substitution for a from equation (3) into equation (4) yields:

$$V_i = \frac{W_r d}{1 + W_r \varrho_{H_2O}}(V_t - V_o). \tag{5}$$

Values for d and W_r are generally specified by the manufacturer. Another method for obtaining
V_i and avoiding the aforementioned objections would be to record the elution volume (V_e) at
which separation of small molecules ceases to occur and subtract this value from V_o (Andrews,
1964).

V_o is generally determined by collecting the volume needed to elute the fully excluded solute
particles. This volume usually amounts to roughly one third of the gel bed; $V_o \approx V_t/3$. This volumetric relationship has been investigated using homogeneous rigid spheres (*e.g.* ball bearings)
packed into a measuring cylinder (Bernal and Mason, 1960; Scott, 1960). The amount of water
occupying the void is between 34 and 38% of the total volume. In practice, the void volume
between the gel particles is somewhat less (and gel volume somewhat higher) due to the hetero-
geneity of the bead sizes and their deformation.

31.4.2 Theoretical Models for Calculating K_d and Similar Constants

Several models have been proposed to describe K_d or a similar measure of solute hindrance in
terms of solute and gel particle dimensions (Laurent and Killander, 1964; Porath, 1963; Squire,
1964). This approach regards the gel as a physical barrier for molecules of certain defined sizes
and is commonly referred to as the steric exclusion approach. All three model descriptions of
solute behavior, mentioned above, assume some structural shape for the gel particles. In contrast, Ackers (1967) defined K_d by assuming that the penetrable volumes (of any shape) within
the gel are distributed randomly with respect to the sizes of molecules they can accommodate.
This assumes that the frequency distribution of sizes follows a normal curve and the region available for penetration is characterized by the radius, a, of the largest molecules it can accommodate. The fraction of the internal volume, V_i, that is penetrable by a solute molecule of radius a is
represented by a Gaussian probability curve. The total fractional volume within the gel occupied
by a given species of radius a is consequently the fraction available to this molecule and all
fractions available to larger molecules. Mathematically, this is expressed as the error function
complement (*erfc*) of the Gaussian distribution in the following manner:

$$K_d = erfc\,[(a-a_o)/b_o] \tag{6}$$

$$\text{or} \quad a = a_o + b_o\,erfc^{-1}K_d \tag{7}$$

where a_o is the position of the maximum value of the distribution and b_o is a measure of the standard deviation. Experimental data are in excellent agreement with a linear relationship between
a and $erfc^{-1}K_d$.

31.4.3 Band Dispersion

A sample containing a solute concentration, C_i, layered on the column as a narrow band is eluted as a widened, bell-shaped band as a result of several diffusional factors. Glueckauf (1955) obtained for practical purposes a Gaussian distribution for the concentration of a solute initially layered as a narrow zone in terms of experimental and theoretical parameters. In obtaining the Gaussian curve, the concept of the height equivalent to a theoretical plate (HETP), first introduced by Martin and Synge (1941), was combined with an appropriate equation for mass conservation. The HETP concept regards the chromatographic column as a series of sections or plates, each having a length such that the mobile phase leaving the plate is in equilibrium with the stationary phase within the plate. With the appropriate differential equation and boundary conditions, the simplified solution obtained for conditions normally encountered in column chromatography is given by:

$$C = \left(\frac{N'}{2\pi}\right)^{1/2} \frac{Q}{(V_e V)^{1/2}} \exp\left[-\frac{N'}{2}\frac{(V_e-V)^2}{VV_e}\right] \tag{8}$$

where C is the sample concentration as a function of the volume of fluid injected, V. V_e is the elution volume, Q is the total mass of solute injected, and N' is the number of theoretical plates from the center of the initial solute zone to the bottom of the column. For large N', $N' \simeq N$, the total number of theoretical plates in the column.

As N' becomes large (above 100, for example, which is usually the case) equation (8) becomes well approximated by the normal distribution, and the techniques commonly applied to normally distributed populations can be applied. The approximate form is:

$$C = \left(\frac{N'}{2\pi}\right)^{1/2} \frac{Q}{V_e} \exp\left[-\frac{N'}{2}\frac{(V_e-V)^2}{V_e^2}\right] \tag{9}$$

The maximum concentration, $\overline{C}$, is given by:

$$\overline{C} = \left(\frac{N'}{2\pi}\right)^{1/2} \frac{Q}{V_e} \tag{10}$$

Equation (9) can be compared with the standard form of the normal distribution:

$$y = \frac{1}{\sigma(2\pi)^{1/2}} \exp\left[\frac{-(x-\mu)^2}{2\sigma^2}\right] \tag{11}$$

where x and y are the generalized abscissa and ordinate, respectively, μ is the average value of x, and σ is the standard deviation.

Comparison of equations (9) and (11) reveals that:

$$y = C/Q \tag{12}$$

and that:

$$\sigma = V_e/\sqrt{N'} \tag{13}$$

Superficially, equation (13) can be read to indicate that the standard deviation (or band spread) increases as V_e increases. However, the reationship between σ and V_e is more complex than indicated because N' is also a function of V_e. Empirically, it has been found that the variance, σ^2, usually goes through a maximum as a function of molecular weight (Ouano, 1977).

Although the value N' can be readily estimated *via* equation (10), measurements of Q, the area under the elution curve, are less convenient. A more general method for evaluating N is given by:

$$N = 16(V_e/W_t)^2 \tag{14}$$

where W_t is defined as the difference between the x-intercepts of the two tangent lines passing through the inflection points of the Gaussian curve.

Although N is a measure of column separating ability, a more widely used expression for column efficiency is HETP (H) which is obtained by dividing the column length by N (the number of plates):

$$H = L/N = L\sigma^2/V_e^2 \tag{15}$$

The smaller the HETP, the more efficient the column.

In order to show the true dependence of σ (or H) on column dimensions, equation (15) can be rewritten in terms of the characteristics of the solute and packing material:

$$H = L/N = \frac{\sigma^2}{LA^2[\varepsilon_o + K_d\varepsilon_i(1 - \varepsilon_o)]^2} \qquad (16)$$

where A = cross-sectional area of column, $\varepsilon_o = V_o/V_t$ = porosity of column external to beads, and $\varepsilon_i = V_i/(V_t - V_o)$ = internal bead porosity.

As mentioned above, the band spread is due to several factors. First, molecular diffusion along the length of the column causes an injected pulse to spread. Second, the flow channels between the packed particles are nonuniform and tortuous, and thus the pulse is spread further. This is referred to as eddy diffusion or axial dispersion. Third, as the solute molecules diffuse into the particles, they must first cross a stagnant liquid film surrounding each particle, a process known as film diffusion. Finally, diffusion within the pore spaces of the particles contributes to the spreading.

Giddings (1965) has shown that the variance of the output peak, σ^2, can be expressed as the sum of contributions from each of these mechanisms, and this in turn is related to the HETP through equation (16). A number of investigators (Giddings, 1965; Horvath and Lin, 1976; van Deemter *et al.*, 1956) have formulated expressions for H; the one by Horvath and Lin (1976) is as follows:

$$H = \frac{2\gamma D_m}{u} + \frac{2\lambda d_p}{1 + \omega \left(\frac{ud_p}{D_m}\right)^{-1/3}} + \frac{\omega K_d^2\varepsilon_o\varepsilon_i^2 d_p}{18[\varepsilon_o + K_d(1 - \varepsilon_o)\varepsilon_i]^2}\left[\frac{ud_p}{D_m}\right]^{2/3} + \frac{K_d(1 - \varepsilon_o)\varepsilon_i d_p^2 u}{30D_s[\varepsilon_o + K_d(1 - \varepsilon_o)\varepsilon_i]^2} \qquad (17)$$

where D_m is the bulk molecular diffusivity, D_s is the diffusivity within the particles, d_p is the particle diameter, and γ, λ and ω are constants for a particular packing. The value u is the interstitial velocity, given by

$$\frac{q}{A\varepsilon_o} \qquad (18)$$

where q is the volumetric flow rate. The terms in equation (17) represent contributions from axial diffusion, axial dispersion, film diffusion and intra particle diffusion, respectively.

One important variable affecting zone spread is the particle diameter, d_p. This can be seen by making a few simplifications in equation (17). First, for large molecules D_m is often of the order 10^{-6} to 10^{-7} cm^2 s^{-1} while u is 1–5 $\times$ 10^{-3} cm s^{-1}, and, therefore, the first term is usually negligible. Second, at high enough velocities, the velocity dependence of axial dispersion (the second term) becomes unimportant. If we further define a relative rate factor, R, by writing

$$R = \frac{V_o}{V_e} = \frac{\varepsilon_o}{\varepsilon_o + K_d(1 - \varepsilon_o)\varepsilon_i} \qquad (19)$$

then equation (17), in somewhat simpler form, becomes

$$H = 2\lambda d_p + \left[\frac{\omega K_d^2\varepsilon_i D_s}{18(1 - \varepsilon_o)(D_m^2 d_p u)^{1/3}} + \frac{K_d\varepsilon_i}{30\varepsilon_o}\right]\frac{R(1 - R)d_p^2 u}{D_s} \qquad (20)$$

This shows more clearly that the smaller the particle diameter is, the smaller is the zone spread.

31.4.4 Column Resolution

The resolution produced on a particular column indicates a column's efficiency, where resolution, R_s, is a measure of zone overlap. For the separation of equal quantities of two components, R_s is given by the following equation:

$$R_s = \frac{V_{e_2} - V_{e_1}}{0.5(W_1 + W_2)} \qquad (21)$$

where W is defined as the width at baseline of the Gaussian curve and the numerals (1 and 2) represent the two components. An $R_s = 1.5$ corresponds to 99.8% of baseline resolution. The σ term mentioned in equation (13) may be substituted for W in equation (21). In terms of the theoretical plate number N, we have:

$$R_s = \frac{1}{4}\left(\frac{\alpha - 1}{\alpha}\right)\frac{K_2'}{1 + K_2'}N^{1/2} = \frac{1}{4}\left(\frac{\alpha - 1}{\alpha}\right)\frac{K_2'}{1 + K_2'}(L/H)^{1/2} \qquad (22)$$

where $\alpha = K_{d_2}/K_{d_1}$ = selectivity or relative retention and $K' = K_d V_i/V_o = (V_e - V_o)/V_o$ = capacity coefficient of column and the number 2 refers to the component with the larger elution volume. With H proportional to d_p (particle diameter), resolution is certainly improved with smaller gel particles.

31.5 OPERATIONS

In discussing column operation, one must distinguish between laboratory-scale and production-scale separations. The principles of operation are the same, but the goals in the two situations differ. In the laboratory, a high-resolution separation in a minimum amount of time is desirable and typically one must trade one off against the other. In the industrial setting, however, minimizing the capital investment and overall operating costs for a given throughput is the goal. High resolution is not a goal in itself, although the degree of resolution achieved can have a considerable effect on the economics of the process.

Because of its importance in both scales of operation, we will briefly review those factors which affect column resolution and then examine how these factors must be balanced in choosing appropriate column operating conditions.

31.5.1 Resolution

Resolution refers to the degree of peak overlap in the product stream. One measure of resolution is R_s, as defined in equation (21). Equation (22), which shows that R_s is proportional to $(L/H)^{1/2}$, evolves because the distance between peaks is proportional to the column length, while peak spreading goes only as $L^{1/2}$.

A number of controllable variables are included in the determination of H. A widely-used simplified form of equation (17) illustrates the salient features:

$$H = \frac{2\gamma D_m}{u} + 2\lambda d_p + \frac{\alpha R(1 - R)d_p^2 u}{D_s} \tag{23}$$

where α is a constant. The first term represents the effect of axial diffusion, which is rarely significant even at the lowest velocities. The second term represents the high-velocity behavior of eddy diffusion, as shown in equation (20). Ordinarily, eddy diffusion (or axial dispersion), is modeled using a dispersion coefficient (D_s) analogous to D_m, the molecular diffusivity. But unlike the diffusivity, which is a fundamental property of the solute in the solvent, the dispersion coefficient is a function of the flow rate and the characteristics of the bed. It is affected most strongly by the particle diameter, and thus the contribution from axial dispersion is proportional to d_p. The third term represents the combined effects of film and intraparticle diffusion at higher velocities. Because a fixed amount of time is required for solutes to diffuse in and out of the particles, there is less time for this process to take place, and the band spread increases. Thus, mass transfer effects are proportional to velocity.

Equation (22) shows that column resolution increases with $L^{1/2}$. Because the first term in equation (23) is usually small, resolution also increases as the velocity decreases. Furthermore, the use of smaller gel particles can improve resolution in two ways. First, axial dispersion, represented by the second term, declines. Second, solute molecules require less time to diffuse into smaller particles and equilibrate with the bulk. The two major terms determining H then decrease, and a more efficient separation results.

The variables which can be adjusted for a given separation are, as shown in equation (23), column length, flow velocity and particle size. Furthermore, because equation (23) is valid only for infinitely small samples, the sample volume also plays a role. How each of these variables affects the design of the separation is considered in the following paragraphs.

31.5.2 Sample Volume

As the sample volume increases, resolution declines. For laboratory work, the sample volume should generally be below 3% of the total column volume (Scopes, 1982). Little improvement is obtained for sample concentrations below 1%. Several output curves are shown in Figure 4 for a given amount of solute applied in different volumes using the method of Scopes (1978).

For production purposes, the optimum sample volume may be 15% or more of total column volume (Delaney, 1980). In this case the central portion of the desired peak is retained and the leading and trailing edges are recycled to the feed.

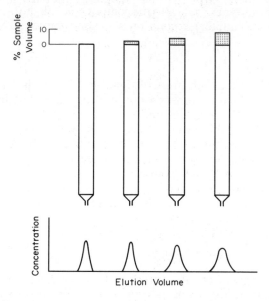

Figure 4 Effect of sample volume on elution pattern (modified from Scopes, 1982)

31.5.3 Flow Rate

Equations (17)–(23) do not reveal the severe limitations on flow rate imposed by the compressible gels normally used. Most exhibit the behavior shown in Figure 5, where beyond a critical imposed pressure drop the bed compresses enough to cause a reduction in flow rate. Often, then, separations are carried out at the maximum feasible flow rate, which is approximately 75% of the theoretical maximum. In small-diameter columns, the column packing material is supported somewhat by the walls. This permits higher flow rates than is possible in larger diameter columns.

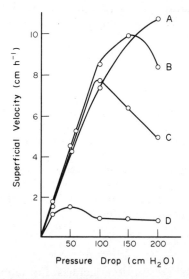

Figure 5 Effect of hydrostatic pressure on flow rate for Sephadex G-150 packed in 30 cm diameter columns of varying heights. A, 8 cm; B, 13 cm; C, 23 cm; D, 52 cm (modified from Janson and Dunnill, 1974)

Unfortunately, complete information on flow rate/pressure drop behavior is not available. This behavior is a complex function of packing material, applied pressure drop, bed height and bed diameter. Hence, the data in Table 1 are intended merely to be representative. They cannot really even be compared because manufacturers have not settled on a single column size or set of sizes as standards for comparative purposes. Such standards are particularly necessary given the current lack of an adequate theory for flow of fluids in compressible porous media. The empirical correlations commonly used in processes involving filtration are completely inadequate because they fail to predict a maximum in flow rate as a function of pressure drop.

One approach may be to employ a differential form of the Blake-Kozeny equation:

$$v = \frac{d_p^2 \varepsilon_o^3}{k\mu(1-\varepsilon_o)^2}\left(-\frac{dp}{dx}\right) \tag{24}$$

where p is the local pressure x is the axial distance along the bed, k is a constant (usually taken to be 150), and ε_o is the external porosity of the bed, equal to V_o/V_t. ε_o is a function of the local mechanical pressure on the beads. Should a rational method be developed for determining its dependence on mechanical pressure this approach would prove valuable. In columns formed with rigid particles, these concerns are moot. Generally, however, resolution declines as flow rate rises, but throughput goes up. Hence there is some optimum flow rate, which in most cases must be found empirically.

31.5.4 Column Length

For a given packing material, the column length is the primary variable controlling resolution. In principle, a column length sufficient for the separation is selected and then the diameter is chosen so that for a given flow rate the velocity is acceptable. However, the resolution is affected strongly by the uniformity of column packing and by the flow distribution at the entrance to the bed. When the packing is uneven or the flow non-uniformly distributed, resolution suffers dramatically.

From a practical standpoint, it is easier to pack a long, narrow column uniformly than a short, wide one. Moreover, in narrow cores any unevenness that exists has a smaller effect on results. Similarly, a uniform flow distribution is easier to achieve with a narrow column. Consequently, on the laboratory scale one manufacturer (Bio-Rad) recommends a length/diameter ratio of at least 5 to 10. This is corroborated by the experience of the chemical industry, where as a rule of thumb, length/diameter ratios greater than 10 are used. Hence, the selection of a column length restricts the diameter somewhat, and this in turn restricts the range of velocities. Therefore, some degree of trade-off between column length and velocity is required. Higher velocities are desirable because they represent higher throughput. However, the higher velocity decreases resolution, which has to be made up with a longer column.

On an industrial scale, such large length/diameter ratios are impossible with the currently available highly compressible packings. Thus, industry has turned to stacks of short columns (Janson and Dunnill, 1974), and has chosen to grapple with the problems of uniform packing and flow distribution. An example of a short column is the module manufactured by Pharmacia Inc., which has a diameter of 37 cm and a length of 15 cm. Design of flow distributors is discussed in more detail by Janson and Dunnill (1974).

31.5.5 Particle Size

To some extent, the trade-off between particle size and column length resembles that between velocity and column length. Smaller particles dramatically improve resolution, but they also dramatically increase pressure drop. This poses no problem on the laboratory scale, so fine particles are routinely used. Industrially, larger particles must be used, but with them come necessarily longer columns and their attendant problems.

31.5.6 Molecular Weight Determination and Column Calibration

Molecular weight estimation is accomplished by first obtaining an elution profile for a set of protein standards of known molecular weight, M. A subsequent plot of V_e vs. log M provides a calibration curve from which molecular weight estimations of protein unknowns can be determined. The calibration curve is nearly linear for the specified molecular weight range of the gel. Resulting V_e vs. log M data points for protein unknowns may deviate from the calibration curve for a number of reasons. Adsorption to the gel may cause considerable error when determining the elution volume, V_e. Also, many protein unknowns have rather a pronounced ellipsoidal shape (as opposed to spherical shape) which can affect the estimated (or measured) molecular weight value. This effect is evident from the following equation.

$$M = \frac{4}{3}\pi a^3 \frac{N}{\bar{v}} \left(\frac{f}{f_o}\right)^{-1}$$

(25)

where N is Avogadro's number, $\bar{v}$ is the partial specific volume, a is the radius of the equivalent hydrodynamic sphere (Stokes' radius), and f/f_o is the frictional ratio which describes the effect of solvation and ellipticity. An f/f_o value of one refers to a hydrated sphere while values greater than one refer to varying degrees of non-sphericity. Usage of V_e vs. M not only implies similar values of f/f_o for different proteins but also similar values for $\bar{v}$. For proteins, $\bar{v}$ is usually between 0.71 and 0.75. Thus, deviations from a calibration curve stem from differences in the shape factor, f/f_o. Using globular proteins with similar frictional ratios and similar specific partial volumes, Determann (1969) determined and illustrated empirical linear curves of V_e/V_o vs. log M for several Sephadex gels.

Column efficiency, represented by the HETP, can be measured simultaneously with calibration. Assuming Gaussian elution profiles, HETP is calculated from the column length, elution volume and peak width for a given solute. Equation (15) can be used ($H = L/N$) if the peak width is measured as the difference between the intersections of tangents drawn through the inflection points on the elution curve at the baseline (for a Gaussian, this difference is 4σ). However, determination of tangents is subject to error. An alternate method is to take W_t as the peak width where C is $0.135\bar{C}$. For perfect Gaussian profiles, the techniques are equivalent.

31.6 FUTURE DIRECTIONS

As one can see, the compressibility of the currently used packing materials imposes constraints on virtually every phase of column operation. This will change as more rigid materials are developed such as agarose–polyacrylamide blend (Ultrogel), dextran cross-linked with N,N'-methylenebisacrylamide (Sephacryl) and the cross-linked synthetics already discussed. The result will be that chromatographic separations will be more practical at the industrial level.

A key drawback of MSC is that it is a batch method. For large-scale production, economic considerations lie greatly in favor of continuous processing. Hence, there has been some effort in developing workable systems. Sussman (1976) has reviewed a number of designs. At present, efforts with MSC have met with only limited success (Delaney, 1980; Fox *et al.*, 1969; Nicholas and Fox, 1969), primarily because of solvent requirements. However, the substantial incentive suggests that there will be further developmental work.

31.7 REFERENCES

Ackers, G. K. (1967). A new calibration procedure for gel filtration columns. *J. Biol. Chem.*, **242**, 3237–3238.
Andrews, P. (1964). Estimation of the molecular weights of proteins by Sephadex gel-filtration. *Biochem. J.*, **91**, 222–233.
Bernal, J. D. and J. Mason (1960). Co-ordination of randomly packed spheres. *Nature (London)*, **188**, 910–911.
Barth, H. G. (1980). A practical approach to steric exclusion chromatography of water soluble polymers. *J. Chromatogr. Sci.*, **18**, 409–429.
Bio-Rad Laboratories (1975). *Gel Chromatography*. Bio-Rad Laboratories, Richmond, CA.
Coupek, J. (1982). Macroporous hydroxyethyl methacrylate copolymers, their properties, activation, and use in high-performance affinity chromatography. In *Affinity Chromatography and Related Techniques*, ed. T. C. J. Gribnau, J. Visser and R. J. F. Nivard. Elsevier, Amsterdam.
Delaney, R. A. M. (1980). Industrial gel filtration of proteins. In *Applied Protein Chemistry*, ed. R. A. Grant. Applied Science, London.
Determann, H. (1969). *Gel Chromatography*. Springer-Verlag, New York.
Dubin, P. L. (1981). Aqueous exclusion chromatography. *Sep. Purif. Methods*, **10**, 287–313.
Dubin, P. L. (1983). Exclusion chromatography of water-soluble polymers. *Am. Lab.*, **15** (1), 62–73.
Fischer, L. (1980). Gel filtration chromatography. In *Laboratory Techniques in Biochemistry and Molecular Biology*, ed. T. S. Work and R. H. Burdon, vol. 1, part 2. Elsevier/North Holland, Amsterdam.
Fox, J. B., R. C. Calhoun and W. J. Eglinton (1969). Continuous chromatography apparatus I. Construction. *J. Chromatogr.*, **43**, 48–54.
Gelotte, B. (1960). Studies on gel filtration: Sorption properties of the bed material Sephadex. *J. Chromatogr.*, **3**, 330–342.
Giddings, J. C. (1965). *Dynamics of Chromatography*. Dekker, New York.
Glueckauf, E. (1955). Theory of chromatography, part 9: the 'theoretical plate' concept in column separations. *Trans. Faraday Soc.*, **51**, 34–44.

Granath, K. A. and P. Flodin (1961). Fractionation of dextran by the gel filtration method. *Makromol. Chem.*, **48**, 160–171.

Gurkin, M. and V. Patel (1982). Aqueous gel filtration chromatography of enzymes, proteins, oligosaccharide and nucleic acids. *Am. Lab.*, **14**, 64–73.

Horvath, C. and H.-J. Lin (1976). Movement and band spreading of unsorbed solutes in liquid chromatography. *J. Chromatogr.*, **126**, 401–420.

Janak, J., J. Coupek, M. Krejci, O. Mikes and J. Turkova (1975). In *Liquid Column Chromatography*, ed. Z. Deyl, K. Macek and J. Janak, p. 187. Elsevier, New York.

Janson, J. C. and P. Dunnill (1974). Factors influencing scale-up of chromatography. In *Industrial Aspects of Biochemistry*, ed. B. Spencer, part I, pp. 81–105. North-Holland, Amsterdam.

Kremmer, T. and L. Boross (1979). *Gel Chromatography*. Wiley, New York.

Laurent, T. C. and J. Killander (1964). A theory of gel filtration and its experimental verification. *J. Chromatogr.*, **14**, 317–330.

Martin, A. J. P. and R. L. M. Synge (1941). A new form of chromatogram employing two liquid phases. *Biochem. J.*, **35**, 1358–1368.

Mikes, O., P. Strop and J. Coupek (1978). Ion-exchange derivatives of Spheron. I. Characterization of polymeric support. *J. Chromatogr.*, **153**, 23–36.

Morris, C. J. O. R. and P. Morris (1976). *Separation Methods in Biochemistry*, 2nd. edn. Wiley, New York.

Nicholas, R. and J. B. Fox (1969). Continuous chromatography apparatus III. Application. *J. Chromatogr.*, **43**, 61–65.

Ouano, A. C. (1977). Kinematics of gel permeation chromatography. *Adv. Chromatogr.*, **15**, 233–271.

Pharmacia Fine Chemicals (1979). *Gel Filtration: Theory and Practice*. Pharmacia, Uppsala, Sweden.

Porath, J. (1963). Some recently developed fractionation procedures and their application to peptide and protein hormones. *Pure Appl. Chem.*, **6**, 233–244.

Porath, J. (1974). General methods and coupling procedures. *Methods Enzymol.*, **34**, 13–30.

Provder, T. (1980). *Size Exclusion Chromatography (GPC)*. American Chemical Society, Washington, DC.

Regnier, F. E. and R. Noel (1976). Glycerolpropylsilane bonded phases in the steric exclusion chromatography of biological macromolecules. *J. Chromatogr. Sci.*, **14**, 316–320.

Scopes, R. K. (1978). Techniques for protein purification. *Techniques in the Life Sciences*, **B101**, 1–42.

Scopes, R. K. (1982) *Protein Purification: Principle and Practice*. Springer-Verlag, New York.

Scott, G. D. (1960). Packing of equal spheres. *Nature (London)*, **188**, 908–909.

Squire, P. G. (1964). A relationship between the molecular weights of macromolecules and their elution volumes based on a model for Sephadex gel filtration. *Arch. Biochem. Biophys.*, **107**, 471–478.

Sussman, M. V. (1976). Continuous chromatography: a status report. *CHEMTECH*, **6**, 260–264.

Van Deemter, J. J., F. J. Zuiderweg and A. Klinkenberg (1956). Longitudinal diffusion and resistance to mass transfer as causes of nonideality in chromatography. *Chem. Eng. Sci.*, **5**, 271–289.

32

Affinity Chromatography

M. L. YARMUSH and C. K. COLTON
Massachusetts Institute of Technology, Cambridge, MA, USA

32.1 INTRODUCTION

Biospecific purification procedures provide an attractive alternative to classical purification techniques which rely on differences in physicochemical properties such as solubility, size and charge. The latter techniques are generally low yield procedures, and, if purification to homogeneity is the desired result, multiple steps must be effected requiring large amounts of starting material and many hours of work. The development of experimental methodology based solely on the biological and functional attributes of the molecules of interest was, however, slow until appropriate support media and simple chemical coupling schemes became available from the preparation of affinity adsorbents. Today, affinity chromatography is an established laboratory tool which has expanded even beyond the molecular domain to include purification of particulate elements such as virus particles, cells and cellular components.

Affinity chromatography is usually realized by covalently linking a binding molecule, called the ligand, to an insoluble support and packing the support into a column as shown in Figure 1. Only compounds with appreciable affinity for the ligand are retained on such a column while others pass through unretarded. Specifically adsorbed compounds (ligates) are then eluted by a variety of means, such as altering the composition of the solvent to favor dissociation. In principle, biospecific adsorption can be applied where any particular ligand interacts specifically with a biomolecule, for example in the purification of enzymes, antibodies, nucleic acids, vaccines and vitamins, to name a few. The method has inherent advantages over the classical techniques of protein and biomolecule purification because it is essentially a single-step procedure which provides with rapidity both high yield and high resolution of the purified species.

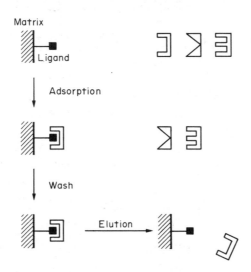

Figure 1 Basic principles of affinity chromatographic separations. A suitable ligand is covalently attached to an insoluble matrix. In the adsorption step, only those molecules with a specific binding site for the ligand bind to the adsorbent; molecules without the proper geometric fit pass through unaffected. In the elution step, the bound molecules are disengaged from the column and collected

Although not very descriptive of the process, the term affinity chromatography was originally proposed by Cuatrecasas *et al.* (1968) for chromatography based on biospecific recognition. Since then, several other terms have been employed including biospecific adsorption, bioselective adsorption, bio-affinity chromatography and ligand-specific chromatography. The more recent terminology has evolved in order to emphasize the dependence of the technique on natural biological interactions. Included in this classification, in addition to systems that offer mono- or limited specificity which, for example, employ enzymes or antibodies, are group specific adsorbents which show specificity for whole classes of molecules. The latter include dye adsorbents, which have a substantial selectivity for nucleotide binding sites on certain enzymes, and lectin adsorbents, which are employed for isolation of glycoproteins, glycolipids and polysaccharides.

This short review will highlight the major components of biospecific adsorption in order to acquaint the reader with the basics of the technique. For more comprehensive treatments, the reader can refer to several excellent detailed monographs and reviews (Gribnau *et al.*, 1982; Scouten, 1981; Lowe and Dean, 1974; Lowe, 1979; Absolom, 1981; Weetall, 1973; Hoffmann-Ostenhof *et al.*, 1978; Porath and Kristiansen, 1975).

32.2 MATRIX

The basic requirements of an ideal general support matrix for affinity chromatography are listed in Table 1. Five types of polymers which have been frequently used as matrix materials are agarose, porous glass or silica, polyacrylamide, methacrylate polymers and cellulose (Figure 2). The properties and suitability of these materials as affinity supports are discussed below.

32.2.1 Agarose

Ever since the pioneering work by Cuatrecasas *et al.* (1968), agarose, a polygalactan obtained from seaweed, has been the most universally employed general matrix for affinity chromatography applications. The main advantage of agarose is its highly porous, hydrophilic structure, combined with relative inertness and chemical stability. The hydroxyl groups on the sugar residues

Commercially available supports

Beaded supports (brand names)	Chemical structure	Suppliers
Agarose (Sepharose, Biogel A, Ultrogel)	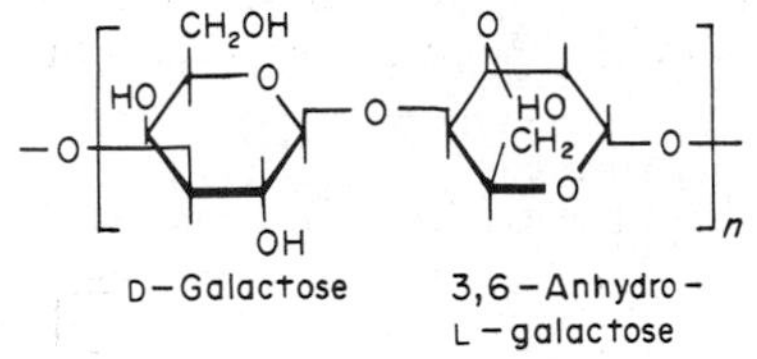	Pharmacia Bio Rad LKB Serva P-L Biochemicals Sigma
Siliceous materials (Controlled Pore Glass, Spherosil)	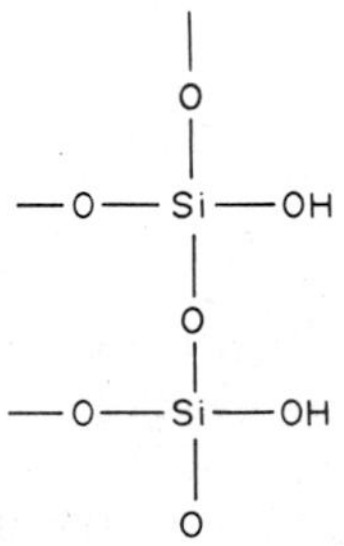	Corning Rhone – Progil Bio Rad Waters Serva Electro – nucleonics
Polyacrylamide (Biogel P)	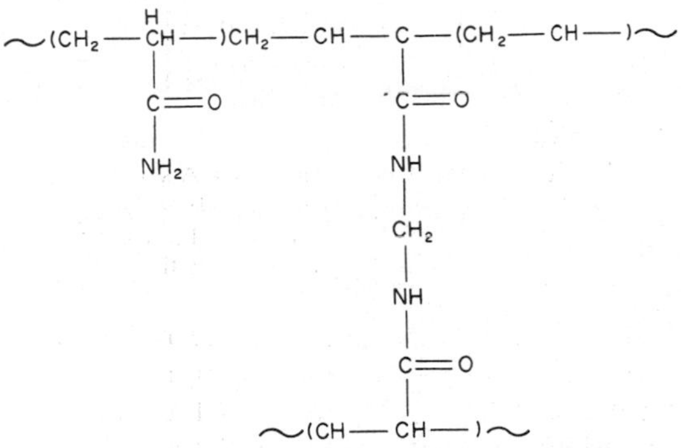	Bio Rad
Hydroxyethyl methacrylate copolymers (Spheron)	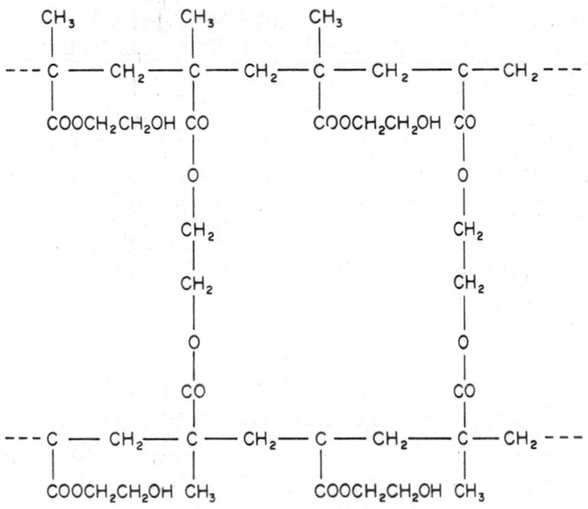	LaChema Koch – Light Knauer – KG
Cellulose	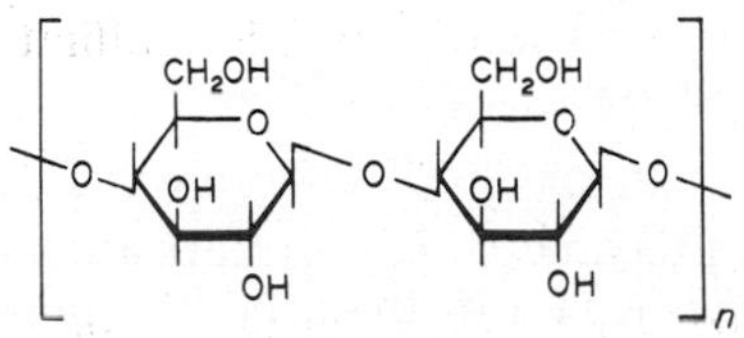	E. Merck Whatman Sigma

Figure 2 Chemical structure of some commercially available support matrices. A wide variety of pre-activated gels with and without spacer arms are also available from the suppliers listed above. Details of these products may be obtained from the suppliers

Table 1 Characteristics of an Ideal Support for
Affinity Chromatographic Separations

(a) Insolubility
(b) Rigidity
(c) Permeability
(d) Macroporosity
(e) Hydrophilicity
(f) Chemical derivatization potential
(g) Minimal non-specific adsorption
(h) Resistance to microbial or enzymatic degradation
(i) Ease of synthesis and/or commercial availability

can be easily derivatized for the covalent attachments of ligands by a variety of methods. In addition, the presence of the sugar unit, 3,6-anhydro-L-galactose, makes agarose exceptionally resistant to degradation. The main drawbacks are modest rigidity, thermolability at high temperatures, fragility with regard to mechanical attrition and relatively high price. Some improvement in the rigidity of agarose, and consequently its ability to withstand increased flow rates, has been attained after crosslinking with epichlorhydrin, divinyl sulfone or 2,3-dibromopropan-1-ol (Porath *et al.*, 1971; Porath and Sundberg, 1972). Since the process of crosslinking reduces the porosity as compared to that of the parent gel, the investigator may wish to purchase commercial crosslinked agarose for which the manufacturer has already determined the porosity. Although crosslinking improves stability toward extremes of pH, temperature and high concentrations of organic solvents and chaotropic agents, there is a compensatory reduction in the number of hydroxyl groups available for covalent attachment. Some reports demonstrate that the loss of OH sites can be reduced by addition of sorbitol or phloroglucinol during the crosslinking process (Coombe and George, 1976; Porath and Sundberg, 1972).

An alternative for increasing maximum flow rates involves increasing the diameter of the beads. An adverse effect accompanying this modification, however, may be a lowering of the efficiency of operation, since the time needed for attainment of equilibrium will be increased in proportion to the square of the diameter.

32.2.2 Porous Glass and Silica

Controlled pore glass (CPG) is the most commonly employed inorganic matrix for immobilization of biological molecules. A wide variety of organic functional groups have been applied to the support, and activation can then be achieved by the same techniques used for other supports. Glass derivatives have outstanding rigidity and are extremely resistant to degradation. In addition, column bed volumes do not change when the aqueous environment is altered. Two serious shortcomings in the use of porous glass are: (1) non-specific adsorption (even with many derivatized surfaces) and (2) slight solubility at alkaline pH. The former difficulty has been significantly reduced by derivitization of the glass with a crosslinked organic coating of dextran or triethoxypropylglycidoxysilane (Weetall, 1973; Regnier and Noel, 1976). One can minimize the solubility problem by using glass coated with other metal oxides such as zirconium or titanium which are available commercially.

A less expensive inorganic alternative to CPG is macroporous silica beads (Spherosil). Spherosil can be readily crosslinked with a variety of hydrophilic copolymers carrying different amounts and types of functional groups linked either directly or indirectly *via* a spacer to the backbone of the polymer. A recent preliminary comparison study suggests that certain Spherosil affinity adsorbents may offer increased adsorption capacity over that obtained with commercial agarose affinity adsorbents (Schutyser *et al.*, 1982).

32.2.3 Polyacrylamide

Polyacrylamide gels are prepared by copolymerization of acrylamide crosslinked with bisacrylamide to provide a polymer with a hydrocarbon framework carrying carboxamide side chains. Advantages of this matrix include stability in a wide pH range and in many organic solvents, lack of ionized groups, biological inertness and resistance to degradation. Despite these favorable properties, polyacrylamide suffers from two major drawbacks: (1) limited mechanical stability (much poorer than agarose) which leads to bead compression under pressure and (2) a reduction in bead porosity concurrent with chemical derivatization. One approach to circumvent the latter problem has been to copolymerize the polyacrylamide gel with a functional group bearing an acrylic or vinyl monomer (Schnaar and Lee, 1975). Details of this procedure will be discussed in the following section.

32.2.4 Hydroxyalkyl Methacrylate

A modified polyacrylamide which has excellent mechanical stability and a range of larger pore sizes is Spheron, a hydrophilic copolymer of 2-hydroxyethyl methacrylate and ethylene dimethacrylate. These supports are chemically and mechanically stable, easily derivatized and only slightly susceptible to non-specific adsorption. The gels contain neutral hydroxyl groups and are thus readily activated by all procedures used to derivatize polysaccharide carriers. Alternatively, the gels may be formed by copolymerization with monomers containing reactive groups by a procedure analogous to that used in polyacrylamide gels (Coupek, 1982). The copolymer can be further modified by covalent attachment of mono- and oligo-saccharides (Filka *et al.*, 1978). These glycosidic derivatives combine the superior mechanical and hydrodynamic properties of the Spheron with the advantageous adsorption characteristics of polysaccharides. Another modified polyacrylamide which has been widely used for enzyme immobilization but infrequently in affinity chromatography applications is Enzacryl, a copolymer of *N*-acryloylmorpholine and *N,N'*-methylenebisacrylamide (Epton *et al.*, 1976).

32.2.5 Cellulose

Before the availability of beaded agarose, cellulose was frequently used to prepare affinity adsorbents. Most commercially available cellulose preparations are in a microcrystalline form and thus have a high flow resistance. Recently, however, several investigators (Chen and Tsao, 1976; Gurvich and Lechtzind, 1982) have prepared cellulose in a beaded form which appears to be highly porous, chemically versatile, and physically rigid. A serious limitation with regard to its uses for large protein purification is its non-uniform character which allows variable penetration of large protein molecules.

32.3 SPACER ARMS

In many instances, the binding site of a particular ligate is located sufficiently deep within the ligate so that the matrix may sterically hinder the binding of the ligate to its ligand. This is often critical in enzyme–substrate systems where the ligand is a small molecule. To overcome this problem, the ligand is generally attached to the matrix *via* a flexible arm or spacer. These spacer molecules are usually linear aliphatic hydrocarbons which may or may not have interposed polar moieties such as secondary amino, hydroxyl and peptide groups. Although the interposition of a spacer may reduce the steric interference generated by the matrix, several studies have suggested that spacers may occasionally not only generate local steric hindrance but also contribute to non-specific adsorption phenomena (O'Carra, 1981). The length of the spacer is most important when the affinity of the immobilized ligand for the ligate is weak (O'Carra *et al.*, 1973; Nishikawa and Bailon, 1975). In general, spacers do not provide any advantage when large molecules such as proteins are the immobilized components. Affinity chromatographic support media containing a variety of spacer arms are available commercially from suppliers listed in Figure 2.

32.4 COUPLING PROCEDURES

Many ligand immobilization procedures have been documented in the literature. The use of a particular procedure depends on the functional group(s) present on the matrix and ligand and the stability (mechanical and chemical) of both matrix and ligand. The chemically reactive groups which are generally used for attachment are hydroxyl, amino and carboxyl functions. Some of these techniques are reviewed briefly below.

A variety of methods are available for derivatization of supports with hydroxyl functions (polysaccharides, glass, hydroxyalkyl methacrylate). The most popular method involves reaction with cyanogen bromide. The chemical reactions of cyanogen bromide are indeed complex; products formed include cyanate esters, cyclic and acyclic imidocarbonates, carbamates and carbonates (Figure 3). The principal reactive component, however, appears to be the cyanate ester (Wilchek *et al.*, 1975). The activated agarose reacts rapidly in moderately alkaline conditions (pH 9–10) with primary amines to yield principally the *N*-substituted isourea derivative. The isourea substitutent is positively charged at neutral pH ($pK_a \sim 10.4$), which introduces an element of anion exchange to the adsorbent. This effect can be circumvented by use of an acid hydrazide instead of an alkylamine to generate an isourea analogue that is not protonated at neutral pH (Lowe, 1979). A residual problem with either method of CNBr activation is the instability of the isourea linkage under alkaline conditions which leads to slow leakage of the ligand from the support.

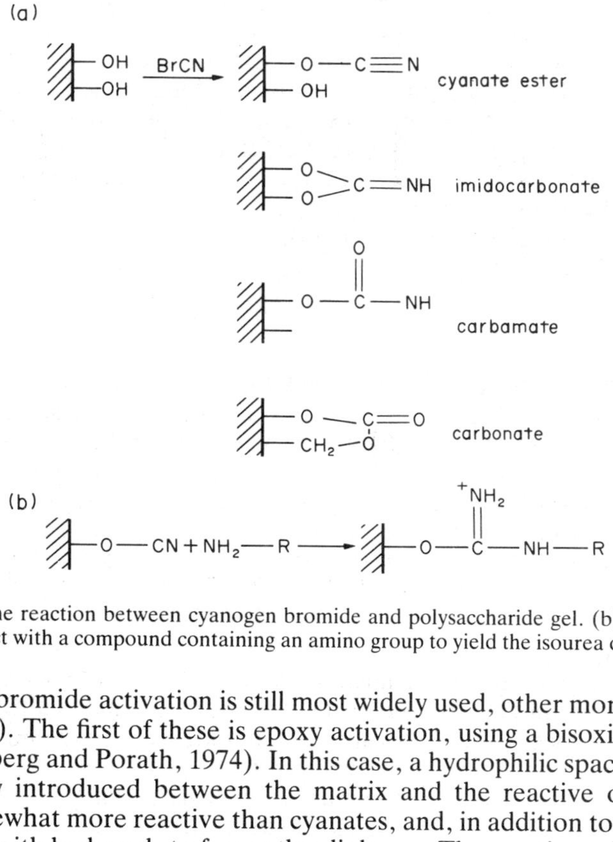

Figure 3 (a) Products of the reaction between cyanogen bromide and polysaccharide gel. (b) Reaction of the cyanate ester product with a compound containing an amino group to yield the isourea derivative

Although cyanogen bromide activation is still most widely used, other more attractive methods are available (Figure 4). The first of these is epoxy activation, using a bisoxirane, 1,3-butanediol diglycidyl ether (Sundberg and Porath, 1974). In this case, a hydrophilic spacer arm containing 11 atoms is automatically introduced between the matrix and the reactive oxirane (Figure 4a). Epoxy groups are somewhat more reactive than cyanates, and, in addition to reactions with primary amines, they react with hydroxyls to form ether linkages. The reaction with hydroxyls is slow, however, requiring higher temperatures (25–40 °C) and more extreme alkaline pH conditions (pH 11–13). The resulting linkage is very stable under alkaline conditions, and the spacer arm, which contains no charged groups, does not generate non-specific ionic adsorption. Elimination of excess unreacted oxirane functions can be difficult requiring either prolonged storage in alkaline solutions or prolonged treatment with small nucleophiles.

A second method involves activation with 1,1'-carbonyldiimidazole (Bethell *et al.*, 1979), which yields a product with comparable reactivity to the cyanate ester (Figure 4b). Subsequent coupling to a primary amine generates an uncharged urethane linkage. The reagent is also less noxious than either cyanogen bromide or bisoxiranes.

The most recent method (Nilsson and Mosbach, 1980), involves the use of toluenesulfonyl chloride (tosyl chloride) or the more reactive 3,3,3-trifluoroethanesulfonyl chloride (tresyl chloride) (Figure 4c). The tosyl group introduced on activation reacts rapidly to give a stable secondary amine. At neutral pH the secondary amine group is charged, however, so that the ultimate charge behavior of the adsorbent may be similar to that obtained with cyanogen bromide activation, but the linkage is more stable.

Chemical derivatization of polyacrylamide can be achieved by primary modification of preformed polyacrylamide beads (Inman and Dintzis, 1969) or by copolymerization of acrylamide

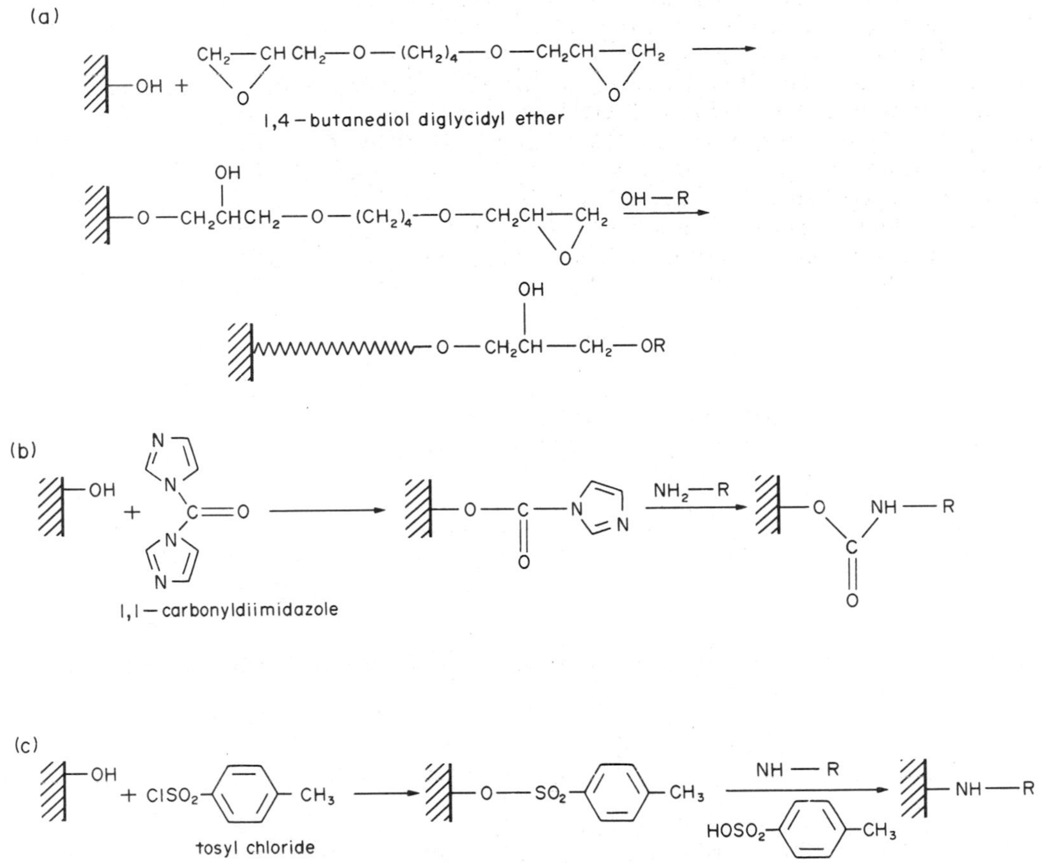

(a)

1,4-butanediol diglycidyl ether

(b)

1,1-carbonyldiimidazole

(c)

tosyl chloride

Figure 4 (a) Activation and reaction of epoxy-activated agarose with a compound containing a hydroxyl residue. Activation and reaction of carbonyldiimidazole-activated (b) and tosyl-activated agarose (c) with a compound containing an amino group

and bisacrylamide with active esters (Schnaar and Lee, 1975; Whitesides *et al.*, 1976). These reactions are summarized in Figure 5. Proteins can be immobilized on derivatized supports containing free amino and carboxyl functions by a variety of methods. A representative listing is shown in Table 2.

32.5 ADSORPTION

Conditions which maximize optimum association between the ligand and ligate and minimize binding of extraneous materials are generally used in the adsorption step. Parameters such as ionic strength, pH and chemical composition of the equilibrium buffer, along with temperature and flow rate, can be adjusted to obtain maximum specific adsorption.

Typically the protein sample in the equilibration buffer is applied to a previously equilibrated affinity column. The column is than washed with several column volumes of buffer. Continued washing should theoretically result in the eventual recovery of the adsorbed material. However, this procedure is generally very time consuming and can result in a very diluted product, so that recovery is usually effected by changing the pH, ionic strength or chemical composition of the buffer (as discussed in the next section).

The apparent kinetics of adsorption are usually slow enough to necessitate sample application at relatively slow flow rates. High sample protein concentrations and/or low concentration of immobilized ligand also demand reduced flow. If the interaction between immobilized ligand and ligate is weak, some preincubation prior to washing may be in order. However, non-specific adsorption may generate similar time-dependent effects and some contamination of the eluted product may result. Batch procedures are most suitable in the case of high affinity systems where

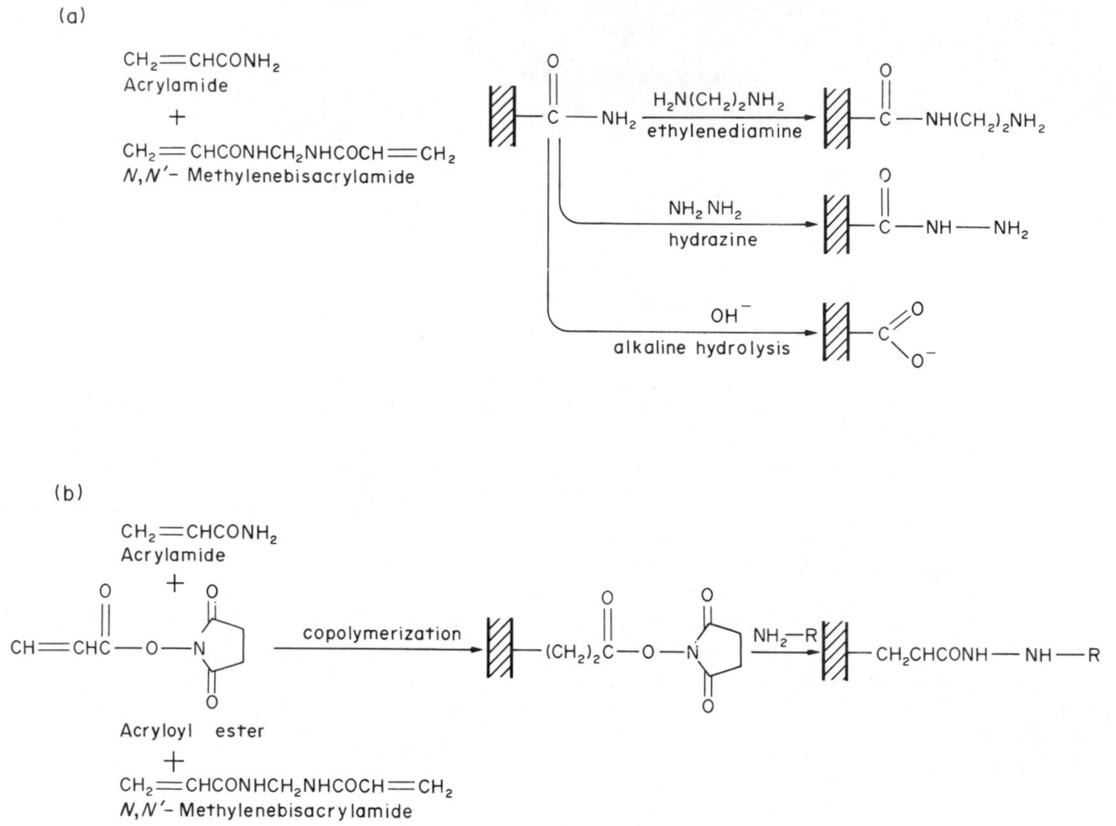

(a)

(b)

Figure 5 (a) Copolymerization of acrylamide and bisacrylamide followed by primary derivatization reactions by aminoethylation, hydrazinolysis and alkaline hydrolysis. (b) Polyacrylamide gel copolymerized with an active ester. Acrylamide and methylenebisacrylamide are copolymerized with an acryloyl ester, *N*-succinimidyl acrylate. Coupling of ligand is achieved by simply displacing the active ester in the gel with an aliphatic amino group

Table 2 Methods of Coupling Ligands to Amino and Carboxyl Functions[a]

Functional group of matrix	Method	Ref.
Amino	Carbodiimides	Weliky *et al.* (1969), Weetall and Weliky (1964)
	Isonitrile	Axen *et al.* (1971), Axen and Vretblad (1971)
	Cyanogen bromide	Axen *et al.* (1967, 1971), March (1974)
	Glutaraldehyde	Weston and Avrameas (1971)
	Diazonium salts	Barker *et al.* (1970) Bar-Eli and Katchalski (1963)
	Isothiocyanate	Barker *et al.* (1970), Axen and Porath (1964)
	Halotriazines	Kay and Lilly (1970)
Carboxyl	Acid anhydride, acid chloride	Levin *et al.* (1964), Isliker (1957)
	Azide	Barker *et al.* (1970), Micheel and Ewers (1949)
	Carbodiimides	Weliky *et al.* (1969), Weetall and Weliky (1964)

[a] Modified from Porath and Kristiansen (1975).

relatively small amounts of specific protein are to be purified from a mixture containing a significant portion of extraneous protein.

A crude extract can often be chromatographed directly on the adsorbent without previous purification so long as particulates which may plug the column are absent. In some cases, a preliminary fractionation by conventional means may be advantageous or even necessary. One obvious example is when the crude sample contains a substance which is capable of degrading the ligand.

32.6 ELUTION

The most critical step in the purification of proteins is often the elution of adsorbed species from the adsorbent. Several general methods of elution have been commonly used for a wide variety of affinity applications. When dealing with high affinity systems none of the general methods described below may be effective in completely liberating the bound molecules. Under these circumstances one may achieve higher yields by employing combinations of these techniques.

32.6.1 Biospecific Elution

Elution of ligate may occasionally be achieved by addition of an excess of a low molecular weight compound which competes for the binding site of the ligate and can be removed from the product stream by subsequent dialysis or gel filtration. This principle has been most successfully used in immunoadsorption with haptens and peptides and in enzymology with cofactors, coenzymes or nucleotides, cosubstrates and competitive inhibitors.

32.6.2 Elution by Reversible Denaturation

Urea and guanidine salts are commonly used to change globular proteins into random-coiling macromolecules that lose their capacity to bind to the ligand. However, these reagents may irreversibly denature the components of the complex during the dissociation process. Several groups have examined the effect of neutral salts, particularly their anionic components, on a variety of macromolecular systems (Dandliker and de Saussure, 1971; Hjerten, 1973). These ions have been termed chaotropic to indicate their ability to disrupt water structure by breaking hydrogen bonds. The elution strength of these ions basically corresponds to the order exhibited in the Hofmeister series which expresses the ability of certain ions to precipitate lyophilic substances from colloidal solution. The decrease in ability to precipitate is accompanied by an increase in ability to promote unfolding of molecules as well as dissociation of complexes. An example of an analogous series which defines the relative effectiveness of a particular ion to induce dissociation of antigen–antibody interactions and thus result in elution is arranged below in order of decreasing elution effectiveness (chaotropicity):

$$CCl_3CO_2^- \geqslant SCN^- > CF_3CO_2^- > ClO_4^- > I^- > Cl^-$$

It is important to note that the order of chaotropicity may vary for other applications. Neutral solutions of chaotropic ions in concentrations up to 3 M have been used as desorbers in a variety of moderate to high affinity systems. Regardless of which ionic species is used, it is advisable to restore the native protein structure promptly by removal of the denaturant through dilution, desalting, or dialysis.

32.6.3 Elution by Dissociation of Bonds from Coulombic and Dispersion Interactions

The noncovalent physicochemical interactions between molecules in affinity chromatography can be classified as emanating from electrostatic forces (including Coulombic forces and forces arising from permanent dipoles), induction forces, and London dispersion forces (also called hydrophobic forces) (Lyklema, 1982). With both small and large molecules, Coulombic and dispersion forces usually predominate (Vilker *et al.*, 1981). The extent of interaction can be influenced by the properties of the solvent, thereby permitting selection of conditions which favor adsorption or desorption. Coulombic forces stem from the attraction between positively and negatively charged groups on the interacting molecules. These attractive forces can be virtually eliminated by increasing the ionic strength of the equilibrium buffer or even made repulsive by altering the pH. From a practical point of view, desorption is often more effective when an ionic or pH gradient is employed rather than by direct application of desorbing medium.

In addition to these Coulombic forces, dispersion (or hydrophobic) forces also participate in the specific interaction between ligand and ligate. These so-called van der Waals interactions can also be reversed. This occurs when the sum of the energies of interactions between two different compounds and a liquid medium is larger than the sum of the energies of interactions between the two compounds and those between the molecules of the liquid medium (van Oss *et al.*, 1980,

1982). In practical terms, the net van der Waals interaction between two different compounds in a liquid is repulsive when the surface tension of the liquid medium has a value intermediate between the values of the interfacial tensions of the two different interacting materials (van Oss, *et al.*, 1980, 1982). In aqueous systems, van der Waals repulsion can be achieved through lowering the surface tension of the aqueous medium by the addition of varying amounts of water-miscible organic solvents such as ethylene glycol, dimethyl sulfoxide or ethanol. This approach has been employed experimentally in the dissociation of antigen–antibody complexes and in certain cases of affinity chromatography (van Oss *et al.*, 1980, 1982). Thus, more effective elution schemes would involve reversal of both Coulombic and van der Waals interactions.

32.6.4   Electrophoretic Desorption

Several reports have appeared on the application of electrophoretic desorption to a variety of interactions (Morgan *et al.*, 1980; Brown *et al.*, 1977). This technique involves direct electrophoresis of proteins from affinity adsorbents using modified gel electrophoresis equipment. In principle, electrophoretic desorption is applicable in all situations where the adsorbed species has an overall net charge. The generated electric field does not actively induce dissociation of ligand–ligate complexes but rather induces migration of the charged ligate after dissociation has occurred as a result of the reversibility of the interactions. The ligand on the other hand does not migrate because it is non-reversibly bound to the support. This technique would appear to be limited in cases of high affinity interactions and with geometries which allow for repeated reassociation of ligate with ligand during migration. The advantages of the technique include mild conditions, high yield (provided adequate time is given for desorption) and small desorbate volume.

32.6.5   Elution by Lowering Ionic Strength

As stated previously, in moderate affinity systems increasing the ionic strength of the eluting buffer may be sufficient to effect desorption. Recently, a hypotonic elution technique (desorption by lowering the ionic strength) has been employed in the immunoadsorbent purification of several enzymes (Danielsen *et al.*, 1982). Indeed, elution with distilled water has also been reported (Vidal *et al.*, 1980). The mechanism of these elution techniques, which may involve both inter- and intra-molecular repulsion, is unclear. Assessment of these techniques will have to await further study.

32.6.6   Elution by Changes in Temperature

The adsorption process in affinity chromatographic separations is generally exothermic (Harvey *et al.*, 1974). According to LeChatelier's principle, elevated temperatures will shift the equilibrium in the direction of heat absorption such that dissociation of the complex is favored. The more exothermic the adsorption process, the more susceptible it will be to elution by an increase in temperature. Advantages of the thermal elution technique are that the purified molecules are recovered under relatively mild conditions and are not mixed with elution reagents such as salts or chaotropic agents. Furthermore, when used as an adjunct to other elution techniques, temperature treatment may allow complex dissociation with relatively milder treatments than that achieved under isothermal adsorption and desorption conditions

32.7   REGENERATION

Adsorbent fouling may present a serious problem when crude extracts and mixtures are used as starting materials. This problem is usually manifested by a reduced adsorbent capacity which may result when denatured material has become tightly bound to the column. One protocol (Lowe, 1979) suggests a routine wash with 2 M KCl/6 M urea after each use. Addition of dioxane, or dimethylformamide, and/or an overnight incubation with pronase (for non-protein adsorbents) may also prove to be beneficial.

32.8 THEORETICAL MODELING

Theoretical models of the process of affinity chromatography have been developed by several authors (Graves and Wu, 1974; Wankat, 1974; Denizot and Delaage, 1975; Dunn and Chaiken, 1975; Kasai and Ishii, 1975; DeLisi and Hethcote, 1982). Of these models, the most instructive for present purposes is that of Wankat (1974) who developed an analysis which included the effect of steric blockage of ligand sites by bound ligand as well as the interactions between an enzyme and a ligand, and between enzyme and washing agent. Overall, column operation was described by a staged, discrete transfer and discrete equilibrium step model that is applicable to counter current distribution, This analysis is briefly described below.

It is assumed that there is specific adsorption of enzyme on the solid surface, that a maximum of a monolayer of enzyme can cover the surface, that free enzyme, available ligand and enzyme–ligand complex are in equilibrium on each stage of the column, and that the enzyme–ligand complex covers a relatively large area and blocks other ligand sites to which it is not attached. Equilibrium and mass balance relations for the reversible enzyme–ligand interaction are combined to yield an expression for the surface concentration of enzyme ligand complex [EB]:

$$[EB] = \frac{B_0[E]}{K_{EB} + [E]A_E B_0} \tag{1}$$

where [E] is the volumetric concentration of free enzyme in solution, [B] is the surface concentration of available ligand, B_0 is the original concentration of ligand, A_E is the surface area covered by a unit mass of EB complex and $K_{EB}=[E][B]/[EB]$, the effective dissociation constant for the enzyme–ligand reaction.

A similar analysis is applied for dissociation of the enzyme–ligand complex by elution with an agent W which does not interact with the ligand but interacts with the enzyme to form a complex EW which also does not bind to the ligand. The volumetric concentration [EW] of the complex is given by:

$$[EW] = \frac{[E](M_{W_{j,S}}/V)}{K_{EW} + [E]} \tag{2}$$

where V is the volume of liquid on each stage, $M_{W_{j,S}}$ is the mass of W (as both free W and that contained in the EW complex) on stage j after transfer step S, and $K_{EW} = [E][W]/[EW]$ is the dissociation constant for enzyme–elution agent reaction.

Lastly, the fraction $f_{E_{j,S}}$ of enzyme in the liquid phase (as E and EW) in stage j after transfer step S is given by:

$$f_{E_{j,S}} = \frac{([E] + [EW])V}{([E] + [EW])V + [EB]A_T} \tag{3}$$

where A_T is the total surface area per stage.

Equations (1) to (3) are used in the staged chromatographic model in which a column of length L is divided into N equilibrium stages, each of which has a length equal to the height equivalent to a theoretical plate (HETP, an experimentally determined parameter). Some fraction, q, of the liquid in each stage is transferred to the next stage at each transfer step. The mass of enzyme in all forms in stage j after transfer step S, $M_{E_{j,S}}$, is given by:

$$M_{E_{j,S}} = qf_{E_{j-1,S-1}}M_{E_{j-1,S-1}} + (1-qf_{E_{j,S-1}})M_{e_{j,S-1}} \tag{4}$$

Similarly, $M_{W_{j,S}}$ is given by:

$$M_{W_{j,S}} = qM_{W_{j-1,S-1}} + (1-q)M_{W_{j,S-1}} \tag{5}$$

Finally, $M_{E_{j,S}}$ can also be defined by:

$$M_{E_{j,S}} = [E]V + [EW]V + [EB]A_T. \tag{6}$$

Wankat (1974) described an iterative technique to solve equations (1) through (6) for successive stages, beginning at the column inlet. Methods were also described for experimentally determining the values of independent variables. A calculation of [E] at the outlet of the column provides a measure of the separation achieved for given inlet, column and operating parameters.

Using the methods outlined, predictions of the performance of an affinity chromatography column may be obtained in both large scale and laboratory scale systems for pulse, step and breakthrough curves. The theory is also flexible and can be modified to account for various phenomena including dispersion, two-enzyme binding and non-specific adsorption.

32.9 LARGE SCALE APPLICATIONS

In principle, affinity chromatography is ideally suited for large-scale industrial separations. High purification can be accomplished in a single step; consequently, yields are high. The nature of the adsorption–desorption process provides high potential product concentration as well. Nonetheless, affinity chromatography has found relatively few large scale applications (Janson, 1982). Most notable are two reports by Robinson *et al.* (1972, 1974), the first of which describes several factors affecting scale-up of affinity purification of β-galactosidase on substituted agarose and the second of which describes a pilot scale system of timed values used to operate a 1.8 l capacity column continuously. From a technological point of view, the main obstacle to development of large scale systems has been the compressibility of the popular matrix materials. Some of the newer matrices discussed earlier, which combine mechanical stability with favorable adsorption charateristics, may prove to be ideally-suited for these purposes.

32.10 IMMUNOADSORPTION

Immunoadsorption is perhaps the most exciting application of biospecific adsorption because of the extraordinary specificity of most immunological interactions. The term immunoadsorption has been used to describe purification of either antigens or antibodies by their immobilized, complementary molecules. This brief discussion will deal with selective adsorption of an antigen to a solid support to which a specific antibody preparation has been covalently attached. The technique is not limited to protein antigens, since antibodies to small molecules (haptens) may be prepared by linking the hapten to macromolecules. Until recently, this approach attracted little attention because the polyclonal antibody preparations used contain a multiplicity of different antibodies. These antibodies have recognition sites for different portions of the antigen (Ag) molecule and a wide distribution of association constants for each site. As a consequence of the presence of high affinity antibodies, harsh conditions are usually required to elute all of the bound antigen which may lead to significant denaturation of antigen and/or antibody. Furthermore, a substantial fraction of the product may become irreversibly bound and therefore unrecoverable on the first separation performed on a virgin column (Eveleigh, 1982). This not only involves loss of material to be isolated, but also loss of adsorbent capacity.

With the advent of hybridoma technology, low to intermediate affinity (10^4–10^7 M^{-1}) monoclonal antibodies (all of which have identical site recognition and affinity characteristics) in theoretically unlimited supply are for the first time available. Laboratory scale systems using immunoadsorption with monoclonal antibodies (Ab) have been reported recently for the purification of several proteins. (Secher and Burke, 1980; Staehelin *et al.*, 1981; Hauri *et al.*, 1980).

Most of the problems encountered using polyclonal antibodies can be avoided with the use of monoclonal antibodies. The facilities needed and the techniques used, although straightforward, are more demanding than those required for ordinary antibody production. However, the advantages over polyclonal antibodies for making immunoadsorbents are considerable. First, the antigen preparation used does not have to be pure; the screening process for detecting suitable clones selects only the cells producing antibody to the antigen of interest and rejects those producing antibody to other molecules. Secondly, as mice are usually used, the amount of antigen required to elicit a response is very small so that less than 100 μg may be sufficient. Antigen preparations for polyclonal antibody production are usually of the order of milligrams. Thirdly, the reproducibility of specificity and affinity is invariant in different batches of a given monoclonal antibody preparation. It is possible to select a monoclonal Ab with an affinity constant in the desirable range of 10^4–10^7 M^{-1}, thus minimizing problems generated by harsh elution conditions. Last, once a suitable clone has been found, the cells can be stored in liquid nitrogen, providing an indefinite supply of that antibody; the supply does not terminate with an animal's death.

One of the drawbacks of immunoadsorbents and affinity adsorbents in general is that, although a high degree of ligand substitution can be achieved using conventional immobilization methods, only a small proportion of the immobilized species attains the appropriate orientation so that typically, only a fraction of potential sites can actually bind the appropriate ligate. Recently, Schneider *et al.* (1982) have described a method whereby the antibody ligand is reacted with Protein A-Sepharose followed by chemical crosslinking with dimethyl pimelimidate. This method allows favorable spatial orientation of antibodies and thus maximizes antigen binding capacity.

Widespread use of immunoadsorption with monoclonal antibodies is presently inhibited by high cost. Two factors largely determine the cost of the purification process: (1) the cost of production of the monoclonal antibody and (2) the number of reuses attainable with the immunoadsorbent in successive adsorption–desorption cycles. One would expect that the cost of monoclonal antibodies will substantially decrease over time. New cell culture techniques are being applied to monoclonal production such as the microencapsulation technique recently announced by Damon Biotech. The combination of these new lower-cost techniques and the expanding markets in clinical assays and medical therapies (Sevier *et al.*, 1981) may reduce the cost of monoclonal antibodies to a level where they can be applied more cost-effectively to the purification of broad classes of biological compounds.

Similar cost benefits would accrue from development of mild desorption methods which would extend the life of immunoadsorbents in addition to preserving the functional activity of the isolated species. Current practice usually involves relatively harsh conditions of extremes of pH, chaotropic salts, or use of concentrated solutions of urea or guanidine hydrochloride, either of which lead to loss of functional activity.

This aspect of the procedure, which has in the majority of cases been governed by arbitrary and empirical criteria, deserves further fundamental study. Advances in this area will be predicated on a basic understanding of the intermolecular forces involved in the immunoadsorption process. This knowledge will ultimately help define specific conditions (and possibly general principles) which on the one hand, maximize the association between antibody and antigen, and, on the other hand, lead to complete dissociation under the most mild conditions attainable.

32.11 REFERENCES

Absolom, D. R. (1981). Affinity chromatography. *Sep. Purif. Methods*, **10**, 239–286.

Arsenis, C. and D. B. McCormick (1964). Purification of liver flavokinase by column chromatography on flavin–cellulose compounds. *J. Biol. Chem.*, **239**, 3093–3097.

Axen, R. and J. Porath (1964). Chemical coupling of amino acids, peptides and proteins to Sephadex. *Acta Chem. Scand.*, **18**, 2193–2195.

Axen, R., J. Porath and S. Ernback (1967). Chemical coupling of peptides and proteins to polysaccharides by means of cyanogen halides. *Nature (London)*, **214**, 1302–1304.

Axen, R., P. Vretblad and J. Porath (1971). The use of isocyanides for the attachment of biologically active substances to polymers. *Acta Chem. Scand.*, **25**, 1129–1132.

Axen, R. and P. Vretblad (1971). Binding of proteins to polysaccharides by means of cyanogen halides. Studies on cyanogen bromide treated Sephadex. *Acta Chem. Scand.*, **25**, 2711–2716.

Axen, R. and S. Ernback (1971). Chemical fixation of enzymes to cyanogen halide activated polysaccharide carriers. *Eur. J. Biochem.*, **18**, 351–360.

Bar-Eli, A. and E. Katchalski (1963). Preparation and properties of water-insoluble derivatives of trypsin. *J. Biol. Chem.*, **238**, 1690–1698.

Barker, S. A., P. J. Somers, R. Epton and J. V. McLaren (1970). Cross-linked polyacrylamide derivatives as water-insoluble carriers of amylolytic enzymes. *Carbohydr. Res.*, **14**, 287–296.

Bethell, G. S., J. S. Ayers, W. S. Hancock and M. T. W. Hearn (1979). A novel method of activation of cross-linked agaroses with 1,1'-carbonyldiimidazole which gives a matrix for affinity chromatography devoid of additional charged groups. *J. Biol. Chem.*, **254**, 2572–2574.

Brown, P. J., M. J. Leyland, J. P. Keenan and P. D. G. Dean (1977). Preparative electrophoretic desorption in the purification of human serum ferritin by immuno-adsorption. *FEBS Lett.*, **83**, 256–259.

Chen, L. F. and G. T. Tsao (1976). Physical characteristics of porous cellulose beads as supporting material for immobilized enzymes. *Biotechnol. Bioeng.*, **18**, 1507–1516.

Coombe, R. G. and A. M. George (1976). An alternative coupling procedure for preparing activated sepharose for affinity chromatography of Penicillinase. *Aust. J. Biol. Sci.*, **29**, 305–316.

Coupek, J. (1982). Macroporous spherical hydroxyethyl methacrylate copolymers, their properties, activation and use in high performance affinity chromatography. In *Affinity Chromatography and Related Techniques*, ed. T. C. J. Gribnau, J. Visser and R. J. F. Nivard, pp. 165–179. Elsevier, Amsterdam.

Cuatrecasas, P., M. Wilchek and C. B. Anfinsen (1968). Selective enzyme purification by affinity chromatography. *Proc. Natl. Acad. Sci. USA*, **61**, 636–643.

Cuatrecasas, P. (1969). Interaction of insulin with the cell membrane: the primary action of insulin. *Proc. Natl. Acad. Sci. USA*, **63**, 450–457.

Dandliker, W. B. and V. A. de Saussure (1971). Stabilization of macromolecules by hydrophobic bonding: role of water structure and of chaotropic ions. In *The Chemistry of Biosurfaces*, ed. M. L. Hair, pp. 1–40. Dekker, New York.

Danielson, E. M., H. Sjostrom and O. Noren (1982). Hypotonic elution, a new desorption principle in immunoadsorbent chromatography. *J. Immunol. Methods*, **52**, 223–232.

De Lisi, C. and H. W. Hethcote (1982). A theory of column chromatography for sequential reactions in heterogenous nonequilibrium systems: applications to antigen–antibody reactions. In *Affinity Chromatography and Related Techniques*, ed. T. C. J. Gribnau, J. Visser and R. J. F. Nivard, pp. 63–78. Elsevier, Amsterdam.

Denizot, F. C. and M. A. Delaage (1975). Statistical theory of chromatography: new outlooks for affinity chromatography. *Proc. Natl. Acad. Sci. USA*, **72**, 4840–4843.

Dunn, B. M. and I. M. Chaiken (1975). Evaluation of quantitative affinity chromatography by comparison with kinetic and equilibrium dialysis methods for analysis of nucleotide binding to Staphlococcal nuclease. *Biochemistry*, **14**, 2343–2349.

Epton, R., B. L. Hibbert and T. H. Thomas (1976). Enzymes covalently bound to polyacrylic and polymethacrylic copolymers. *Methods Enzymol.*, **44**, 84–107.

Eveleigh, J. W. (1982). Practical considerations in the use of immunoadsorbents and associated instrumentation. In *Affinity Chromatography and Related Techniques*, ed. T. C. J. Gribnau, J. Visser and R. J. F. Nivard, pp. 293–303. Elsevier, Amsterdam.

Filka, K., J. Coupek and J. Kocourek (1978). Studies on lectins. XL. *O*-glycosyl derivatives of Spheron in affinity chromatography of lectins. *Biochem. Biophys. Acta*, **539**, 518–528.

Graves, D. J. and Yun-Tai Wu (1974). On predicting the results of affinity procedures. *Methods Enzymol.*, **34**, 140–163.

Gribnau, T. C. J., J. Visser and R. J. F. Nivard (eds.) (1982). *Affinity Chromatography and Related Techniques*. Elsevier, Amsterdam.

Gurvich, A. E. and E. V. Lechtzind (1982). High capacity immunoadsorbents based on preparations of reprecipitated cellulose. *Mol. Immunol.*, **19**, 637–640.

Harvey, M. J., C. R. Lowe and P. D. G. Dean (1974). Affinity chromatography on immobilized adenosine 5'-monophosphate. 5. Some applications of the influence of temperature on the binding of dehydrogenases and kinases. *Eur. J. Biochem.* **41**, 353–357.

Hauri, H. P., A. Quaroni and K. J. Isselbacher (1980). Monoclonal antibodies to sucrose/isomaltase. *Proc. Natl. Acad. Sci. USA* **77**, 6629–6633.

Hjerten, S. (1973). Some general aspects of hydrophobic interaction chromatography. *J. Chromatogr.*, **87**, 325–331.

Hoffmann-Ostenhof, O., M. Breitenbach, F. Koller, D. Kraft and O. Scheiner (1978). *Affinity Chromatography*. Pergamon, Oxford.

Inman, J. K. and H. M. Dintzis (1969). The derivatizion of cross-linked polyacrylamide beads. Controlled introduction of functional groups for the preparation of special-purpose, biochemical adsorbents. *Biochemistry*, **8**, 4074–4082.

Isliker, H. C. (1957). Chemical nature of antibodies. *Adv. Protein Chem.*, **12**, 381–463.

Janson, J. C. (1982). Scaling-up of affinity chromatography, technological and economical aspects. In *Affinity Chromatography and Related Techniques*, ed. T. C. J. Gribnau, J. Visser and R. J. F. Nivard, pp. 503–512. Elsevier, Amsterdam.

Kalderon, N., I. Silman, S. Blumberg and Y. Dudai (1970). A method for the purification of acetylcholinesterase by affinity chromatography. *Biochem. Biophys. Acta*, **207**, 560–565.

Kasai, K. and S. Ishii (1975). Quantitative analysis of affinity chromatography of trypsin. *J. Biochem.*, **77**, 261–264.

Kay, G. and M. D. Lilly (1970). The chemical attachment of chymotrypsin to water-insoluble polymers using 2-amino-4,6-dichloro-*s*-triazine. *Biochim. Biophys. Acta*, **198**, 276–285.

Laas, T. (1975). Improved agarose matrixes for biospecific affinity chromatography. In *Protides of the Biological Fluids*, ed. H. Peeters, pp. 495–503. Pergamon, Oxford.

Levin, Y., M. Pecht, L. Goldstein and E. Katchalski (1964). A water-insoluble polyanionic derivative of trypsin. II. Effect of the polyelectrolyte carrier on the kinetic behavior of the bound trypsin. *Biochemistry*, **3**, 1913–1919.

Lowe, C. R. and P. D. G. Dean (1974). *Affinity Chromatography*. Wiley, New York.

Lowe, C. R. (1979). An introductiuon to affinity chromatography. In *Laboratory Techniques in Biochemistry and Molecular Biology*, ed. T. S. Work and E. Work, vol. 7, pp. 269–510. North Holland, Amsterdam.

Lyklema, J. (1982). Molecular interactions in affinity chromatography and related techniques. In *Affinity Chromatography and Related Techniques*, ed. T. C. J. Gribnau, J. Visser and R. J. F. Nivard, pp. 11–27. Elsevier, Amsterdam.

March, S. C., I. Parikh and P. Cuatrecasas (1974). A simplified method for cyanogen bromide activation of agarose for affinity chromatography. *Anal. Biochem.* **60**, 149–152.

Micheel, F. and J. Ewers (1949). Synthesis of compounds of cellulose with proteins. *Die Makromol. Chem.*, **3**, 200–209.

Morgan, M. R. A., E. George and P. D. G. Dean (1980). Investigations into the controlling factors of electrophoretic desorption: a widely applicable nonchaotropic elution technique in affinity chromatography. *Anal. Biochem.*, **105**, 1–5.

Mosbach, K., H. Builford, R. Ohlsson and M. Scott (1972). General ligands in affinity chromatography. Cofactor–substrate elution of enzymes bound to the immobilized nucleotides adenosine 5'-monophosphate and nicotinamide–adenine dinucleotide. *Biochem J.*, **127**, 625–631.

Nilsson, K. and K. Mosbach (1980). *p*-Toluenesulfonyl chloride as an activating agent of agarose for the preparation of immobilized affinity ligands and proteins. *Eur. J. Biochem.*, **112**, 397–402.

Nishikawa, A. H. and P. Bailon (1975). Lyotropic salt effects in hydrophobic chromatography. *Anal. Biochem.*, **68**, 274–280.

O'Carra, P. (1981). Biospecific binding to immobilized small ligands in affinity chromatography. *Biochem. Soc. Trans.*, **9**, 283.

O'Carra, P., S. Barry and T. Griffin (1973). Spacer-arms in affinity chromatography: the need for a more rigorous approach. *Biochem. Soc. Trans.*, **1**, 289–290.

Porath, J., J. C. Janson and T. Laas (1971). Agar derivatives for chromatography, electrophoresis and gel-bound enzymes. I. Desulfated and reduced cross-linked agar and agarose in spherical bead form. *J. Chromatogr.*, **60**, 167–177.

Porath, J. and L. Sundberg (1972). High capacity chemisorbents for protein immobilization. *Nature, New Biol.*, **238**, 261–262.

Porath, J. and T. Kristiansen (1975). Biospecific affinity chromatography and related methods. In *The Proteins*, ed. H. Neurath and R. L. Hill, vol. 1, pp. 95–178. Academic, New York.

Regnier, F. E. and R. Noel (1976). Glycerolpropylsilane bonded phases in the steric exclusion chromatography of biological macromolecules. *J. Chromatogr. Sci.*, **14**, 316–320.

Robinson, P. J., P. Dunnill and M. D. Lilly (1972). Factors affecting scale up of affinity chromatography of β-galactosidase. *Biochim. Biophys. Acta*, **285**, 28–35.

Robinson, P. J., M. A. Wheatley, J. C. Janson, P. Dunnill and M. D. Lilly (1974). Pilot scale affinity chromatography: purification of β-galactosidase. *Biotechnol. Bioeng.*, **16**, 1103–1112.

Schnaar, R. L. and Y. C. Lee (1975). Polyacrylamide gels copolymerized with active esters. A new medium for affinity systems. *Biochemistry*, **14**, 1535–1541.

Schneider, C., R. A. Newman, D. R. Sutherland, U. Asser and M. F. Greaves (1982). A one-step purification of membrane proteins using a high efficiency immunomatrix. *J. Biol. Chem.*, **257**, 10 766–10 769.

Schutyser, J., T. Buser, D. Van Olden, H. Tolmas, F. Van Houdenhoven and G. Van Dedem (1982). Synthetic polymers applied to macroporous silica beads to form new carriers for industrial affinity chromatography. In *Affinity Chromatography and Related Techniques*, ed. T. C. J. Gribnau, J. Visser and R. J. F. Nivard, pp. 143–153. Elsevier, Amsterdam.

Scouten, W. H. (1981). *Affinity Chromatography*. Wiley, New York.

Secher, D. S. and D. C. Burke (1980). A monoclonal antibody for large-scale purification of human leukocyte interferon. *Nature (London)*, **285**, 446–450.

Sevier, E. D., G. S. David, J. Martinis, W. J. Desmond, R. M. Bartholomew and R. Wang (1981). Monoclonal antibodies in clinical immunology. *Clin. Chem.*, **27**, 1797–1806.

Staehelin, T., D. S. Hobbs, H. Kung, C. Y. Lai and S. Pestka (1981). Purification and characterization of recombinant human leukocyte interferon with monoclonal antibodies. *J. Biol. Chem.*, **256**, 9750–9754.

Sundberg, L. and J. Porath (1974). Preparation of adsorbents for biospecific affinity chromatography. Attachment of group-containing ligands to insoluble polymers by means of bifunctional oxiranes. *J. Chromatogr.*, **90**, 87–98.

van Oss, C. J., D. R. Absolom and A. W. Neumann (1980). Applications of net repulsive van der Waals forces between different particles, macromolecules or biological cells in liquids. *Colloids and Surfaces*, **1**, 45–56.

van Oss, C. J., D. R. Absolom and A. W. Neumann (1982). Role of attractive and repulsive van der Waals forces in affinity and hydrophobic chromatography. In *Affinity Chromatography and Related Techniques*, ed. T. C. J. Gribnau, J. Visser and R. J. F. Nivard, pp. 29–37. Elsevier, Amsterdam.

Vidal, J., G. Godbillon and P. Gadal (1980). Recovery of active, highly purified phosphoenolpyruvate carboxylase from specific immunoadsorbent column. *FEBS Lett.*, **118**, 31–34.

Vilker, V. L., C. K. Colton and K. A. Smith (1981). The osmotic pressure of concentrated protein solutions: effect of concentration of pH in saline solutions of bovine serum albumin. *J. Colloid Interface Sci.*, **79**, 548–566.

Wankat, P. C. (1974). Theory of affinity chromatography separations. *Anal. Chem.*, **46**, 1400–1408.

Weetall, H. H. (1973). Affinity chromatography. *Sep. Purif. Methods*, **2**, 199–229.

Weetall, H. H. and N. Weliky (1964). New cellulose derivatives for the isolation of biologically active molecules. *Nature (London)*, **204**, 896–897.

Weliky, N., F. S. Brown and E. C. Dale (1969). Carrier-bound proteins: properties of peroxidase bound to insoluble carboxymethylcellulose particles. *Arch. Biochem. Biophys.*, **131**, 1–8.

Weston, P. D. and S. Avrameas (1971). Proteins coupled to polyacrylamide beads using glutaraldehyde. *Biochem. Biophys. Res. Commun.*, **45**, 1574–80.

Whitesides, G. M., A. Lamotte, O. Adalsteinsson and C. K. Colton (1976). Covalent immobilization of adenylate kinase and acetate kinase in a polyacrylamide gel: enzymes for ATP regeneration. *Methods Enzymol.*, **56**, 887–897.

Wilchek, M. (1974). Isolation of specific and modified peptides derived from proteins. *Methods Enzymol.*, **34**, 182–195.

Wilchek, M., T. Oka and Y. J. Toper (1975). Structure of a soluble super-active insulin is revealed by the nature of the complex between cyanogen-bromide–activated Sepharose and amines. *Proc. Natl. Acad. Sci. USA*, **72**, 1055–1058.

33

Hydrophobic Chromatography

Z. ER-EL
Hebrew University, Jerusalem, Israel

33.1 INTRODUCTION

Alkylamines react with BrCN-activated agarose to form modified agaroses with new adsorption properties. These modified agaroses discriminate between proteins mainly through hydrophobic interactions of the proteins with the alkyl residues on the support. This discrimination mechanism forms the basis for the development of a new technique called hydrophobic chromatography (Er-el *et al.*, 1972; Shaltiel and Er-el, 1973). The development of this technique in the early seventies was the result of advances in several aspects of column chromatography. First, the development of beaded agarose, a porous and hydrophilic matrix of good stability and minimal non-specific interactions with soluble proteins. Second, the development of several chemical processes for derivatization of the beaded agarose and attachment of ligands (especially the use of BrCN for activation and binding of amines) (Axen *et al.*, 1967). Third, the introduction of the hydrocarbon 'spacer' technique for attachment of ligands to the agarose matrix. Fourth, our observation that the hydrocarbon 'spacers' endow this matrix with the ability to adsorb proteins and that this ability depends on the chain length of the hydrocarbon 'spacer' (Shaltiel, 1974).

By using homologous series of hydrocarbon-coated agaroses varying only in hydrocarbon chain length, it is possible to monitor optimal adsorption and to purify soluble proteins from crude extracts. When such a crude mixture of proteins, *e.g.* a crude muscle extract, is passed through a series of these columns, the number of adsorbed proteins and their affinity to the columns will increase concomitantly with the increase in hydrocarbon chain length. There are proteins that, under given experimental conditions, will not bind even to the long-chain members of the series. Such proteins can be purified extensively merely by passing a crude extract through such a column, *e.g.* decyl- or dodecyl-agarose (Shaltiel, 1974, 1975). This chromatography procedure can be used to separate and resolve isoenzymes, interconvertible enzymes and membrane-bound enzyme complexes. There are preliminary results indicating that hydrophobic chromatography can also be used as a tool for the separation of cells and for the probing of cell surfaces (Shaltiel *et al.*, 1978). Hydrophobic chromatography is used not only for separation purposes but also serves as a tool in the study of the hydrophobicity of biopolymers and in the recognition and organization of proteins.

Hydrophobic chromatography derives its discriminative capacity from the interactions between available hydrophobic 'pockets' or 'patches' on protein surfaces and hydrophobic ligands evenly distributed throughout the hydrophilic, porous matrix. The interactions are primarily determined by hydrophobic effects, yet they may also involve electrostatic interactions, hydrogen bonding or charge-transfer effects, depending on the chemical structure of the support and the chromatographic conditions.

In addition to the activation of agarose by BrCN, other chemical methods of synthesis have been developed for the attachment of hydrophobic ligands to agarose by bonds in which electrostatic charges are not involved. Attempts have been made to develop 'pure' hydrophobic chromatography (Hjertén, 1981). Another approach has been to benefit from the effect of charged groups on the separation process (Yon, 1972). As a result, various names have been given to modifications of the technique, although all of them refer to the concept of hydrophobicity.

Hydrophobic interaction chromatography, hydrophobic salting-out chromatography, phosphate induced protein chromatography, multivalent interaction chromotography and detergent chromatography will all be dealt with in this chapter and the aspects common to all these techniques exemplified. The advantage of using homologous series of columns to optimize the interaction, which is the basis for hydrophobic chromatography, will be demonstrated.

Although agarose or cross-linked agarose is the most common insoluble matrix used in hydrophobic chromatography, there are other matrices as well. The reader is referred to the comprehensive review of Ochoa (1978) for hydrophobic matrices that are not based on agarose. Several comprehensive reviews on hydrophobic chromatography have appeared and the reader may refer to them for further details (Shaltiel, 1978; Srinivasan and Ruckenstein, 1980; Hjertén, 1981).

33.2 STRUCTURE AND SYNTHESIS OF AGAROSE-BASED HYDROPHOBIC MATRICES

Activation of agarose by BrCN, originally developed by Axen *et al.* (1967), was used by Er-el *et al.* (1972) to synthesize alkylagarose

The reaction proceeds as shown in Figure 1, modified after Axen and Ernback (1971), Wilchek *et al.* (1975) and Halperin *et al.* (1981). It has been demonstrated that some of the alkyl side-chains are bound *via* O-agarose-N-substituted isourea (a), which is protonated at neutral pH and affords the column a positive charge. Other alkyl side-chains are bound *via* substituted imidocarbonate (b) and substituted carbonate (c).

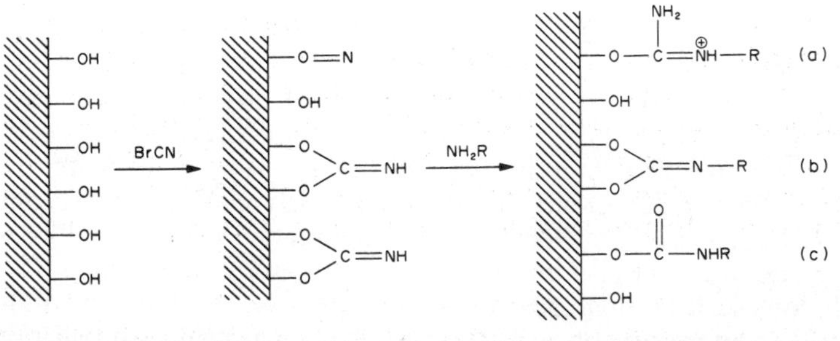

Figure 1 Synthesis of hydrocarbon-coated agaroses; activation of agarose by BrCN

Agarose may be activated by adding BrCN while maintaining the pH close to 11 for several minutes by the addition of 2 M or 5 M NaOH (Er-el *et al.*, 1972; Shaltiel, 1974). The procedure has been simplified by March *et al.* (1974), using a concentrated buffer solution.

The coupling of long-chain alkylamines (longer than octylamine) is cumbersome in aqueous media due to the limited solubility of such amines at room temperature (Shaltiel, 1974). Recently, it was suggested by Halperin *et al.* (1981) to perform the coupling in an organic medium (dioxane–water 95:5 v/v). Special precautions should be taken when transferring the agarose into the organic phase and back to water in order to preserve the porosity of the matrix (Hjertén *et al.*, 1974).

Wilchek and Miron (1976) have suggested the use of acetic anhydride to acetylate the N-substituted isourea and abolish the electrostatic charges (see Figure 2).

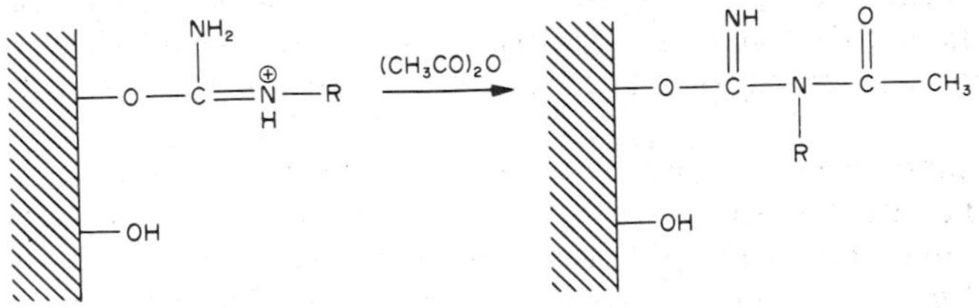

Figure 2 Synthesis of hydrocarbon-coated agaroses; acetylation of N-substituted isourea

The insertion of hydrophobic ligands into the agarose matrix, in the form of uncharged ether bonds, was suggested by Porath *et al.* (1973), who synthesized a benzyl ether of cross-linked agarose. Hjertén *et al.* (1974) prepared glycidyl ethers from alcohols and epichlorhydrin, using boron trifluoride ethyl etherate as catalyst. Glycidyl ethers were coupled to the agarose in an organic solvent (dioxane), according to the overall reaction shown in Figure 3.

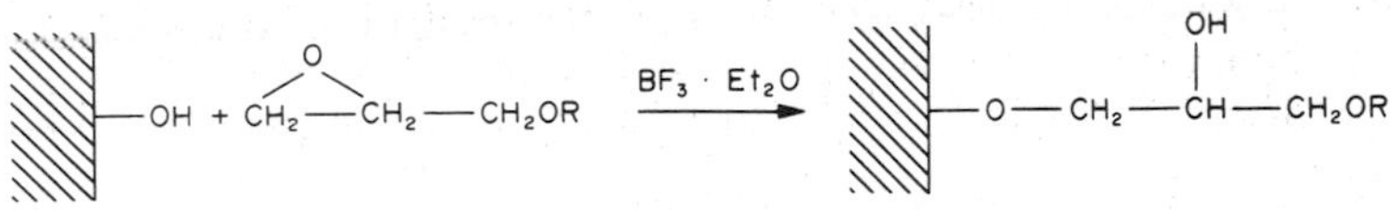

Figure 3 Synthesis of hydrocarbon-coated agaroses; boron trifluoride ethyl etherate catalyzed reaction

Reaction conditions affect the degree of substitution of the agarose gels, namely the concentration of the ligand on the gel, which may be expressed in μmol per ml packed agarose or in mmol per mol galactose (the backbone of the agarose matrix). Jennissen and Heilmeyer (1975) have shown that the ratio of BrCN to agarose determines the degree of substitution of the alkylagaroses. Rosengren *et al.* (1975) found a correlation between the concentration of glycidyl ether and the extent of substitution in a number of ether-linked hydrocarbon-coated agaroses.

The extent of the substitution can be determined directly by including a radioactively labelled ligand in the coupling step, as was demonstrated by Er-el (1975), Jennissen and Heilmeyer (1975), or by nuclear magnetic resonance of the product, as was shown by Rosengren *et al.* (1975). Recently, Halperin *et al.* (1981) observed that the degree of substitution determined by NMR correlates favorably with values obtained from ^{14}C incorporation. The extent of substitution can be estimated indirectly by saturating the column material with a dye, Ponceau S, which in aqueous solution binds irreversibly to the adsorbents (Hofstee, 1973).

Using the above mentioned methods of synthesis and analytical techniques, it became possible to prepare homologous series of columns that are identical in degree of substitution but differ in the attached hydrophobic ligands. The hydrophobic ligands can carry functional groups such as amine, carboxyl or phenyl, as summarized in Table 1. Retention of the protein on these columns will involve, in addition, hydrophobic interactions, charge effects, hydrogen bonding, *etc.* The structure of other important hydrophobic matrices is schematically presented in Table 1.

33.3 THEORETICAL ASPECTS OF HYDROPHOBIC CHROMATOGRAPHY

The name 'hydrophobic chromatography' was given to this chromatographic separation of proteins in view of the large contribution of hydrophobic interactions in the separation process. A detailed discussion on the nature and mechanism of hydrophobic interactions is beyond the scope of this review and the reader is referred to the excellent book by Tanford (1973) or to more recent reviews by Ochoa (1978), Shaltiel (1978) and Srinivasan and Ruckenstein (1980). Briefly, the mechanism of interaction can be explained as follows. Upon introduction of a non-polar solute,

Table 1 Homologous Series of Modified Agaroses Which May be Used for Hydrophobic Chromatography of Proteins and Other Biomolecules

Structure[a]	Abbreviation[b]
(a) *Agarose activated by BrCN*	
$-NH(CH_2)_nH$	Seph - C_n
$-NH(CH_2)_nNH_2$	Seph - C_n - NH_2
$-NH(CH_2)_nOH$	Seph - C_n - OH
$-NH(CH_2)_nPh$	Seph - C_n
$-NH(CH_2)_nNHCO(CH_2)_2 CO_2H$	Seph - C_n - CPA
$-NHCH(CHPh)CO_2H$	Phe - Seph
	A.A. - Seph
(b) *Ether linkages*	
$NOCH_2CH(OH)CH_2O(CH_2)_nH$	Alkyl-ether-Seph
$-OCH_2Ph$	Benzyl-ether-Seph

[a] n denotes the number of carbon atoms in the alkyl chain (generally 1–12). [b] Seph–C_0 = unmodified agarose; C_n-CPA = N–(3-carboxypropionyl)aminoalkyl; Phe = phenylalanyl; A.A. = amino acid residue.

water molecules are forced to form more organized clusters or icelike structures. The association of non-polar solutes in aqueous media is assumed to be driven by entropy since, upon association, the adjacent water molecules are excluded and enter a less ordered state (Ochoa, 1978). Accordingly, the interaction is non-specific and the entropy gain of the water molecules contributes more than other interactions, *e.g.* van der Waals forces. An alternative hypothesis is that the main effect caused by the introduction of non-polar solutes is enthalpic and that it is produced by breakage of the hydrogen bonds of the water molecules. This means that the association of non-polar solutes is driven by enthalpy (Srinivasan and Ruckenstein, 1980). Accordingly, van der Waals forces are important in the process of adsorption of proteins to hydrophobic matrices.

In view of the high performance of hydrophobic chromatography it is considered that the resolving power of the adsorbents cannot be explained on the basis of entropy alone. The involvement of attraction forces such as van der Waals forces may therefore contribute to the discriminating power of the columns.

Protein adsorption to hydrocarbon-coated agaroses is presented as evidence for the existence of hydrophobic 'patches' or 'pockets' on the surface of the proteins. Proteins generally contain amino acids with non-polar chains such as leucine, valine, proline and phenylalanine. When the protein polypeptides are folded into their native tertiary structure, there is a tendency to 'bury' the non-polar side-chains within the macromolecule and expose the hydrophilic chains to the aqueous medium. Several observations serve to point out that many of the non-polar side-chains end up on the surface of the protein where they cluster in the form of hydrophobic 'patches' of different sizes and various degrees of hydrophobicity (Klotz, 1970). It is possible that hydrophobic 'patches' are involved in protein aggregation and biospecific recognition of membranes and other molecules such as lipids.

The study of the mechanism of hydrophobic chromatography is hampered by the complexity of interactions involved in the adsorption and desorption processes. Attempts have been made to simplify the system and present the process in thermodynamic or kinetic models (Morrow *et al.*, 1975; Chang *et al.*, 1980; Srinivasan and Ruckenstein, 1980).

33.3.1 The Agarose Matrix

The ideal support for hydrophobic ligands should be hydrophilic and inert, and should not exert any non-biospecific secondary adsorption effects. Agarose is the most commonly used matrix, although other matrices, synthetic and natural, have also been applied (Ochoa, 1978). The availability of cross-linked agarose beads with improved resistance enables operation under a wider range of experimental conditions. The agarose matrix, however, is not an ideal support since, under certain conditions (*e.g.* very high salt concentrations), it binds halophilic proteins (Ochoa, 1978) or membrane proteins (Hjertén, 1978). Its hydrophobicity stems presumably from the 3–6 methylene diether bridges present in every second galactose residue of the polysaccharide chain.

The main advantages of agarose are its large surface area and porosity. Sepharose 4B, a commercial agarose product, contains beads of 15–100 μm diameter, providing a very large surface area for binding. Also, pores and channels which have openings of up to 0.3 μm in diameter are randomly distributed all over the beads. Agarose swollen in water contains 96% water and offers proteins a hydrophilic environment (Amsterdam *et al.*, 1975).

Activation of agarose with BrCN and subsequent binding of alkylamines to the matrix introduces positive charges in the form of substituted isourea linkages (as shown earlier). The role of electrostatic interactions in hydrophobic chromatography on agarose modified by BrCN created much controversy among scientists in the field. Attempts have been made to attach the hydrophobic ligands *via* non-charged ether linkages and neutral hydrazides, or to acetylate the N-substituted isourea linkages. The rationale was to enable the study of so-called 'pure' hydrophobic effects. The success of this approach will be discussed in the following section.

33.3.2 Hydrophobic *versus* Ionic Interactions

The involvement of electrostatic interactions in hydrophobic chromatography was proposed at an early stage in the development of this method (Shaltiel and Er-el, 1973; Porath *et al.*, 1973; Hjertén, 1973; Hofstee, 1973; Jost *et al.*, 1974). The use of an homologous series of identical alkylagaroses, varying only in the length of the alkyl side-chains, revealed an increasing capacity of the columns to adsorb proteins as a function of increasing alkyl chain length. This served as evidence for the involvement of mainly hydrophobic interactions in the process (Shaltiel, 1974, 1978).

Arguments for the major contribution of electrostatic effects in protein binding to alkylagaroses modified by the CNBr technique depend on the following observations. (1) Elution of a number of proteins has been achieved by increasing salt concentrations (see Table 2). It is known (Tanford, 1973) that salt enhances hydrophobic interactions. (2) Binding of several proteins (*e.g.* α-lactalbumin and ovalbumin) to alkylagaroses was abolished by the use of uncharged alkylagaroses prepared by linking alkyl hydrazides (Jost *et al.*, 1974) or by acetylating the parent, charged aklylagarose (Wilchek and Miron, 1976). Another approach, adopted by Hofstee (1975), was to carry out the procedure in the presence of very high salt concentrations, *e.g.* 3 M NaCl. Under such conditions, ionic effects are nullified (Hofstee and Otillio, 1978).

Halperin *et al.* (1981) made a detailed comparison of charged and electrically neutral column materials. In this comprehensive work, three series of alkyagaroses were compared: alkylagaroses derived by the BrCN technique (Type I), acetylated alkylagaroses (Type I, Acet.) and ether-linked alkylagaroses (Type II). The ligand density of each column in the series was determined by NMR as well as by ^{14}C-labelled ligands and only series with identical ligand density were used. Adsorption profiles of glycogen phosphorylase *b*, hemoglobin, lysozyme and bovine serum albumin were almost identical on Type I and Type I, Acet. columns except for ovalbumin which, similarly to the observation of Wilchek and Miron (1976), was excluded from Seph-C$_n$ ($n = 4$–6) Type I, Acet. columns. By changing the buffer from 50 mM Tris-HCl, pH 8.0 to 100 mM sodium phosphate, pH 7.0, the acetylated columns adsorbed ovalbumin similarly to the charged columns. Halperin *et al.* (1981) concluded that the abnormal adsorption profile observed in the Tris buffer may be associated with buffer-dependent specific ionic effects (Shaltiel and Er-el, 1973; Henderson *et al.*, 1974). They also compared the retention of bovine serum albumin, myoglobin and β-lactoglobulin on homologous series of Seph-C$_n$, Type I and Type II, and obtained very similar adsorption profiles for each of the proteins. When a crude muscle extract was passed on Type I and Type II columns and the excluded fractions subjected to acrylamide gel electrophoresis, there was remarkable similarity between the bank pattern obtained by electrophoresis of the excluded fractions from columns with equally long side-chains. Their conclusion was that hydrophobic, rather than ionic, interactions are the main factors responsible for the adsorption properties of alkylagaroses.

The use of ionic interactions in combination with hydrophobic interactions was suggested by Yon (1972). By planting charged groups at the 'tip' of the alkyl ligand the hydrophobicity of the adsorbents is reduced and thus enables the compounds to be eluted by changing the pH of the medium (Simmonds and Yon, 1976). The ionic effects can augment the resolving power of hydrocarbon-coated agaroses and provide matrices with hydrophobicity intermediate between that of close homologs of the Seph-C$_n$ series (Shaltiel and Er-el, 1973).

33.3.3 Degree of Substitution of the Modified Agaroses

Jennissen and Heilmeyer (1975) have shown that the degree of substitution of alkylagaroses affects protein adsorption. In general, the amount of protein bound per unit volume of adsorbent increases exponentially with ligand density, finally reaching a plateau. They claim that a critical hydrophobicity is necessary for the adsorption of proteins and it can be obtained by increasing the degree of substitution or, alternatively, by elongating the specific alkylamine chain while maintaining a constant degree of substitution. Another observation was that the fraction of enzyme that can be eluted with 50 mM NaCl declines exponentially with increasing hydrophobicity of the gel. Using Hill and Scatchard plots, Jennissen (1978) analyzed the data for phosphorylase *b*, phosphorylase kinase and phosphorylase phosphatase at low and high ionic strengths. He concluded that enzyme adsorption is a result of the simultaneous interaction of a critical number of multiple sites (depending on the degree of substitution) and the affinity of a single binding site (the length of the alkyl residue). If the chain is sufficiently long, interaction with a single alkyl residue may yield the free energy necessary for adsorption.

Multi-point attachment of phycoerythrin to ether-linked alkylagaroses was also observed by Rosengren *et al.* (1975). If substitution is substantial, the gels adsorb less protein due to steric hindrance and loss of porosity. Also, proteins adsorbed to highly substituted gels require drastic elution conditions. Similar results were shown by Er-el (1975).

33.3.4 The Aqueous Milieu (Effect of Salts and Polarity-reducing Agents)

Proteins in an aqueous milieu are not rigid molecules. A change in the environment of the protein, due to the addition of salts or polarity-reducing agents, is likely to affect the conformation of the protein as well as the association of the protein with the hydrophobic ligand (Halperin *et al.*, 1981). Generally, salts enhance hydrophobic interactions. The ions of these salts can be arranged according to their salting-out properties (the Hofmeister, or lyotropic, series). Other ions can destabilize protein structure and are called 'chaotropic' or salting-in ions (Pahlman *et al.*, 1977). When the ionic effect is mild and reversible, the term 'deformer' is used (Er-el *et al.*, 1972).

A number of investigators have reported the effects of salts on protein adsorption and elution (Jennissen and Heilmeyer, 1975; Nishikawa and Bailon, 1975; Pahlman *et al.*, 1977). It was observed that the presence of salting-out ions enhanced the adsorption of proteins to the hydrocarbon-coated agaroses and the effect correlated with the Hofmeister series. Conversely, chaotropic ions were chosen for the elution of strongly adsorbed proteins (see Table 2).

Adsorption in the presence of high concentrations of salting-out ions, *e.g.* $(NH_4)_2SO_4$, has now become common practice in many purification procedures; elution is brought about by a decreasing salt gradient (see Table 2).

A quantitative approach was applied by Morrow *et al.* (1975), to the adsorption of β-galactosidase to Sepharose substituted with 3,3'-diaminodipropylamine. They observed that enzyme adsorption shows a hysteretic effect as the ionic strength is increased and then decreased. By applying a mathematical model similar to the Debye-Hückel theory of protein solubility and by measuring equilibrium constants, they were able to predict the desorption of an enzyme as a function of time, using a decreasing linear gradient of salt in a continuously stirred tank.

Van Oss *et al.* (1979) explained the effects of salts and polarity-reducing agents on the elution process as a result of repulsive van der Waals forces. Proteins begin to be eluted when the surface tension of the eluant is lowered to a value just below that of the protein.

The effects of salts and polarity-reducing agents were dealt with in great detail by Srinivasan and Ruckenstein (1980).

In the aqueous milieu, other factors, such as temperature and pH (Ochoa, 1978; Srinivasan and Ruckenstein, 1980) affect adsorption and elution.

33.4 APPLICATIONS OF HYDROPHOBIC AGAROSES TO PURIFICATION

Hydrophobic chromatography may be included as one of the steps in the purification of proteins. The most suitable column for preparative scale purification is selected by monitoring the adsorption and elution profiles of the crude protein on several short columns (each of about 1 ml settled volume) of a given homologous series (*e.g.* nine columns: Seph-C_1 to Seph-C_8, and Seph-C_0 as control) under similar experimental conditions (pH, ionic strength, buffer, *etc.*).

Table 2 Application of Hydrophobic Chromatography for Purification of Proteins from Crude Extracts (Selected Number of Samples)

No.	Protein and source	H.C. column used	Eluant	Purification achieved	Ref.
1	Phosphorylase b,—rabbit muscle extract	Seph-C_4	Imidazole citrate	100	Er-el *et al.* (1972)
2	Glycogen synthetase—rabbit muscle	Seph-C_4-NH_2	NaCl gradient	25	Shaltiel and Er-el (1973)
3	Histidine binding protein J—*S. typhymurium*	Seph-C_{10}-NH_2	By exclusion	7–10	Shaltiel *et al.* (1973)
4	Several aminoacyl t-RNA synthetases— yellow lupin seeds	Seph-C_6-NH_2	KCl gradient	10–100	Jakubowski and Pawelkiewitz (1973)
5	L-Histidinol P_i amino transferase— *S. typhymurium*	Seph-C_6-NH_2	NaCl gradient	15	Henderson *et al.* (1974)
6	Phosphate kinase and phosphate phos- phatase—muscle extract	Seph-C_1	NaCl	20, 25	Heilmeyer and Jennissen (1975)
7	Maltodextrin phosphorylase—*E. coli*	Seph-C_{10}-NH_2	NaCl gradient	8	Thanner *et al.* (1975)
8	Glutamine synthetase enzymes—*E. coli*	Seph-C_5-NH_2	KCl gradient	10–20	Shaltiel *et al.* (1975)
9	Tryptophanase Aspartokinase I homoserine dehydrogenase β-Galactosidase—*E. coli*	Seph-C_3	D.a.s.g. in different buffers	3 9 5	Raibaud *et al.* (1975)
10	Histones—calf thymus	Seph-C_8	D.a.s.g.	—	Arfman and Shaltiel (1976)
11	Clostripain and collagenase *Clostridium histolyticum*	Seph-C_4-NH_2 Seph-C_7-NH_2	NaCl gradient	9 7	Kula *et al.* (1976)
12	Alcohol dehydrogenase—*S. cerevisiae*	Seph-C_{10}-CO_2H	C_2H_5OH, pH	14	Schopp *et al.* (1976)
13	Alcohol dehydrogenase—*Acinetobacter* sp.	Seph-C_{10}-CO_2H	P_i gradient, pH	22	Schopp *et al.* (1976)
15	c-AMP dep. protein kinase—muscle extract	Seph-C_6	NaCl gradient	30	Hoppe and Wagner (1977)
16	Non-histone high mobility group protein—mouse liver nuclei	Seph-C_4-HN_2	D.a.s.g.	—	Conner and Comings (1981)
17	Nuclear androgen receptor—rat protastic chromatin	Seph-C_6-NH_2	By exclusion	15	Bruchovsky *et al.* (1981)
18	Sphingomyelinase—human placenta	Seph-C_8 and Seph-C_6	TritonX–100 By exclusion	21 (total)	Jones *et al.* (1981)
19	Glycophorin and protein E-erythrocyte membrane 'proteins'	Seph-C_{10}-CPA	SDS + pH change	—	Simmonds and Yon (1976)
20	PEP dep. phosphotransferase E_I—*E. coli*	Octyl-Seph (twice) + phenyl-Seph	Ethylene glycol gradient	870	Robillard *et al.* (1979)
21	Phenylalanine hydroxylase—rat liver	Phenyl-Seph	Remove activator Phe	>200	Shiman *et al.* (1979)
22	Erythropoietin—human urine	Phenyl-Seph	Ethylene glycol	100	Lee Huang (1980)
23	Peroxidase—tomato extract	Phenyl-Seph	D.a.s.g.	3	Jen *et al.* (1980)
24	Heat stable enterotoxin—*E. coli*	Octyl-Seph	5% ethylene glycol	200	Stavric *et al.* (1981)
25	Human prolactin—pituitary glands	Phenyl-Seph	Acetonitrile gradient	85	Hodgkinson and Lowry (1981)
26	Staphylococcal lipase	Octyl-Seph	Triton X-100 gradient	40	Jurgens and Huser (1981)

Each of the columns is loaded with a small volume (50–200 μl) containing several milligrams of the pure protein or of the pure extract, and the columns are then washed with the application buffer (2 ml). The eluates are assayed for the non-adsorbed enzyme or protein. Then elution is attempted with an elution buffer (about 2 ml) and a second fraction is collected and assayed as above. Results, in percent of protein or enzyme applied or in protein concentration of each fraction, are plotted *versus* the number of carbon atoms in the alkyl side-chains. Generally, the adsorption profiles obtained resemble titration curves, whereas the elution profiles have a distinct maximum. The results of such an experiment using a pure protein, glycogen phosphorylase b, are presented in Figure 4 (Er-el *et al.*, 1972; Shaltiel, 1975).

The experiment is repeated under different sets of conditions (*e.g.* changing the buffer compo-

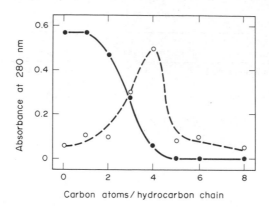

Figure 4   Adsorption profile (●) of glycogen phosphorylase *b* on a Seph-C*n* kit at 22 °C with a buffer composed of sodium β-glycerophosphate (50 mM) and EDTA (1 mM), pH 7.0. Elution (○) is attempted with a 'deforming buffer' composed of imidazole (0.4 M) adjusted to pH 7.0 with citric acid. Aliquots of the AMP-free enzyme (1 mg in 0.1 ml) were applied on each of the columns. The first 2 ml were collected, the deforming buffer applied and then an additional 2 ml were collected and their absorbance at 280 nm was monitored. (Reproduced with permission from *Chromatography of Synthetic and Biological Polymers*, ed. R. Epton, vol. 2 (1978), Ellis Horwood, Chichester)

sition) until reasonable purification of the protein is obtained. If the desired degree of purification cannot be achieved on a given series of homologous columns such as Seph-C*n*, a different set of homologous series of columns can be tried out, *e.g.* Seph-C*n*-NH$_2$ which, at neutral pH, has a positive charge at the 'tip' of the alkyl side-chain or Seph-C*n*-CO$_2$H, which is negatively charged, Seph-C*n*-Ph, *etc.*

The most appropriate column chosen from the homologous series for the large-scale purification is the one that enables total retention of the protein and its elution under mild conditions, with the highest degree of specific activity and the best recovery.

As can be seen in Figure 4, Seph-C$_4$ appears to be the most suitable member of the homologous Seph-C*n* series for the purification of phosphorylase *b* under given chromatographic conditions (see also Table 2).

A 1 ml column typically adsorbs between 1–10 mg of purified protein and, therefore, the purification of many enzymes and proteins in quantities large enough for biochemical research can readily be carried out using a 20–100 ml column. The successful application of a 1000 ml hydrophobic column in the purification of a human placental enzyme was reported by a manufacturer of fine chemicals (personal communication).

Table 2 summarizes the results of a selected list of purification procedures reported in the literature. The first 18 examples included in the table are only those reports in which a specific hydrophobic column was selected, based on the methodology presented above, by comparing adsorption–elution profiles on homologous series of columns. In the last eight examples in the table, no criteria were reported for choosing the particular column used.

The following conclusions may be drawn from the data presented in Table 2. (1) Hydrophobic chromatography may be used to purify enzymes and proteins from various sources: microbial, plant and mammalian. Both extracellular enzymes and membrane-bound proteins can be purified. (2) Soluble enzymes can be adsorbed using buffers commonly used in biochemical studies (see Nos. 1–8, 11, 15, Table 2). It is also possible to carry out the adsorption process in the presence of 'structure-forming salts', *e.g.* (NH$_4$)$_2$SO$_4$ (see Nos. 9, 10, 16, 23, Table 2). (3) Elution can be carried out using salt gradients of increasing or decreasing concentration. Certain salts, such as imidazole citrate, affect and modify protein structure and may be used for elution (No. 1, Table 2). (4) Elution can be achieved with the aid of detergents, *e.g.* Triton X-100 or sodium dodecyl sulfate (Nos. 18, 19, Table 2). Polarity-reducing agents such as ethylene glycol or ethanol are very effective in the elution of proteins (Nos. 12, 20, 22, Table 2). (5) Some hydrophobic columns have charged groups at the 'tip' of the alkyl side-chains and their discriminating power involved charge as well as hydrophobicity. Elution from such columns can be effected by a shift in pH (Nos. 12, 13, 19, Table 2). (6) Hydrophilic proteins can be purified by passing crude extracts through a highly hydrophobic column. Purification is achieved by exclusion of the hydrophilic protein from the column and retention of the more hydrophobic contaminants (Nos. 3, 17, Table 2). (7) Hydrophobic columns with different ligands can be used consecutively in the purification of several proteins, or for obtaining a higher degree of purification of a particular protein (Nos. 11, 20, Table 2). (8) The degree of purification to be expected is between 5–30 fold and recovery is usually better than 50%. Even greater purification and higher recoveries (close to 100%) have been cited (Nos. 1, 4, 20–22, Table 2).

The number of reports, according to which specific columns were chosen systematically according to the adsorption profiles on homologous series of columns, is very small. Only a dozen of the 130 articles published between 1977 and 1982, and bearing the title 'Hydrophobic Chromatography', described such systematic studies in their work. In order to facilitate the methodological approach to the selection of a column for purification purposes, it is crucial to use standardized kits of homologous series of columns (also commercially available).

In spite of all these reservations, the degree of purification reported is often quite impressive (see Table 2, Nos 21, 22).

33.5 THE RESOLVING POWER OF HYDROPHOBIC AGAROSES

The high resolution which can be achieved through the use of hydrophobic chromatography was exemplified earlier in this chapter by the efficient purification of proteins from crude extracts containing many proteins (see Table 2). A number of examples are given below for application of hydrophobic chromatography other than to the recovery of proteins from crude extracts. In each of the examples the investigators emphasized the high resolution obtained.

33.5.1 Resolution of Interconvertible Forms of Enzymes

Er-el and Shaltiel (1974) demonstrated the separation of two metabolically interconvertible forms of rabbit muscle glycogen phosphorylase on Seph-C_n columns. These two forms are identical in their amino acid sequence, except for a unique serine residue in each of the enzyme protomers which becomes O-phosphorylated upon conversion of the b form (a dimer) to the a form (a tetramer). It was found that phosphorylase a is retained on Seph-C_1 and can be released by an NaCl gradient, whereas phosphorylase b is fully adsorbed on Seph-C_4, or higher members of this series. It is possible to separate the two enzyme forms on a Seph-C_4 column; by lowering the pH to 5.8, phosphorylase b is eluted and phosphorylase a is retained and can be eluted subsequently by a gradient of imidazole citrate.

33.5.2 Separation of Protein Subunits and Protein Aggregates

Hydrophobic interactions play a major role in the association of protein subunits. Separation of biologically functional subunits is essential for biochemical studies on protein structure and function. Hydrophobic chromatography has been used by several groups as a tool for separating such subunits and purifying protein aggregates.

Jacobson *et al.* (1978) used pentyl-Seph for the isolation of α and β subunits of human pituitary thyrotropin. The closely associated subunits of the native protein were dissociated by treatment with 1 M propionic acid for 16 h. Using the same solution, two peaks, α and β_1, were resolved; by then passing 0.01 M potassium phosphate, a third peak, β_2 could be eluted. On a larger column only two peaks, α and β, were obtained. Recombination of these fractions resulted in an active compound indicating that the separation process had been sufficiently mild, and was reversible.

Another example of subunit separation was described by Lo *et al.* (1981). Subunits of Pike eel gonadotropin were resolved on phenyl-Seph after dissociation in propionic acid. The reassociated subunits exhibited about 75% of the original activity, showing that the procedure had been sufficiently mild.

Roobol *et al.* (1980) investigated the aggregation of eukaryotic elongation factor eEF-Ts. Initially, it was reported that the elongation factor contains two polypeptide chains of 55 000 daltons and 30 000 daltons. By using phenyl-Seph it was possible to purify eEF-Ts under non-denaturing conditions and to obtain a purified 30 000 dalton protein that exhibited activities usually ascribed to eEF-Ts. Higher molecular weight proteins observed prior to purification were assumed to be elongation factor aggregates with contaminating proteins.

33.5.3 Membrane Proteins and Membrane-bound Proteins

Membrane proteins are generally associated with lipids and organized in a complex quaternary structure. The release of membrane proteins and membrane-bound proteins (which are located at the membrane surface) may require disruption of the membrane by a combination of physical and chemical methods. Detergents are used to extract and solubilize these proteins. The presence of the detergents makes further purification very complicated and therefore an intensive effort was made to use hydrophobic chromatography in such cases.

Simmonds and Yon (1976) solubilized human erythrocyte 'ghosts' in 0.5% (w/v) sodium dodecyl sulfate and subjected the proteins obtained to fractionation, using several charged alkylagaroses. Glycophorin and protein E were efficiently separated on Seph-C_{10}-CPA; spectrin, however, was not fractionated in a well-defined manner.

Hjertén (1978) dealt with the fractionation of erythrocyte membranes and his article includes some general observations on the fractionation of membrane proteins.

Bostwick and Eichberg (1981) attempted to purify phosphatidylinositol kinase from rat liver. In order to purify the enzyme, it was necessary to solubilize it with detergents. However, the presence of the detergent did not permit further purification by gel filtration or affinity chromatography. This might be attributed to association of the enzyme with hydrophobic proteins. By passing the solubilized enzyme preparation on dodecyl-Seph the enzyme was adsorbed and the detergent (Triton X-100) washed off. The enzyme was then eluted using 8 mM deoxycholate and a 2–3 fold increase in specific activity was obtained.

Homcy *et al.* (1977) purified adenylate cyclase, a membrane-bound protein, which had been solubilized by detergent. Following hydrophobic chromatography on dodecyl-Seph, during which the detergent was removed, the protein could be further purified using gel filtration and ion exchange chromatography. The enzyme, which prior to hydrophobic chromatography behaved as an heterogeneous mixture, now revealed itself as a homogeneous species. The length of the alkyl side-chain was crucial in the ability of the column to retain the enzyme and separate it from other hydrophobic proteins in the preparation.

33.5.4 Separation of Peptides and Protein Fragments

Zaidenzaig and Stern (1980) investigated the resolution of diphtheria toxin. The toxin could be obtained by passing *C. diphtheriae* growth medium supernatant through Seph-C_6 and eluting it with 0.1 M NaCl. Alternatively, after mild tryptic digestion, two fragments could be separated on Seph-C_4.

An attempt to use hydrophobic chromatography in the separation of peptides from ampholytes was made by Gelsema *et al.* (1980). The association constants of model peptides and ampholytes to octyl-Seph were determined. It was observed that carrier ampholytes are not retained on octyl-Seph and therefore separation of the peptides is feasible. They concluded that effective separation of ampholytes from peptides requires adsorbents even more hydrophobic than octyl-Seph. A somatomedin-containing peptide mixture served as a model for separation purposes. The elution pattern of the somatomedin-containing peptides, using an ethanol gradient of up to 40% (v/v), could not be explained readily on the basis of hydrophobic bonding alone, since basic peptides eluted late in the chromatogram.

Recently, Sobel *et al.* (1982) used Ph-Seph in combination with other techniques to purify peptides from BrCN digests of Aα chains of human fibrinogen. Two peptides, one consisting of 235 residues and the other of 66 residues, were thus isolated.

33.5.5 Preferential Adsorption of Intact Cells

Halperin and Shaltiel (1976) investigated the adsorption profiles of erythrocytes on homologous series of alkylagarose Seph-C_n. The adsorption of red blood cells from various animal species increased with increasing chain length of these columns. Adsorbed cells could be eluted by physically shaking the beads in the presence of bovine serum albumin (recovery greater than 95%), without significant damage to the cells. The columns could differentiate between mouse and guinea pig and between inbred strains of mice. Halperin and Shaltiel (1978) discussed the mechanism of adsorption to the columns and supplied additional evidence for the involvement of hydrophobic interactions in the process. They compared the adsorption profiles of the erythrocytes on neutral acetylated Seph-C_n and on regular Seph-C_n. They found that the acetylated columns were of lower binding capacity, especially those columns with short, alkyl side-chains. A comparison of the adsorption of guinea pig erythrocytes prior to and after reduction of their surface charge by enzymatic removal of sialic acid residues, revealed very little difference. The authors suggested that, although ionic interactions may be involved in the adsorption process, the predominant effect is a hydrophobic one and, therefore, the columns could prove useful in probing cell-surface properties.

An attempt to use hydrophobic matrices for determining the adhesive properties of intact bacterial cells was made by Smyth *et al.* (1978). This group determined the adsorption properties of porcine enteropathogenic strains of *E. coli* possessing or lacking the K88 antigen. Cells were passed on several hydrophobic agarose gels (*e.g.* phenyl-Seph, octyl-Seph, palmitoyl-Seph). Only those cells that possessed the K88 antigen adsorbed to the columns in the presence of 2 M NaCl or 1 M $(NH_4)_2SO_4$; cells lacking the antigen were not retained. A correlation between adsorption to the column and bacterial adhesion to the target mucosal epithelium was suggested.

33.6 CONCLUSION

Hydrophobic chromatography has now become a common and useful technique in the biochemical laboratory. It is based on the use of hydrocarbon-coated agaroses in which the ligands are attached *via* ionogenic or non-ionogenic bonds. Using this method, in the past decade it has become possible to purify a large number of proteins from crude extracts.

The main controversy on the use of hydrophobic chromatography relates to the role of electrostatic interactions in the adsorption process. Attempts to develop 'pure' hydrophobic chromatography were justified on the basis of the risk of protein denaturation, or the need for better control of the mechanism of adsorption. In the search for 'pure' hydrophobic chromatography, new resins were developed and the general trend was to use high concentrations of salting-out ions to quench any remaining ionic interactions. It is rather doubtful if exposure of sensitive enzymes to salt concentrations greater than 3 M does not increase the risk of denaturation. In addition, it is questionable if attachment of enzymes to highly hydrophobic, non-charged matrices such as octyl-Seph does not result in denaturation.

When using hydrophobic chromatography for purification purposes it is worthwhile to consider the following factors. (a) Adsorption is greatly affected by the hydrophobicity of the hydrocarbonaceous ligand. Therefore, the use of homologous series of columns in which only one parameter is gradually changed is preferable, for example adsorption profiles should be compared using a kit of identical Seph-C_n or Seph-C-$_n$NH$_2$ columns (same size, same degree of substitution, same buffer but different alkyl chain length). (b) Adsorption is influenced by the degree of substitution. The commercially available columns may not be of optimum ligand density. Therefore, after choosing the appropriate column in the kit, it is advisable to study the effect of ligand density on the separation process. (c) The ratio of BrCN used in the activation of agarose to the alkylamine used in the coupling, determines the charge of the column as well as the density of substituted alkyl side-chains. While preparing similar columns with different ligand densities, it is important to use the same molar ratio of BrCN to alkylamine (unpublished data). (d) The most suitable buffers for adsorption and desorption can be chosen only by trial and error. The effects of ions and polarity-reducing agents should be kept in mind. When, in the course of the purification process, an ammonium sulfate precipitate is obtained, it can be directly applied to the hydrophobic column and elution of the desired protein attempted by decreasing salt gradient, or by an increasing ethylene glycol gradient.

Finally, the practical aspects of hydrophobic chromatography are already well documented. Less is known about the mechanism of the adsorption and desorption processes. Only a few attempts have been made to develop kinetic and thermodynamic models and far more work is necessary before proper design of the separation process will replace empiricism. The same is true for industrial scale purifications for which new synthetic hydrophobic matrices will need to be developed.

ACKNOWLEDGMENT

The author wishes to express his gratitude to G. Halperin for fruitful discussions and critical remarks.

33.7 REFERENCES

Amsterdam, A., Z. Er-el and S. Shaltiel (1975). Ultrastructure of beaded agarose. *Arch. Biochem. Biophys.*, **171**, 673–677.

Arfman, H. A. and S. Shaltiel (1976). Resolution and purification of histones on homologous series of hydrocarbon-coated agaroses. *Eur. J. Biochem.*, **70**, 269–274.

Axen, R. and S. Ernback (1971). Chemical fixation of enzymes to cyanogen halide activated polysaccharide carriers. *Eur. J. Biochem.*, **18**, 351–360.

Axen, R., J. Porath and S. Ernback (1967). Chemical coupling of peptides and proteins to polysaccharides by means of cyanogen halides. *Nature (London)*, **214**, 1302–1304.

Bostwick, J. R. and J. Eichberg (1981). Detergent solubilization and hydrophobic chromatography of rat brain phosphatidylinositol kinase. *Neurochem. Res.*, **6**, 1053–1065.

Bruchovsky, N., P. S. Rennie and T. Comeau (1981). Partial purification of nuclear androgen receptor by micrococcal nuclease digestion of chromatin and hydrophobic interaction chromatography. *Eur. J. Biochem.*, **120**, 399–405.

Chang, C. T., B. J. McCoy and R. G. Carbonell (1980). Hydrophobic chromatography of β-galactosidase. *Biotechnol. Bioeng.*, **22**, 377–399.

Conner, B. J. and D. E. Comings (1981). Isolation of a non-histone chromosomal high mobility group protein from mouse liver nuclei by hydrophobic chromotography. *J. Biol. Chem.*, **256**, 3283–3291.

Er-el, Z. (1975). Hydrophobic chromatography in the study of enzymes involved in glycogen metabolism. Ph.D. Thesis submitted to the Weizmann Institute of Science.

Er-el, Z. and S. Shaltiel (1974). Hydrophobic chromatography in the resolution of the interconvertible forms of glycogen phosphorylase. *FEBS Lett.*, **40**, 142–145.

Er-el, Z., Y. Zaidenzaig and S. Shaltiel (1972). Hydrocarbon-coated Sepharoses. Use in the purification of glycogenphosphorylase. *Biochem. Biophys. Res. Commun.*, **49**, 383–390.

Gelsema, W. J., C. L. De Ligny, W. M. Blanken, R. J. Hamer and A. M. P. Roozen (1980). Study of the feasibility of separation of carrier ampholytes from peptides by hydrophobic interaction chromatography on octyl-Sepharose. *J. Chromatogr.*, **196**, 51–58.

Halperin, G., M. Breitenbach, M. Tauber-Finkelstein and S. Shaltiel (1981). Hydrophobic chromatography on homologous series of alkylagaroses. *J. Chromatogr.*, **215**, 211–228.

Halperin, G. and S. Shaltiel (1976). Homologous series of alkyl agaroses discriminate between erythrocytes from different sources. *Biochem. Biophys. Res. Commun.*, **72**, 1497–1503.

Halperin, G. and S. Shaltiel (1978). On the mechanism of adsorption of erythrocytes to hydrocarbon-coated agaroses. In *Affinity Chromatography*, ed. O. Hoffmann-Ostenhof *et al.*, pp. 307–311. Pergamon, Oxford.

Henderson, G. B., S. Shaltiel and E. E. Snell (1974). ω-Aminoalkylagaroses in the purification of L-histidinol-phosphate aminotransferase. *Biochemistry*, **21**, 4335–4338.

Hjertén, S. (1978). Fractionation of membrane proteins by hydrophobic interaction chromatography and by chromatography on agarose equilibriated with a water–alcohol mixture of low or high pH. *J. Chromatogr.*, **159**, 85–91.

Hjertén, S. (1981). Hydrophobic interaction chromatography of proteins, nucleic acids, viruses and cells on noncharged amphiphilic gels. *Methods Biochem. Anal.*, **27**, 89–108.

Hjertén, S., J. Rosengren and S. Påhlman (1974). Hydrophobic interaction chromatography. The synthesis and the use of some alkyl and aryl derivatives of agarose. *J. Chromatogr.*, **101**, 281–288.

Hodgkinson, S. C. and P. J. Lowry (1981). Hydrophobic interaction chromatography and anion-exchange chromatography in the presence of acetonitrile. A two-step purification method for human prolactin. *Biochem. J.*, **199**, 619–627.

Hofstee, B. H. J. (1973). Immobilization of enzymes through non-covalent binding to substituted agaroses. *Biochem. Biophys. Res. Commun.*, **53**, 1137-1144.

Hofstee, B. H. J. (1975). Fractionation of protein mixtures through differential adsorption on a gradient of substituted agaroses of increasing hydrophobicity. *Prep. Biochem.*, **5**, 7–19.

Hofstee, B. H. J. and N. F. Otillio (1978a). Non-ionic adsorption chromatography of proteins. *J. Chromatogr.*, **159**, 57–69.

Hofstee, B. H. J. and N. F. Otillio (1978b). Modifying factors in hydrophobic binding by substituted agaroses. *J. Chromatogr.*, **161**, 153–163.

Homcy, C. J., S. M. Wrenn and E. Haber (1977). Demonstration of the hydrophilic character of adenylate cyclase following hydrophobic resolution on immobilized alkyl residues. *J. Biol. Chem.*, **252**, 8957–8964.

Hoppe, J. and K. G. Wagner (1977). An improved method for the purification of cAMP-dependent protein kinase from rabbit muscle using hydrophobic chromatography. *FEBS Lett.*, **74**, 95–98.

Jacobson, G., P. Roos and L. Wide (1978). Human pituitary thyrotropin. Isolation of the α and β subunits by hydrophobic interaction chromatography. *Biochem. Biophys. Acta*, **536**, 363–375.

Jakubowski, H. and J. Pawelkiewicz (1973). Chromatography of plant amino acyl-tRNA synthetases on ω-amino alkyl Sepharose columns. *FEBS Lett.*, **34**, 150–154.

Jen, J. J., A. Seo and W. H. Flurkey (1980). Tomato peroxidase: Purification *via* hydrophobic chromatography. *J. Food Sci.*, **45**, 60–63.

Jennissen, H. P. (1978). Multivalent interaction chromatography as exemplified by the adsorption and desorption of skeletal muscle enzymes on hydrophobic alkyl-agaroses. *J. Chromatogr.*, **159**, 71–83.

Jennissen, H. P. and L. M. G. Heilmeyer (1975). General aspects of hydrophobic chromatography. Adsorption and elution characteristics of some skeletal muscle enzymes. *Biochemistry*, **14**, 754–759.

Jones, C. S., P. Shakaran and J. W. Callahan (1981). Purification of sphingomyelinase to apparent homogeneity by using hydrophobic chromatography. *Biochem. J.*, **195**, 373–382.

Jost, R., T. Miron and M. Wilchek (1974). The mode of adsorption of proteins to aliphatic and aromatic amines coupled to cyanogen bromide activated agarose. *Biochem. Biophys. Acta*, **362**, 75–82.

Juergens, D. and H. Huser (1981). Large scale purification of staphylococcal lipase by hydrophobic interaction chromatography. *J. Chromatogr.*, **216**, 295–301.

Klotz, I. M. (1970). Comparison of molecular structures of proteins: helix content, distribution of apolar residues. *Arch. Biochem. Biophys.*, **138**, 704–706.

Kula, M. R., D. Hatef-Haghi, M. Tauber-Finkelstein and S. Shaltiel (1976). Consecutive use of ω-amino alkylagaroses. Resolution and purification of clostripain and collagenase from *Clostridium hystolyticum. Biochem. Biophys. Res. Commun.*, **69**, 389–396.

Låås, T. (1975). Agar derivatives for chromatography, electrophoresis and gel bound enzymes. *J. Chromatogr.*, **111**, 373–387.

Lee-Huang, S. (1980). A new preparative method for isolation of human erythropoietin with hydrophobic interaction chromatography. *Blood*, **56**, 620–624.

Lo, T. B., F. L. Huang and G. D. Chang (1981). Separation of subunits of pike eel gonadotropin by hydrophobic interaction chromatography. *J. Chromatogr.*, **215**, 229–233.

March, S. C., I. Parikh and P. Cuatrecasas (1974). A simplified method for cyanogen bromide activation of agarose for affinity chromatography. *Anal. Biochem.*, **60**, 149–152.

Morrow, R. M., R. G. Carbonell and B. J. McCoy (1975). Electrostatic and hydrophobic effects in Affinity chromatography. *Biotechnol. Bioeng.*, **17**, 895–914.

Ochoa, J. L. (1978). Hydrophobic (interaction) chromatography. *Biochimie*, **60**, 1–15.

Pahlman, S., J. Rosengren and S. Hjertén (1977). Hydrophobic interaction chromatography on uncharged Sepharose derivatives. Effect of neutral salts on the adsorption of proteins. *J. Chromatogr.*, **131**, 99–108.

Porath, J., L. Sundberg, N. Fornstedt and I. Olsson (1973). Salting-out in amphiphilic gels as a new approach to hydrophobic adsorption. *Nature (London)*, **245**, 465–466.

Raibaud, O., A. Hogberg-Raibaud and M. E. Goldberg (1975). Purification of *E. coli* enzymes by chromatography on amphiphilic gels. *FEBS Lett.*, **50**, 130–134.

Robillard, G. T., G. Dooijewaard and J. Lolkema (1979). *Escherichia coli* phosphoenolpyruvate dependent phosphotransferase system. Complete purification of Enzyme I by hydrophobic interaction chromatography. *Biochemistry*, **18**, 2984–2989.

Roobol, K., I. Vianden and W. Moller (1980). Aggregation of eucaryotic elongation factor eEF-Ts and its isolation by means of hydrophobic adsorption chromatography. *FEBS Lett.*, **111**, 136–142.

Rosengren, J., S. Påhlman, G. Magnus and S. Hjertén (1975). Hydrophobic interaction chromatography on non-charged Sepharose derivatives. *Biochem. Biophys. Acta*, **412**, 51–61.

Schopp, W., M. Grunow, H. Tauchert and H. Aurich (1976). Specific hydrophobic chromatography of alcohol dehydrogenases on 10-carboxydecyl-Sepharose. *FEBS Lett.*, **68**, 198–202.

Shaltiel, S. (1974). Hydrophobic chromatography. *Methods Enzymol.*, **34**, 126–140.

Shaltiel, S. (1975). Hydrophobic chromatography. Use in the resolution, purification and probing of proteins. *FEBS Proc.*, **40**, 117–127.

Shaltiel, S. (1978). Hydrophobic chromatography. In *Chromatography of Synthetic and Biological Polymers*, ed. R. Epton, vol. 2, pp.13–41. Ellis Horwood, Chichester.

Shaltiel, S., S. P. Adler, D. Purich, C. Caban, P. Senior and E. R. Stadtman (1975). ω-Amino alkyl agaroses in the resolution of enzymes involved in regulation of glutamine metabolism. *Proc. Natl. Acad. Sci. USA*, **72**, 3397-3401.

Shaltiel, S. and Z. Er-el (1973). Hydrophobic chromatography: Use in the purification of glycogen synthetase. *Proc. Natl. Acad. Sci. USA*, **70**, 778–781.

Shaltiel, S., G. Ferro-Luzzi Ames and K. Dale Noel (1973). Hydrophobic chromatography in the purification of the histidine-binding protein J from *Salmonella typhimurium. Arch. Biochem. Biophys.*, **159**, 174–179.

Shaltiel, S., G. Halperin, Z. Er-el, M. Tauber-Finkelstein and A. Amsterdam (1978). Homologous series of hydrocarbon-coated agarose in hydrophobic chromatography. In *Affinity Chromatography*, ed. O. Hoffman-Ostenhof *et al.*, pp. 141–160. Pergamon, Oxford.

Shiman R., D. W. Gray and A. Pater (1979). A simple purification of phenylalanine hydroxylase by substrate-induced hydrophobic chromatography. *J. Biol. Chem.*, **254**, 11300–11306.

Simmonds, R. J. and R. J. Yon (1976). Protein chromatography on adsorbents with hydrophobic and ionic groups. *Biochem. J.*, **157**, 153–159.

Smyth, C. J., P. Jonsson, E. Olsson, O. Soderlind, J. Rosengren, S. Hjertén and T. Wadstrom (1978). Differences in hydrophobic surface characteristics of porcine enteropathogenic *E. coli* with or without K88 antigen as revealed by hydrophobic interaction chromatography. *Infect. Immunol.*, **22**, 462–472.

Sobel, J. H., J. A. Koehn, R. Friedman and R. E. Canfield (1982). Alpha chain crosslinking of human fibrin: purification and radioimmunoassay development for two Aα chain regions involved in cross-linking. *Thrombosis Res.*, **26**, 411–424.

Srinivasan, R. and E. Ruckenstein (1980). Role of physical forces in hydrophobic interaction chromatography. *Sep. Purif. Methods.*, **9**, 267–370.

Stavric, S., N. Dickie, T. M. Gleeson and M. Akhtar (1981). Application of hydrophobic interaction chromatography to purification of *E. coli* heat-stable enterotoxin. *Toxicon*, **19**, 743–747.

Tanford, C. (1973). *The Hydrophobic Effect*. Wiley-Interscience, New York.

Thanner, F., D. Palm and S. Shaltiel (1975). Hydrophobic and biospecific chromatography in the purification of maltodextrin phosphorylase from *E. coli. FEBS Lett.*, **55**, 178–182.

Van Oss, C. J., D. R. Absolom and A. W. Neumann (1979). Repulsive van der Waals forces. II Mechanism of hydrophobic chromatography. *Sep. Sci. Technol.*, **14**, 305–317.

Wilchek, M. and T. Miron (1976). On the mode of adsorption of proteins to 'Hydrophobic Columns'. *Biochem. Biophys. Res. Commun.*, **72**, 108–113.

Wilchek, M., T. Oka and Y. J. Topper (1975). Structure of a soluble super-active Insulin is revealed by the nature of the complex between cyanogen-bromide-activated Sepharose and amines. *Proc. Natl. Acad. Sci. USA*, **72**, 1055–1058.

Yon, R. J. (1972). Chromatography of lipophilic proteins on adsorbents containing mixed hydrophobic and ionic groups. *Biochem. J.*, **126**, 765–767.

Zaidenzaig, Y. N. and T. M. Stern (1980). Simple procedure for purification of diphtheria toxin and separation of its fragments by hydrophobic chromatography. *Anal. Biochem.*, **109**, 71–75.

34

High Performance Liquid Chromatography

C. W. RAUSCH
Waters Associates, Milford, MA, USA
and
A. H. HECKENDORF
Southborough, MA, USA

34.1 INTRODUCTION

'In technology . . . we are entering a period of turbulence, a period of rapid innovation . . . But a time of turbulence is also one of great opportunity for those who can understand, accept, and exploit the new realities. It is, above all, a time of opportunity for leadership.' (Drucker, 1982).

In the chemical process industry today, there has been a need for systems effecting separation of multi-component compounds without affecting each component. This is especially true in the field of biotechnology where the research at present is complicated by engineering difficulties associated with a different group of separation problems. This promises to be a fascinating area as it is now associated with the excitement generated in the field of genetic engineering over the past decade.

Historically, each development in the analytical world has predicated the next. For example, solvent delivery system development required very high pressure pumps, which demanded a change from the septum injector to injectors which could handle higher pressures in an easier fashion. Since the solvent delivery systems were now somewhat pulseless, improved detector design became a central requirement. Once the chromatographic system had been developed in the history of analytical HPLC, automation of that system soon followed with the advent of microprocessor technology. In this progression, the types of problems handled afforded more complex analysis and separation.

With the advent of smaller particle technology in the packing media, lower limits of detection were required. With all the developments in the analytical world there has been a continuing downward trend in particle size technology in microparticulate packings, from fully porous 10 μm spheres to the present 3 μm era of separations where the thrust is for increased resolution, increased speed and increased separating capabilities.

The dilemma now confronting the chemical process industry is which path to take in the development of the technique of preparative chromatography. Each preparative HPLC system is presently really a low-pressure technique. The name HPLC reflects more of an analytical instrumentation aspect *versus* truly open column preparative work that Swett started at the turn of the century. The developments have not been significant in column techniques. Minor enhancements, such as medium pressure or flash chromatography, MPLC techniques, have not really evolved and the resolution power is still limited to alphas of about 1.5 (α = difficulty of separation of components; $\alpha = 1$ = no separation). What is required now is the scale-up of this analytical technique to be used as a visible product process.

34.2 THEORY AND PRACTICE OF HIGH PERFORMANCE LIQUID CHROMATOGRAPHY

All liquid chromatography systems function according to the same basic principles. Rapid, efficient separations distinguish high performance liquid chromatography (LC) from the slower, less efficient open-column techniques. Chromatography simply involves separations due to differencs in the equilibrium distribution of sample components between two different phases. One of these phases is a moving or mobile phase, and the other is a stationary phase.

The sample components move through the chromatographic system only when they are in the mobile phase. The velocity of component migration is in part a function of the equilibrium distribution. The component having a distribution favoring the mobile phase migrates faster than that which has a distribution favoring the stationary phase. Separation results from different velocities of migration as a consequence of the difference in equilibrium distributions (Figure 1).

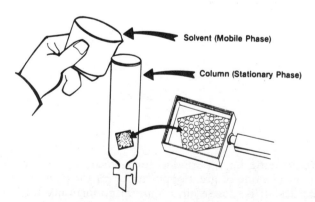

Solvent (Mobile Phase)

Column (Stationary Phase)

Figure 1 Typical open column chromatography system

In a properly selected separation system, there must be an adequate balance in the relative affinities of the sample components for the two phases. Once the correct operating conditions have been established, virtually any sample that can be dissolved can be separated using one or more high performance liquid chromatography (HPLC) methods.

34.2.1 Adsorption Chromatography

Simply stated, when the components of a sample have more relative affinity for the column packing material than the mobile phase, the separatory mechanism is termed 'adsorption'. Adsorption chromatography is a powerful separation mode and currently forms the basis for a large percentage of all high pressure liquid chromatography. An adsorption (process) is any state which affects the solute solubility. Adsorption includes as many as eight different chemical processes such as hydrogen bonding, dipole–dipole interactions, *etc.*, all working in combination to effect the separation.

Manipulation of the composition of mobile phase and packing material allows different adsorptive forces to work within the column. These adsorptive forces cause the molecules of the sample components to reside with the packing material for different periods of time. Those molecules that remain with the column for relatively short periods are apt to emerge from the column more quickly than those molecules which are retained for a longer time.

Two distinct types of adsorption chromatography exist. In normal phase HPLC, the packing material is more polar than the mobile phase. Typically, solvents used in normal phase HPLC are highly nonpolar, such as hexane or heptane, and the packing is polar in nature, such as silica gel or alumina.

In reverse phase HPLC, the packing material is chemically treated to provide a very nonpolar surface. Such chemical treatment may include bonding phenyl, alkylnitrile, octyl, or octadecyl functional groups to the silica, forming very stable siloxane bonds. Highly polar solvents and buffers are generally used for the mobile phase. These include combinations of water-based buffers and a wide variety of organic modifying solvents (for instance methanol, acetonitrile, THF).

Finally, reverse phase and normal phase HPLC are complementary; where one works poorly, the other is likely to provide successful separations. Refer to Table LC Solvents for a comparison of the chromatographic parameters of normal and reverse phase chromatography.

34.2.2 Ion Exchange Chromatography

Ion exchange chromatography utilizes the net positive or negative charge of sample components to achieve a separation. The column packing material used in this type of chromatography is also charged. In general, the packing becomes the fixed anion with a mobile counter ion which exchanges with the mobile counter ion in solution. For example, when an ionic species of a sample passes through an ion (cation) exchange column, the positively charged molecules are adsorbed on to the negatively charged column packing. Weakly charged sample components move rapidly through the column and therefore emerge first, while strongly charged molecules are retained longer until released by the column packing.

The mobile phases typically used in ion exchange chromatography vary from weak to strongly ionic, depending upon the desired separation.

34.2.3 Size Exclusion Chromatography

The difference in the molecular size of sample components in solution is the basis for separation by size exclusion chromatography. Size exclusion is a form of liquid chromatography by which the solute molecules (molecules of a sample in solution) are separated as a result of their permeation into the solvent-filled, porous matrix of the column packing. It may be typically used for molecules from 10 000 MW to 10^6 MW. Large molecules may be excluded from some or all of the pores in the packing by virtue of their physical size and are carried along with the mobile phase. Smaller molecules spend more residence time in the pores and will elute after the large molecules. Molecules will therefore be eluted in order of decreasing molecular size.

33.2.4 Affinity Chromatography

Affinity chromatography separates sample components on the basis of relative preferences for a specific ligand, which is chemically bonded to an inert column packing material. When a sample is passed through a column of this type, those molecules with the greatest affinity for the bound ligand are retained on the column packing material, while all others elute. These retained molecules can be selectively eluted from the column with various buffers, solutions containing the ligand itself or an analog thereof, or with a combination of organic solvents and buffers.

34.3 HIGH PERFORMANCE LIQUID CHROMATOGRAPHY INSTRUMENTATION

Several requirements must be satisfied in designing a high pressure liquid chromatographic system. Mechanisms are needed to: (1) introduce the sample into the mobile phase stream; (2) push the sample and mobile phase through the column; (3) monitor and collect the separated fraction.

HPLC systems are unlike other types of chromatographic systems, because they utilize densely packed columns and relatively high flow rates to effect the separation. These two unique features provide the excellent resolution between sample components for which HPLC is noted. However, to achieve this outstanding separation usually requires the use of high pressure pumps to move the sample/mobile phase mixture through the densely packed column material in a reasonably short amount of time. The liquid chromatographic system must, therefore, contain basic system components designed to work in combination with each other (Figure 2).

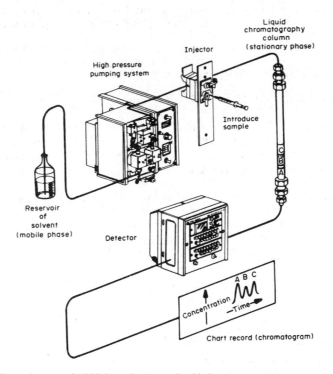

Figure 2   A typical high performance liquid chromatography instrument

A smooth running 'high performance' system requires a pumping system and a system for the introduction of the sample into the mobile phase stream. Another component in a basic HPLC system is the detector. Depending upon the sample chemistry, refractive index (RI), ultraviolet (UV), visible (VIS), or fluorescence detection modes, to name a few, may be employed. The critical point in detector selection is dynamic range and selectivity.

Finally, once the separation is completed and detection has occurred, it is necessary to collect and process all effluent streams appropriately. Equipment used at this stage ranges from a simple rotary evaporator to sophisticated flash evaporators, ultrafiltration systems and reverse osmosis membrane systems.

34.3.1 The Separation Media for the Column

Column selection is the most important consideration of an HPLC system. Different types of columns are available for the separation of a wide variety of complex mixtures and sample loads.

Analytical columns are typically used for methods development and high sensitivity, high resolution applications. These small diameter columns (2–4 mm I.D.) are generally 10–30 cm long. The size of the packed bed within these columns determines the amount of sample material which can be loaded. The sample loading limit is less than 1–10 mg per column depending upon the packing material used and the desired resolution.

The next group, the semi-preparative columns, includes those that prove useful for isolation and separation of larger scale sample loads. Typical dimensions for semi-preparative columns are 6–8 mm I.D. by 25–100 cm long. These columns are used for separations involving sample loads of 10–100 mg.

Preparative columns are the next largest capacity columns. These columns may range from 2–5 cm I.D. and to several feet in length. Typically, the sample load for these columns is approximately gram quantities.

Finally, for large scale production work, process chromatography columns are necessary to separate gram to kilogram quantity samples for commercial biotechnology separation purposes. These columns have a much larger diameter, are longer, and are used for the isolation, separation and high recovery of biologically active compounds.

At present, worldwide application of HPLC separations have been published on thousands of chemical analysis and purification problems. The new areas of interest have included the following: proteins, naturally occurring and synthetic peptides, amino acids, nucleic acids, carbohydrates, organic intermediates, metabolic intermediates, fermentation broths, *etc.*

34.3.2 Fundamentals: Capacity, Selectivity and Resolution

Three terms used frequently in practicing chromatography are fundamental to considering any LC system.

The capacity factor (k') is defined for a single compound in an LC system under a given set of conditions. It is defined as the ratio of the amount of compound on the stationary phase to the amount of compound in the mobile phase during the chromatographic run. The capacity factor may be measured in terms of retention volume of a compound (more specific to adsorption chromatography).

The k' of a compound is simply a measure of its retention in terms of column volumes (Figure 3).

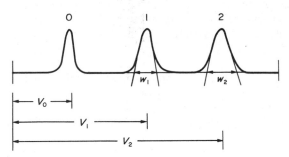

V_0 = retention volume* of unretained compound
   = one column volume

V_1 = retention volume* of compound #1

V_2 = retention volume* of compound #2

w_1 = peak width of compound #1 in terms of volume

w_2 = peak width of compound #2 in terms of volume

*Retention volumes measured at peak apex.

Capacity factor for compound #1 = $k_1' = \dfrac{V_1 - V_0}{V_0}$

Capacity factor for compound #2 = $k_2' = \dfrac{V_2 - V_0}{V_0}$

Theoretical plate number

$= N = 16 \left(\dfrac{V}{w}\right)^2$

$= 16 \left(\dfrac{V_1}{w_1}\right)^2$ for component #1

$= 16 \left(\dfrac{V_2}{w_2}\right)^2$ for components #2

Figure 3 Calculation of capacity factor and theoretical plate number

The selectivity factor (α) indicates the degree to which two sample components separate under a given set of chromatographic conditions. The α is a function of distance between band centers and is measured in terms of k'.

The selectivity factor for compound 1 and compound 2 is deduced using the equation:

$$\alpha_{1,2} = \frac{k_2'}{k_1'} = \frac{V_2 - V_0}{V_1 - V_0}$$

The objective in developing any separation is to obtain resolution (R). The resolution of two components eluting from a chromatographic column is defined as the distance between the peak centers divided by their average peak widths.

Resolution for compounds 1 and 2 is:

$$R = \frac{V_2 - V_1}{1/2(W_1 + W_2)}$$

34.3.3 Throughput, Purity and Load

The ultimate objective of a preparative and process scale chromatographic separation is to obtain a maximum amount of purified material in a minimum amount of time. This is called throughput.

$$\text{Throughput} = \text{amount of purified sample/unit time}$$

The term 'pure', as we shall think of it, is relative and has an arbitrary value, that is a sample is 'pure' when it is sufficiently free of contamination to satisfy some given set of criteria. These criteria and the corresponding purity requirement should be well established before a preparative separation and subsequent scale-up are carried out. Having defined the purity requirements of the finished sample, other factors can be optimized to obtain maximum throughput.

When considering the throughput of a system in terms of the amount of purified sample obtained, it is important to consider load, or the amount of crude sample which is handled in a single run. Obviously, load must be optimized in order to obtain maximum levels of throughput. At times, sample solubility is the sole load-limiting factor. In most cases, solubility presents no insurmountable problem and the loading time is simply the largest amount of crude sample which can be injected into the system without destroying the separation. When the amount of sample injected exceeds the loading limit, the system is in a state of overload (Beverung, 1979).

Purity and load are related in that one may often be further optimized at the expense of the other. To better understand the significance of this relationship, and how it can be used to advantage, it will be useful to become familiar with a few fundamental parameters of a column. A simple model can then be developed, describing load in terms of a group of molecules entering a chromatographic bed. Examining in detail the assumptions made in developing this simple model, those parameters that affect loadability with regard to optimizing throughput will become apparent.

It can be shown that as k' decreases, the loading limit of the system increases, and it generally holds true that systems with smaller k' will have higher loading limits than those with larger k'. Obviously, a smaller k' also means a shorter time period per run. The net result is a sizeable increase in throughput.

In view of this enormous capacity for adsorption or partition, one must explain the fact that typical sample loads for separations are much smaller than our ideal system. To answer this question, we must develop a model further.

Consider two compounds which have been injected into the system, each of which has a k' greater than 0. Each molecule of a single compound will spend a portion of its residence time in the chromatographic bed adsorbed or partitioned onto the stationary phase. A molecule of the second compound will spend either more or less time on the stationary phase, depending upon the relative magnitude of attractive forces of the stationary and mobile phases for that molecule.

The difference between attractive forces of the media surface for two different types of molecules is extremely small. Thus, for one molecule to be selectively retarded and separated from another, it must equilibrate between the media and the mobile phase many times. The minute time lag, thus multiplied, will result in a significant difference in measurable retention time as the molecules approach the end of the column bed.

If it were only necessary to separate two molecules from each other, the problem would be simple. Thus, when a large number of molecules are injected into a chromatographic bed, a fierce competition for adsorptive sites initially exists between different sample molecules and eluent (solvent) molecules (Figure 4). Some molecules may pass a considerable distance down the bed before being adsorbed. Eventually, the sample is sufficiently diluted to allow equal opportunity for each sample molecule to be adsorbed. Herein lies an important load-limiting factor concerned with the equilibrium of the sample between the mobile and stationary phases.

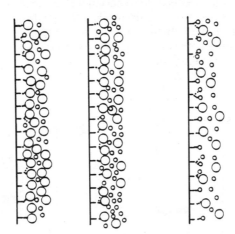

Figure 4 Equilibration in liquid chromatography; increasing dilution from left to right, giving increased chance of molecule adsorption

When a sample is injected, it is introduced into the column bed as a concentrated plug. When the solvent flow begins, the sample plug gradually comes into equilibrium (Figure 5). The volume of column bed required for sample equilibration (volume of equilibration) in a given system depends upon the nature of the sample and the amount injected. When a larger quantity of sample is injected, more volume of column bed will be required for sample equilibration.

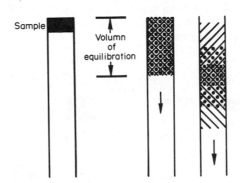

Figure 5 Volume of equilibration

The volume of equilibration was the basis for the original concept of the chromatographic plate. This is not a 'theoretical plate' but a 'real' plate whose size (length, which for a given diameter of cylinder specifies volume) is determined by the quantity of sample injected into the bed. If the size of the plate comprised the entire bed of silica, the column would contain one plate. If the plate occupied only 1/100th of the silica bed, the column would contain one hundred plates.

As mentioned, sample molecules must interact many times with the adsorbent surface in order to achieve separation. It stands to reason that when two compounds differ largely in their affinity for the media, a smaller number of adsorptive interactions are required to achieve adequate separation of the corresponding molecules. As the difference decreases, separation becomes more difficult and a larger number of adsorptive interactions must occur for a separation to take place.

To provide for sufficient interaction with the silica surface of a hard-to-separate sample, the sample size should be decreased appropriately; therefore, the plate volume is decreased, the number of plates increased, and a larger volume of the column bed is available for sample separation.

If, by reducing the effective plate volume in this manner, the desired separation cannot be achieved, the plate number may be further increased by increasing bed length. This is done either by using recycle or by using columns connected in series, or both. The objective is to provide the sample with enough residence time in the system for the required amount of sample transfers to occur between the mobile and stationary phases.

Another way of looking at the effects of load and plate volume on the separation is with respect to the chromatogram. In the case of two compounds differing largely in ther affinities for silica, a correspondingly large α is observed. Increasing the load results in an increase in plate volume (volume of equilibration), and a corresponding increase in peak widths. Large loads are possible wihout significant peak overlap (Figure 6).

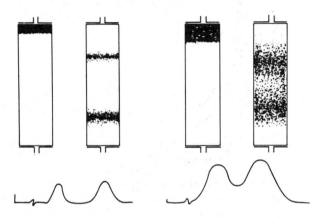

Figure 6   Load and resolution (large α)

When, on the other hand, two compounds differ slightly in their affinity for silica, a small α is observed, and the load must be cut down to avoid peak significant overlap (Figure 7). Increasing length of the column or recycling allows more reasonable load, but the separation takes longer. In summary, a small α results in either smaller loading limits or longer separation times or both, with an ultimate decrease in throughput.

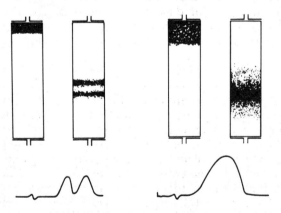

Figure 7   Load and resolution (small α)

It should be apparent now, that large α values are desirable in preparative LC. Achieving maximum α values involves changing the chemical nature of the sample, the stationary phase, or the mobile phase. Most often, there are limitations to the changes which can be made to the sample and stationary phases, so α is generally optimized by making changes in the mobile phase.

For example, there is a dramatic loss of theoretical plates, N, when loading small particle silica columns with compound. From this example, a 20 mm by 30 cm column packed with 10 μm packing was very efficient at low loads but at 50 mg of load, efficiency became less than 1000 theoretical plates. A column of this size is designed for operation in the 100–500 mg range based on a

100:1 silica/sample loading ratio. Substitution of 37–55 μm packing material can give the equivalent performance in the 100–500 mg loading range without the back pressure, flow limitations or cost incurred by use of 10 μm material (Figure 8).

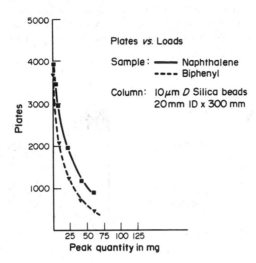

Figure 8   Loss of efficiency with load of compound

34.4   GRADIENT

Gradient elution (Figure 11) in high performance liquid chromatography is best defined as a change in mobile phase (solvent) strength occurring during the course of an analysis. This change can be in the form of solvent polarity, ionic stength, pH or multiple changes when complex interactions exist between column packing and sample components.

Gradient elution is needed when an isocratic (constant) mobile phase composition does not adequately resolve a mixture of components in a reasonable amount of time. The separation of polystyrene oligomers is a good example of such a situation. In Figure 9A, an isocratic mix of 70% tetrahydrofuran (THF) in water eluted all components in 28 minutes, but with poor resolution. In Figure 9B, 60% THF (isocratic) did not elute more than four components. Complete elution with good resolution in 25 minutes is shown in Figure 10, where a gradient from 50–70% THF is used.

The goal of gradient elution, optimum resolution in minimum time, can be achieved by manipulation of several parameters as follows. (1) Curve shape: the rate at which solvent strength changes can occur linearly or nonlinearly throughout the gradient. In Figure 10, the THF content increases faster at the beginning than at the end of the convex gradient, which improves resolution of the latter-eluting oligomers. Other uses of nonlinear curve shapes are shown in Figure 11, where the appropriate solvent profile can enhance the resolution of complex mixtures. (2) Solvent strength range: the strength of initial and final mobile phase composition can be manipulated so that elution of components is over the entire gradient range, thus maximizing resolution per unit time. (3) Run time: the time required to go from initial to final solvent condition has a major effect on resolution, since this controls the rate-of-change of solvent strength. (4) Flow rate: in gradient elution, an increase in flow rate while holding other parameters (run time, *etc.*) constant will usually improve resolution. This is not inconsistent with the fact that an increase in flow rate will reduce resolution in the isocratic mode.

34.4.1   Performance Criteria of a Gradient System

34.4.1.1   *Accurate solvent profiles*

The polystyrene chromatogram in Figure 10 requires a convex curve shape for best overall resolution, meaning that the rate of change for THF is greater at the beginning of the gradient.

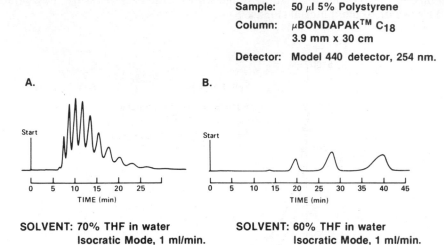

Sample: 50 μl 5% Polystyrene

Column: μBONDAPAK™ C$_{18}$
3.9 mm x 30 cm

Detector: Model 440 detector, 254 nm.

A.

Start

B.

Start

TIME (min)

TIME (min)

SOLVENT: 70% THF in water
Isocratic Mode, 1 ml/min.

SOLVENT: 60% THF in water
Isocratic Mode, 1 ml/min.

Figure 9   Isocratic separation of polystyrene (800 MW)

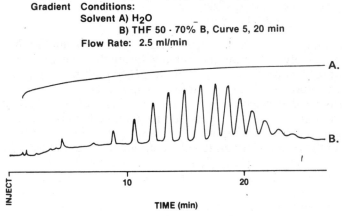

Column: Radial-PAK™ C8, 10 μm, 8 mm ID
Gradient Conditions:
Solvent A) H$_2$O
B) THF 50 - 70% B, Curve 5, 20 min
Flow Rate: 2.5 ml/min

A.

B.

INJECT

TIME (min)

Figure 10   Gradient separation of polystyrene (800 MW)

Generation of a gradient curve occurs in an electronic controller and is transmitted to the solvent delivery system. In order to obtain the expected results, the actual mobile phase composition must follow the electronically-generated profile. This becomes important when transferring methods from one HPLC system to another. Differences in fluid path may cause the actual solvent profiles to vary, even when the electronically-generated curves are the same. Also, when developing methods, if the actual solvent change and the electronic change are different, it becomes difficult to predict accurately the parameters necessary to achieve a gradient separation.

34.4.1.2   *Rapid response to solvent changes*

In addition to accurate solvent profiles, a gradient system should deliver a given change in mobile phase composition to the column as quickly as possible. The delay time between a gradient controller call for a mobile phase change and that change actually reaching the column is mainly a function of the volume, geometry of the mixing device and the volume of the fluid path between the mixer and the column. When developing a separation, rapid mobile phase changes save time by allowing rapid column equilibration with each new mobile phase blend. For repeti-

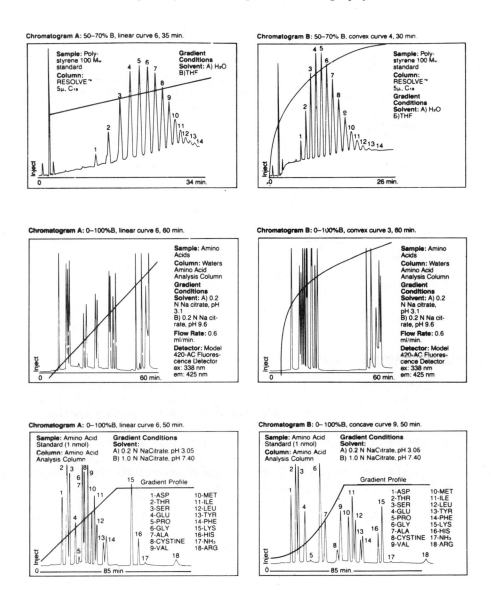

Figure 11 Gradient patterns

tive analysis requiring gradient elution, a rapid return to the initial solvent composition decreases the time required for column equilibration before the next sample can be run. Also, when converting gradient runs to isocratic conditions, minimizing delay simplifies the interpretation of the relationship between a particular point in the gradient and an isocratic solvent composition.

34.4.1.3 Mechanism of gradient formation

Now we shall consider the components of an HPLC system which contribute to accurate solvent profiles and rapid mobile phase changes.

Two approaches commonly used to generate solvent gradients are shown in Figure 12. Method A employs electrically-actuated solenoid valves located before a pump. Gradient precision depends upon the ability of these valves to reproducibly proportion segments of each solvent, consistent with the composition called for by the controller. If reproducible retention times and stable detector outputs are to be obtained, these 'pulses' of solvent must be thoroughly mixed into a homogeneous stream before entering the chromatographic column. Mixing devices in this type of system can often distort the solvent profile and cause slow response to solvent changes.

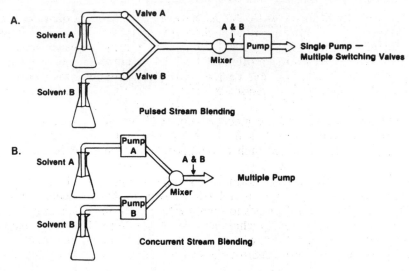

Figure 12 Gradient formation

34.4.1.4 Role of the mixing device

The primary purpose of mixing in gradient LC systems is to smooth the variations in solvent composition caused by the gradient-generating device. These variations can be large when produced by proportioning valves, or small with well-designed solvent delivery systems.

The properties of common mixing devices will vary depending upon design. A stirred chamber is more effective for smoothing large compositional changes than a capillary laminar flow mixer, but it also produces gradient curve distortion and slow mobile phase changes (Figure 13). Curve A is the electronically-generated profile. Curve B is the solvent profile from a stirred chamber, and curve C is the solvent profile from a capillary laminar flow mixer.

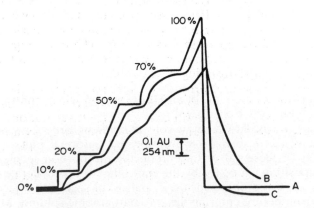

Figure 13 Comparison of mixing devices

34.5 SCALE-UP TO LARGE PROCESS SEPARATIONS

In a general sense, there are three problems associated with traditional chemical engineering unit operations when concerning biotechnology. First, look at the separation unit operations of distillation and evaporation. This is an energy-intensive separation process where extremely dilute fermentation broths will not be concentrated into specific raw material feeds for subsequent purification by this technique. These dilute systems would expend much energy merely in the concentration phase to achieve a separation which would not be cost effective.

Secondly, traditional techniques are mechanically and thermally stressful on the new products that are being generated in this field, and a technique such as chromatography is, indeed, the only technique gentle enough to deal with these types of products, enzymes and other proteins.

The most important role that high performance chromatography can play, once the parameters and understanding for scale-up have been understood, will involve the speed and selectivity that have been used by the analytical chemists to separate the active materials into discrete chemical components. At present, traditional techniques are only marginally acceptable for scale-up. Many of the materials to be separated are similar in nature and must undergo careful scrutiny as to their ability to be separated on a selective basis with older, more traditional and better understood technologies.

Since the introduction of high performance liquid chromatography in the late 1960s, its development has illustrated an efficiency and versatility as an analytical method for separating and determining the number of components that make up various mixtures.

34.5.1 HPLC Scale-up

Until now, many of the advances in this area have been made through a method of increasing pressure and the use of finer particles. To repeat the earlier note, from an analytical point, pressure is not a problem and higher pressures are needed in order to operate columns packed with very fine particles. When undertaking scaling, the analytical separation requires operation at faster flows with the use of very fine particles to run within the optimum performance of the system. The necessity of operating these columns at very high flow velocities is required in order to utilize the efficiency. As is readily apparent, most commercially available equipment is now designed to operate at pressures of 250–400 atm which is not easily applied to a scale-up situation. Use of pressures as high as 4000 atm have been reported and advocated at times from an analytical point of view and at the initial preparative point of view. This is indicative of a general trend for analytical techniques to be applied on a process scale.

It might therefore seem, from a survey of any kind of scientific and commercial literature, that the limit to pressures which can be used should be set only by 'the technology and financial considerations'. If this were true, there would be little point in any exercise of understanding scaling. However, one must not derive such a conclusion. The true essentials for separation scale-up must not lose sight of practicality.

What should be considered in contrast to these general trends from the analytical systems is an understanding of the goal to be achieved in a given separation task and whether the design of the column has been properly thought out. In other words, a reduction of packing particle diameter is not necessarily accompanied by an increase in resolution. The most important force should be dependent upon one's method in separations and purification schemes, not column length or flow velocities, which are only means of achieving a separation.

There have been several theoretical proposals confirming some of the experimental results which demonstrate that the concept of a short column, packed with the right size particles, has kept the scaling parameters constant and will achieve difficult separations in a reasonable time using only moderate pressures. This key point would be an apparent inconsistency with analytical techniques and should raise questions and indicate the need for understanding the pressure required to move the carrier liquid and hence perform a separation in a reasonable amount of time. With large-scale operation, it is necessary that the price required to achieve a separation using high pressure be analyzed so this driving force does not limit the use of liquid chromatography as a unit operations processing tool. What should be considered is scaling around the best use of pressure.

34.5.2 Optimization

The difficulty of optimization from the preparative to a larger scale or process chromatographic separation is due to the fact that the separation of each pair of compounds depends upon many variables. For convenience, according to earlier papers in the area of chromatography, these chromatographic parameters/variables can be separated into main and secondary parameters. The seven main parameters are H as in HETP, the column length L, the number of plates N, the velocity of the mobile phase U, the column pressure drop ΔP, the particle diameter D_p, the retention time of the second solute of a pair of interests T_r and the resolution and the maximum concentration of the peak C-max. The primary equation which describes the resolution of the separation is:

$$N = 16(R_s)^2(\alpha/\alpha-1)^2(K'+1)/K'^2$$

The secondary parameters affecting operating performance of the column are of lesser importance; they are solvent viscosity, the capacity ratio K', the specific column permeability k, or ε, the total external porosity ε_n and ε_e, the temperature T, the diffusion coefficient D_M, the sample size M, the selectivity α and the column diameter. The basis for classification (Martin and Synge, 1941) gives secondary variables that are supposed to be virtually constant for a system and packing method within fixed temperature boundaries. It should be noted that some of the secondary parameters have to be changed into primary parameters in some instances, one of them being preparative and process scale where the mass and column diameter will be key parameters.

Working with these parameters, we can assemble a series of classical equations that will start the process of optimization. Any optimization procedure aims to find the optimum value of one of these parameters as a function of the other seven, in fact, as a function in any three independent variables chosen arbitrarily among the seven factors. The adsorption and desorption of molecules from the liquid phase on the solid support and the liquid solution, which is the carrier fluid passing by the external/internal surface of a pure solid, offers the opportunities for the purification of a process stream. This is reflected in the use of any kind of fluid-contacting device, specifically, for the recovery of valuable components, as in the separation for purifying a pharmaceutical agent or therapeutic agent. In many chemical engineering processes, it is sometimes possible to use countercurrent flow of a fluid in solid phases. More often, the solid particles will be held in a fixed bed and the fluid feed will be run through the bed until it becomes equilibrated with that liquid feed. Then small quantities of material, or larger quantities in the case of preparative chromatography, will be put into this fluid stream and adsorbed onto the surface at various rates until one of them will break through and be part of the column effluent in a different time frame from other components. The bed structure becomes critical to the success of performing an efficient and low cost separation. This semi-batch cycling of injecting sample and selectively collecting the effluent streams is the key to effective separations. It is critical in the case of a gradient system where forward and back mixing would limit the separation power from the fluid changing the separation process. The sample stream as well as the timing for collection of the material at the appropriate peak, utilizing selective recirculation of solvent streams and selective recovery and reuse of solvent streams define operating cost and the operation procedure of scaling-up the chromatographic operations.

These operations are generally referred to as sorption processes where the concentration in the fluid and the solid phase inside the bed depend upon time and a specific component. A steady-state condition for HPLC chromatographic operation is never reached in practice, because the bed structure as well as the bed chemistry may change over time. This adsorptive/desorptive exchange may, at some time, permit a feed to pass through without affording any separation. Thus, the design of fixed-bed devices for chromatographic separations is mathematically more difficult than the design of continuous and steady-state adsorption or distillation equipment. Nevertheless, a rather complete theory of fixed-bed phenomena exists from much of the chemical engineering literature (Bird, 1960). It has largely evolved around the field of ion exchange devices where knowledge of phase equilibria and interphase mass transfer rates are documented and used to predict operation, just as they do in any mass transfer device (Chao, 1982).

The objective of this section is to briefly describe application of the theory to design considerations in scale-up. This HPLC fixed-bed operation is called large scale preparative chromatography (process scale). The first area to be addressed is the effect of high loads on the efficiency of the chromatographic operations. This emerges as a dominating aspect in the discussions of many preparative chromatographic techniques. Numerous studies on the effect of high loads on efficiency with standard columns have been carried out with chromatography in a small scale preparative mode (Jong, 1981), but most of these were carried out with a single solvent and with defined mixtures. The loading of large mass and the subsequent loss of effective resolution severely affects the peak shape. This should not be taken as a negative indication of resolution in the quest for enhanced separation. By increasing the mass of a solute in a constant volume, one is searching for the mass of solute which yields the maximum allowable increase in peak width over the normally observed peak width for a very low mass of solute (and consequently the maximum allowable loss of resolution). Under these conditions, it is also noted that the k' of the peak will shift to an earlier elution time. Despite the poor peak shape, which may exhibit a steep front and broad tail, the material may be fractionated to yield pure compound. Above this mass loading, the peak shape will become badly distorted due to nonlinearity of the isotherm, and this effect will swamp out the column packing dispersive effect, leading to total loss of resolution capability (Jong, 1981). It should be noted that this effect is more related to concentration than volume. This relation demonstrates the need for a good column efficiency, since this will have a direct bearing on the concentration of the elution profile. For a column of the same size, a greater mass can be loaded on a column of high dispersion characteristics (larger particle) than one of low dispersion characteristics (smaller particle size). This effect is seen in Figure 8. The separation and enrichment actually occur in a small part of the column. Under most conditions, the interaction of solutes can be neglected. From the point of view of separation, such experiments and discussions connected with them are somewhat artificial in the practical application of chromatography. It is separation of adjacent peaks which limits the load that can be applied.

In addition, in adsorption chromatography a displacement effect is often observed (frontal displacement). A solute band is accelerated and sometimes contracted to a measurable extent in the presence of a large amount of another solute which is in the column. The competition for adsorption sites in the molecules in the larger band displace those of the first band from the adsorbent. This displacement chromatography caused by an overload of such effects may be as important as the distortion that would occur with a substance on the high load when chromatographed by itself.

34.5.3 Scaling

With the use of this adsorption phenomenon, let us turn to the scale-up. In many cases in engineering, it has been common practice for the current laboratory technique to separate biological compounds to be merely geometrically scaled by using larger dimensions of the laboratory separating device in a process environment. What happens under these single scale-up schemes where various biological mixtures are fractionated is that the packed column is designed for a large volume of material. To make the system scaleable, however, the column and the operation scale require consideration together. From the laboratory information to a pilot scale, a quantifiable and meaningful ratio of length to diameter must be used to achieve economical and reproducible purification. This is especially true in the general area of biological compounds where the proliferation of laboratory techniques has amounted to a proliferation of processing techniques, which most of the time are not meaningful when applied at a routine level of processing operation.

The first criterion for dynamic similarity within the column system is the geometry. This is achieved by making L over D of the packing of each column the same (Charm, 1968). The ratio of the amount of material to be put on to the 'relevant apparent' packing should also be equal in both columns. This means that the loading per gram of packing material/surface gram loading per kilogram of packing must be similar. Dynamic similarity may now be achieved by keeping the Reynolds number ($DV\varrho/\mu$) the same in each case. Since viscosity and porosity of the packing are the same in each column due to the identical nature of the chemistry and the base packing media, the Reynolds number will now be the same if the DV is the same. Thus, when the L over D, DV and the amount of the material for the amount of packing material in the system are the same, dynamic similarity between the columns should exist and the elution patterns will be the same in each column.

Using this information, one should calculate the amount of substance which can be separated per unit time; in other words, the production rate and the peaks eluted. The purity of the isolated compounds will then determine the quantities that will be chromatographed in preparative chromatography (Figure 14).

We must base the design on an operating sequence of the preparative system and on the optimum production rate which will mean optimizing the latter with respect to time, solvent con-

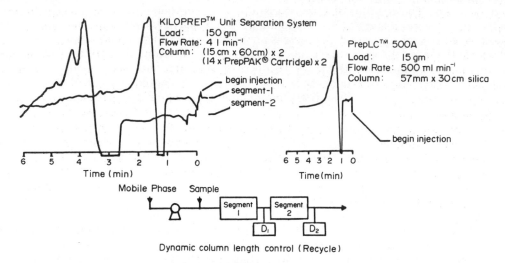

Figure 14 Comparison of column performance at equivalent loading

sumption, the smallest possible solution of the sample and the separate peak in the element material.

For the present exercise, we can design without doing too many calculations and a lot of experimentation. The person involved in scaling can assemble into a reasonable order all the parameters involved in the first section to help derive an optimum design for the system. Fundamentally, the pertinent parameters can be developed into an optimized method which will emphasize operation strategies. It is good to emphasize an operating sequence as well as a strategy in design for the maximizing throughput of a given column (Bailly and Tondeur, 1981).

For the process design, optimization will require a focus on the selection of the column size and ability to operate with the given scale-up Reynolds number which then will encompass a total operating cost to minimize overall costs. Although an engineer should consider each case on its individual merits to get the most economical design, he would be faced with the ability to bracket the economics by approaching either the best use of the column volume available and maximizing the use of the feed with a carrier stream, or looking at maximizing the flow rate through the system with the given column size and varying the amount of material to be put on. In the case of maximizing the column feed or varying the material to be separated with the column, this would drive toward minimizing the column size and shape for the given material to be separated, whereas, in the other case, one would optimize by minimizing the use of the column transport system or feed handling and carrier liquid handling systems to drive toward a very large column. This second case of driving toward a very large column is what has been practiced in the field of biotechnology. A simple laboratory scale-up would drive one to this selected alternative.

Let us present, for both cases, the type of optimization expression needed and fix the system adsorption isotherms with the given column packing material.

Using expressions developed from Timmins (1969) and Hupe (1981) we can express an optimization expression dealing with rapid injections or placements of sample onto the column and a cycle that will be substantially less than the retention time so the column can handle several injections at one time. This leads to a production rate which is related to the column porosity, flow velocity, fluid density and concentration of the feed. The feed time is the amount of mobile phase injected as well as that fraction of injection time which is a part of the total cycle during which injection occurs. The expression would look as follows: A = area, ε = void fraction, V = velocity, e = density, x = mole fraction, t_i = time of injection, t_c = total cycle time. Thus:

$$\text{Production} = A\varepsilon V\varrho x\,(t_i/t_c)$$

Through reformulation of the equations, this expression of production along with the theoretical information about resolution and time can be formulated into an equation which is more convenient for the production rate:

$$\text{Production rate} = 0.4(A\varepsilon Vx)/R_s\left(1 - \frac{L\text{-min}}{L}\right)^{1/2}$$

where L-min is the minimum column length that will still achieve a separation. This relates to the primary parameters discussed earlier where N-minimum $H = L$ is required for the separation.

For the system we consider here, each is obtained from experimentally determined Van Deemter relations (Van Deemter, 1956), and is dependent upon the linear velocity in the column. With this expression, it can be shown that increasing the column length beyond the minimum will permit an increase in throughput. We can reference this by looking at a rearrangement of the above equation to help find the best choice of operating conditions in a column segment arrangement for the case and derive an expression which will relate to the efficiency of the utilization of the mobile phase that is being employed. We will define efficiency as the ratio of the sample injection rate to the amount that is being carried in the flow in the mobile phase (Figure 15). As previously stated, purity requirements will determine the choice of resolution. The percentage of material in the injected sample will also play a part in determining the efficiency of use of the mobile phase. This also relates to the column length that will be required. The ratio is the column length that will be required to the minimum length required to achieve a separation. It is shown here as a simple graph (Figure 16).

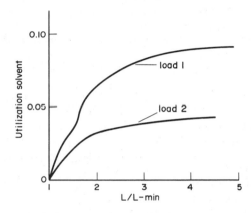

Figure 15   Efficiency of utilization of mobile phase

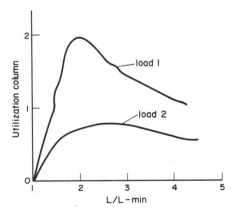

Figure 16   Efficiency of utilization of column

Taking the alternative design selected, conditions can be developed that maximize the utilization of the column. That will indicate a relation between the rate of sample through the column and the volume of the column that is used for the separation. The flow rate P over area and length is equal to an equation.

This equation shows that by increasing the amount of material in the feed stream and keeping the resolution constant, utilization of the column to perform a separation would increase in direct proportion to the density. Velocity over the minimum length required for such separation, which means that columns with good separation capability, *i.e* a low L-min, could operate in relatively high fluid velocity to give the best separation of the system. This is the first example where the soft gel-type material would tend to have drawbacks since it could not operate in a high enough flow rate. Although the utilization of the mobile phase would be independent of the same ratio used for sizing the column, coupling the two together will always demonstrate an increase in the utilization of the column. Figure 16 shows the effect of varying this parameter by increasing the column length above the minimum value. Substantial improvements in this efficiency occur with increasing length until L over L-min is equal to 1.5. Beyond this point, an increase in length actually decreases the ability of the column and the effect of utilization to perform a separation. This would be in contrast to the effect of length on the efficiency where the continued improvement occurs as L over L-min increases. Thus:

$$\text{Column utilization} = \text{production rate}/(\varrho\, U_i A) = 0.4(\chi/R_s).1 - (L\text{-min}/L)^{1/2}$$

This point should be considered rather carefully in dealing with a difficult separation combined with simple parts of a separating requirement. One may want to focus on an L over L-min which would vary, depending upon the separation required of individual components within the system. Developing this thought further leads to the approach of segmenting the column into short segments so that only the column that is necessary would be utilized for separating some parts of the compound mixture. More column could be brought to bear if greater separation were required.

From an engineering point of view, in most cases the best economics will prevail where conditions of the summation of the two curves will maximize either one of the two parameters. Dealing with the liquids for chromatographic separations would indicate a rather high flow rate at rather high values for the mole fraction in the sample injected. In optimizing this to achieve better utilization, mass loading should also be used although it causes extensive peak spreading due to non-linear effects which was discussed earlier in describing mass loadability of the chromatographic column for preparative use.

Given the previously mentioned parameters and some of the design criteria such as Reynolds number and L over D geometric scaling to develop chromatography from a laboratory technique into a practical method for large scale separations, two problems will have to be solved in designing this system.

To obtain good resolution and optimum band profile within the chromatography column, the structure must be maintained and consistent throughout, and the flow being presented to the column and taken away from the column must be managed in such a way as to minimize the effect of the fluid momentum.

Various techniques have been applied to overcome the wall effects and bed shifting which occur in soft and hard materials. To eliminate erosion and packed-bed structure decay resulting from various surface energy fluctuations caused by changing solvent conditions, one must address a method of flow distribution and bed structure stabilization. If one studies the literature, it is indicated that few, if any, attempts have been made to solve this problem, either through a radial pressure technique of conforming the wall to the bed structure or axial pressure technique conforming the packing bed within itself and maintaining the pressure on the bed structure from either end, eliminating the chance for bed degradation. Today most problems are addressed through a structuring of various column lengths and column baffling systems in addition to customized injector systems which try to overcome the inequalities in the bed structure. Despite the great amount of attention devoted to column performance under these different conditions, the extent of the wall effect is still poorly understood. This wall effect, as well as bed changing, which causes packing non-heterogeneities that distort the flow profile, causes the overall axial dispersion to increase and make both isocratic and gradient system impractical. In comparison, short range effects, diffusion and local wall-induced long-range contributions are less than minimal and are difficult to model adequately, as this involves functional relationships between actual diffusivities and velocity within the radial position which, when scaled, is difficult to measure.

Despite the difficulties that seem to be inherent in the scale-up of liquid chromatography, the speed and extremely powerful separation capability it is capable of will find widespread use in the industrial sector, as first seen in its use with high value compounds in the biological environment.

Although every conceivable principle upon which separation could be based has been investigated in the laboratory, interesting new insights from the analytical environment still continue to appear.

High performance liquid chromatography has recently made tremendous impact in the recent environment, with such products as insulin and growth hormone. HPLC has come under scrutiny for scale-up despite the laboratory biochemist's insistence on low pressure techniques.

34.6 SUMMARY

High performance liquid chromatography (HPLC) is recognized as a very effective laboratory technique for the separation and purification of chemical compounds. Its capability to exploit the minute chemical differences between compounds of interest may be the only effective way to disentangle many of the very complex biological mixtures.

The enhancement acquired with gradient chromatography as well as an understanding of load can only assure the technique a successful place in the analytical laboratory.

The challenge to the biochemist and biochemical engineer, who must scale this laboratory technique, lies in knowing the value of fundamental and secondary parameters and assembling them into the appropriate relations (L/D and R_c). The correct design for fluid distribution and column segmentation control, the optimization procedure for correct system design, and correct operational procedure, will all play an important role in the successful scale-up for manufacture of insulin, growth hormone or any other chemical compound of interest.

As in any unit operation, the total system of feed input and feed output, product recovery and by-product waste must be carefully accounted for to determine process economics. In the case of biological compounds, new concentration techniques will play a significant role in the success of process scale chromatography.

34.7 REFERENCES

Bailly, M. and D. Tondeur (1981). Recycle optimization in non-linear productive chromatography. *Chem. Eng. Sci.*, **38**, 1199–1212.

Beverung, W., A. Heckendorf, P. C. Rahn and M. Woodman (1979), *Preparative Liquid Chromatography and Its Relationship to Thin Layer Chromatology*. Waters Associates, Milford, MA.

Bird, R. B., W. E. Stewart and E. N. Lightfoot (1966). *Transport Phenomena*, pp. 196–200. Wiley, New York.

Chao, J. F., J. J. Huang and C. R. Huang (1982). Continuous multi-affinity separation of proteins: cyclic processes. *Recent Adv. Adsorp. Ion Exch.*, **78**, 39–45.

Charm, Stanley E. (1968). The scaling up of elution in chromatography columns. *Chem. Eng. Prog. Symp. Ser.*, **64**, 9–11.

Crane, L. J., M. Zief and J. Horvath (1981). Low pressure preparative liquid chromatography. *American Lab*, **12**, 128–135.

DeJong A. W. J., H. Poppe and J. C. Kraak (1981). Column loadability and particle size in preparative liquid chromatography. *J. Chromatogr.*, **209**, 432–436.

Drucker, P. (1982). Personal communication.

Eon, C. H. (1978). Comparison of broadening patterns in regular and radially compressed large-diameter columns. *J. Chromatogr.* **149**, 29–42.

Hupe, K. P. and H. H. Lauer (1981). Selection of optimal conditions in preparative liquid chromatology. *J. Chromatogr.*, **203**, 41–52.

Jacob, L. and G. Guiochon (1975). A theoretical study of the behavior of the band of solute at its inlet into a Chromatographic Column, *J. Chromatogr. Sci.*, **13**, 18–25.

Janson, J.-C. (1982). Scaling-up of affinity chromatography, technological and economical aspects. *Affinity Chromatography and Related Techniques*, pp. 503–512. Wiley, New York.

Korpi, Jack (10/83, 4/82, 12/81), *Gradient Elution in HPLC*. Waters Associates, Milford, MA.

Martin, M., C. Eon and G. Guiochon (1974). Study of the pertinency of pressure in liquid chromatography. I. Theoretical analysis. *J. Chromatogr.* 3, **99**, 357–376.

Martin, A. J. P. and R. L. M. Synge (1941) *Chem. Eng. Sci.*, 1358–1368.

Mohanraj, S. and W. Herz (1981). Use of the peak shaving-recycle technique for separation of labdadiene and labdatriene isomers by HPLC. *J. Liq. Chromatogr.*, **4**(3), 525–532.

Nettleton, D. E., Jr. (1981). Preparative liquid chromatography. I. Approaches utilizing highly compressed beds. *J. Liq. Chromatogr.*, **4** (Suppl 1), 141–173.

Sheh, C. K., C. N. Snavely, T. E. Molnar, J. L. Meyer, W. B. Caldwell and E. L. Paul (1983). Large scale liquid chromatography system, *Chem. Eng. Prog.*, **6**, 138–139.

Timmins, R. S., L. Mir and J. M. Ryan (1969). Large scale chromatography: a new separation tool. *Chem. Eng.*, May 19, 170–178

Van Deemter, J. J., F. J. Zuiderweg and A. Klinkenberg (1956). Longitudinal diffusion and resistance to mass transfer as causes of nonideality in chromatography. *Chem. Eng. Sci.*, **5**, 271–289.

35

Recovery of Biological Products by Distillation

G. D. MOON, JR
Raphael Katzen Associates International Inc., Cincinnati, OH, USA

35.1 INTRODUCTION

Distillation is a unit operation which has been commercially practiced for many years for the separation and purification of fermentation-produced substances. The products of fermentation can be manifold, depending upon the feed substrate and the conditions under which the fermentation operation is conducted, *e.g.* temperature, pH, fermenting agent, *etc.* The feedstock material is most often biomass or biomass related.

Of greatest long-term commercial importance has been the production of ethanol (ethyl alcohol) from biomass feedstocks containing starch or sugar. The potable alcohol industry (whiskey production) depends upon corn or other feed grains containing starch for feedstock. The production of rum utilizes molasses, a byproduct of the sugar industry, as feedstock. Biomass feedstocks were used for many years for the production of commercial-grade (solvent and chemical intermediate use) ethanol.

With the availability of cheap crude oil and with advances in petroleum chemistry and catalysis in the years following World War II, the production of commercial-grade ethanol by the catalytic hydration of petroleum-derived ethylene became predominant. Synthetically produced ethanol replaced essentially all of the existing market for fermentation alcohol except that devoted to potable alcohol production.

In 1973 the OPEC nations, by unprecedented actions, triggered a world energy crisis. The industrialized nations, including the United States, Western Europe and Japan, were suddenly faced with petroleum shortages caused by the OPEC embargo and a degree of economic destabilization due to the burgeoning cost of energy as it reflected upon petroleum-based economics.

This situation led to a number of remedial actions including self-imposed conservation measures and a search for means of replacing expensive, imported crude oil with cheaper feedstock materials derived from renewable resources. There resulted a reawakened emphasis on the important part that might be played by biomass-derived products, *e.g.* ethanol produced from abundant grain resources as a liquid fuel extender and octane enhancer.

A great deal of time, money and effort has been expended in research leading to the effective commercialization of processes that will convert the cellulosic fraction of biomass materials to simple sugars. Simple sugars, *e.g.* glucose, can then serve as substrate for the production of ethanol by fermentation. Though some of this effort has had a degree of success, no economically viable process has yet been successfully demonstrated at the commercial level.

Other efforts have intensified upon improving existing technology. These have been directed at such things as improving the starch content of abundant biomass feedstocks, fermentation kinetics, thermal stability of yeasts used for fermentation, and overall energy efficiency of the process for producing anhydrous ethanol for fuel use.

35.2 ENERGY REDUCTION

Of the myriad efforts that have been expended to the present time, the greatest success has been achieved in reducing the total energy required to produce motor fuel grade anhydrous ethanol from grain feedstocks. The area of the overall alcohol production process in which the greatest energy reduction has been achieved has been distillation.

There are two primary motivations for reducing energy requirements. These are: (1) to reduce the operating cost component associated with energy required to produce fuel grade ethanol; and (2) to improve the overall energy effectiveness of the fuel alcohol production process.

The overall energy effectiveness of the process is defined as the ratio of the available thermal energy per unit of alcohol product compared to the total thermal energy equivalent required to produce that unit quantity of alcohol product. It is desirable to make this ratio as large as possible. Values in excess of 1.0 are most desirable since they indicate a favorable energy balance, *i.e.* the fuel produced contains energy in excess of the total energy required to produce the fuel. Favorable overall energy effectiveness ratios are being achieved in current commercial practice.

35.3 MOTOR FUEL GRADE ETHANOL

The motor fuel grade ethanol industry that has developed in the United States during the decade following the oil crisis of 1973 bears a great resemblance to the fermentation alcohol industry as it has been practiced for many years. Fundamental processing steps in the overall fuel ethanol production process remain similar, though considerably improved in efficiency. Most motor fuel grade ethanol plants employ grain feedstocks (corn, milo, barley, wheat) and are dry-milling operations. Corn is the predominant feedstock. Typical grain feedstock compositions and ethanol yields are shown in Table 1. A few plants employing molasses have been built. Several wet-milling plants have been built but only in situations where the fuel ethanol plant is subordinate to a wet-milling operation. Wet-milling, in contrast to dry-milling, involves initial removal of some valuable grain byproducts such as corn oil, corn gluten and some starch before the balance of the feedstock enters the fuel ethanol process. Table 2 lists some of the properties of ethanol.

Table 1   Grain Composition and Ethanol Yield Data

Component	Corn[a]	Milo[a]	Wheat[a]	Barley[a]
Starch	72	70	64	67.1
Protein	10	10	14	14.2
Oil, fiber, ash	18	20	22	18.7
Total	100	100	100	100
Moisture (weight percent)	15.5	15.5	14	10.6
Bushel weight (kg)	25.4	25.4	27.2	21.8
Starch weight (kg/bushel)	15.5	15.0	15.0	13.1
Theoretical ethanol yield (l/bushel)	11.1	10.7	10.7	9.3
Practical ethanol yield (l/bushel)	9.5	9.2	9.2	8.0

[a] Weight percent, dry basis.

A typical grain dry-milling plant, producing motor fuel grade ethanol, will contain the following fundamental processing sections (Katzen, 1978): (1) grain storage and preparation; (2) mash preparation and cooking; (3) mash saccharification; (4) fermentation; (5) distillation; (6) stillage separation; (7) thin stillage concentration by evaporation; and (8) drying of concentrated stillage solids.

Table 2 Properties of Ethanol

Molecular formula	C_2H_5OH
Molecular weight	46.07
Specific gravity (at 15.6 °C/15.6 °C)	0.7937
Heat of combustion (kJ kg^{-1})	29 836
Boiling point (at 101.3 kPa; °C)	78.32
Latent heat of vaporization at normal boiling point (kJ kg^{-1})	838.4
Vapor heat capacity (at 25 °C; kJ kg^{-1} °C^{-1})	1.60
Liquid heat capacity (at 25 °C; kJ kg^{-1} °C^{-1})	2.42

Two ancillary operations may be carried out in some plants. These are enzyme (fungal amylase) preparation and yeast (*Saccharomyces cerevisiae*) propagation. The fungal enzymes are used in the mash cooking and saccharification steps. Yeast is employed to accomplish fermentation. Many of the smaller fuel alcohol plants (less than 40×10^{12} l per year) find it more economical to purchase enzymes and yeast from commercial sources and will only require mixing equipment to add these materials at the appropriate point in the process.

35.4 THE DISTILLATION PROCESS

The purpose of this chapter will be to discuss step (5), distillation (Henley and Seader, 1981; McCabe and Thiele, 1925; Smith, 1963).

From a thermodynamic standpoint, the ethanol–water system is interesting. These components form a non-ideal system. Water and ethanol are miscible in all proportions. The system forms a homogeneous, binary, minimum boiling (78.15 °C under atmospheric pressure) azeotrope containing 89.43 mole percent ethanol. As the system operating pressure is reduced, the azeotrope composition approaches pure ethanol. The azeotrope finally disappears under a system operating pressure of 6.7 kPa (Perry, 1941). It is the non-ideal nature of the system which complicates the design of a distillation system to produce anhydrous ethanol from dilute ethanol–water feed mixtures produced in fermentation.

Homogeneous binary azeotropic systems cannot be separated into pure components in a single ordinary distillation tower at normal operating pressures. Only the azeotrope composition and one pure component will result from such an operation. To achieve a complete separation into the pure components, some other means must be used. In the case of ethanol and water, the means most commonly employed is to add a third solvent component to effectively bypass the binary azeotropic composition in an additional distillation step known as azeotropic distillation. In the azeotropic distillation step a ternary minimum boiling azeotrope forms. This ternary azeotrope has an atmospheric boiling temperature (64.9 °C) considerably below the boiling temperature of the atmospheric binary ethanol–water azeotrope (78.15 °C). The ternary azeotrope, upon condensation, forms two phases which can be decanted and separated. This step, in effect, bypasses the binary azeotrope and allows anhydrous ethanol to be produced as one of the terminal products of the azeotropic distillation tower operation.

A typical flowsheet for a distillation system producing 190×10^{12} l per year of motor fuel grade alcohol from corn is shown in Figure 1. Reference to this schematic flow diagram will aid in understanding the following discussion (Katzen, 1980).

Fermented beer is delivered to the distillation system from a 'beer well'. The term 'beer' refers to any stream in which ethanol has been produced by a fermentation process. The beer well is a large accumulator tank normally holding about two days' production of fermented beer. Feed beer will normally contain 8 to 12 volume per cent of ethanol on a solids-free basis; the balance is water. The feed will also contain from 7 to 10 weight percent total solids. These solids are both dissolved and suspended, and the ratio of dissolved to suspended will depend upon the feedstock used. In a typical corn dry-milling operation, the ratio of dissolved to suspended solids is about 0.67. The total solids in the beer feed represents components of the feedstock which are non-fermentable such as oil, fiber, protein material, ash and byproduct yeast propagated in fermentation. A small amount of dissolved carbon dioxide, another fermentation byproduct, will be present in the feed. In addition, the beer feed will contain small quantities of extraneous byproducts of the fermentation operation. These include light ends such as acetaldehyde, and fusel oils (mixtures of higher alcohols such as butanol, pentanol and isopentyl alcohol). There may also be

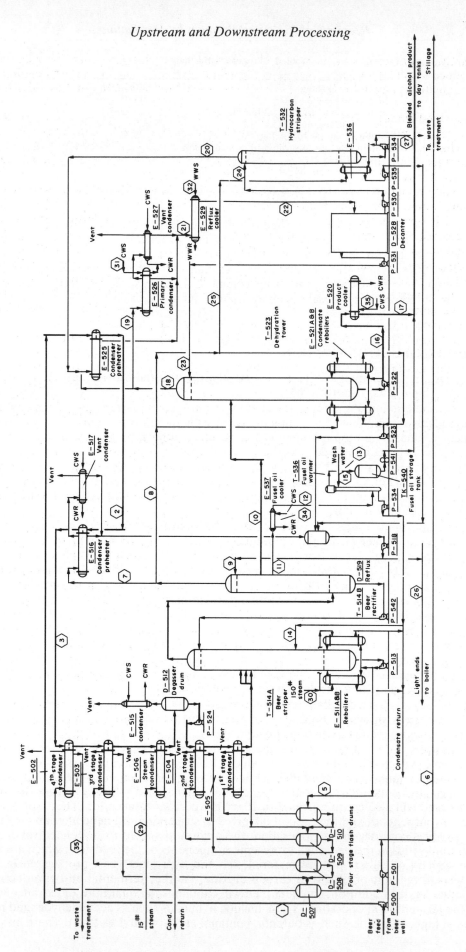

Figure 1 Motor fuel grade anhydrous alcohol system

trace quantities of materials such as glycerol and acetic acid. Grain fermentation operations normally produce about 0.08 kg of yeast per kg of alcohol produced and 4 to 5 l of fusel oil for every 1000 l of anhydrous alcohol produced. Light ends production will be about 1 kg of alcohol produced.

The beer feed from the beer well is pumped by P-500 through a series of heat exchangers in which the feed is preheated to its saturation temperature before the feed enters beer stripper T-514A. The beer feed flows, in succession, through heat exchanger E-525, condenser–preheater E-516, another condenser–preheater E-502, a condenser E-503 and process steam condenser E-506. Partially preheated feed next passes to flash degassing drum D-512. In this operation the carbon dioxide remaining dissolved in the feed following fermentation is flashed-off by reducing the pressure. Some alcohol and water vapor are also removed in the flashing operation. Degasser drum D-512 is surmounted by condenser E-515 which employs cooling water to condense remnants of alcohol and water vapor contained in the carbon dioxide flash vapor. The condensed alcohol and water vapor are returned to the system and carbon dioxide is vented to the atmosphere. The carbon dioxide removal step is necessary to guard against inert gas blanketing that might occur on heat exchange surfaces further along in the separation system.

The partially heated beer feed is pumped by P-524 from the bottom of degasser drum D-512 through two more heat exchangers in series, condensers E-504 and E-505. The beer, at nearly saturation temperature, enters the central portion of beer stripper T-514A.

In tower T-514A the beer feed is partially concentrated to produce an overhead vapor containing about 50 volume percent of ethanol (100° proof ethanol). A term peculiar to the ethanol industry is degrees proof. The 'proof' represents twice the volume percent of contained ethanol. The bottoms stream from beer stripper T-514 contains no more than 0.02 weight percent ethanol (solids-free basis), the balance is water. All of the non-fermentable grain solids present in the original feed are also contained in the T-514A bottoms. This bottoms material with its solids is referred to as 'stillage'.

The hot stillage contains a lot of useful thermal energy. This energy is recovered by subjecting the hot stillage to four successive flash stages in series. The stillage passes, in order, through reduced pressure staging in flash drums D-510, D-509, D-508 and D-507. In each of these flash stages vapor is generated to be passed to condensers E-505, E-504, E-503 and E-502 where the flash vapor (steam) is condensed for the purpose of preheating the beer feed. This preheating step has been referred to earlier in this discussion. Steam condensate from each of these condensers is collected and passed to the next successive flash stage. The stillage finally leaves flash drum D-507 at approximately atmospheric pressure and 100 °C. From there, it is pumped by P-501 to further processing for recovery of the high protein solids which it contains.

The overhead vapor generated in tower T-514A flows to the base of tower T-514B, the beer rectifier. Towers T-514B and T-514A may be thought of as a single tower. Because of the overall height requirement (in excess of 30 m), these towers are physically separated. However, the combination constitutes a single–rectifying operation.

In tower T-514B, the beer rectifier, the alcohol concentration is increased, and the vapors leave the top of T-514B at an alcohol concentration of 95 volume percent (190° proof) or greater.

One of the keys to reducing the energy required for alcohol separation and purification involves the overhead vapors leaving T-514B. These vapors are maintained under a pressure greater than atmospheric pressure. In fact, they may be under a pressure as great as 790 kPa. Normally, tower combination T-514A and T-514B is designed to operate under a head (top) pressure of 446 kPa. This means that the saturation temperature of these vapors (121 °C) can be used to generate additional vapor by condensing them against materials boiling at lower temperatures of around 93 °C. This condensation step is the function that these vapors perform in the final ethanol dehydration.

In the beer stripper–rectifier combination, there must be a source of energy used to generate the vapors developed in this tower combination. That source of energy is process steam at 1136 kPa (185.5 °C). This process steam comes from the plant boiler, or, more usually, has already been used to give up some of its energy in another plant operation. For example, the process steam may be exhaust steam from a back-pressure turbine used to supply shaft work at some other point in the overall alcohol production process.

In any case, the process steam gives up its energy by condensing in parallel flow, forced circulation reboilers E-511A and B located at the base of beer stripper T-514A. The tube-side fluid in the forced circulation reboilers is recycling stillage from which vapors are generated in the base section of tower T-514A. The process steam used as heat source in E-511A and B represents the only energy input required for the entire distillation system. In this typical 190×10^{12} l per year plant, the process steam usage amounts to about 54 000 kg h^{-1} (2.25 kg process steam per liter of anhydrous product alcohol).

Tower combination T-514A and T-514B serves to accomplish two other separation functions. The upper five trays in tower T-514B are used to concentrate light ends (acetaldehyde). These five trays constitute the so-called pasteurizing section. A purge stream taken from the reflux line returning to the top tray of T-514B serves to remove these light ends along with a small amount of alcohol which is lost from the system. Removal of these light ends is necessary since they have a very high volatility. If blended with motor fuel grade alcohol into gasoline, these light end materials could cause serious vapor lock problems, particularly in warm climates. To conserve energy, the light ends purge stream can have its thermal energy content recovered by being burned in the plant boiler.

The other separation function accomplished in tower T-514B involves the concentration and removal of the fusel oils contained in the beer feed. Fortunately, these fusel oil materials (higher alcohols) are more volatile than ethanol when high concentrations of water are present and less volatile than ethanol when low concentrations of water are present. For this reason, these fusel oil materials tend to concentrate on the tower trays located somewhere in the upper portion of the rectifier tower T-514B. In the start up of a system, this location must be determined by a trial-and-error procedure, which is accomplished by providing for multiple liquid side draws spread over a number of trays in the rectifying section. Usually the fusel oils will concentrate at a point in the tower where the tray liquid composition approaches 60 percent ethanol (120° proof).

A continuous liquid draw is made at the fusel oil concentration point. The hot side-draw stream is cooled in fusel oil cooler E-537 and then passed to the base of fusel oil washer T-538. A cold water stream enters the top of the fusel oil washer and countercurrently contacts the cooled fusel oil side-draw stream. In this operation, the ethanol present in the side-draw is extracted into the heavier water phase. The separated fusel oils accumulate as a light layer which can be continuously removed from the top section of the fusel oil washer to be sent to fusel oil storage tank TK-540. The water wash stream containing extracted ethanol is returned to the lower part of beer stripper T-514A for additional alcohol recovery.

If these fusel oil materials had not been removed continuously from tower T-514B, they could have concentrated to such an extent that two liquid phases would separate on the tray corresponding to the fusel oil concentration point. The hydrodynamic stability of the tower operation would be disturbed by such an occurrence, and continuous steady-state operation could not be maintained.

In practice, the separated, washed fusel oil stream is blended with the anhydrous motor fuel grade alcohol product. The fusel oil materials possess a higher unit energy content than ethanol. These, therefore, serve to add fuel value. In addition, these materials tend to increase the water tolerance of gasoline when it is blended with motor fuel grade ethanol.

The mechanical design description for the combination beer stripper–rectifier system can be summarized as follows. Because of the high solids content of the beer feed, the type of contacting stage utilized in the beer stripper portion of the combination tower is important. Ordinary perforated trays have been found to be unsatisfactory due to plugging and solids build-up. The classic bubble-cap tray would be even more unsatisfactory for the same reason. With ordinary trays, maintenance time becomes excessive and turn-around time may become prohibitive. Thus, the ideal type of contactor would be one with reasonable separation efficiency, with solids purging capability, and with relative cleaning ease when maintenance is required. Basically, the type of contactor is a disc-and-donut arrangement. Properly designed, these may be used in service for much longer periods of time than can be realized with such devices as perforated trays. Disc-and-donut contactors have low pressure drop per separation stage and yield a fairly low separation efficiency per stage. This means that more disc-and-donut stages will be required to achieve a given degree of separation than would be the case for perforated trays. However, this does not lead to serious economic consequences since a given separation can be achieved in about the same tower shell height with many actual disc-and-donut trays or the fewer required number of perforated trays.

Because of the possible presence of acetic acid, an extraneous byproduct of the fermentation operation, it is advisable that an acid-resistant stainless steel be employed throughout towers T-514A and T-514B and for pump and heat exchanger components that come in contact with the beer feed or the stillage. Occasionally, carbon steel can be employed with a proper corrosion allowance. If it is possible to insure that all acids present in the beer feed or in subsequent process streams can be satisfactorily neutralized with caustic, then carbon steel components can be specified.

In the 190×10^{12} l per year production facility example, tower T-514A is 3.5 m in diameter and contains 80 disc-and-donut trays. Tower T-514B contains 28 perforated trays and is 2.6 m in diameter.

In Figure 1 the partially concentrated ethanol (190° proof) resulting from the initial distillation step is removed as a liquid side-draw stream from a tray at the bottom of the pasteurizing section atop tower T-514B. This stream becomes the feed stream for dehydration tower T-523.

Dehydration tower T-523 operates at a head pressure corresponding to atmospheric pressure. Tower construction can be of carbon steel since no acids are present in the partially concentrated alcohol feed. Corresponding to a production rate of 190×10^{12} l per year, this tower will be 3.5 m in diameter and will contain 50 to 60 conventional perforated trays. Overall tower height will be in excess of 30 m.

The thermal energy required to generate the necessary vapor traffic in the dehydration tower is supplied by condensing a portion of the alcohol-rich overhead vapor developed in the pressure rectifier tower T-514B. This vapor is condensed on the shell side of parallel condensate reboilers E-521A and B. These reboilers may be of the thermosiphon variety or can be forced circulation and can be of carbon steel construction.

The overhead vapor produced in the dehydration tower will have a composition approaching the atmospheric ternary minimum boiling azeotrope formed by the system ethanol–water–azeotropic solvent. The degree of approach to the azeotropic composition will depend upon the number of trays and reflux ratio used in the dehydration tower. The classic azeotropic solvent to achieve ethanol dehydration is benzene. In fact, it is still the solvent which is most widely used even though it has come under scrutiny because of its carcinogenic properties. Other azeotropic solvents which have been employed include cyclohexane and heptane. Pentane (Black, 1980, 1982) has been suggested as an economically useful azeotropic solvent. Of these other solvents, only cyclohexane has reached any significant commercial usage. Another solvent that has found some usage is diethyl ether.

Tight, well-designed systems employing benzene should pose no real health hazard provided system operation and maintenance receive proper attention. The azeotropic solvent is in total recycle within the anhydrous system.

The bottoms product from dehydration tower T-523 is the desired anhydrous motor fuel grade alcohol. Actually, this bottom product, though called anhydrous, contains about 99.5 volume percent ethanol with the balance being water. This material is suitable for blending with gasoline, and the contained water in the motor fuel grade alcohol product will not give water separation problems in any but the most severely cold climates encountered in the United States. The hot, anhydrous motor fuel grade alcohol product is pumped by P-522 through product cooler E-520 where its temperature is reduced to 37.8 °C. Previously separated fusel oils are continuously blended with the anhydrous motor fuel grade alcohol, and this mixture passes to off-site day tanks where the product may be checked for purity and production quantity.

The overhead vapor from the dehydration tower is divided; a portion is joined by vapor generated in hydrocarbon stripper T-532 to be condensed in condenser preheater E-525. This condenser preheater represents the first stage of beer feed preheating referred to earlier. The balance of the dehydrating tower overhead vapor is condensed in primary condenser E-526, with final remnants of vapor condensed in vent condenser E-527. Adequate condensing surface must be provided in the vent condenser to insure that all benzene and ethanol vapors are condensed to avoid atmospheric discharge.

The condensates developed in E-525, E-526 and E-527 are combined before passing to reflux cooler E-529. In the reflux cooler, the combined condensates are subcooled to about 90 °F to aid in phase separation in decanter D-528.

In decanter D-528, phase separation occurs into a benzene-rich upper layer containing alcohol and some water and a lower layer containing principally water, some alcohol and some benzene.

The layers are continuously decanted with the upper benzene-rich layer being returned by pump P-531, to serve as the reflux stream to the top tray of anhydrous tower T-523. The lower water-rich layer from the decanter is pumped by P-530 to serve as feed entering the top tray of hydrocarbon stripper T-532.

In the hydrocarbon stripper an overhead vapor is developed which will approach the equilibrium vapor composition corresponding to liquid component concentrations present in the feed stream to the hydrocarbon stripper. The last remnants of benzene and alcohol contained in the stripper feed are removed in the overhead. The bottoms product from the hydrocarbon stripper is water. The water removed in this bottoms stream represents the total water entering with the dehydration tower feed less any small amount of water remaining in the anhydrous motor fuel grade alcohol product. The water leaving the base of the hydrocarbon stripper is passed to an off-site waste treatment plant to insure removal of any hazardous material which it might contain.

Energy required to generate vapor traffic necessary for the operation of the hydrocarbon stripper is supplied by an unused fraction of the overhead vapor generated in beer rectifier T-514B. This alcohol-rich vapor fraction is condensed on the shell-side of reboiler E-536. Reboiler E-536 may be of the thermosiphon variety, or it can be designed as a forced circulation unit.

35.4.1 Advantages of the Distillation System

The distillation system for producing anhydrous motor fuel grade alcohol which has been described shows a number of advantages when compared to traditional commercial distillation systems designed according to older technology. Foremost of these advantages is the low energy usage required.

When energy was cheap, little attention was paid to energy recovery and reuse. Distillation system designs were prepared so that system fixed investment was minimized. As a result, the energy usage for obtaining anhydrous ethanol amounted to as much as 4.2 kg of process steam per liter of anhydrous alcohol produced. The system which has been used for illustration requires only 2.25 kg of process steam per liter of anhydrous alcohol produced, a reduction in energy usage amounting to 46%. While it is true that additional heat transfer equipment is necessary to accomplish this saving, the added equipment cost is quickly recovered in the energy saving realized.

Early ethanol distillation systems were mechanically crude compared to those that can be fabricated at present. Corrosion and scaling were always a problem with the tray designs and materials of construction then available. Tray designs which were available could not obtain the separation efficiency currently realized. Equipment maintenance was time consuming and difficult to accomplish.

The use of a pressure cascaded system to create the temperature differential required for heat transfer, and hence to allow for energy reuse, has been commercially demonstrated. The reduction in energy usage needed for ethanol separation and purification contributes significantly to improving the energy effectiveness for the process of converting biomass materials to useful liquid fuels or fuel components.

35.4.2 Investment Cost of the Distillation System

The pressure-cascaded, low energy usage distillation system producing 190×10^{12} l per year of motor fuel grade ethanol from corn, which has been used for illustration, requires a total fixed investment of $\$10 \times 10^{12}$ (1983 US dollars). This figure represents plant-erected cost for the distillation system and includes provision for buildings and structures but does not include land cost.

In general, equipment larger than 3.7 m in diameter cannot be shop fabricated and shipped to site. For this reason, the 190×10^{12} l per year distillation facility represents the upper limit capacity for a shop fabricated, single train facility.

About two years would be required for engineering design and construction of the facility. An additional two months should be allowed for equipment checking and startup.

35.4.3 Operating Factors for the Distillation System

Utility requirements for the 190×10^{12} l per year motor fuel grade ethanol distillation system are as follows: service factor, 0.904; process steam (1136 kPa, 185.5 °C), 54 000 kg h^{-1}; low pressure flash steam (205 kPa, 120.5 °C), 13 100 kg h^{-1}; electric power, 333 kW; process water, 2270 kg h^{-1}; recycle cooking tower water (29.4 °C), 925 000 kg h^{-1}; refrigerated cooling water (10 °C), 284 000 kg h^{-1}.

The labor requirement for around-the-clock operation of the distillation system consists of one operator, one laborer, a half-time laboratory technician and a half-time equipment maintenance person.

35.5 OTHER PROPOSED SYSTEMS

A number of other useful distillation system configurations have been proposed. All of these are aimed at reducing the process energy required to remove the high water content inherent with beer feeds developed in ethanol fermentation.

Among the energy saving methods proposed is that of compression of the overhead vapors from an atmospheric beer stripper–rectifier (Null, 1976). By increasing the temperature and energy content of the overhead vapor sufficiently, these vapors may be used to supply the energy required for the reboiler of the same tower in which they were generated, or they could be used to supply energy for the azeotropic tower or the hydrocarbon stripper tower. In principle, this method would be satisfactory; in practice, the method is limited by the modest compression ratio that compressor manufacturers are willing to use in vapor recompression equipment. At the low compression ratios realized in such units, the temperature difference required for economic heat transfer is not available. Furthermore, there are limitations on the allowable upper limit for vapor temperature leaving a compressor unit which are dictated by materials of construction which are available. In addition, the compression of volatile, flammable substances, such as high concentration ethanol vapor, must always be exercised with great care and attention to safety.

Methods employing vacuum stripping of fermentation beer have been suggested. The purpose of this is to allow for continuous removal of ethanol from fermenting mashes to overcome the fermentation limiting action that ethanol has upon yeast. Vacuum is employed to lower the distillation temperature to a level that can be tolerated by yeast. In a theoretical sense, this method has merit. In fact, at a sufficiently high vacuum (6.7 kPa absolute), the homogeneous binary azeotrope of ethanol and water disappears. Pure anhydrous ethanol could be obtained without resort to azeotropic distillation. The use of vacuum separation has a serious practical limitation due to the size of the distillation equipment required for handling a given vapor throughput. For example, if one halves the absolute operating pressure under which a distillation system is to be operated and a given vapor throughput is desired, the lower pressure tower diameter will have to be 1.4 times the diameter of a tower operated at the higher pressure.

Another useful distillation system configuration that has been proposed involves operating the azeotropic tower under superatmospheric pressure (Messick *et al.*, 1983). The pressurized overhead vapors generated in the azeotropic tower may then be used as the source of energy for a portion of the energy required by the atmospheric stripper–rectifier tower. The energy effectiveness of this configuration will not be as great as that for the system used for illustration. However, grain solids present in the beer feed to the atmospheric stripper–rectifier will be subjected to considerably lower temperatures than occur when the stripper–rectifier unit is operated under pressure. This can have an effect upon the ultimate quality of the protein portion of the grain solids recovered as byproduct.

35.6 SUMMARY

This chapter has focused upon current commercial methods used for the separation and purification of anhydrous motor fuel grade ethanol by distillation; motivation for improving the energy efficiency ratio when using biomass feedstocks to produce liquid fuel substitutes is clearly economic because of the burgeoning cost of energy. A low energy usage distillation system has been used for illustration. System size parameters have been given for a typical system capable of producing 190×10^{12} l per year of anhydrous motor fuel grade ethanol. Basic energy usage, installed cost data and operating factors associated with the distillation system have been stated.

Other proposed distillation system modifications aimed at accomplishing energy reduction have been mentioned.

Though the only biomass-produced chemical (fuel) of current commercial importance is ethanol, there is an ongoing intensification of research effort aimed at producing other biomass-based chemical intermediate materials. These will assume growing importance as non-renewable energy sources become less available, energy costs rise, and the impetus for discovery of substitute energy resources must be found. Many of these will be biomass based and will assume commercial importance when sufficient economic motivation becomes mandated by necessity.

35.7 REFERENCES

Black, C. (1980). Distillation modeling of ethanol recovery and dehydration process for ethanol and gasohol. *Chem. Eng. Prog.*, September, 78.

Black, C. (1982). Simulation applied to special industrial problems. *Chem. Eng. Prog.*, December, 53.

Henley, E. J. and J. D. Seader (1981). *Equilibrium-Stage Separation Operations in Chemical Engineering*. Wiley, New York.

Raphael Katzen Associates (1978). *Grain Motor Fuel Alcohol, Technical and Economic Assessment Study, DOE Contract No. EJ-78-C-01-6639*.

Raphael Katzen Associates (1980). Distillation system for motor fuel grade anhydrous alcohol. *US Pat.* 4 217 178.

McCabe, W. L. and E. W. Thiele (1925). *Ind. Eng. Chem.*, **17**, 605.

Messick, J. R. *et al.* (1983). Anhydrous ethanol distillation method and apparatus. *US Pat.* 4 422 903.

Null, H. R. (1976). Heat pumps in distillation. *Chem. Eng. Prog.*, July, 58.

Perry, J. H. (1941). *Chemical Engineers' Handbook*, 2nd edn., p. 1373. McGraw-Hill, New York.

Smith, B. D. (1963). *Design of Equilibrium Stage Processes*, chapters 5, 6 and 11. McGraw-Hill, New York.

36
Supercritical Fluid Extraction

R. C. WILLSON
Massachusetts Institute of Technology, Cambridge, MA, USA

36.1 INTRODUCTION

The design of biochemical and pharmaceutical separation processes is often constrained by the heat-sensitivity of the products, organisms or enzymes involved and the widespread occurrence of dilute aqueous product streams requiring large energy expenditures for processing. In addition, many biochemical products are intended for use in foods and pharmaceuticals and are therefore subject to regulations which may limit the methods used in their production and recovery.

Use of supercritical or near-critical solvents, which are compounds with critical temperature and pressure near the operating conditions of the process, has been suggested for application to chromatography (Gere, 1983; Giddings *et al.*, 1968), petroleum processing, toxic waste destruction (Josephson, 1982) and a wide variety of other processes reviewed elsewhere (Irani and Funk, 1977; Paulaitis *et al.*, 1980; Williams, 1981). In addition, supercritical fluid extraction is of interest to the biological process industries. This interest stems from the prospect of achieving substantial energy savings while conducting separations at low temperatures using physiologically inert solvents.

36.2   WHAT IS A SUPERCRITICAL FLUID?

36.2.1   Thermodynamic Background

The normal boiling point of a pure compound is the temperature at which liquid and vapor forms of that compound coexist at a pressure of 1 atm. As the pressure varies upward or downward from 1 atm, the boiling temperature increases or decreases respectively. The locus of boiling points is called the boiling line (Figure 1). For common substances, the liquid and vapor phases which coexist at a pressure of 1 atm differ substantially in physical properties such as density, viscosity, *etc*. At higher pressures, however, the coexisting phases become gradually less different. Figure 1 illustrates this trend in the case of the density of carbon dioxide. At 10 °C the densities of liquid and gaseous carbon dioxide are 0.86 g ml^{-1} and 0.14 g ml^{-1} respectively. At 20 °C, the figures are 0.77 g ml^{-1} and 0.19 g ml^{-1} and at 30 °C the liquid and gas densities are 0.59 and 0.35. At some temperature and pressure, the 'vapor' and 'liquid' phases become identical. The point at which the two-phase system collapses to a single phase is called the critical point. The critical point constants of a given substance are its critical temperature (31.1 °C for carbon dioxide), critical pressure (73 atm or slightly over 1000 p.s.i. for carbon dioxide) and critical density (0.47 g ml^{-1} for carbon dioxide, which is comparable to the densities of conventional liquid solvents). At temperatures and pressures above their critical values, the substance exists in a form with properties intermediate between those of a liquid and those of a gas. This phase is known as a supercritical fluid.

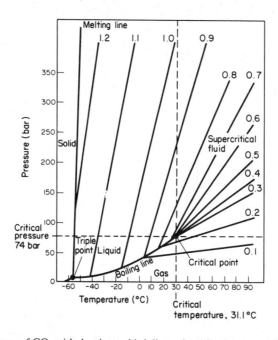

Figure 1   Phase diagram of CO_2 with density as third dimension; densities given from 0.1 to 1.2 g ml^{-1}

36.2.2   Properties of Supercritical Fluids

As shown in Table 1, the supercritical fluid state is reached by a variety of common substances at temperatures which are ideal for the processing of many biological products and at pressures which, though high, are within the range of normal industrial practice. The physical and transport properties of supercritical fluids generally lie between those of liquids and gases and are generally favorable for use as extraction solvents. In particular, the viscosities of supercritical fluids are substantially below the range of values for liquids, leading to the facilitation of diffusional transport and reduced energy requirements for pumping.

Table 1 Possible Supercritical Solvents for Biochemical Products

Compound	Critical temperature (°C)	Critical pressure (atm)
Carbon dioxide	31.1	73
Ethane	32.3	48
Ethylene	9.5	49
Nitrous oxide	36.5	70
Trifluoromethane	25.9	46

36.2.3 Solvent Regeneration

The unique property of supercritical fluids which makes them of industrial interest is the strong dependence of their densities and solvent properties on temperature and pressure in the vicinity of the critical point. The practical importance of this phenomenon is that it can reduce the process of solvent regeneration (recovery of the extracted product from the extraction solvent) to a small temperature or pressure shift. This effect can be used to reduce the energy needed to carry out the overall extraction process and also allows the entire process to be carried out at ambient temperatures because no secondary distillation is required to separate product and solvent.

36.3 SOLVENT PROPERTIES OF SUPERCRITICAL FLUIDS

36.3.1 General Trends

The solubility of most substances in supercritical fluids is generally lower than their solubilities in conventional liquid solvents. Observed solubilities, however, range from negligible to at least 30 wt %. Supercritical fluids are generally non-polar solvents and so are better solvents for non-polar compounds than for polar compounds (Brogle, 1982; Stahl and Quirin, 1983).

36.3.2 Ionized Compounds

Ionized compounds are essentially insoluble in supercritical fluids. This includes compounds possessing dissociated acid groups or protonated basic groups, as well as inorganic salts. An example of the effect of ionized groups on solubility in supercritical fluids is the negligible extraction of light carboxylic acids from aqueous solutions at pH values above their pK_a values as contrasted with their solubility in supercritical solvents in the unionized form. Other compounds which possess ionized groups include amino acids, proteins and nucleic acids and these materials are poorly dissolved by supercritical fluids.

36.3.3 Polar Compounds

Unionized but highly polar compounds are also less soluble in supercritical fluids than are less polar compounds of comparable size. In general, the addition of hydroxyl, amino or carboxyl groups to a compound produces a compound of substantially lower solubility in supercritical fluids, while carbonyl groups have a somewhat smaller effect on solubility. Detailed treatments of the effect of molecular structure on solubility in supercritical fluids have been given in works by Stahl *et al.* (1978) and Stahl and Quirin (1983) which also describe successful extractions of a variety of alkaloids and other biochemicals. Examples of compounds which are generally too rich in polar substituents to be soluble in supercritical solvents include polysaccharides and sugars.

36.3.4 Easily Extracted Compounds

Hydrocarbons and lipid-soluble organic compounds of moderate polarity are, in general, readily soluble in supercritical fluids. Such compounds include paraffins, many esters, ethers and lactones, and glycerides (Williams, 1981). A variety of fairly complex organic molecules, such as chemotherapeutic drugs (Krukonis *et al.*, 1979), caffeine (Studiengesellschaft Kohle, 1981), α-tocopherol (vitamin E), nicotine (Hag AG., 1974), and many steroids and alkaloids (Stahl *et al.*, 1978), have been successfully extracted with supercritical fluids. In addition, many flavor and aroma elements (Hag AG., 1973), including those of cloves, pepper, cinnamon, vanilla, nutmeg and hops (Gardner, 1982; Hag AG., 1971), have been recovered.

36.4 EXTRACTION FROM AQUEOUS SOLUTIONS

The bulk of the reported measurements of supercritical fluid solubilities have been for dry, solid materials in equilibrium with a pure supercritical fluid. Recently, however, increased attention has been given to the use of supercritical fluids to extract dissolved substances from dilute aqueous solutions such as might be produced by fermentation. Compounds of recent interest have included ethylene oxide (Bhise, 1983), light organic acids such as acetic and butyric acids (Busche *et al.*, 1982; Shimshick, 1981, 1983; Yates, 1981), and light alcohols such as (particularly) ethanol, isopropanol and butanol (Kuk and Montagna, 1981; McHugh *et al.*, 1981; Moses and deFilippi, 1983). Potential applications of supercritical fluid extraction are summarized in Table 2.

Table 2 Potential Applications of Supercritical Fluid Extraction

---

Recovery of light organic acids (Busche *et al.*, 1982) and alcohols (deFilippi and Moses, 1982)

Removal of alkanes and alcohols used as carbon sources from fermentation products (Bott, 1980)

Recovery of products such as fat-soluble vitamins (A, D, E, K), alkaloids, naphthoquinone (Ensley *et al.*, 1983) and some antibiotics

Removal of conventional solvent residues

Recovery of polyethylene glycol from aqueous two-phase systems (Kula *et al.*, 1979), or products produced using these systems

Recovery of fluorocarbons used in aqueous media (Chibata, 1974) to enhance oxygen transport

Drying of and/or lipid removal from biomass

36.5 INDUSTRIAL APPLICATIONS

36.5.1 Decaffeination

The first, and most cited, industrial application of supercritical fluid extraction is in the decaffeination of coffee, the first US patent for which was issued in 1974 (Roselius *et al.*, 1974). In the process described in this patent, the coffee beans are roasted and ground to a maximum particle diameter of about 5 mm and the coffee beans are stripped of aroma components with dry supercritical carbon dioxide, which extracts about 10–12 wt % of oil from the coffee. The aromatic oil is recovered from the carbon dioxide stream by lowering the pressure to below the critical pressure and is returned to the product at a later step. Caffeine is then extracted from the remaining coffee solids with moist supercritical carbon dioxide containing 0.5 to 1.0% water by weight. The aroma oil is then mixed either with the water-soluble substances remaining in the coffee after aroma and caffeine extraction to produce an instant coffee, or with the coffee solids to produce a decaffeinated ground coffee for brewing.

Although details of the commercial process have not been made public, it is believed that commercial coffee decaffeination has been carried out in Germany since 1978 in a manner similar to that described in US patent 4 260 639 (Studiengesellschaft Kohle, 1981). This process involves contacting wet, green coffee beans with supercritical carbon dioxide, which selectively removes caffeine from the beans while leaving the aroma constituents behind. The caffeine is removed from the solvent stream either by pressure reduction, which decreases the solvent power of the carbon dioxide, or by adsorbing the caffeine onto activated carbon, which minimizes the process energy consumption by avoiding the need for recompression of the regenerated solvent. It appears that this process, which involves processing raw coffee beans, avoids degradation of the subtle flavor and aroma elements which are produced by roasting and so can produce a superior caffeine-free product. The raw beans are extracted while swollen with water to reduce the diffusional mass transfer resistance inside the beans.

36.5.2 Hop Extraction

Another commercial process of interest is the extraction of flavor essences from hops with liquid carbon dioxide, which has been carried out at Reigate, England by Pauls and Whites International, Ltd. (Gardner, 1982). Liquid carbon dioxide has been found to be an effective solvent for many compounds (Francis, 1954; Schultz and Randell, 1970; Zosel, 1978). The advantage of its use in this process is that at ambient temperature, carbon dioxide is near its critical point, so that the properties of the liquid and vapor phases are relatively similar. An important consequence of this similarity is that the difference in specific enthalpies of the two phases, which is the amount of energy which must be added to evaporate the liquid into the vapor phase, is lower than it would be farther from the critical point. In particular, it is much lower than the value for conventional liquid solvents, which at ambient temperatures are far from their critical points.

The process, then, involves contacting pelleted milled hops with liquid carbon dioxide, which extracts the flavor constituents. The carbon dioxide is then passed to the shell side of a heat exchanger, where heat from the recompression of recycled solvent is used to evaporate carbon dioxide into the vapor phase, concentrating and precipitating the hop resins. The carbon dioxide vapor is then recompressed to a liquid state, cooled in the heat exchanger and recycled to the extractors. Additional plants for hop extraction are currently under construction.

36.5.3 Other Processes

Several processes based on supercritical extraction are believed to be under development. Among these are methods of extraction from aqueous solutions, including recovery of isopropanol from pharmaceutical processing streams (Moses and deFilippi, 1983), recovery of ethanol from dilute (typically 10% by weight) solutions (deFilippi and Moses, 1982), and production of ethylene glycol by extraction of ethylene oxide from water with supercritical carbon dioxide, followed by catalytic reaction of the ethylene oxide and carbon dioxide (Bhise, 1983). Apparatus for extraction from liquids is shown in Figure 2. This illustration, which is a diagram of pilot-scale equipment constructed by Arthur D. Little, Inc. (Moses *et al.*, 1982), includes an ambient-temperature distillation column suitable for recovery of dissolved products from a near-critical liquid. If a supercritical fluid were used as the solvent, the solvent would be flashed to a lower pressure to cause precipitation of products and then recompressed and recycled as shown in the figure.

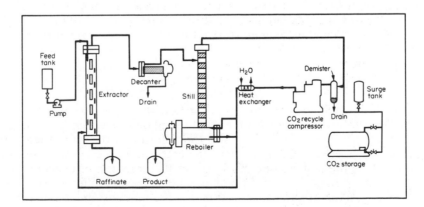

Figure 2 Pilot scale near-critical liquid extraction apparatus

36.6 ENTRAINERS AND MIXTURES

36.6.1 Background

Aside from the high capital cost of high-pressure equipment, the major economic drawback to the use of supercritical fluid extraction in the recovery of biological products is the generally low solubilities of many biochemicals in supercritical solvents, leading to the use of large solvent:feed ratios. Another potentially significant difficulty in the use of supercritical fluids is the narrow range of temperature for a given solvent in which the dependence of solvent power on pressure and temperature is very strong. While this difficulty is minimized in practice by the existence of a number of attractive potential solvents with critical temperatures in the range of 9–35 °C, it would be desirable to allow greater flexibility in the selection of operating temperatures. In particular, it would be desirable to have available a supercritical fluid extractant with a critical temperature in the vicinity of the optimal processing temperature of any system of interest.

36.6.2 Entrainers

The solubilities of many substances in supercritical fluids have been found to be increased by the addition of small quantities of a component intermediate in volatility between the supercritical fluid and the compound to be extracted (Brunner, 1983; Brunner and Peter, 1981). Such compounds are known as 'modifiers', or, more commonly, as 'entrainers' after their ability to 'entrain' the desired substance into the supercritical phase. The extraction of polar compounds may be especially improved by the addition of substances able to form hydrogen, Lewis acid–base or other intermolecular bonds with the solute. An example of the use of entrainers is given by Shimshick (1983) who describes the enhancement of the extractability of carboxylic acids with carbon dioxide from 1.0 M sodium salt solutions at 90 °C and 220 bar by addition of 12 wt % dimethyl ether to the supercritical extractant. The supercritical fluid phase concentration of acetic acid was 0.05 wt % without dimethyl ether and 0.17% in the presence of the entrainer. For propanoic acid, the figures were 0.15% without and 0.45% with the entrainer. Finally, the solubility of butanoic acid was increased from 0.80% to 1.55% by the addition of the entrainer. In addition to the effect of the entrainer, the influence of the increased size of the non-polar fraction of the molecule on extractability is evident in these figures.

Another important example of the potential usefulness of entrainers is the proposed coffee decaffeination process described above. The process involves the initial use of dry carbon dioxide as an extractant to remove flavor and aroma constituents from the coffee beans while leaving the caffeine behind. A second extraction would then be carried out in which carbon dioxide moistened with water as an entrainer would be used to remove the caffeine. The ability of water to enhance the solvent properties of supercritical fluids may also facilitate the recovery of fermentation products from aqueous solutions.

A final effect of the addition of an entrainer to a supercritical fluid extractant can be to increase the temperature dependence of the solubility of non-volatile components in the supercritical fluid. In some cases this effect might allow the use of an increase in temperature to precipitate the dissolved compounds out of the supercritical fluid (Brunner and Peter, 1981). This process could be more energy-efficient than the use of pressure cycling for solvent regeneration.

36.6.3 Solvent Mixtures

The introduction of an additional component of similar volatility to a supercritical fluid extractant can also be used to give mixtures with critical temperatures which lie between the critical temperatures of the components. This can allow the tailoring of the critical temperature of the extractant (and so the operating temperature of the extraction process) to meet other temperature constraints such as the temperature sensitivity of the organism or enzyme used in the process, heat lability of the product to be recovered, melting point of a solid to be recovered, *etc.* For biochemical processes, likely components for use in minor amounts in mixtures include propane (critical temperature, 97 °C), Frean 22 (96 °C) and ammonia (132 °C). Temperature-tailoring components may also be selected on the basis of their ability to serve as entrainers.

36.7 ECONOMICS

The economics of a supercritical-fluid extraction process, like those of any solvent extraction process, depend strongly on the solvent to feed ratio used in the process, which will in turn be directly proportional to the solvent power of the extractant. Both fixed costs, such as depreciation of capital investment, and variable costs, such as electrical power for compressor operation, are approximately proportional to the solvent flow required. The lack of detailed equilibrium data for many systems of interest, therefore, makes it difficult to evaluate the economic viability of candidate processes.

A detailed treatment of one potentially important process, the recovery of isopropanol from pharmaceutical processing streams with near-critical carbon dioxide, has been published by Arthur D. Little, Inc. (Moses and deFilippi, 1983). This study used as its design basis the recovery of 20 million gallons per year of anhydrous (98.5 wt %) alcohol from a 40 wt % aqueous solution. The use of conventional distillation was also analyzed for comparison. While the initial capital cost of the carbon dioxide extraction plant was about 40% higher than that of the distillation equipment, this cost was more than compensated for by the savings due to reduced utility consumption. The annual cost of cooling water and steam for the distillation plant was more than seven times the cost of electricity for carbon dioxide compression in the extraction plant. With all costs considered, the incremental capital cost of the extraction plant had a payback period of about six months, while the cost of a retrofit to existing equipment would be paid back in about two years.

36.8 PROSPECTS

The use of supercritical and near-critical solvents is likely to be of increasing importance in the future, both in the recovery of biological products and in a variety of other fields such as chromatography and petroleum processing. Substantial amounts of equilibrium data must be collected to support process development and design. Some other areas in which additional research and development are needed have been pointed out by King and Bott (1982). A summary of the advantages and disadvantages of this technology is presented in Table 3.

Table 3 General Characteristics of Supercritical Fluid Extraction

Advantages	Disadvantages
High energy efficiency	High pressure equipment needed
Low temperatures	Relatively low solvent power
Non-toxic, inexpensive solvents	Poor solvents for polar compounds
Low viscosity, high diffusivity	Lack of data on which to base designs
Adjustable solvent power	

36.9 REFERENCES

Bhise, V. S. (1983). Process for preparing ethylene glycol. *US Pat.* 4 400 559.
Bott, T. R. (1980). Supercritical gas extraction. *Chem. Ind.*, 15 March, p. 228.
Brogle, H. (1982). Carbon dioxide as a solvent—its properties and applications. *Carbon Dioxide in Solvent Extraction—Proceedings of SCI Food Group Food Engineering Panel Meeting, London.* SCI, London.
Brunner, G. and S. Peter (1981). Gas extraction in the presence of an entrainer. *Proceedings of Second World Conference of Chemical Engineers, Montreal, Canada*, p. 81.
Brunner, G. (1983). Selectivity of supercritical compounds and entrainers with respect to model substances. *Fluid Phase Equilib.*, **10**, 289.
Busche, R. M., E. J. Shimshick and R. A. Yates (1982). Recovery of acetic acid from dilute aqueous solution. *Biotechnol. Bioeng. Symp.*, **12**, 249.
Chibata, I. (1974). Cultivation of aerobic microbes using perfluorotributylamine. *US Pat.* 3 850 753.
deFilippi, R. P. and J. M. Moses (1982). Extraction of organics from aqueous solutions using critical-fluid carbon dioxide. *Biotechnol. Bioeng. Symp.*, **12**, 205.
Ensley, B. D., B. J. Ratzkin, T. D. Osslund, M. J. Simon, L. P. Wackett and D. T. Gibson (1983). Expression of naphthalene oxidation genes in *Escherichia coli* results in the biosynthesis of indigo. *Science*, **222**, 167.
Francis, A. W. (1954). Ternary systems of liquid carbon dioxide. *J. Phys. Chem.*, **58**, 1099.
Gardner, D. S. (1982). Industrial scale hop extraction with liquid carbon dioxide. *Carbon Dioxide in Solvent Extraction—Proceedings of SCI Food Group Food Engineering Panel Meeting, London.* SCI, London.
Gere, D. R. (1983). Supercritical fluid chromatography. *Science*, **222**, 253.

Giddings, J. C., M. N. Myers, L. McLaren and R. A. Keller (1968). High pressure gas chromatography of nonvolatile species. *Science*, **162**, 67.

Hag AG (1971). Method of producing hop extracts. *Br. Pat.* 1 388 581.

Hag AG (1973). Method of producing spice extracts with a natural composition. *Br. Pat.* 1 336 511.

Hag AG (1974). Method of extracting nicotine from tobacco. *Br. Pat.* 1 357 645.

Irani, C. A. and E. W. Funk (1977). Separations using supercritical gases. *CRC Handbook: Recent Developments in Separation Science*, vol. III, part A, p. 171. CRC Press, Boca Raton, FL.

Josephson, J. (1982). Supercritical fluids. *Environ. Sci. Technol.*, **16**, 548a.

King, M. B. and T. R. Bott (1982). Problems associated with the development of gas extraction and similar processes. *Sep. Sci. Technol.*, **17**, 119.

Krukonis, V. J., A. R. Branfman and M. G. Broome (1979). Supercritical extraction of plant materials containing chemotherapeutic drugs. *Proceedings of the 87th AIChE National Meeting, Boston, Massachusetts*. AIChE, New York.

Kuk, M. S. and J. C. Montagna (1981). Solubility of oxygenated hydrocarbons in supercritical carbon dioxide. *Proceedings of the AIChE Annual Meeting, New Orleans, Louisiana*. AIChE, New York.

Kula, M.-R., K.-H. Kroner, W. Stach, H. Hustedt, A. Durekovic and S. Grandja (1979). Process for the separation of enzymes. *US Pat.* 4 144 130.

McHugh, M. A., M. W. Mallett and J. P. Kohn (1981). High pressure fluid phase equilibria of alcohol–water–supercritical solvent mixtures. *Proceedings of the AIChE Annual Meeting, New Orleans, Louisiana*. AIChE, New York.

Moses, J. M. and R. P. deFilippi (1983). Critical fluid extraction processes for high-volume organic chemicals. *Proceedings of the International Solvent Extraction Conference, Denver, Colorado*, Paper 42c. AIChE, New York.

Moses, J. M., K. E. Goklen and R. P. deFilippi (1982). Pilot plant critical fluid extraction of organics from water. *Proceedings of the AIChE Annual Meeting, Los Angeles, California*, Paper 127c. AIChE, New York.

Paulaitis, M. E., V. J. Krukonis, R. T. Kurnik and R. C. Reid (1980). Supercritical fluid extraction. *Rev. Chem. Eng.*, **1**, 179.

Roselius, W., O. Vitzthum and P. Hubert (1974). Method for the production of caffeine-free coffee extract. *US Pat.* 3 843 824.

Schultz, W. G. and J. M. Randall (1970). Liquid carbon dioxide for selective aroma extraction. *Food Technol.*, **24** (11), 94.

Shimshick, E. J. (1981). Recovery of organic acids from dilute aqueous solutions of salts of organic acids by supercritical fluids. *US Pat.* 4 250 331.

Shimshick, E. J. (1983). Extraction with supercritical carbon dioxide. *Chemtech*, June, 374.

Stahl, E., W. Schilz, E. Schutz and E. Willing (1978). A quick method for the microanalytical evaluation of the dissolving power of supercritical gases. *Angew. Chem., Int. Ed. Engl.*, **17**, 731.

Stahl, E. and K. W. Quirin (1983). Dense gas extraction on a laboratory scale—a survey of some recent results. *Fluid Phase Equilib.*, **10**, 269.

Studiengesellschaft Kohle (1981). Process for the decaffeination of coffee. *US Pat.* 4 260 639.

Williams, D. F. (1981). Extraction with supercritical gases. *Chem. Eng. Sci.*, **36**, 1769.

Yates, R. A. (1981). Removal and concentration of low molecular weight organic acids from dilute solutions. *US Pat.* 4 282 323.

Zosel, K. (1978). Separation with supercritical gases: practical applications. *Angew. Chem., Int. Ed. Engl.*, **17**, 702.

37

Electrodialysis

S. M. JAIN and P. B. REED
Ionics Inc., Watertown, MA, USA

37.1 INTRODUCTION

Electrodialysis (ED) is a rapid process for altering the composition and/or concentration of electrolytes in aqueous solutions by transferring ions across semi-permeable membranes, under the influence of a direct electric current. The membranes, containing ion-exchange groups, have a fixed electrical charge. Membranes with positively fixed charge groups (anion membranes) will be selective to the passage of anions and will repel cations. Negatively charged (cation) membranes will conversely be selective to pass cations and will repel anions. By alternating these membranes it is possible to construct a cell which will allow the dilution or concentration of the electrolytes in the original solution being processed. The solution will become further concentrated by the transfer of water through the membrane with the salt ions.

As will be explained further in this chapter these characteristics are used in the desalting and concentration of proteins from fermentation broths, selected fractionation and purification of proteins, and the recovery and reuse of valuable reagents often used in bioprocesses.

Traditionally ED is used to desalt brackish water to render it potable, or to produce deionized water for industrial use. Over 1000 plants have been installed throughout the world with an installed capacity of over 340 000 m^3 per day. Large scale ED plants are also used world-wide for desalting cheese whey protein solutions, thus converting a former waste product into a useful food source used in the making of infant formulas, bakery products, ice creams, *etc.* Other applications of ED in industry include the recovery of plating wastes, deashing of sugar solutions, conversion of alkali salts of organic acids to the acids, generation of caustic and chlorine and the purification of fermentation broths for recovering genetically engineered proteins. The process can also be used for recovery and purification of useful reagents such as GuHCl, SDS, $(NH_4)_2SO_4$, ampholytes, urea, *etc.*

The ED process is essentially independent of diffusion, being electrically driven, hence it can be used to transfer salts from a lower concentration to one at a higher concentration and can be used either to add or remove salts. This distinguishes ED from other known methods for desalting. Dialysis, for instance, commonly used at the laboratory level to remove salt from a higher to a lower concentration, depends on diffusion as the driving force and is inherently slower than ED. In ED there is also no dilution of the product even when desalting to very low salt levels.

Electrodialysis is also distinguished from ion exchange resins even though the membranes used in an ED system are chemically equivalent. In ion exchange, charged species are removed by exchanging ionic charge groups on the resins and hence resins require periodic regeneration. In ED the regeneration is not required since the ionic species is electrically driven out of the solution.

In reverse osmosis (RO) the application of pressure greater than the osmotic pressure over the concentrated solution causes the flow of water to the less concentrated solution across a semi-permeable membrane. The amount of pressure varies depending upon the concentration of the solutions. Thus for sea water containing about 0.6 M salt, the pressures required are 30–40 atm to produce the flow of water. In bioprocessing the salt levels are much higher, making the pressures required for RO impractical. Since both the proteins and the salt would be rejected by the RO membrane this process is extremely limited.

Ultrafiltration (UF) is somewhat akin to RO in that it filters on the basis of molecular size, but differs in that the salts and peptides will pass through. UF operates at moderate pressures of the order of 3–4 atm. The UF membranes are rated on the basis of nominal molecular weight cut off and usually develop a concentrated layer of macromolecules at the membrane surface causing concentration polarization and thereby reducing flow rates. Many techniques such as crossflow, stirring, or addition of solvent (in diafiltration) are used to minimize this effect. For desalting protein solutions (especially those with a high concentration) filtration is slow compared to ED, and non-specific adsorption of proteins which occurs on UF membranes at low ionic strengths causes a lower yield. Thus UF and diafiltration can be effective methods for separation of molecules with a large size differential but are not efficient for desalting.

37.2 BRIEF HISTORY

The earliest work on ED can be traced to 1869 when Maignot and Sabates received a German patent on a three compartment cell device for carrying out ED. The center compartment was bound by two semi-permeable membranes. In this simplest application of ED, the membranes function as a barrier resisting the mixing of products of electrolysis but do not exert any selective action with regard to passing ions. Figure 1 illustrates the demineralization of a sodium chloride solution in this type of apparatus. Membranes M_1 and M_2 are freely permeable to all ions and resist concentration gradients. With the imposition of an electric current sodium ions pass from compartment 1 into compartment 2 and an equal number from compartment 2 into 3; chloride ions simultaneously proceed in the reverse direction. At the electrodes hydrogen and hydroxyl ions are generated in compartments 1 and 3 and combine in compartment 2 as H_2O. At this stage, the rate of influx *vs.* the rate of efflux of Na^+ and Cl^- is less, due to the retention of Na^+ and Cl^- in compartments 3 and 1 respectively, resulting in a decrease of NaCl content in the center compartment 2, production of NaOH in 3 and HCl in compartment 1. Because the H^+ ion is more mobile than the OH^- ion, compartment 2 becomes acidic. This was a major disadvantage of a three compartment cell containing non-selective membranes. In addition, the three compartment cell had a low current efficiency (defined as equivalent of salts removed to Faradays of electricity passed). It was pointed out by Meyer and Strauss (1937) that the ion selective membranes would overcome this difficulty.

Today, ED with ion selective membranes provides greater versatility and relative freedom from the complications mentioned above. The most significant advance in commercialization of

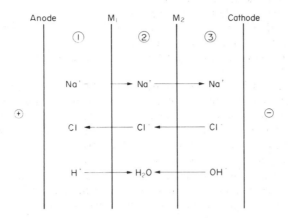

Figure 1 Three compartment electrodialysis cell

ED was the realization that the ions removed from demineralizing compartments need not be eliminated at the electrodes, but could be received into an adjacent cell or brine compartment. However, the use of neutral membranes with their inherent high electrical resistance kept ED too expensive. A major breakthrough in the 1950s came when Juda and McRae (1950) in the US and Kressman (1950) in the UK were successful in making synthetic ion exchange membranes with low electrical resistance. The commercial availability of these membranes led to significant advances in ED particularly with regard to its application in desalting brackish water with 3000–10 000 p.p.m. salt. Since then there has been an increasing commercial acceptance of ED.

By the late 1970s and early 1980s the application of ED to biotechnology was developed in order to selectively remove or purify protein solutions by the use of either the desalting or salting out mode of ED (Jain, 1983). The known precipitation characteristics of certain euglobulins permit selective protein fractionation as the ionic concentration is altered. By lowering the salt level by ED and simultaneously controlling the pH to the pI of impurities, it is possible to remove the impurities and leave the target protein intact. Contrastingly, salting out with the direct addition of saturated solutions of $(NH_4)_2SO_4$ or Na_2SO_4 is a well known technique in protein purification schemes. Salting out by ED though is a unique process in that it requires a lower bulk concentration of the salting out agent and offers the economic advantage of its subsequent recovery. These applications will be discussed in more detail in the following sections.

37.3 ION EXCHANGE MEMBRANES

Ion exchange (or ion selective) membranes are the heart of the ED system and are basically ion exchange resins in sheet form. The chemical structure of typical modern membrane resins is shown schematically in Figure 2.

There are two membrane types: cationic and anionic. The key to their ability to act as a carrier of electric current is the presence of a high concentration of mobile counterions. The cation exchange or (cation selective) membrane consists of polystyrene with negatively charged sulfonate groups bonded to most of the phenyl groups in the polystyrene. The negative charges are electrically balanced by positively charged cations (counterions). Within the imbibed water the positively charged counterions M^+ are strongly dissociated from the bound or fixed negatively charged sulfonic groups and hence are free to migrate, *i.e.* they can be exchanged with other cations from an ambient solution. Electrical neutrality is preserved during this exchange. Anion membranes on the other hand have positively charged quaternary ammonium groups $(-N_3R)^+$ chemically bonded to the phenyl groups in the polystyrene whereas X^- forms the mobile counterion and hence the principal carrier of electric current. While these mobile counterions account for low electrical resistance of these membranes, the high concentration of fixed charged groups tends to exclude charged co-ions from the ambient solution. This negligible permeability of ions of the same sign as the fixed ion of the membrane is then responsible for the high selectivity of these membranes, *i.e.* cation selective membranes (having fixed negative charges) tend to

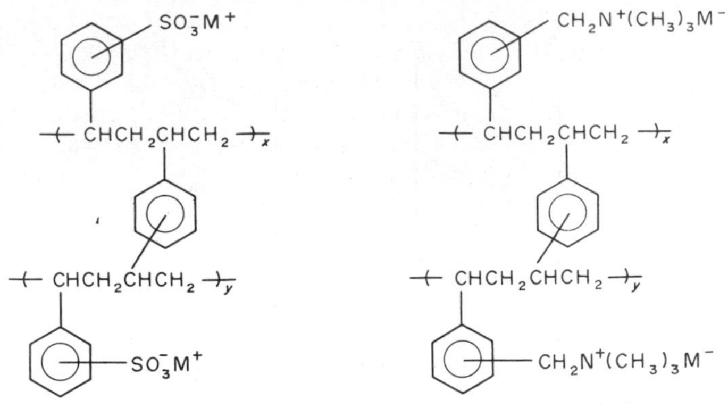

Figure 2 Schematic representation of ion-exchange resins

exclude anions having a negative charge whereas anion selective membranes with a fixed positive charge similarly exclude cations with a positive charge and allow the mobile anions to pass through. In order to control the large swelling tendency of sulfonated polystyrene when in water, cross-linking agents such as divinylbenzene (as shown in Figure 2) are usually included in the polymer.

The pore size of commercially manufactured ED membranes may vary from 10–100 Å and can be made 'loose' or 'tight' depending upon the application. The membrane is usually reinforced with an inert synthetic fabric to provide mechanical strength. A typical membrane for ED use has a pore size of 10–20 Å with a capacity of 1.6–3.0 meq per dry gram of resin. Water content may range from 20% to 70% of resin. Membrane thickness of 0.15–0.6 mm and electrical resistance of *ca.* 3–30 ohm cm^{-2} at room temperature when in equilibrium with a 0.5 N sodium chloride solution would be common values for an ion exchange membrane used in an efficient ED cell.

The efficiency of demineralization in the dilute (or dialysate) compartment is strongly influenced by selection of membranes since it affects counter transfer processes such as co-ion transfer, diffusion and water transport associated with counterion movement and osmosis. Hence, besides having such qualities as mechanical strength, chemical inertness and dimensional stability, an ion selective membrane should have the following characteristics (McRae, 1979): (1) High ion selectivity. This negligible permeability towards co-ions (ions of the same sign as the fixed charges) is usually expressed in terms of a transport number of the counterions in the membrane and is defined as the fraction of the current carried through the membrane by counterions. If the co-ions penetrate into a membrane reducing the transport number of the counterions, the coulomb efficiency of the ED process is decreased. Thus if t^c is the anion transport number in the cation permeable membrane and t^a is the cation transport number in the anion permeable membrane, the efficiency, η of an ED process is given by:

$$\eta = 1 - (t^c + t^a) \tag{1}$$

i.e. the efficiency is decreased by an amount equal to the combined co-ion transfer of the cation and anion selective membranes resulting in a lower net transport of counterions. Most of the commercial membranes available have ion transport numbers of *ca.* 0.85–0.95 in dilute solutions, but these values are lower if the salt concentration in the ambient solution is high. (2) High electrical conductance. This factor influences the ohmic resistance and hence power requirements of an ED unit. The membrane resistance is the controlling factor for solutions having a concentration in excess of 0.1 N; for more dilute solutions however, the resistance is mainly controlled by the solution contribution. (3) Negligible rate of free electrolyte diffusion. Salt diffusion through the membrane counteracts the electrodialytic salt transport. Unless the membrane is highly selective, the diffusion of the salt tends to increase across the membrane as the concentration differential increases. Designers of ED apparatus therefore, try to minimize concentration differences across membranes and also utilize highly selective membranes to counteract this back diffusion. (4) Low osmotic permeability. Ion exchange membranes tend to transport high anomalous osmotic flow of water as the electrolytes are transported. This transport of water through the membranes with the ions is known as endosmotic water transfer. With the high concentration of counterions in ion selective membranes, the transport of ions and water are closely coupled and the internal solution moves in a piston-like flow. This endosmotic water transfer is about 100–200 ml per equivalent of ions transferred. While the endosmotic water transfer is not significant for dilute solutions, it becomes an important advantage for desalting of concentrated solutions, such as most biological solutions, since the water loss tends to concentrate the protein solutions being desalted. This water loss can be minimized if desired by proper choice of membranes (membranes with low water content and high concentration of fixed charged groups have low endosmotic water transfer).

37.4 PRINCIPLE AND THEORY OF OPERATION

37.4.1 Major Embodiments

37.4.1.1 Concentration and dilution of salts

This is the most widely used embodiment of the ED process on a commercial scale, where a stream (called the 'dilute' stream) needs to be desalted, by decreasing the salt concentration, and simultaneously the adjacent stream (called the 'conc' stream) has its salt concentration increased as a result of salt being transferred across the membranes. Figure 3 shows schematically the concentrating–diluting ED process principle. A number of compartments or chambers called diluting (D^*) and concentrating (C^*) cells separated by alternating anion and cation membranes are bound by two electrode stream chambers and two electrodes, cathode and anode. These C^* and D^* compartments are made of thin plastic materials to support the membrane and allow the liquid to be pumped in and out. The diluting cells are filled with a liquid containing a salt with M^+ as cations and X^- as anions. With the imposition of an electric potential the cations M^+, being positively charged, move toward the anode. The cation selective membrane C allows M^+ ions to pass through from the diluting compartment to the concentrating compartment. However the further progress of M^+ and X^- out of the 'conc' compartment is blocked by the fixed charge groups in the anion and cation selective membranes and hence this compartment C^* gets enriched and compartment D^* gets depleted in the salt M^+X^-.

One concentrating and one diluting cell separated by a cation and anion membrane constitute a cell pair. Several repeating cell pairs arranged between two electrode compartments and electrodes form a membrane stack. The end electrode compartments are separated by a special electrode membrane from the other cell pairs and hence the incidental electrode reaction products are kept separated from the process stream. It should be noted that since the process is electrically driven, this arrangement of cells will continuously desalt the process stream (and enrich the other streams), with no need for chemical regeneration such as is required in ion exchange columns.

37.4.1.2 Ion substitution

A second embodiment of ED is the ion substituting mode where a process stream is modified in its ionic composition. The liquid to be modified is passed between membranes of like charge (*e.g.* two cation or anion transfer membranes), ions will leave the solution in one direction through one membrane, and simultaneously equivalent ions of like charge will enter the solution through the opposite membrane from a make-up solution flowing on the other side of the latter membrane. Figure 4 illustrates such an arrangement. The process stream contains an electrolyte M^+X^-; the object is to replace M^+ with L^+. On application of a potential the cations L^+ of the make-up stream move into the process stream and M^+ of the process stream leaves and enters the waste stream where they combine with the X^- ions transferred through the anion membrane. Thus, as a result, the process stream's M^+ is replaced by L^+, M^+ being rejected to the waste stream; simultaneously the L^+X^- content of the make-up stream is depleted. In such an arrangements the repeat unit is a cell trio instead of a cell pair and consists of process, make-up and waste compartments. In the case where cation substitution is the object, the cell trio will require two cation membranes and one anion membrane. Where anion substitution is desired the cell trio will consist of two anion membranes and one cation membrane.

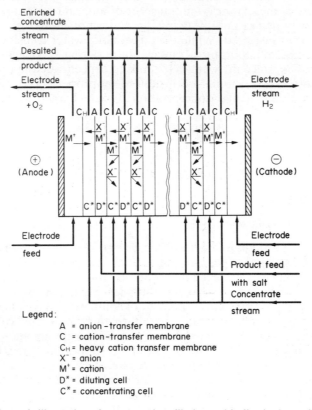

Figure 3 Schematic illustration of concentrating–diluting multicell pair electrodialysis process

37.4.1.3 *Electrolytic ED*

In the above two modes the electrode reactions are not important but in this third embodiment, the electrodes play a fundamental part in the process. Such an embodiment is shown in Figure 5. A membrane separates the two electrode compartments and their reaction products. A salt feed M^+X^- is fed to the anode compartment. On the imposition of an electrical potential the cation M^+ moves out through the cation membrane towards the cathode and there combines with OH^- to give MOH whereas X^- becomes X at the anode and is given out as a gaseous product. A familiar example of an electrolytic cell is the chlor–alkali cell where an NaCl brine feed is converted to NaOH (at cathode) and Cl_2 (at anode).

It is apparent that by a suitable choice of embodiments of an ED stack a variety of processes can be carried out in a simple and controllable manner.

37.4.2 Operation

The operation (Mason and Kirkham, 1959, Jain, 1979) of a cell stack shown in Figure 3 is governed by well-known electrochemical principles. According to Faraday's law, 96 500 C (1 faraday) or 26.8 A h are required theoretically for electrolytic transfer of one gram equivalent. In a stack containing n cell pairs, this current will theoretically transfer n gram equivalents. Therefore, the demineralizing capacity of a stack can be calculated from the following equation:

$$\text{g-equivalents/min} = 60neiA_p/F \qquad (2)$$

where n = number of cell pairs; e = fractional current efficiency; i = current density, amp cm^{-2}; A_p = available transfer area per cell pair, cm^2; and F = Faraday's constant, 96 500 C/g-equivalent.

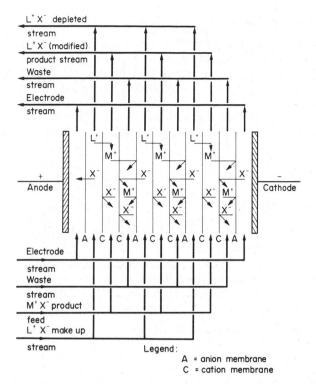

Figure 4 Schematic illustration of ion substitution electrodialysis process

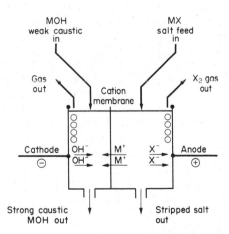

Figure 5 Schematic representation of an electrodialysis cell in electrolyte mode

For a batch process, the differential demineralization rate is approximated by assuming perfect mixing:

$$- V_D \, dN = \frac{eIn}{F} \, d\theta \tag{3}$$

where V_D = batch volume, l; N = normality of electrolyte in batch, g equivalents per l; I = current, A or C/s; and θ = time, s.

But at constant i/N:

$$I = (i/N)NA_p \tag{4}$$

substituting in equation (3), integrating and rearranging gives:

$$P \equiv V_D / [(nA_p)\theta] = [e(i/N)] \ln(N_i/N_f) \tag{5}$$

where N_i and N_f are the initial and final normalities of the diluting stream, and P the production rate of desalted product in l/s/cm². Equation (5) can be used to calculate the rate or time required to desalt a batch volume at a certain operating i/N. Equation (5) also shows that the production rate is directly proportional to i/N and inversely proportional to $\ln[N_i/N_f]$. As high a value of i/N is chosen as is permitted by polarization considerations determined for a particular solution.

As the current is carried through solutions by the ions, the current-carrying capacity (*i.e.* conductivity) of a solution depends upon the number of ions present, or the solution normality. Thus, the ratio of current density to normality (i/N) is an important variable in membrane system design. As i/N increases, the concentration of ions at the membrane–solution interface in the diluting cells decreases. If i/N becomes excessive there will not be enough ions to carry the desired current, and a limiting i/N will be reached where film depletion or concentration polarization occurs (Wilson, 1960).

Polarization can be minimized and allowable i/N maximized by creating rapid stirring at the membrane solution interface through proper design of the spacers. Local spot polarization can be minimized by avoiding stagnant areas exposed to current. Polarization generally is not encountered in high concentration solutions (greater than 0.1 N).

In dilute solution, the avoidance of local or general polarization is the most crucial task of cell designers. Ionics' tortuous path spacers, from the small laboratory cell to the large commercial size, are designed with this in mind and have been successsful in operating at high i/N values. At high concentrations, factors other than i/N limit the current density. These factors are high voltage drop and excessive heating and power consumption.

37.5　EQUIPMENT

Electrodialysis systems consist of three basic building blocks: a membrane stack, hydraulic system and a rectifier as the source of direct current. Figure 6 shows a schematic ED system.

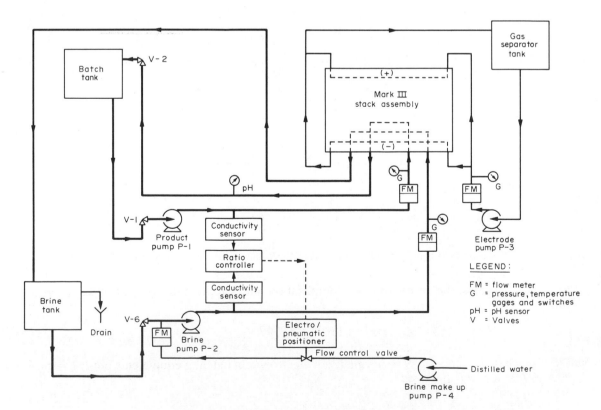

Figure 6　Schematic representation of an ED system

37.5.1 Membrane Stack

A desalting membrane stack consists of cation and anion selective membranes arranged alternately between intermembrane spacers in a stacked configuration with a positive and negative electrode at the ends. A typical membrane and spacer manufactured by Ionics is shown in Figure 7. The spacers are designed to promote turbulence and mixing so that the boundary layer effects at the membrane surface are minimized and the process is made essentially diffusion free. Spacers manufactured by Ionics incorporate a tortuous path flow configuration between the inlet and exit manifolds. The intermembrane spacers which form the demineralizing or dilute compartments in the stack are fed in parallel from a common manifold. Similarly concentrating compartments are also fed from a separate common manifold. The electrodes spacers are fed by a separate stream and the reaction products are isolated from the process streams. Usually a concentrate stream at each end separates the electrode stream from the dilute stream. A long tortuous path in the compartments allows a high percentage (35–50%) of salt to be removed per pass as the solution is pumped through the demineralizing compartment. To enhance salt removal of the total solution, electrodialysis stacks are often fed in parallel and manifolded in series with other sections of spacers in the stack. In a batch process the solution is recirculated several times through the stack until the desired level of total desalting is achieved. Alternatively, several stacks may be operated in series to reach the desired desalting.

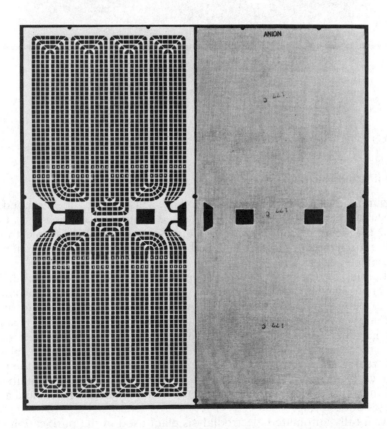

Figure 7 Ionics' mark III-4 membranes and tortuous path spacers

Electrodialysis stacks can be equipped with any number of cell pairs but typical stacks range from 100 cell pairs, with effective membrane area of approximately 30 m^2 for high salinity solutions, to 700 cell pairs, with effective membrane area of 210 m^2 for brackish water. The inlet and outlet manifolds for dilute and concentrate compartments are located at top and bottom of the stack. Electrical connections of the electrodes to power source are made through Joy connectors and the entire stack is protected by a safety shield to protect the operator from accidental electrical shock. A typical stack is shown in Figure 8, without its protective shield.

Figure 8   Ionics' Bio-Prep™ Membrane stack (without protective shield)

37.5.2   Hydraulics and Rectifier

The piping, pumps, valves and gauges, controllers, sensors, *etc.* are mounted on a skid and usually form a separate section. The stack has flanged connections which end at the side of the skid. Usually the electrode stream requires smaller amounts of solution (typically 40–60 l for one stack of 100 cell pairs) and a degassifier tank with a blower to exhaust the small volume of gases generated at the electrode streams.

The choice of piping for the product stream depends upon the type of end use. Thus for ordinary desalting of brackish water, PVC piping is used whereas for pharmaceutical and food industry use, the piping is usually stainless steel and all valves and connections are of quick disconnect sanitary type. In most biotechnology applications, the brine line is also made of stainless steel.

The rectifier is generally in a separate cabinet along with various electrical accessories, switches, alarms, recorders, *etc.* housed in the same unit. In most pharmaceutical and food industry applications, rectification control of the current and voltage is critical to the process. A conductivity voltage/current program is incorporated to prevent exceeding the limiting current for a particular solution concentration, preventing electrical shorting in the stack, and at the same time maximize the rates of desalting.

Figure 9 shows a fully automated electrodialysis plant used in the purification of a genetically engineered interferon solution. The three basic building blocks can be seen in this overall picture.

37.5.3   Considerations in Scale-up from Laboratory to Pilot and to Industrial Scale Electrodialysis Plants

Desalting of solutions in biotechnology is a common problem. Since the composition differs widely, it is essential to test the behavior on a laboratory scale and that proper conditions for flow rate, suitable current density/voltage, pressure drops, power consumption, *etc.* be obtained with the laboratory size equipment. Long term behavior of the membrane stacks can be studied with a

Figure 9 Ionics' Bio-Prep™ Mark III electrodialysis unit for interferon purification

laboratory or a pilot plant system and then dedicated commercial size plants can be designed directly from data which incorporates the experience gained with these laboratory and pilot size units. Ionics' BioPrep™ system (Figure 9) with effective membrane area of 30 m² has been scaled up from data obtained on MediMat™ laboratory units and the pilot scale Stackpack® unit (Figure 10) having an effective membrane area of 0.4 m². This capability for direct correlation in scale-up, regardless of batch size, is unique to electrodialysis as has been demonstrated in many standard batch or continuous electrodialysis systems for food and pharmaceutical industry use throughout the world. For example, a typical Electromat® standard whey line processes 125 000–440 000 kg per day of raw whey depending upon 50–90% deashing. Such a line has about 180 m² of effective cell pair area and processes 20–30% preconcentrated whey solids solutions. Operating rates for calculating scale-up size for these units were also directly derived from the data obtained in the lab with a Stackpack®. While these BioPrep™ and Electromat® systems are available, for both batch and continuous processing, most food and pharmaceutical products normally operate by a batch process. The major reason for this is that in cases such as whey deashing, the high salt content of the whey would require 30–35 passes to obtain 90% deashing depending upon type of whey, hence for continuous processing it would require 30 or more stages in series with each stage at a different DC voltage and different current density. Interstage pumping would also be required. This makes the process control complicated and expensive. Another disadvantage of continuous process is operating with different feeds of differing salt content. To maintain a constant product composition in a continuous demineralization process, it will be necessary to adjust the operating conditions for the individual feed. In a batch process constant product composition is maintained at the same operating conditions merely by shortening or lengthening the batch cycle time.

37.5.4 Clean-in-place (CIP) Techniques

The electrodialysis equipment used in food and pharmaceutical industry requires very strict sanitary conditions. Usually after each 4–5 hour batch, a cleaning cycle is instituted for 1–2 hours. This consists of flushing the process streams with water, followed by 0.1 N caustic solution, water, 0.1 N hydrochloric acid solution and a final water rinse. Separate tanks are used to hold these

Figure 10   Ionics' Stackpack®3K-laboratory scale ED unit

cleaning solutions to flush the system without necessitating opening of the stacks. In cases where there is a gap of a few days between batches, the system can be left filled with a 25% solution of ethanol or similar sterilant.

Periodically (depending upon the extent of use), electrodialysis stacks are removed from the line and disassembled for cleaning and inspection. A stand-by stack can quickly be connected to the hydraulic system so that production schedules remain uninterrupted.

37.5.5   Automation of Industrial and Pilot Plants

In most of the modern pilot and industrial scale electrodialysis systems used in the food and pharmaceutical industries there is extensive automation. Thus, for example, the BioPrep system is fully automatic and process control is effected through changes in the electrical conductivity of the product solution during desalting. The recirculating brine solution used to carry the salts removed from the product is automatically diluted with fresh water to maintain a certain conductivity ratio between the product and brine streams, thus minimizing diffusion between the two streams. A variable DC voltage program automatically controlled by product and brine solution conductivities provides maximum salt removal rates throughout the batch. Batch end point is preset by a desired conductivity. The unit is shut off automatically once the desired end point is reached. After removal of the batch, the system is flushed with CIP solutions automatically with the operator needing merely to change the appropriate cleaning solutions in the CIP tank and switch to appropriate flush cycles. Thus a minimum of labor is involved in operating these plants.

37.6   APPLICATIONS OF ELECTRODIALYSIS IN BIOTECHNOLOGY

Applications of electrodialysis in biotechnology can be grouped in four different classes: (1) conventional, these applications include desalting, pH adjustment, ion substitution, electrolytic conversion and waste treatment; (2) pretreatment, prior to electrophoretic separations and ion exchange columns; (3) protein fractionation, novel use of ED either for gross purification or selective removal of proteins by desalting or salting out; (4) recovery/purification of reagents, ED used to recover ionic reagents such as GuHCl, $(NH_4)_2SO_4$ or purify the non-ionic agents such as urea by removing the ionic impurities.

37.6.1 Conventional Applications

The most widely used application of ED is the desalting of brackish water to make it potable. Another application is desalting of whey containing about 12% proteins, 75% lactose and 8% ash or salts.

An interesting example of ED is in the manufacture of a high purity ion complexed dextrin product used as injectable iron solution for pigs (Meinhold, 1966). Production of ion dextrin prior to ED was a slow process requiring a dozen steps consisting of washing, precipitation and pH adjustment and ion exchange. A single batch took as long as four days; the batch size was limited and there was an inherent danger of contamination because of excessive handling. The use of ED greatly reduced the number of steps (therefore the cost) by eliminating repeated washings of the precipitate, pH adjustment with ion exchange resins and completed the process cycle in 6–8 hours (compared with 4 days).

For similar reasons, ED as a unit operation in genetically engineered product solutions for desalting/purification is now looked at very favorably. The first commercial application of ED to a genetically engineered product was to reduce the high salt concentration (2 N NaCl) from α-interferon after elution from a column. The electrodialysis process operates with minimum loss of the protein even at very high desalting levels required by this application since there is negligible non-specific adsorption of the protein of interest.

In biotechnology, process schemes are enhanced by a desalting step after fermentation, column elution or salting out, to purify the product by reducing the salt. ED can serve as an exquisite technique since the process is gentle to the protein, very fast, does not produce any dilution of product with desalting, the degree of desalting can be easily and specifically controlled and high yields of proteins are possible. No regeneration such as with ion exchange columns is needed.

Increasing attention is being given to two areas of application involving ion substitution, adjustment of pH solutions and ionic modification of natural biological solutions such as milk or drugs to contain low sodium. Adjustment of pH by ED can be done by a process shown in Figure 4, by having the make-up stream consist of an acid solution and replacing the M^+ ions with H^+ ions. This method of pH adjustment as opposed to direct acid addition has the advantage that H^+ is introduced into the product without the addition of an anion, so that the salt level in the product is decreased instead of increasing. The same holds true for alkalinization of the product. The pH control is very precise and readily effected.

The other application concerns ion modification of biological solutions or drugs without change of pH. Examples include replacing sodium with calcium or potassium in milk or drugs which are useful for persons on low sodium diets.

37.6.2 Pretreatment

In the pharmaceutical industry certain drugs such as antibiotics are purified by adsorption on an ion exchange column. Often the feed contains salts which compete with the ion exchange sites of the resin thereby reducing the capacity of the ion exchange column and requiring frequent regeneration. If the feed is desalted by ED prior to the ion exchange column adsorption step, the process not only results in a cleaner operation but increases the capacity of ion exchange columns for adsorption of the protein. Thus smaller columns can be used or, conversely, larger batches can be processed.

Similarly, efficiency of electrophoretic separations such as isoelectric focusing or electrophoresis can be greatly increased if the solutions are desalted by ED to a controlled low level prior to applying these techniques, since with low salt higher voltages can be applied safely without causing overheating.

37.6.3 Protein Purification/Fractionation

In ED, ions are electrically transported through semi-permeable anion and cation selective membranes. The ED process is essentially independent of diffusion and hence can be used either to selectively remove or add salts. These two characteristics of ED, *i.e.* (a) reducing ionic concentration, desalting and (b) increasing ionic concentration, salting out, form the basis of novel fractionation techniques for proteins and can serve as a tool either for broad-range purification methods of removing impurities that become insoluble at low or high ionic strengths or simply to selectively recover the protein of interest. The separations can be further enhanced by combining ED with a pH adjustment step to approach the pI of the proteins to be precipitated.

Such a desalting process has been used for fractionation of plasma proteins by ED desalting. As the salt level of the plasma is decreased, most of the non-albumin proteins precipitate out. By continuing desalting, it is possible (Jain, 1983) to obtain, for example, an IgA-rich precipitate and an albumin-rich solution. This technique is also applicable for broad purification or impurity removal or concentration of genetically-engineered protein solutions.

ED desalting for protein fractionation has many advantages over conventional methods. It effects a fast and controlled removal of salts, without dilution of the product. Because of low membrane area requirements, there is negligible non-specific adsorption of protein resulting in high yields of proteins. Very low molecular weight proteins and peptides can be rapidly desalted because of the small pore size of the membrane (10–100 Å) and the process is easy to scale up.

Efficient separations of proteins have also been effected by ED salting out (Jain, 1983). In the separation of plasma protein albumin from IgG by this method, ED effects about 80% removal of IgG at a bulk concentration of only about 1.1 N Na_2SO_4, while the same separation by direct addition requires about 1.8–2.0 N Na_2SO_4. Another advantage of the ED process for fractionation is the ability to recycle the salting out agent; ED can salt out one stream and simultaneously desalt the other. This obviates the necessity of a waste disposal or recovery of the salting out agent.

37.6.4 Recovery and Purification of Reagents

At present the majority of reagents used in biotechnology processing such as GuHCl used as a chaotropic agent or $(NH_4)_2SO_4$ for salting out, are discarded after single use by dialyzing them out. This may be acceptable if the operation is on a laboratory scale, but the expense of these reagents can be considerable for large scale operation. ED can be successfully used to recover these reagents in a useful concentration of 3–5 M as opposed to a very dilute solution obtained as a result of dialyzing them out. With ED it has been shown that over 99% of GuHCl can be removed from 65 l of 7.0 M solution in approximately 2.5 h using an ED stack with 10 m^2 effective cell pair area giving a processing rate of 2.5 l h^{-1} m^{-2}. Similarly $(NH_4)_2SO_4$ can be recovered and concentrated for reuse by ED. This not only solves a waste problem but also recovers an expensive reagent.

Non-ionic reagents such as urea are often used to solubilize proteins in biotechnology/biosynthesis schemes. After separation of large molecular weight proteins by ultrafiltration, the urea solution is contaminated with salts and smaller molecular weight proteins not removed by ultrafiltration. Large quantities of concentrated urea solution are generated as a waste stream requiring considerable expense for waste disposal. This waste stream can be cleaned by first desalting the urea to a low salt level by electrodialysis and later the small molecular weight proteins can be adsorbed in special ion exchange columns thereby rendering the urea solution free of salts and proteins to recycle. Here dialysis would not work since urea has a low molecular weight and hence passes through a dialysis membrane along with the salt. However, with ED urea, being non-polar, stays behind in the dilute stream, only a negligible amount being lost to the 'conc' stream.

With the use of special 'open pore' membranes, use of ED can be extended to remove other reagents such as ampholytes (which have higher molecular weight than the ordinary inorganic salts) from biological solutions.

37.7 ECONOMIC CONSIDERATIONS

Operating costs for ED, when used on a large scale for desalting brackish water, will generally range from $0.26 to $0.52 per m^3. These costs will vary with the feed salinity, plant capacity and the unit cost of electric power in the area.

In biotechnology, where salinity levels are considerably higher, desalting levels are significantly greater and product yield is more important, the operating cost per m^3 cannot be comparable to that of water. However, extensive experience is available in large scale desalting of protein solutions such as whey and one can use that experience as a guide for economic considerations of an ED plant for biotechnology use. Table 1 shows the chemical composition of dried whey before and after ED demineralization.

Table 1 Chemical Composition (%) of Dried Whey Before and After Electrodemineralization

	Before	After		
	Whole whey	25% De-ashed whey	50% De-ashed whey	90% De-ashed whey
Protein	12.0	12.5	13.0	13.5
Lactose, hydrate	74.8	76.3	77.8	80.5
Butterfat	0.7	0.7	0.7	0.7
Moisture[a]	4.5	4.5	4.5	4.5
Ash	8.0	6.0	4.0	0.8

[a] Total moisture including water of crystallization of lactose.

The plant capacity, product yield and processing costs of whey vary considerably with percent of ash which is to be removed (see Table 2). The production capacity is doubled when demineralizing to 75% rather than 90% ash removal and doubles again when demineralizing to only 50%. Operating and maintenance costs are about 11.6 ¢ kg^{-1} for 90% ash removal, 6.4 ¢ kg^{-1} for 75% and only 3.0 ¢ kg^{-1} for 50% ash removal.

Table 2 Plant Capacity and Typical Operating and Maintenance Costs[a]

	50% De-ashed	75% De-ashed	90% De-ashed
Production capacity (kg of feed solids per day)	27 800	13 600	7900
Overall plant yield on feed solids	90%	85%	80%
Operating and maintenance cost breakdown (¢/kg)			
Electrical energy	0.37	0.84	1.39
Steam at $11/1000 kg	0.24	0.66	1.72
Chemicals and other utilities	0.18	0.29	0.53
Waste treatment 11¢/kg BOD	0.48	0.84	1.23
Operating labor	0.48	1.06	1.91
Replacement of stack components	1.14	2.53	4.58
Other maintenance materials	0.07	0.13	0.24
Total operating and maintenance costs (¢/kg product solids)	2.96	6.35	11.6

[a] Based on 6000 operating hours per year; electrical energy at 3.5 ¢ kWh^{-1}; labor at $15 per work hour.

Because of frequent chemical usage for clean-in-place maintenance (see Section 37.5.4) the life of membranes in whey plants is of the order of 6 months to 12 months (compared with 5–7 years for brackish water desalting).

The capital depreciation factor used in calculating return on investment is not shown in the table since this will vary depending on extent of automation required by the end user.

37.8 SUMMARY

Electrodialysis is a very powerful tool for the separation and purification of genetically engineered proteins. By electrically driving salt ions out of solution through highly efficient ion-exchange membranes, the proteins are rapidly desalted and concentrated with nominal loss. ED can be equally effective when applied to the recovery of GuHCl, $(NH_4)_2SO_4$, urea and other valuable reagents.

The unique design and configuration of the membrane cells allow accurate prediction of size requirements for commercial scale-up from the rate data and yields obtained from the bench top units at the research level. Economics for production systems can be calculated for any size plant. The systems can be self-contained, fully automated, of sanitary type construction and readily sterilized between production batches through full clean-in-place capabilities.

ED can be used in several modes such as desalting–concentrating, ion-substitution and electrolytic. To date the most common use of ED in biotechnology has been the desalting mode applied to the purification of several genetically engineered pharmaceutical products (such as interferon). The relatively low operating costs for ED should make it an equally economically viable process for non-drug and other commodity priced products produced in large volumes. Application of other modes of ED, *e.g.* ion substitution, electrolytic, salting out for precipitation of proteins, *etc.*, are expected to gain popularity in the field of biotechnology as the workers in the field become more familiar with the versatile nature of the process and its economic advantages.

37.9 REFERENCES

Jain, S. M. (1983). Electric membrane process for protein recovery. In *Proc. Biochem. Eng. Conf., 3rd, 1982*. Reprinted from *Ann. N.Y. Acad. Sci.*
Jain, S. M. (1979). Electrodialysis and its applications. *Am. Lab.*, 65–76.
Juda, W. and W. A. McRae (1950). *J. Am. Chem. Soc.*, **72**, 1044.
Kressman, T.R.E. (1950). *Nature (London)*, **165**, 568.
McRae, W. A. (1979). Electrodialysis. *Encyclopedia of Chemical Technology*, 3rd edn., vol. 8, pp. 726–738. Wiley, New York.
Mason, E. A. and T. A. Kirkham (1959). Design of electrodialysis equipment. *Chem. Eng. Prog.*, **55**, 173–189.
Meinhold, T. F. (1966). Drug output spurts tenfold *via* electrodialysis route. *Chemical Processing*.
Meyer, K. H. and W. Strauss (1937). *Helv. Chim. Acta*, **23**, 795.
Reed, P. B. (1984). Electrodialysis for the purification of protein solutions. *Chem. Eng. Prog.*, December 1984.
Wilson, J. R. (ed.) (1960). *Demineralization by Electrodialysis*. Butterworth, London.

Appendix 1: Glossary of Terms

Because of the broad multidisciplinary nature of biotechnology, both beginners and specialists in this field may find this glossary useful. It covers terms often used in the relevant areas of the biological, chemical and engineering sciences. It is not intended to be exhaustive. The material was generated primarily from four sources: 'Commercial Biotechnology', Office of Technology Assessment, Congress of the US (1984); 'Advances in Biotechnology', M Moo-Young *et al.* (1981); *Pure and Applied Chemistry*, Vol. 54, No. 9, pp. 1743–1749 (1982); and 'Dictionary of Biochemistry', J. Stenish (1975).

Acclimatization: The biological process whereby an organism adapts to a new environment. For example, it describes the process of developing microorganisms that degrade toxic wastes in the environment.

Activated sludge: Biological growth that occurs in aerobic, organic-containing systems. These growths develop into suspensions that settle and possess clarification and oxidative properties.

Activation energy: The difference in energy between that of the activated complex and that of the reactants; the energy that must be supplied to the reactants before they can undergo transformation to products.

Active immunity: Disease resistance in a person or animal due to antibody production after exposure to a microbial antigen following disease, inapparent infection or inoculation. Active immunity is usually long-lasting.

Active transport: The movement of a solute across a biological membrane such that the movement is directed against the concentration gradient and requires the expenditure of energy.

Activity: A measure of the effective concentration of an enzyme, drug, hormone or some other substance. It is also the product of the molar concentration of an ionic solute and its activity coefficient.

Adsorption: The taking up of molecules of gases, dissolved substances or liquids at the surfaces of solids or liquids with which they are in contact.

Aerobic: Living or acting only in the presence of free-form oxygen, as in air.

Affinity chromatography: The use of compounds, such as antibodies, bound to an immobile matrix to 'capture' other compounds as a highly specific means of separation and purification.

Airlift fermenter: Vessel in which a bioconversion process takes place; the sparged gas is the only source of agitation. The presence of a draft tube inside a fermenter distinguishes this type of fermenter from a bubble column.

Alga (pl. algae): A chlorophyll-containing, photosynthetic protist; algae are unicellular or multicellular, generally aquatic and either eukaryotic or prokaryotic.

Amino acids: The building blocks of proteins. There are 20 common amino acids.

Amino acid sequence: The linear order of amino acids in a protein.

Amylase: An enzyme that catalyzes the hydrolysis of starch.

Anabolism: The phase of intermediary metabolism that encompasses the biosynthetic and energy-requiring reactions whereby cell components are produced. Also, the cellular assimilation of macromolecules and complex substances from low-molecular weight precursors (*cf.* Catabolism).

Anaerobic: Living or acting in the absence of free-form oxygen.

Anaerobic digestion: The energy-yielding metabolic breakdown of organic compounds by microorganisms that generally proceeds under anaerobic conditions and with the evolution of gas. The term is most often used to describe the reduction of waste sludges to less solid mass and for methane gas production.

Antibiotic: A specific type of chemical substance that is administered to fight infections, usually bacterial infections, in humans or animals. Many antibiotics are produced by using microorganisms; others are produced synthetically.

Antibody: A protein (immunoglobulin) produced by humans or higher animals in response to

591

exposure to a specific antigen and characterized by specific reactivity with its complementary antigen (see also Monoclonal antibodies).

Antigen: A substance, usually a protein or carbohydrate, which, when introduced into the body of a human or higher animal, stimulates the production of an antibody that will react specifically with it.

Antiserum: Blood serum containing antibodies from animals that have been inoculated with an antigen. When administered to other animals or humans, antiserum produces passive immunity.

Aromatic compound: A compound containing a benzene ring. Many speciality and commodity chemicals are aromatic compounds.

Ascites: Liquid accumulations in the peritoneal cavity. Used as a method for producing monoclonal antibodies.

Asepsis: The prevention of access of microorganisms causing disease, decay or putrefaction to the site of a potential infection.

Aseptic: Of, or pertaining to, asepsis. In directed fermentation processes, the exclusion of unwanted (contaminating) organisms.

Assay: A technique that measures a biological response.

Attenuated vaccine: Whole, pathogenic organisms that are treated with chemical, radioactive, or other means to render them incapable of producing infection. Attenuated vaccines are injected into the body, which then produces protective antibodies against the pathogen to protect against disease.

Autolysis: The self-destruction of a cell as a result of the action of its own hydrolytic enzymes.

Autotrophic: Capable of self-nourishment (opposed to heterotrophic).

Axial dispersion: Mixing along the flow path of fluids during processing as in a bioreactor.

Bacteria: Any of a large group of microscopic organisms having round, rodlike, spiral or filamentous unicellular or noncellular bodies that are often aggregated into colonies, are enclosed by a cell wall or membrane, and lack fully differentiated nuclei. Bacteria may exist as free-living organisms in soil, water, organic matter, or as parasites in the live bodies of plants and animals.

Bacteriophage (or **phage**)/**bacterial virus**: A virus that multiplies in bacteria. Bacteriophage lambda is commonly used as a vector in rDNA experiments.

Batch processing: A method of processing in which a bioreactor, for example, is loaded with raw materials and microorganisms, and the process is run to completion, at which time products are removed (*cf.* Continuous processing).

Binary fission: Asexual division in which a cell divides into two, approximately equal, parts, as in the growth method of some single cells.

Biocatalyst: An enzyme, in cell-free or whole-cell forms, that plays a fundamental role in living organisms or industrially by activating or accelerating a process.

Biochemical: Characterized by, produced by, or involving chemical reactions in living organisms; a product produced by chemical reactions in living organisms.

Biochip: An electronic device that uses biological molecules as the framework for molecules which act as semiconductors and functions as an integrated circuit.

Bioconversion: A chemical conversion using a biocatalyst.

Biodegradation: The breakdown of substances by biological agents, especially microbes.

Biological oxygen demand (BOD): The oxygen used in meeting the metabolic needs of aerobic organisms in water containing organic compounds.

Biological response modifier: Generic term for hormones, neuroactive compounds and immunoactive compounds that act at the cellular level; many are possible targets for production with biotechnology.

Biologics: Vaccines, therapeutic serums, toxoids, antitoxins and analogous biological products used to induce immunity to infectious diseases or harmful substances of biological origin.

Biomass: Organic matter of biological origin such as microbial and plant material.

Biooxidation: Oxidation (the loss of electrons) catalyzed by a biocatalyst.

Biopolymers: Naturally occurring macromolecules that include proteins, nucleic acids and polysaccharides.

Bioprocess: Any process that uses complete living cells or their components (*e.g.* enzymes, chloroplasts) to effect desired physical or chemical changes.

Bioreactor: Vessel in which a bioprocess takes place; examples include fermenter, enzyme reactor.

Biosensor: A device, usually electronic, that uses biological molecules to detect specific compounds.

Biosurfactant: A compound produced by living organisms that helps solubilize compounds such as organic molecules (*e.g.* oil and tar) by reducing surface tension between the compound and liquid.

Biosynthesis: Production, by synthesis or degradation, of a chemical compound by a living organism.

Biotechnology: Use of biological agents or materials to produce goods or services for industry, trade and commerce; a multidisciplinary field.

Bubble column: A gas–liquid contacting vessel in which sparged gas is the only source of agitation.

Bubbly flow: Type of two-phase flow in which the gas phase is distributed in the liquid in the form of bubbles whose dimensions are small compared to the characteristic dimension of the flow cross-section, as in certain designs of bioreactor geometrics.

Budding: A form of asexual reproduction typical of yeast, in which a new cell is formed as an outgrowth from the parent cell.

Buffer: A solution containing a mixture of a weak acid and its conjugate weak base that is capable of resisting substantial changes in pH upon the addition of small amounts of acidic or basic substances, as used for some fermentation media.

Bulking: Increase in volumetric solids as in certain waste treatment bioreactors, which limits the weight concentration that a clarifier can handle.

Callus: An undifferentiated cluster of plant cells that is a first step in regeneration of plants from tissue culture.

Carbohydrate: An aldehyde or a ketone derivative of a polyhydroxy alcohol that is synthesized by living cells. Carbohydrates may be classified either on the basis of their size into mono-, oligo- and poly-saccharides, or on the basis of their functional group into aldehyde or ketone derivatives.

Carboxylation: The addition of an organic acid group (COOH) to a molecule.

Catabolism: The phase of intermediary metabolism that encompasses the degradative and energy-yielding reactions whereby nutrients are metabolized. Also, the cellular breakdown of complex substances and macromolecules to low-molecular weight compounds (*cf.* Anabolism).

Catalysis: A modification, especially an increase, in the rate of a chemical reaction induced by a material (*e.g.* enzyme) that is chemically unchanged at the end of the reaction.

Catalyst: A substance that induces catalysis; an agent that enables a chemical reaction to proceed under milder conditions (*e.g.* at a lower temperature) than otherwise possible. Biological catalysts are enzymes; some nonbiological catalysts include metallic complexes.

Cell: The smallest structural unit of living matter capable of functioning independently; a microscopic mass of protoplasm surrounded by a semipermeable membrane, usually including one or more nuclei and various nonliving products, capable alone, or interacting with other cells, of performing all the fundamental functions of life.

Cell culture: The *in vitro* growth of cells usually isolated from a mixture of organisms. These cells are usually of one type.

Cell differentiation: The process whereby descendants of a common parental cell achieve and maintain specialization of structure and function.

Cell fusion: Formation of a single hybrid cell with nuclei and cytoplasm from different cells (as in 'cell fusion technology'; see also Hybridoma.)

Cell line: Cells that acquire the ability to multiply indefinitely *in vitro* (especially in plant cell tissues).

Cellulase: The enzyme that digests cellulose to sugars.

Cellulose: A polymer of six carbon sugars found in all plant matter; the most abundant biological compound on earth.

Chemical clarification: Characterization of a wastewater process involving distinct operations: coagulation, flocculation and sedimentation.

Chemostat: An apparatus for maintaining microorganisms in the (exponential) phase of growth over prolonged periods of time. This is achieved by the continuous addition of fresh medium and the continuous removal of effluent, so that the volume of the growing culture remains constant.

Chemostat selection: Screening process used to identify microorganisms with desired properties, such as microorganisms that degrade toxic chemicals (see Acclimatization).

Chemotherapeutic agent: A chemical that interferes with the growth of either microorganisms or cancer cells at concentrations at which it is tolerated by the host cells.

Chemotherapy: The treatment of a disease by means of chemotherapeutic agents.

Chitin: A homopolysaccharide of *N*-acetyl-D-glucosamine that is a major constituent of the hard, horny exoskeleton of insects and crustaceans.

Chlorophyll: The green pigment that occurs in plants and functions in photosynthesis by absorbing and utilizing the radiant energy of the sun.

Chloroplasts: Cellular organelles where photosynthesis occurs.

Chromosome: A structure in the nucleus of eukaryotic cells that consists of one or more large double-helical DNA molecules that are associated with RNA and histones; the DNA of the chromosome contains the genes and functions in the storage and in the transmission of the genetic information of the organism.

Chromatography: A process of separating gases, liquids or solids in a mixture or solution by adsorption as the mixture or solution flows over the adsorbent medium, often in a column. The substances are separated because of their differing chemical interaction with the adsorbent medium.

Chromosomes: The rodlike structures of a cell's nucleus that store and transmit genetic information; the physical structures that contain genes. Chromosomes are composed mostly of DNA and protein and contain most of the cell's DNA. Each species has a characteristic number of chromosomes.

Clinical trial: One of the final stages in the collection of data for drug approval where the drug is tested in humans.

Clone: A group of genetically identical cells or organisms produced asexually from a common ancestor.

Cloning: The amplification of segments of DNA, usually genes.

Coagulation: The process whereby chemicals are added to wastewater resulting in a reduction of the forces tending to keep suspended particles apart.

Coagulation–flocculation aids: Materials used in relatively small concentrations which are added either to the coagulation and/or flocculation basins and may be classified as: oxidants, such as chlorine or ozone; weighting agents, such as bentomite clay; activated silica; and polyelectrolytes.

Coding sequence: The region of a gene (DNA) that encodes the amino acid sequence of a protein.

Codon: The sequence of three adjacent nucleotides that occurs in messenger RNA (mRNA) and that functions as a coding unit for a specific amino acid in protein synthesis. The codon determines which amino acid will be incorporated into the protein at a particular position in the polypeptide chain.

Coefficient of thermal conductivity: A physical parameter characterizing intensity of heat conduction in a substance; it is numerically equal to the conductive heat flux density due to a temperature gradient of unity.

Coenzyme: The organic molecule that functions as a cofactor of an enzyme.

Cofactor: The nonprotein component that is required by an enzyme for its activity. The cofactor may be either a metal ion (activator) or an organic molecule (coenzyme) and it may be attached either loosely or tightly to the enzyme; a tightly attached cofactor is known as a prosthetic group.

Colony: A group of contiguous cells that grow in or on a solid medium and are derived from a single cell.

Complementary DNA (cDNA): DNA that is complementary to messenger RNA; used for cloning or as a probe in DNA hybridization studies.

Compulsory licensing: Laws that require the licensing of patents, presumably to ensure early application of a technology and to diffuse control over a technology.

Constitutive enzyme: An enzyme that is present in a given cell in nearly constant amounts regardless of the composition of either the tissue or the medium in which the cell is contained.

Conjugation: The covalent or noncovalent combination of a large molecule, such as a protein or a bile acid, with another molecule. Also, the alternating sequence of single and double bonds in a molecule. Also, the genetic recombination in bacteria and in other unicellular organisms that resemble sexual reproduction and that entails a transfer of DNA between two cells of opposite mating type which are associated side by side.

Continuous processing: Method of processing in which raw materials are supplied and products are removed continuously, at volumetrically equal rates (*cf.* Batch processing).

Continuum: A medium whose discrete heterogeneous structure can be neglected, as in certain fluid flow problems.

Convective mass transfer: Mass transfer produced by simultaneous convection and molecular diffusion. The term is usually used to describe mass transfer associated with fluid flow and involves the mass transfer between a moving fluid and a boundary surface or between two immiscible moving fluids.

Convective transfer: The transfer of mass, heat or momentum in a medium with a nonhomogeneous distribution of velocity, temperature, or concentration; it is accompanied by the displacement of macroscopic elements through the medium.

Cosmid: A DNA cloning vector consisting of plasmid and phage sequences.

Corporate venture capital: Capital provided by major corporations exclusively for high-risk investments.

Crabtree effect: The inhibition of oxygen consumption in cellular respiration that is produced by increasing concentrations of glucose (see also Pasteur effect).

Critical dilution rate: Dilution rate at which wash-out conditions of cells in a bioreactor occurs (see Chemostat).

Cross flow filtration: Method of operating a filtration device whereby the processed material prevents undue build-up of filtered material on filter.

Crystalloid: A noncolloidal low-molecular weight substance.

Culture medium: Any nutrient system for the artificial cultivation of bacteria or other cells; usually a complex mixture of organic and inorganic materials.

Cyclic batch culture: Method of operating a bioreactor whereby a fill-and-dump approach retains enough biocatalyst to avoid need for re-inoculation. In cell cultures, relatively insignificant growth between fill and dump stages occurs.

Cytoplasm: The 'liquid' portion of a cell outside and surrounding the nucleus.

Cytotoxic: Damaging to cells.

Declining phase (or **Death phase**): The phase of growth of a culture of cells that follows the stationary phase and during which there is a decrease in the number (or the mass) of the cells.

Denitrification: The formation of molecular nitrogen from nitrate by way of nitrite.

Deoxyribonucleic acid (DNA): A linear polymer, made up of deoxyribonucleotide repeating units, that is the carrier of genetic information; present in chromosomes and chromosomal material of cell organelles such as mitochondria and chloroplasts, and also present in some viruses. The genetic material found in all living organisms. Every inherited characteristic has its origin somewhere in the code of each individual's DNA.

Diagnostic products: Products that recognize molecules associated with disease or other biological conditions and are used to diagnose these conditions.

Dialysis: The separation of macromolecules from ions and low-molecular weight compounds by means of a semipermeable membrane that is impermeable to (colloidal) macromolecules but is freely permeable to crystalloids and liquid medium.

Dicots (dicotyledons): Plants with two first embryonic leaves and nonparallel veined mature leaves. Examples are soybean and most flowering plants.

Diffusion boundary layer: Characterized by a transverse concentration gradient of a given component in a mixture; the effect of the gradient produces a transverse (mass transfer) of this component.

Diffusion coefficient: A physical parameter that appears as a proportionality coefficient with the gradient of concentration of a specified component in the equation which establishes the dependence of the mass diffusion flux density of the given component on the concentration gradients of all the components in the mixture. 'Self-diffusion coefficient' denotes a physical parameter which characterizes the diffusion of some molecules in the same medium with respect to others for a single-component medium.

Diffusional mass flux: Mass flux due to molecular diffusion.

Dilution rate: Reciprocal of the residence time of a culture in a bioreactor; given by the flow rate divided by bioreactor volume.

Dimensional analysis: Method of determining the number and structure of dimensionless groups consisting of variables essential to a given process on the basis of a comparison of the dimensions of these variables.

Diploidy (or **Diploid state**): The chromosome state in which each of the various chromosomes, except the sex chromosome, is represented twice.

Disclosure requirements: A patent requirement for adequate public disclosure of an invention that enables other people to build and use the invention without 'undue' experimentation.

Distal: Remote from a particular location or from a point of attachment.

DNA: Deoxyribonucleic acid (see above).

DNA base pair: A pair of DNA nucleotide bases. Nucleotide bases pair across the double helix in a very specific way: adenine can only pair with thymine; cytosine can only pair with guanine.

DNA probe: A sequence of DNA that is used to detect the presence of a particular nucleotide sequence.

DNA sequence: The order of nucleotide bases in the DNA helix; the DNA sequence is essential to the storage of genetic information.

DNA synthesis: The synthesis of DNA in the laboratory by the sequential addition of nucleotide bases.

Doubling time: The observed time required for a cell population to double in either the number of cells or the cell mass; it is equal to the generation time only if all the cells in the population are capable of doubling, have the same generation time, and do not undergo lysis (see Generation time).

Downstream processing: After bioconversion of materials in a bioreactor, the separation and purification of the product(s).

Drug: Any chemical compound that may be administered to humans or animals as an aid in the treatment of disease.

Dry weight: The weight of a sample from which liquid has been removed, usually by drying of filtered material.

Eddy mass diffusivity: A quantity characterizing the intensity of turbulent mass transfer of a particular component, as in intensely mixed bioreactors.

Electrophoresis: The movement of charged particles through a stationary liquid under the influence of an electric field. Electrophoresis is a tool for the separation of particles in both preparative and analytical studies of macromolecules. Separation is achieved primarily on the basis of the charge on the particles and to a lesser extent on the basis of the size and shape of the particles. Potentially useful in downstream processing.

Elution: The removal of adsorbed material from an adsorbent, such as the removal of a product from an enzyme bound in a chromatography column.

Emulsification: The process of making lipids, oils, fats, more soluble in water.

Endoenzyme: An enzyme that acts at random in cleaving molecules of substrate.

Endorphins: Opiate-like, naturally occurring peptides with a variety of analgesic effects throughout the endocrine and nervous systems.

Enrichment culture: A culture used for the selection of specific strains of an organism from among a mixture; such a culture favors the growth of the desired strain under the conditions used.

Enzyme: Any of a group of catalytic proteins that are produced by living cells and that mediate and promote the chemical processes of life without themselves being altered or destroyed.

Enzyme induction: The process whereby an inducible enzyme is synthesized in response to an inducer. The inducer combines with a repressor and thereby prevents the blocking of an operator by the repressor.

***Escherichia coli* (*E. coli*)**: A species of bacteria that inhabits the intestinal tract of most vertebrates. Some strains are pathogenic to humans and animals. Many nonpathogenic strains are used experimentally as hosts for rDNA.

Eukaryote: A cell or organism with membrane-bound, structurally discrete nuclei and well-developed cell organelles. Eukaryotes include all organisms except viruses, bacteria and blue-green algae (*cf*. Prokaryote).

Exoenzyme: An enzyme that acts by cleaving the ends of molecular chains in a substrate.

Exponential growth: The growth of cells in which the number of cells (or the cell mass) increases exponentially.

Export controls: Laws that restrict technology transfer and trade for reasons of national security, foreign policy or economic policy.

Facultative: Capable of living under more than one set of conditions, usually with respect to aerobic or anaerobic conditions (*e.g.* a facultative anaerobe is an organism or a cell that can grow either in the absence, or in the presence, of molecular oxygen).

Fatty acids: Organic acids with long carbon chains. Fatty acids are abundant in cell membranes and are widely used as industrial emulsifiers.

Fed-batch culture: As in 'cyclic batch culture' except that significant changes in the medium (*e.g.* cell growth) occur during the addition and/or removal of materials from bioreactor.

Feedback inhibition: A negative feedback mechanism in which a product of an enzymatic reaction inhibits the activity of an enzyme that functions in the synthesis of this product.

Feedstocks: Raw materials used for the production of chemicals.

Fermentation: A bioprocess. Fermentation is carried out in bioreactors and is used in various industrial processes for the manufacture of products such as antibiotics, alcohols, acids and vaccines by the action of living organisms (strictly speaking, anaerobically).

Film boiling: Boiling in which a continuous film of vapor that collapses periodically into the bulk of the liquid is formed on the heated surface, as in certain evaporation processes.

Flagellum: (pl. **flagella**): A threadlike, cellular extension that functions in the locomotion of bacterial cells and of unicellular eukaryotic organisms.

Flavin adenine dinucleotide (FAD): The flavin nucleotide, riboflavin adenosine diphosphate, which is a coenzyme form of the vitamin riboflavin, and which functions in dehydrogenation reactions catalyzed by flavoproteins.

Flocculating agent: A reagent added to a dispersion of solids in a liquid to bring together the fine particles into larger masses.

Flocculation: The agglomeration of suspended material to form particles that will settle by gravity, as in the 'tertiary' treatment of waste materials.

Food additive (or **Food ingredient**): A substance that becomes a component of food or affects the characteristics of food and, as such, is regulated, *e.g.* by the US Food and Drug Administration.

Forced convection: Motion of fluid elements induced by external forces, *e.g.* in a bioreactor by a mechanical stirrer.

Free convection: Motion of fluid elements induced by 'natural' forces, *e.g.* by density differences caused by concentration or temperature gradients.

Free-living organism: An organism that does not depend on other organisms for survival.

Fruiting body: A mass of vegetative cells which swarm together at the same stage of growth.

Fungus: Any of a major group of saprophytic and parasitic plants that lack chlorophyll, including molds, rusts, mildews, smuts and mushrooms.

Gamma globulin (GG): A protein component of blood that contains antibodies and confers passive immunity.

Gene: The basic unit of heredity; an ordered sequence of nucleotide bases, comprising a segment of DNA. A gene contains the sequence of DNA that encodes one polypeptide chain (*via* RNA).

Gene amplification: In biotechnology, an increase in gene number for a certain protein so that the protein is produced at elevated levels.

Gene expression: The mechanism whereby the genetic directions in any particular cell are decoded and processed into the final functioning product, usually a protein (see also Transcription and Translation).

Gene transfer: The use of genetic or physical manipulation to introduce foreign genes into host cells to achieve desired characteristics in progeny.

Generation time: The time required by a cell for the completion of one cycle of growth (see also Doubling time).

Genetic engineering: Loose term used to describe any gene manipulative technique, especially recombinant DNA techniques.

Genome: The genetic endowment of an organism or individual.

Genus: A taxonomic category that includes groups of closely related species.

Germ cell: The male and female reproductive cells; egg and sperm.

Germplasm: The total genetic variability available to a species.

Glycoproteins: Proteins with attached sugar groups.

Glycoside: A mixed acetal (or ketal) derived from the cyclic hemiacetal (or hemiketal) form of an aldose (or a ketose); a compound formed by replacing the hydrogen or the hydroxyl group of the anomeric carbon of the carbohydrate with an alkyl or aryl group.

Glycosylation: The attachment of sugar groups to a molecule, such as a protein.

Gram negative: Designating a bacterium that does not retain the initial Gram stain but retains the counterstain. Gram-negative bacteria possess a relatively thin cell wall that is not readily digested by the enzyme lysozyme, and in which the peptidoglycan layer is covered with lipopolysaccharide.

Gram positive: Designating a bacterium that retains the initial Gram stain and is not stained by the counterstain. Gram-positive bacteria generally possess a relatively thick and rigid cell wall that is readily digested by the enzyme lysozyme, and that consists of a layer of peptidoglycan.

Gram stain: A set of two stains (chemicals) that are used to stain bacteria; the staining depends on the composition and the structure of the bacterial cell wall.

Growth hormone (GH): A group of peptides involved in regulating growth in higher animals.

Heat flux: The quantity of heat that passes through an arbitrary surface per unit time.

Heat flux density: Heat flux per unit area.

Heat transfer: Spontaneous irreversible process of heat transmission in a space with a nonisothermal temperature field, as in the cooling or heating of bioreactors and ancilliary equipment.

Hemicellulose: A polymer of D-xylose that contains side chains of other sugars and that serves to cement plant cellulose fibers together.

Herbicide: An agent (*e.g.* a chemical) used to destroy or inhibit plant growth; specifically, a selective weed killer that is not injurious to crop plants.

Heterofermentative lactic acid bacteria: Lactic acid bacteria that produce in fermentation less than 1.8 moles of lactic acid per mole of glucose; in addition to lactic acid, these organisms produce ethanol, acetate, glycerol, mannitol and carbon dioxide (see also Homofermentative lactic acid bacteria).

Heterotroph: A cell or organism that requires a variety of carbon-containing compounds from animals and plants as its source of carbon, and that synthesizes all of its carbon-containing biomolecules from these compounds and from small inorganic molecules.

Heterotrophic: Pertaining to a regulatory enzyme in which the effector is a metabolite other than the substance of the enzyme.

High performance liquid chromatography (HPLC): A recently developed type of chromatography that is potentially important in downstream processing of bioreactor products.

Histone: A basic, globular, and simple protein that is characterized by its high content of arginine and lysine. Histones are found in association with nucleic acids in the nuclei of many eukaryotic cells.

Homofermentative lactic acid bacteria: Lactic acid bacteria that produce 1.8–2.0 moles of lactic acid per mole of glucose during fermentation.

Hormone: A chemical messenger found in the circulation of higher organisms that transmits regulatory messages to cells.

Host: A cell whose metabolism is used for growth and reproduction of a virus, plasmid or other form of foreign DNA.

Host–vector system: Compatible combinations of host (*e.g.* bacterium) and vector (*e.g.* plasmid) that allow stable introduction of foreign DNA into cells.

Hybrid: The offspring of two genetically dissimilar parents (*e.g.* a new variety of plant or animal that results from cross-breeding two different existing varieties; a cell derived from two different cultured cell lines that have fused).

Hybridization: The act or process of producing hybrids.

Hybridoma: Product of fusion between myeloma cell (which divides continuously in culture and is 'immortal') and lymphocyte (antibody-producing cell); the resulting cell grows in culture and produces monoclonal antibodies.

Hybridoma technology: See Monoclonal antibody technology.

Hydrolysis: Chemical reaction involving addition of water to break bonds.

Hydroxylation: Chemical reaction involving the addition of hydroxyl (OH) groups to chemical compounds.

Hypha (pl. **hyphae**): The filamentous and branched tube that forms the network which contains the cytoplasm of the mycelium of a fungus.

Immobilized enzyme or cell techniques: Techniques used for the fixation of enzymes or cells on to solid supports. Immobilized cells and enzymes are used in continuous bioprocessing in bioreactors and upstream or downstream processing of materials.

Immune response: The reaction of an organism to invasion by a foreign substance. Immune responses are often complex, and may involve the production of antibodies from special cells (lymphocytes), as well as the removal of the foreign substance by other cells.

Immunization: The administration of an antigen to an animal organism to stimulate the production of antibodies by that organism. Also, the administration of antigens, antibodies or lymphocytes to an animal organism to produce the corresponding active, passive or adoptive immunity.

Immunoassay: The use of antibodies to identify and quantify substances. The binding of antibodies to antigen, the substance being measured, is often followed by tracers such as radioisotopes.

Immunogenic: Capable of causing an immune response (see also Antigen).

Immunotoxin: A molecule attached to an antibody capable of killing cells that display the antigen to which the antibody binds.

Inducible enzyme: An enzyme that is normally either absent from a cell or present in very small amounts, but that is synthesized in appreciable amounts in response to an inducer in the process medium.

Interface: The boundary between two phases.

Interferons (Ifns): A class of glycoproteins (proteins with sugar groups attached at specific locations) important in immune function and thought to inhibit viral infections.

In vitro: Literally, in glass; pertaining to a biological reaction taking place in an artificial apparatus; sometimes used to include the growth of cells from multicellular organisms under cell culture conditions. *In vitro* diagnostic products are products used to diagnose disease outside of the body after a sample has been taken from the body.

In vivo: Literally, in life; pertaining to a biological reaction taking place in a living cell or organism. *In vivo* products are products used within the body.

Ionic strength: A measure of the ionic concentration of a solution.

Laminar flow: Fluid motion in which the existence of steady fluid particle trajectories can exist; in processing equipment, it represents relatively low levels of mixing intensities.

Isoelectric pH (isoelectric point): The pH at which a molecule has a net zero charge; the pH at which the molecule has an equal number of positive and negative charges, which includes those due to any ions bound by the molecule.

Isoelectrophoretic pH (isoelectrophoretic point): The pH at which the electrophoretic mobility of a protein is zero; this pH may coincide with the theoretical isoelectric pH of the protein, depending on the surface structure of the protein, the ionic strength, and the nature of the ionic double layer around the protein.

Lag phase: That phase of growth of a cell that precedes the exponential phase and during which there is only little or no growth.

Leaching: The removal of a soluble compound such as an ore from a solid mixture by washing or percolating.

Lignocellulose: The composition of woody biomass, including lignin and cellulose.

Lignolytic: Pertaining to the breakdown of lignin.

Lime: Various natural forms of the chemical compound CaO, as in hydrated lime and dolomitic lime.

Linker: A small fragment of synthetic DNA that has a restriction site useful for gene cloning, which is used for joining DNA strands together.

Lipase: An enzyme that catalyzes the hydrolysis of fats.

Lipids: A large, varied class of water-insoluble fat-based organic molecules; includes steroids, fatty acids, prostaglandins, terpenes and waxes.

Lipopolysaccharide: A water-soluble lipid–polysaccharide complex.

Liposome transfer: The process of enclosing biological compounds inside a lipid membrane and allowing the complex to be taken up by a cell.

Lymphocytes: Specialized white blood cells involved in the immune response; B lymphocytes produce antibodies.

Lymphokines: Proteins that mediate interactions among lymphocytes and are vital to proper immune function.

Lyophilization: The removal of water under vacuum from a frozen sample; a relatively gentle process in which water sublimes.

Lysis: The rupture and dissolution of cells.

Mass exchange: Mass transfer across an interface or a permeable wall (membrane) between two phases.

Mass flux: The mass of a given mixture component passing per unit time across any surface.

Mass flux density: Mass flux per unit area of surface.

Mass transfer: Spontaneous irreversible process of transfer of mass of a given component in a space with a nonhomogenous field of the chemical potential of the component. In the simplest case, the driving force is the difference in concentration (in liquids) or partial pressure (in gases) of the component. Other physical quantities, *e.g.* temperature difference (thermal diffusion), can also induce mass transfer.

Mass transfer coefficient: A quantity characterizing the intensity of mass transfer; it is numerically equal to the ratio of the mass flux to the difference of its mass fractions. For the case of mass transfer between a liquid medium and a gas, the mass fraction of a given component in the liquid is determined by phase equilibrium parameters (distribution coefficient) with allowance, if necessary, for resistance to transfer at the phase boundary *per se*.

Mass velocity: Mass flow rate across a unit area perpendicular to the direction of the velocity vector.

Mesophile: An organism that grows at moderate temperatures in the range 20–45 °C, and that has an optimum growth temperature in the range 30–39 °C.

Mesophilic: Of, or pertaining to, mesophiles.

Messenger RNA (mRNA): RNA that serves as the template for protein synthesis; it carries the transcribed genetic code from the DNA to the protein synthesizing complex to direct protein synthesis.

Metabolism: The physical and chemical processes by which chemical components are synthesized into complex elements, complex substances are transformed into simpler ones, and energy is made available for use by an organism.

Metabolite: Any reactant, intermediate or product in the reactions of metabolism.

Metallothioneins: Proteins, found in higher organisms, that have a high affinity for heavy metals.

Methanogens: Bacteria that produce methane as a metabolic product.

Microorganisms: Microscopic living entities; microorganisms can be viruses, prokaryotes (*e.g.* bacteria) or eukaryotes (*e.g.* fungi). Also referred to as microbes.

Microencapsulation: The process of surrounding cells with a permeable membrane.

Mitochondrion (pl. **mitochondria**): A subcellular organelle in aerobic eukaryotic cells that is the site of cellular respiration and that carries out the reactions of the citric acid cycle, electron transport, and oxidative phosphorylation. Mitochondria contain DNA and ribosomes, carry out protein synthesis, and are capable of self-replication.

Mixed culture: Culture containing two or more types of microorganisms.

Molecular diffusion: Mass transfer resulting from thermal motion. Concentration diffusion refers to molecular diffusion resulting from a nonhomogenous distribution of concentrations of components of a mixture.

Monoclonal antibodies (MAbs): Homogeneous antibodies derived from a single clone of cells; MAbs recognize only one chemical structure. MAbs are useful in a variety of industrial and medical capacities since they are easily produced in large quantities and have remarkable specificity.

Monoclonal antibody technology: The use of hybridomas that produce monoclonal antibodies for a variety of purposes. Hybridomas are maintained in cell culture or, on a larger scale, as tumors (ascites) in mice. Also referred to as 'hybridoma' technology.

Monocots (monocotyledons): Plants with single first embryonic leaves, parallel-veined leaves, and simple stems and roots. Examples are cereal grains such as corn, wheat, rye, barley and rice.

Monosaccharide: A polyhydroxy alcohol containing either an aldehyde or a ketone group; a simple sugar.

Multigenic: A trait specialized by several genes.

Multi-phase medium: A medium consisting of two or more single-phase portions with physical properties changing discontinuously (stepwise) at the boundaries of the medium, as in a gas–liquid dispersion used in aerobic fermentations.

Mutagenesis: The induction of mutation in the genetic material of an organism; researchers may use physical or chemical means to cause mutations that improve the production of capabilities of organisms.

Mutagen: An agent that causes mutation.

Mutant: An organism with one or more DNA mutations, making its genetic function or structure different from that of a corresponding wild-type organism.

Mutation: A permanent change in a DNA sequence.

Mycelium (pl. **mycelia**): The vegetative structure of a fungus that consists of a multinucleate mass of cytoplasm, enclosed within a branched network of filamentous tubes known as hyphae.

Myeloma: Antibody-producing tumor cells.

Myeloma cell line: Myeloma cells established in culture.

Natural convection: Free motion due to gravitational forces in a system with a non-homogeneous density distribution (see also Free convection).

Neurotransmitters: Small molecules found at nerve junctions that transmit signals across those junctions.

Newtonian fluid: A fluid, the viscosity of which is independent of the rate and/or duration of shear.

NIH Guidelines: Guidelines, established by the US National Institutes of Health, on the safety of research involving recombinant DNA.

Nitrogen fixation: The conversion of atmospheric nitrogen gas to a chemically combined form, ammonia (NH_3), which is essential to growth. Only a limited number of microorganisms can fix nitrogen.

Nodule: The anatomical part of a plant root in which nitrogen-fixing bacteria are maintained in a symbiotic relationship with the plant.

Nodulins: Proteins, possibly enzymes, present in nodules; function unknown.

Non-Newtonian fluid: A fluid, the viscosity of which depends on the rate and/or duration of shear.

Nucleate boiling: Boiling in which vapor is generated in the form of periodically produced and growing discrete bubbles.

Nucleic acids: Macromolecules composed of sequences of nucleotide bases. There are two kinds of nucleic acids: DNA, which contains the sugar deoxyribose, and RNA, which contains the sugar ribose.

Nucleoside: A glycoside composed of D-ribose or 2-deoxy-D-ribose and either a purine or a pyrimidine.

Nucleotide base: A structural unit of nucleic acid. The bases present in DNA are adenine, cytosine, guanine and thymine. In RNA, uracil substitutes for thymine.

Nucleus: In the biological sciences, a relatively large spherical body inside a cell that contains the chromosomes.

Oligomer: A molecule that consists of two or more monomers linked together, covalently or non-covalently.

Oligonucleotides: Short segments of DNA or RNA.

Optical density (absorbance): A measure of the light absorbed by a solution.

Organelle: A specialized part of a cell that conducts certain functions. Examples are nuclei, chloroplasts and mitochondria, which contain most of the genetic material, conduct photosynthesis and provide energy, respectively.

Organic compounds: Molecules that contain carbon.

Organic micropollutant: Low molecular weight organic compounds considered hazardous to humans or the environment.

Osmosis: The movement of water or another solvent across a semipermeable membrane from a region of low solute concentration to one of a higher solute concentration.

Osmotic pressure: The pressure that causes water or another solvent to move in osmosis from a solution having a low solute concentration to one having a high solute concentration; it is equal to the hydrostatic pressure that has to be applied to the more concentrated solution to prevent the movement of water (solvent) into it.

Oxidation ponds: Quiescent earthen basins that provide sufficient hydraulic hold-up time for the natural processes to effect removal and stabilization of organic matter.

Oxygen transfer rate (OTR): Mass transfer for oxygen solute as in fermentation medium.

Parasite: An organism that lives in or upon another organism from which it derives some or all of its nutrients.

Passive immunity: Disease resistance in a person or animal due to the injection of antibodies from another person or animal. Passive immunity is usually short-lasting (*cf.* Active immunity).

Pasteur effect: The inhibition of glycolysis and the decrease of lactic acid accumulation that is produced by increasing concentrations of oxygen.

Patent: A limiting property right granted to inventors by a government allowing the inventor of a new invention the right to exclude all others from making, using or selling the invention unless specifically approved by the inventor, for a specified time period in return for full disclosure by the inventor about the invention.

Pathogen: A disease-producing agent, usually restricted to a living agent such as a bacterium or virus.

Pectin: A polysaccharide that occurs in fruits and that consists of a form of pectic acid in which many of its carboxyl groups have been methylated.

Peptide: A linear polymer of amino acids. A polymer of numerous amino acids is called a *polypeptide*. Polypeptides may be grouped by function, such as 'neuroactive' polypeptides.

Permease: An enzyme that is instrumental in transporting material across a biological membrane or within a biological fluid.

Pharmaceuticals: Products intended for use in humans, as well as *in vitro* applications to humans, including drugs, vaccines, diagnostics and biological response modifiers.

Photorespiration: Reaction in plants that competes with the photosynthetic process. Instead of fixing CO_2, RuBPCase can utilize oxygen, which results in a net loss of fixed CO_2.

Photosynthesis: The reaction carried out by plants where carbon dioxide from the atmosphere is fixed into sugars in the presence of sunlight; the transformation of solar energy into biological energy.

Plasma: The liquid (noncellular) fraction of blood. In vertebrates, it contains many important proteins (*e.g.* fibrinogen, responsible for clotting).

Plasmid: An extrachromosomal, self-replicating, circular segment of DNA; plasmids (and some viruses) are used as 'vectors' for cloning DNA in bacterial 'host' cells.

Plug flow: Flow of materials in which there is no mixing in the direction of flow (see Axial dispersion).

Polymer: A linear or branched molecule of repeating subunits.

Polypeptide: A long peptide, which consists of amino acids.

Polysaccharide: A polymer of sugars.

Pool boiling: Boiling with convective (free) motion in a liquid volume whose dimensions in all directions are large compared to the breakaway diameter of the bubble.

Primary metabolite: Metabolite that is required for the function of the organism's life support system.

Proinsulin: A precursor protein of insulin.

Prokaryote: A cell or organism lacking membrane-bound, structurally discrete nuclei and organelles. Prokaryotes include bacteria and the blue-green algae. (*cf.* eukaryote).

Promoter: A DNA sequence in front of a gene that controls the initiation of 'transcription' (see below).

Prophylaxis: Prevention of disease.

Protease: Protein-digesting enzyme.

Protein: A polypeptide consisting of amino acids. In their biologically active states, proteins function as catalysts in metabolism and, to some extent, as structural elements of cells and tissues.

Protist: A unicellular or multicellular organism that lacks the tissue differentiation and the elaborate organization that is characteristic of plants and animals.

Protoplast fusion: The joining of two cells in the laboratory to achieve desired results, such as increased viability of antibiotic-producing cells.

Protozoa: Diverse forms of eukaryotic microorganisms; structure varies from simple single cells to colonial forms; some protozoa are pathogenic.

Psychrophile: An organism that grows at low temperatures in the range of 0–25 °C, and that has an optimum growth temperature in the range 20–25 °C.

Pure culture: A culture containing only microorganisms from one species.

Pyrogenicity: The tendency for some bacterial cells or parts of cells to cause inflammatory reactions in the body, which may detract from their usefulness as pharmaceutical products.

Recarbonation: Unit water treatment process in which carbon dioxide is added to a lime-treated water. Basic purpose is the downward adjustment of the pH of the water.

Recombinant DNA (rDNA): The hybrid DNA produced by joining pieces of DNA from different organisms together *in vitro* (*i.e.* in an artificial apparatus).

Recombinant DNA technology: The use of recombinant DNA for a specific purpose, such as the formation of a product or the study of a gene.

Recombination: Formation of a new association of genes or DNA sequences from different parental origins.

Reducing sugar: A sugar that will reduce certain inorganic ions in solution, such as the copper(II) ions of Fehling's or Benedict's reagent; the reducing property of the sugar is due to its aldehyde or potential aldehyde group.

Regeneration: In biological sciences, the laboratory process of growing a whole plant from a single cell or small clump of cells.

Regulatory sequence: A DNA sequence involved in regulating the expression of a gene.

Repressible enzyme: An enzyme, the synthesis of which is decreased when the intracellular concentration of specific metabolites reaches a certain level.

Resistance gene: Gene that provides resistance to an environmental stress such as an antibiotic or other chemical compound.

Respiration: The cellular oxidative reactions of metabolism, particularly the terminal steps, by which nutrients are broken down; the reactions which require oxygen as the terminal electron acceptor, produce carbon dioxide as a waste product, and yield utilizable energy.

Restriction enzymes: Enzymes that cut DNA at specific DNA sequences.

Ribosome: One of a large number of subcellular, nucleoprotein particles that are composed of approximately equal amounts of RNA and protein and that are the sites of protein synthesis in the cell.

Ribosomal RNA: The RNA that is linked noncovalently to the ribosomal proteins in the two ribosomal subunits and that constitutes about 80% of the total cellular RNA.

RNA: Ribonucleic acid (see also Messenger RNA).

RuBPCase (ribulosebisphosphate carboxylase): An enzyme that catalyzes the critical step of the photosynthetic CO_2 cycle.

Salting out: The decrease in the solubility of a protein that is produced in solutions of high ionic strength by an increase of the concentrations of neutral salts.

Saccharification: The degradation of polysaccharides to sugars.

Scale-up: The transition of a process from research laboratory bench scale to engineering pilot plant or industrial scale.

Secondary metabolite: Metabolite that is not required by the producing organism for its life-support system.

Semiconductor: A material such as silicon or germanium with electrical conductivities intermediate between good conductors such as copper wire and insulators such as glass.

Shake flask: A laboratory flask for culturing microorganisms in a shaker–incubator which provides mixing and aeration.

Shear rate: The variation in velocity within a flowing material, as in a bioreactor.

Single cell protein (SCP): Cells, or protein extracts, of microorganisms grown in large quantities for use as human or animal protein supplements. A misnomer for multicellular SCP products.

Single-phase medium: Continuous single- or multi-component medium whose properties in space can vary in a continuous manner with no phase boundaries, *e.g.* a gas or a liquid.

Slaking: Process of adding water to quicklime, or recalcined lime, to produce a slurry of hydrated lime.

Slant culture: A culture grown in a tube that contains a solid nutrient medium which was solidified while the tube was kept in a slanted position.

Slimes: Aggregations of microbial cells that pose environmental and industrial problems; may be amenable to biological control.

Sludge: Precipitated solid matter produced by water and sewage treatment or industrial problems; may be amenable to biological control.

Slug flow: Type of two-phase flow in which the gas phase flows in the form of large bubbles whose transverse dimensions are commensurate with the characteristic dimension of the flow cross section as in some pipeline operations.

Somaclonal variation: Genetic variation produced from the culture of plant cells from a pure breeding strain; the source of the variation is not known.

Species: A taxonomic subdivision of a genus. A group of closely related, morphologically similar individuals which actually or potentially interbreed.

Specific growth rate: The rate of growth of a population of microorganisms, per unit mass of cells.

Spectrometer: An instrument used for analyzing the structure of compounds on the basis of their light-absorbing properties.

Spheroplast: A bacterial cell that is largely, but not entirely, freed of its cell wall.

Spore: A dormant cellular form, derived from a bacterial or a fungal cell, that is devoid of metabolic activity and that can give rise to a vegetative cell upon germination; it is dehydrated and can survive for prolonged periods of time under drastic environmental conditions.

Starch: The major form of storage carbohydrates in plants. It is a homopolysaccharide, composed of D-glucose units, that occurs in two forms: amylose, which consists of straight chains, and in which the glucose residues are linked by means of alpha(1–4) glycosidic bonds; and amylopectin, which consists of branched chains, and in which the glucose residues are linked by means of both alpha(1–4) and alpha(1–6) glycosidic bonds.

Stationary phase: The phase of growth of a culture of microorganisms that follows the exponential phase and in which there is little or no growth.

Sterile: Free from viable microorganisms.

Sterilization: The complete destruction of all viable microorganisms in a material by physical and/or chemical means.

Steroid: A group of organic compounds, some of which act as hormones to stimulate cell growth in higher animals and humans.

Stirred tank bioreactor: Agitated vessel in which a bioprocess takes place; mixing is provided by the mechanical action of an impeller/agitator.

Storage protein genes: Genes coding for the major proteins found in plant seeds.

Strain: A group of organisms of the same species having distinctive characteristics but not usually considered a separate breed or variety. A genetically homogeneous population of organisms at a subspecies level that can be differentiated by a biochemical, pathogenic or other taxonomic feature.

Substrate: A substance acted upon, for example, by an enzyme.

Subunit vaccine: A vaccine that contains only portions of a surface molecule of a pathogen. Subunit vaccines can be prepared by using rDNA technology to produce all or part of the surface protein molecule or by artificial (chemical) synthesis of short peptides.

Surfactant: A substance that alters the surface tension of a liquid, generally lowering it; detergents and soaps are typical examples.

Symbiont: An organism living in symbiosis, usually the smaller member of a symbiotic pair of dissimilar size.

Symbiosis: In the biological sciences, the living together of two dissimilar organisms in mutually beneficial relationships.

Synchronous growth: Growth in which all of the cells are at the same stage in cell division at any given time.

Tangential flow filtration: See Cross-flow filtration.

Taxis: The movement of an organism in response to a stimulus.

Taxonomy: The scientific classification of plants and animals that is based on their natural relationships; includes the systematic grouping, ordering and naming of the organisms.

Therapeutics: Pharmaceutical products used in the treatment of disease.

Thermal diffusivity: Numerically equal to the ratio of the coefficient of thermal conductivity to the volumetric specific heat of a substance (see also Heat transfer).

Thermophile: An organism that grows at high temperatures in the range 45–70 °C (or higher temperatures) and that has an optimum growth temperature in the range 50–55 °C.

Thermophilic: Heat loving. Usually refers to microorganisms that are capable of surviving at elevated temperatures; this capability may make them more compatible with industrial biotechnology schemes.

Thermotolerant: Capable of withstanding relatively high temperatures (45–70 °C).

Thrombolytic enzymes: Enzymes such as streptokinase and urokinase that initiate the dissolution of blood clots.

Ti plasmid: Plasmid from *Agrobacterium tumefacciens*, used as a plant vector.

Tissue plasminogen activator (TPA): A hormone that selectively dissolves blood clots that cause heart attacks and strokes.

Totipotency: The capacity of a higher organism cell to differentiate into an entire organism. A totipotent cell contains all the genetic information necessary for complete development.

Toxicity: The ability of a substance to produce a harmful effect on an organism by physical contact, ingestion or inhalation.

Toxin: A substance, produced in some cases by disease-causing microorganisms, which is toxic to other living organisms.

Toxoid: Detoxified toxin, but with antigenic properties intact.

Transcription: The synthesis of messenger RNA on a DNA template; the resulting RNA sequence is complementary to the DNA sequence. This is the first step in gene expression (see also Translation).

Transfer RNA (tRNA): A low-molecular weight RNA molecule, containing about 70–80 nucleotides, that binds an amino acid and transfers it to the ribosomes for incorporation into a polypeptide chain during translation.

Transformation: In the biological sciences, the introduction of new genetic information into a cell using naked DNA.

Transistor: An active component of an electrical circuit consisting of semiconductor material to which at least three electrical contacts are made so that it acts as an amplifier, detector or switch.

Translation: In the biological sciences, the process in which the genetic code contained in the nucleotide base sequence of messenger RNA directs the synthesis of a specific order of amino acids to produce a protein. This is the second step in gene expression (see also Transcription).

Transposable element: Segment of DNA which moves from one location to another among or within chromosomes in possibly a predetermined fashion, causing genetic change; may be useful as a vector for manipulating DNA.

Turbulent flow: In the engineering sciences, fluid motion with particle trajectories varying chaotically (randomly) with time; irregular fluctuations of velocity, pressure and other parameters, non-uniformly distributed in the flow; indicates relatively high levels of mixing (*cf.* Laminar flow).

Vaccine: A suspension of attenuated or killed bacteria or viruses, or portions thereof, injected to produce active immunity (see also Subunit vaccine).

Vector: DNA molecule used to introduce foreign DNA into host cells. Vectors include plasmids, bacteriophages (virus) and other forms of DNA. A vector must be capable of replicating autonomously and must have cloning sites for the introduction of foreign DNA.

Viable: Describing a cell or an organism that is alive and capable of reproduction.

Viable count: The number of viable cells in a culture of microorganisms.

Virus: Any of a large group of submicroscopic agents infecting plants, animals and bacteria and unable to reproduce outside the tissues of the host. A fully formed virus consists of nucleic acid (DNA or RNA) surrounded by a protein or protein and lipid coat.

Viscosity: A measure of a liquid's resistance to flow.

Volatile fatty acids (VFAs): Mixture of acids, primarily acetic and propionic, produced by acidogenic microorganisms during anaerobic digestion.

Volatile organic compounds (VOCs): Group of toxic compounds found in ground water and that pose environmental hazards; their destruction during water purification may be done biologically.

Wash out: Condition at which the critical dilution rate is exceeded in a chemostat.

Wild-type: The most frequently encountered phenotype in natural breeding populations.

Yeast: A fungus of the family Saccharomycetacea that is used especially in the making of alcoholic liquors and fodder yeast for animal feeds, and as leavening in baking. Yeasts are also commonly used in bioprocesses.

Yield: For a general chemical reaction: the weight of product obtained divided by the theoretical amount expected. Also, for the isolation of an enzyme: the total activity at a given step in the isolation divided by the total activity at a reference step.

Appendix 2:
Nomenclature Guidelines

Provisional list of symbols and units recommended for use in biotechnology by the IUPAC Commission on Biotechnology as reported in *Pure Appl. Chem.*, **54**, 1743–1749 (1982). This list is intended for use in conjunction with other recommendations on symbols, in particular *Manual of Symbols and Terminology for Physicochemical Quantities and Units* (Pergamon, 1979) and 'Letter Symbols for Chemical Engineering', *Chem. Eng. Prog.*, 73–80 (1978).

A2.1 GENERAL CONCEPTS

	Symbol	SI units	Other units
Activation energy	E	$J\,mol^{-1}$	$cal\,mol^{-1}$
for growth	E_g		
for death	E_d		
Area dimensions			
area per unit volume	a	m^{-1}	cm^{-1}
Linear dimension			
impeller diameter	D_i	m	m
tank diameter	D_t	m	m
liquid depth	D_l	m	m
width of baffle	D_b	m	m
Pressure	p	Pa	atm, bar
denote partial pressure with appropriate subscript, *e.g.* P_{O_2} for partial pressure of oxygen			
Ratio, in general	R		
for stoichiometric mass ratio, *e.g.* mass of substrate A consumed per mass of substrate B consumed	$R_{A/B}$		
for stoichiometric molar ratio, *e.g.* mole of substrate A consumed per mole of B consumed	$R_{MA/B}$		
Temperature			
absolute	T	K	K
general	t, T	°C	°C
Time	t	s	min, h
identify specific time periods by appropriate subscripts, *e.g.* t_d for			

	Symbol	SI units	Other units

doubling time, t_l for lag time, and t_r for replacement or mean residence time

Volume dimensions

	Symbol	SI units	Other units
volume	V	m^3, L	L

identify by subscript, *e.g.* V_1 for volume of stage 1, *etc.*

	Symbol	SI units	Other units
Yield, general mass ratio expressing output over input	Y		

without further definition, Y refers to the mass conversion ratio in terms of g dry weight biomass per g mass of substrate used. It should be further defined by subscripts to denote other ratios, *e.g.* $Y_{P/S}$ and $Y_{P/X}$ for g mass of product per g mass of substrate and per g dry weight of biomass, respectively

	Symbol	SI units	Other units
Yield, growth mass ratio corrected for maintenance, where	Y_G		

$$\frac{1}{Y} = \frac{1}{Y_G} + \frac{m}{\mu}$$

or

$$q_s = \frac{\mu}{Y_G} + m$$

	Symbol	SI units	Other units
Yield, molar growth	Y_{GM}	$kg\,mol^{-1}$	$g\,mol^{-1}$

kg biomass formed per mole of mass used, or further defined as above to denote other molar yields

A2.2   CONCENTRATIONS AND AMOUNTS

Concentration

Biomass*

	Symbol	SI units	Other units
total mass (dry wt. basis)	x	kg	g
mass concentration (dry wt. basis)	X	$kg\,m^{-3}$	$g\,L^{-1}$
volume fraction	ϕ		
total number	N		
number concentration	n	m^{-3}	
Substrate concentration mass or moles per unit volume	C_S	$kg\,m^{-3}$, $kmol\,m^{-3}$	$mg\,L^{-1}$, $mmol\,L^{-1}$
Product concentration mass or moles per unit volume	C_P	$kg\,m^{-3}$, $kmol\,m^{-3}$	$mg\,L^{-1}$, $mmol\,L^{-1}$
Gas hold-up volume of gas per volume of dispersion	ε_G		

* Note: because of the difficulty in expressing biomass (cells) in molar terms, a separate symbol (other than C) is recommended.

	Symbol	SI units	Other units
Inhibitor concentration mass or moles per unit volume	C_i	$kg\ m^{-3}$, $kmol\ m^{-3}$	$mg\ L^{-1}$, $mmol\ L^{-1}$
Inhibitor constant dissociation constant of inhibitor–biomass complex	K_i	$kg\ m^{-3}$	$g\ L^{-1}$
Saturation constant as in the growth rate expression $\mu = \mu_m C_S/(K_s + C_S)$	K_s	$kmol\ m^{-3}$	$g\ L^{-1}$, $mmol\ L^{-1}$
Total amount, *e.g.* mass or moles	C	kg, kmol	g, mol

A2.3 INTENSIVE PROPERTIES

	Symbol	SI units	Other units
Density, mass	ϱ	$kg\ m^{-3}$	$g\ L^{-1}$
Diffusivity, molecular, volumetric	D_v	$m^2\ s^{-1}$	$cm^2\ s^{-1}$
Enthalpy, mass, of growth heat produced per unit of dry weight biomass formed	H_X	$J\ kg^{-1}$	$J\ g^{-1}$
Enthalpy, molar, of substrate consumption or of product formation	H_S, H_P	$J\ mol^{-1}$	$J\ mol^{-1}$
Vapor pressure denote with appropriate subscript, *e.g.* $p_i{}^*$ = vapor pressure of material i	p^*	Pa	atm, bar
Viscosity, absolute	μ	Pa s	poise
Viscosity, kinematic	v	$m^2\ s^{-1}$	$cm^2\ s^{-1}$

A2.4 RATE CONCEPTS

	Symbol	SI units	Other units
Death rate, specific $\delta = -(dn/dt)/n$	δ	s^{-1}	s^{-1}, h^{-1}
Dilution rate volume flow rate/culture volume	D	s^{-1}	h^{-1}, d^{-1}
Dilution rate, critical value at which biomass washout occurs in continuous flow culture	D_c	s^{-1}	h^{-1}, d^{-1}
Doubling time, biomass $t_d = (\ln 2)/\mu$	t_d	s	min, h
Flow rate, volumetric identify stream by appropriate subscript, *e.g.* A for air, G for gas, L for liquid, *etc.*	F	$m^3\ s^{-1}$	$L\ h^{-1}$
Growth rate, colony radial rate of extension of biomass colony on a surface	K_r	$m\ s^{-1}$	$\mu m\ h^{-1}$
Growth rate, maximum specific	μ_m	s^{-1}	h^{-1}, d^{-1}
Growth rate, specific $\mu = (dx/dt)/x$	μ	s^{-1}	h^{-1}, d^{-1}

	Symbol	*SI units*	*Other units*
Heat transfer coefficient			
individual	h	$\text{W m}^{-2}\,\text{K}^{-1}$	$\text{cal h}^{-1}\,\text{cm}^{-2}\,^{\circ}\text{C}^{-1}$
overall	U	$\text{W m}^{-2}\,\text{K}^{-1}$	$\text{cal h}^{-1}\,\text{cm}^{-2}\,^{\circ}\text{C}^{-1}$
Maintenance coefficient, substrate or non-growth term associated with substrate consumption as defined in yield relationship (see yield term)	m	s^{-1}	h^{-1}
Mass transfer coefficient (molar basis)			
Individual, area basis	k	$\text{kmol m}^{-2}\,\text{s}^{-1}\,(\text{driving force})^{-1}$	$\text{gmol h}^{-1}\,\text{cm}^{-2}\,(\text{driving force})^{-1}$
gas film	k_G	$\text{kmol m}^{-2}\,\text{s}^{-1}\,\text{kPa}^{-1}$	"
liquid film	k_L	m s^{-1}	"
Overall, area basis	K	"	"
gas film	K_G	"	"
liquid film	K_L	"	"
Individual, volumetric basis	ka	$\text{kmol m}^{-3}\,\text{s}^{-1}\,\text{kPa}^{-1}$	h^{-1}
gas film	$k_\text{G}a$	"	h^{-1}
liquid film	$k_\text{L}a$	s^{-1}	h^{-1}
Metabolic rate, maximum specific	q_m	s^{-1}	h^{-1}
Metabolic rate, specific	q	s^{-1}	h^{-1}

$q = (\text{d}C/\text{d}t)/X$
where C may be a substrate or product mass concentration. Subscripts may further define the rates, *e.g.* q_S, q_P, q_{O_2}, which are substrate utilization, product formation, and oxygen uptake rates, respectively

Mutation rate	w	s^{-1}	h^{-1}
Power	P	W	W
Productivity, mass concentration rate basis, use appropriate subscripts, *e.g.* r_X for biomass productivity and r_P for product productivity	r	$\text{kg m}^{-3}\,\text{s}^{-1}$	$\text{kg m}^{-3}\,\text{h}^{-1}$
Revolutions per unit time or stirring speed	N	s^{-1}	s^{-1}
Velocity	V	m s^{-1}	cm s^{-1}

V_s for superficial gas velocity = $F_\text{G}/\pi D_\text{t}^2$

V_i for impeller tip velocity = $\pi N D_\text{i}$

Subject Index